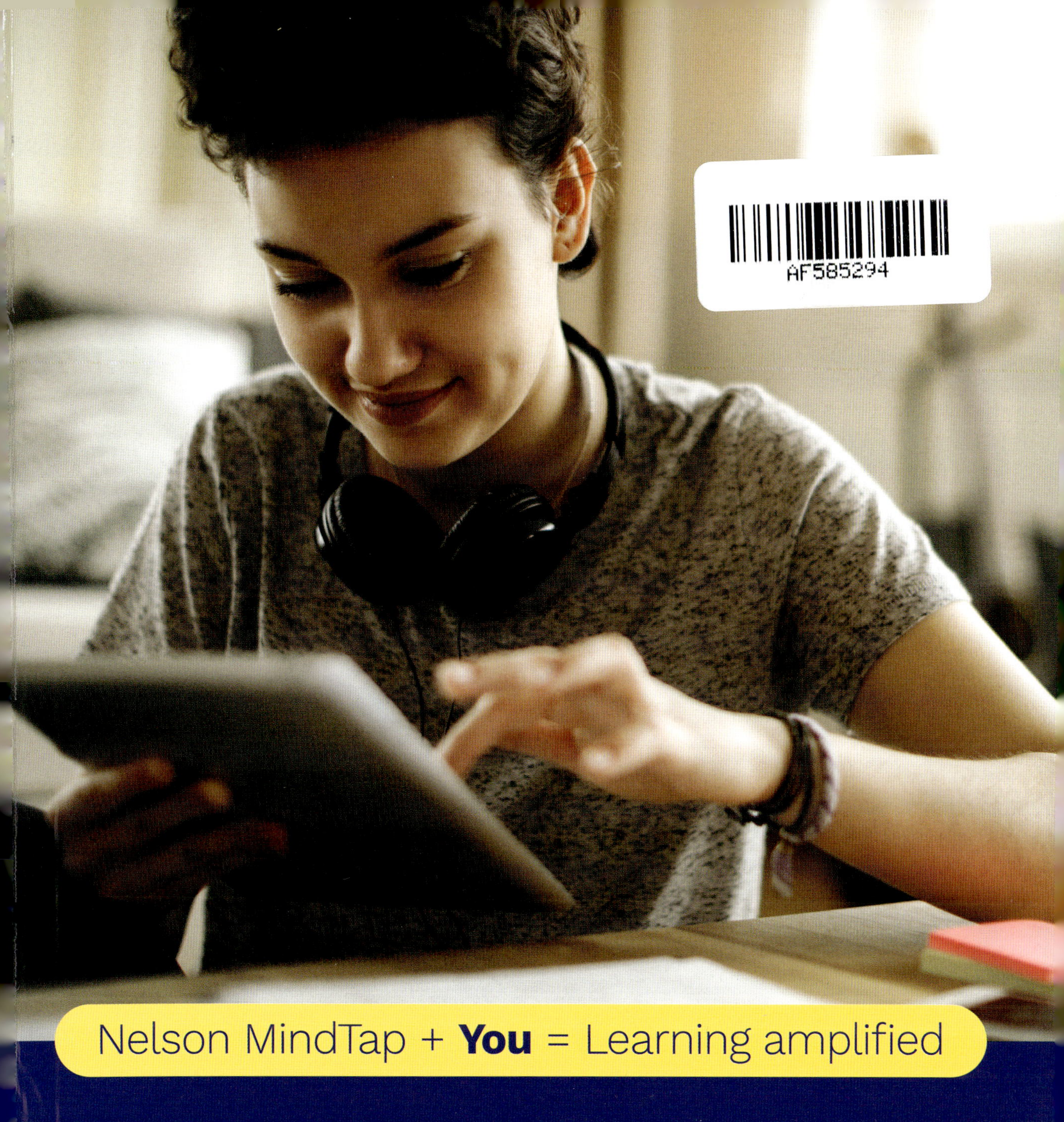

Nelson MindTap + **You** = Learning amplified

"I love that everything is interconnected, relevant and that there is a clear learning sequence. I have the tools to create a learning experience that meets the needs of all my students and can easily see how they're progressing."

— **Sarah,** Secondary School Teacher

nelson maths.

8

Rachel Theunissen
Rashmi Bhagwati
Judy Binns
Gaspare Carrozza
Series editor
Robert Yen

Nelson Maths 8 for the Australian Curriculum WA
1st Edition
Rachel Theunissen
Rashmi Bhagwati
Judy Binns
Gaspare Carrozza
ISBN 9780170465601

Publisher: Robert Yen
Project editor: Alan Stewart
Series cover design: Leigh Ashforth (Watershed Art & Design) and Cengage Creative Studio
Cover image: Shutterstock.com/IKT COLLECTIONS
Series text design: Leigh Ashforth (Watershed Art & Design)
Series designer: Danielle Maccarone
Permissions researcher: Catherine Kerstjens
Production controller: Karen Young
Proofreader / Answer checker: MPS Limited
Typesetter: KGL

Any URLs contained in this publication were checked for currency during the production process. Note, however, that the publisher cannot vouch for the ongoing currency of URLs.

For product information and technology assistance,
in Australia call **1300 790 853**;
in New Zealand call **0800 449 725**

For permission to use material from this text or product, please email **aust.permissions@cengage.com**

ISBN 978 0 17 046560 1

Cengage Learning Australia
Level 7, 80 Dorcas Street
South Melbourne, Victoria Australia 3205

Cengage Learning New Zealand
Unit 4B Rosedale Office Park
331 Rosedale Road, Albany, North Shore 0632, NZ

For learning solutions, visit **cengage.com.au**

Printed in Singapore by C.O.S. Printers Pte Ltd.
1 2 3 4 5 6 7 26 25 24 23 22

ACKNOWLEDGEMENT OF COUNTRY

Nelson acknowledges the Traditional Owners and Custodians of the lands of all First Nations Peoples of Australia. We pay respect to their Elders past and present.

We recognise the continuing connection of First Nations Peoples to the land, air and waters, and thank them for protecting these lands, waters and ecosystems since time immemorial.

Warning – First Nations Australians are advised that this book and associated learning materials may contain images, videos or voices of deceased persons.

Preface

Nelson Maths 7–10 has been designed for the 2020s classroom, focussing on core skills with explicit grading of exercise questions, 'flipped classroom' video tutorials, online interactivity, more applications and problem-solving questions, and worked solutions to every question.

This new series is carefully mapped to the Australian Curriculum and built on solid pedagogical foundations that integrate into every chapter practical classroom activities, engaging investigations, problem-solving, reasoning, communicating, reflecting, summarising, extension, revision, mental calculation, technology, numeracy and literacy.

This book, *Nelson Maths 8*, has been designed for Year 8 students, but includes Year 7 revision and Year 9 extension work.

The *Nelson MindTap* online learning platform contains print and multimedia content: worksheets, videos, quizzes, interactives, topic tests, worked solutions and much more. We have provided an abundance of resources for teachers to plan and teach for a variety of pathways. *Nelson Maths* is clear, concise, fresh and smart. We have designed this series to be user-friendly and uncomplicated so that teachers and students everywhere can pick it up and use straight away. So let's get started.

About the authors

Rashmi Bhagwati teaches at a high school in southeast Brisbane and has been a mathematics educator for over 20 years. She has been involved in external exam marking and assessment processes, and was a panellist for Mathematics A (senior General Mathematics). Rashmi wrote topic tests for the *Nelson Senior Maths Essential Mathematics* series for Queensland and WA.

Rachel Theunissen is the Head of Mathematics at Lesmurdie Senior High School in Perth, where she also leads the Literacy and Numeracy team. She is a Curriculum Support Teacher with the Department of Education and a Volunteer of MAWA and the Australian Mathematics Trust (AMT). Rachel has been teaching for over 25 years and presents regularly at conferences on mathematics and problem-solving.

Judy Binns was the Head Teacher of Mathematics at Mulwaree High School in Goulburn, NSW. She has an interest in motivating students with learning difficulties and wide experience in teaching senior practical mathematics courses.

Gaspare Carrozza was the Head Teacher of Mathematics at Homebush Boys High School in Sydney. He has conducted training workshops in both mathematics and computing.

Series editor **Robert Yen** taught at Hurlstone Agricultural High School in Sydney and works for Cengage as the Mathematics Publisher.

Contributing authors

Sarah Hamper wrote the technology sections.

Kahlia Dreyer, Lachlan Grierson, Brook Johns, Tracey Johnson, Kuldip Khehra, Judy Binns, Gaspare Carrozza and **Robert Yen** wrote many worksheets and topic tests.

Rashmi Bhagwati, Jade Lori, Deborah Da Cruz, Monique Ellement, John Drake, Katie Jackson, Joanne Magner, Scott Smith and **Robert Yen** created the video tutorials.

With thanks to **Dr Caty Morris** on behalf of ATSIMA (Aboriginal and Torres Strait Islander Mathematics Alliance) for her content review and advice.

Table of contents

Curriculum grids

Australian Curriculum

Strand	*Nelson Maths 7* chapter	*Nelson Maths 8* chapter
NUMBER	1 Integers	1 Pythagoras' theorem
	3 Whole numbers	2 Working with numbers
	4 Fractions and percentages	5 Area and volume
	7 Decimals	6 Fractions and percentages
	9 Formulas and graphs	
	12 Ratios and time	
ALGEBRA	5 Algebra and equations	2 Working with numbers
	9 Formulas and graphs	3 Algebra
	12 Ratios and time	10 Equations and inequalities
		11 Ratios, rates and time
		12 Graphing linear functions
MEASUREMENT	2 Angles	1 Pythagoras' theorem
	6 Geometrical figures	4 Area and volume
	8 Area and volume	11 Ratios, rates and time
	12 Ratios and time	
SPACE	6 Geometrical figures	4 Geometry
		8 Congruent and similar figures
		12 Graphing linear functions
STATISTICS	10 Analysing data	7 Investigating data
PROBABILITY	11 Probability	9 Probability

Year 8 content descriptions

Australian Curriculum descriptions (© ACARA 2022).

Content description	*Nelson Maths 8* chapter
NUMBER (N)	
AC9M8N01: recognise irrational numbers in applied contexts, including square roots and π.	1 Pythagoras' theorem
	2 Working with numbers
	5 Area and volume
AC9M8N02: establish and apply the exponent laws with positive integer exponents and the zero-exponent, using exponent notation with numbers	2 Working with numbers
AC9M8N03: recognise terminating and recurring decimals, using digital tools as appropriate	2 Working with numbers

Content description	Nelson Maths 8 chapter	
AC9M8N04: use the 4 operations with integers and with rational numbers, choosing and using efficient strategies and digital tools where appropriate	2	Working with numbers
	6	Fractions and percentages
AC9M8N05: use mathematical modelling to solve practical problems involving rational numbers and percentages, including financial contexts; formulate problems, choosing efficient calculation strategies and using digital tools where appropriate; interpret and communicate solutions in terms of the situation, reviewing the appropriateness of the model	6	Fractions and percentages
	7	Decimals
ALGEBRA (A)		
AC9M8A01: create, expand, factorise, rearrange and simplify linear expressions, applying the associative, commutative, identity, distributive and inverse properties	3	Algebra
AC9M8A02: graph linear relations on the Cartesian plane using digital tools where appropriate; solve linear equations and one-variable inequalities using graphical and algebraic techniques; verify solutions by substitution	3	Algebra
	10	Equations and inequalities
	12	Graphing linear functions
(EXTENSION) YEAR 9 AC9M9A01: apply the exponent laws to numerical expressions with integer exponents and extend to variables	3	Algebra
(EXTENSION) YEAR 9 AC9M9A02: simplify algebraic expressions, expand binomial products and factorise monic quadratic expressions	3	Algebra
(EXTENSION) YEAR 9 AC9M9A04: identify and graph quadratic functions, solve quadratic equations graphically and numerically, and solve monic quadratic equations with integer roots algebraically, using graphing software and digital tools as appropriate	10	Equations and inequalities
AC9M8A03: use mathematical modelling to solve applied problems involving linear relations, including financial contexts; formulate problems with linear functions, choosing a representation; interpret and communicate solutions in terms of the situation, reviewing the appropriateness of the model	12	Graphing linear functions
AC9M8A04: experiment with linear functions and relations using digital tools, making and testing conjectures and generalising emerging patterns	12	Graphing linear functions
MEASUREMENT (M)		
AC9M8M01: solve problems involving the area and perimeter of irregular and composite shapes using appropriate units	5	Area and volume
AC9M8M02: solve problems involving the volume and capacity of right prisms using appropriate units	5	Area and volume
AC9M8M03: solve problems involving the circumference and area of a circle using formulas and appropriate units	5	Area and volume
(EXTENSION) YEAR 9 AC9M9M01: solve problems involving the volume and surface area of right prisms and cylinders using appropriate units	5	Area and volume
AC9M8M04: solve problems involving duration, including using 12- and 24-hour time across multiple time zones	11	Ratios, rates and time

Content description	*Nelson Maths 8* chapter	
AC9M8M05: recognise and use rates to solve problems involving the comparison of 2 related quantities of different units of measure	11	Ratios, rates and time
AC9M8M06: use Pythagoras' theorem to solve problems involving the side lengths of right-angled triangles	1	Pythagoras' theorem

SPACE (SP)

Content description	Chapter	
AC9M8SP01: identify the conditions for congruence and similarity of triangles and explain the conditions for other sets of common shapes to be congruent or similar, including those formed by transformations	8	Congruent and similar figures
AC9M8SP02: establish properties of quadrilaterals using congruent triangles and angle properties, and solve related numerical problems using reasoning	8	Congruent and similar figures
AC9M8SP03: describe the position and location of objects in 3 dimensions in different ways, including using a three-dimensional coordinate system with the use of dynamic geometric software and other digital tools	12	Graphing linear functions
AC9M8SP04: design, create and test algorithms involving a sequence of steps and decisions that identify congruency or similarity of shapes, and describe how the algorithm works	8	Congruent and similar figures

STATISTICS (ST)

Content description	Chapter	
AC9M8ST01: investigate techniques for data collection including census, sampling, experiment and observation, and explain the practicalities and implications of obtaining data through these techniques	7	Investigating data
AC9M8ST02: analyse and report on the distribution of data from primary and secondary sources using random and non-random sampling techniques to select and study samples	7	Investigating data
AC9M8ST03: compare variations in distributions and proportions obtained from random samples of the same size drawn from a population and recognise the effect of sample size on this variation	7	Investigating data
AC9M8ST04: plan and conduct statistical investigations involving samples of a population; use ethical and fair methods to make inferences about the population and report findings, acknowledging uncertainty	7	Investigating data

PROBABILITY (P)

Content description	Chapter	
AC9M8P01: recognise that complementary events have a combined probability of 1; use this relationship to calculate probabilities inn applied contexts	9	Probability
AC9M8P02: determine all possible combinations for 2 events, using two-way tables, tree diagrams and Venn diagrams, and use these to determine probabilities of specific outcomes in practical situations	9	Probability
AC9M8P03: conduct repeated chance experiments and simulations, using digital tools to determine probabilities for compound events, and describe results	9	Probability

About this book

Coverage of the Australian Curriculum

- *Nelson Maths 8* covers the Australian curriculum, as shown by the table of contents and curriculum grid on the previous pages.
- This book contains Year 8 content. It also contains revision of some Year 7 content and some Year 9 or extension work marked in **red** and by *.
- Each chapter begins with a **chapter outline** that includes the curriculum proficiencies covered in each section.

U = UNDERSTANDING	F = FLUENCY
Understanding is 'knowing and relating' maths. It is more than just learning facts. It's deep understanding, seeing how mathematical content is interconnected, knowing 'why' as well as 'how'.	**Fluency** is 'applying' maths. It is being able to use mathematics competently and effectively. When you are fluent in a language, you have mastered it so that you can improvise and confidently use the correct word or phrase. Fluency in maths is choosing an appropriate skill, method or formula to use at the right place and time.
PS = PROBLEM SOLVING	**R = REASONING**
Problem-solving is 'modelling and investigating' with maths. It involves interpreting a rich, elaborate problem, selecting an appropriate strategy or model, solving the problem, then evaluating, communicating and justifying the solution.	**Reasoning** is 'generalising and proving' with maths, using higher-order thinking to connect specific facts to general principles, using algebra, logic, proof and justification.

To these proficiencies, we have added C = **COMMUNICATING.**

Communicating is 'describing and explaining' maths, representing mathematical theory and solutions in words, algebraic symbols, special notations, diagrams, graphs and tables.

Understanding and **Fluency** can be found in every exercise and activity, while **Problem-solving**, **Reasoning** and **Communicating** are found in the **Investigations**, **Technology**, **Mental skills**, **Language of Maths** and **Topic summary** activities, and explicitly labelled in every exercise (see below).

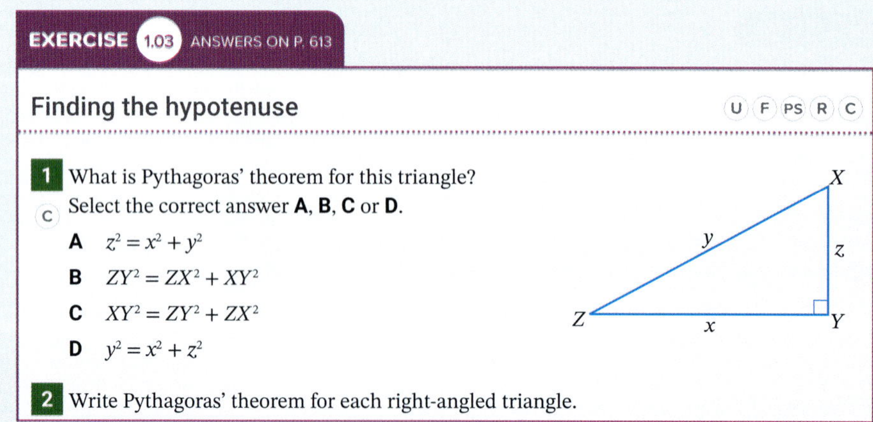

At the beginning of each chapter

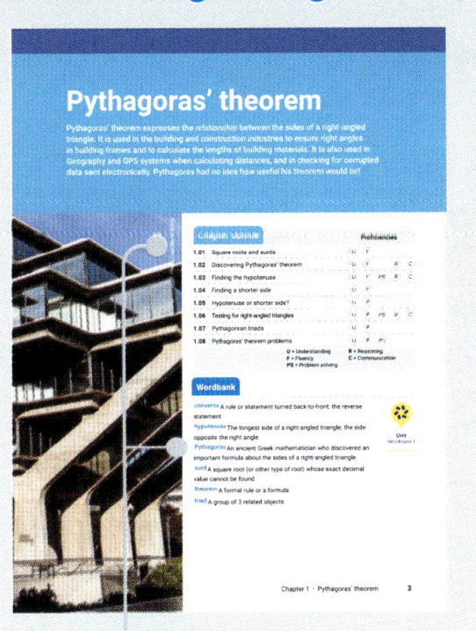

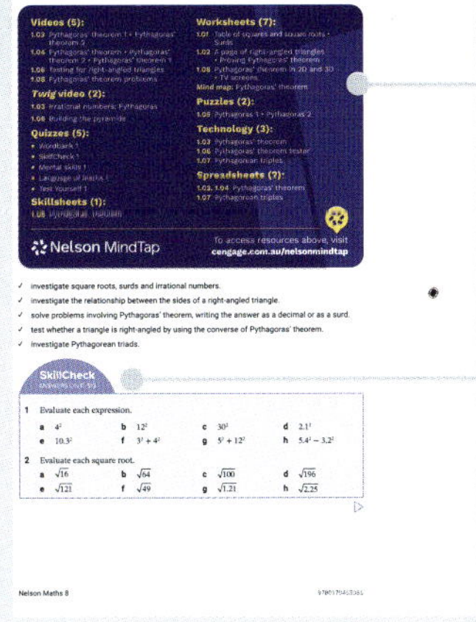

A listing of **Nelson MindTap** chapter resources

SkillCheck reviews prerequisite skills and knowledge for the chapter

A **Chapter outline** and a **Wordbank** chapter glossary (a full glossary appears at the back of the book)

In each chapter

Important facts and formulas are highlighted in a shaded box.

Glossary terms are printed in **blue**.

Graded exercises are linked to worked examples and include multiple-choice questions, exam-style problems and realistic applications.

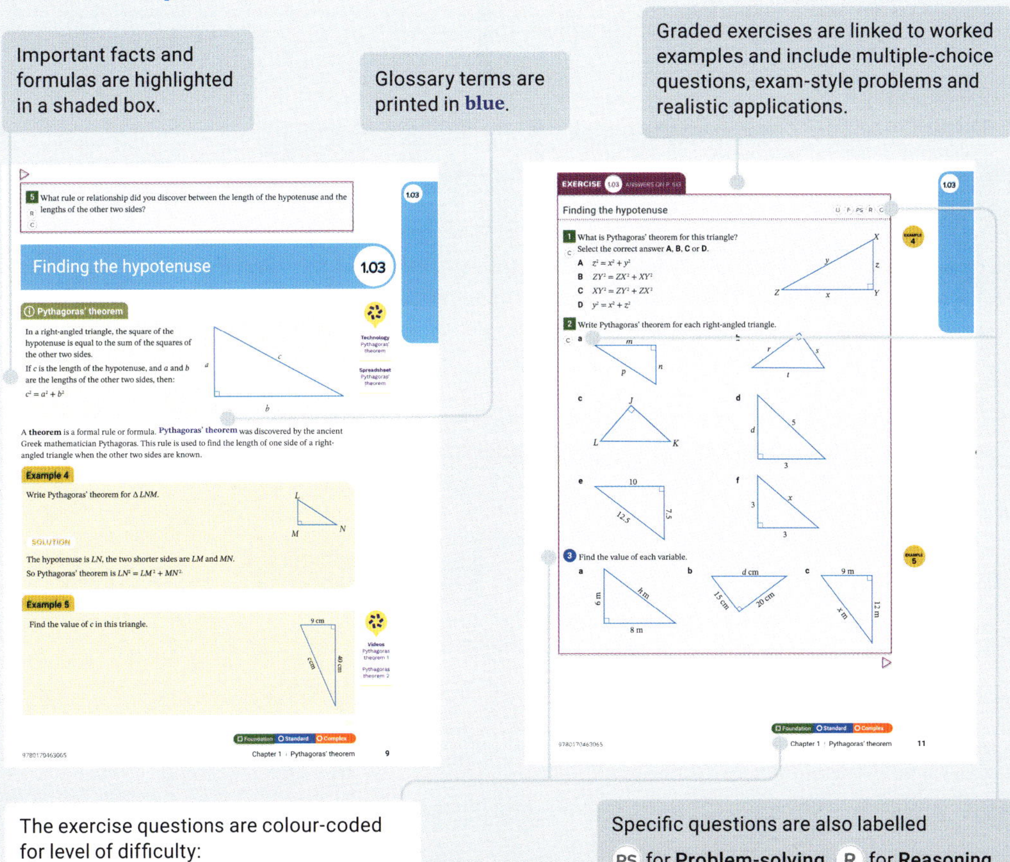

The exercise questions are colour-coded for level of difficulty:

Foundation Standard Complex

Specific questions are also labelled PS for **Problem-solving,** R for **Reasoning** and C for **Communicating**.

MENTAL SKILLS

Mental skills reinforce mental calculation strategies ('calculator-free maths').

TECHNOLOGY

Technology includes spreadsheets, dynamic geometry software and the Internet.

INVESTIGATION

Investigations explore the syllabus in more detail, through group work, discovery and modelling activities.

DID YOU KNOW?

Did you know? contains interesting facts and applications of the mathematics learned in the chapter.

At the end of each chapter

Power plus is an extension/ challenge exercise.

Language of maths has a chapter word list and literacy questions.

Topic summary has a mind map activity with downloadable solutions.

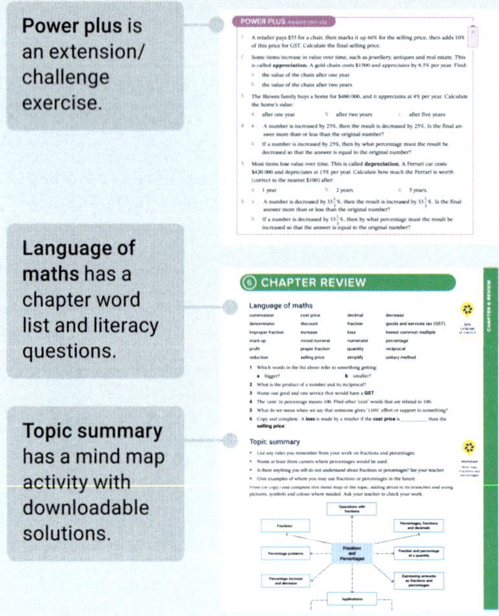

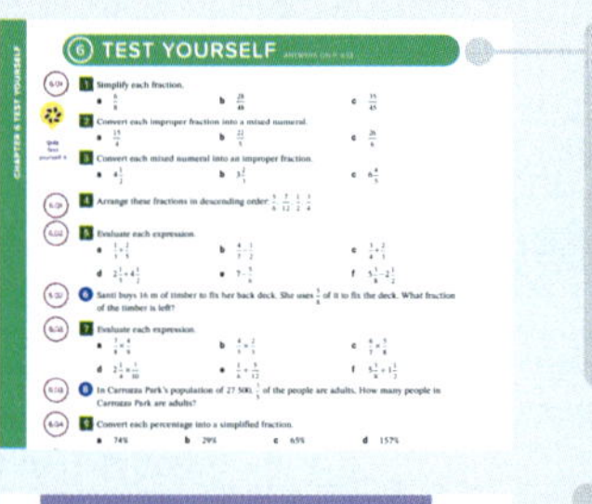

Test Yourself contains chapter revision linked to the relevant exercise set.

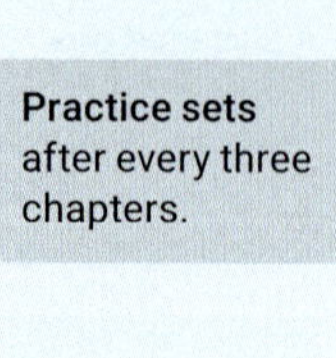

Practice sets after every three chapters.

At the end of the book

General practice exercise

Answers (worked solutions are available for teachers).

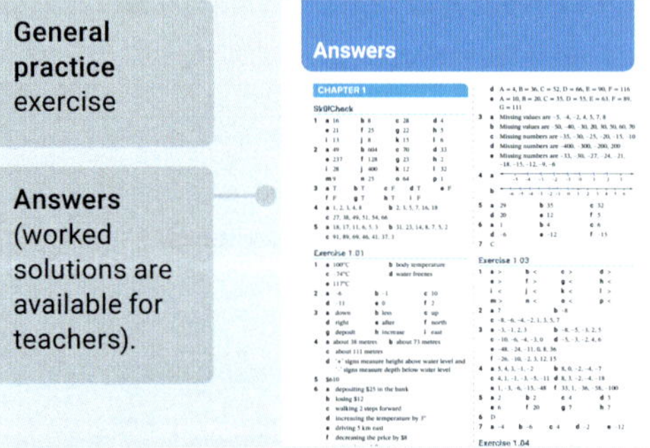

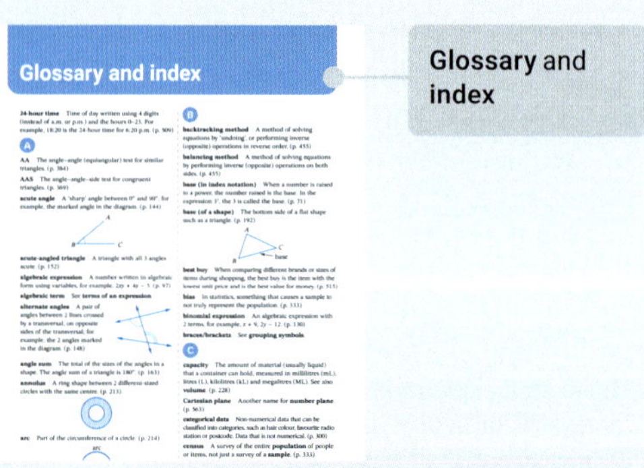

Glossary and **index**

Nelson MindTap

An online learning space that provides students with tailored learning experiences.

- Access tools and content that make learning simpler yet smarter to help you achieve maths mastery.
- Includes an eText with integrated interactives and online assessment.
- Margin links in the student book signpost multimedia student resources found on MindTap.

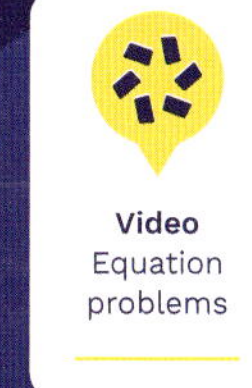

Video
Equation problems

For students:

- **Watch** video tutorials featuring expert teacher advice to unpack new concepts and develop your understanding.
- **Revise** using quizzes, worksheets and skillsheets to practise your skills and build your confidence.
- **Navigate** your own path, accessing the content, analytics and support as you need it.
- ***Twig* mini-documentaries** showing the background or context of mathematics
- ***PhET* interactives**: maths simulations

For teachers*:

- Tailor content to different learning needs – assign directly to the student, or the whole class.
- Monitor progress using assessment tools like Gradebook and Reports.
- Integrate content and assessment directly within your school's LMS for ease of access.
- Access topic tests, teaching plans and worked solutions to each exercise set.

* Complimentary access to these resources is only available to teachers who use this book as part of a class set, book hire or booklist. Contact your Cengage Education Consultant for information about access and conditions

Nelson Maths 7–10 series

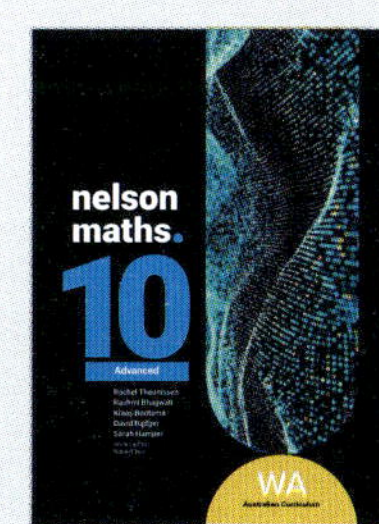

Symbols and abbreviations

$=$	is equal to	$\parallel$	is parallel to
$\neq$	is not equal to	$\perp$	is perpendicular to
$\approx$	is approximately equal to	$\therefore$	therefore
$<$	is less than	x^2	x squared, $x \times x$
$>$	is greater than	x^3	x cubed, $x \times x \times x$
$\leq$	is less than or equal to	$\sqrt{\ }$	square root, radical sign
$\geq$	is greater than or equal to	$\sqrt[3]{\ }$	cube root
( )	parentheses, round brackets	$P(E)$	the probability of event E
[ ]	(square) brackets	LHS	left-hand side
{ }	braces	RHS	right-hand side
$\pm$	plus or minus	%	percentage
-3	negative 3	p.a.	per annum (per year)
$0.\dot{1}5\dot{2}$	the recurring decimal 0.152152 ...	$\overline{x}$	the mean (average)
$^\circ$	degree	μ	micro-, mu
$\angle A$	angle A		
$\triangle ABC$	triangle ABC		

Mathematical verbs

A glossary of 'doing words' commonly found in mathematics problems

analyse: study in detail the parts of a situation

bisect: cut in half

calculate: *see* **evaluate**

classify, identify: state the type, category or feature of an item or situation

comment: express an observation or opinion about a result

complete: fill in detail to make a statement, diagram or table correct or finished

compare: show how two or more things are similar or different

construct: draw an accurate diagram

convert: change from one form to another, for example, from a fraction to a decimal, or from kilograms to grams

decrease: make smaller

describe: state the features of a situation

estimate: make an educated guess for a number, measurement or solution, to find roughly or approximately

evaluate, calculate: find the value of a numerical expression, for example, 3×8^2 or $4x + 1$ when $x = 5$

expand: remove brackets in an algebraic expression by multiplying, for example, expanding $3(2y + 1)$ gives $6y + 3$

explain: describe why or how

factorise: take out the highest common factor (HCF) of an expression and insert brackets, for example, factorising $5x - 20$ gives $5(x - 4)$. The opposite of **expand**.

give reasons: show the rules or thinking used when solving a problem. See also **justify**.

graph: display as a number line, number plane or statistical graph.

hence find/prove: find an answer or prove a result using previous answers or information supplied

identify: *see* **classify**

increase: make larger

interpret: find meaning in an answer or result

justify: give reasons or evidence to support your argument or conclusion. See also **give reasons**.

measure: use an instrument to find the size of something, for example, use a ruler to determine the length of a pen.

prove that: *see* **show that**

recall: remember and state

reduce (a fraction) to its lowest terms: see **simplify (a fraction)**

round (a number): find the nearest approximation of a number. For example, 4.3 rounded to the nearest whole number is 4, \$12.9598 rounded to the nearest cent is \$12.96, 0.166 66 rounded to three decimal places is 0.167

show that, prove that: (in questions where the answer is given) use calculation, procedure or reasoning to prove that an answer or result is true

show working: show the steps you used to find the answer

simplify: give a result in its most basic, shortest, neatest form, for example, simplifying a ratio or algebraic expression

simplify (a fraction): reduce the numerator and denominator of a fraction by dividing by their HCF, for example, $\frac{16}{20}$ simplified is $\frac{4}{5}$

simplify (a ratio or rate): reduce the terms or units of a ratio or rate by dividing by their HCF, for example, 10 : 4 simplified is 5 : 2

sketch: draw a rough diagram that shows the general shape or idea, less accurate than **construct**

solve: find the value(s) of an unknown variable in an equation or inequality

state: *see* **write**.

substitute: replace a variable by a number and evaluate

verify: check that a solution or result is correct, usually by substituting back into the equation or referring back to the problem

write correct to: See **round (a number)**.

write, state: give the answer, formula or result without showing any working or explanation (this usually means that the answer can be found mentally, or in one step)

1

MEASUREMENT, NUMBER

Pythagoras' theorem

Pythagoras' theorem expresses the relationship between the sides of a right-angled triangle. It is used in the building and construction industries to ensure right angles in building frames and to calculate the lengths of building materials. It is also used in Geography and GPS systems when calculating distances, and in checking for corrupted data sent electronically. Pythagoras had no idea how useful his theorem would be!

iStock.com/Sherry Smith

Chapter outline

		Proficiencies				
1.01	Square roots and surds	U	F			
1.02	Discovering Pythagoras' theorem	U	F		R	C
1.03	Finding the hypotenuse	U	F	PS	R	C
1.04	Finding a shorter side	U	F			
1.05	Hypotenuse or shorter side?	U	F			
1.06	Testing for right-angled triangles	U	F	PS	R	C
1.07	Pythagorean triads	U	F			
1.08	Pythagoras' theorem problems	U	F	PS		

U = Understanding
F = Fluency
PS = Problem solving
R = Reasoning
C = Communication

Wordbank

converse A rule or statement turned back-to-front; the reverse statement

hypotenuse The longest side of a right-angled triangle; the side opposite the right angle

Pythagoras An ancient Greek mathematician who discovered an important formula about the sides of a right-angled triangle

surd A square root (or other type of root) whose exact decimal value cannot be found

theorem A formal rule or a formula

triad A group of 3 related objects

Quiz
Wordbank 1

Videos (5):

1.03 Pythagoras' theorem 1 • Pythagoras' theorem 2
1.04 Pythagoras' theorem • Pythagoras' theorem 2 • Pythagoras' theorem 1
1.06 Testing for right-angled triangles
1.08 Pythagoras' theorem problems

Twig videos (2):

1.03 Irrational numbers: Pythagoras
1.06 Building the pyramids

Quizzes (5):

- Wordbank 1
- SkillCheck 1
- Mental skills 1
- Language of Maths 1
- Test Yourself 1

Skillsheet (1):

1.05 Pythagoras' theorem

Worksheets (7):

1.01 Table of squares and square roots • Surds
1.02 A page of right-angled triangles • Proving Pythagoras' theorem
1.08 Pythagoras' theorem in 2D and 3D • TV screens
Mind map: Pythagoras' theorem

Puzzles (2):

1.05 Pythagoras 1 • Pythagoras 2

Technology (3):

1.03 Pythagoras' theorem
1.06 Pythagoras' theorem tester
1.07 Pythagorean triples

Spreadsheets (2):

1.03, 1.04 Pythagoras' theorem
1.07 Pythagorean triples

Nelson MindTap

To access resources above, visit **cengage.com.au/nelsonmindtap**

1 In this chapter you will:

- ✓ investigate square roots, surds and irrational numbers.
- ✓ investigate the relationship between the sides of a right-angled triangle.
- ✓ solve problems involving Pythagoras' theorem, writing the answer as a decimal or as a surd.
- ✓ test whether a triangle is right-angled by using the converse of Pythagoras' theorem.
- ✓ investigate Pythagorean triads.

SkillCheck

ANSWERS ON P. 613

Quiz SkillCheck 1

1 Evaluate each expression.

a 4^2 **b** 12^2 **c** 30^2 **d** 2.1^2
e 10.3^2 **f** $3^2 + 4^2$ **g** $5^2 + 12^2$ **h** $5.4^2 - 3.2^2$

2 Evaluate each square root.

a $\sqrt{16}$ **b** $\sqrt{64}$ **c** $\sqrt{100}$ **d** $\sqrt{196}$
e $\sqrt{121}$ **f** $\sqrt{49}$ **g** $\sqrt{1.21}$ **h** $\sqrt{2.25}$

3 Find the perimeter of each shape.

a

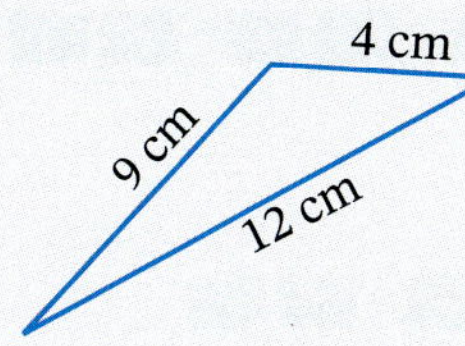

b

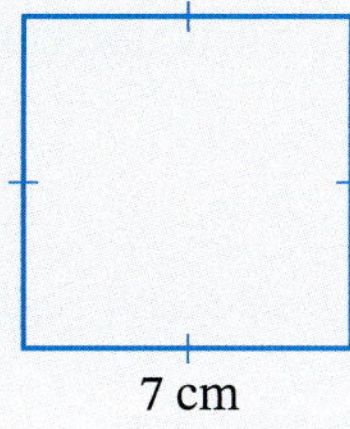

c

4 Find the area of each shape.

a

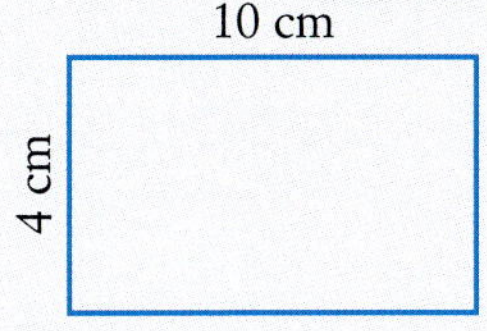

b

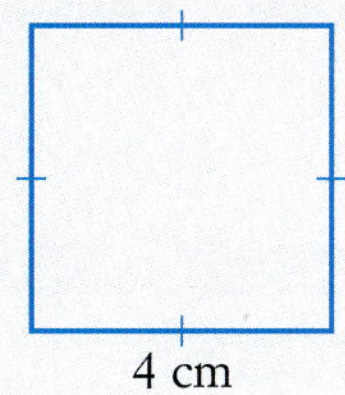

c

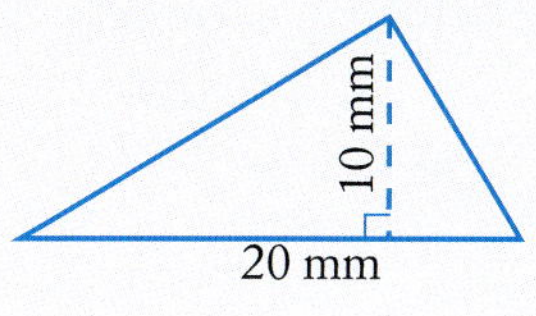

5 Write each number correct to one decimal place.

a 10.33 **b** 4.67 **c** 7.654 **d** 0.8888

6 Evaluate each expression correct to one decimal place.

a $\sqrt{3^2+7^2}$ **b** $\sqrt{5.6^2+9.2^2}$ **c** $\sqrt{15^2-8^2}$

Square roots and surds 1.01

The **square root** ($\sqrt{\ }$) of a number is the **positive** value which, if squared, will give that number. For example:

- $\sqrt{25}=5$ because $5^2=25$ 'the square root of 25'.

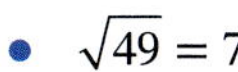

- $\sqrt{49}=7$ because $7^2=49$ 'the square root of 49'.

Worksheets
Table of squares and square roots
Surds

Most square roots do not give exact answers like the ones above. For example, $\sqrt{10}=3.16227766\ldots \approx 3.2$. Such square roots are called **surds**. A surd is a square root ($\sqrt{\ }$), cube root ($\sqrt[3]{\ }$) or any other type of root whose exact decimal or fraction value cannot be found. As a decimal, its digits run endlessly *without repeating*, so they are *neither* terminating nor recurring decimals. A surd cannot be written in fraction form $\frac{a}{b}$ so it is also called an **irrational** number.

Example 1

Evaluate each expression correct to two decimal places.

a $\sqrt{7^2+9^2}$ **b** $\sqrt{12^2-10^2}$

SOLUTION

a $\sqrt{7^2+9^2} = 11.40175\ldots$ On a calculator, enter √ (7 x^2 + 9 x^2) =

≈ 11.40

OR:

$7^2 + 9^2 = 130$ On a calculator, enter 7 x^2 + 9 x^2 =

$\sqrt{130} = 11.40175\ldots$ √ Ans =

≈ 11.40

b $\sqrt{12^2-10^2} = 6.63324\ldots$ √ (12 x^2 − 10 x^2) =

≈ 6.63

OR:

$12^2 - 10^2 = 44$ 12 x^2 − 10 x^2 =

$\sqrt{44} = 6.63324\ldots$ √ Ans =

≈ 6.63

Example 2

Select the surds from this list of square roots:

$\sqrt{72}$ $\sqrt{121}$ $\sqrt{64}$ $\sqrt{90}$ $\sqrt{28}$

SOLUTION

$\sqrt{72} = 8.4852...$ $\sqrt{121} = 11$ $\sqrt{64} = 8$

$\sqrt{90} = 9.4868...$ $\sqrt{28} = 5.2915...$

So the surds are $\sqrt{72}$, $\sqrt{90}$ and $\sqrt{28}$.

EXERCISE 1.01 ANSWERS ON P. 613

Square roots and surds

U F

1 Evaluate each square root.

R **a** $\sqrt{196}$ **b** $\sqrt{900}$ **c** $\sqrt{64}$

d $\sqrt{625}$ **e** $\sqrt{121}$ **f** $\sqrt{361}$

2 Evaluate each square root, correct to two decimal places.

R **a** $\sqrt{12}$ **b** $\sqrt{45}$ **c** $\sqrt{1001}$ **d** $\sqrt{325}$

e $\sqrt{153}$ **f** $\sqrt{207}$ **g** $\sqrt{98}$ **h** $\sqrt{888}$

i $\sqrt{24}$ **j** $\sqrt{110}$ **k** $\sqrt{297}$ **l** $\sqrt{689}$

☐ Foundation ○ Standard ⬡ Complex

3 Evaluate each expression, correct to two decimal places.

Ⓡ **a** $\sqrt{12^2 - 5^2}$ **b** $\sqrt{5^2 + 7^2}$ **c** $\sqrt{6^2 + 11^2}$

d $\sqrt{17^2 - 12^2}$ **e** $\sqrt{8^2 - 3^2}$ **f** $\sqrt{4^2 + 9^2}$

g $\sqrt{(1.5)^2 + (4.2)^2}$ **h** $\sqrt{(12.5)^2 - (7.1)^2}$ **i** $\sqrt{(25.7)^2 + (18.2)^2}$

4 Which of the following is **not** a surd? Select the correct answer **A**, **B**, **C** or **D**.

Ⓡ **A** $\sqrt{10.24}$ **B** $\sqrt{24}$ **C** $\sqrt{1000}$ **D** $\sqrt{3}$

5 Select all the surds from this list of square roots.

Ⓡ $\sqrt{98}$ $\sqrt{9}$ $\sqrt{225}$ $\sqrt{160}$ $\sqrt{36}$ $\sqrt{52}$

$\sqrt{144}$ $\sqrt{77}$ $\sqrt{18}$ $\sqrt{196}$ $\sqrt{200}$ $\sqrt{81}$

Discovering Pythagoras' theorem 1.02

A **right-angled triangle** has one right angle (90°) and two smaller angles. The side opposite the right angle is the *longest* side and is called the **hypotenuse**.

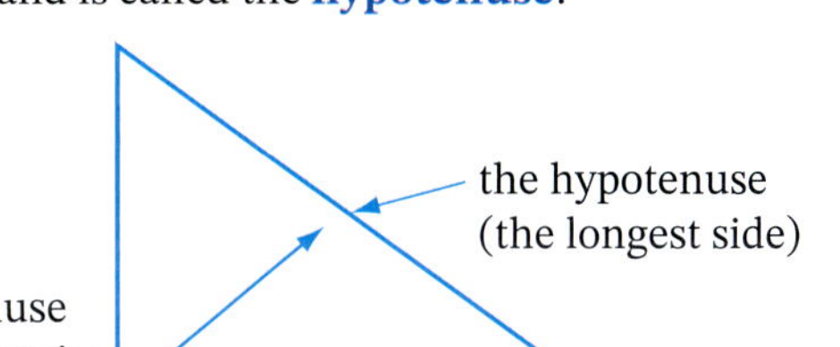

Worksheets
A page of right-angled triangles
Proving Pythagoras' theorem

Pythagoras' theorem is a rule that describes the relationship between the hypotenuse and the other two (shorter) sides.

Example 3

a Construct a right-angled triangle with the two shorter sides being of length 3 cm and 4 cm.

b Measure the length of the hypotenuse.

c Evaluate $3^2 + 4^2$.

d If c is the length of the hypotenuse in centimetres, evaluate c^2.

SOLUTION

a

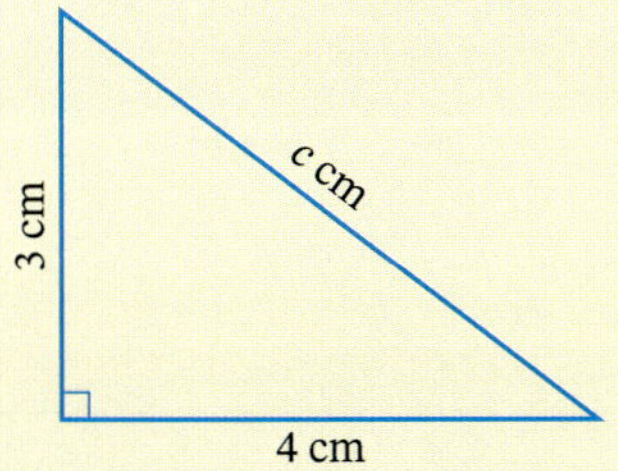

b Measuring with a ruler, the hypotenuse is 5 cm long.

c $3^2 + 4^2 = 25$

d $c^2 = 5^2 = 25$

☐ Foundation ○ Standard ⬡ Complex

EXERCISE 1.02 ANSWERS ON P. 613

Discovering Pythagoras' theorem

U F R C

1 Name the hypotenuse in each triangle.

C **a**

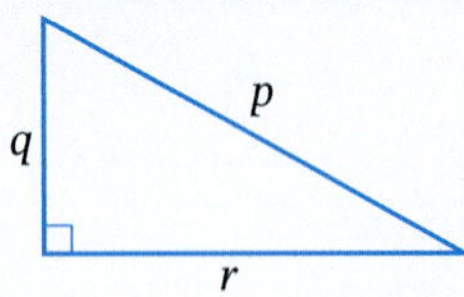

b

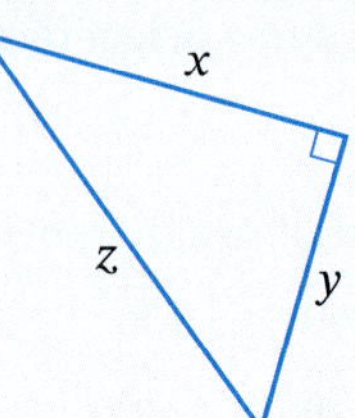

c

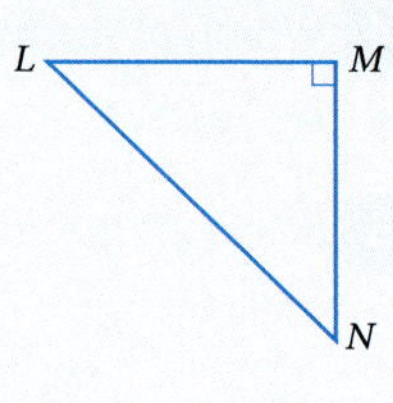

EXAMPLE 3

2 **a** Draw a right-angled triangle with short sides measuring 5 cm and 12 cm.

C **b** Measure the length of the hypotenuse.

c Evaluate $5^2 + 12^2$.

d If c is the length of the hypotenuse in centimetres, evaluate c^2.

e Copy and complete each statement.

The square of the hypotenuse = $\_\_\_\_^2$ = ____

The sum of the squares of the two shorter sides = $\_\_\_\_^2 + \_\_\_\_^2$ = ____

3 Which equation is true for the right-angled triangle shown? Select the correct answer **A**, **B**, **C** or **D**.

C

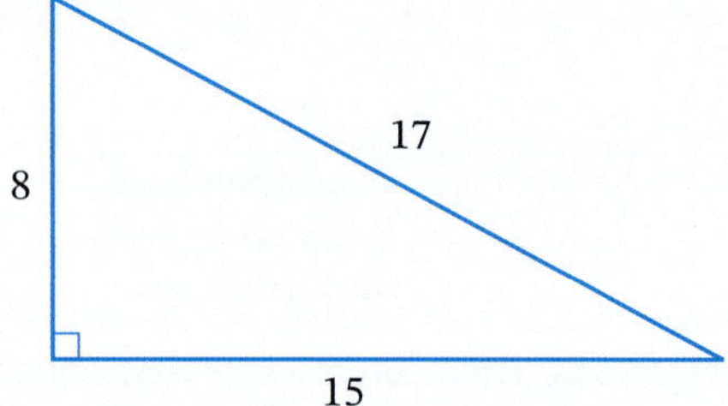

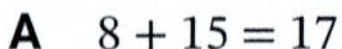

A $8 + 15 = 17$

B $8^2 + 15^2 = 17^2$

C $8^2 + 17^2 = 15^2$

D $17^2 + 15^2 = 8^2$

4 **a** Use dynamic geometry software (or pencil, ruler and 5 mm grid paper) to draw a right-angled triangle with the shorter sides of length 6 cm and 8 cm, then measure the length of the hypotenuse.

C **b** Copy and complete this table by:

- constructing each right-angled triangle with the measurements given for the two shorter sides a and b (as you did in part a).
- measuring the length, c, of the hypotenuse.
- evaluating the values of a^2, b^2, $a^2 + b^2$, c^2.

Shorter sides (cm)		Hypotenuse (cm)				
a	b	c	a^2	b^2	$a^2 + b^2$	c^2
6	8					
2.5	6					
1.5	2					
4	7.5					

▷

☐ Foundation ◯ Standard ⬡ Complex

5 What rule or relationship did you discover between the length of the hypotenuse and the lengths of the other two sides?

R

C

Finding the hypotenuse 1.03

Pythagoras' theorem

In a right-angled triangle, the square of the hypotenuse is equal to the sum of the squares of the other two sides.

If c is the length of the hypotenuse, and a and b are the lengths of the other two sides, then:

$c^2 = a^2 + b^2$

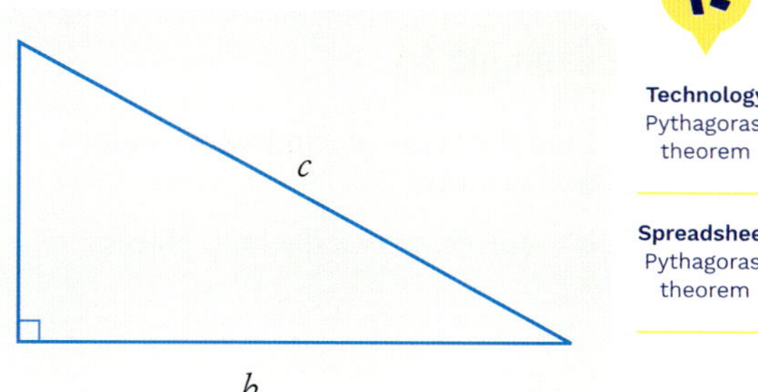

Technology Pythagoras' theorem

Spreadsheet Pythagoras' theorem

A **theorem** is a formal rule or formula. **Pythagoras' theorem** was discovered by the ancient Greek mathematician Pythagoras. This rule is used to find the length of one side of a right-angled triangle when the other two sides are known.

Example 4

Write Pythagoras' theorem for $\triangle LNM$.

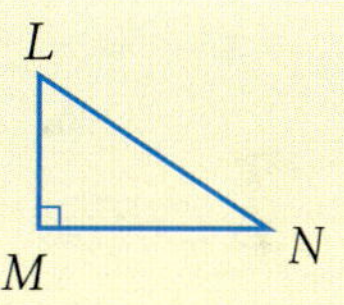

SOLUTION

The hypotenuse is LN, the two shorter sides are LM and MN.

So Pythagoras' theorem is $LN^2 = LM^2 + MN^2$.

Example 5

Find the value of c in this triangle.

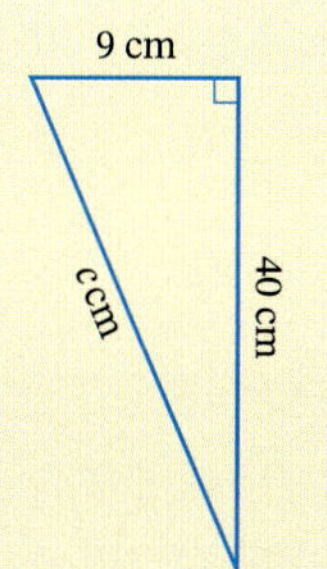

Videos Pythagoras theorem 1

Pythagoras theorem 2

Foundation Standard Complex

SOLUTION

c cm is the length of the hypotenuse. Using Pythagoras' theorem:

$c^2 = a^2 + b^2$

$c^2 = 9^2 + 40^2$ Substitute the lengths of the two shorter sides.

$= 1681$

$c = \sqrt{1681}$ Use the square root to find c.

$= 41$

Check: A hypotenuse of 41 cm looks reasonable from the diagram: it is the longest side and a little longer than 40 cm.

Example 6

Find the value of c in this triangle.

a as a surd.

b correct to two decimal places.

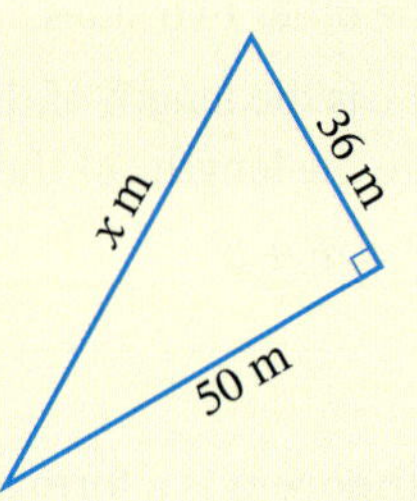

SOLUTION

a Write Pythagoras' theorem.

$c^2 = a^2 + b^2$

$x^2 = 50^2 + 36^2$ The hypotenuse in this question is x.

$= 3796$ Substitute the lengths of the two shorter sides.

$x = \sqrt{3796}$ Because the question asks for a **surd**, leave the answer in square root ($\sqrt{\ }$) form.

b $x = \sqrt{3796}$ From part **a**.

$= 61.6116\ldots$

≈ 61.61 *Check*: From the diagram, this answer looks reasonable.

Note: The surd answer is more exact than the decimal answer because it is not rounded.

EXERCISE 1.03 ANSWERS ON P. 613

Finding the hypotenuse

U F PS R C

1 What is Pythagoras' theorem for this triangle?

C Select the correct answer **A**, **B**, **C** or **D**.

EXAMPLE 4

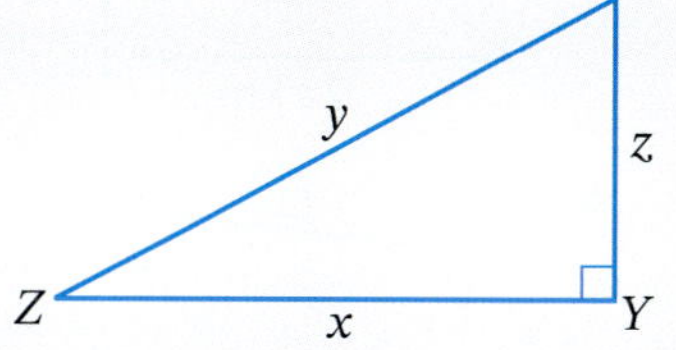

A $z^2 = x^2 + y^2$

B $ZY^2 = ZX^2 + XY^2$

C $XY^2 = ZY^2 + ZX^2$

D $y^2 = x^2 + z^2$

2 Write Pythagoras' theorem for each right-angled triangle.

C **a**

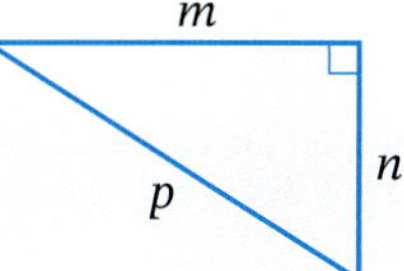

b

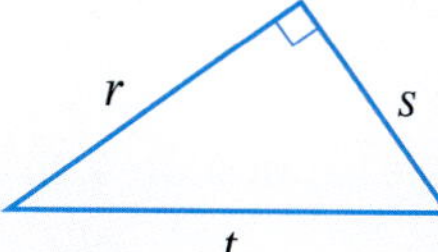

c

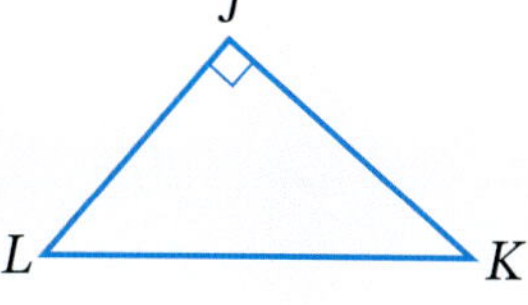

d

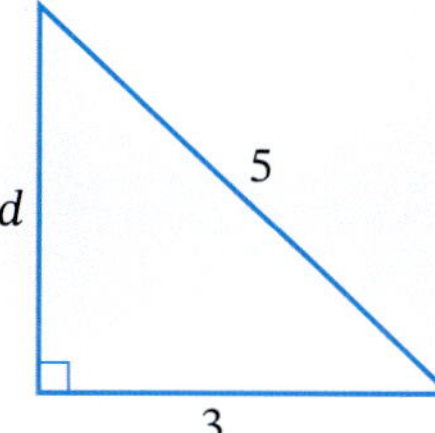

e

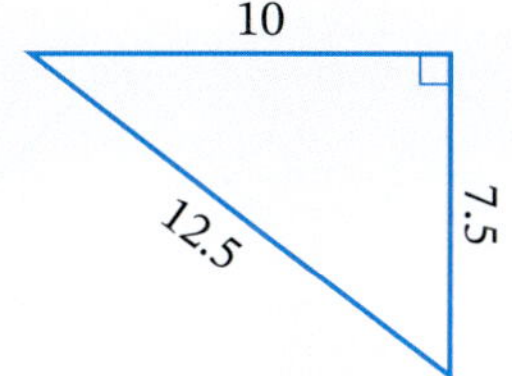

f

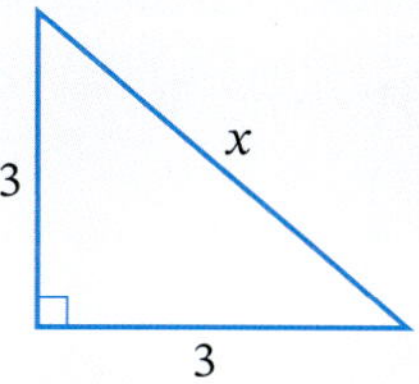

3 Find the value of each variable.

a

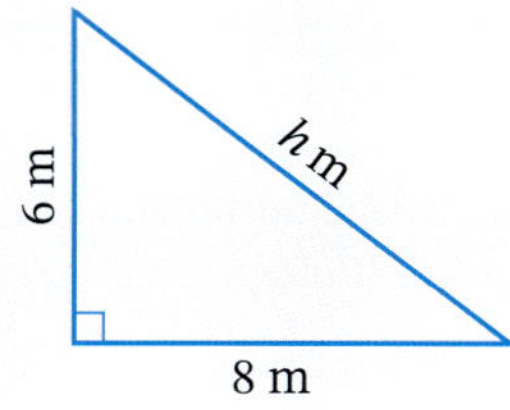

b

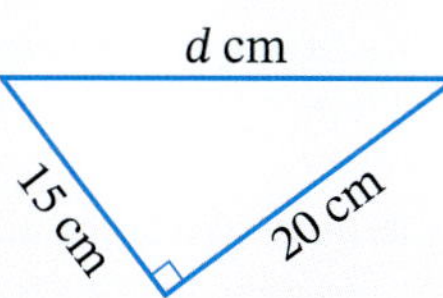

c

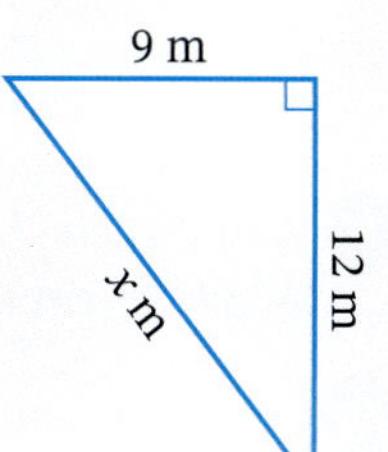

☐ Foundation ○ Standard ○ Complex

4 Find the value of each variable as a surd.

a
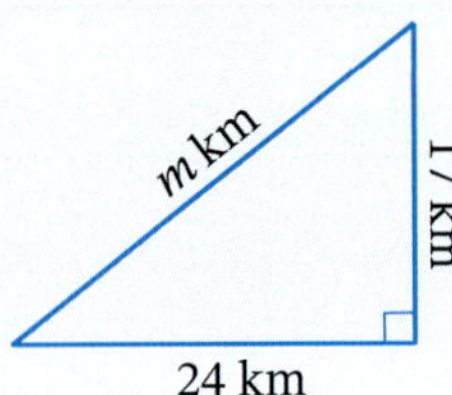

b
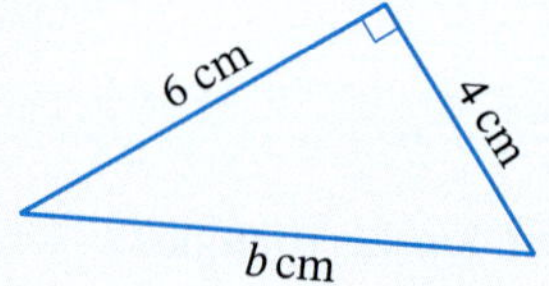

c
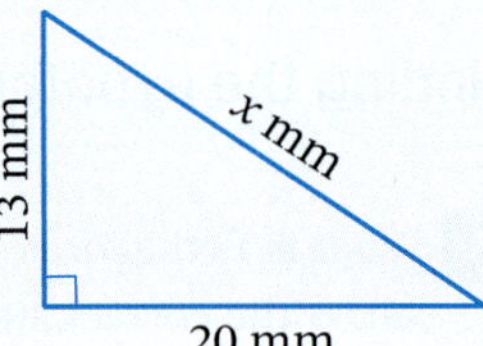

d
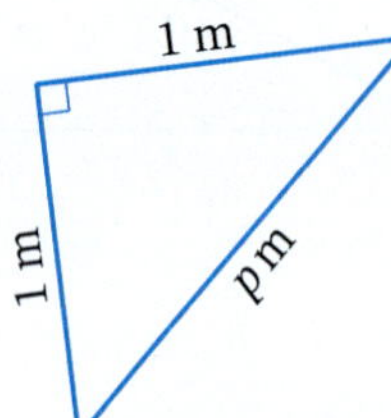

e
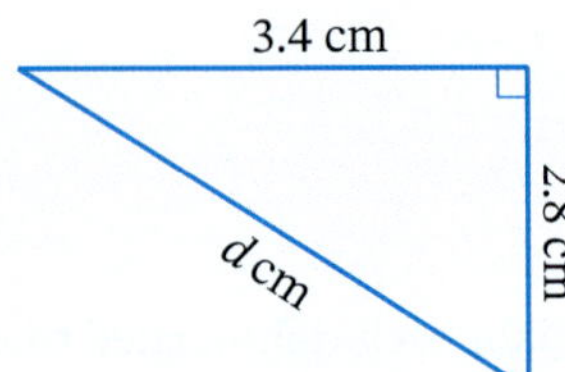

5 For each triangle in question **4**, find correct to one decimal place the value of each variable.

6 In this triangle, what is the value of d? Select **A**, **B**, **C** or **D**.

A 22.8 **B** 12.9

C 34.5 **D** 25.5

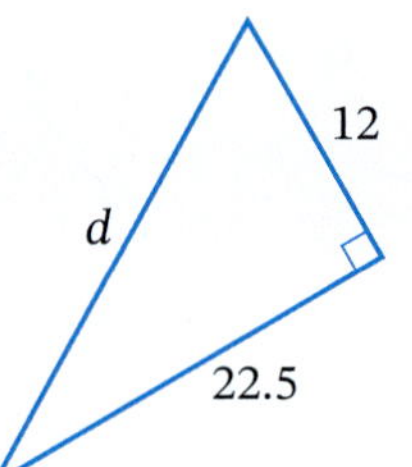

7 Find the value of each variable, correct to two decimal places.

a
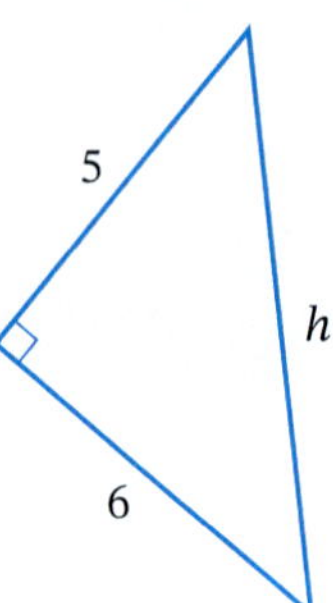

b
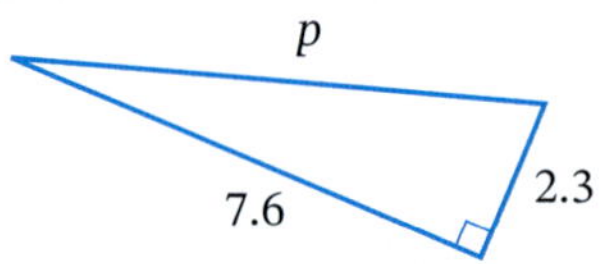

c
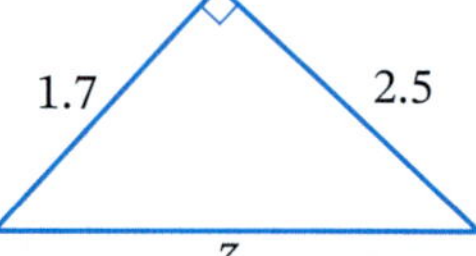

8 Find the length of the hypotenuse of a right-angled triangle with shorter sides of length 7.5 cm and 10 cm.

9 Copy and complete each equation.

a $24^2 + 32^2 = \_\_\_\_^2$

b $36^2 + 15^2 = \_\_\_\_^2$

c $9^2 + 40^2 = \_\_\_\_^2$

d $21^2 + \_\_\_\_^2 = 35^2$

e $60^2 + \_\_\_\_^2 = 61^2$

f $\_\_\_\_^2 + 15^2 = 17^2$

g $\_\_\_\_^2 + 24^2 = 51^2$

h $\_\_\_\_^2 + \_\_\_\_^2 = 10^2$

i $\_\_\_\_^2 + \_\_\_\_^2 = 15^2$

□ Foundation ○ Standard ⬡ Complex

10 Iluka left his campsite and walked directly East for 800 m to a water hole and then directly North for 1500 m to the meeting place.

PS R C

a If he had been able to walk in a straight line from his campsite to the meeting place, what distance would he have walked?

b How much less walking would he have done?

c What is one reason that Iluka might not have been able to walk directly to the meeting place?

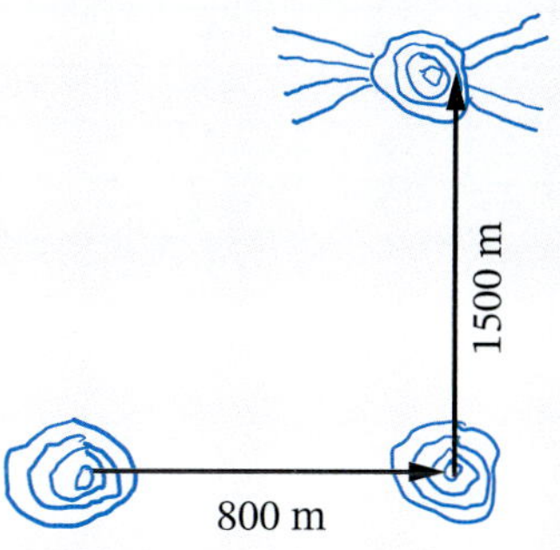

DID YOU KNOW?

Pythagoras

Pythagoras was a Greek philosopher and mathematician. He was born on the Island of Samos, part of ancient Greece, and lived from 570 BCE to 490 BCE. He was the first person to set down a mathematical proof for his famous theorem, even though it was known and used long before his time.

The Pythagoreans were a group of men who followed Pythagoras' philosophical ideas and worked on mathematical problems. They formed a secret society known as the Pythagoreans. Apparently, they were so upset about the discovery of surds that they tried to keep it a secret. Hippasus, one of the Pythagoreans, was drowned for revealing the secret to outsiders.

Which American president published a proof of Pythagoras' theorem? When did he do this?

Shutterstock.com/OmoPhoto

Video Irrational numbers: Pythagoras

TECHNOLOGY

Finding the hypotenuse

In this activity, we will create a spreadsheet to calculate the length of the hypotenuse, c, of a right-angled triangle, given the lengths of the other two sides (a and b).

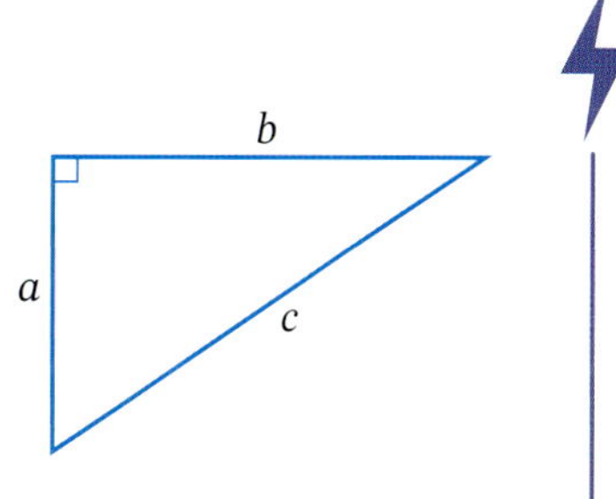

1 Enter the headings and values shown.

☐ Foundation ○ Standard ⬡ Complex

	A	B	C	D	E
1	a	b	a^2+b^2	c	
2	3	4			
3	5	12			
4	6	8			
5	7	24			
6					
7					

Remember: a^2 + b^2 means $a^2 + b^2$.

2 Enter into cell C2 the formula **=A2^2+B2^2**, and use **Fill Down** to copy this formula into cells C3 to C5.

3 To calculate the length of each hypotenuse in column D, we need to find the square root of the values in column C, so in cell D2, enter **=sqrt(C2)**. Then **Fill Down** to copy this formula into cells D3 to D5.

4 Use your spreadsheet to check your answers to questions **3** to **6** of Exercise 1.03.

1.04 Finding a shorter side

Video
Pythagoras' theorem

Spreadsheet
Pythagoras' theorem

Pythagoras' theorem can also be used to find the length of one of the shorter sides in a right-angled triangle if the hypotenuse and the other side are given. In this case, we *subtract* the square of the length of the known shorter side from the square of the length of the hypotenuse.

Finding a shorter side

To find the length of a shorter side, a, in a right-angled triangle with hypotenuse c and other shorter side b, use Pythagoras' theorem in the form:

$$a^2 = c^2 - b^2$$

Remember the correct order for $c^2 - b^2$, the hypotenuse is squared first, otherwise we will get a negative number.

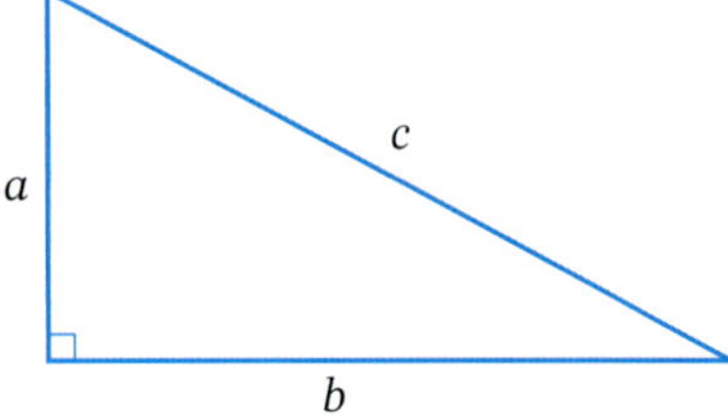

Example 7

Find the value of d in this triangle.

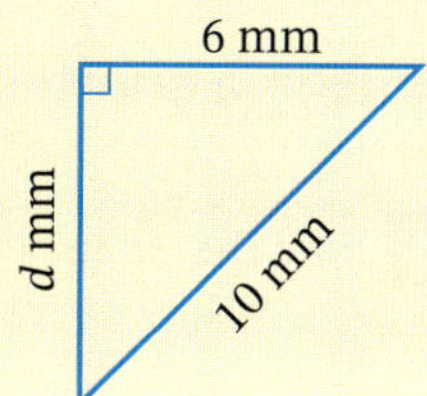

Video
Pythagoras' theorem 2

SOLUTION

d is one of the shorter sides.

$a^2 = c^2 - b^2$

$d^2 = 10^2 - 6^2$ (Hypotenuse)^2 – (other side)^2

$= 64$

$d = \sqrt{64}$ Take the square root to find d.

$= 8$ *Check*: From the diagram, the answer $d = 8$ looks reasonable.

Example 8

Find the value of y in this triangle:

a as a surd

b correct to two decimal places.

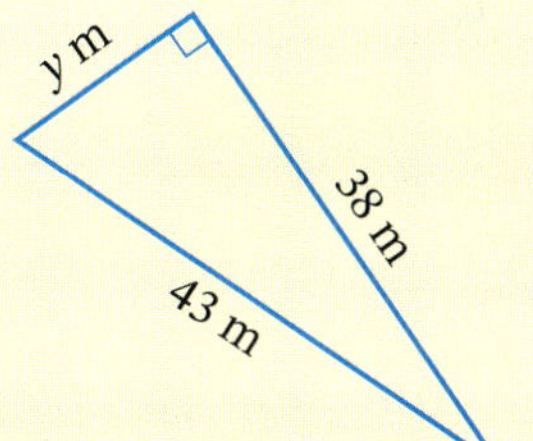

Video
Pythagoras' theorem 1

SOLUTION

a $a^2 = c^2 - b^2$

$y^2 = 43^2 - 38^2$ (Hypotenuse)^2 – (other side)^2

$= 405$

$y = \sqrt{405}$ Surd answer

b $y = \sqrt{405}$ from part **a**

$= 20.1246\ldots$

≈ 20.12

Check: From the diagram, this answer appears to be reasonable.

EXERCISE 1.04 ANSWERS ON P. 613

Finding a shorter side

U F R

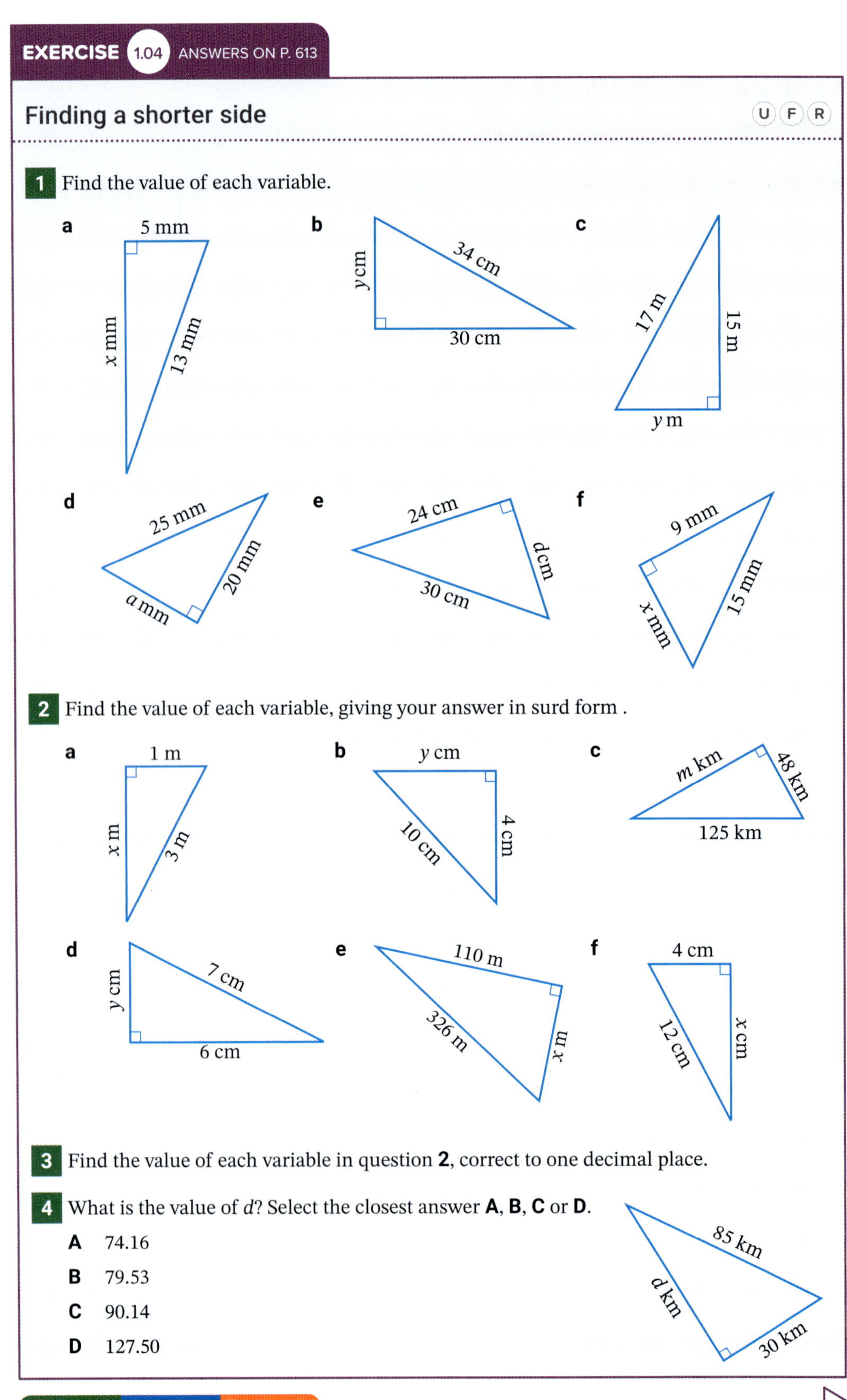

1 Find the value of each variable.

2 Find the value of each variable, giving your answer in surd form .

3 Find the value of each variable in question **2**, correct to one decimal place.

4 What is the value of d? Select the closest answer **A**, **B**, **C** or **D**.

A 74.16

B 79.53

C 90.14

D 127.50

☐ Foundation ○ Standard ○ Complex

5 Find the value of each variable, correct to two decimal places.

a

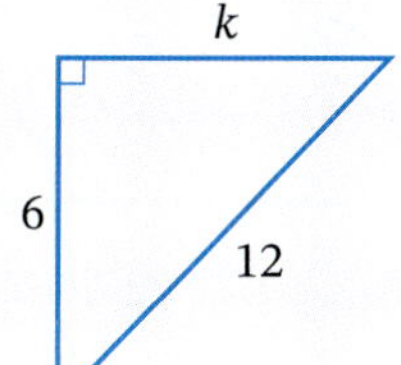

b

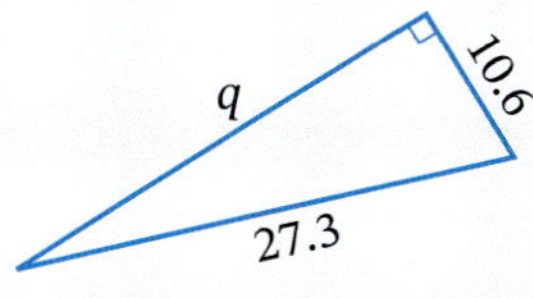

c

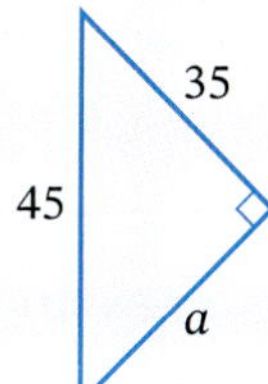

6 Find the length of the unknown side of a right-angled triangle with a hypotenuse of length 32.5 cm and one of the short sides of length 12.5 cm.

7 Birrani walked 1300 m in a north-easterly direction from his campsite to a meeting place 1200 m directly north of a waterhole. How far was his original campsite from the waterhole?

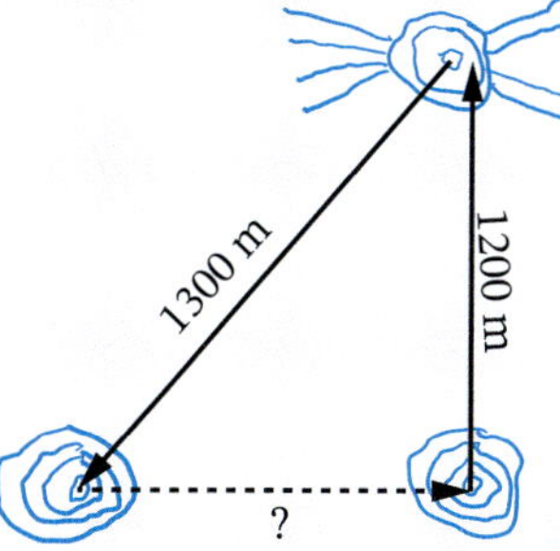

8 What could be the lengths of the other two sides of a right-angled triangle if:

R **a** one of the shorter sides is $\sqrt{7}$ cm long?

b one of the shorter sides is 7 cm long?

c the hypotenuse is $\sqrt{20}$ cm long?

d the hypotenuse is 20 cm long?

TECHNOLOGY

Finding a shorter side

In this activity, we will create a spreadsheet to calculate the length of a shorter side, a, of a right-angled triangle, given the lengths of the hypotenuse (c) and the other side (b).

b

a

c

1 Enter the headings and values shown below.

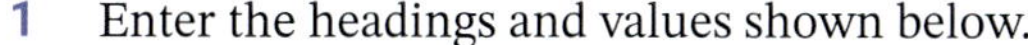

	A	B	C	D
1	c	b	c^2 - b^2	a
2	25	20		
3	37	12		
4	11	7		
5	29	21		

2 Enter a formula into cell C2 to calculate $c^2 - b^2$, which is 225, and use **Fill Down** to copy this formula into cells C3 to C5.

Foundation Standard Complex

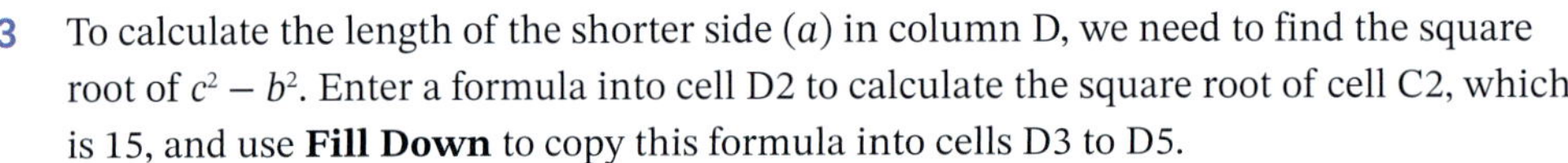

3 To calculate the length of the shorter side (a) in column D, we need to find the square root of $c^2 - b^2$. Enter a formula into cell D2 to calculate the square root of cell C2, which is 15, and use **Fill Down** to copy this formula into cells D3 to D5.

4 Use your spreadsheet to check your answers to Exercise 1.04.

Quiz
Mental skills 1

☆ **MENTAL SKILLS** 1 ANSWERS ON P. 613 **Maths without calculators**

Squaring a number ending in 5 or 1

The square of a number ending in 5 always ends in 25.

For example, $35^2 = 1225$ and $105^2 = 11\,025$.

A mental calculation trick requires three easy steps:

- delete 5 from the number.
- multiply the remaining number by the next consecutive number.
- write '25' at the end of the product.

1 Study each example.

a To calculate 35^2:

- deleting the 5 from 35 leaves 3.
- multiply 3 by the next consecutive number: $3 \times 4 = 12$.
- write '25' at the end: 1225 $35^2 = 1225$.

b To calculate 105^2:

- deleting the 5 from 105 leaves 10.
- multiply 10 by the next consecutive number: $10 \times 11 = 110$.
- write 25 at the end: 11 025 $105^2 = 11\,025$.

2 Now calculate each square number.

a	25^2	**b**	55^2	**c**	45^2	**d**	85^2
e	115^2	**f**	7.5^2	**g**	95^2	**h**	195^2
i	1.5^2	**j**	65^2	**k**	155^2	**l**	245^2

The square of a number ending in 1 always ends in 1.

For example, $41^2 = 1681$ and $71^2 = 5041$.

A mental calculation trick requires three steps:

- round the number down to the nearest 10 (by subtracting 1) to make a new number.
- square the new number.
- to your answer, add the new number and its next consecutive number.

3 Study each example.

a To calculate 41^2:
- round 41 down to 40
- square 40: $40^2 = 1600$
- add 40 and 41: $1600 + 40 + 41 = 1681$

$41^2 = 1681$

b To calculate 71^2:
- round 71 down to 70
- square 70: $70^2 = 4900$
- $4900 + 70 + 71 = 5041$

$71^2 = 5041$

4 Now calculate each square number.

a 21^2 **b** 101^2 **c** 31^2 **d** 91^2

e 5.1^2 **f** 81^2 **g** 61^2 **h** 201^2

i 1.1^2 **j** 4.1^2

Hypotenuse or shorter side? 1.05

Let's solve some mixed problems that require using Pythagoras' theorem. To find the length of any unknown side in a right-angled triangle, follow these steps.

1 Decide whether the unknown side is the **hypotenuse** or one of the **shorter sides**.

2 To find the hypotenuse, use $c^2 = a^2 + b^2$ and *add*.

To find a shorter side, use $a^2 = c^2 - b^2$ and *subtract*.

3 If a **surd answer** is required, leave the answer in square root ($\sqrt{\ }$) form. Otherwise, calculate the value correct to the required number of decimal places.

Skillsheet Pythagoras' theorem

Puzzles Pythagoras 1

Pythagoras 2

Example 9

Find the value of x as a surd.

SOLUTION

x is the hypotenuse.

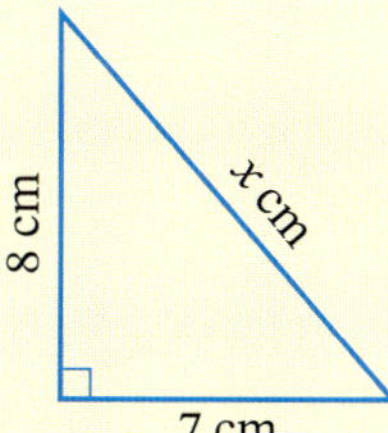

$c^2 = a^2 + b^2$

$x^2 = 7^2 + 8^2$

$= 113$

$x = \sqrt{113}$ As a surd.

Example 10

Find the value of n, correct to one decimal place.

SOLUTION

n is a shorter side.

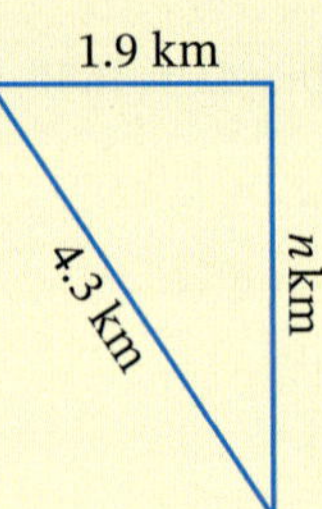

$$a^2 = c^2 - b^2$$

$$n^2 = 4.3^2 - 1.9^2$$ (Hypotenuse)^2 – (other side)^2

$$= 14.88$$

$$n = \sqrt{14.88}$$ Find the square root.

$$= 3.85746\ldots$$

$$\approx 3.9$$

EXERCISE 1.05 ANSWERS ON P. 614

Hypotenuse or shorter side?

U F R C

1 For each triangle, state whether the unknown side is the hypotenuse (H) or one of the shorter sides (S).

R C

a

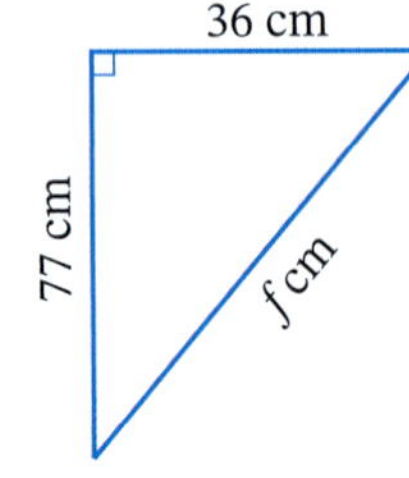

b

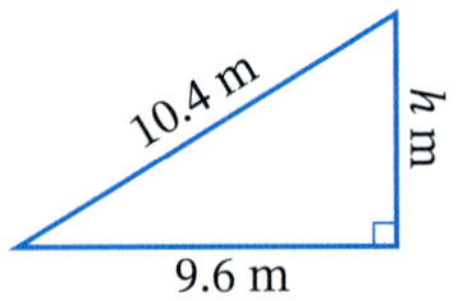

c

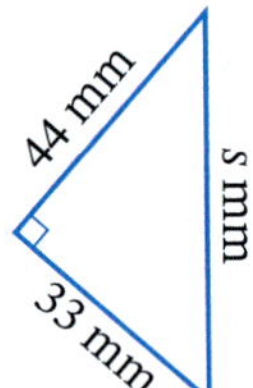

d

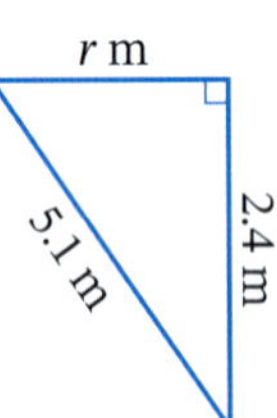

e

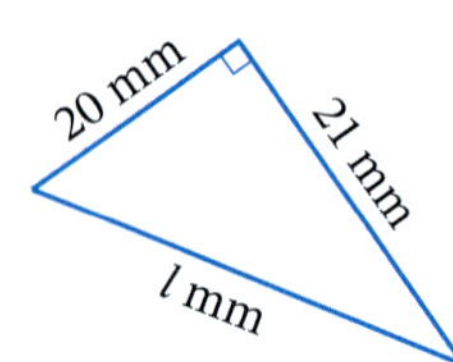

f

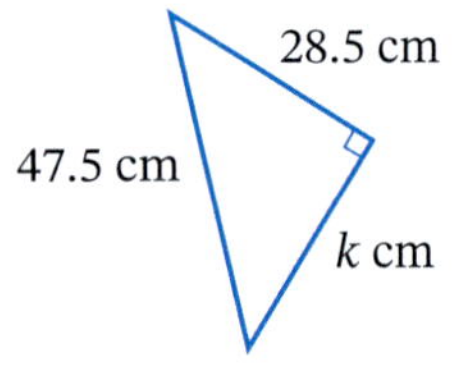

▷

□ Foundation ○ Standard ○ Complex

2 Find the value of each variable in each triangle in question **1**.

3 Find the value of each variable as a surd.

a

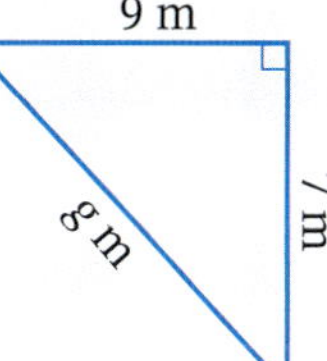

b

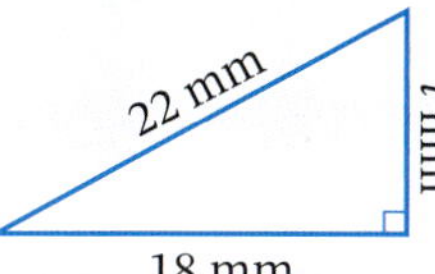

c

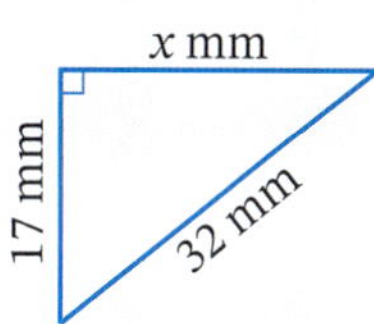

4 Find the value of each variable correct to one decimal place.

a

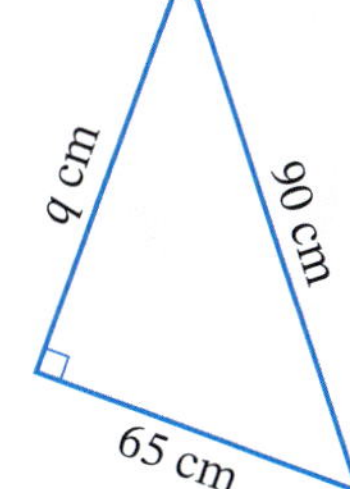

b

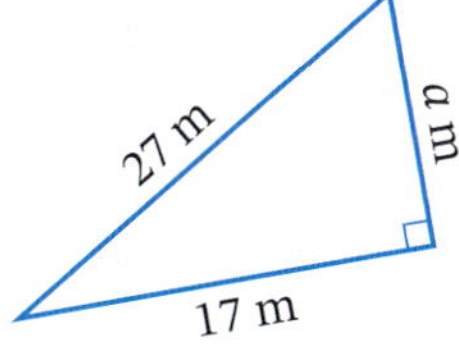

c

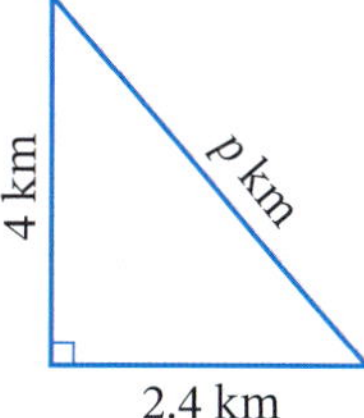

5 What is the value of y? Select the closest answer **A**, **B**, **C** or **D**.

A 25.29 **B** 25.30

C 30.47 **D** 30.46

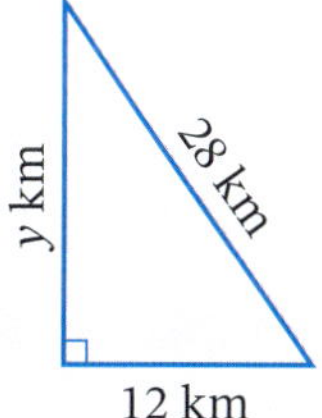

6 Find the value of each variable correct to two decimal places.

a

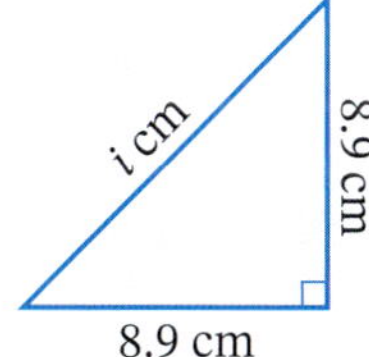

b

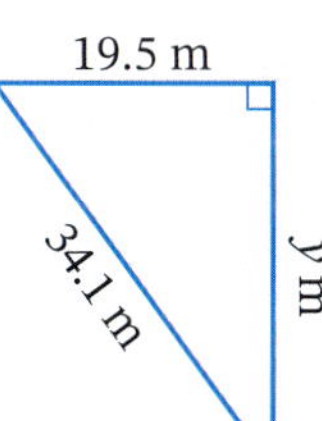

c

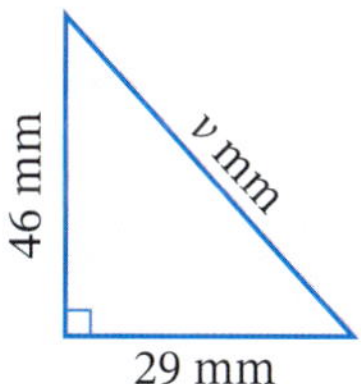

7 Find the value of each variable as a surd.

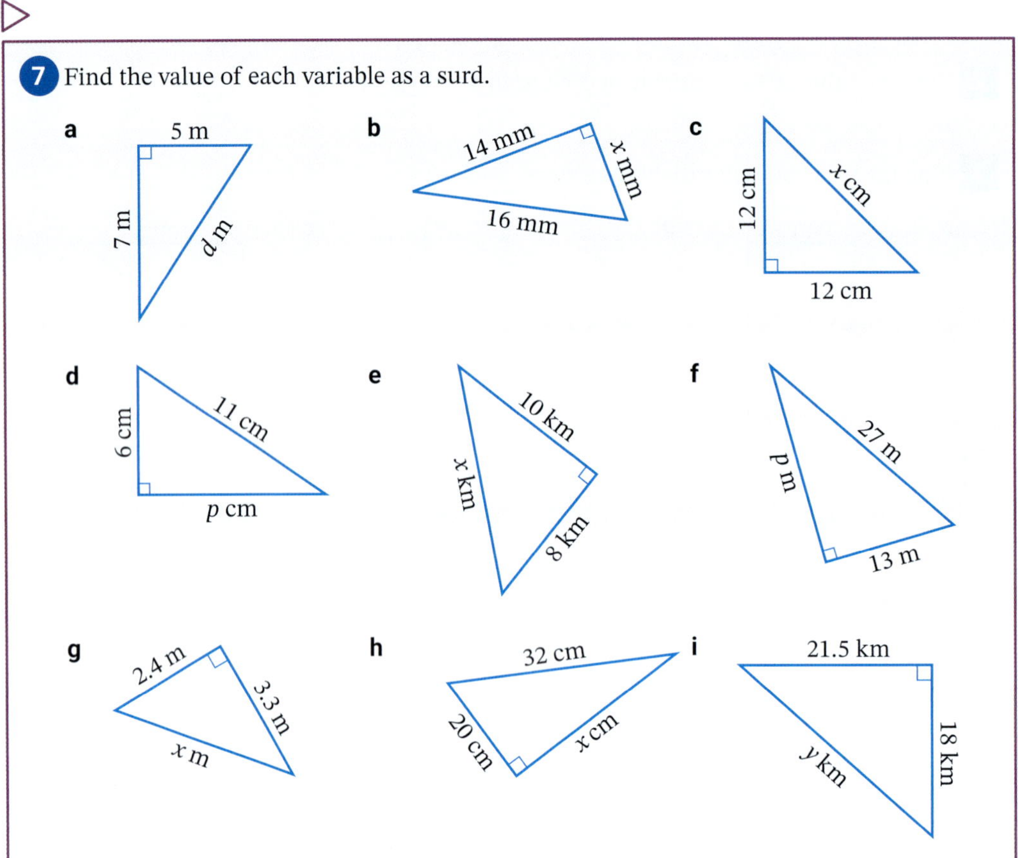

1.06 Testing for right-angled triangles

Technology
Pythagoras' theorem tester

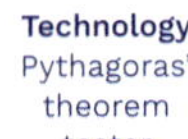

Video
Building the pyramids

Pythagoras' theorem can be used to test whether a triangle is right-angled.

Testing for right-angled triangles

If the sides of a triangle have lengths a, b and c, where c is the largest, and they follow the formula $c^2 = a^2 + b^2$, then the triangle must be right-angled.

The right angle is opposite the longest side, c.

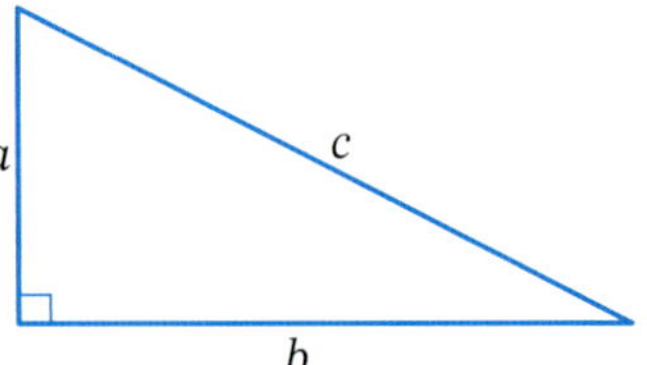

This is called the **converse** of Pythagoras' theorem. The word 'converse' means to 'turn around' or 'reverse'. So the converse of a theorem is the theorem written 'back-to-front'.

Example 11

Test whether each triangle is right-angled.

a

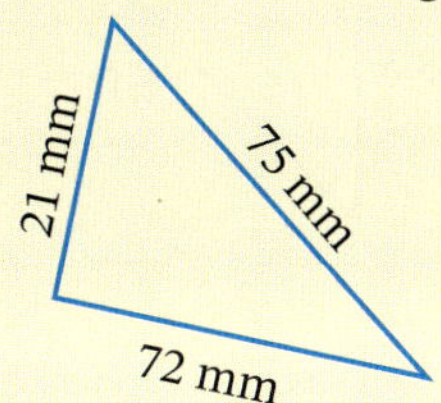

b

15 m
14 m
6 m

Video Testing for right-angled triangles

SOLUTION

a $75^2 = 5625$ — Squaring the longest side.

$21^2 + 72^2 = 5625$ — Squaring the two shorter sides, and adding.

$\therefore 75^2 = 21^2 + 72^2$ — The three sides follow $c^2 = a^2 + b^2$.

$\therefore$ The triangle is right-angled (with the right angle opposite the 75 mm side).

$\therefore$ is the symbol for 'therefore'

b $15^2 = 225$ — Squaring the longest side.

$6^2 + 14^2 = 232$ — Squaring the two shorter sides, and adding.

$\therefore 15^2 \neq 6^2 + 14^2$ — The three sides do not follow $c^2 = a^2 + b^2$.

$\therefore$ The triangle is not right-angled.

EXERCISE 1.06 ANSWERS ON P. 614

Testing for right-angled triangles

U F PS R C

1 Copy each triangle, test whether it is right-angled, then mark the right angle on the triangle. EXAMPLE 11

R **a**

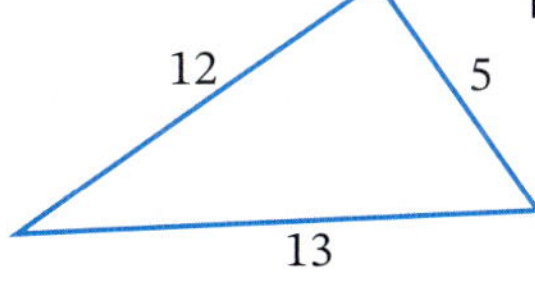

b

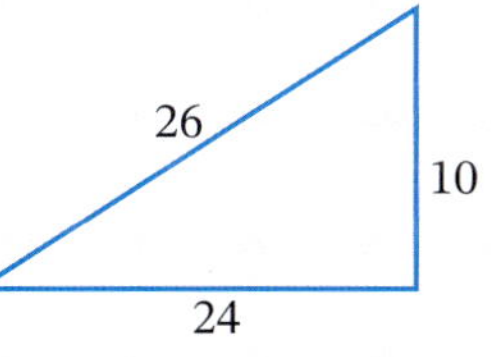

c

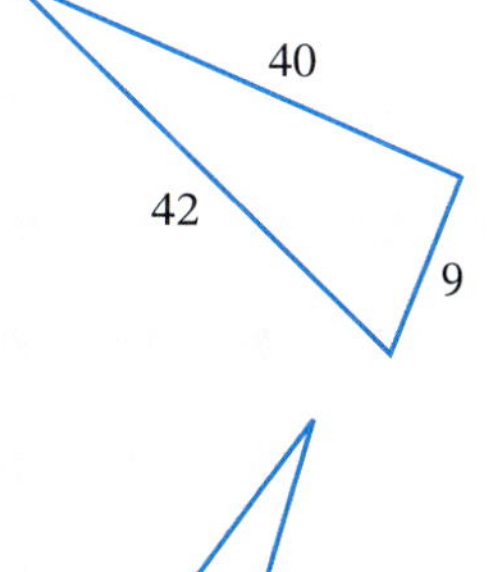

d

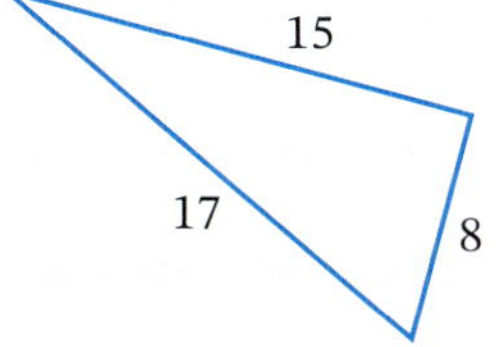

e

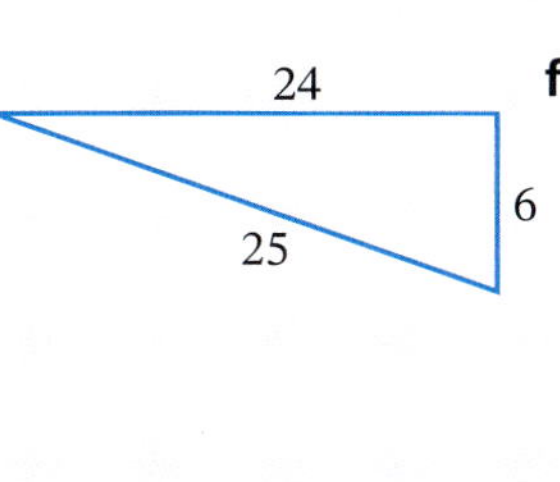

f

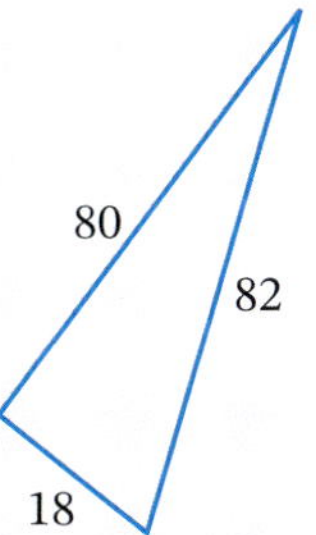

Foundation Standard Complex

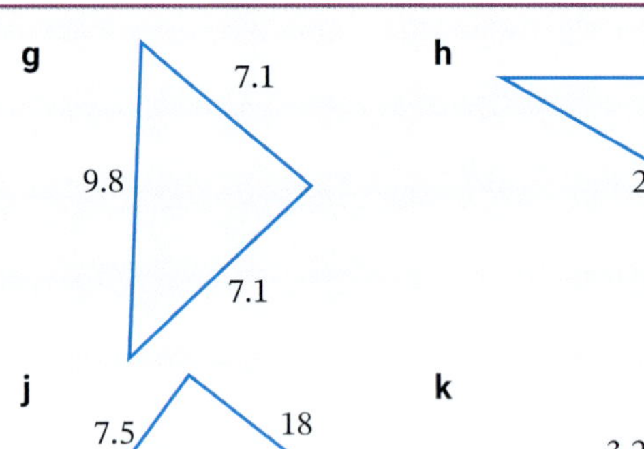

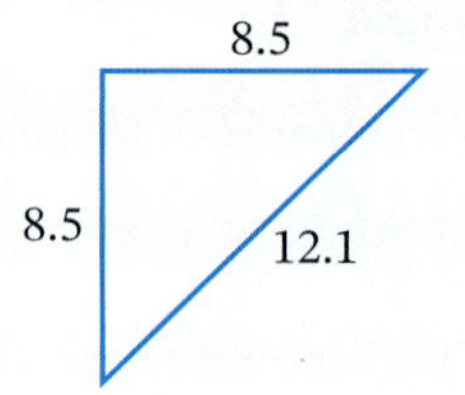

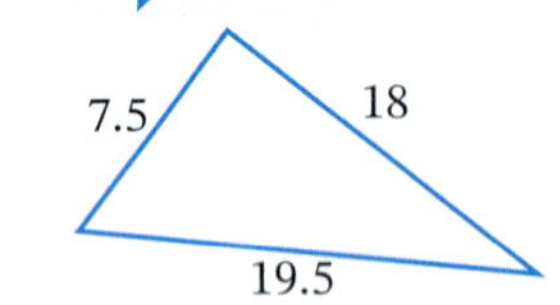

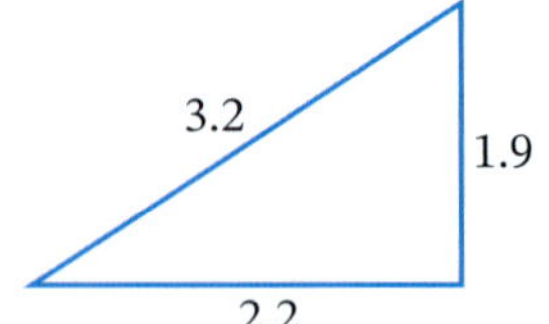

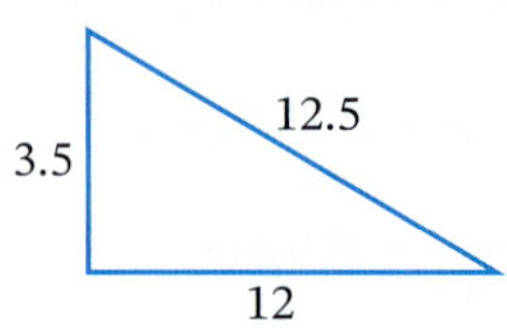

2 PS R C Jane and Mark are building the framework for a rectangular concrete slab. They want to check if it has right angles at the corners. They measure the sides and the diagonal. The sides are 3.6 m and 4.8 m. The diagonal is 6 m.

a Draw a diagram showing this information.

b Have they constructed their framework with right angles at the corners? Justify your answer with a calculation.

TECHNOLOGY

Pythagoras' theorem tester

In this activity, we will create a spreadsheet to calculate the square of the longest side of a triangle (c), then calculate the sum of the squares of the two shorter sides (a and b) of the triangle and test to see whether the two values are the same.

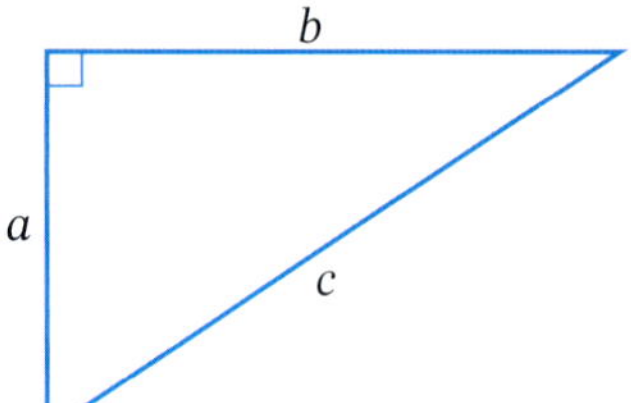

1 Enter the headings and values shown below.

	A	B	C	D	E
1	c	c^2	a	b	a^2+b^2
2	13		5	12	
3	26		10	24	
4	42		9	40	
5	17		8	15	
6	25		6	24	
7	82		18	80	
8	9.8		7.1	7.1	
9					

2 Enter a formula into cell B2 to calculate c^2, which is 169, and use **Fill Down** to copy this formula into cells B3 to B8.

3 To calculate the sum of the squares of the other two sides (a *and* b) in column E, we need to find $a^2 + b^2$. Enter a formula into cell E2 to calculate the square of cell C2 added to the square of cell D2, which is 169, and use **Fill Down** to copy this formula into cells E3 to E8.

4 If the number in column B matches the number in column E the triangle is a right-angled.

You can enter an 'if' rule to check this for you in column F.

5 Use your spreadsheet to check your answers to Exercise 1.06.

Pythagorean triads 1.07

A **Pythagorean triad** or **Pythagorean triple** is any group of three numbers that follow Pythagoras' theorem, for example, (3, 4, 5) or (2.5, 6, 6.5). The word **triad** means a group of three related items ('tri-' means 3).

Technology
Pythagorean triples

Spreadsheet
Pythagorean triples

ⓘ Pythagorean triads

(a, b, c) is a **Pythagorean triad** if $c^2 = a^2 + b^2$.

Any multiple of (a, b, c) is also a Pythagorean triad.

Example 12

Test whether (5, 12, 13) is a Pythagorean triad.

SOLUTION

$13^2 = 169$	Squaring the largest number.
$5^2 + 12^2 = 169$	Squaring the 2 smaller numbers and adding them.
$\therefore 13^2 = 5^2 + 12^2$	These 3 numbers follow Pythagoras' theorem.
$\therefore$ (5, 12, 13) is a Pythagorean triad.	

Example 13

{3, 4, 5} is a Pythagorean triad. Create other Pythagorean triads by multiplying (3, 4, 5) by:

a 2 **b** 9 **c** $\frac{1}{2}$

SOLUTION

a $2 \times (3, 4, 5) = (6, 8, 10)$

Checking: $10^2 = 100$

$6^2 + 8^2 = 100$

$\therefore 10^2 = 6^2 + 8^2$

$\therefore$ (6, 8, 10) is a Pythagorean triad.

b $9 \times (3, 4, 5) = (27, 36, 45)$

Checking: $45^2 = 2025$

$27^2 + 36^2 = 2025$

$\therefore 45^2 = 27^2 + 36^2$

$\therefore$ (27, 36, 45) is a Pythagorean triad.

c $\frac{1}{2} \times (3, 4, 5) = (1.5, 2, 2.5)$

Checking: $2.5^2 = 6.25$

$1.5^2 + 2^2 = 6.25$

$\therefore 2.5^2 = 1.5^2 + 2^2$

$\therefore$ (1.5, 2, 2.5) is a Pythagorean triad.

EXERCISE 1.07 ANSWERS ON P. 614

Pythagorean triads

1 Test whether each triad is a Pythagorean triad.

R **a** (8, 15, 17) **b** (10, 24, 26) **c** (30, 40, 50)

d (5, 7, 9) **e** (9, 40, 41) **f** (4, 5, 9)

g (11, 60, 61) **h** (7, 24, 25) **i** (15, 114, 115)

2 Which of the following is a Pythagorean triad? Select the correct answer **A**, **B**, **C** or **D**.

R **A** (4, 6, 8) **B** (5, 10, 12) **C** (6, 7, 10) **D** (20, 48, 52)

3 If you used the technology worksheet **Pythagoras' theorem tester** on page 20 to create a spreadsheet, then use it to check your answers to questions **1** and **2**.

4 For each Pythagorean triad, create another Pythagorean triad by multiplying each number in the triad by:

R

i a whole number **ii** a fraction **iii** a decimal

a (5, 12, 13) **b** (8, 15, 17) **c** (30, 40, 50) **d** (7, 24, 25)

Check that each answer follows Pythagoras' theorem.

Foundation Standard Complex

5 (R) Pythagoras developed a formula for finding Pythagorean triads (a, b, c). If one number in the triad is a, the formulas for the other two numbers are $b = \frac{1}{2}(a^2 - 1)$ and $c = \frac{1}{2}(a^2 + 1)$.

a If $a = 5$, use the formulas to find the values of b and c.

b Hence show that (a, b, c) is a Pythagorean triad.

6 Use the formulas to find Pythagorean triads for each value of a.

a $a = 7$ **b** $a = 11$ **c** $a = 15$ **d** $a = 4$

e $a = 9$ **f** $a = 19$ **g** $a = 10$ **h** $a = 51$

7 There are many other formulas for creating Pythagorean triads. Use the Internet to search for some of them and test that they are correct.

Pythagoras' theorem problems 1.08

Pythagoras' theorem has many practical applications, from finding the diagonal length of a TV screen to calculating the distance a ship has sailed.

Worksheets
Pythagoras' theorem in 2D and 3D

TV screens

Example 14

The size of a TV screen is described by the length of its diagonal. If a flat-screen TV is 89 cm wide and 50 cm high, what is the size of its screen?

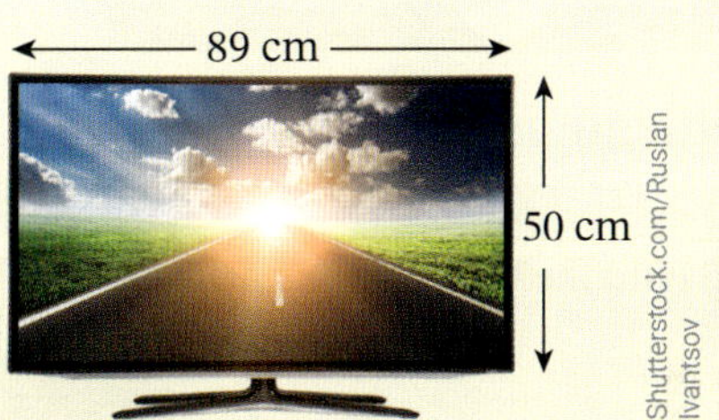

Shutterstock.com/Ruslan Ivantsov

Answer to the nearest centimetre.

SOLUTION

Let the diagonal length be d cm.

$d^2 = 89^2 + 50^2$ — d is the hypotenuse.

$= 10\,421$

$d = \sqrt{10\,421}$

$= 102.0832\ldots$

≈ 102

The size of the screen is 102 cm.

Foundation Standard Complex

Video
Pythagoras' theorem problems

Example 15

A tower is supported by a wire that is 20 m long and attached to the ground 10 m from the base of the tower. How high does the wire reach up the tower? Answer correct to the nearest 0.1 m.

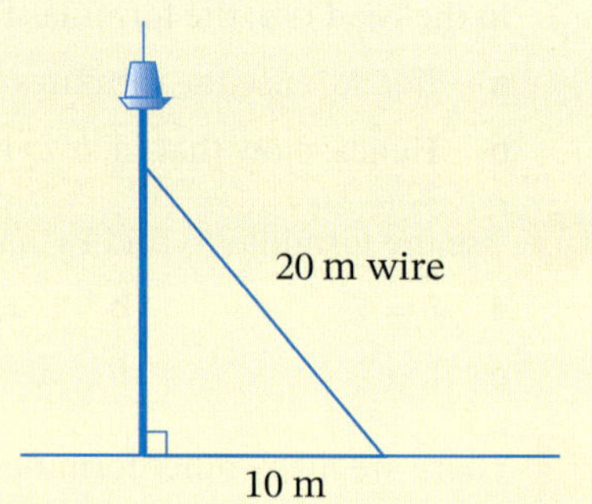

SOLUTION

Let h m be how high the wire reaches up the tower.

$h^2 = 20^2 - 10^2$ h is a shorter side.

$= 300$

$h = \sqrt{300}$

$= 17.3205$

≈ 17.3

The wire reaches 17.3 m up the tower.

EXERCISE 1.08 ANSWERS ON P. 614

Pythagoras' theorem problems

U F PS R

EXAMPLE 14

1. Find the length of this playground slide, correct to two decimal places.

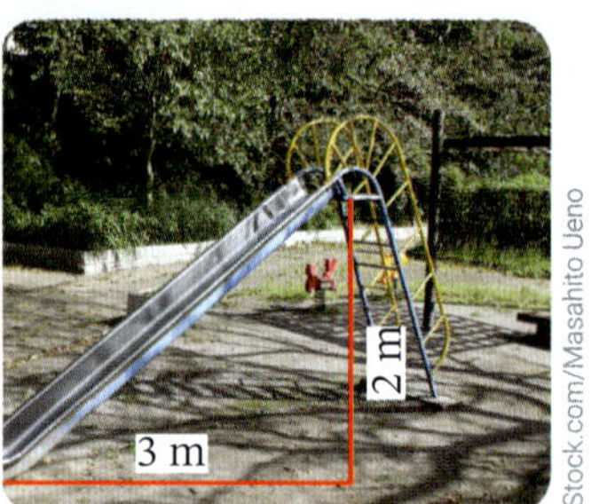

iStock.com/Masahito Ueno

2. A ship sails 60 km south and then 70 km east. How far is it from its starting point, correct to one decimal place?

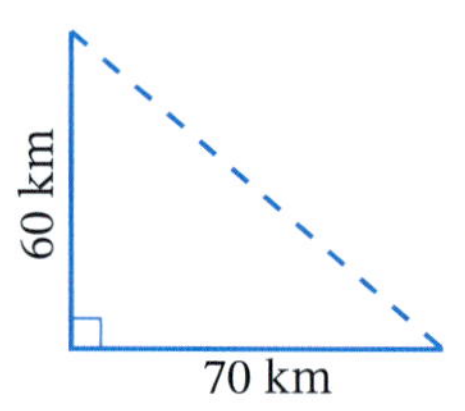

3 What is the length of a cable used to stabilise a flagpole that is 12 m high, if the cable is secured to the ground 4 m from the base of the flagpole? Select the closest answer **A**, **B**, **C** or **D**.

A 11.3 m **B** 12.6 m

C 16 m **D** 80 m

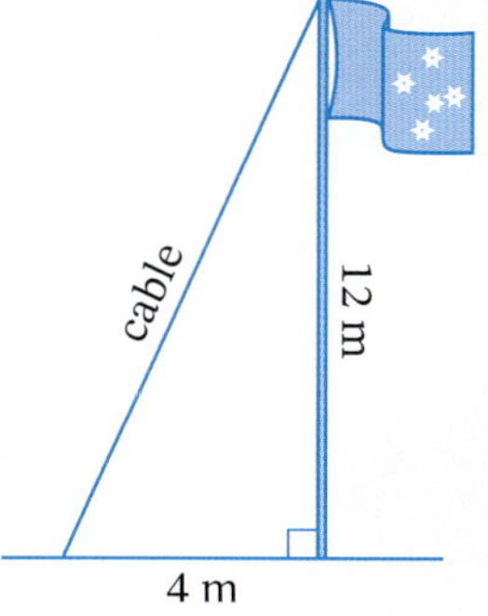

4 Find the length of the diagonal of a square with sides of length 6 m. Give your answer as a surd.

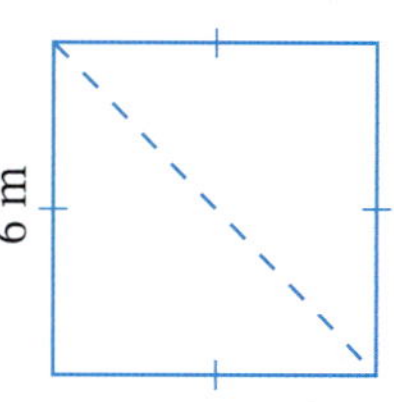

5 A firefighter places a ladder on a window sill 9.5 m above the ground. If the foot of the ladder is 1.8 m from the wall, how long is the ladder, correct to one decimal place?

Alamy Stock Photo/Willows Photos UK

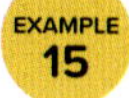
EXAMPLE 15

6 Find the height of a TV screen with a 140 cm diagonal if its length is 122 cm. Give your answer correct to one decimal place.
PS

Foundation Standard Complex

7 By first using Pythagoras' theorem to find the length of the unknown side, find the perimeter of each shape below, correct to two decimal places where necessary.

a

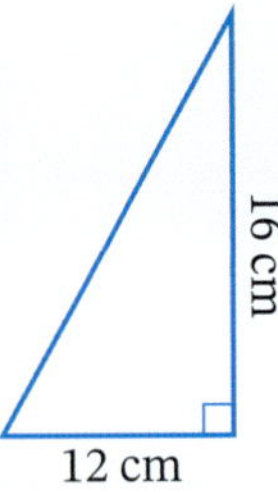

b

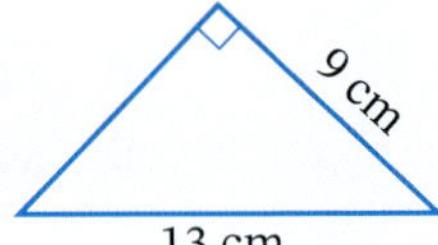

c

d

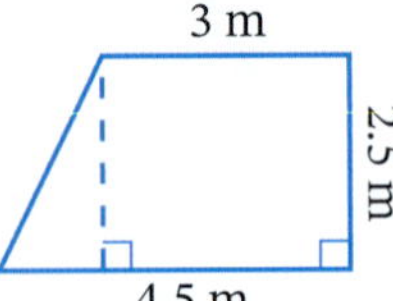

e

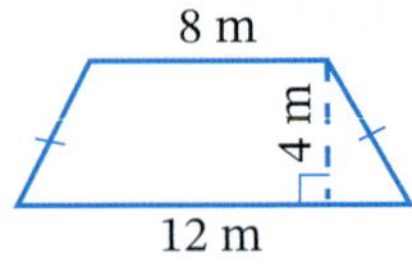

8 PS R An equilateral triangle has sides of length 12 cm. Find its perpendicular height, h, correct to two decimal places.

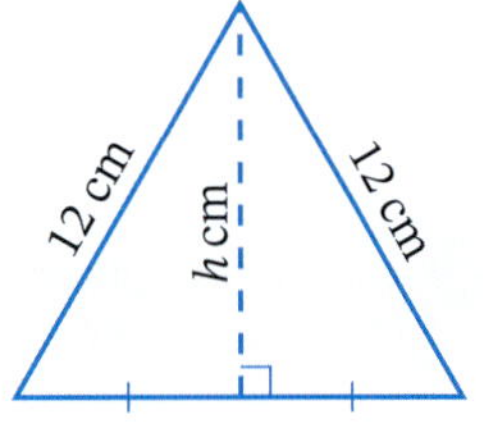

9 PS R A rope is tied to the top of the 6 m wall of a tent and tied to a peg in the ground. The peg is 2 m from the bottom of the tent. How long is the rope, correct to two decimal places?

10 PS R Olga holds a kite string 1.2 m above the ground. How high is the kite above the ground, to the nearest metre?

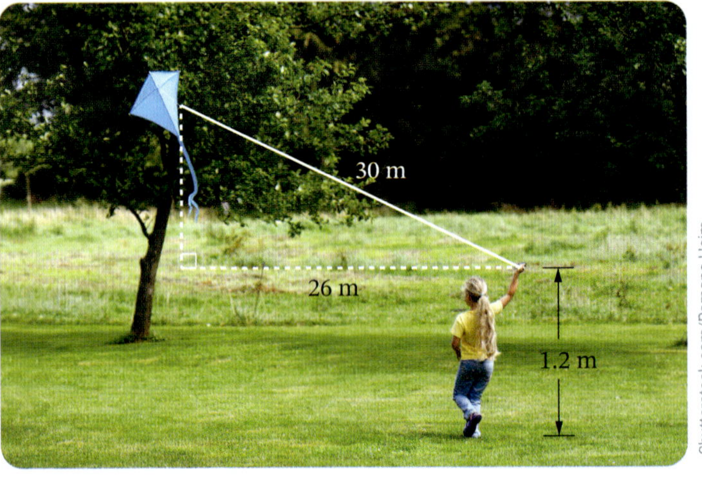

Shutterstock.com/Ramona Heim

Foundation Standard Complex

11 Find the area of each shape below, correct to one decimal place where necessary.

R

a

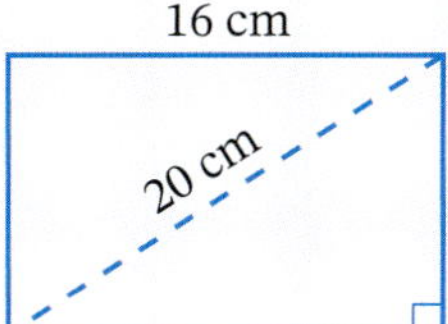

b

c

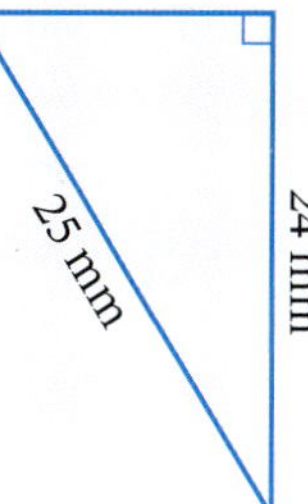

d

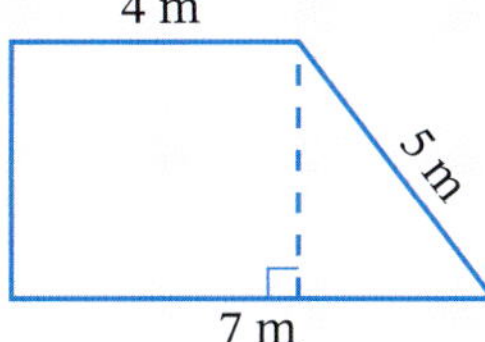

e

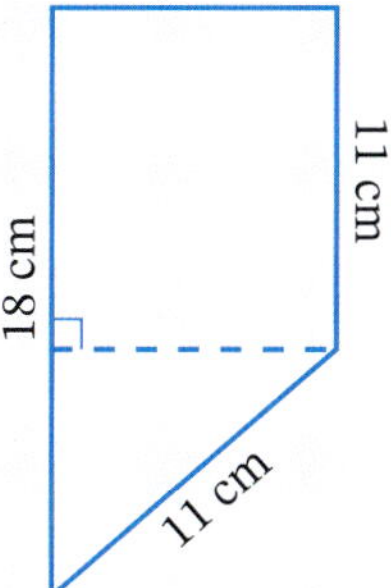

12 This baseball diamond is a square shape of length 27 m. What is the distance from the home plate to second base, correct to the nearest metre?

Shutterstock.com/Creative Droneworks

13 The slant height of a cone is 27 cm and its vertical height is 14 cm. Find the radius of the base circle, correct to the nearest centimetre.

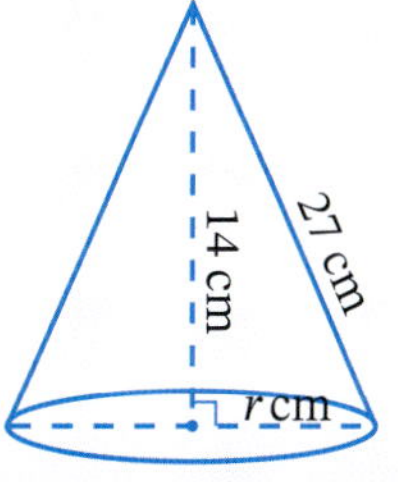

14 A ladder 5 m long leans against a wall, with its base 2 m from the bottom of the wall. How far does the ladder reach up the wall, correct to nearest centimetre?

PS

R

☐ Foundation ○ Standard ⬡ Complex

POWER PLUS ANSWERS ON P. 614

1 Find the value of x in each triangle, correct to one decimal place.

a
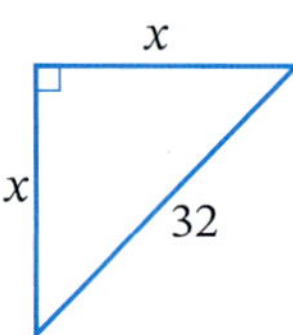

b
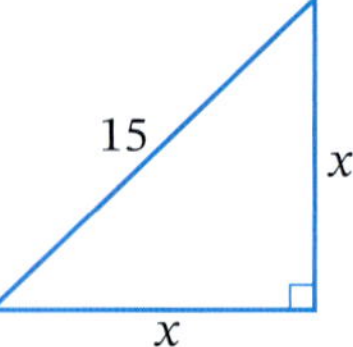

c
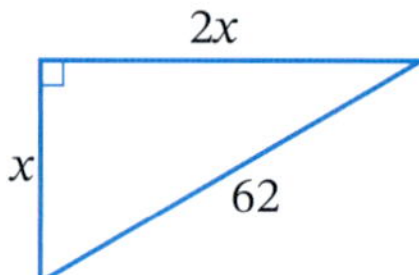

d
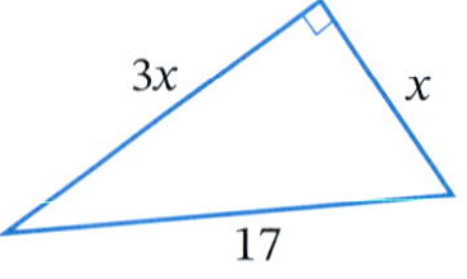

e
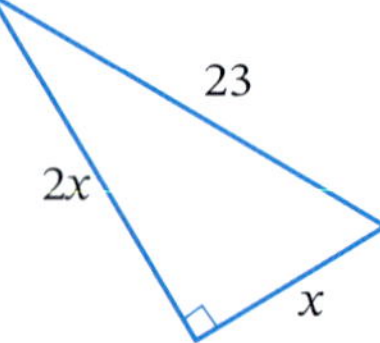

f
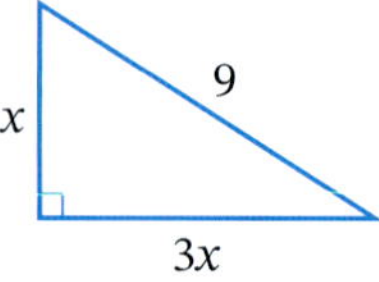

2 Find, correct to one decimal place, the length of:

a HD

b DE

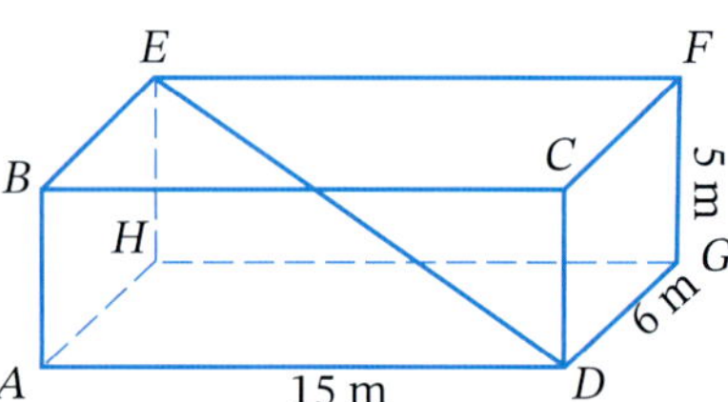

3 For this cube of side length 15 cm, find, correct to one decimal place, the length of:

a QS, the diagonal of the base.

b QT, the diagonal of the cube.

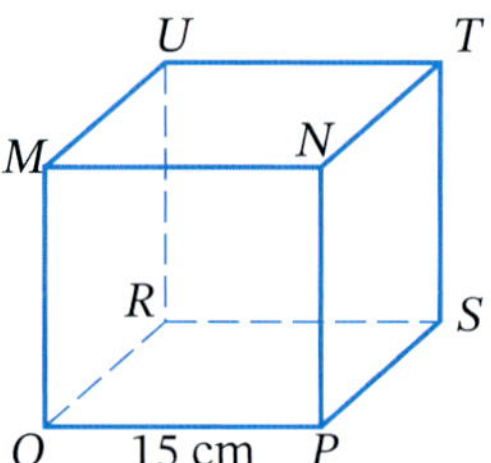

4 For this square pyramid, find the slant height EF correct to one decimal place.

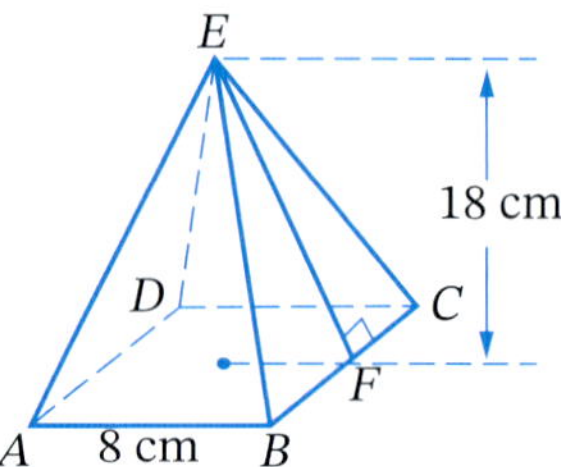

① CHAPTER REVIEW

Language of maths

Quiz Language of maths 1

area	converse	diagonal	formula
hypotenuse	irrational	perimeter	Pythagoras
right-angled	shorter	side	square root
surd	theorem	triad	unknown

1 Who was **Pythagoras** and which country did he come from?

2 Describe the **hypotenuse** of a right-angled triangle in two ways.

3 What is another word for 'theorem'?

4 For what type of triangle is **Pythagoras' theorem** used?

5 What is a **surd**?

6 What is the name given to a set of three numbers that follows Pythagoras' theorem?

Topic summary

Worksheet Mind map: Pythagoras' theorem

- How relevant do you think Pythagoras' theorem is to our world? Give reasons for your answer.

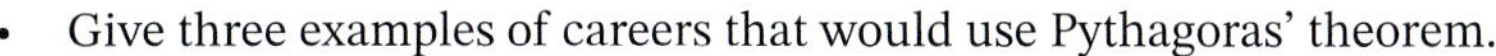

- Give three examples of careers that would use Pythagoras' theorem.

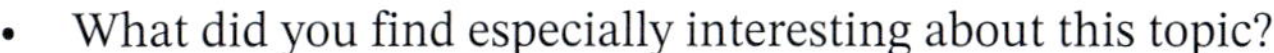

- What did you find especially interesting about this topic?

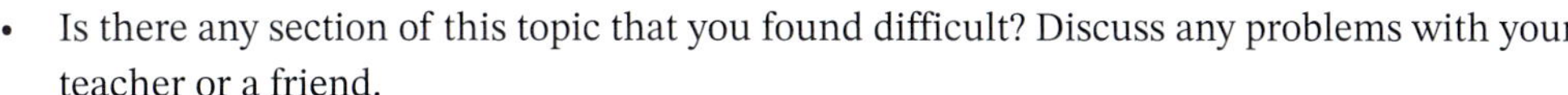

- Is there any section of this topic that you found difficult? Discuss any problems with your teacher or a friend.

Print (or copy) and complete this mind map of the topic, adding detail to its branches and using pictures, symbols and colour where needed. Ask your teacher to check your work.

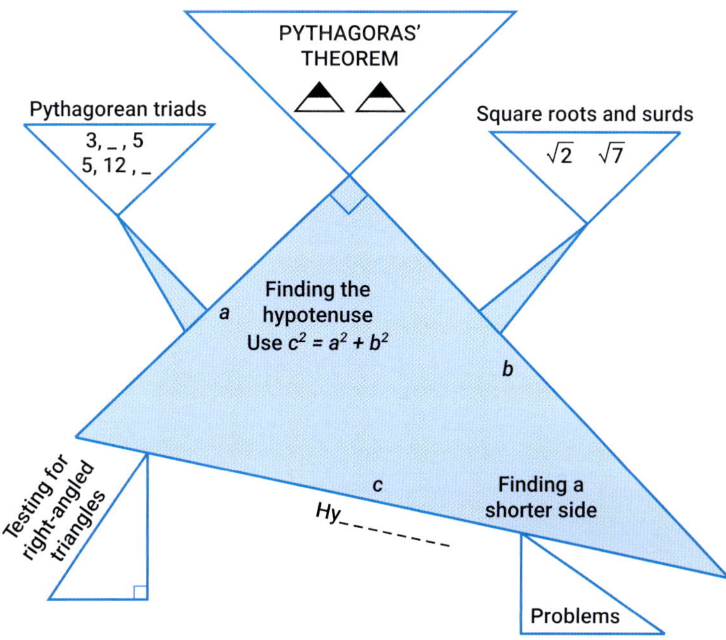

1 TEST YOURSELF

ANSWERS ON P. 614

1 Evaluate each expression, correct to two decimal places.

a $\sqrt{203}$ **b** $\sqrt{15^2+6^2}$ **c** $\sqrt{7^2-4^2}$ **d** $\sqrt{3.4^2+5.2^2}$

Quiz Test yourself 1

2 Select all the surds from this list of square roots.

$\sqrt{25}$ $\sqrt{104}$ $\sqrt{96}$ $\sqrt{256}$ $\sqrt{12}$

$\sqrt{169}$ $\sqrt{121}$ $\sqrt{45}$ $\sqrt{729}$ $\sqrt{88}$

3 For each right-angled triangle, name the hypotenuse and write Pythagoras' theorem.

a

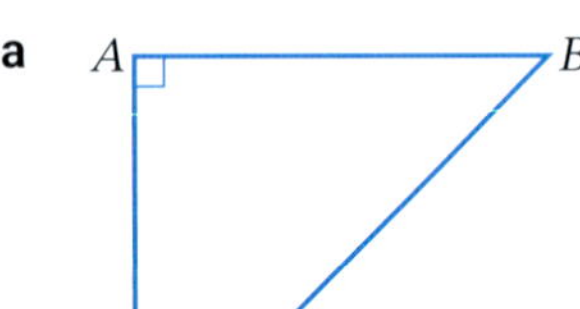

b

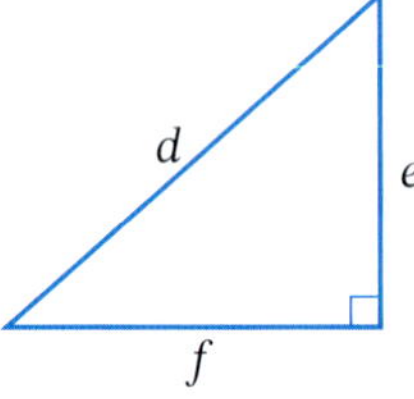

c

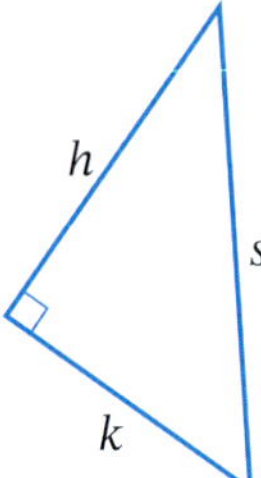

1.03

4 Find the value of each variable, giving your answer as a surd.

a

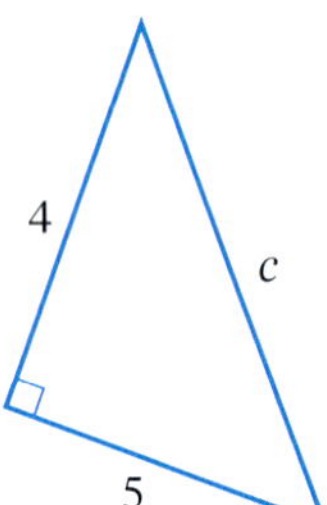

b

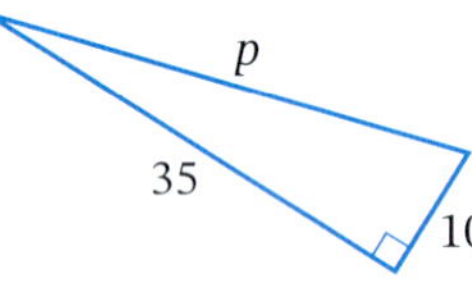

c

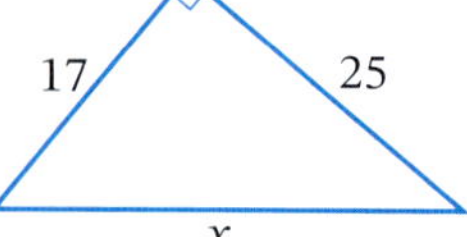

1.04

5 Find the value of each variable, giving your answer correct to two decimal places.

a

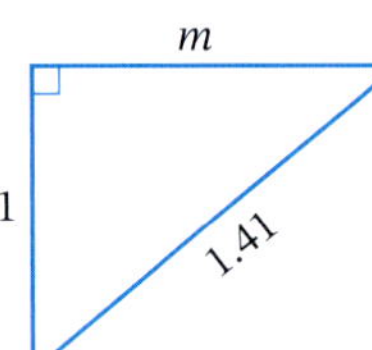

b

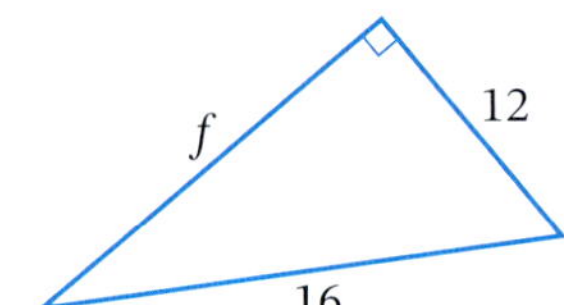

c

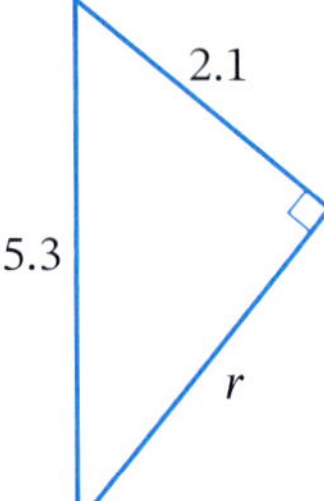

☐ Foundation ○ Standard ⬡ Complex

6 Find the value of each variable, correct to one decimal place. 

a 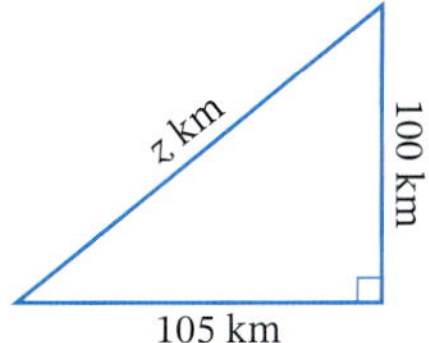

b 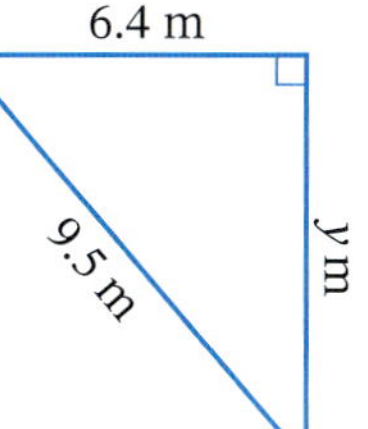

c 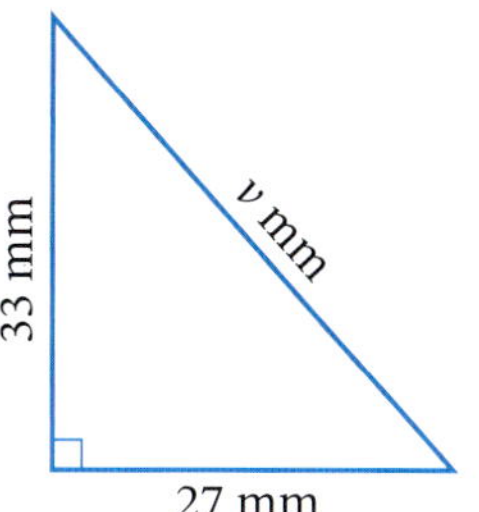

7 Test whether each triangle is right-angled. If the triangle is right-angled, sketch it showing the right angle. 

a 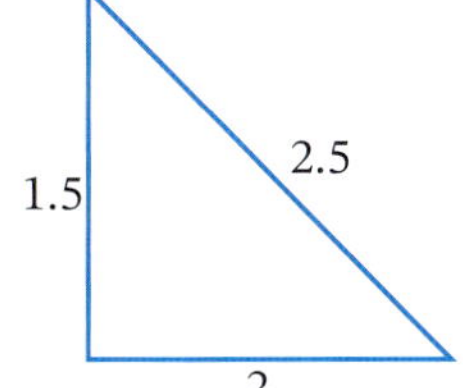

b 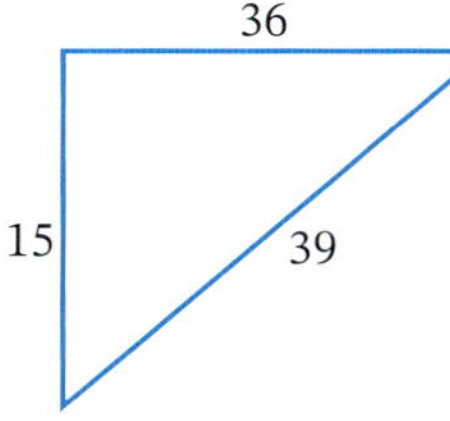

c 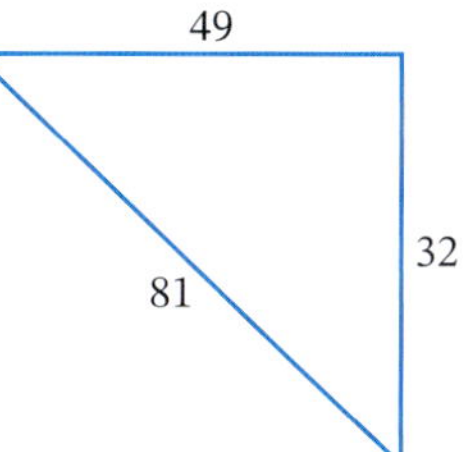

8 Test whether each triad is a Pythagorean triad. 

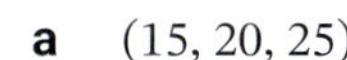 **a** (15, 20, 25) **b** (11, 14, 20)  **c** (20, 21, 29) **d** (2.5, 6, 6.5)

9 Find the length of the longest pencil that can fit inside this pencil case. Answer to the nearest millimetre. 

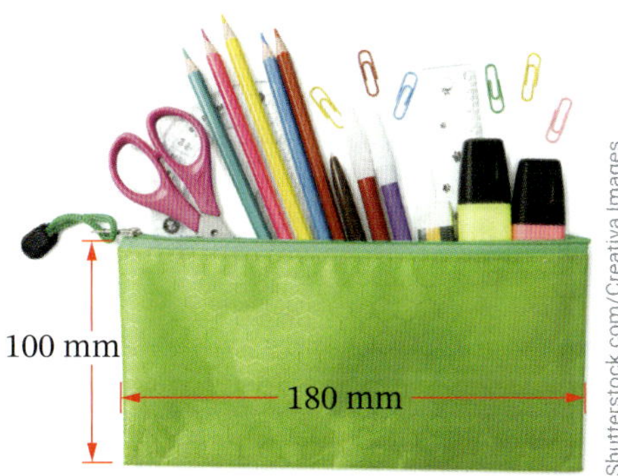

Shutterstock.com/Creativa Images

10 Find the perimeter of each shape, correct to one decimal place. 

a 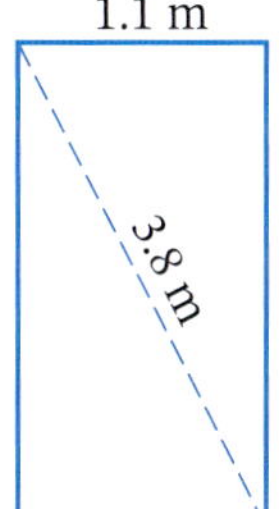

b 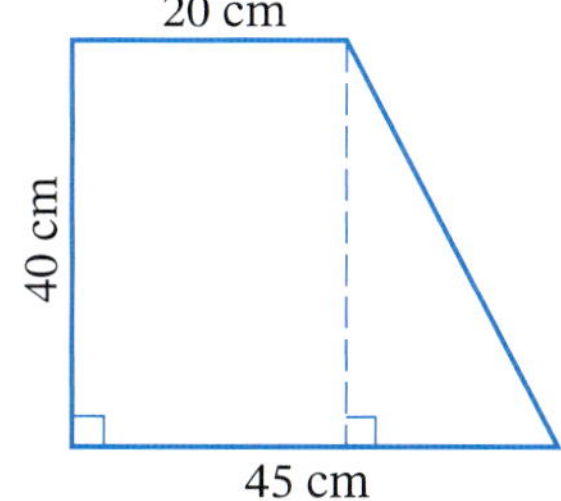

□ Foundation ○ Standard ⬡ Complex

2
NUMBER

Working with numbers

Numbers are interesting! You already know about odd and even numbers, prime and composite numbers and integers. There are also triangular and square numbers, Fibonacci numbers, perfect and amicable numbers, and complex numbers. Numbers can form special patterns such as magic squares, Pythagorean triads and Pascal's triangle. These patterns can be found in biology, probability, digital technology and software design.

In this chapter, we will revise and extend our number skills and examine the patterns involved when calculating with powers, using mental calculation, pen-and-paper methods and calculators.

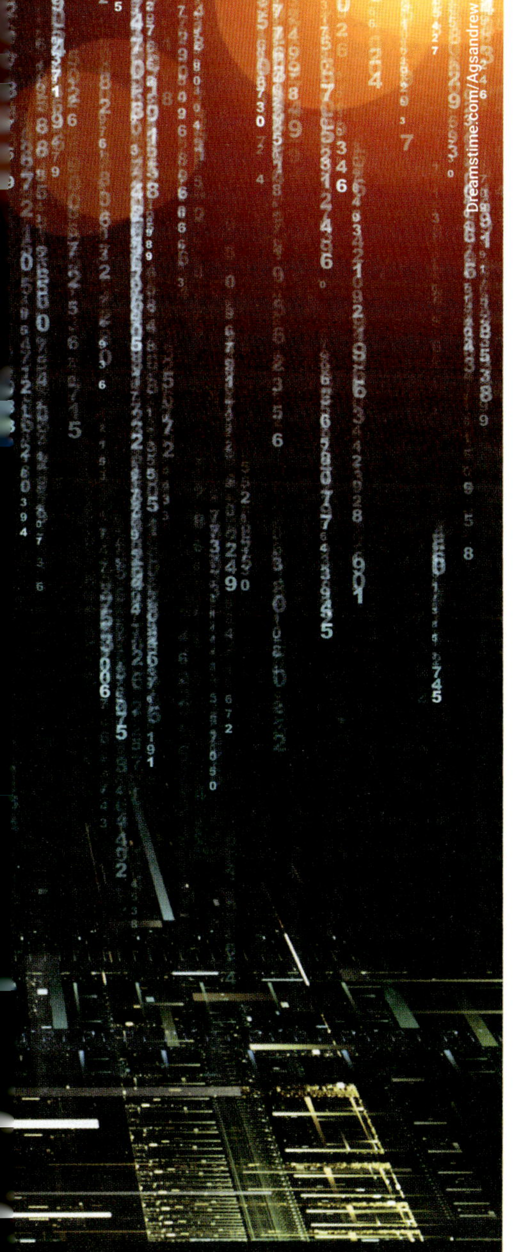

Chapter outline

		Proficiencies				
2.01	Mental calculation	U	F	PS	R	
2.02	Adding and subtracting integers	U	F	PS	R	
2.03	Multiplying integers	U	F		R	C
2.04	Dividing integers	U	F		R	C
2.05	Order of operations	U	F	PS	R	C
2.06	Decimals	U	F	PS	R	
2.07	Multiplying and dividing decimals	U	F	PS	R	
2.08	Terminating and recurring decimals	U	F		R	C
2.09	Powers and roots	U	F		R	C
2.10	Factor trees	U	F		R	C
2.11	Index laws for multiplying and dividing	U	F		R	C
2.12	More index laws	U	F		R	C
2.13	Investigating patterns in fractions and decimals	U	F		R	C

U = Understanding
F = Fluency
PS = Problem solving
R = Reasoning
C = Communication

Wordbank

Quiz
Wordbank 2

base (in index notation) A number raised to a power, for example, in 3^5, 3 is the base

cube root The value which, if cubed, will give the number required, for example, $\sqrt[3]{8} = 2$ because $2^3 = 8$

factor tree A diagram that lists the prime factors of a number

index notation A way of writing powers for the repeated multiplication of a number, for example, 3^5

mental calculation To operate with numbers 'in your head', without using a pen or calculator

order of operations The rules for calculating an expression involving mixed operations, such as $14 - 2 \times 4 + 1$

recurring decimal A decimal that has one or more digits that repeat endlessly

terminating decimal A decimal that is not recurring, but comes to an end

Videos (19):

2.01 Mental multiplication • Dividing numbers • Multiplying by 8, 9, 11 and 12
2.02 Adding integers • Subtracting integers
2.03, 2.04 Multiplying and dividing integers
2.05 Order of operations 1 • Order of operations 2 • BODMAS
2.06 Rounding decimals • Adding and subtracting decimals
2.07 Multiplying decimals • Dividing decimals
2.08 Converting fractions to decimals • Recurring decimals
2.10 Factor trees • Highest common factor • Using factor trees to find square roots
2.11 Powers, indices and exponents • The prime number code • Index laws
2.12 Index laws

***Twig* videos (7):**

2.01 Number theory: Gauss • Chinese development of maths • The Fibonacci sequence
2.02 India and negative numbers
2.04 Chinese development of maths
2.06 Rounding: Snails vs rockets
2.09 Binary: the computer language
2.10 The prime number code

***PhET* interactives (5):**

SkillCheck Number line: Integers
2.01 Arithmetic
2.02 Number line: Operations • Number line: Distance
2.07 Area model Decimals

Quizzes (6):

- Wordbank 2
- SkillCheck 2
- Mental skills 2A
- Mental skills 2B
- Language of Maths 2
- Test Yourself 2

Skillsheets (13):

SkillCheck Decimals • Fractions and decimals
2.02 Integers • Integers using diagrams • Integers using coloured squares
2.03 Integers
2.05 Order of operations • Spreadsheets
2.06 Decimals
2.08 Fractions and decimals
2.09 Index notation • Square roots and cube roots
2.10 Factors and divisibility • Prime factors by repeated division

Worksheets (21):

SkillCheck Calculation aids • Number grids
2.02 A page of number lines
2.03 Integers writing activity
2.04 What is the integer equation? • Integer review • Integers writing activity
2.06 Decimals wall • Dewey decimals • Decimals 1 • Decimals 10
2.07 Decimal cards • Shopping and change • What's the point? • Decimals 7
2.08 Fraction families
2.09 Powers and roots • Big numbers
2.10 Perfect and amicable numbers
2.12 What is the power question? • Power calculations
2.13 Fraction families
Mind map: Working with numbers

Puzzles (14):

SkillCheck Magic squares
2.01 The accidental detective
2.02 Integer Snap
2.03 Multiplying integers game
2.04 Integers group clues
2.05 Directed numbers • Order of operations puzzle
2.07 Operations with decimals • Which decimals? • Decimal number grids
2.08 Decimal squaresaw 2
2.09 Square root Snap
2.10 Crossnumber puzzles • Crossnumber challenges
2.13 Crossnumber challenges

Technology (2):

2.05 Review of spreadsheets
2.07 Movie night

Presentations (4):

2.02 Number lines • Adding directed numbers • Subtracting directed numbers
2.10 Highest common factors

Spreadsheet (2):

2.03 Integer quiz: Multiplication

To access resources above, visit
cengage.com.au/nelsonmindtap

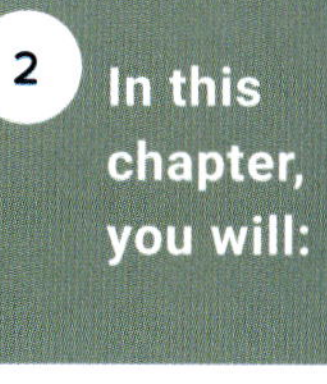

- ✓ apply number laws to help mental calculation.
- ✓ calculate with integers, decimals and fractions, using mental and written strategies and technology.
- ✓ apply the order of operations to evaluate mixed expressions.
- ✓ investigate terminating and recurring decimals.
- ✓ use computational thinking to describe number patterns.
- ✓ evaluate expressions involving powers, square root and cube root, after first estimating.
- ✓ find square roots and cube roots of any non-square whole number using a calculator, after first estimating.
- ✓ explore the properties of squares and square roots of products: $(ab)^2$ and $\sqrt{ab}$.
- ✓ use a factor tree to write a number as a product of its prime factors.
- ✓ use factor trees to find the highest common factor (HCF) and lowest common multiple (LCM) of 2 or more numbers, and to find the square root or cube root of a square number or cube number, respectively.
- ✓ explore the index laws with positive and zero powers.

SkillCheck

ANSWERS ON P. 614

1 Copy and complete each number line.

a

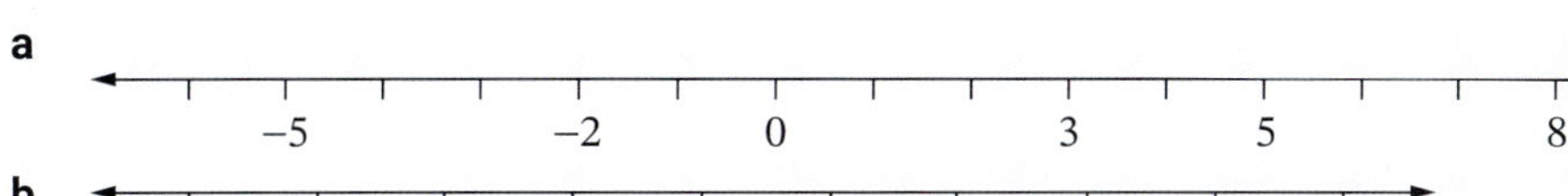

b

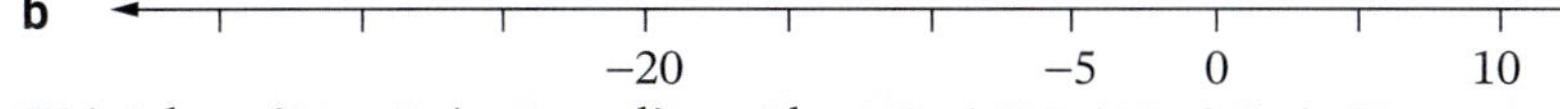

2 Write these integers in ascending order: −7, 6, −5, 11, −2, 1, 0, 7.

3 Write these integers in descending order: 12, −3, 5, −1, 2, −9, 4, 8.

4 In the decimal 2.718, name the digit in the:

a tenths place **b** thousandths place **c** hundredths place

5 Convert each decimal to a fraction.

a 0.03 **b** 0.007 **c** 0.49 **d** 0.9

6 How many decimal places have each number?

a 14.451 **b** 86.6 **c** 28.04 **d** 3.141 59

7 a List the factors of 15.

b List the first 5 multiples of 15.

8 State whether each number is prime or composite.

a 2 **b** 17 **c** 25 **d** 27

Quiz
SkillCheck 2

Skillsheets
Decimals

Fractions and decimals

Worksheets
Calculation aids

Number grids

Puzzle
Magic squares

Interactive
Number line: Integers

2.01 Mental calculation

Videos
Mental multiplication

Dividing numbers

Number theory: Gauss

Chinese development of maths

Puzzle
The accidental detective

Interactive
Arithmetic

The word 'mental' means 'using the mind' and mental calculation is the skill of working with numbers 'in your head', without using a pen or calculator. In the 'Mental skills' sections of *New Nelson Maths 7*, you learnt many mental strategies for calculating with numbers. Let's revise them now.

Example 1

Estimating answers by rounding numbers

a $15 + 37 + 18 + 45 + 22 \approx 20 + 40 + 20 + 40 + 20$ (exact answer = 137)
$= 140$

b $18 \times 12 \approx 20 \times 11$ (exact answer = 216)
$= 220$

c $504 \div 8 \approx 500 \div 10$ (exact answer = 63)
$= 50$

Example 2

Adding and multiplying numbers in any order

- The commutative laws: $a + b = b + a$ and $a \times b = b \times a$
- The associative laws: $(a + b) + c = a + (b + c)$ and $(a \times b) \times c = a \times (b \times c)$

a $15 + 37 + 18 + 45 + 22$ Pair numbers that add to multiples of 10.
$= (15 + 45) + (18 + 22) + 37$
$= 60 + 40 + 37$
$= 137$

b $7 \times 4 \times 5 = 7 \times (4 \times 5)$ Pair numbers that multiply to multiples of 10.
$= 7 \times 20$
$= 140$

Example 3

Adding and subtracting 8 or 9

a $43 + 9 = 43 + 10 - 1$ Add 10, count back 1.
$= 53 - 1$
$= 52$

b $97 + 8 = 97 + 10 - 2$ Add 10, count back 2.
$= 107 - 2$
$= 105$

c $61 - 8 = 61 - 10 + 2$ Subtract 10, count forward 2.

$= 51 + 2$

$= 53$

Example 4

Calculating differences using a number line

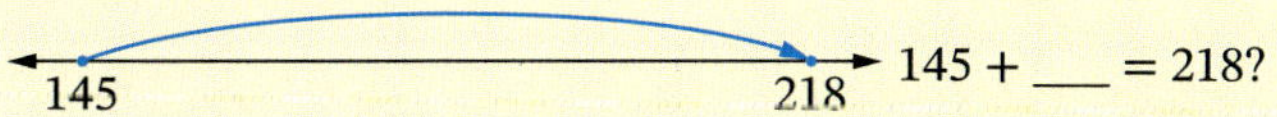

$145 + \_\_\_ = 218?$

Think of this as the difference or 'gap' between 145 and 218 on a number line.

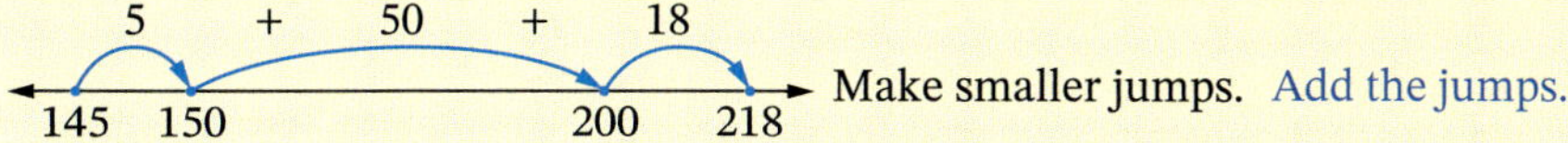

Make smaller jumps. Add the jumps.

$5 + 50 + 18 = 73$

So $218 - 145 = 73$

This is like counting out change with money.

Example 5

Multiplying and dividing by a multiple of 10

a $9 \times 60 = 9 \times 6 \times 10$

$= 54 \times 10$

$= 540$

b $2700 \div 30 = 270\cancel{0} \div 3\cancel{0}$

$= 270 \div 3$

$= 90$

Example 6

Doubling and halving numbers, Multiplying and dividing by 4

a $47 \times 2 = 40 \times 2 + 7 \times 2$

$= 80 + 14$

$= 94$

b $\frac{1}{2} \times 68 = \frac{1}{2} \times 60 + \frac{1}{2} \times 8$

$= 30 + 4$

$= 34$

c $\frac{1}{2} \times 192 = \frac{1}{2} \times 180 + \frac{1}{2} \times 12$

$= 90 + 6$

$= 96$

d $29 \times 4 = 29 \times 2 \times 2$ Double twice

Double $29 = 58$, double $58 = 116$

$29 \times 4 = 116$

e $16 \times 8 = 16 \times 2 \times 2 \times 2$ Double 3 times

Double $16 = 32$, double $32 = 64$, double $64 = 128$

$16 \times 8 = 128$

f $560 \div 4 = 560 \div 2 \div 2$ Halve twice

$\frac{1}{2} \times 560 = 280$, $\frac{1}{2} \times 280 = 140$

$560 \div 4 = 140$

Example 7

Simplifying multiplication by factorising

a $45 \times 16 = 9 \times 5 \times 4 \times 4$
$= 5 \times 4 \times 4 \times 9$
$= 20 \times 36$
$= 720$

b $12 \times 18 = 6 \times 2 \times 9 \times 2$
$= 6 \times 9 \times 2 \times 2$
$= 54 \times 4$
$= 216$

Example 8

Multiplying and dividing by 5, 15, 25, 50

a $16 \times 15 = 16 \times \frac{1}{2} \times 30$ or $8 \times 2 \times 15$
$= 8 \times 30$
$= 240$

b $24 \times 25 = 24 \times \frac{1}{4} \times 100$ or $6 \times 4 \times 25$
$= 6 \times 100$
$= 600$

c $420 \div 5 = 420 \div 10 \times 2$
$= 42 \times 2$
$= 84$

d $300 \div 25 = 300 \div 100 \times 4$
$= 3 \times 4$
$= 12$

Video
Multiplying by 8, 9, 11 and 12

Example 9

Multiplying by 8, 9, 11, 12

The distributive laws:

$a \times (b + c) = a \times b + a \times c$ and $a \times (b - c) = a \times b - a \times c$

a $25 \times 9 = 25 \times (10 - 1)$
$= 25 \times 10 - 25 \times 1$
$= 250 - 25$
$= 225$

b $18 \times 8 = 18 \times (10 - 2)$
$= 18 \times 10 - 18 \times 2$
$= 180 - 36$
$= 144$

c $13 \times 11 = 13 \times (10 + 1)$
$= 13 \times 10 + 13 \times 1$
$= 130 + 13$
$= 143$

d $27 \times 12 = 27 \times (10 + 2)$
$= 27 \times 10 + 27 \times 2$
$= 270 + 54$
$= 324$

ⓘ Short and long division

Short division is a technique for dividing by one-digit numbers (whole numbers less than 10).

Long division is a technique for dividing by numbers with 2 or more digits (whole numbers greater than 10).

Example 10

Evaluate each quotient.

a $520 \div 8$ **b** $945 \div 21$

SOLUTION

a By short division:

$8\overline{)520}$ 8 into 5 goes 0.

$\begin{array}{r} 6 \\ 8\overline{)52^{4}0} \end{array}$ 8 into 52 goes 6, remainder 4.

$\begin{array}{r} 6\;5 \\ 8\overline{)52^{4}0} \end{array}$ 8 into 40 goes 5 (exactly).

$520 \div 8 = 65$ Check by estimating: $520 \div 8 \approx 500 \div 10 = 50$. (65 is close to 50)

OR:

$520 \div 8 = 520 \div 2 \div 2 \div 2$ Halve three times.

$520 \div 2 = 260$, $260 \div 2 = 130$,
$130 \div 2 = 65$

$520 \div 8 = 65$

b By long division:

$$\begin{array}{r} 45 \\ 21\overline{)\;945} \\ \underline{-84}\downarrow \\ 105 \\ \underline{-105} \\ 0 \end{array}$$

21 into 94 is 4, remainder 10.

21 into 105 is 5.

$945 \div 21 = 45$ Check by estimating: $945 \div 21 \approx 900 \div 20 = 45$.

OR:

Guessing with 'easy' multiples of 21.

$21\overline{)\;945}$	
$\underline{-420}$	20 times
525	
$\underline{-420}$	20 times
105	
$\underline{-84}$	4 times
21	
$\underline{-21}$	1 time
0	45 times

EXERCISE 2.01 ANSWERS ON P. 614

Mental calculation

EXAMPLE 1

1 Estimate the value of each expression, and then check your estimate with a calculator.

a $27 + 11 + 87 + 142 + 64$ **b** $55 + 34 - 22 - 46 + 136$
c $684 + 903$ **d** $35 + 81 + 110 + 22 + 7$
e $517 - 96$ **f** $766 - 353$
g 46×8 **h** 367×2
i 51×12 **j** $245 \div 7$
k $828 \div 3$ **l** $564 \div 12$

EXAMPLE 2

2 Evaluate each sum.

PS R

a $55 + 18 + 25 + 9 + 12$ **b** $23 + 42 + 16 + 24 + 7$
c $140 + 33 + 12 + 20 + 28$ **d** $37 + 11 + 29 + 5 + 13$
e $45 + 49 + 121 + 25 + 10$ **f** $74 + 99 + 21 + 32 + 16$

3 What is the value of $720 + 1456$? Select the correct answer **A**, **B**, **C** or **D**.

A 2176 **B** 2206 **C** 2806 **D** 8656

4 Evaluate each expression.

R

a $27 + 9$ **b** $45 + 9$ **c** $16 + 8$ **d** $34 + 28$
e $17 + 11$ **f** $22 + 21$ **g** $69 + 12$ **h** $114 + 32$
i $45 - 9$ **j** $27 - 9$ **k** $76 - 8$ **l** $32 - 18$
m $55 - 12$ **n** $183 - 11$ **o** $67 - 41$ **p** $121 - 22$

5 Evaluate each difference.

PS R

a $560 - 327$ **b** $1614 - 239$ **c** $452 - 367$

6 Write the answer to each multiplication.

a 3×8 **b** 4×4 **c** 7×5 **d** 6×9
e 4×9 **f** 8×5 **g** 8×8 **h** 7×3
i 4×6 **j** 6×6 **k** 9×3 **l** 7×4

7 Write the answer to each division.

a $49 \div 7$ **b** $20 \div 4$ **c** $36 \div 9$ **d** $27 \div 3$
e $54 \div 6$ **f** $28 \div 7$ **g** $15 \div 5$ **h** $72 \div 8$
i $16 \div 4$ **j** $45 \div 5$ **k** $24 \div 8$ **l** $40 \div 8$

▷

Foundation Standard Complex

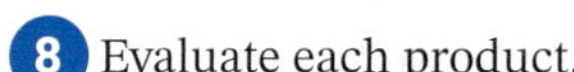

8 Evaluate each product.

PS R

a $8 \times 2 \times 50$	**b** $6 \times 5 \times 3$	**c** $25 \times 7 \times 4$	**d** $4 \times 8 \times 5$
e $5 \times 2 \times 10$	**f** $10 \times 4 \times 25$	**g** $15 \times 50 \times 2$	**h** $3 \times 4 \times 20$
i $6 \times 6 \times 5$	**j** $2 \times 5 \times 32$	**k** $20 \times 17 \times 5$	**l** $10 \times 7 \times 20$

9 Evaluate each expression.

R

a 8×20	**b** 7×500	**c** 2×9000	**d** 15×20
e 3×70	**f** 6×300	**g** 9×4000	**h** 12×200
i $900 \div 30$	**j** $1200 \div 200$	**k** $250 \div 50$	**l** $320 \div 40$
m $1800 \div 90$	**n** $2000 \div 20$	**o** $42\,000 \div 6000$	**p** $240 \div 80$

10 How many hours are there in one week? Select the correct answer **A**, **B**, **C** or **D**.

A 84 **B** 151 **C** 168 **D** 240

11 Evaluate each expression.

R

a 85×2	**b** 39×2	**c** 57×4	**d** 28×4
e 16×8	**f** 33×8	**g** $\frac{1}{2} \times 648$	**h** $\frac{1}{2} \times 232$
i $216 \div 4$	**j** $488 \div 4$	**k** $872 \div 8$	**l** $208 \div 8$

12 Evaluate each product by factorising.

PS R

a 7×40	**b** 25×16	**c** 22×6	**d** 13×20
e 35×6	**f** 21×3	**g** 16×12	**h** 18×18

13 Evaluate each expression.

PS R

a 22×5	**b** 14×5	**c** 28×50	**d** 36×15
e 44×25	**f** 52×50	**g** $800 \div 50$	**h** $90 \div 5$
i $120 \div 15$	**j** $1000 \div 25$	**k** $410 \div 5$	**l** $700 \div 50$

14 Nathan can type 76 words per minute. How many words can he type in 15 minutes?

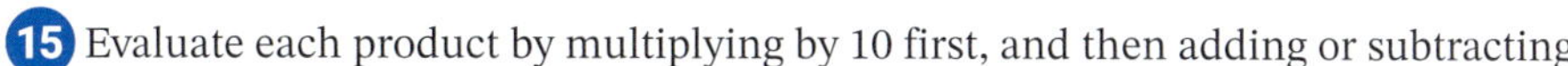

15 Evaluate each product by multiplying by 10 first, and then adding or subtracting.

R

a 34×9	**b** 51×9	**c** 27×8	**d** 72×8
e 36×11	**f** 41×11	**g** 64×12	**h** 22×12

16 Evaluate each quotient.

a $2640 \div 8$	**b** $1638 \div 9$	**c** $616 \div 7$	**d** $1584 \div 4$
e $675 \div 15$	**f** $756 \div 14$	**g** $806 \div 31$	**h** $729 \div 27$

17 There are 135 students in Year 8. If they are divided evenly into five classes, how many students will be in each class? Select the correct answer **A**, **B**, **C** or **D**.

A 21 **B** 25 **C** 27 **D** 29

18 Divide a restaurant bill of $204 evenly among six people.

Foundation Standard Complex

TECHNOLOGY

Sum, difference, product and quotient

In this activity, we will create a spreadsheet to calculate the sum, difference, product or quotient of two numbers. This spreadsheet can be used to check the answers to many of the questions in the Mental calculation section you have just completed.

1 Enter the headings and values shown below.

	A	B	C	D
1	Sum	Difference	Product	Quotient
2	27	560	3	49
3	9	327	8	7
4				

2 Enter a formula into cell A4 to calculate A2+A3, which is 36.

3 Enter a formula into cell B4 to calculate B2-B3, which is 233.

4 Enter a formula into cell C4 to calculate C2×C3, which is 24.

5 Enter a formula into cell D4 to calculate D2÷D3, which is 7.

6 Use your spreadsheet to check your answers to the Mental calculations.

Video
The Fibonacci sequence

DID YOU KNOW?

Fibonacci numbers

Fibonacci numbers are a number pattern 1, 1, 2, 3, 5, 8, ... named after the Italian mathematician Fibonacci (pronounced 'fibbon-archie'), also known as Leonardo of Pisa. He introduced this number pattern to Europe in 1202, although it was known earlier in India.

iStock.com/Manakin

The Fibonacci sequence of numbers is calculated as follows:

Start with 1 and 1.

$1 + 1 = 2$	1, 1, 2
$1 + 2 = 3$	1, 1, 2, 3
$2 + 3 = 5$	1, 1, 2, 3, 5
$3 + 5 = 8$	1, 1, 2, 3, 5, 8

Each number is the sum of the previous two numbers.

Surprisingly, Fibonacci numbers have proved useful in computer algorithms, the growth of plants, software development and optics, even though they were first calculated hundreds of years ago.

Calculate the next five Fibonacci numbers.

Adding and subtracting integers

2.02

Integers are all the positive and negative whole numbers and zero. Integers can be added and subtracted on a number line. A negative number can be entered into a calculator using the (−) or +/− key.

Skillsheets
Integers
Integers using diagrams
Integers using coloured squares

Presentations
Number lines
Adding directed numbers
Subtracting directed numbers

Worksheet
A page of number lines

Puzzle
Integer Snap

Videos
Adding integers
Subtracting integers
India and negative numbers

Interactives
Number line: Operations
Number line: Distance

ⓘ Adding and subtracting integers

Adding a negative number is the same as subtracting its opposite: $7 + (-4) = 7 - 4 = 3$.

Subtracting a negative number is the same as adding its opposite: $7 - (-4) = 7 + 4 = 11$.

Example 11

Evaluate each expression.

a $-8 + 3$ **b** $-8 - 3$ **c** $2 + (-6)$ **d** $2 - (-6)$

SOLUTION

a $-8 + 3 = -5$

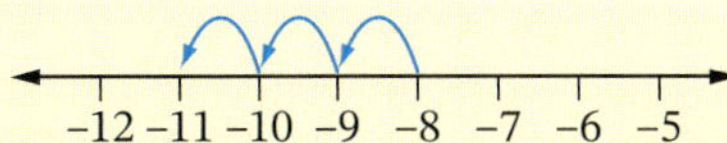

Start at −8, go forward 3

On a calculator, enter: (−) 8 + 3 =.

b $-8 - 3 = -11$

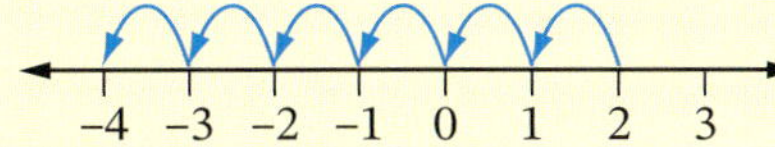

Start at −8, go back 3.

(−) 8 − 3 =.

c $2 + (-6) = 2 - 6 = -4$

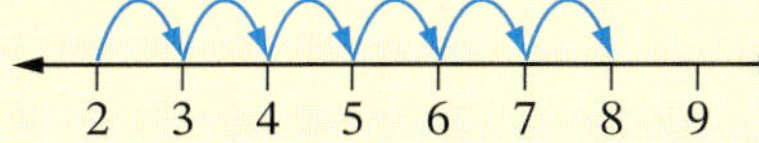

Start at 2, go back 6.

2 + (−) 6 =.

d $2 - (-6) = 2 + 6 = 8$

2 3 4 5 6 7 8 9

Start at 2, go forward 6.

2 − (−) 6 =.

EXERCISE 2.02 ANSWERS ON P. 615

Adding and subtracting integers

U F PS R

1 Evaluate each sum and check your answer using a calculator.

A **sum** is the answer to an addition.

a $-3+1$	**b** $-5+10$	**c** $-2+4$	**d** $-1+7$
e $-10+14$	**f** $-9+6$	**g** $-4+8$	**h** $-5+5$
i $9+(-3)$	**j** $4+(-1)$	**k** $6+(-4)$	**l** $2+(-2)$
m $3+(-8)$	**n** $7+(-10)$	**o** $5+(-6)$	**p** $1+(-9)$

2 Evaluate each difference and check your answer using a calculator.

A **difference** is the answer to a subtraction.

a $7-9$	**b** $5-11$	**c** $3-10$	**d** $2-6$
e $-4-4$	**f** $-8-2$	**g** $-1-5$	**h** $-3-9$
i $6-(-6)$	**j** $4-(-1)$	**k** $7-(-5)$	**l** $10-(-8)$
m $-5-(-4)$	**n** $-6-(-9)$	**o** $-3-(-3)$	**p** $-1-(-2)$

3 R At the ski resort, the temperature rose from −2 to 7°C. What was the increase in temperature?

4 PS R Rhys' bank account balance was −$44, meaning he was $44 in debt. He deposited $120 into his account. What is his balance now? Select the correct answer **A**, **B**, **C** or **D**.

A $64 **B** $74 **C** $76 **D** $164

5 Evaluate each expression.

a $12-3-11$	**b** $6-10+4$	**c** $-18+10-3$
d $-7+3+8$	**e** $8-15+(-6)$	**f** $-2-12+20$
g $3+(-5)+1$	**h** $12-14-(-4)$	**i** $-9+6-(-1)$
j $7-7+(-7)$	**k** $-5+(-5)+(-5)$	**l** $8-10+(-3)$

6 R The temperature changed from 6 to −8°C overnight. What was the change in temperature? Select the correct answer **A**, **B**, **C** or **D**.

A 2°C decrease **B** 14°C decrease **C** 14°C increase **D** 2°C increase

7 PS R Copy and fill in the blank for each equation. Use a number line to help you if you need.

a $5+\_\_\_=-4$	**b** $-2+\_\_\_=1$	**c** $-4+\_\_\_=-9$
d $\_\_\_+(-7)=-11$	**e** $3-\_\_\_=-3$	**f** $-6-\_\_\_=-7$
g $-2-\_\_\_=5$	**h** $\_\_\_-2=-6$	**i** $\_\_\_-(-8)=0$

8 PS R The table shows the temperature in four towns at two different times of the morning. Which town had the smallest change in temperature?

Town	5 a.m.	8 a.m.
Jackson Creek	-6°C	-3°C
Smith Valley	-2°C	3°C
Lake Drake	2°C	6°C
Mount Magner	-3°C	4°C

☐ Foundation ○ Standard ⬡ Complex

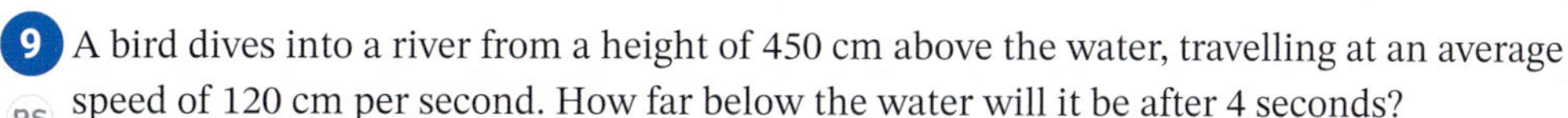

9 A bird dives into a river from a height of 450 cm above the water, travelling at an average speed of 120 cm per second. How far below the water will it be after 4 seconds?

PS

10 Copy and complete the blanks for each equation, where at least one of the numbers must be negative.

R

a ___ + ___ = 4 **b** ___ + ___ = −1 **c** ___ + ___ = 0

d ___ + ___ = −9 **e** ___ + ___ = 2 **f** ___ − ___ = 4

g ___ − ___ = −8 **h** ___ − ___ = 9 **i** ___ − ___ = −1

11 If a and b are integers, test whether each equation is true or false.

R

a $a + b = b + a$ **b** $a - b = b - a$

12 Simplify each algebraic expression by using integers to test them.

R

a $a + (-a)$ **b** $a - (-a)$ **c** $a + a$ **d** $a - a$

13 Investigate whether each sum or difference is always positive (+), always negative (−) or could be either positive or negative (E).

R

a A larger integer minus a smaller integer.

b A positive integer minus a positive integer.

c A positive integer plus a positive integer.

d A smaller integer minus a larger integer.

e A positive integer plus a negative integer.

f A negative integer plus a negative integer.

g A positive integer minus a negative integer.

h A negative integer minus a negative integer.

INVESTIGATION

Multiplying integers

1 Copy and complete each multiplication.

a $3 \times (-7) = (-7) + (-7) + (-7) =$ ____

b $2 \times (-5) = (-5) + (-5) =$ ____

c $3 \times (-4) = (-4) + (-4) + (-4) =$ ____

d $4 \times (-2) = (-2) + (-2) + (-2) + (-2) =$ ____

e $2 \times (-6) =$ ____

f $3 \times (-3) =$ ____

2 a In pairs or as a group activity, copy the multiplication grid onto a large piece of paper and complete the shaded section.

b Complete the first six rows.

c Continue the pattern for each column.

□ Foundation ○ Standard ⬡ Complex

×	5	4	3	2	1	0	−1	−2	−3	−4	−5
5											
4											
3											
2											
1											
0											
−1											
−2											
−3											
−4											
−5											

3 Use your completed table to simplify each product.

a $4 \times (-3)$ **b** -3×5

c $-4 \times (-2)$ **d** $5 \times (-1)$

e $0 \times (-4)$ **f** -3×3

g $2 \times (-4)$ **h** $-5 \times (-5)$

4 How do the signs (positive or negative) of the integers in the question affect the sign of the product (answer)? Try to express the rules for multiplying integers in words.

2.03 Multiplying integers

Worksheet Integers writing activity

Skillsheet Integers

Puzzle Multiplying integers game

Spreadsheet Integer quiz: Multiplication

Video Multiplying and dividing integers

ⓘ Multiplying integers

positive × positive = positive

positive × negative = negative

negative × positive = negative

negative × negative = positive

or

If both numbers have the **same** sign, the answer is **positive**.

If both numbers have **different** signs, the answer is **negative**.

×	+	−
+	+	−
−	−	+

Example 12

Evaluate each product.

A **product** is the answer to a multiplication.

a -3×5 **b** $-6 \times (-9)$ **c** $4 \times (-10)$ **d** $(-7)^2$

SOLUTION

a $-3 \times 5 = -15$

negative × positive = negative

On a calculator, enter: (–) 3 × 5 =

b $-6 \times (-9) = 54$

negative × negative = positive

(–) 6 × (–) 9 =

c $4 \times (-10) = -40$

positive × negative = negative

d $(-7)^2 = (-7) \times (-7) = 49$

negative × negative = positive

$(-7)^2$ is '−7 squared' and means multiplying −7 by itself

((–) 7) x^2 =

EXERCISE 2.03 ANSWERS ON P. 615

Multiplying integers

EXAMPLE 12

1 Evaluate each product and check your answer using a calculator.

a -5×4 **b** $3 \times (-6)$ **c** $-4 \times (-8)$ **d** -9×5

e $10 \times (-7)$ **f** $-15 \times (-2)$ **g** $(-3)^2$ **h** $6 \times (-4)$

i -7×3 **j** $(-10)^2$ **k** $9 \times (-4)$ **l** $-8 \times (-5)$

m $4 \times (-11)$ **n** -20×3 **o** $(-6)^2$ **p** $-7 \times (-9)$

q -11×5 **r** $(-9)^2$ **s** $10 \times (-10)$ **t** $-6 \times (-7)$

2 For each product, select the correct answer **A**, **B**, **C** or **D**.

a $12 \times (-3)$

A 15 **B** −15 **C** −36 **D** 36

b $-3 \times 6 \times 2$

A 0 **B** −16 **C** −18 **D** −36

c $5 \times 3 \times (-2)$

A −15 **B** −30 **C** 15 **D** 30

3 Copy and complete each multiplication grid.

a

×	−3	5
−3		
5		

b

×	−5	5
5		
−5		

c

×	6	−4
−4		
6		

d

×	-1	1
-1		
1		

e

×	7	-3
4		
-6		

×	−3	−6
8		
2		

Foundation  Standard Complex

4 Copy and complete each equation.

a $7 \times \_\_\_ = -14$ **b** $-3 \times \_\_\_ = 12$ **c** $-5 \times \_\_\_ = -25$

d $\_\_\_ \times 3 = -21$ **e** $\_\_\_ \times (-4) = 28$ **f** $\_\_\_ \times 9 = -63$

5 Evaluate each product and check your answer using a calculator.

a $-2 \times (-2) \times 7$ **b** $6 \times 5 \times (-3)$ **c** $4 \times (-1) \times 10$

d $3 \times (-5)^2$ **e** $-5 \times 2 \times (-2)$ **f** $(-2)^3$

6 Evaluate $-2 + 7 \times (-3) + 12$. Select the correct answer **A**, **B**, **C** or **D.**

A -3 **B** -11 **C** 31 **D** 61

7 Investigate whether each product is always positive (+), always negative (–) or could be either positive or negative (E).

a The product of two positive integers.

b The product of two negative integers.

c The product of a positive integer and a negative integer.

d A positive integer squared. **e** A negative integer squared.

8 If a and b are integers, test whether $a \times b = b \times a$ is true or false.

9 What happens when you multiply an integer by its opposite?

10 Copy and complete for each equation, where at least one of the numbers must be negative.

a $\_\_\_ \times \_\_\_ = 20$ **b** $\_\_\_ \times \_\_\_ = -16$ **c** $\_\_\_ \times \_\_\_ = 36$

d $\_\_\_ \times \_\_\_ = -5$ **e** $\_\_\_ \times \_\_\_ = 9$ **f** $\_\_\_ \times \_\_\_ = -22$

11 Simplify each algebraic expression by using integers to test them.

a $a \times a$ **b** $(-a) \times (-a)$

INVESTIGATION

Dividing integers

1 In pairs or as a group activity, copy and complete each question.

a $5 \times \_\_\_ = -20$ b $-6 \times \_\_\_ = 12$

c $-3 \times \_\_\_ = 27$ d $\_\_\_ \times (-4) = -16$

e $\_\_\_ \times 2 = -10$ f $\_\_\_ \times (-5) = 15$

g $-20 \div 5 = \_\_\_$ h $27 \div (-3) = \_\_\_$

i $-16 \div (-4) = \_\_\_$ j $-14 \div 2 = \_\_\_$

k $-12 \div (-12) = \_\_\_$ l $21 \div (-3) = \_\_\_$

2 Write down the rules for dividing integers in words.

Foundation Standard Complex

Dividing integers 2.04

Because division is the opposite of multiplication, the rules for dividing integers are the same.

Worksheets
What is the integer question?
Integer review
Integers writing activity

Video
Multiplying and dividing integers

Puzzle
Integer group clues

ⓘ Dividing integers

positive ÷ positive = positive
positive ÷ negative = negative
negative ÷ positive = negative
negative ÷ negative = positive

÷	+	−
+	+	−
−	−	+

Example 13

Evaluate each quotient.

A quotient is the answer to a division.

a $16 \div (-2)$ **b** $-20 \div (-10)$ **c** $-24 \div 4$ **d** $\frac{-27}{9}$

SOLUTION

a $16 \div (-2) = -8$

positive ÷ negative = negative
On a calculator, enter: 16 [÷] [(−)] 2 [=].

b $-20 \div (-10) = 2$

negative ÷ negative = positive
[(−)] 20 [÷] [(−)] 10 [=]

c $-24 \div 4 = -6$

negative ÷ positive = negative
[(−)] 24 [÷] 4 [=].

d $\frac{-27}{9} = -27 \div 9 = -3$

negative ÷ positive = negative
[(−)] 27 [÷] 9 [=].

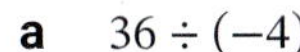

Dividing integers

U F R C

1 Evaluate each quotient and check your answer using a calculator.

a	$36 \div (-4)$	**b**	$-15 \div (-3)$	**c**	$-14 \div 2$	**d**	$60 \div (-10)$
e	$-25 \div (-5)$	**f**	$-28 \div 4$	**g**	$45 \div (-9)$	**h**	$-32 \div 16$
i	$-36 \div (-6)$	**j**	$56 \div (-7)$	**k**	$-24 \div 3$	**l**	$-42 \div (-6)$
m	$40 \div (-8)$	**n**	$-20 \div (-1)$	**o**	$-81 \div 9$	**p**	$24 \div (-4)$

☐ Foundation ○ Standard ⬡ Complex

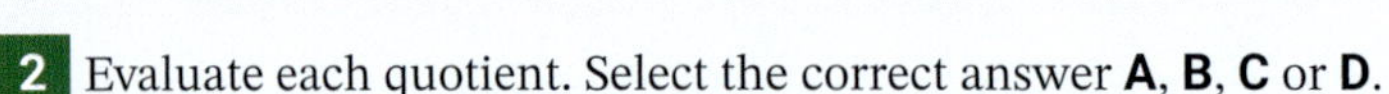

2 Evaluate each quotient. Select the correct answer **A**, **B**, **C** or **D**.

a $-45 \div (-5)$

A 9 **B** -9 **C** -50 **D** -40

b $\frac{42}{-6}$

A 36 **B** -7 **C** 7 **D** -48

3 Evaluate each quotient.

a $\frac{-18}{2}$ **b** $\frac{16}{-4}$ **c** $\frac{-30}{-5}$ **d** $\frac{21}{-3}$

e $\frac{-32}{-8}$ **f** $\frac{-100}{25}$ **g** $\frac{-81}{-9}$ **h** $\frac{56}{-8}$

4 Divide the top row by the left-hand column to complete each division grid.

a

÷	−6	−3	3	6
−6				
−3				

b

÷	8	−8	32	−32
−2				
4				

5 Evaluate each expression and check your answer using a calculator.

a $-28 \div 2 \div 7$ **b** $45 \div (-3) \div (-5)$ **c** $36 \div (-9) \div 2$

d $-90 \div 3 \div (-1)$ **e** $100 \div 2 \div (-10)$ **f** $-24 \div (-6) \div (-4)$

g $8 \times (-5) \div 2$ **h** $-4 \times 7 \div 14$ **i** $35 \div (-5) \times 10$

6 Copy and complete each equation.

R **a** $-12 \div$ ___ $= 2$ **b** $15 \div$ ___ $= -3$ **c** $-9 \div$ ___ $= -1$

d ___ $\div (-4) = -8$ **e** ___ $\div 12 = -2$ **f** ___ $\div (-10) = -10$

7 Evaluate $4 - 24 \div (-6) + 7$. Select the correct answer **A**, **B**, **C** or **D**.

A -20 **B** 7 **C** 10 **D** 15

8 For each quotient, state whether it is always positive (+), always negative (−) or could be either positive or negative (E).

R C

a The quotient of two positive integers.

b The quotient of two negative integers.

c The quotient of a positive integer and a negative integer.

9 If a and b are integers, test whether $a \div b = b \div a$ is true or false.

R

10 What happens when you divide an integer:

R C

a by itself? **b** by its opposite?

11 Simplify each algebraic expression by using integers to test them.

a $a \div a$ **b** $(-a) \div a$

Foundation Standard Complex

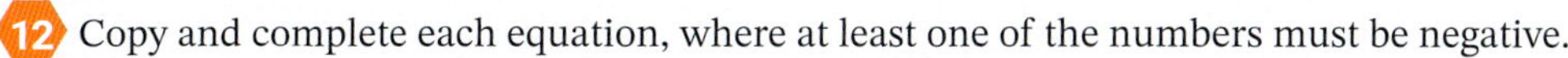

12 Copy and complete each equation, where at least one of the numbers must be negative.

R

a ___ ÷ ___ = 9   b ___ ÷ ___ = −5   c ___ ÷ ___ = 3

d ___ ÷ ___ = −4   e ___ ÷ ___ = 10   f ___ ÷ ___ = −7

DID YOU KNOW?

Pascal's triangle

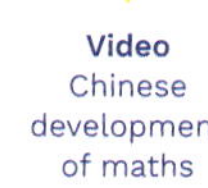

Video
Chinese development of maths

Pascal's triangle is a number pattern named after the French mathematician Blaise Pascal (1623–1662) but it was known about much earlier.

Each number is found by adding the two numbers just to the left and right of it in the previous line.

There are many patterns in this triangle:

- The sum of each row are the powers of 2.
- The Fibonacci numbers are found by summing minor diagonals.

```
            1
          1   1
        1   2   1
      1   3   3   1
    1   4   6   4   1
  1   5  10  10   5   1
1   6  15  20  15   6   1
1  7  21  35  35  21  7  1
```

- The counting numbers and triangular numbers are found along major diagonals.

The triangle has these simple patterns but also provides much more complex patterns, such as binomial expansions in algebra and probability.

Use the Internet to research the pattern in the triangle that gives the powers of 11.

Find two other patterns in Pascal's triangle.

INVESTIGATION

Multiplying decimals

1 Copy and complete each multiplication.

a $1000 \times 6 =$ ____

b $100 \times 6 =$ ____

c $10 \times 6 =$ ____

d $1 \times 6 =$ ____

e $0.1 \times 6 =$ ____

f $0.01 \times 6 =$ ____

Foundation | Standard | Complex

2 a In pairs or as a group activity, copy the multiplication grid below onto a large piece of paper and complete.

×	10	9	8	7	6	5	4	3	2	1	0
1000											
100											
10											
1											
0.1											
0.01											
0.001											

3 Use your completed table to simplify each product.

a 10×0.1 b 9×100 c 8×0.01 d 7×10

e 0×1 f 1×0 g 4×0.001 h 3×1000

4 a When a number is multiplied by 1 what happens to the number?

b Copy and complete:

When a number is multiplied by a number greater than 1 it becomes _____________ (bigger/smaller).

c Copy and complete:

When a number is multiplied by a number between 0 and 1 it becomes _____________ (bigger/smaller).

Quiz
Mental skills 2A

☆ MENTAL SKILLS 2A ANSWERS ON P. 616 Maths without calculators

Multiplying decimals

1 Study each example.

a $3 \times 8 = 24$, so $3 \times 0.8 = 2.4$

0 dp + 1 dp = 1 dp (dp = decimal places)

The number of decimal places in the answer is equal to the total number of decimal places in the question. Also, the answer sounds reasonable because, by estimation: $3 \times 0.8 \approx 3 \times 1 = 3$ $(2.4 \approx 3)$

b $6 \times 5 = 30$, so $0.6 \times 0.5 = 0.30 = 0.3$

1 dp + 1 dp = 2 dp

By estimation, $0.6 \times 0.5 \approx 0.5 \times 0.5 = \frac{1}{2} \times \frac{1}{2} = \frac{1}{4} = 0.25$ $(0.3 \approx 0.25)$

c $7 \times 3 = 21$, so $0.07 \times 0.3 = 0.021$

2 dp + 1 dp = 3 dp

By estimation, $0.07 \times 0.3 \approx 0.07 \times \frac{1}{3} \approx 0.02$ $(0.021 \approx 0.02)$

2 Now evaluate each product.

a 0.7×5 b 12×0.2 c 0.4×0.3 d $(0.6)^2$

e 8×0.1 f 0.03×0.9 g 4×0.05 h 1.1×8

i 0.3×0.8 j 0.2×0.06 k 9×0.2 l 0.07×0.4

3 Study each example.

Given that $15 \times 23 = 345$, evaluate each product.

a $1.5 \times 2.3 = 3.45$

1 dp + 1 dp = 2 dp (*Estimate*: $1.5 \times 2.3 \approx 2 \times 2 = 4$)

b $150 \times 0.23 = 15 \times 10 \times 0.23 = 15 \times 0.23 \times 10 = 3.45 \times 10 = 34.5$

0 dp + 2 dp = 2 dp

(*Estimate*: $150 \times 0.23 \approx 150 \times 0.2 = 150 \times \frac{1}{5} = 30$)

c $0.15 \times 2300 = 0.15 \times 23 \times 100 = 3.45 \times 100 = 345$

2 dp + 0 dp = 2 dp

(*Estimate*: $0.15 \times 2300 \approx 0.2 \times 2300 = \frac{1}{5} \times 2300 = 460$)

4 Now given that $39 \times 17 = 663$, evaluate each product.

a 3.9×17 b 39×170 c 39×0.17 d 0.39×1.7

e 3.9×1.7 f 390×1.7 g 3.9×0.17 h 3.9×170

i 3900×1.7 j 39×1.7 k 39×0.017 l 0.39×0.17

Order of operations 2.05

When evaluating a mixed expression, such as $18 \div (-2 + 1) \times 2$, there is a specific order in which the different operations are performed.

Skillsheet Order of operations

Puzzles Directed numbers

Order of operations puzzle

Videos Order of operations 1

Order of operations 2

BODMAS

ⓘ Order of operations

1 Simplify any expression inside grouping symbols (brackets): start with the innermost brackets first.
2 Simplify any multiplication (×) and division (÷), from left to right.
3 Simplify any addition (+) and subtraction (−), from left to right.

Beware of cheap calculators that do not follow the 'order of operations' rules!

Example 14

Evaluate each expression.

a $5 \times 2 + 3 \times (-9)$ **b** $2 \times [(25 - 4) \div 3]$

SOLUTION

a $5 \times 2 + 3 \times (-9) = 10 + (-27)$

$= -17$

Multiply first: $5 \times 2 = 10$ and $3 \times (-9) = -27$

On a calculator, enter:

5 [×] 2 [+] 3 [×] [(−)] 9 [=].

b $2 \times [(25 - 4) \div 3] = 2 \times [21 \div 3]$

$= 2 \times 7$

$= 14$

Innermost brackets first: $25 - 4 = 21$

Next brackets: $21 \div 3 = 7$

2 [×] [(] [(] 25 [−] 4 [)] [÷] 3 [)] [=].

Example 15

Simplify each fraction.

a $\frac{8+16}{40 \div 10}$ **b** $\frac{10-30}{6^2}$

SOLUTION

To simplify each fraction, evaluate the numerator and denominator separately, and then divide the numerator by the denominator.

a $\frac{8+16}{40 \div 10} = \frac{24}{4}$

$= 6$

On a calculator, enter: [(] 8 [+] 16 [)] [▭/▭] [(] 40

[÷] 10 [)] [=]

OR if using MATH Mode, press [▭/▭] and enter 8 + 16 and 40 ÷ 10 separately into the blank spaces.

b $\frac{10-30}{6^2} = \frac{-20}{36}$

$= -\frac{5}{9}$

Simplifying the fraction.

EXERCISE 2.05 ANSWERS ON P. 616

Order of operations

1 Evaluate each expression and check your answer on a calculator.

a $15 \div (7 - 4)$	**b** $[14 + (-9)] \times 6$	**c** $20 - (5 - 7)$
d $8 \times (-3 + 5 - 6)$	**e** $-2 \times (10 - 9) + 28$	**f** $(16 + 8) \div 2 - 4$
g $18 \div (5 - 2) \times (-2)$	**h** $[4 - (-7)] \times 3 - 10$	**i** $-26 \div (14 + 12)$
j $(7 - 10) \times 10 \div 6$	**k** $5 \times [(22 - 10) \div (-3)] + 1$	**l** $30 + [7 \times (2 - 6)] - 2$

▷

☐ Foundation ◯ Standard ⬡ Complex

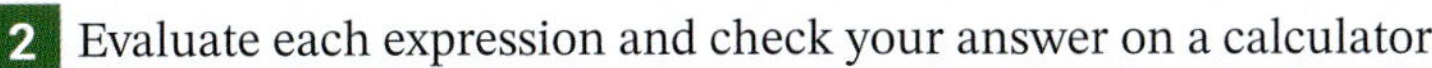

2.05

2 Evaluate each expression and check your answer on a calculator.

a $8 + 5 \times (-2)$ **b** $7 - 2 \times 3$ **c** $6 \times 5 - (-1)$

d $12 + (-6) \div 3$ **e** $3 \times 6 + 2 \times 5$ **f** $9 - 11 + 15 \div 3$

g $-5 \times 10 + 16 \div 2$ **h** $-3 \times 6 - (-2) \times 5$ **i** $42 \div 6 + 9$

j $17 + 8 - 3 \times 2$ **k** $4 \times 3 - (-7) + 5$ **l** $16 - 3 \times (-4) \div 2$

3 Simplify each fraction.

EXAMPLE 15

a $\frac{5\times2}{16+4}$ **b** $\frac{19+5}{18-6}$ **c** $\frac{9\times3}{40-10}$ **d** $\frac{58+8}{-7-4}$

e $\frac{4^2}{16\div(-2)}$ **f** $\frac{5\times3-1}{16+10\times4}$ **g** $\frac{28-5\times3}{(56-30)\div2}$ **h** $\frac{25\div5+13}{\sqrt{36}}$

4 Which expression is equal to 4? Select the correct answer **A**, **B**, **C** or **D**.

A $20 \div 4 + 6 \times 2$ **B** $20 \div (4 + 6) \times 2$

C $20 \div [(4 + 6) \times 2]$ **D** $(20 \div 4 + 6) \times 2$

5 Ms Ferme, the Mathematics Head Teacher, has a parent complaining about the marking of his daughter's maths exam. He claims his daughter's correct answer was marked wrong.

This is the question: $-9 + 13 \times 4$

R C His daughter's answer was 16.

a Why was the daughter's answer wrong?

b How could Ms Ferme explain why the daughter's answer is wrong?

c Put brackets in $-9 + 13 \times 4$ to make his daughter's answer correct.

6 Copy and complete each equation.

R

a $-8 \times 3 - \_\_\_ = -32$ **b** $\_\_\_ \times (5 - 9) = -20$

c $(8 - \_\_\_) \times 7 = 35$ **d** $9 \times \_\_\_ + (-2) = -20$

e $24 \div (-3) \times \_\_\_ = 8$ **f** $\_\_\_ + 5 \times 3 = -5$

g $(-2 - \_\_\_) \times (-4) = -16$ **h** $(-3 - 9) \div \_\_\_ = -2$

i $\_\_\_ \div (-4) + 6 = 1$

7 Copy each equation and insert grouping symbols in the correct places to make the equation true.

R

a $8 - 3 + 7 = -2$ **b** $40 - 10 \times 5 = 150$

c $27 \div 9 \div 3 = 9$ **d** $8 + 4 - 3 \times 2 = 10$

e $8 + 4 - 3 \times 2 = 18$ **f** $6 + 4 \times (-1) - 1 = -11$

g $13 + 3 \div 4 - 6 = -8$ **h** $100 \div 10 + 10 + 5 = 10$

i $100 \div 10 + 10 + 5 = 4$

8 Use all the integers 5, −2 and 10 and grouping symbols to complete each equation.

PS R

a $\_\_\_\_\_\_ = 30$ **b** $\_\_\_\_\_\_ = -1$ **c** $\_\_\_\_\_\_ = -60$

Foundation Standard Complex

Skillsheet Spreadsheets

Technology Review of spreadsheets

TECHNOLOGY

Order of operations

1 Enter these seven values in column A of a new spreadsheet: 36, 7, 8.5, 18, 50, 2 and −3.

2 In column B, write an appropriate formula to evaluate each expression in the given cell. Try to predict the answer before you enter each formula. The first one has been done for you.

B1: $36 + 7 - 2$ In cell B1, enter **=A1+A2−A6**

B2: $(7 + 8.5) \div 2$ B3: $7 + 8.5 \div 2$ B4: $50 - 2 \times (-3)$

B5: $18 \times 7 + 36 \times 8.5$ B6: $18 \times (7 + 36) \times 8.5$ B7: $\frac{50}{2+18}$

B8: $50 + 8.5 - 2 - (-3)$ B9: $7 \times 2 + 50 \div 2$ B10: $\frac{50\times(2+18+8.5)}{-3}$

2.06 Decimals

Skillsheet Decimals

Worksheets Decimals wall

Dewey decimals

Decimals 1

Decimals 10

ⓘ Ordering decimals

- List them in a column with the decimal points in line so that the place values can be compared.
- Fill any gaps at **the** end with 0s.
- **'Ascending order'** means going up, from smallest to largest.
- **'Descending order'** means going down, from largest to smallest.

Example 16

Write these decimals in ascending order: 3.5, 3.751, 3.15, 3.157.

SOLUTION

3.500 List the numbers in a column with the decimal points aligned.
3.751 Fill any gaps at the end with zeros.
3.150 Order the decimals from smallest to largest by comparing the place values.
3.157 The smallest is 3.150, next is 3.157, next is 3.500, the largest is 3.751.

In ascending order, the decimals are 3.15, 3.157, 3.5, 3.751.

ⓘ Rounding decimals

To **round a decimal**, cut it at the required decimal place and look at the digit in the next place:

- if that digit is less than 5 (i.e. 0, 1, 2, 3 or 4), **round down**.
- if the digit is 5 or more (i.e. 5, 6,7, 8 or 9), **round up**.

Example 17

a Round 5.261 to the nearest tenth.

b Write 14.8239 correct to two decimal places.

SOLUTION

a 'nearest tenth' = to one decimal place.

5.2 | 61

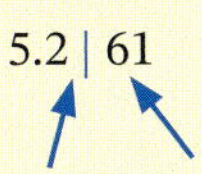

cut the next digit is 6 (>5), so round **up** to 5.3.

b 14.82 | 39

cut the next digit is 3 (<5), so round **down** to 14.82.

$14.8239 \approx 14.82$

ⓘ Adding and subtracting decimals

When **adding and subtracting decimals**, keep decimal points below one another. Fill any gaps with 0s.

Example 18

Evaluate each expression.

a $2.7 + 44.3 + 13.25$

b $23.8 - 5.65$

SOLUTION

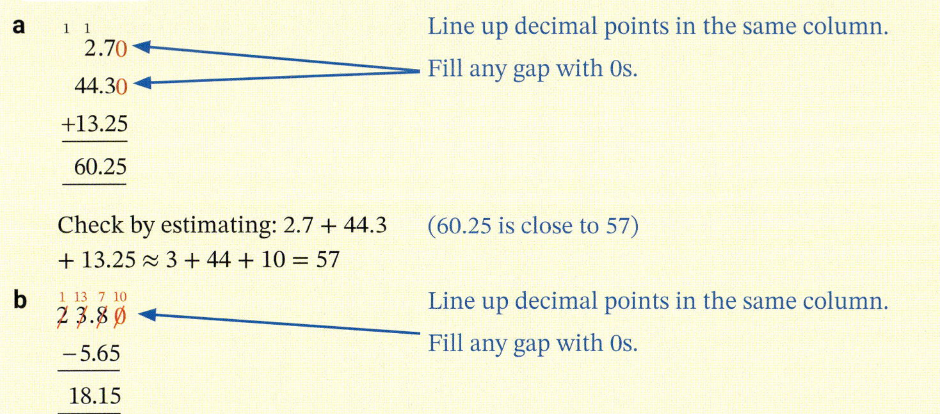

a Line up decimal points in the same column. Fill any gap with 0s.

Check by estimating: $2.7 + 44.3 + 13.25 \approx 3 + 44 + 10 = 57$ (60.25 is close to 57)

b Line up decimal points in the same column. Fill any gap with 0s.

Check by estimating:

$23.8 - 5.65 \approx 24 - 6$

$= 18$ (18.15 is close to 18)

EXERCISE 2.06 ANSWERS ON P. 616

Decimals

1 Which decimal is the largest? Select the correct answer **A**, **B**, **C** or **D**.

A 6.408 **B** 6.48 **C** 6.048 **D** 6.4

2 Write each set of decimals in ascending order.

a 4.35, 4.3, 4.05, 4.035
b 17.2103, 17.12, 12.173, 17.231
c 14.8, 14.08, 14.801, 14.81
d 0.3, 0.295, 0.032, 0.2

3 Write each set of decimals in descending order.

a 1.612, 1.621, 1.61, 1.16
b 3.05, 3.053, 3.035, 3.305

4 How many decimal places have each decimal?

a 3.824 **b** 1.09 **c** 51.6 **d** 22
e 12.573 **f** 450.304 **g** 788.5 **h** 0.45

EXAMPLE 17

5 Write each decimal correct to one decimal place.

a 3.851 **b** 4.0736 **c** 0.3333 **d** 7.34
e 15.0801 **f** 2.976 **g** 2.048 **h** 16.1919

6 What is 12.3752 rounded to the nearest hundredth? Select the correct answer **A**, **B**, **C** or **D**.

A 12.8 **B** 12.37 **C** 12.38 **D** 12.375

7 Round each decimal to the nearest hundredth.

a 68.9109 **b** 107.0594 **c** 3.5963 **d** 4.7077
e 3.198 **f** 32.999 **g** 19.7291 **h** 11.254

8 Write each decimal correct to three decimal places.

a 9.7043 **b** 13.167 54 **c** 0.08281 **d** 53.094 23

9 Round $460.39512 to the nearest:

a cent **b** dollar **c** tenth **d** 5 cents **e** thousandth

EXAMPLE 18

10 Evaluate each expression, and then check your answer with an estimate.

a 28.51 + 136.4
b 7.2 + 18.16
c 34.7 + 29.4 + 8.5
d 127.81 − 36.2
e 46.5 − 30.8
f 100.87 − 23.6
g 20.03 − 1.59
h 12.56 − 9.88
i 71.5 − 4.82
j 65.21 − 13.6
k 9 − 4.36
l 3.671 − 1.28

11 Evaluate 0.52 + 14.025. Select the correct answer **A**, **B**, **C** or **D**.

A 65.025 **B** 19.225 **C** 14.545 **D** 14.077

Foundation | Standard | Complex

12 Yesterday, the temperature dropped from 24.1 to 16.5°C. What was the difference between the two temperatures?

13 Monique built a wooden frame with dimensions 0.8 m by 0.5 m. How much wood will be left from a 3.4 m length of timber?

PS

14 Tahir bought the following items: an exercise book for \$2.70, two pens for \$1.60 each, a drink for \$1.50 and a packet of chips for \$2.65.

PS

a How much did Tahir spend in total?

b If he paid with a \$20 note, how much change did he receive?

15 Copy and complete each equation with the correct decimals.

R

a 2.8 + _____ = 5.84

b _____ − 7.6 = 5.14

c 18.75 − ____ = 13.3

d _____ + 0.83 = 1.719

e _____ − 1.07 = 3.256

f _____ − _____ = 4.89

INVESTIGATION

Rounding up or down?

Sometimes, the decision on whether to round an answer up or down depends on the situation. In groups 2–4, discuss each of the following situations and decide whether it is more appropriate to round up or to round down. You must give a reason for your choice.

1 Seven friends have dinner at a restaurant and the total bill is \$206. They decide to share the bill evenly: \$206 ÷ 7 = \$29.428 57...

How much should each friend pay, to the nearest dollar: \$29 or \$30?

Shutterstock.com/G-Stock Studio

2 You have a budget of \$65 for buying drinks for a party. One can of drink costs \$1.74 at the supermarket: \$65 ÷ \$1.74 = 37.356 32...

How many whole cans of drink can you buy: 37 or 38?

3 You need to find the average number of people living in each home in your street. You survey 48 homes and count a total of 166 people: 166 ÷ 48 = 3.458 33...

What is the average whole number of people living in each house: 3 or 4?

4 You need to paint the walls of a house with a total surface area of 334 m^2. One tin of paint covers 64 m^2: 334 ÷ 64 = 5.218 75.

How many whole tins of paint do you need for the job: 5 or 6?

5 Jodie has \$2098 in her bank account and the bank pays her 4.71% in interest:

4.71% × \$2098 = \$98.8158...

How much interest will the bank pay Jodie: \$98.81 or \$98.82?

TECHNOLOGY

The FIX mode on a calculator

Most scientific calculators have a FIX mode that rounds a displayed answer to a given number of decimal places. However, the exact answer is still stored in the calculator's memory to make sure that any further calculations involving this answer are accurate.
Find out how to use the FIX mode on your calculator.
The FIX mode has been used to set this calculator to display answers to five decimal places.

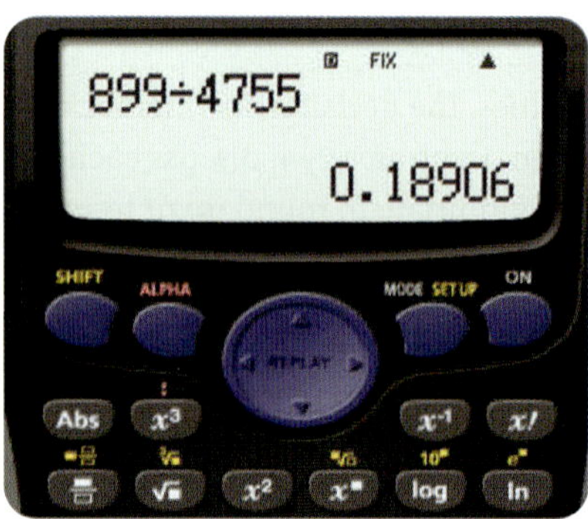

2.07 Multiplying and dividing decimals

Video
Multiplying decimals

Interactive
Area model
Decimals

ⓘ Multiplying decimals

When **multiplying decimals,** the number of decimal places in the answer equals the *total* number of decimal places in the question.

Example 19

Evaluate each product.

a 0.8×0.3 **b** 7.2×5 **c** 5.31×1.3

SOLUTION

a $8 \times 3 = 24$ Multiply without decimal points first.

Count the number of decimal places in the question. 0.8 has one decimal place, 0.3 has one decimal place.

$0.8 \times 0.3 = 0.24$ Write the answer with $1 + 1 = 2$ decimal places.

Check by estimating: $0.8 \times 0.3 \approx 1 \times 0.3 = 0.3$ $(0.24 \approx 0.3)$.

b $72 \times 5 = 360$ Multiply without decimal points first.

7.2 has one decimal place, 5 has no decimal place.

$7.2 \times 5 = 36.0$

$= 36$ Write the answer with $1 + 0 = 1$ decimal place.

Check by estimating: $7.2 \times 5 \approx 7 \times 5 = 35$ $(36 \approx 35)$.

c
```
  531
× 13
 1593
 5310
 6903
```
Multiply without decimal points first.

5.31 has two decimal places and 1.3 has one decimal place.

$5.31 \times 1.3 = 6.903$ Write the answer with $2 + 1 = 3$ decimal places.

Check by estimating: $5.31 \times 1.3 \approx 5 \times 1 = 5$ $(6.903 \approx 5)$.

Worksheets
Decimal cards
Shopping and change
What's the point?
Decimals 7

Puzzles
Operations with decimals
Which decimals?
Decimal number grids

Technology
Movie night

2.07

ⓘ Dividing a decimal by a whole number

- Rewrite the question in 'short division' form.
- Make the decimal point in the answer line up with the decimal point in the question.
- Add 0s to the end of the decimal being divided, if needed.

ⓘ Dividing a decimal by another decimal

- Make the second decimal a whole number by moving the decimal point the required number of places to the right.
- Move the point in the first decimal the same number of places to the right.
- Divide the new first number by the whole number.

This works because we are multiplying both decimals by the same power of 10 before dividing.

Video Dividing decimals

Example 20

Evaluate each quotient.

a $6.28 \div 5$ **b** $12.4 \div 0.04$ **c** $4.98 \div 1.2$

SOLUTION

a

$$5\overline{)6.{}^{1}2{}^{2}8{}^{3}0}\quad = 1.256$$

Write $6.28 \div 5$ as a short division.

- 5 into 6 goes 1, remainder 1.
- Make the decimal point in the answer line up with the decimal point in the question.
- 5 into 12 goes 2, remainder 2.
- 5 into 28 goes 5, remainder 3.
- Add 0 to 6.28 so that you can complete the division.
- 5 into 30 goes 6 exactly.

$6.28 \div 5 = 1.256.$

Check by estimating: $6.28 \div 5 \approx 6 \div 5 \approx 1$ $(1.256 \approx 1)$.

b $12.4 \div 0.04 = 1240 \div 004$

$= 1240 \div 4$

$= 310$

We added a 0 to the end of 12.4 so that we can move the point two places to the right.

Move both decimal points two places to the right so that 0.04 becomes a whole number.

Check by estimating: $1240 \div 4 \approx 1200 \div 4 \approx 300$ $(310 \approx 300)$.

c $4.98 \div 1.2 = 49.8 \div 12$

Move both decimal points one place to the right so that 1.2 becomes a whole number.

```
    4.15
12)49.80
  -48↓
     18
    -12
     60
    -60
      0
```

Write $49.8 \div 12$ as a long division:

12 into 49 is 4, remainder 1.

12 into 18 is 1, remainder 6.

12 into 60 is 5 exactly.

$4.98 \div 1.2 = 4.15$

Check by estimating: $4.98 \div 1.2 \approx 5 \div 1 = 5$ $(4.15 \approx 5)$.

EXERCISE 2.07 ANSWERS ON P. 616

Multiplying and dividing decimals

U F PS R

EXAMPLE 19

1 Evaluate each product, and then check your answer with an estimate.

a	6.5×4	**b**	0.45×9	**c**	2.3×5	**d**	7×3.9
e	0.7×0.3	**f**	0.8×0.11	**g**	0.05×0.4	**h**	4.2×0.3
i	6.3×4.5	**j**	28.7×3.1	**k**	1.3×0.62	**l**	0.72×0.51

2 Felix was building a fence. He needed 9.4 metres of wire at $5.80 per metre. What was the cost of the wire? Select the correct answer **A**, **B**, **C** or **D**.

PS

A $15.20 **B** $54.52 **C** $152.00 **D** $545.20

3 Thao's mobile phone plan charges $20 per month plus $0.18 for each phone call. How much will Thao need to pay if she made 112 calls in one month?

PS

EXAMPLE 20

4 Evaluate each quotient, and then check your answer with an estimate.

a	$24.8 \div 4$	**b**	$114.1 \div 7$	**c**	$7.83 \div 3$
d	$0.695 \div 5$	**e**	$63.3 \div 0.2$	**f**	$4.173 \div 0.06$
g	$18.5 \div 0.05$	**h**	$23.22 \div 0.4$	**i**	$14.5 \div 25$
j	$6.24 \div 1.2$	**k**	$238 \div 1.4$	**l**	$0.252 \div 0.21$

5 What is the value of $12.92 \div 0.04$? Select the correct answer **A**, **B**, **C** or **D**.

A 32.3 **B** 323 **C** 3.23 **D** 3230

6 Brad uses 7.24 metres of wood to make a garden bed fence.

PS R

Shutterstock.com/Paul Maguire

a How much wood does he need to make six garden bed fences?

b If one of the garden beds is square-shaped, how long is each side?

7 Evaluate each expression.

a	$0.3 \times 0.7 + 5.7$	**b**	$2.8 \div (7.4 - 7.2)$	**c**	$3.62 + 4.08 \div 2$

Foundation Standard Complex

8 Jordan's car holds 45 litres of petrol. If the price of petrol is 149.9 cents per litre, how much will Jordan need to pay to fill the tank? Give your answer to the nearest 5 cents.

PS

9 Copy and complete each equation with the correct decimals.

R

a $0.3 \times$ ______ $= 2.4$ **b** ______ $\times 0.6 = 1.2$ **c** ______ $\times 0.7 = 0.028$

d $3.1 \times$ ______ $= 2.79$ **e** $83.4 \div$ ______ $= 6$ **f** ______ $\div 8.1 = 3.2$

10 A car travels 145.8 kilometres in 9 litres of petrol. How many kilometres could it travel in one litre of petrol?

PS

11 Anja earned $169.20 for working a 12-hour shift. How much was earned each hour?

PS

12 Henry has a faulty calculator that does not show the decimal point. For each calculation in the table shown, write the correct answer.

PS R

	Calculation	Answer with missing decimal point
a	3.42 × 12	4104
b	4.145 × 0.2	829
c	37.3 × 8.8	32824
d	0.03 × 157.64	47292
e	8.3902 × 0.3	251706

2.08 Terminating and recurring decimals

Skillsheet Fractions and decimals

Worksheet Fraction families

Puzzle Decimal squaresaw 2

Video Converting fractions to decimals

When converting a fraction to a decimal, or dividing two numbers, the decimal answer can be **terminating** or **recurring**.

Terminating decimals, such as 0.625, have a definite number of decimal places, while **recurring decimals**, such as 0.272727..., written as $0.\dot{2}\dot{7}$ or $0.\overline{27}$, have one or more digits that repeat endlessly. We use dots or lines to mark the repeating section. For example, 0.259 259 259... $= 0.\dot{2}5\dot{9}$ or $0.\overline{259}$.

'Terminate' means 'to stop', while 'recurring' means 'repeating'.

☐ Foundation ◯ Standard ⬡ Complex

Example 21

Convert each fraction to a recurring decimal.

a $\frac{5}{6}$ **b** $\frac{2}{11}$

SOLUTION

a $\frac{5}{6}$ means $5 \div 6$

$$6\overline{)5.0000\ldots} = 0.8333\ldots$$

Add 0s to complete the division.

A recurring decimal.

$\frac{5}{6} = 0.8333... = 0.8\dot{3}$ or $0.8\overline{3}$

b $\frac{2}{11}$ means $2 \div 11$

$$11\overline{)2.00000\ldots} = 0.18181\ldots$$

A recurring decimal.

$\frac{2}{11} = 0.181\ 818... = 0.\dot{1}\dot{8}$ or $0.\overline{18}$

Video Recurring decimals

Example 22

Evaluate each quotient.

a $2.8 \div 9$ **b** $0.7 \div 0.12$

SOLUTION

a $2.8 \div 9$

$$9\overline{)2.8000\ldots} = 0.3111\ldots$$

A recurring decimal.

$2.8 \div 9 = 0.3111... = 0.3\dot{1}$ or $0.3\overline{1}$

b $0.7 \div 0.12 = 70 \div 12$

$$12\overline{)70.0000\ldots} = 5.8333\ldots$$

A recurring decimal.

$0.7 \div 0.12 = 5.83333... = 5.8\dot{3}$ or $5.8\overline{3}$

EXERCISE 2.08 ANSWERS ON P. 616

Terminating and recurring decimals

U F R C

1 Convert each fraction to a terminating decimal.

a $\frac{2}{5}$	**b** $\frac{3}{8}$	**c** $\frac{3}{4}$	**d** $\frac{2}{2}$
e $\frac{2}{4}$	**f** $\frac{6}{8}$	**g** $\frac{3}{5}$	**h** $\frac{2}{8}$
i $\frac{5}{8}$	**j** $\frac{7}{8}$	**k** $\frac{4}{8}$	**l** $\frac{5}{5}$

EXAMPLE 21

Foundation Standard Complex

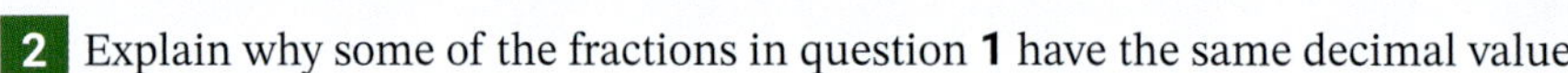

2 Explain why some of the fractions in question **1** have the same decimal value.

R C

3 Rewrite each recurring decimal using dot notation.

C
a 0.6666... **b** 4.2727... **c** 0.2828... **d** 3.8333...
e 9.607607... **f** 0.15191519... **g** 0.05252... **h** 12.23444...

4 Write each recurring decimal showing the repeated pattern.

R C
a $0.858\dot{3}$ **b** $0.1\dot{6}\dot{5}$ **c** $0.\dot{2}\dot{7}$ **d** $0.\dot{4}6153\dot{8}$

5 Convert each fraction to a recurring decimal.

C
a $\frac{1}{9}$ **b** $\frac{1}{6}$ **c** $\frac{5}{6}$ **d** $\frac{1}{7}$ **e** $\frac{2}{3}$
f $\frac{2}{7}$ **g** $\frac{2}{9}$ **h** $\frac{3}{7}$ **i** $\frac{4}{7}$ **j** $\frac{4}{9}$
k $\frac{4}{6}$ **l** $\frac{5}{9}$ **m** $\frac{6}{7}$ **n** $\frac{7}{9}$ **o** $\frac{5}{7}$

6 Copy and complete the table and note the pattern.

Fraction	$\frac{1}{7}$	$\frac{2}{7}$	$\frac{3}{7}$	$\frac{4}{7}$	$\frac{5}{7}$	$\frac{6}{7}$
Decimal						

EXAMPLE 22

7 Evaluate each expression as a recurring decimal.

a $11 \div 3$ **b** $15.4 \div 9$ **c** $58.43 \div 0.11$ **d** $1.96 \div 0.6$

Quiz Mental skills 2B

☆ MENTAL SKILLS 2B ANSWERS ON P. 617 Maths without calculators

Dividing decimals

To divide one decimal by another, first move the decimal points in **both** decimals the same number of places to the right so that the second decimal is a **whole number**.

1 Study each example.

a $0.24 \div 0.06 = 24 \div 6 = 4$
b $0.45 \div 0.5 = 4.5 \div 5 = 0.9$
c $0.006 \div 0.3 = 0.06 \div 3 = 0.02$
d $27 \div 0.9 = 270 \div 9 = 30$
e $1.6 \div 0.4 = 16 \div 4 = 4$
f $5.6 \div 0.07 = 560 \div 7 = 80$

2 Now evaluate each quotient.

a $0.25 \div 0.5$ **b** $63 \div 0.7$ **c** $3.2 \div 0.4$
d $0.18 \div 0.2$ **e** $2.7 \div 0.03$ **f** $0.042 \div 0.06$
g $4 \div 0.5$ **h** $1.2 \div 0.04$ **i** $0.072 \div 0.9$
j $0.35 \div 0.1$ **k** $0.28 \div 0.07$ **l** $0.033 \div 0.11$

☐ Foundation ○ Standard ○ Complex

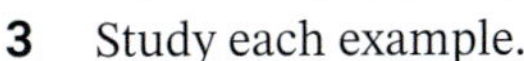

3 Study each example.

Given that $112 \div 14 = 8$, evaluate each expression.

a $112 \div 1.4 = 112.0 \div 1.4$
$= 1120 \div 14$
$= 112 \times 10 \div 14$
$= 112 \div 14 \times 10$
$= 8 \times 10$
$= 80$

Estimate: $112 \div 1.4 \approx 112 \div 1 = 112$

b $0.112 \div 0.14 = 0.112 \div 0.14$
$= 11.2 \div 14$
$= 112 \div 10 \div 14$
$= 112 \div 14 \div 10$
$= 8 \div 10$
$= 0.8$

Estimate: $0.112 \div 0.14 \approx 0.1 \div 0.1 = 1$

c $1120 \div 1.4 = 11200 \div 14$
$= 112 \times 100 \div 14$
$= 112 \div 14 \times 100$
$= 8 \times 100$
$= 800$

Estimate: $1120 \div 1.4 \approx 1120 \div 1 = 1120$

d $1.12 \div 14 = 112 \div 100 \div 14$
$= 112 \div 14 \div 100$
$= 8 \div 100$
$= 0.08$

Estimate: $1.12 \div 14 \approx 1.12 \div 10 = 0.112$

4 Now given that $368 \div 23 = 16$, evaluate each quotient.

a $36.8 \div 2.3$ **b** $368 \div 2.3$ **c** $3.68 \div 2.3$
d $0.368 \div 0.23$ **e** $36.8 \div 23$ **f** $3.68 \div 0.23$
g $36.8 \div 0.23$ **h** $0.368 \div 2.3$ **i** $0.368 \div 23$
j $3.68 \div 0.023$ **k** $3.68 \div 23$ **l** $0.368 \div 230$

Powers and roots 2.09

Powers and the use of **index notation** allow us to write repeated multiplication in a shorter way.

$4^6 = 4 \times 4 \times 4 \times 4 \times 4 \times 4$ '4 to the power of 6'

In 4^6, the number 4 is called the **base** and is the number that is repeated in the multiplication. The small raised number 6 is called the **power** or **index**.

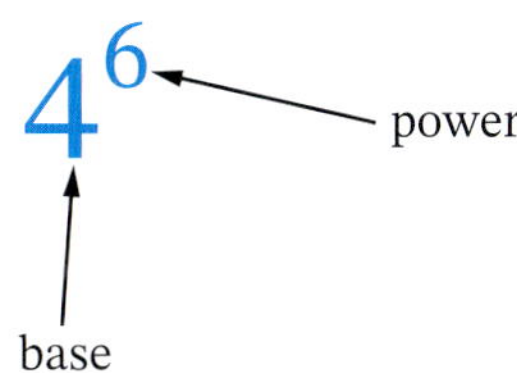

Skillsheets
Index notation
Square roots and cube roots

Worksheets
Powers and roots
Big numbers

Puzzle
Square root Snap

Video
Binary: the computer language

ⓘ The power keys on a calculator

- The square of a number is found using the x^2 key
- The cube of a number is found using the x^3 key
- Any power of a number can be found using the $x^{■}$ or y^x key.

Example 23

Evaluate each power.

a 11^3 **b** $(-8)^4$

SOLUTION

a $11^3 = 1331$ On a calculator, enter: 11 x^3 =

b $(-8)^4 = 4096$ ((−) 8) $x^{■}$ 4 =

ⓘ Square root and cube root

- The **square root** ($\sqrt{}$) of a number is the **positive** value which, if squared, will give that number.
- The **cube root** ($\sqrt[3]{}$) of a given number is the **positive or negative** value which, if cubed, will give that number.

For example:

$\sqrt{36} = 6$ because $6^2 = 36$ 'the square root of 36'

$\sqrt[3]{125} = 5$ because $5^3 = 125$ 'the cube root of 125'

$\sqrt[3]{-8} = -2$ because $(-2)^3 = -8$ 'the cube root of − 8'

Most roots do not give exact answers like the ones above, and are called **surds**. For example, $\sqrt{7} = 2.645751311... \approx 2.6$. A surd is a square root ($\sqrt{}$), cube root ($\sqrt[3]{}$) or any other type of root whose exact decimal or fraction value cannot be found. As a decimal, its digits run endlessly *without repeating*, so they are *neither* terminating nor recurring decimals. A surd cannot be written in fraction form $\frac{a}{b}$, so it is also called an **irrational** number.

Example 24

Evaluate each cube root, correct to two decimal places where necessary.

a $\sqrt[3]{729}$ **b** $\sqrt[3]{100}$

SOLUTION

a $\sqrt[3]{729} = 9$ On a calculator, enter: $\sqrt[3]{}$ 729 =.

b $\sqrt[3]{100} = 4.6415 ... \approx 4.64$ $\sqrt[3]{}$ 100 =

Example 25

Estimate the value of $\sqrt{50}$, correct to one decimal place.

SOLUTION

There is no *exact* answer for the square root of 50, because there isn't a number which, if squared, equals 50 exactly. However, we can find a decimal whose square is close to 50.

Noting that:

$6^2 = 36$

$7^2 = 49$

$8^2 = 64$

we can tell that $\sqrt{50}$ must lie somewhere between 7 and 8. Also, it is much closer to 7 because 50 is just over 49.

So we can estimate that $\sqrt{50} \approx 7.1$.

(In fact, $7.1^2 = 50.41$.)

EXERCISE 2.09 ANSWERS ON P. 617

Powers and roots

1 Write each expression using index notation.

C
- **a** $7 \times 7 \times 7$
- **b** $4 \times 4 \times 4 \times 4 \times 4$
- **c** $5 \times 5 \times 5 \times 5 \times 5 \times 5 \times 5 \times 5 \times 5 \times 5$
- **d** $-8 \times (-8) \times (-8) \times (-8) \times (-8) \times (-8) \times (-8)$

2 Evaluate each expression. EXAMPLE 23

- **a** $(-8)^2$
- **b** 2^3
- **c** $(-6)^3$
- **d** 3^4
- **e** 7^1
- **f** $(-2)^5$
- **g** 10^3
- **h** 4^4
- **i** $(-11)^6$
- **j** $3^9 - 9^3$
- **k** 4×3^5
- **l** $7^4 + 2 \times 8^2$
- **m** $16^4 \div 2^{10} - 5^2$
- **n** $(-4)^6 - 18 \div 3^2$

3 Find the value of $\square$ to make each equation true.

R
- **a** $2^{\square} = 8$
- **b** $3^{\square} = 27$
- **c** $(-10)^{\square} = 100$
- **d** $4^{\square} = 4096$
- **e** $5^{\square} = 125$
- **f** $(-3)^{\square} = -243$

4 What is the value of $(1.3)^2$? Select the correct answer **A**, **B**, **C** or **D**.

- **A** 1.9
- **B** 1.69
- **C** 2.6
- **D** 1.09

5 Which is smaller: 100^3 or 3^{100}?

6 Which is larger: 10^2 or 2^{10}?

Foundation Standard Complex

7 a Evaluate $(2 \times 3)^2$ without using a calculator.

C b Evaluate $2^2 \times 3^2$ without using a calculator.

c Does $(2 \times 3)^2 = 2^2 \times 3^2$? Explain your answer.

8 a Evaluate $(4 \times 5)^2$.

C b Evaluate $4^2 \times 5^2$.

c Does $(4 \times 5)^2 = 4^2 \times 5^2$? Explain your answer.

9 Use the pattern you found in questions **7** and **8** to complete each equation.

R C a $(3 \times 8)^2 =$ ______ $\times$ ______ b $(a \times b)^2 =$ ______ $\times$ ______

10 Copy and complete each pattern.

R a $18^2 = (6 \times 3)^2 = 6^2 \times$ ______ $=$ ______

C b $22^2 = (2 \times 11)^2 =$ ____ $\times$ ____ $=$ ____

c $16^2 = (2 \times$ ___$)^2 =$ ____ $\times$ ____ $=$ _____

d $15^2 = ($___ $\times 5)^2 =$ ____ $\times$ ____ $=$ ____

e $30^2 = (10 \times$ ___$)^2 =$ ___ $\times$ ____ $=$ _____

f $28^2 = ($___ $\times$ ___$)^2 =$ ____ $\times$ ____ $=$ ____

EXAMPLE 24

11 Evaluate each root.

a $\sqrt{784}$ b $\sqrt{256}$ c $\sqrt{289}$ d $\sqrt{1089}$

e $\sqrt[3]{8}$ f $\sqrt[3]{343}$ g $\sqrt[3]{2197}$ h $\sqrt[3]{216}$

12 Evaluate each root, correct to one decimal place.

a $\sqrt{37}$ b $\sqrt[3]{900}$ c $\sqrt{502}$ d $\sqrt[3]{6.5}$

e $\sqrt[3]{-495}$ f $\sqrt{2000}$ g $\sqrt{1.1}$ h $\sqrt[3]{1103}$

13 Copy and complete this table.

Number, x	1	2	3	4	5	6	7	8	9	10	11	12
Number squared, x^2				16			49					
Number cubed, x^3								512		1000		

Use the table from Question **13** to answer Questions **14–17**.

EXAMPLE 25

14 Between which two consecutive whole numbers does $\sqrt{80}$ lie? Select the correct answer **A**, **B**, **C** or **D**.

A 40 and 41 **B** 9 and 10 **C** 79 and 81 **D** 8 and 9

15 Between which two consecutive whole numbers does $\sqrt[3]{130}$ lie? Select the correct answer **A**, **B**, **C** or **D**.

A 4 and 5 **B** 10 and 11 **C** 11 and 12 **D** 5 and 6

16 Between which two consecutive whole numbers does $\sqrt[3]{31}$ lie?

Foundation Standard Complex

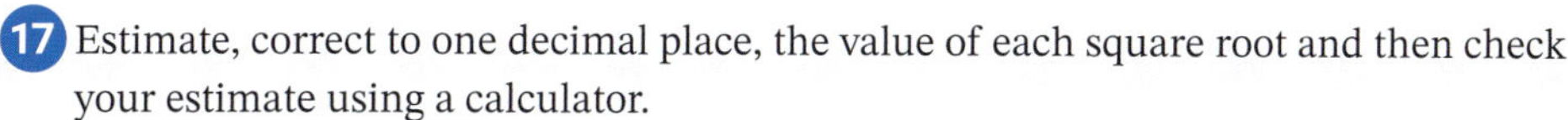

17 Estimate, correct to one decimal place, the value of each square root and then check your estimate using a calculator.

a $\sqrt{33}$ **b** $\sqrt{105}$ **c** $\sqrt{68}$

18 Select all the surds from this list of roots.

$\sqrt{6}$ $\sqrt{2025}$ $\sqrt{62}$ $\sqrt{961}$ $\sqrt{14}$ $\sqrt[3]{181}$ $\sqrt[3]{49}$ $\sqrt[3]{1000}$ $\sqrt[3]{674}$ $\sqrt[3]{1331}$

19 **a** Evaluate $\sqrt{4\times 25}$.

C **b** Evaluate $\sqrt{4}\times\sqrt{25}$.

c What do you notice about your answers to parts **a** and **b**?

20 **a** Evaluate $\sqrt{27\times 3}$.

C **b** Evaluate $\sqrt{27}\times\sqrt{3}$.

c What do you notice about your answers to parts **a** and **b**?

21 Use the pattern you found in questions **19** and **20** to complete each equation.

R C **a** $\sqrt{16\times 9}=$ ____ $\times$ ____ **b** $\sqrt{a\times b}=$ ____ $\times$ ____

22 Copy and complete each pattern.

R **a** $\sqrt{64}=\sqrt{4}\times\sqrt{16}=$ ____ $\times$ ____ $=$ ____

C **b** $\sqrt{441}=\sqrt{9}\times\sqrt{\_\_\_\_}=$ ____ $\times$ ____ $=$ ____

c $\sqrt{144}=\sqrt{36}\times\sqrt{\_\_\_\_}=$ ____ $\times$ ____ $=$ ____

d $\sqrt{2025}=\sqrt{81}\times\sqrt{\_\_\_\_}=$ ____ $\times$ ____ $=$ ____

TECHNOLOGY

Powers and roots

1 Start a new spreadsheet and enter the label **Powers** in cell A1.

2 In cell A2, enter **=2^3** to evaluate 2^3. The caret symbol, ^ (shift 6), is the symbol for power on a spreadsheet.

3 In cell A3, enter **=3^5** to evaluate 3^5.

4 In cell A4, enter **=4^2** to evaluate 4^2.

5 Write an appropriate formula to evaluate each expression in the given cell.

A5: 6.1^2 A6: 4.25^2 A7: $5^2\times 3^2$ A8: $(5\times 3)^2$

6 Does your answer in cell A7 equal your answer in cell A8?

7 Write an appropriate formula to evaluate each expression in the given cell.

A9: 2.9^3 A10: 15^4 A11: 8.3^5 A12: $(-1)^3+(-2)^4$ A13: $-1^3+(-2^4)$

8 Does your answer in cell A12 equal your answer in cell A13?

9 In cell B1, enter **Square roots**.

Foundation Standard Complex

10 In cell B2, enter **=sqrt(49)** to evaluate $\sqrt{49}$ (**=sqrt** stands for 'square root').

11 Write an appropriate formula to evaluate each square root in the given cell.

B3: $\sqrt{81}$ B4: $\sqrt{625}$ B5: $\sqrt{10000}$ B6: $\sqrt{71289}$ B7: $\sqrt{0.064}$ B8: $\sqrt{2.89}$

B9: $\sqrt{16.9}$ (Use **Format Cells** to round to two decimal places.)

B10: $\sqrt{\frac{81}{25}}$

2.10 Factor trees

Video
Factor trees

The prime number code

Skillsheets
Factors and divisibility

Prime factors by repeated division

Worksheet
Perfect and amicable numbers

Puzzles
Crossnumber puzzles

Crossnumber challenges

A **factor tree** can be used to find the prime factors of a number. Every number can be written as a product of its prime factors.

Example 26

Write 60 as a product of its prime factors.

SOLUTION

Draw a factor tree for 60.

Stop when all of the factors are prime.

So 60 as a product of its prime factors is:

$60 = 2 \times 2 \times 3 \times 5 = 2^2 \times 3 \times 5$.

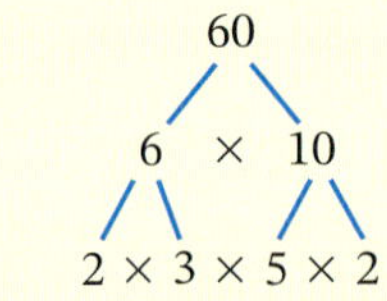

Note: It is possible to draw different factor trees using different factors for the same number, but the final list of prime factors should still be the same.

Here is another factor tree for 60:

$60 = 2 \times 2 \times 3 \times 5 = 2^2 \times 3 \times 5$.

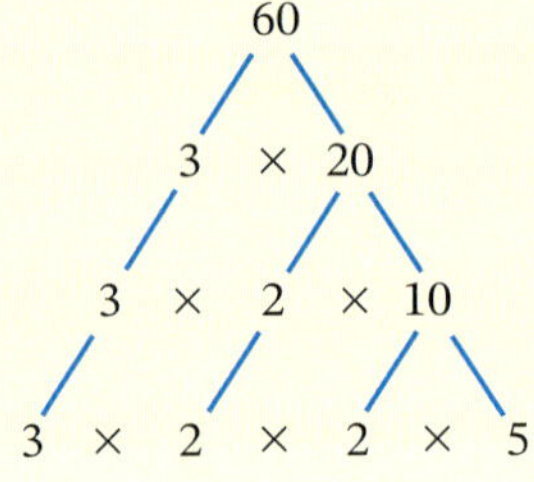

Using prime factors to find the HCF

Presentation
Highest common factors

Video
Highest common factor

Highest common factor

The **highest common factor** or **greatest common divisor** (GCD) of two (or more) numbers is the largest number that is a factor of **both (or all)** of these numbers.

Divisor is just another name for **factor**.

To find the HCF of two numbers using their prime factors:

1 circle common prime factors.

2 multiply them together.

Example 27

Find the HCF of 36 and 45:

a by listing factors. **b** using factor trees.

SOLUTION

a Factors of 36 = **1**, 2, **3**, 4, 6, **9**, 12, 18, 36.

Factors of 45 = **1**, **3**, 5, **9**, 15, 45.

The common factors are 1, 3, 9.

The HCF is **9**.

b Draw factor trees for 36 and 45.

Circle common prime factors: 3 and 3.

Multiply 3 and 3 to calculate the HCF:

HCF = $3 \times 3 =$ **9**.

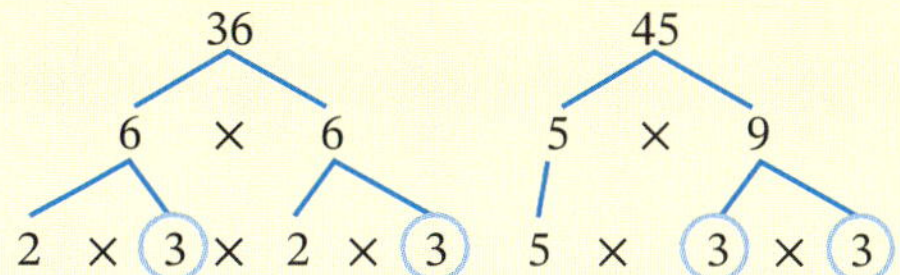

Video
HCF and factor trees

2.10

Using prime factors to find the LCM

Lowest common multiple

Also called **least** common multiple.

The **lowest common multiple** of two (or more) numbers is the smallest number that is a multiple of **both (or all)** of these numbers.

To find the LCM of two numbers using their prime factors:

1 circle common prime factors.

2 cross out **one** of each pair of common factors.

3 multiply the remaining factors.

Example 28

Find the LCM of 36 and 45:

a by listing multiples. **b** using factor trees.

SOLUTION

a Multiples of 36 = 36, 72, 108, 144, **180**, ...

Multiples of 45 = 45, 90, 135, **180**, 225, ...

The LCM is **180**.

b Draw factor trees for 36 and 45.

Circle common prime factors: 3 and 3.

Cross out the two 3s in the second factor tree.

Multiply the remaining factors to calculate the LCM.

LCM = $(2 \times 3 \times 2 \times 3) \times 5 =$ **180.**

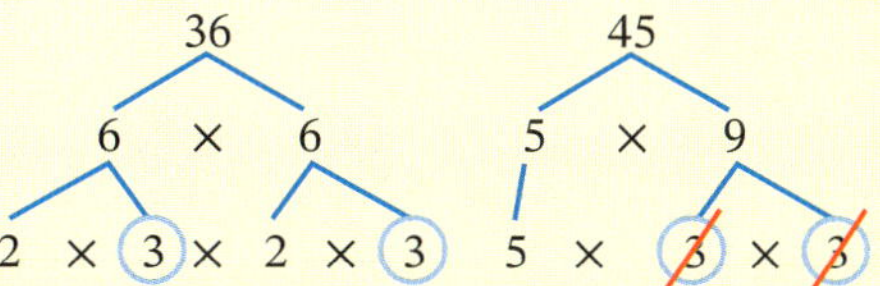

Finding square and cube roots using prime factors

Video Using factor trees to find square roots

Example 29

Use a factor tree to find the value of:

a $\sqrt{196}$ **b** $\sqrt[3]{216}$

SOLUTION

a Draw a factor tree for 196.

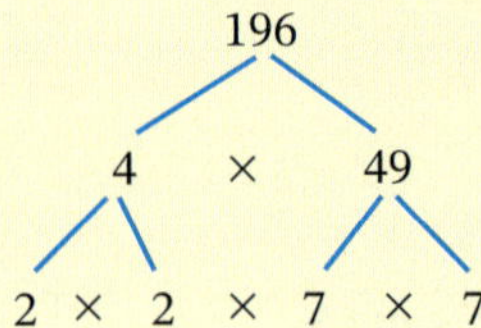

So $196 = 2 \times 2 \times 7 \times 7$

$$\therefore \sqrt{196} = \sqrt{2 \times 2 \times 7 \times 7}$$
$$= 2 \times 7$$
$$= 14$$

$\sqrt{2 \times 2} = 2$, $\sqrt{7 \times 7} = 7$

b Draw a factor tree for 216.

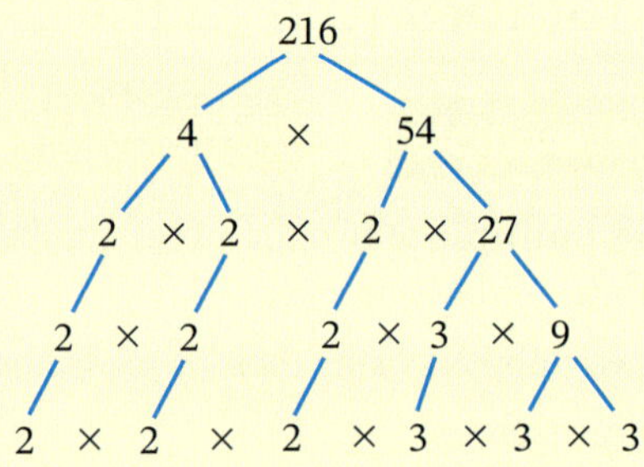

So $216 = 2 \times 2 \times 2 \times 3 \times 3 \times 3$

$$\therefore \sqrt[3]{216} = \sqrt[3]{2 \times 2 \times 2 \times 3 \times 3 \times 3}$$
$$= 2 \times 3$$
$$= 6$$

$\sqrt[3]{2 \times 2 \times 2} = 2$, $\sqrt[3]{3 \times 3 \times 3} = 3$

EXERCISE 2.10 ANSWERS ON P. 617

Factor trees

U F R C

1 Use a factor tree to write each number as a product of its prime factors in index notation.

R C

a 27	**b** 108	**c** 588	**d** 84
e 600	**f** 51	**g** 125	**h** 432

2 What is 42 expressed as a product of its prime factors? Select the correct answer **A**, **B**, **C** or **D**.

A 2×21	**B** $3 \times 3 \times 7$	**C** $2 \times 3 \times 7$	**D** 3×14

☐ Foundation ○ Standard ○ Complex

3 Draw two different factor trees for 80 and show that both give the same prime factors.

EXAMPLE 27

4 Find the HCF of each pair of numbers.

a 8, 28 **b** 14, 35 **c** 20, 5 **d** 45, 60
e 24, 90 **f** 72, 24 **g** 16, 48 **h** 30, 18

5 Use factor trees to find the HCF of each pair of numbers.

a 45, 60 (check against answer to Question **4d**)
b 208, 78 **c** 33, 176

EXAMPLE 28

6 Find the LCM of each pair of numbers.

a 9, 8 **b** 4, 6 **c** 10, 22 **d** 7, 4
e 6, 10 **f** 12, 8 **g** 9, 45 **h** 11, 10

7 Use factor trees to find the LCM of each pair of numbers.

a 90, 15 **b** 28, 36 **c** 16, 20

EXAMPLE 29

8 Use a factor tree to evaluate each root.

a $\sqrt{2025}$ **b** $\sqrt{441}$ **c** $\sqrt{256}$
d $\sqrt[3]{4096}$ **e** $\sqrt[3]{729}$ **f** $\sqrt[3]{5832}$

9 Explain why the 'factor trees' method for finding the LCM works, using 28 and 36 from question **7b** as an example.

10 **a** Find the HCF of 28 and 76.

b Use the formula (rule):

LCM of a and $b = \frac{a \times b}{\text{HCF of } a \text{ and } b}$ to find the LCM of 28 and 76.

INVESTIGATION

Index laws for multiplying and dividing

The terms 3^6 and 3^2 have the same base, 3.

$3^6 = 3 \times 3 \times 3 \times 3 \times 3 \times 3$ and $3^2 = 3 \times 3$.

What happens when we multiply them?

$3^6 \times 3^2 = (3 \times 3 \times 3 \times 3 \times 3 \times 3) \times (3 \times 3) = 3^8 = 6561$.

What happens when we divide 3^6 by 3^2?

$3^6 \div 3^2 = \frac{3 \times 3 \times 3 \times 3 \times 3^1 \times 3^1}{3^1 \times 3^1} = 3 \times 3 \times 3 \times 3 = 3^4 = 81$.

1 Copy and complete each pattern.

a $2^4 \times 2^3 = (2 \times \_\_ \times \_\_ \times \_\_) \times (2 \times \_\_ \times \_\_) = 2^{\square} = \_\_\_\_$

b $7^2 \times 7^2 = (7 \times \_\_) \times (\_\_ \times \_\_) = 7^{\square} = \_\_\_\_$

Foundation Standard Complex

c $10^2 \times 10^3 = (\_\_ \times \_\_) \times (\_\_ \times \_\_ \times \_\_) = 10^{\square} = \_\_\_\_$

d $4^3 \times 4^5 = (\_\_\_\_\_\_\_\_) \times (\_\_\_\_\_\_\_\_\_\_\_\_) = 4^{\square} = \_\_\_\_$

e $5^6 \times 5 = (\_\_\_\_\_\_\_\_\_\_) \times \_\_\_\_ = 5^{\square} = \_\_\_\_$

2 Copy and complete each pattern.

a $3^3 \times 3^7 = 3^{\square}$

b $2^5 \times 2^3 = 2^{\square}$

c $9^6 \times 9^4 = 9^{\square}$

d $6^3 \times 6^3 = 6^{\square}$

e $8^8 \times 8 = 8^{\square}$

3 Copy and complete each statement.

a $a^4 \times a^5 = \_\_\_\_\_\_$

b $a^m \times a^n = \_\_\_\_\_\_$

c When multiplying terms with the same base, ______________ the powers.

4 Copy and complete each pattern.

a $2^5 \div 2^3 = \frac{2\times2\times2\times2\times2}{2\times2\times2} = \_\_\_ \times \_\_\_ = 2^{\square} = \_\_\_\_$

b $7^8 \div 7^4 = \frac{7\times7\times7\times\_\_\times\_\_\times\_\_\times\_\_\times\_\_}{7\times7\times7\times\_\_} = \_\_\_ \times \_\_\_ \times \_\_\_ \times \_\_\_ = 7^{\square} = \_\_\_\_$

c $4^6 \div 4^3 = \frac{\_\_\times\_\_\times\_\_\times\_\_\times\_\_\times\_\_}{\_\_\times\_\_\times\_\_} = \_\_\_ \times \_\_\_ \times \_\_\_ = 4^{\square} = \_\_\_\_$

d $3^6 \div 3 = \_\_\_\_\_\_\_\_\_\_ = \_\_\_\_\_\_\_\_\_\_ = 3^{\square} = \_\_\_\_$

e $10^7 \div 10^2 = \_\_\_\_\_\_\_\_\_\_ = \_\_\_\_\_\_\_\_\_\_ = 10^{\square} = \_\_\_\_$

5 Copy and complete each pattern.

a $3^5 \div 3^2 = 3^{\square}$

b $10^9 \div 10^5 = 10^{\square}$

c $8^6 \div 8^4 = 8^{\square}$

d $6^7 \div 6 = 6^{\square}$

e $2^6 \div 2^5 = 2^{\square}$

6 Copy and complete each statement.

a $a^5 \div a^3 = \_\_\_\_\_\_$

b $a^m \div a^n = \_\_\_\_\_\_$

c When dividing terms with the same base, ______________ the powers.

2.11 Index laws for multiplying and dividing

Video
Powers, indices and exponents

Index laws

When **multiplying terms with the same base**, *add* the powers.

$$a^m \times a^n = a^{m+n}$$

When **dividing terms with the same base**, *subtract* the powers.

$$a^m \div a^n = a^{m-n}$$

These are called **index laws** because **index** is another name for 'power'. These can also be called **exponent laws** because **exponent** is also another name for 'power'.

2.11

Video
Index laws

Example 30

Use index notation to simplify each product.

a $5^3 \times 5^4$ **b** $6^4 \times 6^5$ **c** $2 \times 2^3 \times 2^4$

SOLUTION

a $5^3 \times 5^4 = 5^{3+4} = 5^7$ Adding the powers.

b $6^4 \times 6^5 = 6^{4+5} = 6^9$ Note: Do **not** multiply the bases.

$2 \times 2^3 \times 2^4 = 2^{1+3+4} = 2^8$ Note: $2 = 2^1$.

Video
Index laws

Example 31

Use index notation to simplify each quotient.

a $5^8 \div 5^3$ **b** $\frac{7^5}{7^2}$ **c** $6^{11} \div 6^{10}$

SOLUTION

a $5^8 \div 5^3 = 5^{8-3} = 5^5$ Subtracting the powers.

b $7^5 \div 7^2 = 7^{5-2} = 7^3$ Note: Do **not** divide the bases.

b $6^{11} \div 6^{10} = 6^{11-10} = 6^1 = 6$

EXERCISE 2.11 ANSWERS ON P. 618

Index laws for multiplying and dividing

U F R C

EXAMPLE 30

1 Use index notation to simplify each product.

R C

a $6^3 \times 6^2$ **b** $4^3 \times 4^4$ **c** $2^5 \times 2^6$ **d** 8×8^9

e $11^4 \times 11^4$ **f** $7^3 \times 7^7$ **g** 5×5^6 **h** $6^6 \times 6^{10}$

i $2^3 \times 2^2 \times 2^4$ **j** $5^3 \times 5^4 \times 5^2$ **k** $2^4 \times 2^4 \times 2^6$ **l** $3^2 \times 3^9 \times 3^3 \times 3^2$

2 Which term is equal to $2^3 \times 2^3 \times 2$? Select the correct answer **A**, **B**, **C** or **D**.

A 2^6 **B** 2^9 **C** 8^7 **D** 2^7

3 Simplify each product, and then evaluate it as a whole number.

a $5^2 \times 5^4$ **b** $7^6 \times 7^2$ **c** $3^4 \times 3^9$ **d** $4^3 \times 4^3$

e $9^2 \times 9^5$ **f** $6^5 \times 6^4$ **g** $8^4 \times 8^6$ **h** 7×7^3

EXAMPLE 31

4 Use index notation to simplify each quotient.

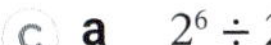

C

a $2^6 \div 2^2$ **b** $3^9 \div 3^7$ **c** $7^8 \div 7^2$ **d** $\frac{2^9}{2^4}$

e $5^{11} \div 5^9$ **f** $10^{11} \div 10^7$ **g** $11^{12} \div 11^5$ **h** $6^{18} \div 6^{11}$

i $2^5 \div 2^4$ **j** $\frac{16^{13}}{16^7}$ **k** $35^{12} \div 35$ **l** $\frac{20^7}{20^6}$

▷

Foundation Standard Complex

5 Which term is equal to $5^8 \div 5^2$? Select the correct answer **A**, **B**, **C** or **D**.

C **A** 1^6 **B** 5^4 **C** 1^5 **D** 5^6

6 Simplify each quotient, and then evaluate it as a whole number.

a $2^6 \div 2^2$ **b** $5^9 \div 5^7$ **c** $4^5 \div 4^2$ **d** $10^9 \div 10^4$

e $\frac{8^{11}}{8^8}$ **f** $6^{10} \div 6^2$ **g** $11^4 \div 11^3$ **h** $\frac{3^7}{3}$

7 Copy and complete this table of powers.

Number, n	n^2	n^3	n^4	n^5	n^6
2	4	8			
3		27		243	
4	16		256		
5		125			15 625

8 Use your answers from question **7** to evaluate each expression.

a $2^3 \times 2^2$ **b** $3^2 \times 3^2$ **c** $5^2 \times 5$ **d** $5^3 \times 5^2$

e $4^4 \times 4^2$ **f** $2^2 \times 3^2 \times 3^3 \times 2$ **g** $4^5 \div 4^2$ **h** $5^6 \div 5$

i $\frac{3^6}{3^4}$ **j** $2^5 \div 2^4$ **k** $\frac{4^4}{4^3}$ **l** $3^3 \div 3^3$

9 Copy and complete each equation.

R **a** $5^2 \times$ ____ $= 5^6$ **b** ____ $\div 3^4 = 3^3$ **c** $2^7 \div$ ____ $= 2^4$

10 Simplify each algebraic expression.

R **a** $a^3 \times a^7$ **b** $x^2 \times x^4$ **c** $b^4 \times b$ **d** $y^3 \times y^2$

C **e** $d^6 \div d^2$ **f** $x^5 \div x^3$ **g** $\frac{a^8}{a^7}$ **h** $\frac{r^5}{r}$

INVESTIGATION

More index laws

What happens when we raise 2^3 to a power of 4?
How do we simplify $(2^3)^4$ using index notation?

$(2^3)^4 = 2^3 \times 2^3 \times 2^3 \times 2^3 = 2^{3+3+3+3} = 2^{12}$

1 Copy and complete each pattern.

a $(32)^5 =$ __ × __ × __ × __ × __ = 3 ______ = 3__

b $(74)^2 =$ __ × __ = 7 ___ = 7__

c $(16^3)^3 =$ ____ × ____ × ____ = 16 ______ = 16__

d $(2^6)^4 =$ __ × __ × __ × __ = 2 ______ = 2__

e $(5^2)^3 =$ __ × __ × __ = 5______ = 5__

2 Copy and complete each statement.

a $(a^5)^2 =$ ______

b $(a^m)^n =$ ______

c When a term with a power is raised to another power, __________ the powers.

Foundation Standard Complex

3 We know that $3^2 = 3 \times 3$ and $3^1 = 3$, but what is the value of 3^0 (3 to the power of 0)?

a Copy and complete this pattern.

$$3^5 = 3 \times 3 \times 3 \times 3 \times 3 = 243$$
$$3^4 = 3 \times 3 \times 3 \times 3 = 81$$
$$3^3 = 3 \times \_\_\_ \times \_\_\_ = \_\_\_$$
$$3^2 = \_\_\_\_\_\_\_\_ = \_\_\_$$
$$3^1 = \_\_\_\_\_\_\_\_ = \_\_\_$$

b As the powers of 3 decrease by 1, what happens to its calculated value?

c According to the pattern, what should be the value of 3^0?

4 a Use index notation to simplify $2^5 \div 2^5$.

b But $2^5 \div 2^5 = \frac{2\times2\times2\times2\times2}{2\times2\times2\times2\times2}$. What is the answer to this?

c What is any number divided by itself equal to?

d What is the value of 2^0?

5 a Use index notation to simplify $5^2 \times 5^0$. What do you notice about the answer?

b If a number multiplied by 5^0 equals itself, what must be the value of 50?

c Use index notation to simplify $10^6 \div 10^0$. What do you notice about the answer?

d If a number divided by 10^0 equals itself, what must be the value of 100?

6 Copy and complete each statement.

a $4^0 = \_\_\_\_\_\_$

b $a^0 = \_\_\_\_\_\_$

c When any number is raised to a power of zero, the answer is always _________.

More index laws 2.12

ⓘ More index laws

When a **term with a power** is raised to **another power**, **multiply** the powers.

$$(a^m)^n = a^{m \times n}$$

Any number raised to the **power of zero** is equal to **1**.

$$a^0 = 1$$

Worksheets
What is the power question?
Power calculations

Video
Index laws

Example 32

Use index notation to simplify each expression.

a $(6^2)^5$ **b** $(7^3)^3$

SOLUTION

a $(6^2)^5 = 6^{2\times5} = 6^{10}$ Multiply the powers.

b $(7^3)^3 = 7^{3\times3} = 7^9$

Video
Index laws

Example 33

Evaluate each expression.

a 6^0 **b** $(3 \times 4)^0$ **c** 3×4^0

SOLUTION

a $6^0 = 1$ Any number raised to the power of 0 equals 1.

b $(3 \times 4)^0 = 12^0 = 1$

c $3 \times 40 = 3 \times 1 = 3$

EXERCISE 2.12 ANSWERS ON P. 618

More index laws

EXAMPLE 32

1 Use index notation to simplify each expression.

R C

a $(2^9)^2$	**b** $(5^2)^3$	**c** $(7^4)^5$	**d** $(8^2)^7$
e $(3^8)^3$	**f** $(9^6)^4$	**g** $(6^7)^2$	**h** $(11^5)^5$
i $(8^{10})^3$	**j** $(4^9)^5$	**k** $[(-2)^4]^7$	**l** $[(-7)^3]^6$

2 Which term is equal to $(3^4)^2$? Select the correct answer **A**, **B**, **C** or **D**.

A 9^6	**B** 6^4	**C** 3^6	**D** 3^8

3 Simplify each expression, and then evaluate it as a whole number.

a $(5^2)^4$	**b** $(2^5)^3$	**c** $(3^9)^2$	**d** $(6^3)^4$
e $(7^2)^3$	**f** $(4^6)^2$	**g** $(8^0)^5$	**h** $(4^3)^3$

EXAMPLE 33

4 Evaluate each expression.

a 9^0	**b** 3^0	**c** 10^0	**d** 16^0
e $(-2)^0$	**f** 6^0	**g** $(7-3)^0$	**h** $7-3^0$
i $(-5)^0$	**j** 7×3^0	**k** $(7 \times 3)^0$	**l** $2+3^0$
m $(6+6)^0$	**n** 6^0+6^0	**o** $6 \div 6^0$	**p** $10-10^0$

▷

Foundation Standard Complex

5 Evaluate $6^2 \div 2^0$. Select the correct answer **A**, **B**, **C** or **D**.

A 3 **B** 36 **C** 18 **D** 6

6 Copy and complete this table of powers of 2.

2^0	2^1	2^2	2^3	2^4	2^5	2^6	2^7	2^8	2^9	2^{10}	2^{11}	2^{12}
		4			32					1024		

7 Use your answers from question **6** to evaluate each expression.

R **a** $(2^2)^3$ **b** $(2^6)^2$ **c** $(2^2)^2$ **d** $(2^3)^3$

e $(2^0)^4$ **f** $(2^2)^4$ **g** $(2^{10})^0$ **h** $(2^7)^1$

8 Copy and complete each equation.

R **a** $(6^2)^{-} = 6^8$ **b** $(5^{-})^3 = 5^9$ **c** $(10^{-})^{-} = 10^{18}$

9 Simplify each algebraic expression.

R **a** $(a^4)^2$ **b** $(x^6)^6$ **c** $(d^2)^5$ **d** $(n^3)^6$

C **e** p^0 **f** $4 \times y^0$ **g** $(4 \times y)^0$ **h** $a^0 - b^0$

Investigating patterns in fractions and decimals 2.13

Computational thinking describes a set of techniques which you can use to solve a problem. Some of the techniques are:

- breaking complex problems into smaller, simpler problems (decomposition).
- recognising patterns and making connections between similar problems (pattern recognition).
- filtering out the parts of the pattern that are not needed to solve the problem (abstraction).
- developing a step-by-step process to solve the problem which can be used by computers or humans (algorithms).

Worksheet Fraction families

EXERCISE 2.13 ANSWERS ON P. 618

Investigating patterns with fractions and decimals

U F R C

In the following exercise, we will use our computational thinking and reasoning skills to investigate patterns with fractions and decimals.

Foundation Standard Complex

1 What types of fractions convert to terminating decimals and what types convert to recurring decimals? For each fraction presented in the table below, convert it to a decimal and also find the prime factors of its denominator. Copy and complete the table.

Fraction	Decimal	Terminating or recurring?	Prime factors of denominator
$\frac{1}{2}$	0.5	Terminating	2
$\frac{1}{3}$	$0.\dot{3}$	Recurring	3
$\frac{1}{4}$			
$\frac{1}{5}$			
$\frac{1}{6}$			
$\frac{1}{7}$			
$\frac{1}{8}$			
$\frac{1}{9}$	$0.\dot{1}$	Recurring	3
$\frac{1}{10}$			
$\frac{1}{11}$			
$\frac{1}{12}$			
$\frac{1}{13}$			
$\frac{1}{14}$			
$\frac{1}{15}$			
$\frac{1}{16}$			

2 Using the results from question **1**, determine what types of fractions convert to terminating decimals.

Foundation Standard Complex

3 If a fraction has a large denominator, what does its decimal form look like? What happens as the denominator of $\frac{1}{n}$ increases, where n is a positive number? Copy and complete this table.

Fraction	Decimal
$\frac{1}{20}$	
$\frac{1}{50}$	
$\frac{1}{100}$	
$\frac{1}{500}$	
$\frac{1}{1000}$	
$\frac{1}{2000}$	
$\frac{1}{5000}$	

4 Using the results from question **3**, write down what you notice about the value of the decimal as the denominator of the fraction gets larger.

5 For the fraction $\frac{n-1}{n}$, whose numerator is 1 less than its denominator, what does its decimal form look like? What happens when the value of the denominator n increases? Copy and complete this table.

Fraction	Decimal
$\frac{19}{20}$	
$\frac{49}{50}$	
$\frac{99}{100}$	
$\frac{499}{500}$	
$\frac{999}{1000}$	
$\frac{1999}{2000}$	
$\frac{4999}{5000}$	

6 Using the results from question **5**, write down what you notice about the decimal value of $\frac{n-1}{n}$ when the value of n increases.

☐ Foundation ○ Standard ⬡ Complex

7 What is the sum of $\frac{1}{n}$ and $\frac{1}{n+1}$, where n is any number? Copy and complete this table.

n	$\frac{1}{n}+\frac{1}{n+1}$	Fraction
1	$\frac{1}{1}+\frac{1}{2}$	$\frac{3}{2}$
2		
3		
4		
5		
6		
7		
8		
9		
10		
11		

8 Using the results from question **7**, write down what patterns you notice about the value of $\frac{1}{n}+\frac{1}{n+1}$.

☐ Foundation ◯ Standard ⬡ Complex

POWER PLUS ANSWERS ON P. 619

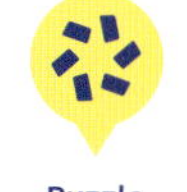

2.13

1 Even if we don't know the exact value of a surd, we can use the rule $\sqrt{ab} = \sqrt{a}\times\sqrt{b}$ to simplify a surd if one of its factors is a square number. Study each example:

- $\sqrt{40} = \sqrt{4\times10} = \sqrt{4}\times\sqrt{10} = 2\sqrt{10}$
- $\sqrt{32} = \sqrt{16\times2} = \sqrt{16}\times\sqrt{2} = 4\sqrt{2}$
- $\sqrt{36} = \sqrt{9\times4} = \sqrt{9}\times\sqrt{4} = 3\times2 = 6$
- $\sqrt{125} = \sqrt{25\times5} = \sqrt{25}\times\sqrt{5} = 5\sqrt{5}$

Simplify each surd.

a $\sqrt{81}$	b $\sqrt{8}$	c $\sqrt{27}$	d $\sqrt{50}$
e $\sqrt{72}$	f $\sqrt{45}$	g $\sqrt{180}$	h $\sqrt{128}$
i $\sqrt{28}$	j $\sqrt{200}$	k $\sqrt{48}$	l $\sqrt{245}$

2 a Use index notation to simplify $2^4 \div 2^5$.

b But $2^4 \div 2^5 = \frac{2\times2\times2\times2}{2\times2\times2\times2\times2}$. What is the answer to this?

c What is the value of 2^{-1}?

d What is the value of 2^{-2}?

3 a Does $(-2)^5 = -2^5$?

b Does $(-3)^4 = -3^4$?

c For what values of n does $(-1)^n = -1$?

d For what value of x is x^0 not defined?

4 Scientific notation is a special way of writing very large or very small numbers using powers of 10. For example:

- 25 600 can be written as $2.56 \times 10 \times 10 \times 10 \times 10 = 2.56 \times 10^4$ because $2.56 \times 10^4 = 2.56 \times 10 \times 10 \times 10 \times 10$.
- 0.001678 can be written as $1.678 \div 10 \div 10 \div 10 = 1.678 \times 10^{-3}$ because $1.678 \times 10^{-3} = 1.678 \div 10 \div 10 \div 10$.

Note that the first number is a decimal between 1 and 10 and the second number is a power of 10.

Convert each number written in scientific notation to an ordinary number.

a 2.4×10^4	b 4.55×10^5	c 9.33×10^{-2}
d 6.38×10^{-2}	e 8.7×10^6	f 5.82×10^{-3}
g 1.26×10^{-1}	h 2.69×10^3	i 3.14×10^8

Can you see a quick way of writing the answer each time?

5 Write each number in scientific notation.

a 12 000	b 345 000 000	c 0.007
d 4000	e 0.0005	f 0.00041
g 1 920 000	h 0.000 361	i 0.000 000 063 7

6 a What is the largest number that can be displayed on your calculator?

b What is the smallest?

2 CHAPTER REVIEW

Language of maths

Quiz
Language of maths 2

base	consecutive	cube root	decimal places
estimate	evaluate	factor tree	highest common factor
grouping symbol	index laws	index notation	integer
long division	lowest common multiple	mental calculation	order of operations
prime factors	product	quotient	recurring decimal
round	square root	surd	terminating decimal

1 If you round a decimal to the nearest hundredth, how many decimal places is this?

2 Write a numerical expression that requires the use of:

a 'order of operations' rules **b** long division

3 Write 'the cube root of -64' using mathematical symbols, and then evaluate it.

4 In 8^5, what are the names given to the '8' and the '5'?

5 What is another name for:

a grouping symbols? **b** index?

6 What diagram is used to write a number as a product of its prime factors?

Topic summary

Worksheet
Mind map: Working with numbers

- What parts of this chapter do you remember from last year?
- Are there any parts of this chapter that you do not understand?
- Discuss any problems with your teacher or a friend.

Print (or copy) and complete this mind map of the topic, adding detail to its branches and using pictures, symbols and colour where needed. Ask your teacher to check your work.

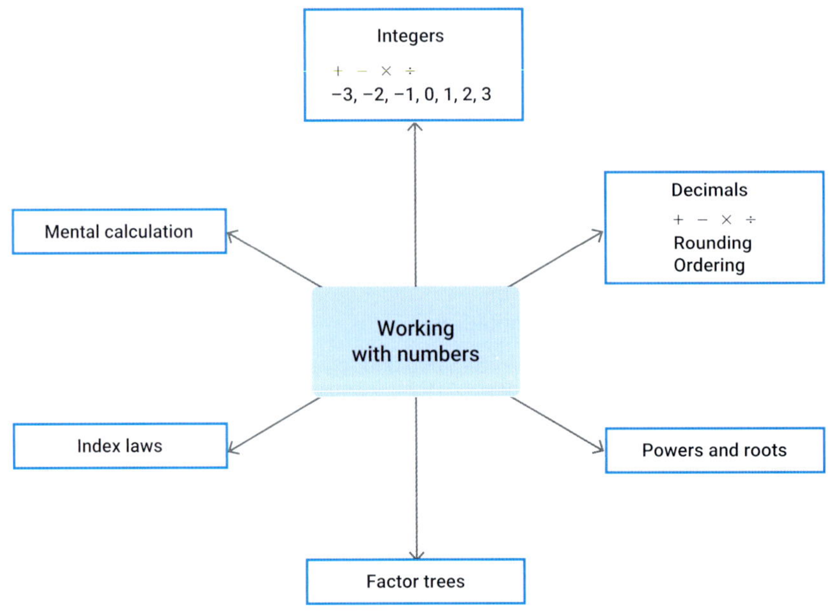

2 TEST YOURSELF

ANSWERS ON P. 619

1 Use mental calculation strategies to evaluate each expression. 2.01

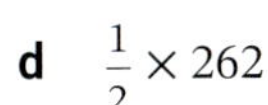

a 7×200 **b** 14×15 **c** $8000 \div 50$

d $\frac{1}{2} \times 262$ **e** $68 + 19$ **f** 35×11

g 65×4 **h** 8×60 **i** $26 + 61 + 156 + 19 + 14$

j $8 \times 25 \times 4$ **k** 23×12 **l** $168 \div 4$

m $2700 \div 300$ **n** 40×70 **o** 37×9

p 18×5 **q** $384 \div 8$ **r** $970 - 245$

s $4 \times 8 \times 5$

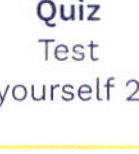

Quiz
Test yourself 2

2 Evaluate each quotient. 2.01

a $812 \div 7$ **b** $846 \div 9$ **c** $396 \div 18$ **d** $2139 \div 23$

3 Evaluate each expression. 2.02

a $-4 + 6$ **b** $5 - (-2)$ **c** $-3 - 8$ **d** $-2 - (-2)$

4 Death Valley in the USA is 397 m below sea level and Mt Everest is 8840 m above sea level. What is the vertical distance between the bottom of Death Valley and the top of Mt Everest? 2.02

5 On 17 July, the maximum temperature in Armidale was 17°C and the minimum temperature was −6°C. By how much had the temperature risen during the day? 2.02

6 Evaluate each product. 2.03

a -2×7 **b** $-3 \times (-8)$ **c** $(-7)^2$ **d** $16 \times (-5)$

7 Evaluate each quotient. 2.04

a $36 \div (-4)$ **b** $-28 \div 7$ **c** $24 \div (-2)$ **d** $-15 \div (-5)$

8 Evaluate each expression. 2.05

a $2 + (13 - 8) \times 11$ **b** $14 - 5 \times (-3)$ **c** $15 - 18 \div 3 \times 2$

d $80 - [(7 - 10) \times (-6)]$ **e** $\frac{8-3}{100 \div 4}$ **f** $\frac{15 \times (-6+8)}{24 \div 2^2}$

9 Write these decimals in descending order: 0.417, 0.47, 0.147, 0.471. 2.06

10 Write each number correct to the number of decimal places shown in the brackets. 2.06

a 0.473 [1] **b** 13.1051 [2] **c** 98.4273 [3]

11 To complete the wiring of a house, Joe, the electrician, needed these lengths of cables: 2.3 m, 1.9 m, 4.2 m and 3.8 m. How much cable did he need altogether? 2.07

12 Evaluate each expression. 2.07

a $37.4 - 6.9$ **b** $13.3 + 0.82 + 5.6$ **c** 2.6×4 **d** 5.5×0.6

e 0.71×1.3 **f** $(2.5)^2$ **g** $9.57 \div 4$ **h** $8.12 \div 0.7$

13 A drink bottle holds 0.5 L. How many drink bottles can be filled from a container that holds 7.5 L? 2.08

□ Foundation ○ Standard ⬡ Complex

CHAPTER 2 TEST YOURSELF

2.09

14 Convert each fraction to a decimal.

a $\frac{2}{9}$ **b** $\frac{7}{8}$ **c** $\frac{1}{6}$

15 Evaluate each expression.

a 7^3 **b** 11^5 **c** $(-5)^4$

d $\sqrt{400}$ **e** $\sqrt[3]{27}$ **f** $\sqrt[3]{-125}$

16 Without using a calculator, estimate $\sqrt{31}$, correct to one decimal place.

17 Copy and complete:

a
$$\begin{aligned} 20^2 &= (\_\_ \times 10)^2 \\ &= \_\_^2 \times 10^2 \\ &= \_\_ \times \_\_ \\ &= \_\_ \end{aligned}$$

b
$$\begin{aligned} \sqrt{196} &= \sqrt{} \times \sqrt{49} \\ &= \_\_ \times \_\_ \\ &= \_\_ \end{aligned}$$

18 **a** Find the HCF of 36 and 42.

b Find the LCM of 20 and 14.

19 Use factor trees to write 132 and 88 as products of their prime factors, and then use them to find the HCF of 132 and 88.

20 Use factor trees to write 12 and 45 as products of their prime factors, and then use them to find the LCM of 12 and 45.

21 Use a factor tree to write 784 as a product of its prime factors, and then use it to evaluate $\sqrt{784}$.

22 Simplify each expression using index notation.

a $4^2 \times 4^5$ **b** $6^4 \times 6^7$ **c** $12^4 \times 12^4$ **d** $9^3 \times 9$

e $2^7 \div 2^3$ **f** $5^9 \div 5^6$ **g** $3^8 \div 3^7$ **h** $6^{11} \div 6^4 \times 6^5$

23 Simplify each expression using index notation.

a $(3^4)^2$ **b** $(7^2)^5$ **c** $(6^3)^3$ **d** $(3^5)^3$

24 Evaluate each expression.

a 3^0 **b** 6^0 **c** 4×4^0 **d** $9 - 9^0$

e 10^0 **f** $(12 \times 3)^0$ **g** $(5 - 3)^0$ **h** $5^0 - 3^0$

☐ Foundation ○ Standard ⬡ Complex

25 What is the difference between $\frac{1}{n}$ and $\frac{1}{n+1}$, where n is any number? 

Copy and complete this table.

n	$\frac{1}{n}-\frac{1}{n+1}$	Fraction
1	$\frac{1}{1}-\frac{1}{2}$	$\frac{1}{2}$
2	$\frac{1}{2}-\frac{1}{3}$	$\frac{1}{6}$
3		
4		
5		
6		
7		
8		
9		
10		
11		

26 Using the results from question **25**, write down what patterns you notice about the value of $\frac{1}{n}-\frac{1}{n+1}$. 

☐ Foundation ○ Standard ○ Complex

3

ALGEBRA

Algebra

Algebra is the study of mathematical patterns and rules. The subject can be traced back over 4000 years ago to the ancient Babylonians. Today, all modern technology relies on algebra – the internet, satellites, computers, TVs and all other digital devices rely on the application of algebra.

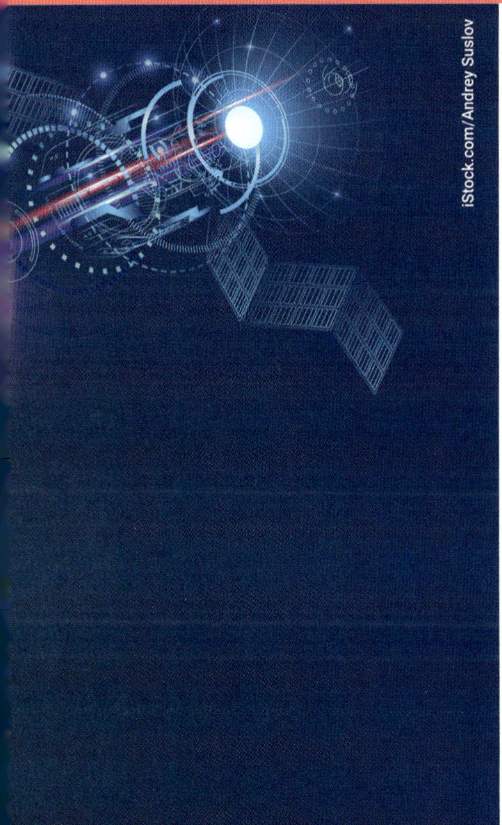

iStock.com/Andrey Suslov

Chapter outline

		Proficiencies				
3.01	Variables	U	F		R	C
3.02	Substitution	U	F	PS		
3.03	Creating algebraic expressions	U	F		R	C
3.04	The laws of arithmetic	U	F		R	C
3.05	Adding and subtracting terms	U	F		R	C
3.06	Multiplying and dividing terms	U	F		R	C
3.07	Extension: The index laws*	U	F			
3.08	Expanding expressions	U	F		R	C
3.09	Factorising algebraic terms	U	F		R	
3.10	Factorising expressions	U	F		R	C
3.11	Factorising with negative terms	U	F		R	C
3.12	Extension: Expanding binomial products*	U	F		R	C

***YEAR 9 EXTENSION**

U = Understanding
F = Fluency
PS = Problem solving
R = Reasoning
C = Communication

Wordbank

algebraic expression An expression that describes a quantity by using variables and numerals, such as $3x + 4y - 8$

algebraic term A part of an algebraic expression, such as $3x$ in $4x^2 + 3x - 11$

evaluate To find the value of an expression

expand To rewrite an expression such as $4(2k + 4)$ without brackets

factorise The opposite of expand; to rewrite an expression, such as $8k + 16$, with brackets

like terms Terms that have exactly the same variables, for example, $5p$ and $2p$

substitute To replace a variable with a number

variable (or **pronumeral**) A quantity that can take on different values, represented by a symbol such as a letter of the alphabet

Quiz
Wordbank 3

Videos (15):

3.01 Algebraic expressions • Simplifying algebraic expressions 2
3.02 Substitution 2 • Substitution 1
3.03 From words to algebraic expressions
3.05 Adding and subtracting like terms
3.06 Multiplying terms • Dividing terms
3.07 Powers, indices and exponents • Simplifying with the index laws • Index laws 1 • Index laws 2
3.08 Expanding expressions • Expanding expressions 2
3.09 HCF of algebraic terms
3.10, 3.11 Factorising expressions
3.12 Expanding binomial products

Twig videos (2):

3.02 The heartbeat formula • Heptathlon

PhET interactives (2):

3.05 Expression exchange
3.08 Area model Algebra

Quizzes (6):

- Wordbank 3
- SkillCheck 3
- Mental skills 3A
- Mental skills 3B
- Language of Maths 3
- Test Yourself 3

Skillsheets (5):

3.01 Order of operations
3.03 Algebraic expressions
3.05, 3.08 Algebra using diagrams
3.09, 3.10, 3.11 Algebra using diagrams • Factorising using diagrams
3.12 Area diagrams

Worksheets (11):

SkillCheck What does the symbol mean?
3.02 Substitution
3.03 Writing algebraic expressions • What does the symbol mean?
3.06 Simplifying algebraic expressions • Perimeter and area • Why aren't they the same?
3.08 Expanding brackets
3.11 Algebra review • Algebraic expressions review
3.12 Binomial products
Mind map Algebra

Puzzles (15):

3.01 Algebra bingo game
3.02 Formula 1 game • Substitution game
3.03 Generalised arithmetic
3.05 Algebra bingo game • Adding and subtracting like terms
3.07 Indices puzzle
3.08 Expandominoes • Expanding • Expand the brackets • Expand negative brackets
3.10 Factorising
3.11 Expandominoes • Factorising puzzle • Factorising with negative terms
3.12 Expanding brackets

Technology (1):

3.12 Expanding binomials

Presentations (1):

3.05 Collecting like terms

Spreadsheets (1):

3.12 Expanding binomials

Nelson MindTap

To access resources above, visit **cengage.com.au/nelsonmindtap**

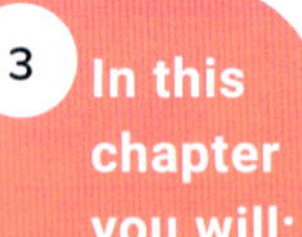

3 **In this chapter you will:**

- ✓ learn about variables as a way of representing numbers using letters.
- ✓ generalise number laws and properties to algebraic terms and expressions.
- ✓ evaluate algebraic expressions by substituting values for each variable.
- ✓ convert between word and algebraic expressions.
- ✓ simplify algebraic expressions using the four arithmetic operations.
- ✓ (YEAR 9 EXTENSION) apply index laws to algebraic expressions with whole number and zero powers.
- ✓ use the distributive law to expand algebraic expressions with grouping symbols (brackets).
- ✓ factorise algebraic terms and expressions.
- ✓ (YEAR 9 EXTENSION) expand binomial products.

SkillCheck

ANSWERS ON P. 620

Quiz
SkillCheck 3

1 Evaluate each expression.

a $3 - 4$ **b** $-5 + 9$ **c** $-8 - 2$

d $3 - (-2)$ **e** -3×6 **f** $-12 \div (-4)$

g $-12 \div 4 - 10$ **h** $-4 \times (-8)$ **i** $(-6)^2$

j 2^3 **k** $(-1)^3$ **l** 2×3^2

2 Simplify each algebraic expression.

a $7 \times p$ **b** $a \times b$ **c** $k \times k$

d $n \times n \times n$ **e** $2 \times p \times m$ **f** $p \times k \times d$

g $y \times 6$ **h** $t \times 3 \times q$ **i** $6d - 7d$

j $q + q + q$ **k** $a + a + a + a$ **l** $1x$

3 Choose which operation (+,–, × or ÷) is related to each word.

a sum **b** product **c** difference **d** quotient

e decrease **f** increase **g** more than **h** less than

4 If $x = 8$ and $y = 2$, evaluate each expression.

a $x + y$ **b** $3x$ **c** $x - y$ **d** $5y$

e xy **f** $9 - y$ **g** $x + 17$ **h** $2x + 4$

i $3y - 2$ **j** $x \div y$ **k** $5x \div y$ **l** $4x + 5y$

5 Find:

a the sum of 8 and 5 **b** double 9

c 6 increased by 10 **d** the difference between 14 and 11

e the product of 3 and 7 **f** how many times 5 divides into 40

Worksheet
What does the symbol mean?

Variables 3.01

In algebra, a **variable** or **pronumeral** is a symbol, usually a letter of the alphabet, that stands for a number. It is called a variable because its value can vary (change).

When writing **algebraic expressions**, we use the following abbreviations.

- $3 \times k = 3k$ for multiplication, we leave out the '×' symbol.
- $m \div 4 = \dfrac{m}{4}$ for division, we can write in fraction form.
- $r \times r = r^2$ for powers.
- $1x = x$ for multiplying by 1.
- $c \times a \times 6 = 6ac$ write the number first, then the variables in alphabetical order.

Video
Algebraic expressions

Puzzle
Algebra bingo game

Video
Simplifying algebraic expressions 2

Example 1

Simplify each expression.

a $p+p+p+p$ **b** $d \times 5$ **c** $r \div 7$

d $m \times n \times 6$ **e** $4v + 2v$ **f** $5 \times k \times 2 \times h$

g $y \times y \times y$ **h** $5a - 4a$ **i** $b \times b \times 15$

SOLUTION

a $p+p+p+p=4p$ — 4 lots of p.

b $d \times 5 = 5d$ — Write the number first.

c $r \div 7 = \frac{r}{7}$

d $m \times n \times 6 = 6mn$ — Number first, then variables in alphabetical order.

e $4v + 2v = 6v$ — $(v+v+v+v)+(v+v) = 6$ lots of v

f $5 \times k \times 2 \times h = 5 \times 2 \times k \times h$
$= 10hk$

g $y \times y \times y = y^3$

h $5a - 4a = 1a$ — $(a+a+a+a+a)-(a+a+a+a) =$ one a left
$= a$ — Just a: no need to write the '1'.

i $b \times b \times 15 = 15b^2$ — The '2' (squared) belongs to the b only.

Example 2

Skillsheet
Order of operations

Use order of operations to simplify each expression.

a $2 \times m - 4$ **b** $15 - k \div 3$ **c** $13 \div (d \times d) + 4$

SOLUTION

a $2 \times m - 4 = 2m - 4$ — $\times$ first, then $-$

b $15 - k \div 3 = 15 - (k \div 3)$ — $\div$ first, then $-$
$= 15 - \frac{k}{3}$

c $13 \div (d \times d) + 4 = 13 \div d^2 + 4$ — ( ) first, then $\div$, then $+$
$= \frac{13}{d^2} + 4$

Example 3

Write each expression in expanded form.

Note that expanding is the opposite of simplifying.

a $5bc$ **b** $-4kr^2$ **c** $x^2 - \frac{t}{11}$

SOLUTION

a $5bc = 5 \times b \times c$

b $-4kr^2 = -4 \times k \times r \times r$

c $x^2 - \frac{t}{11} = x \times x - t \div 11$

EXERCISE 3.01 ANSWERS ON P. 620

Variables

1 Simplify each expression.

a	$w \times 5$	**b**	$m \times m$	**c**	$3 \times c \times b \times a$
d	$k + k + k + k + k + k$	**e**	$m \div 8$	**f**	$f + f + f$
g	$12m + 4m$	**h**	$7 \times w \times 3 \times x$	**i**	$11p - 2p$
j	$4 \times t \times t$	**k**	$6d - 5d$	**l**	$b \times c \times d \times 16$
m	$26 \div n$	**n**	$h \times 9 \times h$	**o**	$9x + 11x$
p	$a + a - a$	**q**	$12q - 4q$	**r**	$2 \times c \times 6 \times n$
s	$3 \times r \times d \times (-2) \times d$	**t**	$a + a + a + c + c$	**u**	$a \times a \times a \times x \times x$

2 Use order of operations to simplify each expression.

a	$13 \div (a \times 2)$	**b**	$9 + 4 \times m$	**c**	$(e - 6) \div 5$
d	$4 \times t - 8$	**e**	$y \times y + z \times z$	**f**	$k \div 9 - n$
g	$p \div (9 + n)$	**h**	$22 - e \div 2$	**i**	$11 + u \times u \times 11$

EXAMPLE 2

3 Which expression is equal to $m \times m \times 5 + 2$? Select the correct answer **A**, **B**, **C** or **D**.

A $5m^2 + 2$ **B** $5m^3 + 2$ **C** $5m + 2$ **D** $m^2 + 7$

4 Explain the meaning of:

a pq **b** $\frac{3y}{x}$ **c** $1x = x$

5 Which expression is **not** equal to $4p$? Select the correct answer **A**, **B**, **C** or **D**.

A $5p - p$ **B** p^4 **C** $p + p + p + p$ **D** $p \times 4$

EXAMPLE 3

6 Write each expression in expanded form.

a	$8st$	**b**	$-2y^2$	**c**	$\frac{8}{f}$	**d**	$r^2 + t^2$
e	$4b^2d$	**f**	$5xy - 2a$	**g**	$\frac{x-2}{3}$	**h**	q^3
i	$16 - 2d^2$	**j**	$\frac{-4d}{3}$	**k**	h^2i^2	**l**	$4j - \frac{9}{d}$

7 What is $7e^3$ in expanded form? Select the correct answer **A**, **B**, **C** or **D**.

A $7e \times 7e \times 7e$ **B** $7e + 7e + 7e$ **C** $7 \times e + e + e$ **D** $7 \times e \times e \times e$

Foundation Standard Complex

INVESTIGATION

The laws of arithmetic

We can use algebraic symbols to describe general laws about arithmetic.

For example, if we add zero to any number, the answer is still that number.

If we let N stand for any number: $N + 0 = N$

In groups of 2–4, answer the following questions.

1 Describe in words what each number rule below means.

a $N \times 1 = N$ b $N \times 0 = 0$ c $a - b \neq b - a$

d $a + b + c = b + a + c$ e $N - 0 = N$ f $ab = ba$

2 Write each rule algebraically using variables.

a Any number divided by 1 equals itself.

b Multiplying a number by 8 is the same as doubling it three times.

c Any three numbers can be multiplied together in any order.

d Any number added to itself is the same as multiplying that number by 2.

e Any number subtracted from itself equals 0.

f Any number multiplied by its reciprocal equals 1.

3 Is each equation true or false? Test your answer by substituting a number for each variable and checking.

a $a \div b = b \div a$ b $N \div N = 1$

c $4a - a = 4$ d a is a factor of a

e If N is even, then $N + 3$ is odd f $\frac{1}{2}N = N - 2$

g $a + (-a) = 2a$ h $0 \div N = 0$

i $N \div 0 = 0$ j $N \div 1 = N$

4 If k is an odd number, what is an expression for:

a the previous odd number? b the next even number?

DID YOU KNOW?

The X factor

In algebra the letter x is used quite frequently as a variable. During the 9th century, the Arabic word *al–jabr* ('algebra') described the process of moving variables from one side of an equation to the other to find the value of an unknown. The word for 'thing' or 'object' (the unknown number) in Arabic was *shei*, translated into Greek as *xei* and shortened to x. *Xenos* is also the Greek word for unknown, stranger or foreigner, which could explain why early European mathematicians used the letter x in algebra.

One further explanation is that ancient scholars could only write using feathers dipped in ink. The two easiest letters to write were x and y. Mathematicians needed variables they could write quickly without mixing them up. As a result, x and y are still the two most commonly used variables in algebra.

Find examples of X–words that involve something unknown, such as X–ray, Brand X or X factor.

Substitution

3.02

The word **substitute** means replacing one thing with another. In algebra, substitution means to replace a variable with a value to evaluate an algebraic expression.

Puzzles
Formula 1 game
Substitution game

Worksheet
Substitution

Videos
Substitution 2
Substitution 1
The heartbeat formula
Heptathlon

Substituting into formulas

A **formula** is a general rule that is written as an algebraic equation and shows the relationship between variables. Solving mathematical problems often involves substituting values into formulas.

Example 4

If $p = 3$ and $q = 14$, evaluate each expression.

a $2p + q$ **b** $\dfrac{pq+2}{4}$ **c** $5q - p^2$

SOLUTION

a $2p+q = 2\times3+14$
$= 20$

b $\dfrac{pq+2}{4} = \dfrac{3\times14+2}{4}$
$= \dfrac{44}{4}$
$= 11$

c $5q - p^2 = 5\times14 - 3^2$
$= 61$

Example 5

The formula for the perimeter of a rectangle is $P = 2l + 2w$, where P is the perimeter, l is the length and w is the width. Use this formula to calculate the perimeter of a rectangle with length 12 m and width 9.5 m.

SOLUTION

When length $l = 12$ and width $w = 9.5$,

$P = 2l + 2w$
$= 2\times12 + 2\times9.5$
$= 43$ m

 **EXERCISE 3.02** ANSWERS ON P. 620

Substitution

U F PS 

1 If $x = 5$ and $y = 6$, what is the value of $2x - y$? Select the correct answer **A**, **B**, **C** or **D**.

A 19 **B** -2 **C** 4 **D** 16

EXAMPLE 4

Foundation Standard Complex

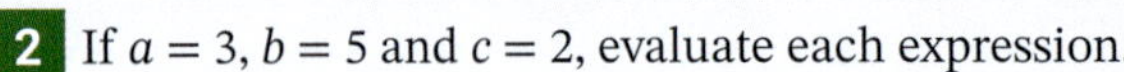

2 If $a = 3$, $b = 5$ and $c = 2$, evaluate each expression.

a $a + b$ **b** $4a$ **c** c^2 **d** $a + b + c$

e abc **f** $3b - a$ **g** $c + a - b$ **h** $4a - 2c$

i $10 - b$ **j** $ab + c^2$ **k** $c - b$ **l** $\frac{2a}{c}$

3 If $d = 10$, $e = -3$ and $f = 5$, evaluate each expression.

a def **b** f^2 **c** e^3 **d** $d \div f$

e $de \div 2$ **f** $d^2 - f^2$ **g** $7e$ **h** $3e + 2d - f$

i $f + d - e$ **j** $\frac{d}{2} - 3f$ **k** $6ef$ **l** $\frac{e+f}{4}$

4 If $p = -3$, what is the value of $2p^2$? Select the correct answer **A**, **B**, **C** or **D**.

A 36 **B** 18 **C** −18 **D** −36

5 If $y = -1$ and $z = 4$, evaluate each expression.

a $3 - 4z$ **b** $-3y^2$ **c** $5(z - 2)$ **d** $y(z + 1)$

e $\frac{y}{z}$ **f** $\frac{z}{4}$ **g** $\frac{y+z}{9}$ **h** $\frac{10}{3+y}$

6 If $n = 6$, which of the following expressions is equal to 22? Select the correct answer **A**, **B**, **C** or **D**.

A $\frac{n+4}{2}$ **B** $\frac{2n+4}{2}$ **C** $\frac{n^2}{2}+4$ **D** $\frac{n^2+4}{2}$

EXAMPLE 5

7 The formula for the area of a triangle is $A = \frac{1}{2}bh$, where A is the area of the triangle, b is the length of the base and h is the height.

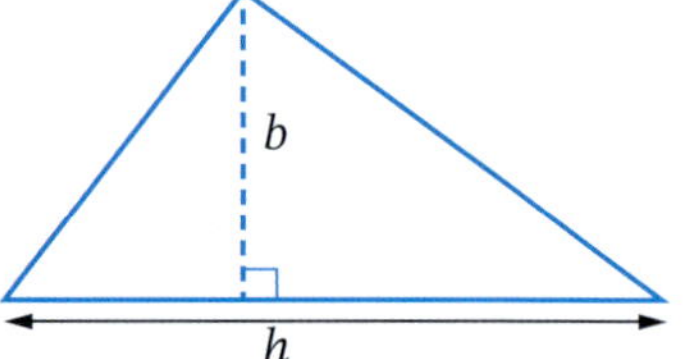

Find the area of a triangle with base length 12 m and height 7 m.

8 A plumber charges according to the formula $C = 42h + 65$, where C is the charge in dollars and h is the number of hours worked. How much does the plumber charge for working for $3\frac{1}{2}$ hours?

Shutterstock.com/kurhan

9 PS The formula for converting Celsius temperatures to Fahrenheit temperatures is $F = \frac{9}{5}C + 32$, where F is the temperature in Fahrenheit and C is the temperature in celsius. Use the formula to convert 26°C to the Fahrenheit.

Foundation Standard Complex

10 The number of hours of sleep recommended for children has the formula $H = 17 - \frac{A}{2}$, where H is the number of hours and A is the age of the child.

PS Find the number of hours of sleep recommended for a six years old.

11 The formula for the cost of a party is $C = 45n + 500$, where C is the cost in dollars and n is the number of guests. Use this formula to calculate the cost of a party for 120 guests.

PS

12 The time (t minutes) taken to travel to school depends on how far the student walks and how far the student travels on the bus. The formula for the time taken is $t = 11w + 4b$, where w is the walking distance and b is the distance travelled by bus.

PS

a Find the time taken for each student to travel to school:

i Michael, who walks 1 km and goes 5 km by bus.

ii Vittoria, who walks 4 km and goes 3 km by bus.

b If school starts at 9 a.m., what is the latest time Wendy can leave home if she needs to walk 2 km and go by bus for 7 km and still reach school on time?

Creating algebraic expressions 3.03

Example 6

If x is a variable representing any number, write an algebraic expression for:

a 4 times the number.

b the number divided by 6.

c the previous consecutive number.

d one-third of the number.

consecutive means 'following in order': for example, 7, 8, 9 are consecutive numbers

Video From words to algebraic expressions

Skillsheet Algebraic expressions

Worksheets Writing algebraic expressions

What does the symbol mean?

Puzzle Generalised arithmetic

Video From words to algebraic expressions

SOLUTION

a $4 \times x = 4x$

b $x \div 6$ or $\frac{x}{6}$

c $x - 1$

d $\frac{1}{3}x = \frac{x}{3}$

The number before x is $x - 1$ (for example, the number before 4 is $4 - 1 = 3$).

Example 7

Think of a number, y, multiply it by 3 and subtract 7. Write an algebraic expression for the answer.

SOLUTION

First, we multiply the number y by 3 to get $y \times 3 = 3y$.

Then, we subtract 7 from our result to get the answer, $3y - 7$.

☐ Foundation ○ Standard ⬡ Complex

Video
From words to algebraic expressions

Example 8

Describe the algebraic expression $\frac{a+b}{2}$ in words.

SOLUTION

$\frac{a+b}{2}$ is the sum of a and b, divided by 2, or the average of a and b.

EXERCISE 3.03 ANSWERS ON P. 621

Creating algebraic expressions

U F R C

EXAMPLE 6

1 If n represents any number, write an algebraic expression for:

R **a** five times the number
b the number plus 6
C **c** the number divided by 10
d the next consecutive number
e half of the number
f 20 less than the number
g the number multiplied by itself
h 20 minus the number
i the previous consecutive number
j double the number
k 6 divided by the number
l the difference between the number and 8
m the product of the number and (−4)
n the square root of the number

2 If a, b and c represent any three numbers, write an algebraic expression for:

R **a** the sum of a, b and c
C **b** the product of a and b
c the difference between b and c, where b is greater than c
d the product of a and c, plus b
e three times b, plus a
f c divided by b
g how much a is greater than b
h 5 more than (b times c)
i c less than b
j a times b divided by c
k the square root of the sum of a and c
l the product of a and b, decreased by 7 times c

3 A cricketer scored a total of x runs in 6 innings. What was her average score?
R Select the correct answer **A**, **B**, **C** or **D**.
C **A** $6x$ **B** $6 \div x$ **C** x^6 **D** $x \div 6$

 Foundation  Standard Complex

4 Write an algebraic expression for the answer to each instruction. EXAMPLE 7

C **a** Think of a number (y), multiply it by 3 and add 8.

b Think of a number (p), divide it by 3 and then subtract 6.

c Think of a number (M), multiply it by 4 and then add 27.

d Think of a number (k), multiply it by 5 and then divide it by 2.

e Think of a number (x), divide it by 6 and add 13.

f Think of a number (B), halve it and then subtract 2.

g Think of a number (R), divide it by 9 and then add 10.

h Think of a number (n), multiply it by 7 and then multiply it by 3.

5 Describe each algebraic expression in words. EXAMPLE 8

C **a** $x + y - z$ **b** $5K + L$ **c** $xy - 3$ **d** $\frac{W}{5}$

e $3x - 1$ **f** $\frac{y-x}{10}$ **g** $2N + 21$ **h** $7 - e$

6 Write an expression for:

R **a** the number of girls in a class if there are b boys and a total of t students.

C **b** the number of bottles of drink needed for a party if each bottle can be shared by five guests and there are n guests.

c the amount of change in dollars that Sandra receives if she paid for three hamburgers costing \$$p$ each with \$$y$ in cash.

d the perimeter of an isosceles triangle with equal sides of length r and an uneven side of length 9.

e the cost of one egg if a carton of a dozen eggs costs \$$a$.

f the perimeter of a square of length x.

g the number of rows of chairs needed to place 160 chairs in the hall if each row contains r chairs.

h the total cost of attending the pool for x adults and y children, where admission is \$5 per adult and \$3 per child.

TECHNOLOGY

Substitution

1 Enter the variables and values shown into a new spreadsheet.

This means $m = 10$.

	A	B
1	**variable**	**value**
2	m	10
3	n	12
4	p	4
5	q	0
6	r	2.7
7	s	-8
8	t	25
9	u	-5

2 In cell D1, enter the formula **=B4–B9** to evaluate $p - u$. You should get 9 because $4 - (-5) = 9$.

3 In cell D2, enter the formula **=B2+B3** to evaluate $m + n$. You should get 22 because $10 + 12 = 22$.

4 For each cell below, enter a formula to evaluate the given expression.

Foundation Standard Complex

D3: pr

D4: $\frac{t}{u}$

D5: $n + t - p$

D6: $n - \frac{m}{u}$

D7: $\frac{n - m}{u}$

D8: $\frac{s}{p} \times r$

D9: p^3

D10: $(s + n) \times (p + r)$

D11: $t + pq + s$

D12: $\sqrt{t}$

D13: $mnpq$

D14: the average of r, s, t and u

D15: $\frac{6m - n}{p + s}$

D16: $\sqrt{n + p}$

D17: ts^2

D18: $n^3 - r^2$

3.04 The laws of arithmetic

Let the variable x stand for an unknown number. Let x be represented by a cup that holds the unknown number of balls.

Then the total number of balls represented in the diagram below can be written using the algebraic expression $3 \times x + 5 = 3x + 5$.

Note that even though we don't know how many balls are in each cup, it is the same number of balls, x, and we can still describe the total number of balls algebraically.

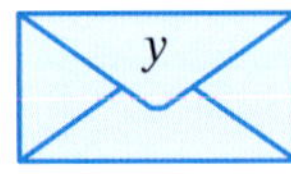

Let y stand for an unknown number that is different from x. Let y be represented by an envelope that contains this number of balls.

Then the total number of balls represented in the diagram below can be written using the algebraic expression $4x + 3y + 1$.

EXERCISE 3.04 ANSWERS ON P. 621

Collecting variables

U F R C

1 Write an algebraic expression for the number of balls represented in each diagram, if each cup holds x balls.

R C

a 

b 

c 

d

e 

f

g

h

i 

j

2 Write an algebraic expression for the number of balls represented in each diagram, if each cup holds x balls and each envelope holds y balls.

R C

a 

b

c

d

e

f

g

h

3 Draw cups and balls to represent each algebraic expression, if each cup holds x balls.

C

a $x + 3$

b $x + x + 1$

c $4 + 2x$

d $2 + x + 2$

e $x + 5 + 2x$

f $1 + 2x + 3 + x$

4 If N stands for the number of paperclips in each envelope, which diagram represents $3N + 2$ paperclips? Select the correct answer **A**, **B**, **C** or **D**.

R C

A

B

C

D

Foundation Standard Complex

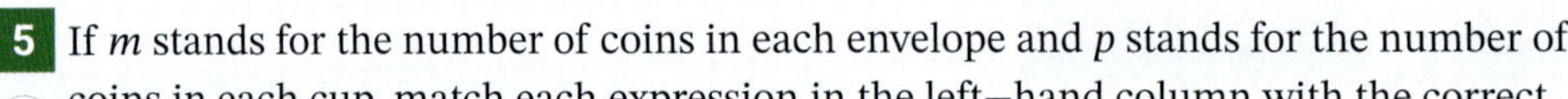

5 If m stands for the number of coins in each envelope and p stands for the number of coins in each cup, match each expression in the left–hand column with the correct expression in the right–hand column.

R C

a 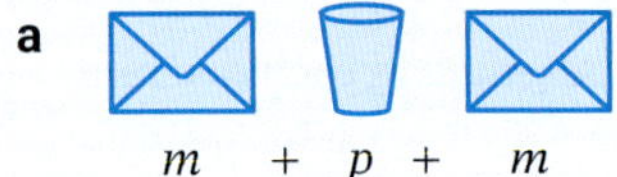
$m + p + m$

A 
$2m + 2p$

b 
$p + m + m + p$

B 
$2p + 3$

c 
$p + m + p + m + p$

C 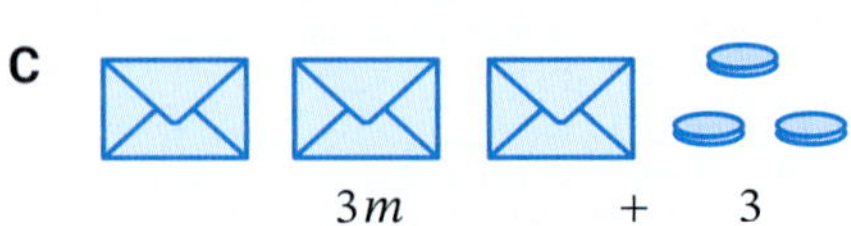
$3m + 3$

d 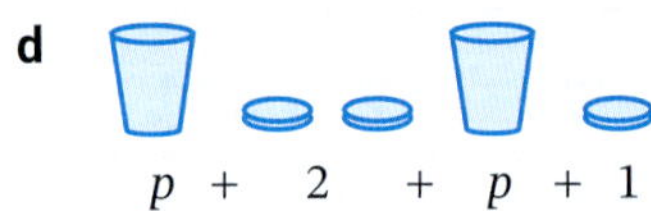
$p + 2 + p + 1$

D 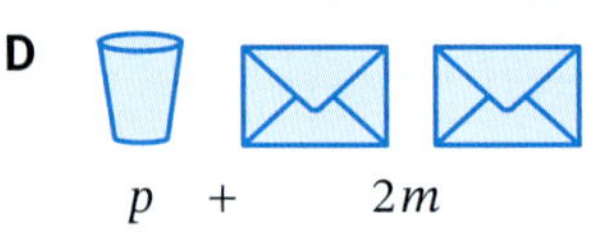
$p + 2m$

e 
$2m + 2 + m + 1$

E 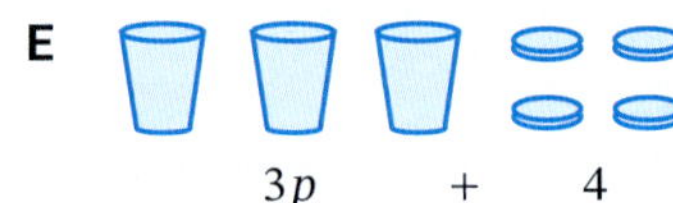
$3p + 4$

f 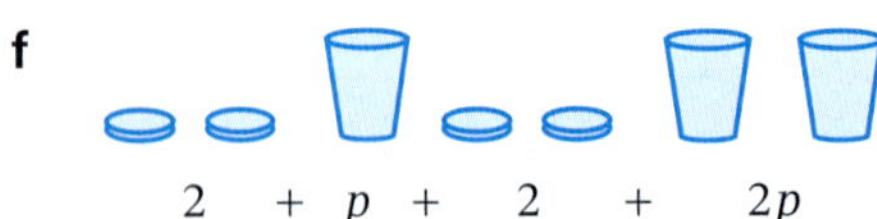
$2 + p + 2 + 2p$

F 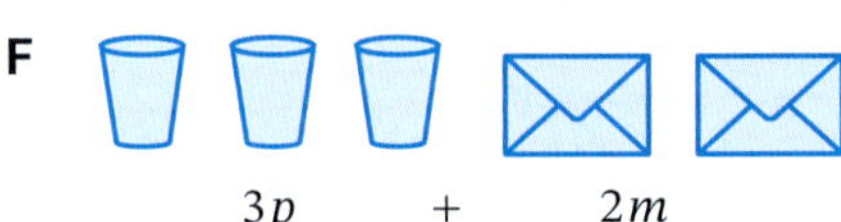
$3p + 2m$

6 Match each expression in the left column with the correct expression on the right.

R C

a $m + p - m$

A $2m$

b $m + 4 - 2$

B $2p + 2$

c $2m + p - p$

C $2p$

d $2p + 6 - 4$

D p

Foundation Standard Complex

e / E

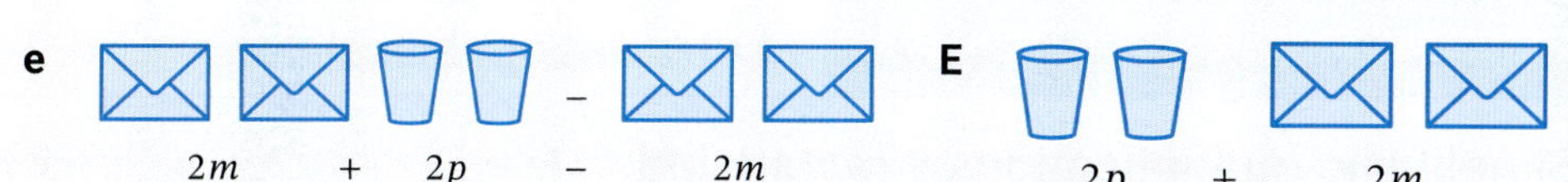

f / F

7 Simplify each expression.

R C

a

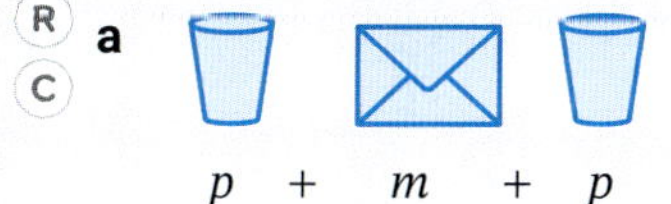

b

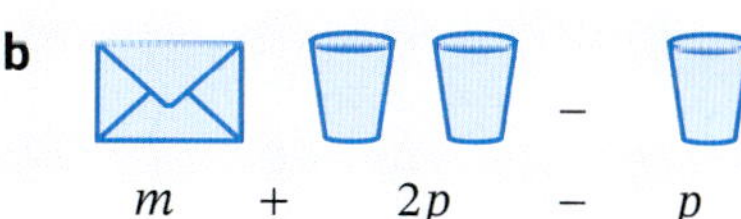

c

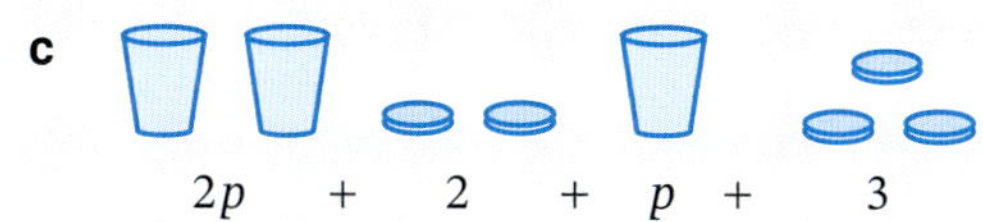

d

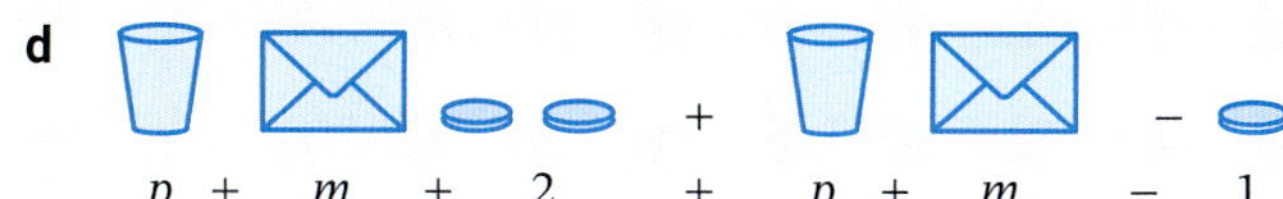

e

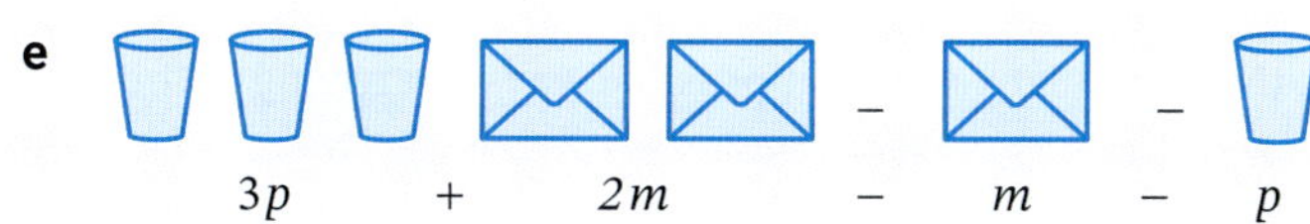

8 Simplify $9a + 4b - 5a + 2b$. Select the correct answer **A**, **B**, **C** or **D**.

R **A** $14a + 6b$ **B** $10ab$ **C** $4a + 6b$ **D** $14a + 11b$

9 Simplify each expression.

R

a $3d + 2d + 4a$ **b** $6h + 4r + 5r - h$ **c** $e + e - e$

d $2k + 3 + 2k - 6$ **e** $3x + 4z - x + 2z$ **f** $2s - p - s + 3p$

g $6u + 4 - 2u + 8$ **h** $7n - 6 + 5 - 3n$ **i** $20 + 6i - 3i - 20$

Quiz
Mental skills 3A

☆ MENTAL SKILLS 3A ANSWERS ON P. 621 **Maths without calculators**

Simplifying multiplication by factorising

Sometimes, one big multiplication may be made into many simpler little multiplications if we break one or both terms into two 'easier' factors. Then we use the property that numbers can be multiplied in any order.

1 Study each example.

a $24 \times 8 = 6 \times 4 \times 2 \times 4$
$= 4 \times 4 \times 6 \times 2$
$= 16 \times 6 \times 2$
$= 96 \times 2$
$= 192$

Look for pairs of numbers that you can multiply easily and rearrange.

b $15 \times 6 = 3 \times 5 \times 6$
$= 3 \times 30$
$= 90$

c $14 \times 36 = 7 \times 2 \times 6 \times 6$
$= 6 \times 7 \times 6 \times 2$
$= 42 \times 6 \times 2$
$= 252 \times 2$
$= 504$

d $30 \times 22 = 10 \times 3 \times 22$
$= 10 \times 66$
$= 660$

Note: For each question, there are many different possible ways of arriving at the correct result.

2 Now evaluate each product by factorising first.

a	30×24	**b**	18×27	**c**	25×33	**d**	16×8
e	28×20	**f**	12×18	**g**	21×9	**h**	36×12
i	16×35	**j**	22×28	**k**	9×15	**l**	27×25

3.05 Adding and subtracting terms

Skillsheet Algebra using diagrams

Puzzles Algebra bingo game

Adding and subtracting like terms

Interactive Expression exchange

What is the fastest way to mentally calculate the sum 3 + 3 + 5 + 3 + 5 + 5 + 3 + 3 + 5?

Do you think of collecting all the 3s together and all the 5s together like this?

$$\begin{aligned}3+3+5+3+5+5+3+3+5 &= (3+3+3+3+3)+(5+5+5+5)\\ &= (5\times 3)+(4\times 5)\\ &= 15+20\\ &= 35\end{aligned}$$

We can do the same thing algebraically.

Suppose an unknown number p is represented by a cup 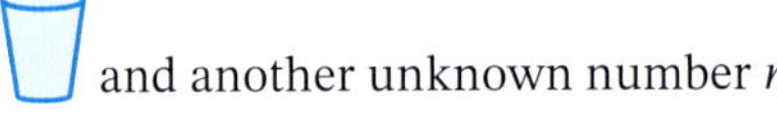 and another unknown number m is represented by an envelope.

Now consider these two sums.

a $p + m + p + m = 2p + 2m$

b $m + p + m + p + m = 3m + 2p$

Note that we can add all the p **terms** together, and we can add all the m terms together, but we cannot add the ps and ms together because p and m represent different numbers. We cannot simplify the answers $2p + 2m$ and $3m + 2p$ because p and m are different.

These are examples of collecting **like terms**, terms that contain exactly the same variable(s), such as $4h$, $5h$ and h. Here are some more examples.

$3x$ and $-5x$	$4mw$ and $2mw$	$12m$ and m
$-6d$ and $10d$	$5ab$ and $2ba$	xyz and $2yzx$

Note: $ab = ba$

Here are some examples of **unlike terms**, terms that have different variables.

$3x$ and $5m$	$4mw$ and $2mn$	$-6d$ and $10d^2$

We can add or subtract like terms because the variables in them represent the same number. We can only add or subtract things that are the same. For example:

$3m + 2p - 2m = m + 2p$

$$(3 \text{ lots of } m)+(2 \text{ lots of } p)-(2 \text{ lots of } m)=(3 \text{ lots of } m-2 \text{ lots of } m)+(2 \text{ lots of } p)$$
$$=(1 \text{ lot of } m)+(2 \text{ lots of } p)$$

This makes sense because if $m = 5$ and $p = 8$, then

$$3m+2p-2m=(3\times5)+(2\times8)-(2\times5)$$
$$=(3\times5)-(2\times5)+(2\times8) \quad \text{grouping like terms}$$
$$=(1\times5)+(2\times8) \quad 3 \text{ lots of } 5-2 \text{ lots of } 5=1 \text{ lot of } 5$$

ⓘ Adding and subtracting terms

When **adding and subtracting terms**, only **like terms** can be collected.

Like terms have exactly the same variables.

Example 9

Select the like terms from each list below.

a $5x, -x, x^2, xy, \frac{1}{2}x, -8x, \frac{4x}{3}$

b $3ab, 5, a, ba, -2b, 4ab^2, -ab$

SOLUTION

a $5x, -x, \frac{1}{2}x, -8x, \frac{4x}{3}$ are like terms because they each have a single x in them.

b $3ab$, ba and $-ab$ are like terms because each contains ab.

Video
Adding and subtracting like terms

Presentation
Collecting like terms

Example 10

Simplify each expression by collecting like terms.

a $7d + 3e + 3d + e$

b $6m - 8 + 3m + 6$

c $6xy + 7x^2 - 3yx - 12x^2$

d $14 - 4t - 9t + 2$

SOLUTION

a $7d + 3e + 3d + e = 7d + 3d + 3e + e$
$= 10d + 4e$

Grouping like terms.

b $6m - 8 + 3m + 6 = 6m + 3m - 8 + 6$
$= 9m - 2$

The + or − sign goes with the term after it.

c $6xy + 7x^2 - 3yx - 12x^2 = 6xy - 3xy + 7x^2 - 12x^2$
$= 3xy - 5x^2$

$3yx = 3xy$

d $14 - 4t - 9t + 2 = 14 + 2 - 4t - 9t$
$= 16 - 13t$

EXERCISE 3.05 ANSWERS ON P. 621

Adding and subtracting terms

EXAMPLE 9

1 Select the like terms from each list.

a $4p, 2y, 2p, 5z, p$

b $m^2, m, n, 3m^2, mn$

c $5ac, 4a, 7ca, 5$

d $vw, 5v, 9v, w, v^2, 2wv$

e $p, 6pq, 2qp, 3pq, 7q$

f $2d, 3d^3, 7d^2, 9, 3d$

g $4mn, 3m, 4, 2nm, mn, 2n$

h $x^2y, 2x, 3y, 4x^2y$

i $7, 2a, 4b, 5ba, ab$

2 What are the like terms in $3ab, 2a^2, 2b^2a, 2p^2, 2ba$? Select the correct answer **A**, **B**, **C** or **D**.

A $2a^2, 2p^2$

B $2a^2, 2b^2a, 2p^2, 2ba$

C $3ab, 2ba$

D $2b^2a, 2ba$

☐ Foundation ○ Standard ⬡ Complex

3 Simplify each expression by collecting like terms.

a $4k + 7k$ **b** $5mn + 2nm$ **c** $xy + xy$
d $3abc + 4abc + 2bac$ **e** $3x^2 + 2x^2 + x^2$ **f** $ef^2 + 4ef^2$
g $8d - 3d$ **h** $12mk - 7mk$ **i** $12xy - 7yx - xy$
j $8de - 12de$ **k** $6k^2 - 3k^2$ **l** $4w^2 - w^2$

4 Which expression is $2m + 4p - 5m - 2p$ simplified? Select the correct answer **A**, **B**, **C** or **D**.

EXAMPLE 10

A $-mp$ **B** $7m + 6p$ **C** $-3m + 2p$ **D** $-3p$

5 Simplify each expression.

a $5k - 2j + 6j$ **b** $10ab + 2 - ab$ **c** $12m^2 - 5m^2 + 2m$
d $12s^2 - 7st - 2s^2$ **e** $4mn + 7mn - mn$ **f** $14k + 3 - 5k$
g $y^2 + 2y^2 - y^2$ **h** $9d - 4e + 3d$ **i** $7 - 2x - 3$
j $15gb - 8gb + gb$ **k** $4 - 2q + 5$ **l** $5a + 6c - 4c$

6 Simplify each expression.

a $4x - y - x - 2y$ **b** $7ab + 2bc - 3ab + bc$ **c** $4k^2 + 3 + 5k^2 + 8$
d $4bc - 8a + 2bc - a$ **e** $x^2 - 6x - x + 7$ **f** $p^2 + q^2 + 3p^2 - q^2$
g $3 + 4y + 8y + 11$ **h** $11r - 3s - 4r + 6s$ **i** $7g + 8b - b - g$
j $6x^2 - 4x - 9x + 6$ **k** $20a + 11m + 32a - 11m$ **l** $3q - 7 - 8q + 12q$
m $5 - 2y - 5 - 4y$ **n** $s^2 + 4s^2 + 12 - 5s^2$

7 Which expression is $a^5 + a^5$ simplified? Select the correct answer **A**, **B**, **C** or **D**.

A a^{10} **B** a^{25} **C** $2a^5$ **D** $2a^{10}$

8 Write an algebraic expression involving a sum that simplifies to:

R C

a $2x + 6$ **b** $6m + 12p$ **c** $x^2 - x$

9 Write an algebraic expression involving a difference that simplifies to:

a $2x + 6$ **b** $6m + 12p$ **c** $x^2 - x$

Foundation Standard Complex

3.06 Multiplying and dividing terms

Worksheets
Simplifying algebraic expressions

Perimeter and area

Why aren't they the same?

Video
Multiplying terms

The associative law of multiplication tells us that to calculate the product $6 \times 4 \times 5 \times 3$ mentally, we can rearrange the numbers to pair convenient factors like this:

$$\begin{aligned}6\times4\times5\times3&=(4\times5)\times(6\times3)\\&=20\times18\\&=360\end{aligned}$$

We can do the same algebraically. As numbers can be multiplied in any order, we can multiply like *and* unlike terms.

Example 11

Simplify each product.

a $7b \times 3a$ **b** $8x \times (-2y)$ **c** $5d \times 6d \times 4$

d $9m^2 \times 4p$ **e** $4a \times 3b \times (-2a)$

SOLUTION

a $\begin{aligned}7b \times 3a &= 7 \times b \times 3 \times a\\&= 7 \times 3 \times a \times b\\&= 21ab\end{aligned}$

Group numbers and variables separately.

b $\begin{aligned}8x \times (-2y) &= 8 \times (-2) \times x \times y\\&= -16xy\end{aligned}$

c $\begin{aligned}5d \times 6d \times 4 &= 5 \times 6 \times 4 \times d \times d\\&= 120d^2\end{aligned}$

d $\begin{aligned}9m^2 \times 4p &= 9 \times 4 \times m^2 \times p\\&= 36m^2p\end{aligned}$

e $\begin{aligned}4a \times 3b \times (-2a) &= 4 \times 3 \times (-2) \times a \times a \times b\\&= -24a^2b\end{aligned}$

ⓘ Multiplying terms

When **multiplying algebraic terms**, multiply the numbers first, then multiply the variables. Write all variables in alphabetical order.

Video
Multiplying terms

Example 12

Use index notation to simplify each expression.

a $k \times k \times k \times k$ **b** $j \times j \times j \times j \times j \times j$

SOLUTION

a $k \times k \times k \times k = k^4$ **b** $j \times j \times j \times j \times j \times j = j^6$

ⓘ Dividing terms

When **dividing algebraic terms**, first divide the numbers and then divide the variables. Write all variables in the alphabetical order.

Example 13

Simplify each quotient.

a $20p \div 5$ **b** $\frac{24y}{8y}$ **c** $\frac{-2m}{4mn}$ **d** $12a^2b \div 3a$

Video Dividing terms

SOLUTION

a $20p \div 5 = \left(\frac{20}{5}\right)p$
$= 4p$

Group numbers and variables separately.

b $\frac{24y}{8y} = \left(\frac{24}{8}\right)\frac{\cancel{y}}{\cancel{y}}$
$= 3$

c $\frac{-2m}{4mn} = \left(\frac{-2}{4}\right)\frac{\cancel{m}}{\cancel{m}n}$
$= \left(\frac{-1}{2}\right)\frac{1}{n}$
$= -\frac{1}{2n}$

d $12a^2b \div 3a = \left(\frac{12}{3}\right)\frac{\cancel{a}ab}{\cancel{a}}$
$= 4ab$

EXERCISE 3.06 ANSWERS ON P. 621

Multiplying and dividing terms

U F R C

1 Simplify $2p \times 5pq$. Select the correct answer **A**, **B**, **C** or **D**. EXAMPLE 11

A $10pq$ **B** $7p^2q$ **C** $7pq$ **D** $10p^2q$

2 Simplify each expression.

a $5 \times 3y$ **b** $8 \times 2m$ **c** $4k \times m$ **d** $4t \times 5$
e $2 \times f \times 6$ **f** $2x \times 8y$ **g** $10p \times 3m$ **h** $3b \times 6d$
i $3d \times 4d$ **j** $5 \times 4m \times 6n$ **k** $2y \times 3y \times 8$ **l** $3r \times 3 \times 3r$
m $2 \times 4a \times 9b$ **n** $x \times xy \times y$ **o** $2j \times 3k \times 7$ **p** $5rst \times 2rs$
q $3ac \times 6cd$ **r** $hjk \times hj \times jk$ **s** $4q^2 \times 6p^2$ **t** $2m \times 3n \times mn$

3 Simplify each expression.

a $-4b \times 2d$ **b** $6m \times (-4m)$ **c** $-2 \times 5k \times 9l$ **d** $20 \times (-3p)$
e $9k \times (-7)$ **f** $-15b \times (-5b)$ **g** $-3m \times mn \times 4n$ **h** $6abc \times (-2ab)$
i $5r \times (-2r) \times (-4)$ **j** $9d \times (-4e) \times 6d$ **k** $-8x \times 9x \times (-2)$ **l** $5m \times 4 \times (-2m)$

4 Which expression simplifies to $14a^2n^2$? Select the correct answer **A**, **B**, **C** or **D**.

A $2a \times an \times 7$ **B** $2a \times 7n \times an$ **C** $7a^2 \times 2n$ **D** $7a^2n^2 \times 2a^2n^2$

5 Simplify each expression. EXAMPLE 12

a $y \times y \times y \times y$ **b** $a \times a \times a$ **c** $2 \times c \times c$
d $t \times t \times t \times t \times t$ **e** $q \times q \times q \times 7$ **f** $k \times k \times k \times l \times l$
g $m \times m \times m \times n \times n$ **h** $-5 \times p \times p \times p \times p \times k \times k \times k$
i $-3 \times y \times y \times y$ **j** $d \times d \times (-1) \times e \times e \times f$
k $q \times q \times r \times r \times r \times r \times 4$

EXAMPLE 13

6 Simplify $\frac{55m^2n}{5m}$. Select the correct answer **A**, **B**, **C** or **D**.

A $11mn$ **B** $11m^2n$ **C** $50mn$ **D** $11n$

7 Simplify each quotient.

a $12m \div 3$ **b** $36e \div 4$ **c** $21p \div (-3)$ **d** $\frac{81x}{9}$

e $\frac{60w}{15}$ **f** $\frac{8m}{m}$ **g** $-25k \div 5k$ **h** $30y \div 40y$

i $28pq \div 14q$ **j** $\frac{63a}{-9a}$ **k** $\frac{4pq}{20q}$ **l** $\frac{-56mn}{8m^2}$

8 Simplify each quotient.

a $14m^2 \div 7$ **b** $6e^2 \div (-2e)$ **c** $-5p^2 \div p$ **d** $\frac{14x^2}{2x}$

e $\frac{6wm}{-3m}$ **f** $\frac{8m^2n}{2mn}$ **g** $32k^2y \div (-4k)$ **h** $9jk^2 \div 12jk^2$

i $\frac{15d^2e^2}{3de}$ **j** $\frac{3a^2b^2c}{3ab^2c}$ **k** $\frac{-24acb}{3bc}$ **l** $\frac{-10m^2n}{5mn}$

9 Simplify $12x \div (-3) \times 2x$. Select the correct answer **A**, **B**, **C** or **D**.

A $-2x$ **B** -2 **C** $-8x$ **D** $-8x^2$

10 Simplify each expression.

a $-18pq \div 6pq$ **b** $48p \div (-8) \div 2p$ **c** $y \times 3y \times (-5y)$

d $9d \times 4e \div 6d$ **e** $8x \times 9x \div 12$ **f** $48m^2 \div 12m \times 2n$

g $\frac{15c}{5c}$ **h** $\frac{a \times 3ab}{ab}$ **i** $\frac{2ab}{10bc}$

j $\frac{3d \times 4e}{6e}$ **k** $\frac{15mn^2p}{3m^2n^2}$ **l** $\frac{-4x^2y \times 3y}{2xy \times (-6xy)}$

11 Write an algebraic expression involving a product that simplifies to:

R C **a** $18de$ **b** $-20a^2y$ **c** $5pr^2$

12 Write an algebraic expression involving a quotient that simplifies to:

R C **a** $7m$ **b** $-3ab$ **c** $5pr^2$

INVESTIGATION

Equivalent expressions

State whether each statement below is true or false, then check by substituting a number for the variable and testing whether the left–hand side equals the right–hand side.

a $6 + 4m + 4 = 4m + 10$ **b** $5x + 3x = 8x^2$ **c** $4k - 7k = -3k$

d $7g + 8h - h - g = 7h + 6g$ **e** $3 + 7n + 13 + 2n = 25n$ **f** $3p \times 7q = 21pq$

g $21m \div 3m = 7m$ **h** $5y - 4y = y$ **i** $\frac{-16a}{4} = -4a$

Foundation Standard Complex

Extension: The index laws

3.07

In the previous chapter, *Working with numbers*, we discovered properties about powers, called **index laws**.

Here are some examples:

- $2^3 \times 2^4 = 2^{3+4} = 2^7$
- $7^5 \times 7 = 7^{5+1} = 7^6$
- $4^6 \div 4^2 = 4^{6-2} = 4^4$
- $\dfrac{10^3}{10^2} = 10^{3-2} = 10^1$
- $(3^4)^2 = 3^{4\times2} = 3^8$
- $(8^5)^3 = 8^{5\times3} = 8^{15}$
- $5^0 = 1$
- $8^0 = 1$

Now we can write these index laws algebraically and use them to simplify algebraic expressions.

YEAR 9 EXTENSION

Puzzle Indices puzzle

Videos Powers, indices and exponents

Simplifying with the index laws

Index laws 1

Index laws

When **multiplying terms with powers** of the same base, **add** the powers.

$$a^m \times a^n = a^{m+n}$$

When **dividing terms with powers** of the same base, **subtract** the powers.

$$a^m \div a^n = \frac{a^m}{a^n} = a^{m-n}$$

When **raising a term with a power** to another power, **multiply** the powers.

$$(a^m)^n = a^{m\times n}$$

Any number raised to the **power of zero** equals **1**.

$$a^0 = 1$$

Now we can apply these index laws algebraically to variables.

Example 14

Video Index laws 2

Simplify each expression.

a $p^6 \times p^2$ **b** $3m^2n \times 7m^3n^5$ **c** $a^{12} \div a^4$ **d** $\dfrac{42y^6}{7y^2}$

e $(d^4)^5$ **f** $(4m^2)^3$ **g** $2k^0$ **h** $(2k)^0$

SOLUTION

a $p^6 \times p^2 = p^{6+2}$ Adding the powers.

$= p^8$

YEAR 9 EXTENSION

b $3m^2n\times7m^3n^5=3\times7\times m^{2+3}\times n^{1+5}$
$=21m^5n^6$

c $a^{12}\div a^4=a^{12-4}$ Subtracting the powers.
$=a^8$

d $\dfrac{42y^6}{7y^2}=\dfrac{42}{7}\times y^{6-2}$
$=6y^4$

e $(d^4)^5=d^{4\times5}=d^{20}$ Multiplying the powers.

f $(4m^2)^3=4^3\times(m^2)^3$ because $(4m^2)^3=4m^2\times4m^2\times4m^2$.
$=64m^{2\times3}$
$=64m^6$

g $2k^0=2\times k^0$ The zero power belongs to k only.
$=2\times1$
$=2$

h $(2k)^0=1$ The zero power belongs to $(2k)$.

EXERCISE 3.07 ANSWERS ON P. 622

The Index laws

EXAMPLE 14

1 Simplify $a^2b\times a^4b^3$. Select the correct answer **A**, **B**, **C** or **D**.

A a^8b^3 **B** a^8b^4 **C** a^6b^3 **D** a^6b^4

2 Simplify each product.

a $m^5\times m^2$ **b** $k^8\times k$ **c** $y^4\times y^6$ **d** $x^7\times x$
e $n^4\times n^4$ **f** $q^2\times q^8$ **g** $b^4\times b^{10}\times b^2$ **h** $s^3\times s^5\times s^7$
i $6k^3\times7k^3$ **j** $5y^3\times4y$ **k** $d^4\times3d^5$ **l** $11e^3\times5e$
m $h^4n^2\times h^6n^5$ **n** $10c^3d^4\times(-3c^3d)$ **o** $f^5g^4\times f^3g^7$ **p** $-5p^3q\times2p^4q$

3 Simplify $\dfrac{36d^6e^4}{12d^3e^4}$. Select the correct answer **A**, **B**, **C** or **D**.

A $3d^3$ **B** $3d^3e$ **C** $3d^2e$ **D** $3d^2$

4 Simplify each quotient.

a $y^6\div y^5$ **b** $p^4\div p$ **c** $\dfrac{q^5}{q^2}$ **d** $n^8\div n^3$
e $\dfrac{k^{11}}{k^9}$ **f** $x^7\div x^5$ **g** $r^8\div r^8$ **h** $m^{12}\div m^5\div m^2$
i $q^5\times q^3\div q^7$ **j** $45g^6\div5g^2$ **k** $18p^7\div6p^3$ **l** $a^{12}\div a^6\div a$
m $66x^4y^2\div6x^3y$ **n** $\dfrac{x^4\times x^2}{x^3}$ **o** $\dfrac{70k^5l^6}{10k^2l}$ **p** $\dfrac{15e^3f}{21ef^2}$

Foundation Standard Complex

YEAR 9 EXTENSION

3.07

5 Simplify $(a^4b^2)^3$. Select the correct answer **A**, **B**, **C** or **D**.

A a^7b^5 **B** $a^{12}b^6$ **C** $a^{12}b^5$ **D** a^7b^6

6 Simplify each expression.

a $(y^8)^2$ **b** $(x^2)^5$ **c** $(y^3)^3$ **d** $(m^2)^9$

e $(k^7)^3$ **f** $(t^{20})^2$ **g** $(3m)^2$ **h** $(2m^4)^3$

i $(xy)^3$ **j** $(p^3q)^4$ **k** $(3h^4)^4$ **l** $(k^3z)^5$

m $(2kj)^4$ **n** $(4c^3d)^2$ **o** $(5x^2y^8)^3$ **p** $(2m^4p^2)^6$

7 Simplify each expression.

a y^0 **b** $(xy)^0$ **c** $(2r-r)^0$ **d** $5u^0$

e $\left(\frac{x}{2}\right)^0$ **f** $(2k^7)^0$ **g** $m^8 \div m^8$ **h** $h^2 \div (h^0)^3$

i $(b^3)^2 \div 2b^0$ **j** $(3x)^0 \times (4x)^2$ **k** $(d^0)^5 \times (d^8)^0$ **l** $26x^4y^3 \div 2x^3y^3$

m $4m^0 + 3m - 3k^0$ **n** $\frac{27b^3d}{9bd}$ **o** $(8h)^2 \times 2h^0 \div 4h^2$ **p** $\frac{6ef^4}{12ef^4}$

8 Simplify each expression.

a $\frac{3pq}{9p^2q^2}$ **b** $\frac{2m^3n^3}{m^2n^2}$ **c** $\frac{8xy}{12xy^2}$

☆ **MENTAL SKILLS 3B** ANSWERS ON P. 622 **Maths without calculators**

Quiz Mental skills 3B

Multiplying by 9, 11, 99 or 101

We can use expanding when multiplying by a number near 10 or near 100.

1 Study each example.

a
$$\begin{aligned}25\times11&=25\times(10+1)\\&=25\times10+25\times1\\&=250+25\\&=275\end{aligned}$$

b
$$\begin{aligned}14\times9&=14\times(10-1)\\&=14\times10-14\times1\\&=140-14\\&=126\end{aligned}$$

c
$$\begin{aligned}32\times12&=32\times(10+2)\\&=32\times10+32\times2\\&=320+64\\&=384\end{aligned}$$

d
$$\begin{aligned}7\times99&=7\times(100-1)\\&=7\times100-7\times1\\&=700-7\\&=693\end{aligned}$$

e
$$\begin{aligned}27\times101&=27\times(100+1)\\&=27\times100+27\times1\\&=2700+27\\&=2727\end{aligned}$$

f
$$\begin{aligned}18\times8&=18\times(10-2)\\&=18\times10-18\times2\\&=180-36\\&=144\end{aligned}$$

☐ Foundation ○ Standard ⬡ Complex

2 Now evaluate each product.

a	16×11	**b**	33×11	**c**	29×9	**d**	45×9
e	62×11	**f**	7×101	**g**	18×101	**h**	36×99
i	19×8	**j**	45×12	**k**	21×102	**l**	6×98

3.08 Expanding expressions

Skillsheet Algebra using diagrams

Worksheet Expanding brackets

Puzzles Expandominoes

Expanding

Expand the brackets

Expand negative brackets

Interactive Area model Algebra

The **distributive law** says that you can multiply by a number by splitting it into the sum or difference of two other numbers. Look at these two examples.

a $23 \times 12 = 23 \times (10 + 2)$ — 23 times

$= (10 + 2) + (10 + 2) + (10 + 2) + ...$ — 23 times each

$= 10 + 10 + 10 + \ldots + 2 + 2 + 2 + \ldots$

$= (23 \times 10) + (23 \times 2)$

$= 230 + 46$

$= 276$

So, $23 \times (10 + 2) = 23 \times 10 + 23 \times 2$

b $35 \times 9 = 35 \times (10 - 1)$ — 35 times

$= (10 - 1) + (10 - 1) + (10 - 1) + ...$ — 35 times each

$= 10 + 10 + 10 + \ldots - 1 - 1 - 1 -$

$= (35 \times 10) - (35 \times 1)$

$= 350 - 35$

$= 315$

So, $35 \times (10 - 1) = 35 \times 10 - 35 \times 1$

ⓘ The distributive law for expanding algebraic expressions

If a, b and c stand for numbers, then:

$a(b + c)$ means $a \times (b + c)$.

$$a(b + c) = ab + ac$$

$$a(b - c) = ab - ac$$

We can apply the distributive law to algebraic expressions. Suppose an unknown number m is represented by an envelope [m] that contains m coins.

Then $2(m + 3)$ or '2 lots of $(m + 3)$', can be represented by this diagram.

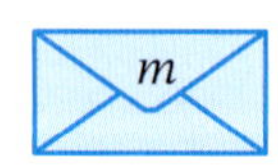

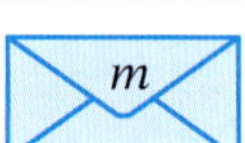

It can be seen that $2(m + 3) = 2 \times m + 2 \times 3 = 2m + 6$.

The number outside the brackets is *multiplied* by each term inside.

Here is another example for $3(2m + 1)$.

=
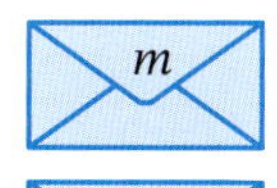
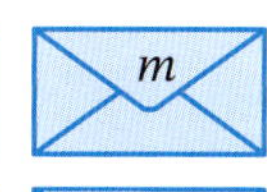

It can be seen that $3(2m + 1) = 3 \times 2m + 3 \times 1 = 6m + 3$.

Removing the brackets in an algebraic expression by multiplying each term inside by the term outside is called **expanding** the expression.

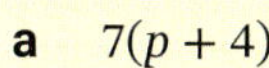

Example 15

Expand each expression.

a $7(p + 4)$ **b** $3(x - 5)$ **c** $5(2m - 7)$

d $-2(a + 3)$ **e** $-(4d - 2)$ **f** $w(9w - 1)$

SOLUTION

a $7(p + 4) = 7 \times p + 7 \times 4$
$= 7p + 28$

7 is multiplied by each term inside the brackets separately.

b $3(x - 5) = 3 \times x - 3 \times 5$
$= 3x - 15$

c $5(2m - 7) = 5 \times 2m - 5 \times 7$
$= 10m - 35$

d $-2(a + 3) = -2 \times a + (-2) \times 3$
$= -2a + (-6)$
$= -2a - 6$

−2 is multiplied by each term inside the brackets separately.

e $-(4d - 2) = -1(4d - 2)$
$= -1 \times 4d - (-1) \times 2$
$= -4d + 2$

$-(4d - 2)$ is the same as $-1 \times (4d - 2)$.

f $w(9w - 1) = w \times 9w - w \times 1$
$= 9w^2 - w$

Video
Expanding expressions 2

Example 16

Expand and simplify each expression.

a $2(x + 5) + 6(x - 4)$ **b** $9k - 4(k - 3)$ **c** $3p - (p + 6)$

SOLUTION

a $2(x+5)+6(x-4)=2\times x+2\times 5+6\times x+6\times(-4)$ Expanding

$=2x+10+6x-24$ Collecting like terms

$= 8x - 14$

b $9k-4(k-3)=9k-4\times k-4\times(-3)$

$=9k-4k+12$

$=5k+12$

c $3p-(p+6)=3p-1(p+6)$

$=3p-1\times p-1\times 6$

$=3p-p-6$

$=2p-6$

EXERCISE 3.08 ANSWERS ON P. 622

Expanding expressions

U F R C

EXAMPLE 15

1 Expand each expression.

a $5(m + 3)$ **b** $8(m - 2)$ **c** $2(j + 10)$ **d** $3(a + 2)$
e $4(h - 7)$ **f** $6(x + y)$ **g** $9(4 - b)$ **h** $11(y - 6)$
i $4(2k + 5)$ **j** $3(5m - 2)$ **k** $6(3x + 4)$ **l** $10(2a + 5b)$

2 Expand each expression.

a $-3(n + 4)$ **b** $-4(v - 5)$ **c** $-7(d + 3)$ **d** $-10(k - 2)$
e $-2(d + e)$ **f** $-5(5 - z)$ **g** $-6(h + 1)$ **h** $-(m - 1)$
i $-9(5 + 2x)$ **j** $-(2f + 3)$ **k** $-3(u - v)$ **l** $-8(4 - k)$

3 Expand $-3(a - 2b)$. Select the correct answer **A**, **B**, **C** or **D**.

A $-3a - 2b$ **B** $-3a + 6b$ **C** $-3a - 6b$ **D** $-3a + 5b$

4 Expand each expression.

a $x(y + 6)$ **b** $p(q - 2)$ **c** $a(a + 4)$ **d** $b(3b - 1)$
e $d(d + 3e)$ **f** $2(4m + 5n)$ **g** $5k(3k - 2)$ **h** $7j(3j + ik)$
i $x(4x + y)$ **j** $-2(4p - 3q)$ **k** $k(k - 12)$ **l** $-m(4 - m)$
m $4n(2n + 5m)$ **n** $2x(1 - 7x)$ **o** $-f(2f + 4)$ **p** $-(5j - 7k)$

5 Expand $3y(y - 4)$. Select the correct answer **A**, **B**, **C** or **D**.

A $3y + 12$ **B** $3y^2 - 7y$ **C** $3y^2 - 12y$ **D** $3y + 7$

6 Explain the difference between $2a + 1$ and $2(a + 1)$.

R C

7 Samir says that, to find the perimeter of a rectangle, you double its length, double its width and add the answers together. Tash says that you add the length and width first, then double the answer. Who is correct? Can you write each of their answers algebraically?

R C

8 Expand $12 - 8(x + 2)$. Select the correct answer **A**, **B**, **C** or **D**.

EXAMPLE 16

A $4(x + 2)$ **B** $4x + 8$ **C** $-8x + 14$ **D** $-8x - 4$

9 Expand and simplify each expression.

a $6(a + 2) + 14$ **b** $4k + 2(k + 3)$ **c** $3(q - 3m) + 5m$

d $8 + 6(p - 5)$ **e** $18 - (2y + 14)$ **f** $31 - 7(t + 2)$

g $6(2h - 1) + 11$ **h** $4x - 4(3x - 2)$ **i** $2(9n - 2) + 8$

10 Check that $8(2b - 3) = 16b - 24$ by substituting any value for b and showing that the answers to both sides are equal.

R

11 Expand and simplify each expression.

a $3(a + 4) + 2(a + 1)$ **b** $2(x + 5) + 6(x - 1)$

c $7(4 + k) + 5(2k - 2)$ **d** $5(2t + 1) + 3(4t + 2)$

e $3(n + 5) - 2(n + 4)$ **f** $10(d + 2) - 6(d + 3)$

g $8(4y + 1) - 5(2y - 3)$ **h** $3(p - 4) - 5(p - 1)$

i $m(m + 1) + 3(m + 1)$ **j** $4(d + e) + 3(d - e)$

k $2f(3f + 4) - 4(2f - 1)$ **l** $5e(1 - 7e) - e(3e + 4)$

10 Copy and complete each equation.

R

a $2(\_\_\_\_\_) = 2x + 10$ **b** $4(\_\_\_\_\_) = 4r - 12$

c $3(\_\_\_\_\_) = 24k - 9$ **d** $5(\_\_\_\_\_) = 10a + 35$

e $-2(\_\_\_\_\_) = -2y - 18$ **f** $-3(\_\_\_\_\_) = -6t + 21$

INVESTIGATION

Expanding with numbers

Expansions can be used with numbers to break a multiplication into parts, for example:

$$\begin{aligned} 9 \times 356 &= 9 \times (300 + 50 + 6) \\ &= 9 \times 300 + 9 \times 50 + 9 \times 6 \\ &= 2700 + 450 + 54 \\ &= 3204 \end{aligned}$$

Foundation Standard Complex

1 Copy this table and complete it by finding the place value of each digit in each number.

Number	1000s	100s	10s	1s
274	0	2	7	4
1234				
5306				
9870				
4025				

2 Now use the information in the table to copy and complete the numerical expansions below.

a $5 \times 274 = 5 \times (0 + 200 + 70 + 4)$
$= 0 + 1000 + 350 + 20$
$= 1370$

b $7 \times 1234 = 7 \times (\_\_\_\_ + \_\_\_\_ + \_\_\_\_ + \_\_\_\_)$
$=$
$=$

c $3 \times 5306 = 3 \times (\_\_\_\_ + \_\_\_\_ + \_\_\_\_ + \_\_\_\_)$
$=$
$=$

d $6 \times 9870 = 6 \times (\_\_\_\_ + \_\_\_\_ + \_\_\_\_ + \_\_\_\_)$
$=$
$=$

e $8 \times 4025 = 8 \times (\_\_\_\_ + \_\_\_\_ + \_\_\_\_ + \_\_\_\_)$
$=$
$=$

Check your solutions using a calculator.

DID YOU KNOW?

Algebraic notation

Algebraic notation is the mathematical language used to describe patterns and rules. It has its own grammar and terminology. For example, the expression $3x^4 - 5x + 7$ has the following features:

exponent: power or index, 4

coefficient: a number that multiplies a variable, 3 and -5

term: group of numbers and variables, $3x^4$, $-5x$, 7

operator: addition or subtraction

constant: the term with no variable, 7

For the expression $-2x^6 + x^3 - 4$, find the:

a exponent(s) **b** coefficient of x^6 **c** term containing x^3 **d** constant

Factorising algebraic terms

3.09

The **factors** of a number are those numbers that divide into it evenly. For example, the factors of 12 are 1, 2, 3, 4, 6 and 12.

Algebraic terms also have factors. The **factors** of an algebraic term are those numbers or terms that divide into it evenly. For example, $10xy = 2 \times 5 \times x \times y$, so *some* of the factors of $10xy$ are 2, 5, x, y, 10, $2y$ and $10x$. Often there are too many to list them all easily.

Skillsheets
Algebra using diagrams

Factorising using diagrams

Example 17

List 4 factors of $40a^2b$.

SOLUTION

There are many ways of multiplying terms to get $40a^2b$, such as:

$8 \times 5 \times a \times a \times b$

$4 \times 10 \times a^2 \times b$

$40 \times a^2 \times b$

So, four factors of $40a^2b$ are: 4, 5, a and a^2b. (*Note*: It is impractical to list them all.)

The highest common factor of algebraic terms

The **highest common factor (HCF)** (or **greatest common divisor (GCD)**) of two numbers is the largest number that divides into both numbers evenly. For example, the HCF of 18 and 30 is 6.

The **highest common factor** of two algebraic terms is the largest term that divides into both terms evenly. For example, the HCF of $18ab$ and $30b$ is $6b$.

Highest common factor

To find the **HCF** of two or more algebraic terms:

- find the HCF of the numbers,
- find the HCF of the variables, and
- multiply them together to make the highest common algebraic factor.

Example 18

Find the HCF of each pair of terms.

a $16x$ and $40xy$ **b** $12r$ and 28

Video
HCF of algebraic terms

SOLUTION

a First, find the HCF of the numbers.

The factors of 16 are **1**, **2**, **4**, **8** and 16.

The factors of 40 are **1**, **2**, **4**, 5, **8**, 10, 20 and 40.

So the HCF of 16 and 40 is **8**.

Then find the HCF of the variables:

The HCF of x and xy is x.

$\therefore$ The HCF of $16x$ and $40xy$ is $8 \times x = 8x$.

b First, find the HCF of the numbers:

The factors of 12 are: **1**, **2**, 3, **4**, 6 and 12.

The factors of 28 are: **1**, **2**, **4**, 7, 14 and 28.

So the HCF of 12 and 28 is **4**.

Then find the HCF of the variables:

The second term (28) has no variable so the HCF of the variables is 1.

$\therefore$ The HCF of $12r$ and 28 is $4 \times 1 = 4$.

EXERCISE 3.09 ANSWERS ON P. 622

Factorising algebraic terms

U F R

1 List the factors of each number.

a 15 **b** 8 **c** 17

d 25 **e** 55 **f** 42

2 Find the HCF of each pair of numbers.

R **a** 9 and 6 **b** 6 and 14 **c** 18 and 24

d 16 and 12 **e** 33 and 22 **f** 15 and 25

g 21 and 9 **h** 45 and 30 **i** 16 and 48

EXAMPLE 17

3 List four factors of each term.

a $5m$ **b** $7pq$ **c** $10x$ **d** $18d$

e $4k^2$ **f** $21bc$ **g** $32x^2y$ **h** $15ab$

4 Which term is **NOT** a factor of $36m^2n$? Select the correct answer **A**, **B**, **C** or **D**.

A $4m$ **B** $9n$ **C** $8mn$ **D** $12m^2n$

EXAMPLE 18

5 Find the HCF of each pair of terms.

R **a** $3xy$ and $12x$ **b** $4j$ and $24j$ **c** $6x$ and $4ax$

d $21kj$ and $14k$ **e** 5 and $10p$ **f** $16m$ and $12m^2$

g $48cd$ and $32bc$ **h** de and de^2 **i** 9 and $27x^2$

j $3m$ and $6n$ **k** $22m$ and $4mn$ **l** $16y$ and $4y$

m a^2x and ay **n** $64j$ and 32 **o** $15w^2$ and $20wy$

p $22x^2y$ and $36xy^2$ **q** cd and ad **r** $36t$ and $12ty^2$

6 Which term is the HCF of $8x^2$ and $12xy$? Select the correct answer **A**, **B**, **C** or **D**.

A 4 **B** $2x$ **C** $4x$ **D** $24x^2y$

Foundation Standard Complex

Factorising expressions

3.10

Expanding an algebraic expression means removing its brackets by multiplying terms.

Factorising is the opposite of expanding. To factorise an algebraic expression, find the HCF of all terms and insert brackets.

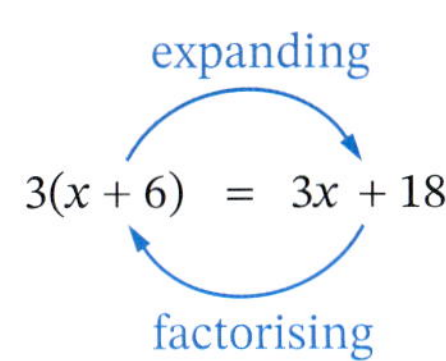

Skillsheets Algebra using diagrams

Factorising using diagrams

Puzzle Factorising

ⓘ Factorising algebraic expressions

- Find the HCF of all terms and write it outside the brackets.
- Divide each term by the HCF and write them inside the brackets.
- $ab + ac = a(b + c)$
- $ab - ac = a(b - c)$

To check that the answer is correct, expand it.

Factorising algebraic expressions

Example 19

Factorise each expression.

a $2x + 6$ **b** $7 - 7y$ **c** $3ab - 5b$ **d** $r^2 + 7rs$

Video Factorising expressions

SOLUTION

a $2x + 6 = 2 \times x + 2 \times 3$
$= 2(x + 3)$

HCF of $2x$ and 6 is 2.
Think: $2 \times$ ___ $= 2x$, $2 \times$ ___ $= 6$.
Check by expanding: $2(x + 3) = 2x + 6$.

b $7 - 7y = 7 \times 1 - 7 \times y$
$= 7(1 - y)$

HCF of 7 and $7y$ is 7.
Think: $7 \times$ ___ $= 7$, $7 \times$ ___ $= 7y$.
Check by expanding: $7(1 - y) = 7 - 7y$.

c $3ab - 5b = b \times 3a - b \times 5$
$= b(3a - 5)$

HCF of $3ab$ and $5b$ is b.
Think: $b \times$ ___ $= 3ab$, $b \times$ ___ $= 5b$.
Check by expanding: $b(3a - 5) = 3ab - 5b$.

d $r^2 + 7rs = r \times r + r \times 7s$
$= r(r + 7s)$

HCF of r^2 and $7rs$ is r.
Think: $r \times$ ___ $= r^2$, $r \times$ ___ $= 7rs$.
Check by expanding: $r(r + 7s) = r^2 + 7rs$.

EXERCISE 3.10 ANSWERS ON P. 622

Factorising expressions

U F R C

1 Find the HCF of the terms in each expression.

R C

a $10a + 2b$ **b** $4 - 12x$ **c** $ax + xy$ **d** $y^2 - y$

2 Copy and complete each factorisation.

R C

a $2x + 4 = 2(\_\_\_\_)$ **b** $4p - 16 = 4(\_\_\_\_)$

c $12k - 15 = 3(\_\_\_\_)$ **d** $5j - 20k = 5(\_\_\_\_)$

e $3x^2 + 3y^2 = 3(\_\_\_\_)$ **f** $5x - 15 = \_\_\_\_(x - 3)$

g $12x - 6 = \_\_\_\_(2x - 1)$ **h** $2a + 8b = \_\_\_\_(a + 4b)$

i $4pq + 10 = \_\_\_\_(2pq + 5)$ **j** $5 + 15x = 5(\_\_\_\_)$

k $xy + xz = x(\_\_\_\_)$ **l** $ab - ac = a(\_\_\_\_)$

m $pq - p = p(\_\_\_\_)$ **n** $cd + bc = c(\_\_\_\_)$

o $a^2 + 2a = a(\_\_\_\_)$ **p** $5hk - 8h = \_\_\_\_(5k - 8)$

q $24mn + 11n = \_\_\_\_(24m + 11)$ **r** $xy + y^2 = \_\_\_\_(x + y)$

s $3k^2 + 4jk = \_\_\_\_(3k + 4j)$ **t** $ef - 2f^2 = \_\_\_\_(e - 2f)$

EXAMPLE 19

3 Factorise each expression.

R C

a $4x + 8$ **b** $3m + 6$ **c** $5a - 10$ **d** $15m - 10a$

e $8x - 4y$ **f** $12m - 16w$ **g** $hk - h$ **h** $2xy - xm$

i $12p + 5pr$ **j** $x^2 + 5x$ **k** $4m - m^2$ **l** $y^2 - 11y$

4 Factorise $16k^2 + 18k$. Select the correct answer **A**, **B**, **C** or **D**.

A $2(8k^2 + 9k)$ **B** $k(16k + 18)$ **C** $2k(8k + 9)$ **D** $4k(4k + 9)$

5 Copy and complete each factorisation.

R C

a $2am + 2an = 2a(\_\_\_\_)$ **b** $6xy - 3x = 3x(\_\_\_\_)$

c $4ap + 2a = 2a(\_\_\_\_)$ **d** $5mn - 5mp = 5m(\_\_\_\_)$

e $3xy - 12x = 3x(\_\_\_\_)$ **f** $2pq + 2pr = 2p(\_\_\_\_)$

g $14x - 2xy = 2x(\_\_\_\_)$ **h** $20kj + 5j = 5j(\_\_\_\_)$

i $9xy - 12yz = 3y(\_\_\_\_)$

6 Factorise $3d^2 + 21d$. Select the correct answer **A**, **B**, **C** or **D**.

A $3(d^2 + 7d)$ **B** $d(3d + 21)$ **C** $d(3d + 7)$ **D** $3d(d + 7)$

Foundation Standard Complex

7 Factorise each expression.

(R) **a**	$7ab + 14a$	**b**	$2x - 6xy$	**c**	$3m - 9mw$	**d**	$10xy - 25xw$
(C) **e**	$8am - 12mp$	**f**	$9xy + 6my$	**g**	$3x + 9xy$	**h**	$ab + abc$
i	$xyz + xya$	**j**	$18m - 24mn$	**k**	$2ac + 3abc$	**l**	$pqr + mpq$
m	$rs + 7r$	**n**	$4s^2 + 16$	**o**	$p^2 - 2p$	**p**	$6k^3 - 2k$
q	$3j + j^2$	**r**	$7n - n^2$	**s**	$7e + 21e^2$	**t**	$6q - 12q^2$
u	$a^4 + a^2$	**v**	$x^2y + xy^2$	**w**	$3m^2n + 9m$	**x**	$12a^2b^2c^2 + 3abc$

Factorising with negative terms 3.11

ⓘ Factorising with negative terms

When **factorising an algebraic expression whose first term is** *negative*, include the negative sign in the factor.

$$-ab - ac = -a(b + c)$$
$$-ab + ac = -a(b - c)$$

Skillsheets
Algebra using diagrams
Factorising using diagrams

Worksheets
Algebra review
Algebraic expressions review

Video
Factorising expressions

Puzzles
Expandominoes
Factorising puzzle
Factorising with negative terms

Example 20

Factorise each expression.

a $-5x - 10$ **b** $-9n + 21$ **c** $-gh - h$ **d** $-mn + n^2$

SOLUTION

a $-5x - 10 = -5 \times x + (-5) \times 2$ HCF of $-5x$ and -10 is -5.
$= -5(x + 2)$ Think: $-5 \times \_\_\_ = -5x$, $-5 \times \_\_\_ = -10$.
Check by expanding: $-5(x + 2) = -5x - 10$.

b $-9n + 21 = -3 \times 3n + (-3) \times (-7)$ HCF of $-9n$ and 21 is -3.
$= -3(3n + (-7))$ Think: $-3 \times \_\_\_ = -9n$, $-3 \times \_\_\_ = 21$.
$= -3(3n - 7)$ Check by expanding: $-3(3n - 7) = -9n + 21$.

c $-gh - h = -h \times g + (-h) \times 1$ HCF of $-gh$ and $-h$ is $-h$.
$= -h(g + 1)$ Think: $-h \times \_\_\_ = -gh$, $-h \times \_\_\_ = -h$.
Check by expanding: $-h(g + 1) = -gh - h$.

d $-mn + n^2 = -n \times m + (-n) \times (-n)$ HCF of $-mn$ and n^2 is $-n$.
$= -n(m + (-n))$ Think: $-n \times \_\_\_ = -mn$, $-n \times \_\_\_ = n^2$.
$= -n(m - n)$ Check by expanding: $-n(m - n) = -mn + n^2$.

Foundation Standard Complex

EXERCISE 3.11 ANSWERS ON P. 623

Factorising with negative terms

U F R C

1 Copy and complete each factorisation.

R C

a $-6k - 6 = -6(\_\_\_\_\_\_\_\_\_)$ **b** $-mn - mp = -m(\_\_\_\_\_\_\_\_\_\_)$

c $-5r + 10 = -5(\_\_\_\_\_\_\_\_)$ **d** $-3k + 27 = -3(\_\_\_\_\_\_\_\_\_\_)$

e $-ac + ad = \_\_\_\_\_\_\_\_\_(c - d)$ **f** $-y^2 - 7y = \_\_\_\_\_\_\_\_\_(y + 7)$

g $-12y - 30 = \_\_\_\_\_\_\_\_\_(2y + 5)$ **h** $-4f + f^2 = \_\_\_\_\_\_\_\_\_(4 - f)$

2 Factorise each expression.

R C

a $-4x + 12$ **b** $-3m - 9$ **c** $-10x + 20$ **d** $-15x + 10m$

e $-6x - 8m$ **f** $-5x - 30$ **g** $-a - ab$ **h** $-4xyz + 8abc$

i $-20 + 4x$ **j** $-dx - dy$ **k** $-2m + 5mp$ **l** $-24x - 16y$

3 Factorise $-14wy - 28wz$. Select the correct answer **A**, **B**, **C** or **D**.

A $-2(7wy + 14wz)$ **B** $-14w(y + 2z)$ **C** $14w(-y - 2z)$ **D** $-2w(7y + 14z)$

4 Factorise each expression.

R C

a $-16xy - 24xk$ **b** $-20x + 12xy$ **c** $-8mw + 12m$

d $5e - 15ef$ **e** $-ab - abc$ **f** $xya + xyw$

g $-12c - 18cw$ **h** $-2rt + 7rst$ **i** $20fh + 5h$

j $-19bw - 38w$ **k** $-abc + abcd$ **l** $-abc - abcd$

5 **a** Substitute $w = 2$ into the expressions $-w^2 + 7w$ and $-w(w - 7)$.

R C

b Is $-w^2 + 7w = -w(w - 7)$ for any value of w?

6 Factorise $-ab + a^2$. Select the correct answer **A**, **B**, **C** or **D**.

A $-a(b - a)$ **B** $a(-b - a)$ **C** $-a(b + a)$ **D** $-a(b + a^2)$

7 Factorise each expression.

R C

a $a^4bc - a^3b^2c$ **b** $2\pi r^2 + 2\pi rh$

3.12 Extension: Expanding binomial products

YEAR 9 EXTENSION

$(x + 4)$ and $(x - 7)$ are called **binomial expressions,** because they each have exactly two terms.

$(x + 3)(x - 1)$ is called a **binomial product,** because it is a product of two binomial expressions.

binomial = 'two terms'

Example 21

Expand $(a + 2)(a + 3)$ using an area diagram.

SOLUTION

Draw an area diagram (rectangle) with length $(a + 2)$ and width $(a + 3)$ and divide the diagram into four smaller rectangles.

Find the area of each smaller rectangle.

Expand $(a + 2)(a + 3)$ by adding the areas of the four rectangles.

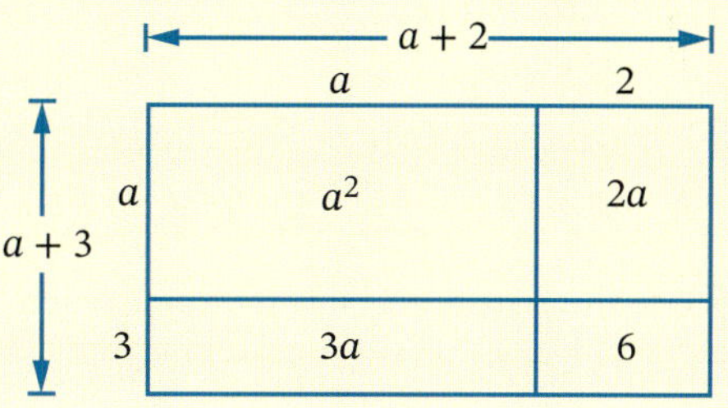

$(a + 2)(a + 3) = a^2 + 2a + 3a + 6$ Expanding

$\qquad\qquad\qquad\;\; = a^2 + 5a + 6$ Simplifying by collecting like terms.

Another way of expanding binomial products is to use the distributive law. Each term in the first binomial is multiplied by each term in the second binomial to give **four terms**, which are collected and simplified.

Example 22

Expand each binomial product.

a $(x + 5)(x + 9)$ **b** $(k + 3)(k - 7)$ **c** $(x - 6)(4x + 2)$

d $(3t - 1)(2t - 5)$ **e** $(a - 6)^2$

SOLUTION

a $(x + 5)(x + 9) = x(x + 9) + 5(x + 9)$ Expanding to make four terms.

$= x^2 + 9x + 5x + 45$ Simplifying.

$= x^2 + 14x + 45$

b $(k + 3)(k - 7) = k(k - 7) + 3(k - 7)$

$= k^2 - 7k + 3k - 21$

$= k^2 - 4k - 21$

c $(x - 6)(4x + 2) = x(4x + 2) - 6(4x + 2)$

$= 4x^2 + 2x - 24x - 12$

$= 4x^2 - 22x - 12$

d $(3t - 1)(2t - 5) = 3t(2t - 5) - 1(2t - 5)$

$= 6t^2 - 15t - 2t + 5$

$= 6t^2 - 17t + 5$

e $(a - 6)^2 = (a - 6)(a - 6)$

$= a(a - 6) - 6(a - 6)$

$= a^2 - 6a - 6a + 36$

$= a^2 - 12a + 36$

Skillsheet Area diagrams

Worksheet Binomial products

Puzzle Expanding brackets

Technology Expanding binomials

Spreadsheet Expanding binomials

Video Expanding binomial products

YEAR 9 EXTENSION

EXERCISE 3.12 ANSWERS ON P. 623

Expanding binomial products

U F R C

EXAMPLE 21

1 Expand each binomial product by copying and completing the area diagram.

a $(x+2)(x+7)$

	x	2
x	---	---
7	---	---

b $(x+5)(x+1)$

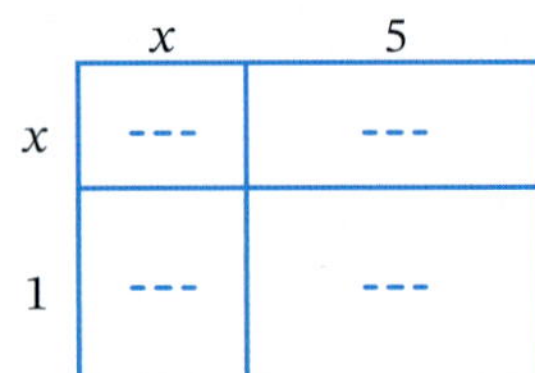

2 Expand $(p+6)(p+2)$. Select the correct answer **A**, **B**, **C** or **D**.

A p^2+12 **B** $p^2+6p+12$ **C** p^2+8p+8 **D** $p^2+8p+12$

EXAMPLE 22

3 Use the distributive law to expand each binomial product algebraically.

R
a $(x+1)(x+4)$ **b** $(x+6)(x+3)$ **c** $(x+5)(x+4)$ **d** $(x+2)(x+2)$
e $(x+2)(x+6)$ **f** $(x+10)(x+3)$ **g** $(x+8)(x+1)$ **h** $(x+7)(x+7)$
i $(x-1)(x+4)$ **j** $(x-3)(x+2)$ **k** $(x+1)(x-4)$ **l** $(x-5)(x+6)$
m $(x-4)(x-5)$ **n** $(x+9)(x-3)$ **o** $(x-5)(x-3)$ **p** $(x-7)(x-7)$

4 R C
a Expand each binomial product.
i $(x+3)(x-3)$ **ii** $(x+4)(x-4)$
iii $(x-2)(x+2)$ **iv** $(x-7)(x+7)$

b Can you see a simpler way of expanding these products? Explain how you would expand $(x+8)(x-8)$.

5 Expand each binomial product.

a $(x+1)(2x+4)$ **b** $(3x-1)(x+2)$ **c** $(2x+3)(x+1)$
d $(x-3)(5x+2)$ **e** $(4x-3)(x-2)$ **f** $(5x+2)(2x-3)$
g $(x-3)(3x-1)$ **h** $(4x+7)(x-3)$

6 Expand each binomial square.

a $(a+5)^2$ **b** $(x+2)^2$ **c** $(y+7)^2$
d $(x-3)^2$ **e** $(2a+1)^2$ **f** $(3a-5)^2$

7 Expand:

a $(a+b)^2$ **b** $(a+b)(a-b)$

8 Expand $\left(4x+\frac{1}{2}\right)^2$. Select the correct answer **A**, **B**, **C** or **D**.

A $16x^2+4x+\frac{1}{4}$ **B** $16x^2+\frac{1}{4}$ **C** $4x^2+2x+\frac{1}{4}$ **D** $16x^2+2x+\frac{1}{4}$

☐ Foundation ○ Standard ⬡ Complex

9780170465601

POWER PLUS ANSWERS ON P. 623

1 Simplify each expression.

a $-7p - 8q + 3p - 5q + 4q$

b $-m - 3n - 5m - 7mn$

c $3k - 2j - 5 - 4k + 6j + 9$

d $a^2 + 2b^2 + 3b^2 - a^2$

e $4x^2 + 6x - x^2 - 12x$

f $xy + yz + xz + 4xy + 3xz + yz$

g $-d^2 + 7 + 6e^2 - 3d^2 - 2e^2 + 9$

h $5f^2 + 6f + 2 + 8f^2 - 1$

2 For each shape, write an algebraic expression for:

i its perimeter

ii its area

(Give your answer in simplest form.)

a

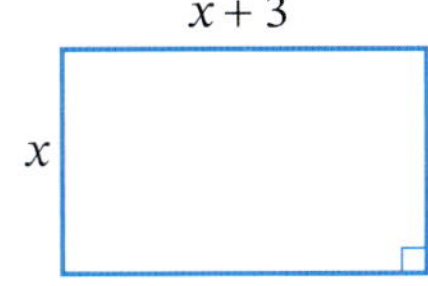

b

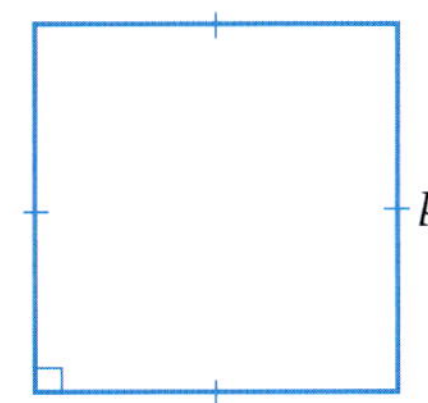

c

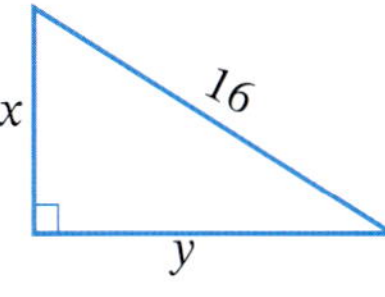

d

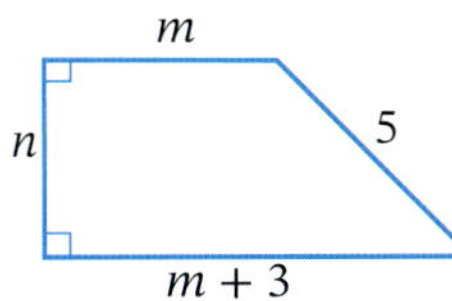

e

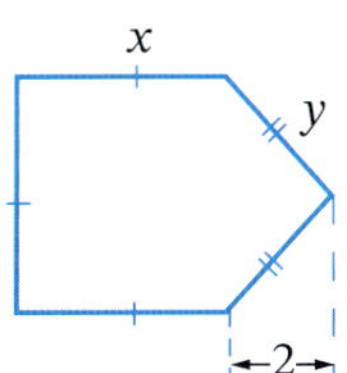

f

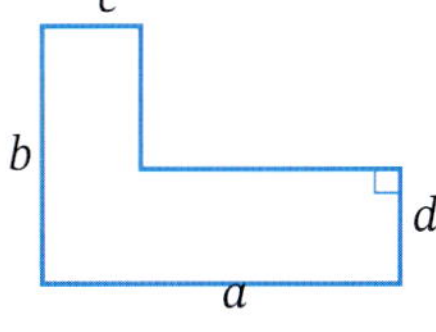

3 Expand and simplify each expression.

a $x(x - 3) + 5(x - 3)$

b $m(a + 4) + m(a - 7)$

c $-3(y - 4) + 7(y - 3)$

d $4(2a + b + 3c) - 2(a + 3b - c)$

e $p(p - 3q + 2r) + 2p(4r - 3p - 2q)$

f $7k^2 + 3(k^2 - 3k - 1) - (6k^2 - 4k + 2)$

g $12 - 5(4f - 6g - 3)$

h $5(2x^2 - 5x + 3) - 4(3x^2 - 8x - 25)$

i $(x + 3)(x + 5)$

j $(p - 1)(p + 10)$

4 Evaluate each expression if $p = 4$, $q = -6$, $r = 3$ and $t = -1$.

a $p + q$

b pq

c $r^2 + t^2$

d $q^2 - 6p + rt$

e $\frac{pq}{rt}$

f $6 - 2r + p$

g $2p + 3q + 4r - 5t$

h $\frac{p+4}{2}$

i $\frac{q+r}{p+t}$

j $\frac{r+7}{5} - \frac{q-t}{5}$

k $\frac{p+3}{10} + \frac{r-2}{2}$

l $q + pr - qt$

5 Factorise each expression.

a $9pq + 5p - 2p^2$

b $44k - 33l - 11m$

c $2x - 7x^2 + 3xy$

d $x(y + 2) + 3(y + 2)$

e $a(b - 3) - 7(b - 3)$

f $4(k + 1) - j(k + 1)$

6 Simplify each expression.

a $\frac{m}{2} + \frac{m}{2}$

b $\frac{7n}{10} - \frac{3n}{10}$

c $\frac{2k}{3} + \frac{k}{3}$

d $\frac{3d}{5} - \frac{d}{5}$

e $\frac{5f}{8} + \frac{7f}{8}$

f $\frac{3c}{10} + \frac{4c}{10}$

g $\frac{4}{x} + \frac{2}{x}$

h $\frac{7}{a} + \frac{5}{b}$

3 CHAPTER REVIEW

Quiz
Language of maths 3

Language of maths

base	brackets	collecting	consecutive
distributive law	evaluate	expand	expanded form
expression	factor	factorise	formula
highest common factor	index laws	indices	like terms
power	pronumeral	substitute	term
unknown	unlike terms	variable	

1 Why does the word 'variable' have that name? Look up its meaning in the dictionary.

2 What are **unlike terms**?

3 How do you **expand** an algebraic expression?

4 What is the name given to the largest algebraic term that divides evenly into two or more given terms?

5 What is the opposite of **expanding** an algebraic expression?

6 Look up some non–mathematical meanings of the word 'factor'. Use 'factor' in a sentence to show one of these meanings.

Worksheet
Mind map: Algebra

Topic summary

- What parts of algebra did you remember from Year 7? What new rules have you learnt?
- Are there parts of this chapter that you still do not understand? Talk to your teacher.
- Give two examples of jobs where algebra is needed.

Print (or copy) and complete this mind map of the topic, adding detail to its branches and using pictures, symbols and colour where needed. Ask your teacher to check your work.

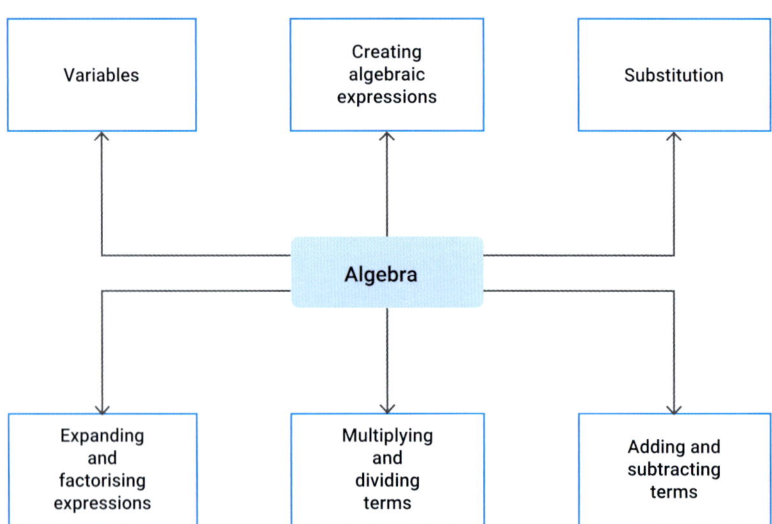

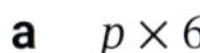

3 TEST YOURSELF

ANSWERS ON P. 623

Quiz Test yourself 3

1 Simplify each expression. 3.01

a $p \times 6$ **b** $r \times (-1)$ **c** $4 \times d \times g$

d $t + t + t + t + t$ **e** $14m + 6m$ **f** $11k - 6k$

g $6 \times y \times y$ **h** $16 \div a$ **i** $5 \times c \times 4 \times d$

2 Write each expression in expanded form. 3.01

a cd **b** $3y^2$ **c** $\frac{w}{6}$ **d** $c^2 + y$

3 Evaluate $2n - p$ when $n = 11$ and $p = -3$. Select the correct answer **A**, **B**, **C** or **D**. 3.02

A 19 **B** 25 **C** 208 **D** 214

4 If $x = 5$ and $y = -2$, evaluate each expression. 3.02

a $x + y$ **b** xy **c** $3x - 2y$

d $6y^2$ **e** $6x \div y$ **f** $x^2 + y^2$

5 Write each statement as an algebraic expression. 3.03

a 4 more than y **b** 6 times b minus 3 **c** p divided by q

6 Simplify each expression. 3.05

a $5k + 7k$ **b** $29d - 12d$

c $4m + 18m - 5m$ **d** $13m - 3m + 8m$

e $4ac + 4ac - 5ca$ **f** $11cd + 7cd - 9cd$

g $7x + 3y + 3x + 7y$ **h** $13r + 4s + 11s + 2r$

i $33h + 6g - 5h + g$

7 Simplify each expression. 3.06

a $5 \times 2x$ **b** $2b \times 7d$ **c** $11m \times 6$

d $3p \times 8q$ **e** $4 \times 2w \times 4w$ **f** $15j \times 5$

g $3m \times 6m \times 2$ **h** $4p \times 3p \times 5p$

8 Simplify each expression. 3.06

a $50x \div 2$ **b** $28b \div 7d$ **c** $66m \div 6$

d $\frac{32rs}{4}$ **e** $\frac{15p}{3p}$ **f** $14d^2 \div 2$

g $14e^2 \div 2e$ **h** $13b^2 \div b$ **i** $\frac{18x^2}{9x}$

j $\frac{44m^2n^2}{11m}$ **k** $\frac{16e^2g}{4eg}$ **l** $\frac{35a^2e^2}{7ae}$

Foundation Standard Complex

9 Expand and simplify each expression.

- **a** $6(n + 2)$
- **b** $-5(2b + 7)$
- **c** $j(9j - 3)$
- **d** $10 + 2(n + 4)$
- **e** $22v - 3(v + 5)$
- **f** $5(y + 2) + 4(y - 7)$

10 Which expression is $5(c - d) - 2c + 3d$ simplified? Select the correct answer **A**, **B**, **C** or **D**.

- **A** $3c - 2d$
- **B** $3c + 8d$
- **C** $3c - 8d$
- **D** $5c - 5d - 2c + 3d$

11 Find the HCF of each pair of terms.

- **a** $6a$ and 12
- **b** $3p^2$ and $18p$
- **c** $7t$ and 14

12 Factorise each expression.

- **a** $3m + 9$
- **b** $5h - 35$
- **c** $6a + 9ab$
- **d** $10h - 12hm$
- **e** $xy + xyz$
- **f** $cd - de$
- **g** $a^2 - 6a$
- **h** $8kj + 24j^2$

13 Factorise each expression.

- **a** $-2 - 4q$
- **b** $-4f - 16$
- **c** $-12 + 15d$
- **d** $-ab + b$
- **e** $-16b - 24bd$
- **f** $-4k^2 - k$
- **g** $-x^2 + 7x$
- **h** $-h^2i - hi^2$

9780170465601

Practice set 1 ANSWERS ON P. 624

1 Evaluate each expression. 2.09

a 7^3 b $(-35)^2$ c 4^4

d $\sqrt{400}$ e $\sqrt[3]{2744}$ f $\sqrt{2.25}$

2 Evaluate each expression without using a calculator. 2.01

a 28×5 b 15×9 c $8 \times 3 \times 5$

d 16×30 e $720 \div 4$ f $300 \div 20$

3 Find the value of each variable, correct to one decimal place where necessary.

a
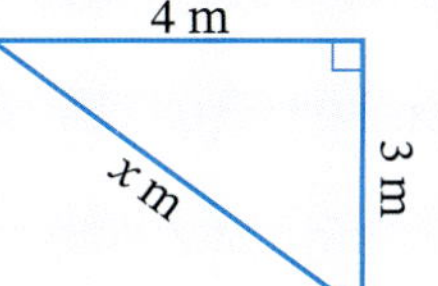

b
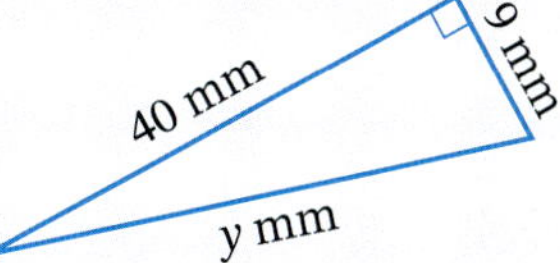

c
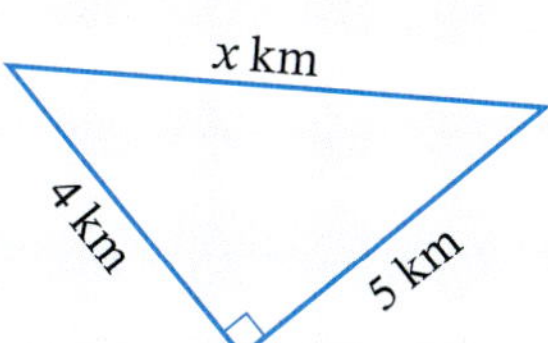

d
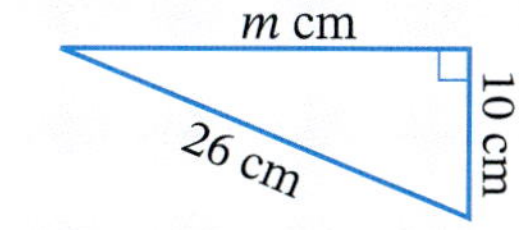

e
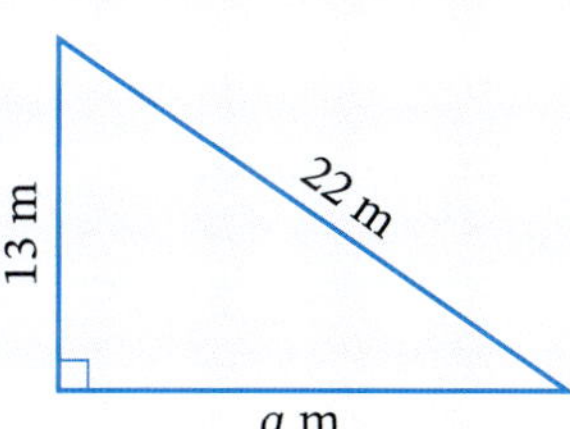

f
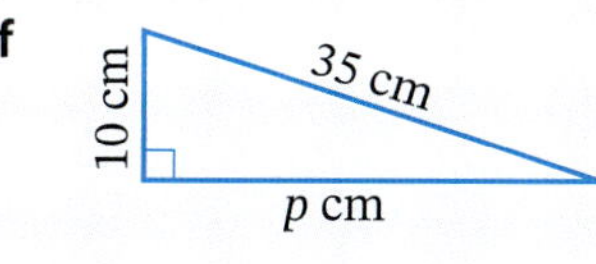

4 Evaluate each expression. 2.02–2.04

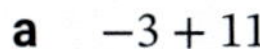

a $-3 + 11$ b $6 \times (-5)$ c $-11 - 9$

d $3 - (-2)$ e $-7 \times (-4)$ f $\frac{42}{-7}$

g $7 - 9$ h $-4 - (-6)$ i $-6 - 10$

j -4×7 k $-54 \div (-9)$ l $75 \div (-5)$

5 Evaluate each expression. 2.05

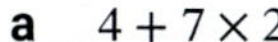

a $4 + 7 \times 2$ b $16 - 28 \div (-4)$ c $(-32 + 11) \div 7$

d $32 \div (-8) - 7 \times 1$ e $(3 + 7 \times 5) \div 2$ f $(17 - 6) \times (3 - 12)$

g $\frac{12+7}{4}$ h $\frac{8.8}{6.5-5.4}$ i $\frac{64-12}{15+11}$

☐ Foundation ○ Standard ○ Complex

6 If $a = 3$ and $b = 5$, then evaluate each expression.

a $2b + a$ **b** a^2 **c** $5a - b$ **d** $ab - b$

7 Simplify each expression.

a $3k + 2j - k + 2j$ **b** $5p - p$

c $4p + 3q + 7p + 2q$ **d** $10x + 7y - 2x - y$

8 Evaluate each expression.

a $1.35 + 0.2 - 5.78$ **b** $35.3 - 12.91$

9 Evaluate each expression.

a 0.8×9 **b** 1.5×0.3 **c** $2.1 \div 5$ **d** $5.52 \div 0.3$

10 Write each statement as an algebraic expression.

a 5 more than k **b** 20 times k

c half of k **d** 3 lots of k increased by 10

e the difference between k and 12 where k is larger

f k multiplied by itself

11 Simplify each expression.

a $5 \times 2p$ **b** $12m \div 2m$ **c** $-4x \times 3xy$

d $3a \times 2b \times 4$ **e** $\frac{14xy}{7xy}$ **f** $\frac{5a \times 4a}{10a}$

12 Use a factor tree to write each number as a product of its prime factors, in index notation.

a 180 **b** 875

13 Use factor trees to find the highest common factor of each pair of numbers.

a 28 and 72 **b** 36 and 90

14 Use index notation to simplify each expression.

a $2^4 \times 2^3$ **b** $7^6 \times 7^5$ **c** $5^8 \div 5^3$ **d** $8^5 \div 8^4$

e $(3^4)^2$ **f** $(11^3)^5$ **g** 7^0 **h** $(8 - 5)^0$

15 Find the height, h cm, of an equilateral triangle with sides 8 cm long. Write your answer:

a as a surd

b correct to two decimal places.

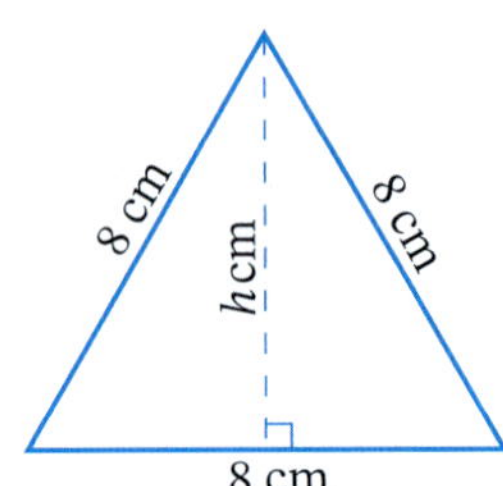

16 Test whether this triangle is right-angled. If it is, copy the triangle and draw the right angle. 

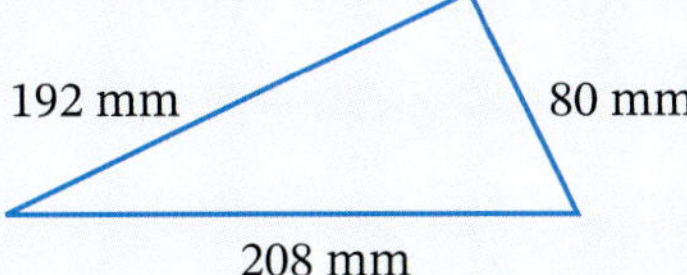

17 Use factor trees to find the lowest common multiple of each pair of numbers. 

a 12 and 18 **b** 18 and 27

18 Expand and simplify each expression. 

a $4(x + 2)$ **b** $6(m - 4)$ **c** $-7(2k + 1)$

d $-(3p + q)$ **e** $2(m + 2) + 4m$ **f** $8 + 4(p - 2)$

19 Factorise each expression. 3.10, 3.11

a $2m + 4$ **b** $d^2 - d$ **c** $6p + 12q$

d $-12x - 24$ **e** $2ab + 3a$ **f** $-2x^2 + 6x$

PRACTICE SET 1

☐ Foundation ○ Standard ⬡ Complex

4

MEASUREMENT, SPACE

Geometry

Geometry is all around us. Designs and patterns involving angles, parallel lines, triangles and quadrilaterals can be found everywhere, in our homes, on transport, in construction, art and nature. This building construction scene shows the importance of angles, lines and shapes in architecture and design.

Alamy Stock Photo/EyeEm

Chapter outline

		Proficiencies			
4.01	Angle geometry	U	F	R	C
4.02	Angles on parallel lines	U	F	R	C
4.03	Classifying triangles	U	F	R	C
4.04	Classifying quadrilaterals	U	F	R	C
4.05	Properties of quadrilaterals	U	F	R	C
4.06	Angle sums of triangles and quadrilaterals	U	F	R	C
4.07	Extension: Angle sum of a polygon*	U	F	R	C

*Extension

U = Understanding
F = Fluency
PS = Problem solving
R = Reasoning
C = Communication

Wordbank

angle sum The total size of the angles in a shape, such as a triangle

bisect To cut in half

convex quadrilateral A quadrilateral whose vertices all point outwards.

diagonal An interval joining two non-adjacent vertices of a shape

exterior angle of a triangle An 'outside' angle of a triangle formed after extending one of the sides of the triangle

scalene triangle A triangle with no equal sides

supplementary angles Two angles that add to 180°

Quiz
Wordbank 4

Videos (10):

4.01 Angles at a point and vertically opposite angles • Angle geometry

4.02 Angles on parallel lines • Angle relationships • Angle geometry

4.03 Properties of triangles • Classifying triangles

4.04, 4.05 Properties of quadrilaterals • Classifying quadrilaterals

4.06 Angle sums of triangles and quadrilaterals • Angles in polygons • Exterior angle of a triangle

Quizzes (6):

- Wordbank 4
- SkillCheck 4
- Mental skills 4A
- Mental skills 4B
- Language of Maths 4
- Test Yourself 4

Skillsheets (3):

SkillCheck Types of angles

4.02 Angles and parallel lines

4.03 Naming shapes

Worksheets (20):

SkillCheck A page of angles

4.01 Straight angles, right angles and revolutions

4.02 Investigating angles on parallel lines • Angles in parallel lines • What is the diagram? • Matching angles

4.03 Properties of triangles • Triangle geometry

4.04 Properties of quadrilaterals • Always, sometimes, never true • Classifying triangles and quadrilaterals

4.05 Properties of quadrilaterals • Diagonal properties of quadrilaterals • Always, sometimes, never true

4.06 Angles in triangles and quadrilaterals • Find the unknown angle 2 • Deductive geometry • Shapes and angles review

4.07 Angle sum of a polygon • Angles in regular polygons • Equal angles

Mind map: Geometry

Puzzles (10):

SkillCheck Angle cards

4.01 Angles: A dog day • Angles

4.02 Angles in parallel lines

4.04 What shape am I? • Singing in the car

4.06 Mixed angles • Angles in triangles • Triangles and quadrilaterals

4.07 Shape puzzle 1

Presentation (1):

4.05 Angles and shapes

Nelson MindTap

To access resources above, visit **cengage.com.au/nelsonmindtap**

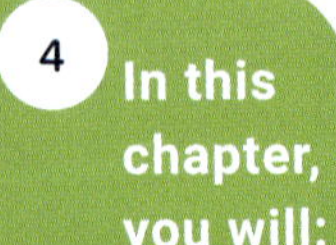

In this chapter, you will:

- ✓ solve geometrical problems involving angles on a straight line, angles at a point and vertically opposite angles.
- ✓ solve geometrical problems involving corresponding, alternate and co-interior angles on parallel lines crossed by a transversal.
- ✓ classify triangles according to their side and angle properties.
- ✓ identify convex and non-convex quadrilaterals.
- ✓ classify squares, rectangles, rhombuses, parallelograms, kites and trapeziums.
- ✓ solve geometrical problems involving the angle sum of a triangle and quadrilateral and the exterior angle of a triangle.
- ✓ (EXTENSION) solve geometrical problems involving the angle sum of a polygon.

SkillCheck

ANSWERS ON P. 624

Quiz
SkillCheck 4

Skillsheet
Types of angles

Worksheet
A page of angles

Puzzle
Angle cards

1 Draw two different examples of each type of angle.

a acute angle **b** right angle
c obtuse angle **d** reflex angle

2 Classify each angle size as being acute, obtuse, right, reflex, straight or a revolution.

a $25°$ **b** $100°$ **c** $300°$ **d** $128°$
e $90°$ **f** $360°$ **g** $286°$ **h** $180°$

3 Name each type of angle(s) marked.

a

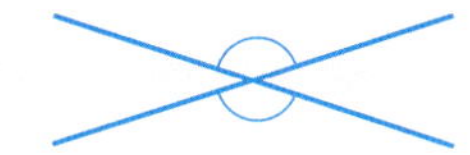

b

c

4 Find the value of x in each equation.

a $x + 30 = 90$ **b** $x + 57 = 180$ **c** $x + 121 + 77 = 360$

5 Copy each shape, name it and draw all axes of symmetry.

a

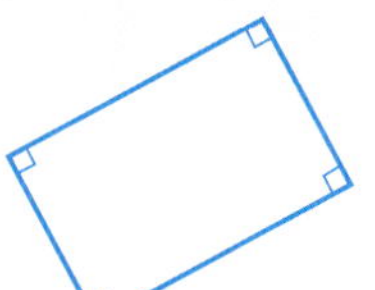

b

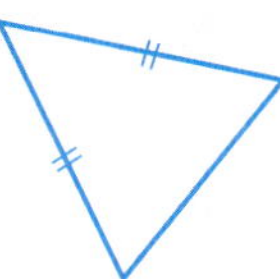

c

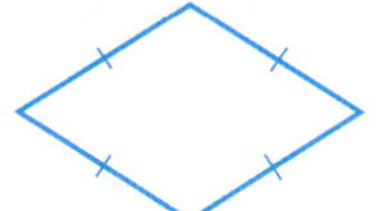

d

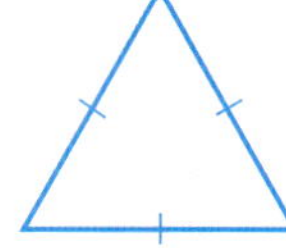

6 Write the order of rotational symmetry of each shape in question **5**.

7 Which quadrilateral has all four sides equal and all four angles equal? Select the correct answer **A**, **B**, **C** or **D**.

A parallelogram **B** rhombus **C** rectangle **D** square

8 Draw a scalene triangle.

9 **a** Draw a parallelogram and its diagonals.

b Are the lengths of the diagonals of a parallelogram equal?

4.01 Angle geometry

Classifying angles

Worksheet
Straight angles, right angles and revolutions

Puzzles
Angles: A dog day

Angles

Right angle	Straight angle	Revolution
90° (quarter-turn)	180° (half-turn)	360° (complete turn)

Acute angle	Obtuse angle	Reflex angle
Less than 90°	Between 90° and 180°	Between 180° and 360°

Angle facts

Video
Angles at a point and vertically opposite angles

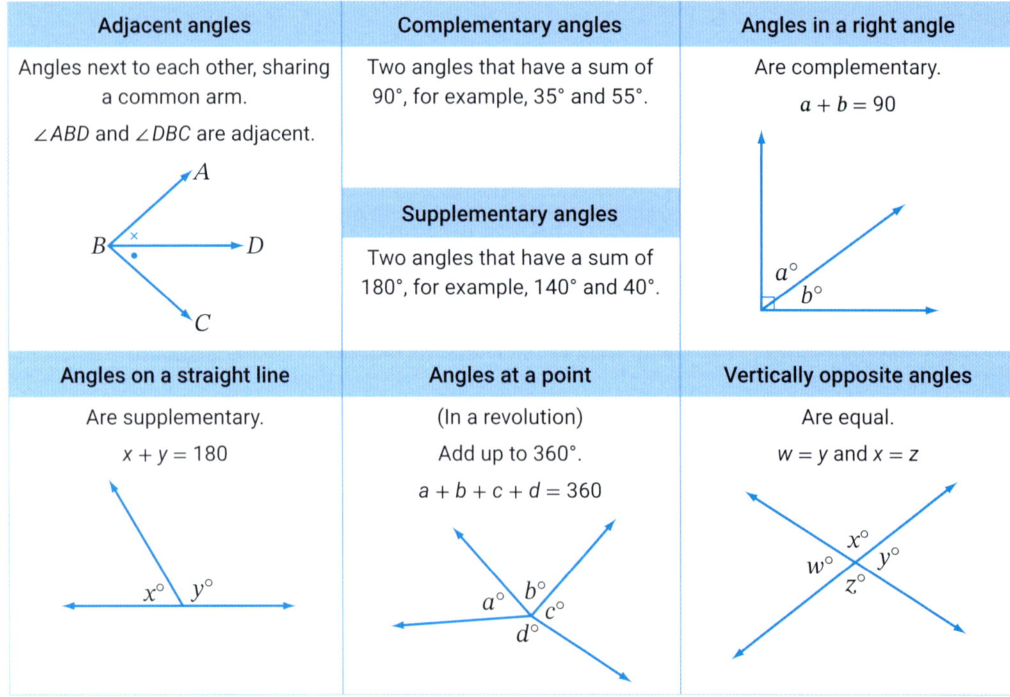

Adjacent angles	Complementary angles	Angles in a right angle
Angles next to each other, sharing a common arm. $\angle ABD$ and $\angle DBC$ are adjacent.	Two angles that have a sum of 90°, for example, 35° and 55°.	Are complementary. $a + b = 90$
	Supplementary angles Two angles that have a sum of 180°, for example, 140° and 40°.	

Angles on a straight line	Angles at a point	Vertically opposite angles
Are supplementary. $x + y = 180$	(In a revolution) Add up to 360°. $a + b + c + d = 360$	Are equal. $w = y$ and $x = z$

Example 1

Find the value of each variable, giving reasons.

a

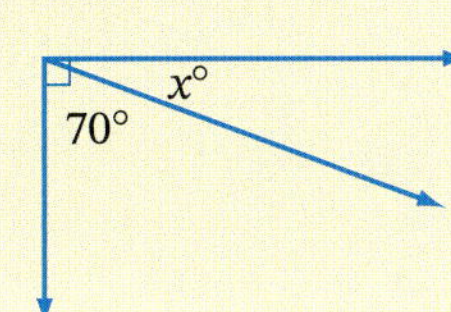

b

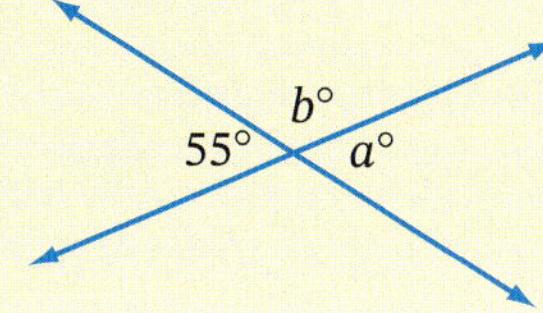

c

134°
72°
y°

Video Angle geometry

SOLUTION

a $x + 70 = 90$ (Angles in a right angle)

$x = 90 - 70$

$= 20$

b $a = 55$ (Vertically opposite angles.)

$b + 55 = 180$ (Angles in a straight line)

$b = 180 - 55$

$= 125$

c $y + 72 + 134 + 90 = 360$ (Angles at a point)

$y + 296 = 360$

$y = 360 - 296$

$= 64$

The reason is written inside brackets.

EXERCISE 4.01 ANSWERS ON P. 624

Angle geometry

U F R C

1 Classify each type of angle(s).

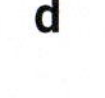

a

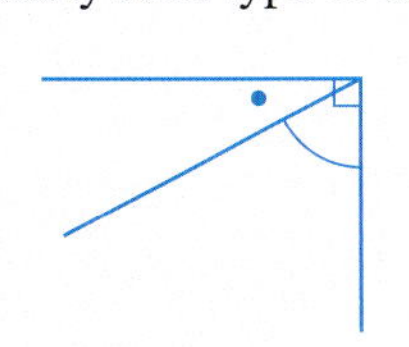

b

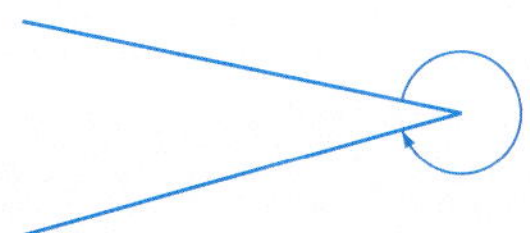

c

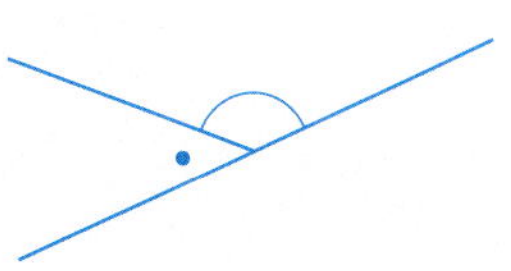

d

e

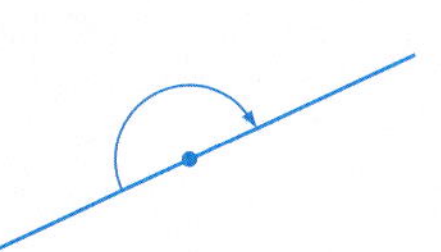

f

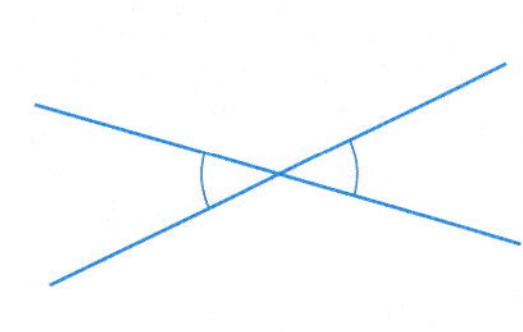

g

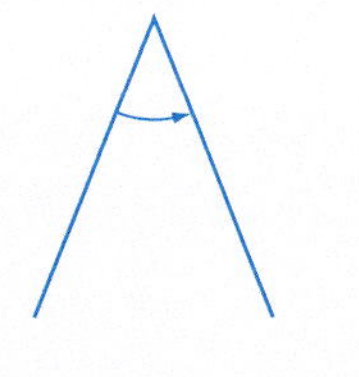

h

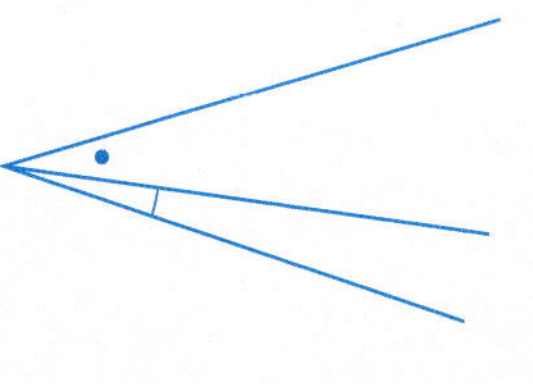

i

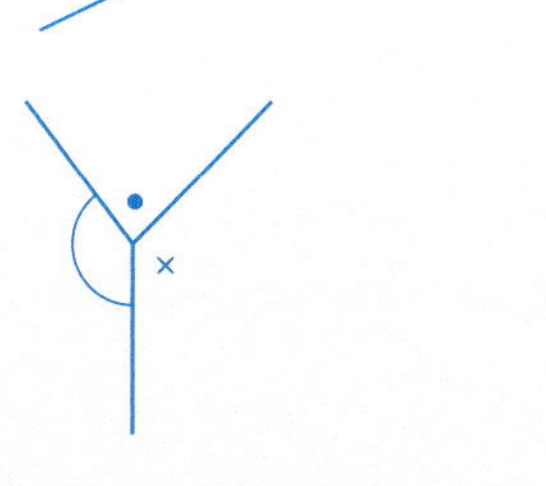

Foundation Standard Complex

2 For each angle size, find the complementary angle.

a 16° **b** 33° **c** 71° **d** 2°

3 For each angle size, find the supplementary angle.

a 25° **b** 149° **c** 107° **d** 48°

4 What is the sum of the angles at a point? Select the correct answer **A**, **B**, **C** or **D**.

A 90° **B** 180° **C** 270° **D** 360°

EXAMPLE 1

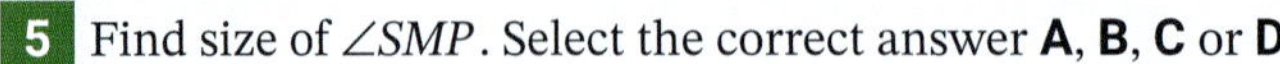

5 Find size of $\angle SMP$. Select the correct answer **A**, **B**, **C** or **D**.

A 80° **B** 280°

C 100° **D** 10°

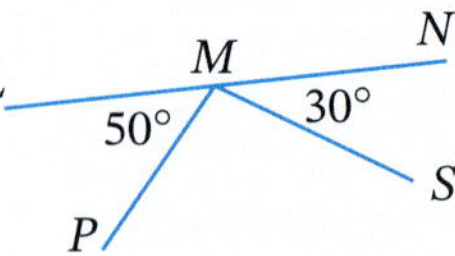

6 Find the value of each variable, giving reasons.

a

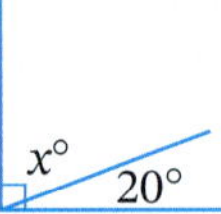

b

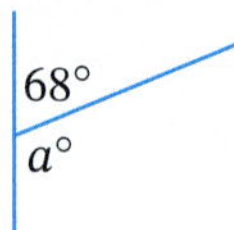

c

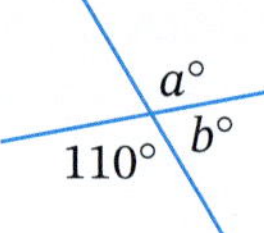

d

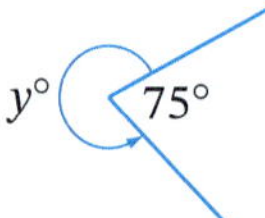

e

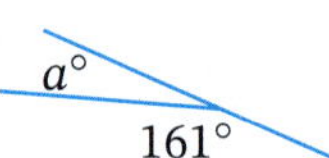

f

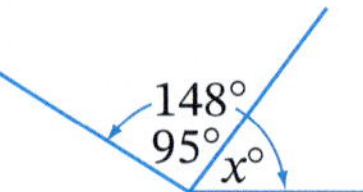

g

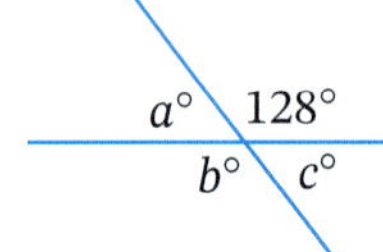

h

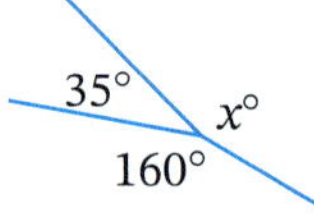

i

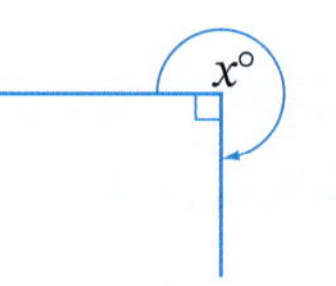

j

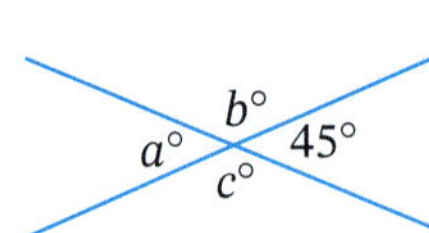

k

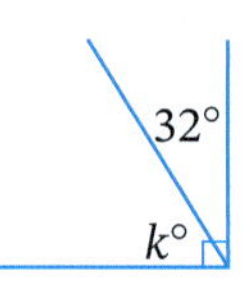

l

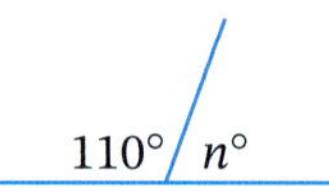

m

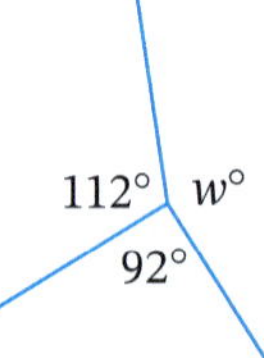

n

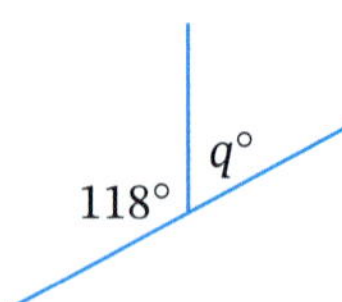

o

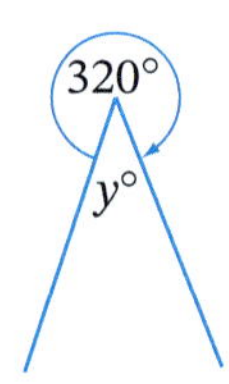

Foundation Standard Complex

p

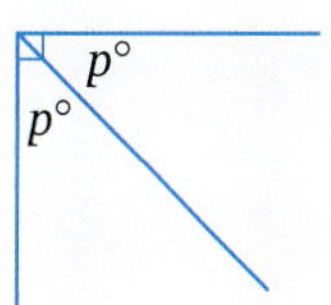

q

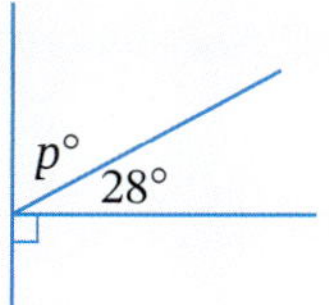

r

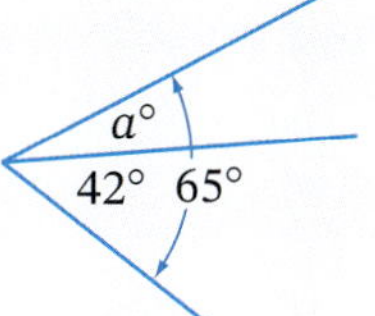

7 Find the value of each variable, giving reasons.

a

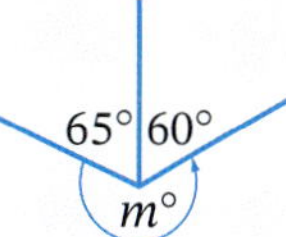

b

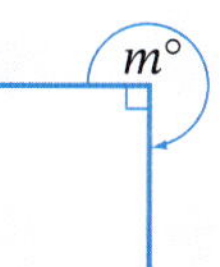

c

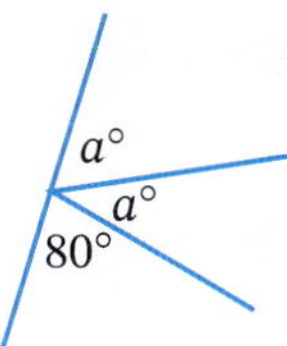

d

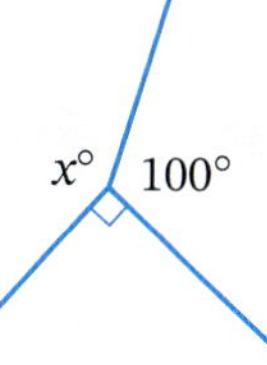

e

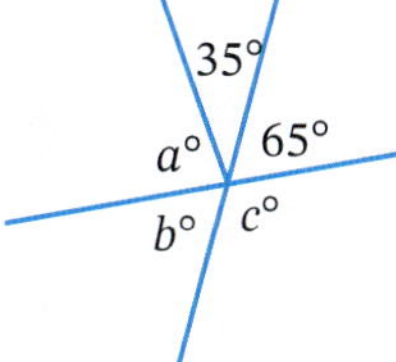

f

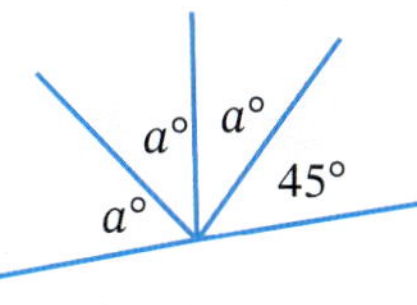

g

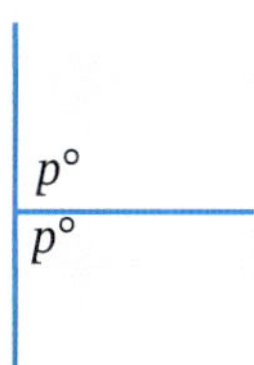

h

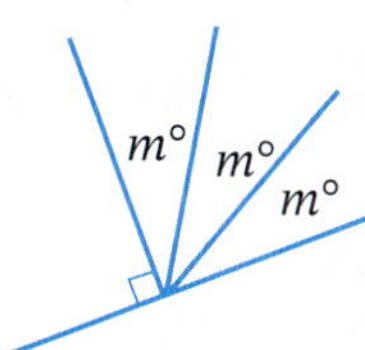

8 *ABC* is a straight line and *EB* bisects ∠*DBF*. Find the size of ∠*EBC*. Select the correct answer **A**, **B**, **C** or **D**.

A 95° **B** 110°

C 85° **D** 140°

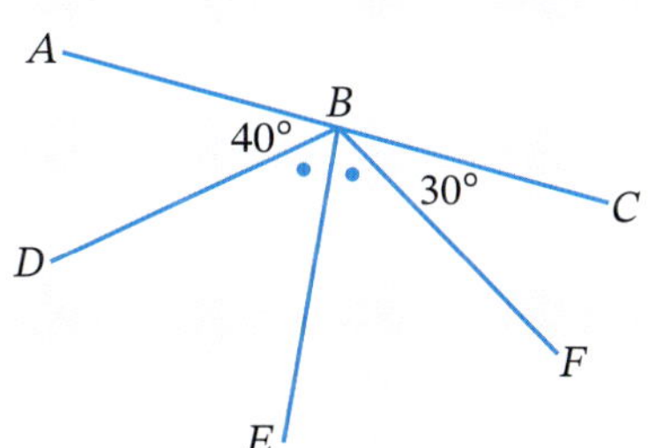

4.01

Foundation Standard Complex

4.02 Angles on parallel lines

Skillsheet
Angles and parallel lines

Worksheets
Investigating angles on parallel lines

Angles in parallel lines

What is the diagram?

Matching angles

Puzzle
Angles in parallel lines

Videos
Angles on parallel lines

Angle relationships

When **parallel lines** are crossed by another line (called a **transversal**), special pairs of angles are formed.

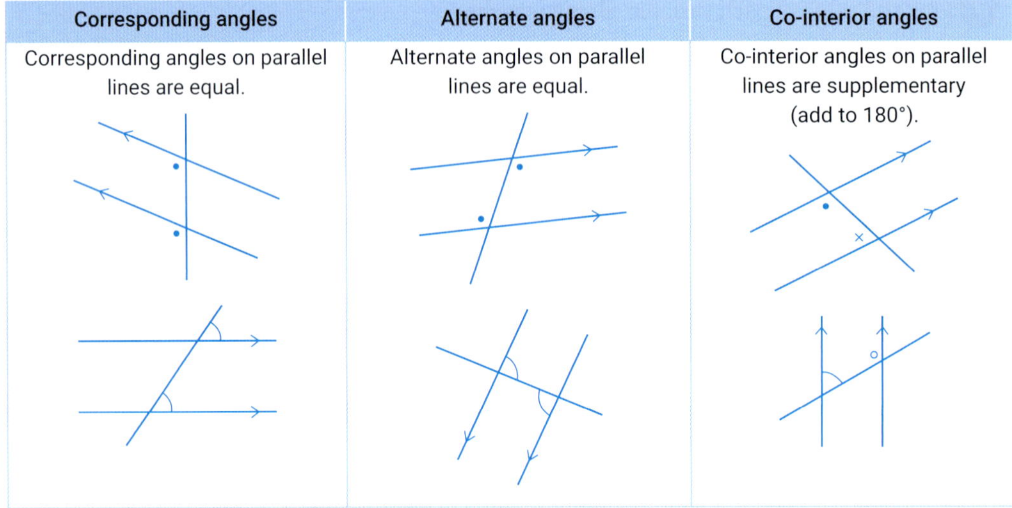

Corresponding angles	Alternate angles	Co-interior angles
Corresponding angles on parallel lines are equal.	Alternate angles on parallel lines are equal.	Co-interior angles on parallel lines are supplementary (add to 180°).

- Corresponding angles are in 'matching' positions on the *same* side of the transversal: 'corresponding' means 'matching'.
- Alternate angles are between the parallel lines on *opposite* sides of the transversal: 'alternate' means 'going back and forth between two sides'.
- Co-interior angles are between the parallel lines on the *same* side of the transversal: 'co-interior' means 'together inside'.

The positions of corresponding angles will form the letter 'F', the positions of alternate angles form the letter 'Z' and the position of co-interior angles form the letter 'C'.

Corresponding angles	Alternate angles	Co-interior angles
Angles are equal in size	Angles are equal in size	Angles add to 180°

You may need to rotate or reflect the diagrams in order to see the letters 'F', 'Z' or 'C'.

Example 2

Find the value of each variable, giving reasons.

a

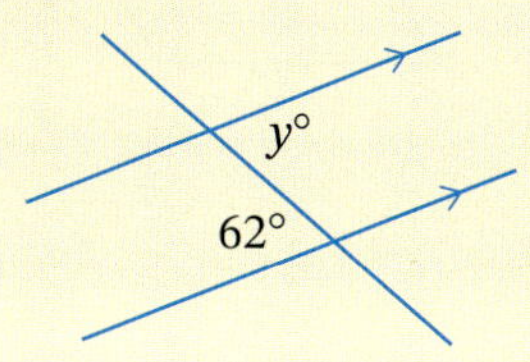

b

$a°$ 110°

SOLUTION

a $y = 62$ (Alternate angles on parallel lines)

b $a + 110 = 180$ (Co-interior angles on parallel lines)

$a = 180 - 110$

$= 70$

Video
Angle geometry

4.02

Example 3

Prove that the lines AB and CD are parallel.

A E B 76° 76° C F D

SOLUTION

$\angle AEF$ and $\angle EFD$ are alternate angles.

$\angle AEF = \angle EFD = 76°$

$\therefore AB \parallel CD$ (Alternate angles are equal)

'$\therefore AB \parallel CD$' means 'Therefore line AB is parallel to line CD'

EXERCISE 4.02 ANSWERS ON P. 625

Angles on parallel lines

U F R C

1 Is each marked pair of angles corresponding, alternate or co-interior?

a

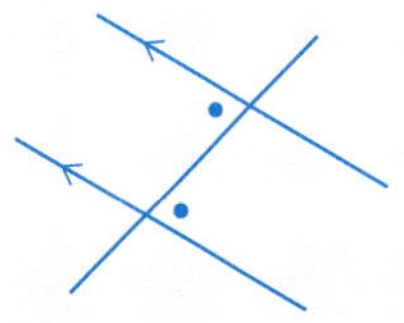

b

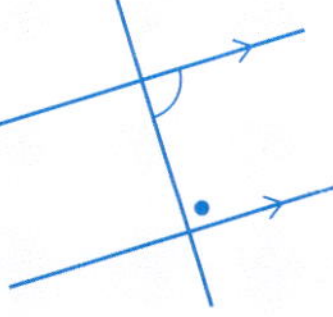

c

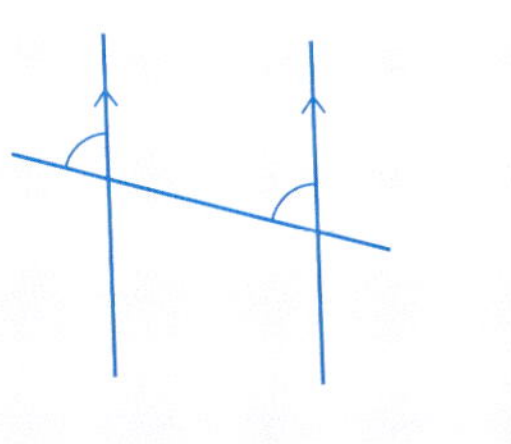

Foundation Standard Complex

d

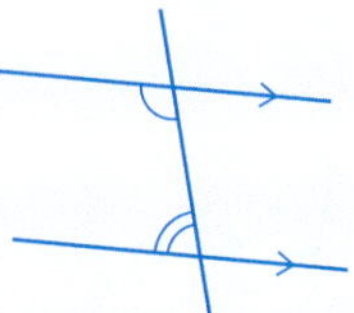

e

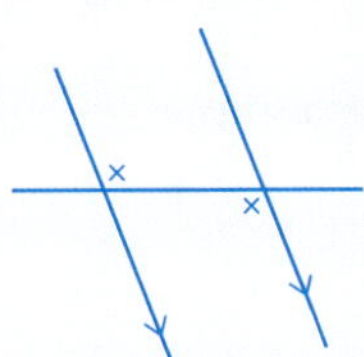

f

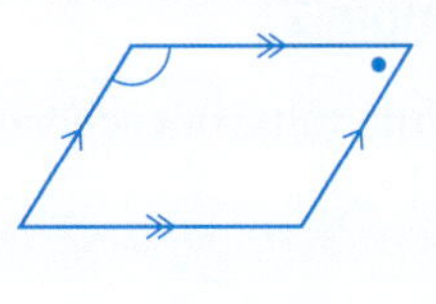

2 Which angle is corresponding to the angle marked g°? Select the correct answer *A*, *B*, *C* or *D*.

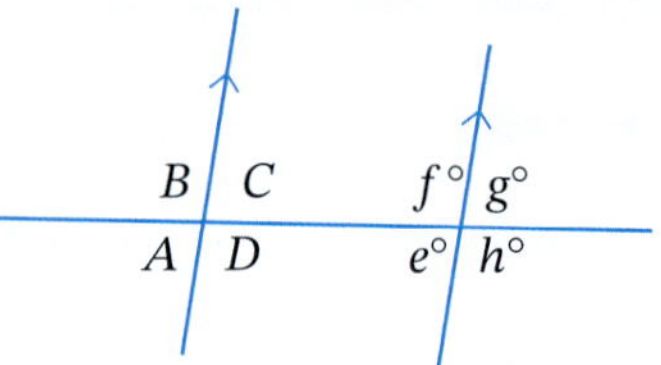

3 For the diagram in question **2**, name an angle that is:

a co-interior to *D*.

b alternate to the angle marked *e*°.

c equal to *C*.

d corresponding to *B*.

e supplementary to the angle marked *e*°.

EXAMPLE 2

4 Find the value of the variable in each diagram, giving reasons for your answers.

R C

a

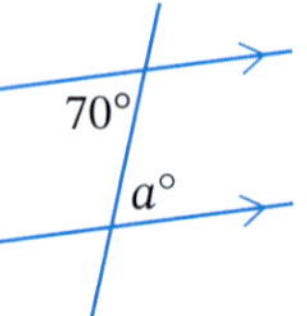

b

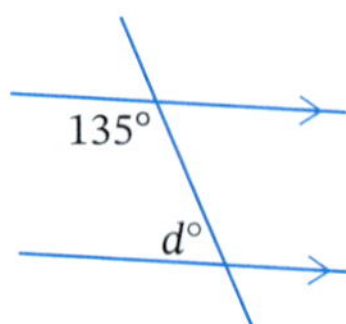

c

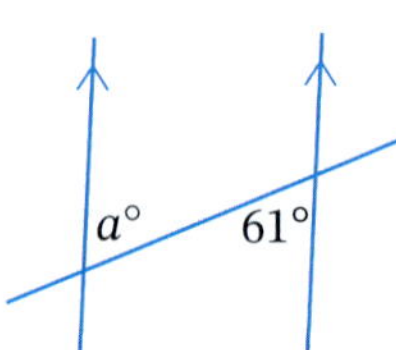

d

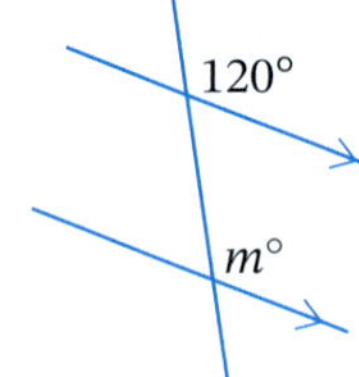

e

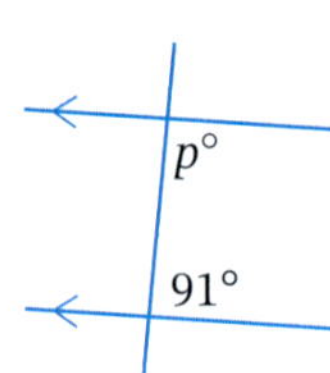

f

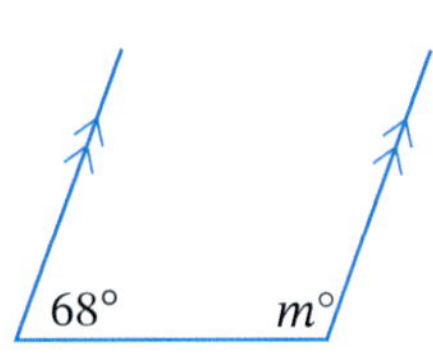

5 Find the value of each variable, giving reasons.

R C

a

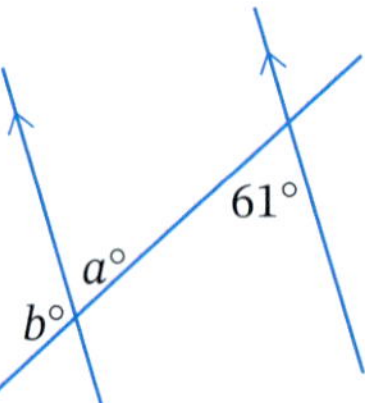

b

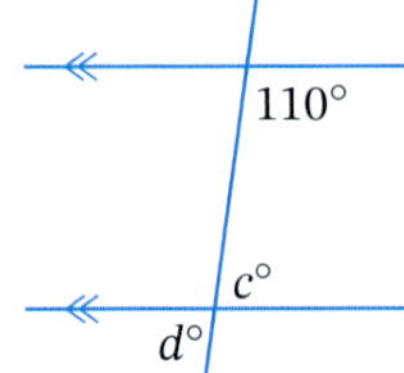

c

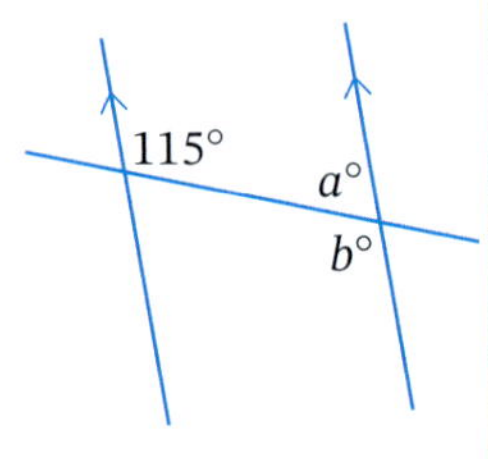

☐ Foundation ○ Standard ⬡ Complex

d

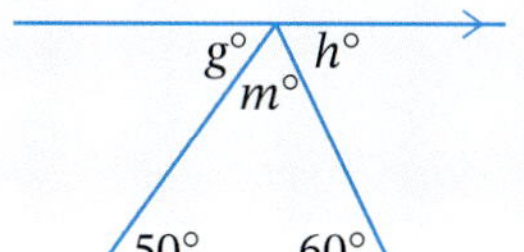

e

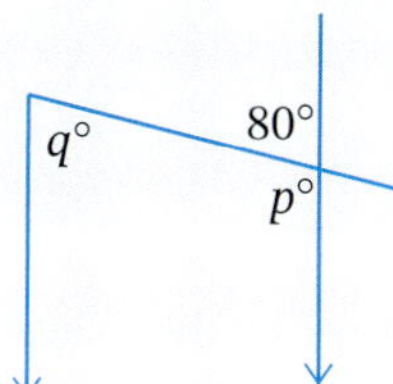

f

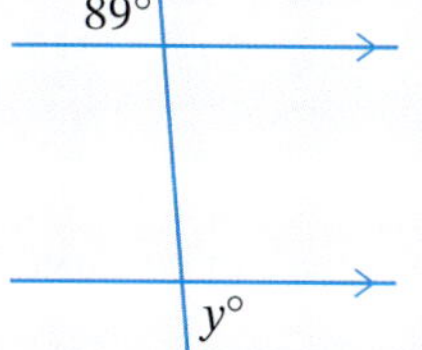

g

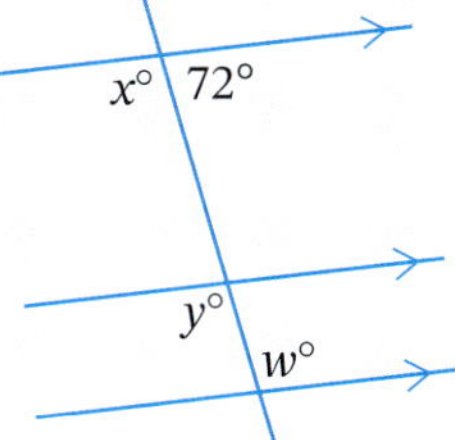

h

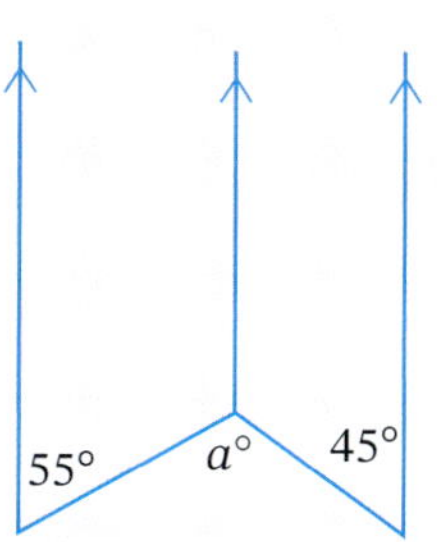

i

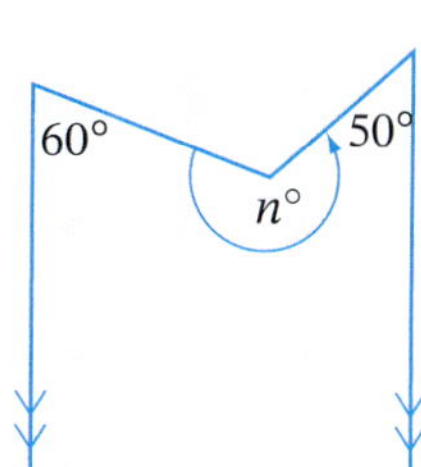

6 What is the value of y in this diagram?

R Select the correct answer **A**, **B**, **C** or **D**.

A 85 **B** 40

C 35 **D** 65

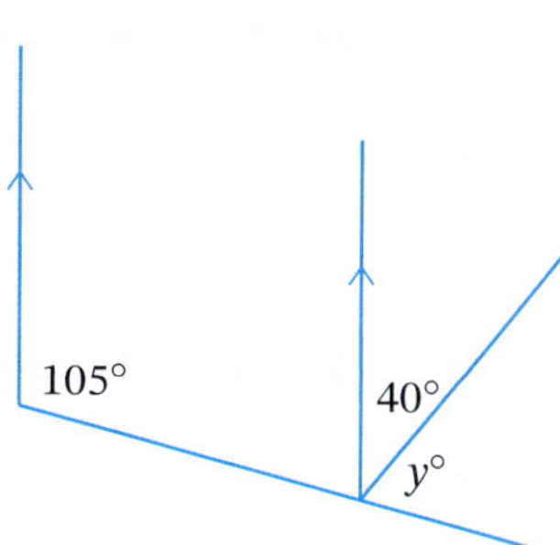

7 For each diagram, decide whether AB is parallel to CD.

R If it is, then prove it.

C **a**

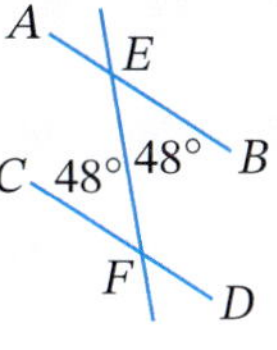

b

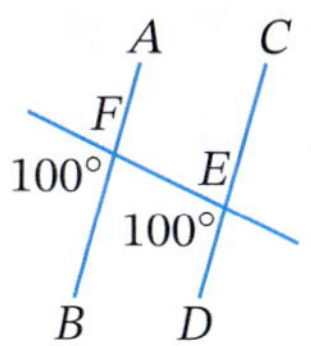

c

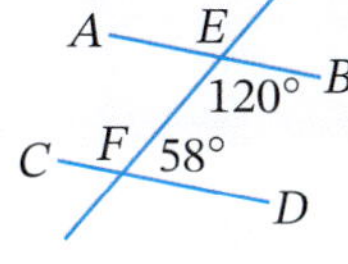

d

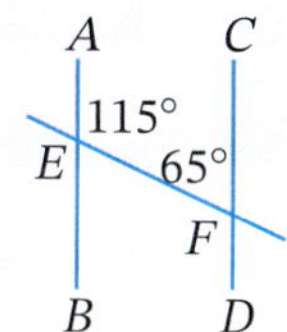

EXAMPLE 2

8 Find the value of m in each diagram.

R **a**

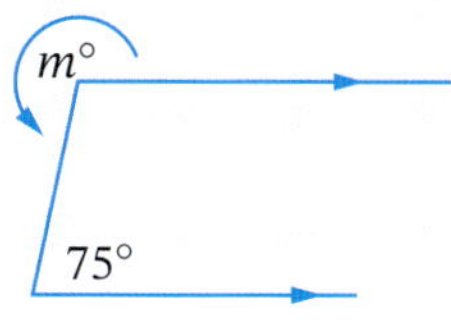

b

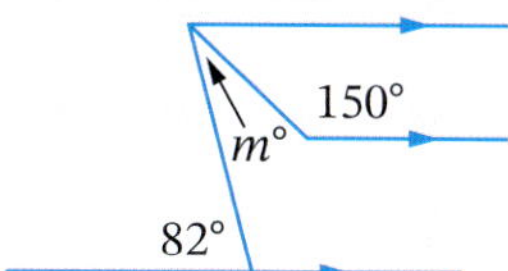

Foundation Standard Complex

4.03 Classifying triangles

Skillsheet
Naming shapes

Videos
Properties of triangles

Classifying triangles

Worksheets
Properties of triangles

Triangle geometry

Triangles can be classified in two ways: by their sides or by their angles.

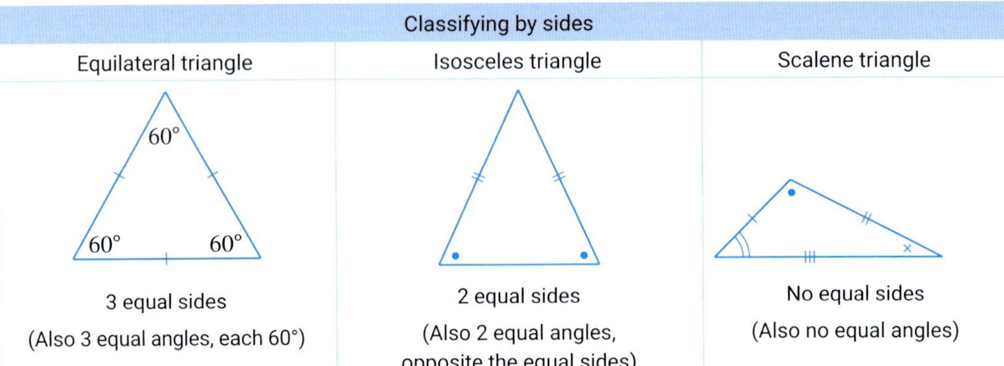

Classifying by sides		
Equilateral triangle	Isosceles triangle	Scalene triangle
3 equal sides (Also 3 equal angles, each 60°)	2 equal sides (Also 2 equal angles, opposite the equal sides)	No equal sides (Also no equal angles)

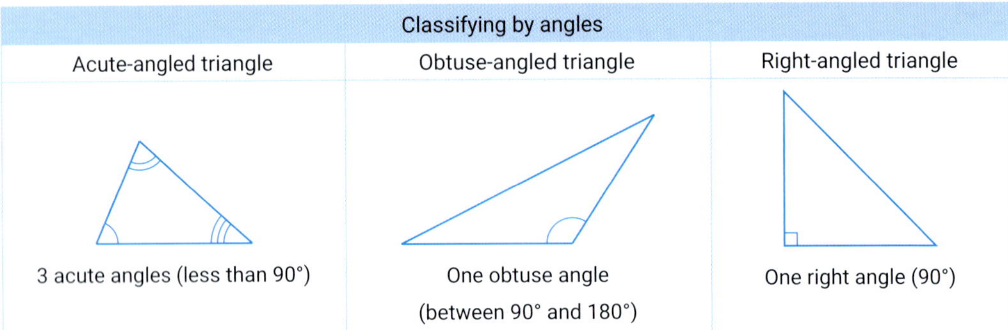

Classifying by angles		
Acute-angled triangle	Obtuse-angled triangle	Right-angled triangle
3 acute angles (less than 90°)	One obtuse angle (between 90° and 180°)	One right angle (90°)

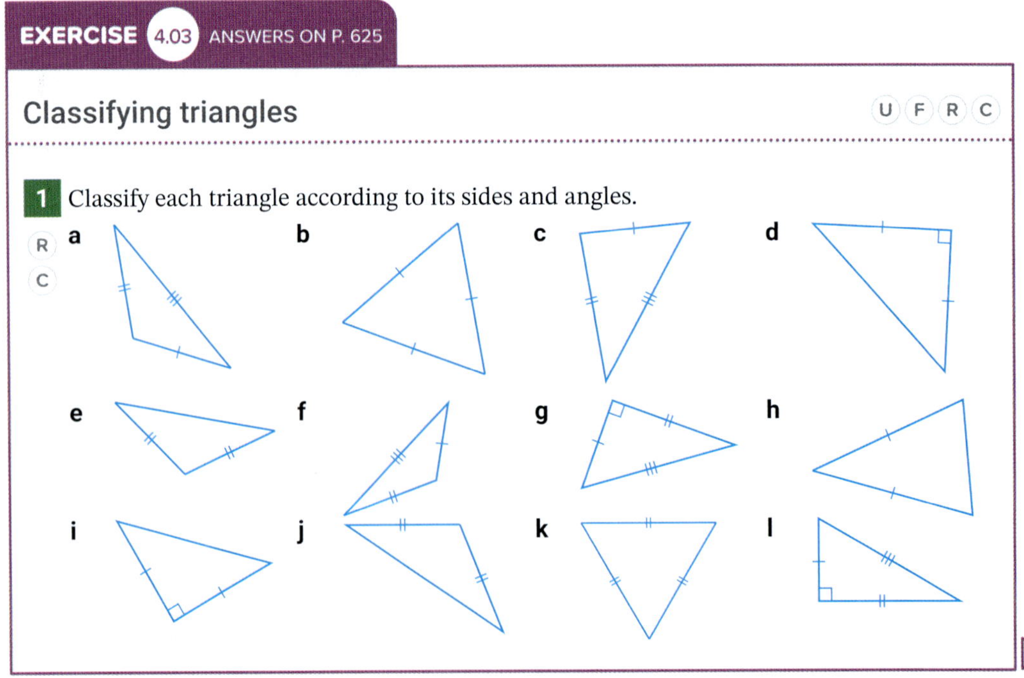

EXERCISE 4.03 ANSWERS ON P. 625

Classifying triangles

U F R C

1 Classify each triangle according to its sides and angles.

R C

9780170465601

2 Sketch a triangle that is:

R C

a right-angled and isosceles **b** equilateral

c scalene and obtuse-angled **d** acute-angled and scalene

e right-angled and scalene **f** acute-angled and isosceles

3 Which type of triangle is this? Select the correct answer **A**, **B**, **C** or **D**.

35° 35°

R C

A isosceles and obtuse-angled

B isosceles and acute-angled

C scalene and obtuse-angled

D scalene and acute-angled

4 Is it possible to draw an obtuse-angled equilateral triangle? Justify your answer.

R C

5 Which triangles in question **1** have:

a line symmetry? **b** rotational symmetry?

6 Find the value of each variable, giving a reason.

R C

a

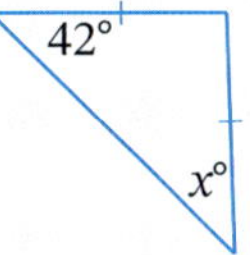

b

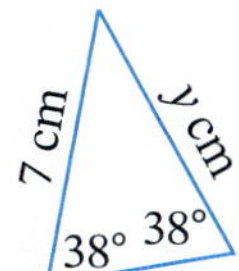

c

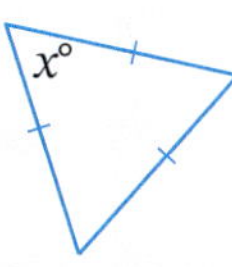

d

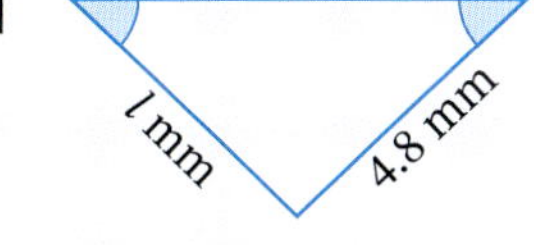

e

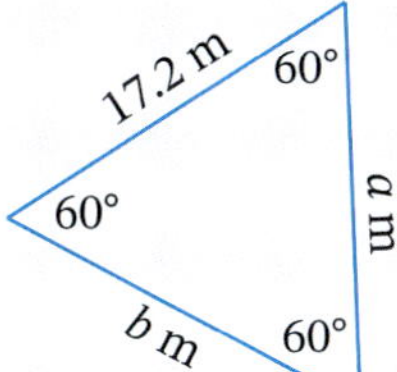

f

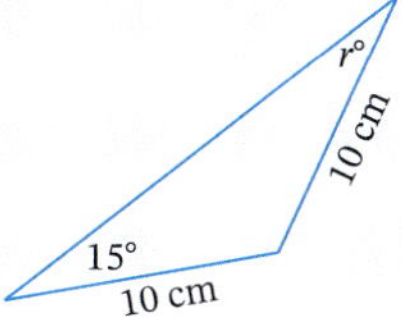

7 Is it possible to draw a triangle with two obtuse angles? Why?

R C

8 Which triangle is both obtuse-angled and scalene? Select the correct answer **A**, **B**, **C** or **D**.

R C

a

b

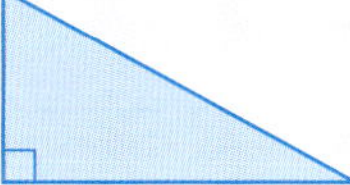

c

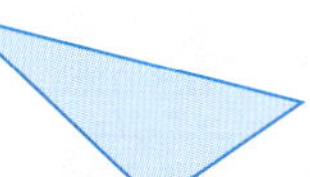

d

9 Copy and complete this table.

R C

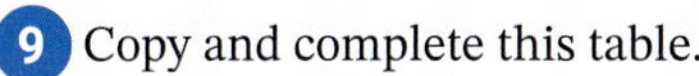

Triangle	Number of axes of symmetry	Order of rotational symmetry
Equilateral triangle		
Isosceles triangle		No rotational symmetry
Scalene triangle		

☐ Foundation ○ Standard ⬡ Complex

DID YOU KNOW?

It's all Greek or Latin to me!

Many of our words in geometry come from Greek or Latin. Latin was the language of the ancient Roman Empire.

Word	Origin	Meaning
Equilateral	Latin: aequus latus	Equal sides
Equiangular	Latin: aequus angulus	Equal corners
Isosceles	Greek: isos skelos	Equal legs
Scalene	Greek: skalenos	Uneven leg
Acute	Latin: acutus	Sharp
Obtuse	Latin: obtusus	Dull or blunt
Reflex	Latin: reflexus	Bent back
Triangle	Latin: tri angulus	Three corners
Rectangle	Latin: rectus angulus	Right corners
Quadrilateral	Latin: quadri latus	Four sides
Polygon	Greek: poly gonon	Many angles
Diagonal	Greek: dia gonios	From angle to angle
Trapezium/Trapezoid	Latin/Greek: trapeza	Small table

Explain what this sentence means and illustrate with a diagram: 'A rhombus is equilateral but not equiangular'.

INVESTIGATION

The perpendicular bisector in an isosceles triangle

$\triangle ABC$ is an isosceles triangle with $AC = AB$. It has one axis of symmetry, AD.

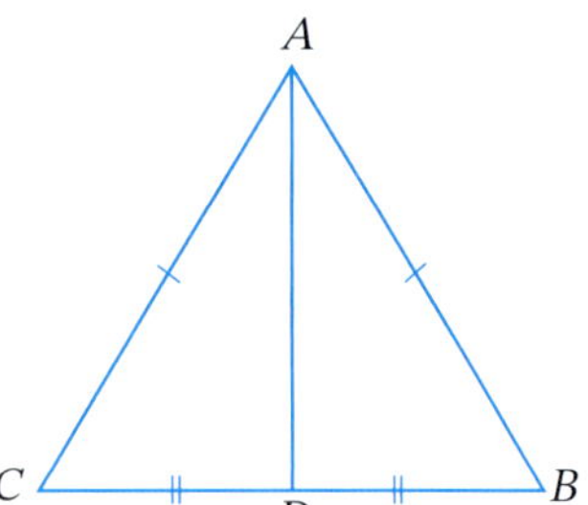

1 Why is $CD = DB$?

2 Why is $\angle ADC = \angle ADB$?

3 What is the size of $\angle ADC$ and $\angle ADB$?

4 AD bisects side CB. What does 'bisect' mean?

5 $AD \perp CB$. What does '$\perp$' mean?

6 'In an isosceles triangle, the axis of symmetry is the perpendicular bisector of the uneven side'. Explain what this means in your own words.

☆ MENTAL SKILLS 4A ANSWERS ON P. 626

Maths without calculators

Quiz
Mental skills 4A

Converting fractions and decimals to percentages

To convert a fraction or decimal into a percentage, multiply it by 100%.

1 Study each example.

a $\frac{2}{5} = \frac{2}{5} \times 100\% = \frac{2}{\cancel{5}_1} \times \cancel{100}^{20}\% = 2 \times 20\% = 40\%.$

b $\frac{24}{40} = \frac{24}{40} \times 100\% = \frac{\cancel{24}^3}{\cancel{40}_5} \times 100\% = \frac{3}{\cancel{5}_1} \times \cancel{100}^{20}\% = 3 \times 20\% = 60\%$

2 Now convert each fraction to a percentage.

a	$\frac{7}{10}$	**b**	$\frac{33}{50}$	**c**	$\frac{27}{60}$	**d**	$\frac{22}{25}$	**e**	$\frac{24}{32}$
f	$\frac{30}{40}$	**g**	$\frac{60}{75}$	**h**	$\frac{4}{5}$	**i**	$\frac{11}{20}$	**j**	$\frac{28}{80}$
k	$\frac{15}{50}$	**l**	$\frac{16}{20}$	**m**	$\frac{54}{60}$	**n**	$\frac{18}{40}$	**o**	$\frac{13}{25}$

3 Study each example.

a $0.41 = 0.41 \times 100\%$
$= 0.41$
$= 41\%$

b $0.08 = 0.08 \times 100\%$
$= 0.08$
$= 8\%$

c $0.9 = 0.9 \times 100\%$
$= 0.90$
$= 90\%$

d $0.375 = 0.375 \times 100\%$
$= 0.375$
$= 37.5\%$

4 Now convert each decimal to a percentage.

a	0.25	**b**	0.68	**c**	0.17	**d**	0.6	**e**	0.1
f	0.333	**g**	0.59	**h**	0.702	**i**	0.84	**j**	0.7
k	0.428	**l**	0.055	**m**	0.91	**n**	0.7825	**o**	0.314

4.04 Classifying quadrilaterals

Videos
Properties of quadrilaterals
Classifying quadrilaterals

Worksheets
Properties of quadrilaterals
Always, sometimes, never true
Classifying triangles and quadrilaterals

Puzzles
What shape am I?
Singing in the car

A **quadrilateral** is any shape with four straight sides. A quadrilateral may be **convex** or non-convex.

Convex quadrilateral	Non-convex quadrilateral
• All vertices (corners) point outwards. • All diagonals lie within the shape. • All angles are less than 180°.	• One vertex points inwards • One diagonal lies outside the shape. • One angle is more than 180° (reflex angle).

There are six special types of quadrilaterals.

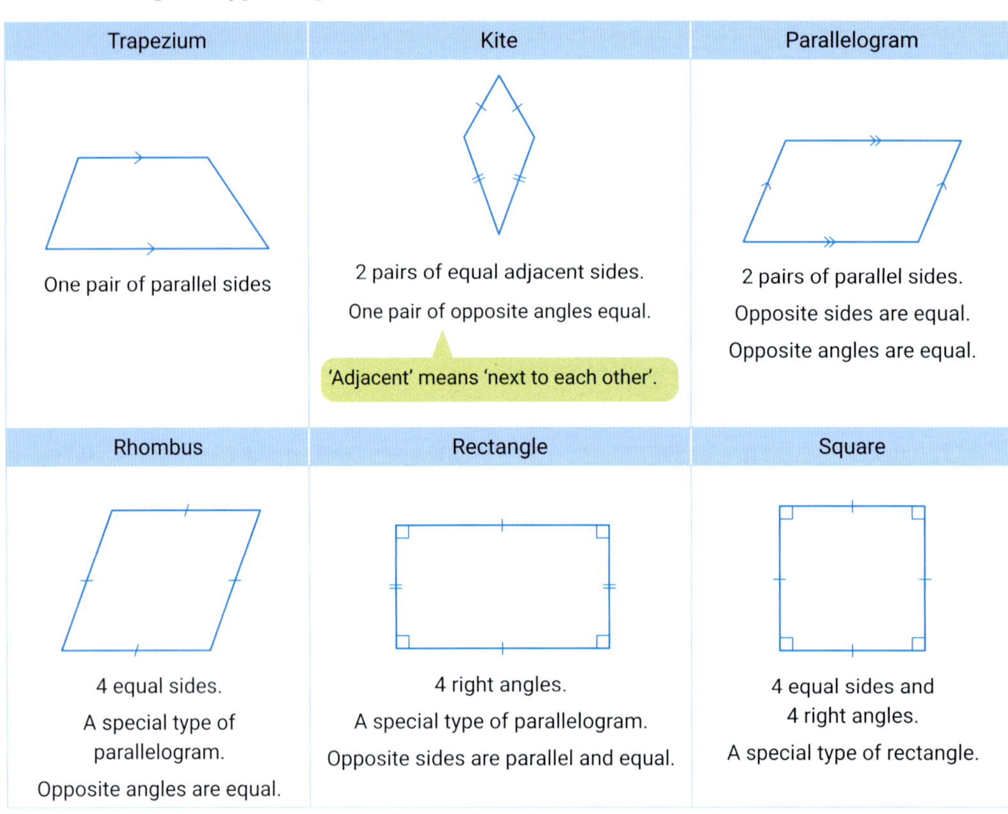

Trapezium	Kite	Parallelogram
One pair of parallel sides	2 pairs of equal adjacent sides. One pair of opposite angles equal. 'Adjacent' means 'next to each other'.	2 pairs of parallel sides. Opposite sides are equal. Opposite angles are equal.

Rhombus	Rectangle	Square
4 equal sides. A special type of parallelogram. Opposite angles are equal.	4 right angles. A special type of parallelogram. Opposite sides are parallel and equal.	4 equal sides and 4 right angles. A special type of rectangle.

EXERCISE 4.04 ANSWERS ON P. 626

Classifying quadrilaterals

U F R C

1 Is each quadrilateral convex or non-convex?

a

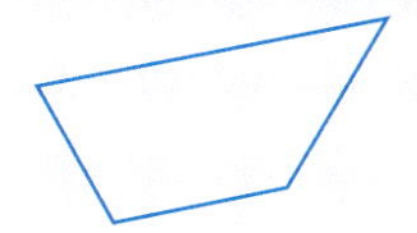

b

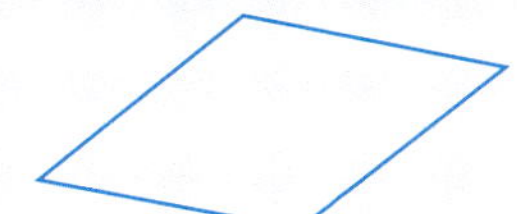

c

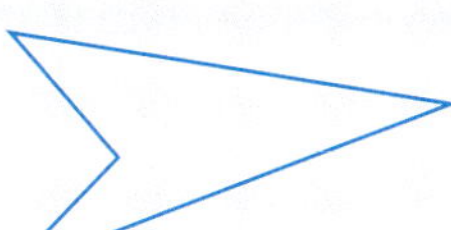

d

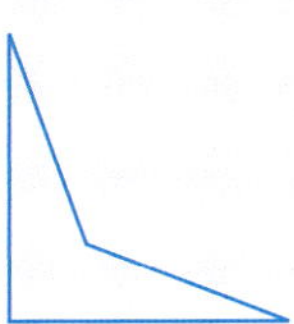

e

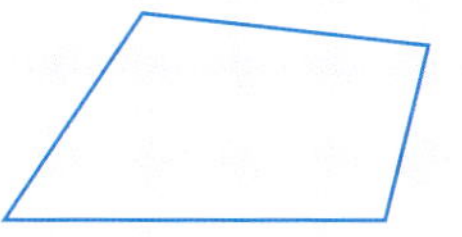

f

2 Name each quadrilateral.

C **a**

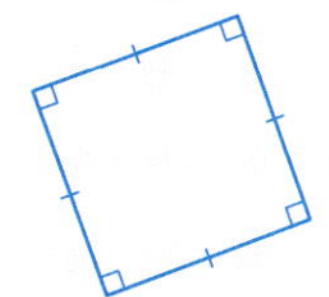

b

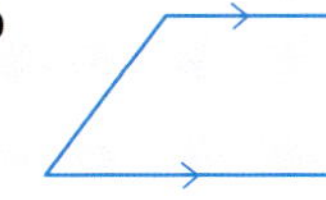

c

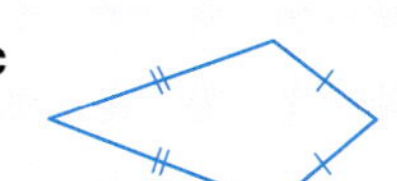

d

3 Name all the quadrilaterals that have:

R **a** four right angles

b exactly one pair of parallel sides

C **c** four equal sides

d opposite sides equal

e opposite sides parallel

f two pairs of equal adjacent sides

4 Which quadrilateral is also called a 'diamond'?

R C

5 When a shape with line symmetry is folded along an axis of symmetry, the two halves of the shape fit exactly on top of each other. For a shape with rotational symmetry, the number of times a shape fits on itself in one revolution is called its order of rotational symmetry.

Copy and complete this table.

Quadrilateral	Number of axes of symmetry	Order of rotational symmetry
Rectangle		
Parallelogram		
Trapezium		No rotational symmetry
Rhombus		
Square		
Kite		

Foundation Standard Complex

6 For this diagram, what type of quadrilateral is:

C **a** *ACDF*? **b** *FBCD*?

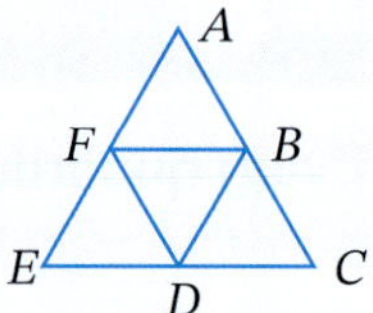

7 A parallelogram is any quadrilateral with both pairs of opposite sides parallel.
R Which of the following is **not** a special type of parallelogram? Select the correct answer
C **A**, **B**, **C** or **D**.

A square **B** kite **C** rectangle **D** rhombus

8 Sketch each quadrilateral, showing its main features.

C **a** rectangle **b** trapezium **c** rhombus **d** kite

9 True or false? (Explain your answers).

R **a** A rhombus is a square.

C **b** A square is a rhombus.

c A rectangle is a parallelogram.

d A parallelogram is a quadrilateral with its opposite sides parallel and equal.

e The diagonals of a parallelogram meet at right angles.

f A square is a rectangle.

g A rectangle is a square.

TECHNOLOGY

Properties of quadrilaterals

In this activity, you will use your dynamic geometry software to construct quadrilaterals.

Square

1. Draw a square of length 7.5 cm.
2. Construct the two diagonals for the square. Measure the length of each diagonal in the square. What do you notice?
3. Now measure the **size** of each angle of the square. What do you notice?
4. Draw another square with side length 9 cm. Repeat steps **2** and **3**. List two properties of the square.

Rectangle

1. Draw a rectangle with sides 6 cm by 3 cm.
2. Measure the length of the diagonals of the rectangle.

3 Now measure the **size** of each angle to check your accuracy. What should be the size of each angle?

4 Draw another rectangle with length 5 cm and width 8.4 cm. Measure the length of each diagonal and the size of each angle.

5 Copy and complete this property: The ____________ in a rectangle are ________.

6 In one rectangle, measure the distance from each vertex to the point of intersection of the two diagonals. Repeat for the second rectangle. What do you notice?

7 Copy and complete this property: The ____________ of a rectangle _____________ each other.

Parallelogram

1 Draw an interval 6 cm long. Label it AB.

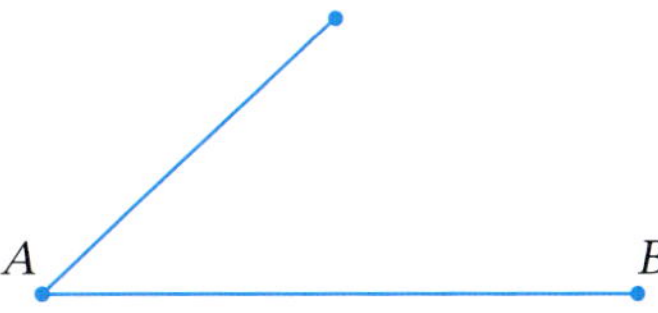

2 From A, draw the interval AC. Make the interval 4 cm.

3 Complete the parallelogram by drawing an interval from C to D and from B to D.

4 Copy and complete: The ___________ sides of a parallelogram are _________.

5 Now measure the size of $\angle CAB$ and $\angle CDB$. Repeat for $\angle ACD$ and $\angle ABD$. What do you notice?

6 Copy and complete: The ___________ angles of a parallelogram are _________.

7 Now draw the diagonals of the parallelogram. Measure the length of each diagonal. What do you notice?

8 Copy and complete: The diagonals of a parallelogram are _________.

9 Do the diagonals of a parallelogram bisect each other? Repeat step **6** from 'Rectangle' to help you.

10 Copy and complete: The diagonals of a parallelogram _________ bisect each other.

11 Drag any vertices of the parallelogram that you can. Is it possible to draw other parallelograms with the same dimensions, 6 cm by 4 cm? What do you notice?

12 Use your dynamic geometry software to **accurately** construct other quadrilaterals such as a rhombus, kite or trapezium.

4.05 Properties of quadrilaterals

Worksheets
Properties of quadrilaterals
Diagonal properties of quadrilaterals
Always, sometimes, never true

Videos
Properties of quadrilaterals
Classifying quadrilaterals

Presentation
Angles and shapes

EXERCISE 4.05 ANSWERS ON P. 626

Properties of quadrilaterals

U F R C

1 R C Copy the table below or use the link to print one out. Accurately draw each of the six special quadrilaterals below on a sheet of paper, then cut them out. Use your shapes to help you complete the table of quadrilateral properties.

a	Trapezium	• One pair of opposite sides are ______
b	Kite	• Two pairs of adjacent sides are ______ • One pair of opposite ______ are equal. • Diagonals intersect at ______
c	Parallelogram	• ______ sides are equal. • Opposite ______ are parallel. • Opposite angles are ______ • Diagonals ______ each other.
d	Rhombus	• All ______ are equal. • ______ sides are ______ • ______ angles are ______ • Diagonals bisect each other at ______ angles. • Diagonals ______ the angles of the rhombus.
e	Rectangle	• Opposite sides are ______ • Opposite sides are also ______ • All angles are ______ • Diagonals are ______ • Diagonals ______ each other.
f	Square	• All sides are ______ • All angles are ______ • Opposite ______ are parallel. • Diagonals are ______ • Diagonals bisect each other at ______

9780170465601

2 Name all special quadrilaterals that have each property.

R C

a Opposite sides are equal.
b Diagonals cross at right angles.
c Opposite angles are equal.
d One pair of opposite sides are parallel.
e Diagonals bisect each other.
f Opposite sides are parallel.
g Adjacent sides are of different lengths.
h Two equal diagonals.
i All angles are equal.
j All sides are equal.

3 Copy and complete the working to find the values of a and b.

R C

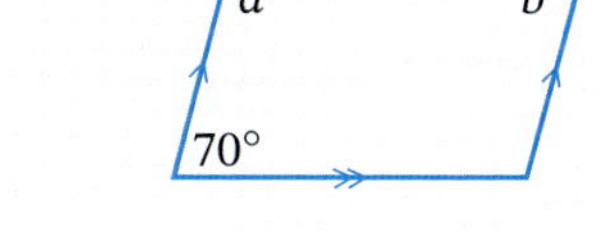

$a + 70 = 180$ (______ angles on ______ lines)

$a =$ ______

$b =$ ______ (opposite ______ of a parallelogram)

4 Find the value of each variable, giving reasons.

R C

a

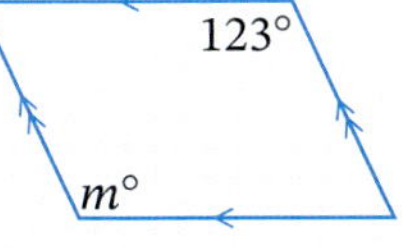

b

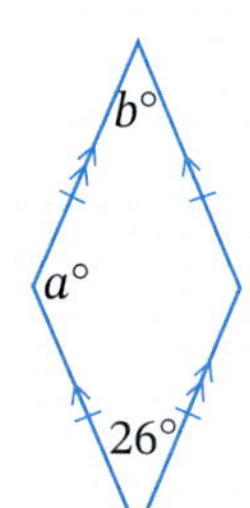

c

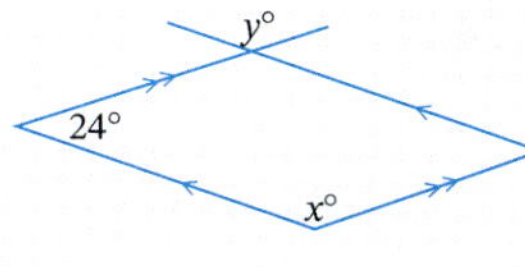

d

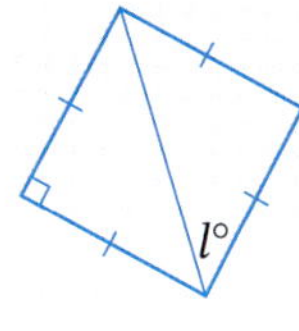

e

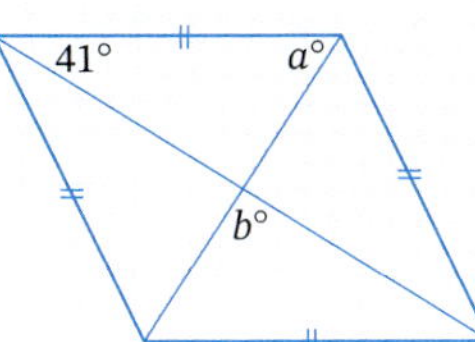

f

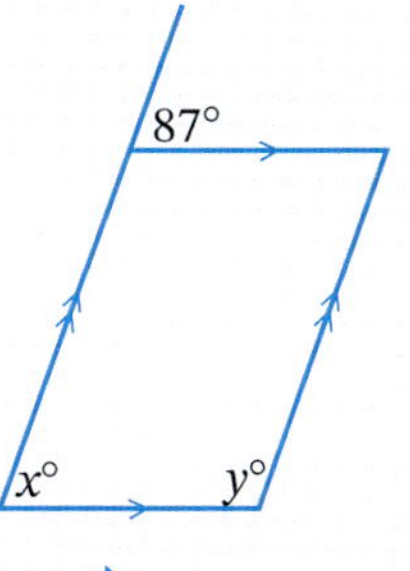

g

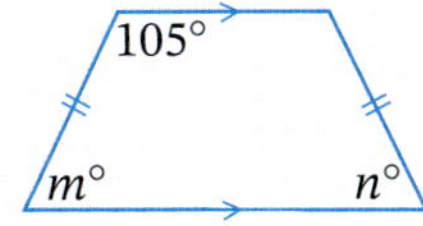

h

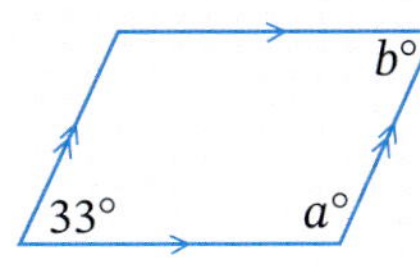

i

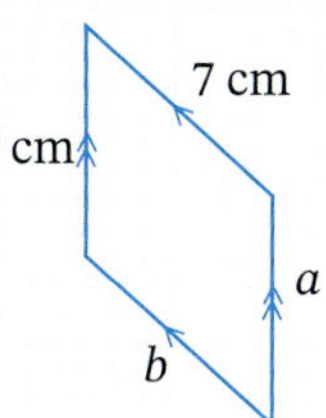

j

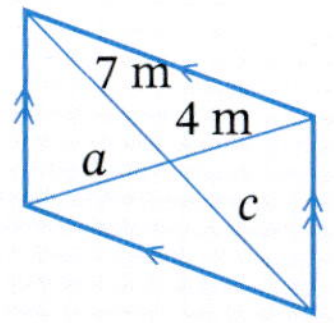

k

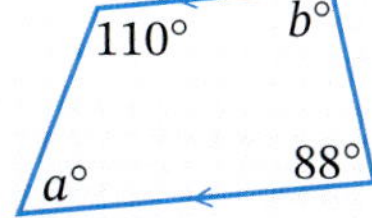

◻ Foundation ◯ Standard ⬡ Complex

5 I am a quadrilateral with opposite sides equal and parallel.

R C My diagonals are equal and I have two axes of symmetry.

Which quadrilateral am I? Select the correct answer **A**, **B**, **C** or **D**.

A parallelogram **B** rectangle **C** square **D** rhombus

6 I am a quadrilateral with opposite sides equal.

R C My diagonals bisect each other and meet at right angles.

Which quadrilateral am I? Select the correct answer **A**, **B**, **C** or **D**.

A parallelogram **B** trapezium **C** rectangle **D** rhombus

7 A rectangle is a quadrilateral with four right angles.

R C Which one of the following is a special type of rectangle? Select the correct answer **A**, **B**, **C** or **D**.

A square **B** kite **C** parallelogram **D** rhombus

8 A rhombus is a quadrilateral with all sides equal.

R C Which one of the following is a special type of rhombus? Select the correct answer **A**, **B**, **C** or **D**.

A square **B** kite **C** parallelogram **D** rectangle

9 The quadrilateral $WXYZ$ is shown.

R C **a** If $WX = YZ$ and $WZ = XY$, must $WXYZ$ be a rectangle? Why?

b If $\angle WZY = 90°$, must $WXYZ$ be a rectangle? Why?

c If the information in parts **a** and **b** are both true, must $WXYZ$ be a rectangle? Why?

W X Z Y

TECHNOLOGY

Exterior angle of a triangle

In this activity, you will use dynamic geometry software to discover an important property about the interior and exterior angles of any triangle.

1 Draw a triangle and label each vertex, A, B and C, as shown.

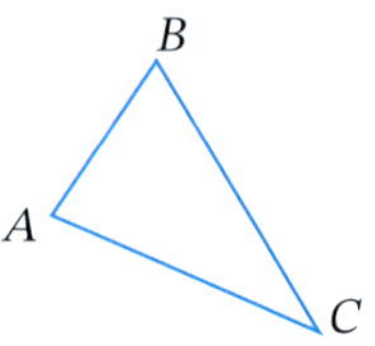

2 Draw a ray from A through C to point D.

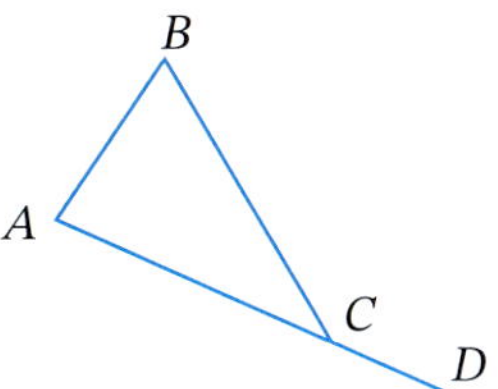

3 Find the size of $\angle ACB$, $\angle BAC$ and $\angle BCD$.

4 Calculate $\angle ACB + \angle BAC$.

5 Compare your answer from question **4** with the size of $\angle BCD$. What do you notice?

6 Repeat steps **1** to **5** for three other triangles.

7 Copy and complete: The __________ angle of a triangle is _______ to the sum of the _____ opposite _________ angles.

Angle sums of triangles and quadrilaterals 4.06

Angle sum of a triangle

The **angle sum of a triangle** is 180°.

$$a + b + c = 180$$

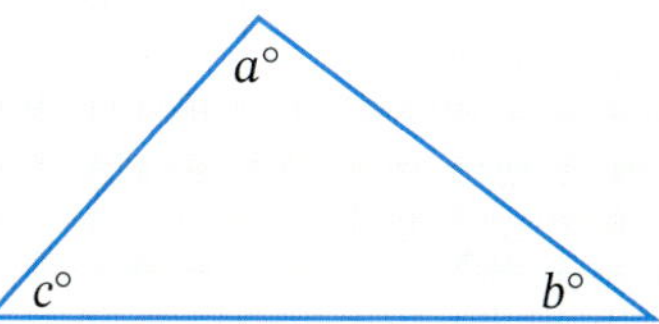

Videos
Angle sums of triangles and quadrilaterals
Angles in polygons

Worksheets
Angles in triangles
Find the unknown angle 2
Deductive geometry
Shapes and angles review

Puzzles
Mixed angles
Angles in triangles
Triangles and quadrilaterals

Example 4

Find the value of each variable, giving reasons.

a

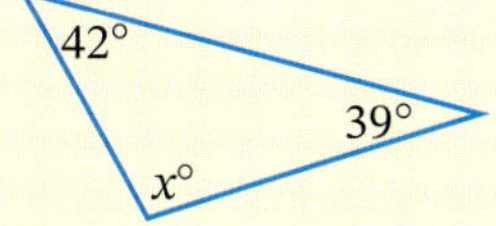

b

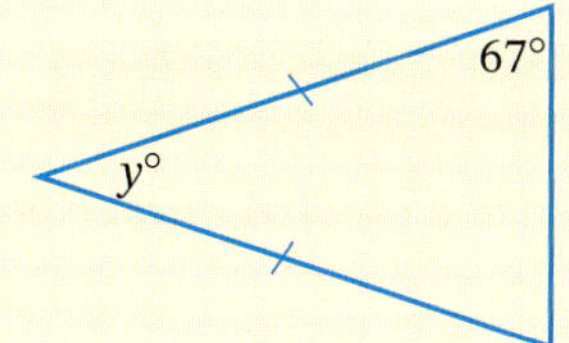

SOLUTION

a $x + 42 + 39 = 180$ (angle sum of a triangle)

$x = 180 - 42 - 39$

$= 99$

b $y + 67 + 67 = 180$ (angle sum of an isosceles triangle)

$y = 180 - 134$

$= 46$

ⓘ Exterior angle of a triangle

The **exterior angle of a triangle** is equal to the sum of the two interior opposite angles.

$$z = x + y$$

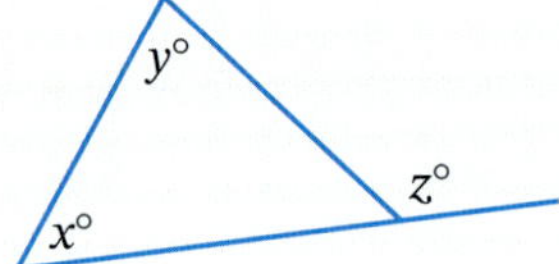

Example 5

Video
Exterior angle of a triangle

Find the value of each variable, giving reasons.

a

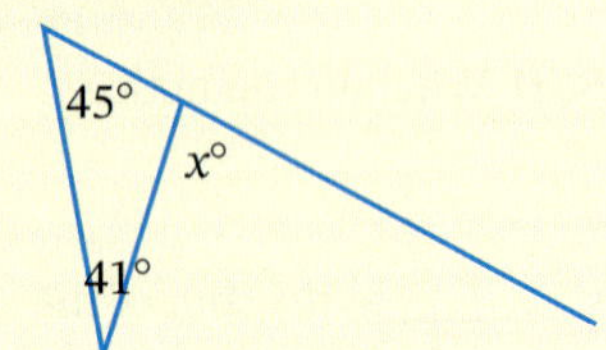

b

m°
116°
54°

SOLUTION

a $x = 45 + 41$ (exterior angle of a triangle)

$= 86$

b $m + 54 = 116$ (exterior angle of a triangle)

$m = 116 - 54$

$= 62$

To prove the angle sum of a quadrilateral, we can divide a quadrilateral into two triangles along one of its diagonals. Because the angles in each triangle add to 180°, the angles in both triangles add to $2 \times 180° = 360°$.

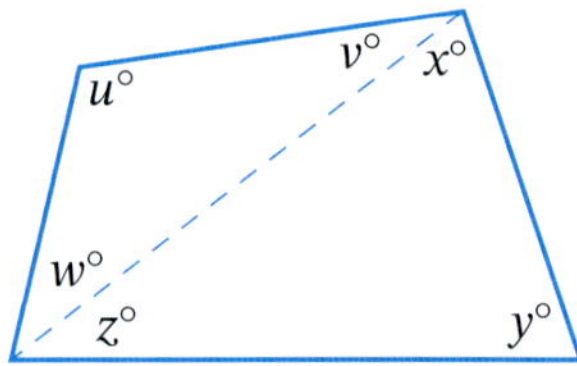

$u° + v° + w° = 180°$ and $x° + y° + z° = 180°$

$\therefore$ Angle sum of quadrilateral $= 180° + 180°$

$= 360°$

ⓘ Angle sum of a quadrilateral

The **angle sum of a quadrilateral** is 360°.

$$a + b + c + d = 360$$

This property is true for both convex and non-convex quadrilaterals.

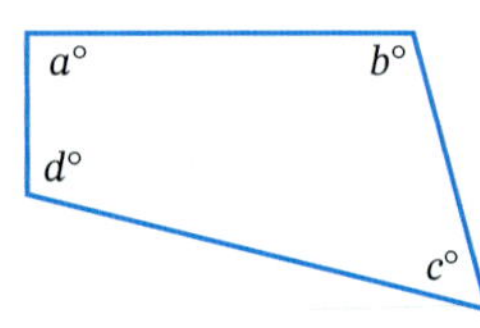

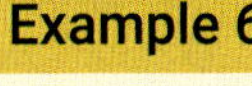

Example 6

Find the value of each variable, giving reasons.

a

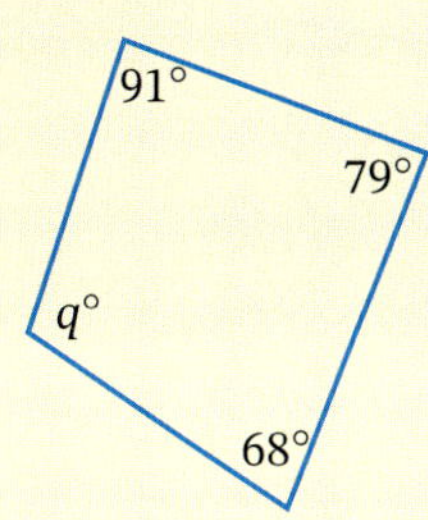

b

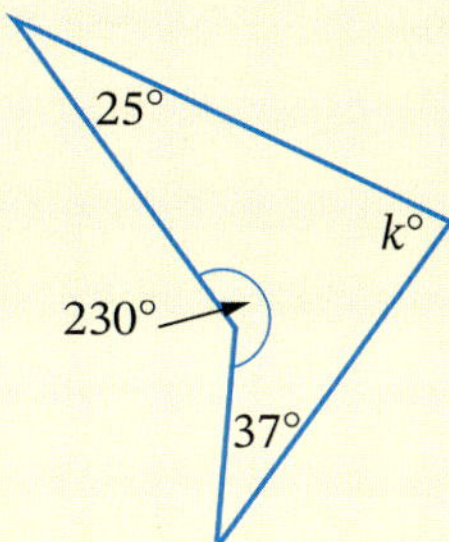

SOLUTION

a $q + 91 + 79 + 68 = 360$ (angle sum of a quadrilateral)

$q + 238 = 360$

$q = 360 - 238$

$= 122$

b $k + 25 + 37 + 230 = 360$ (angle sum of a quadrilateral)

$k + 292 = 360$

$k = 360 - 292$

$= 68$

Video
Angle sums of triangles and quadrilaterals

4.06

EXERCISE 4.06 ANSWERS ON P. 626

Angle sums of triangles and quadrilaterals

U F R C

1 Find the value of each variable.

EXAMPLE 4

R **a**

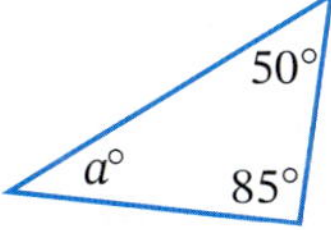

b

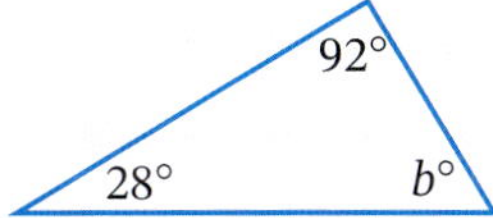

c

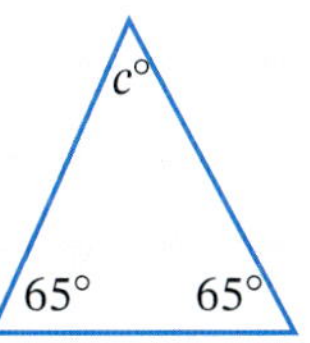

d

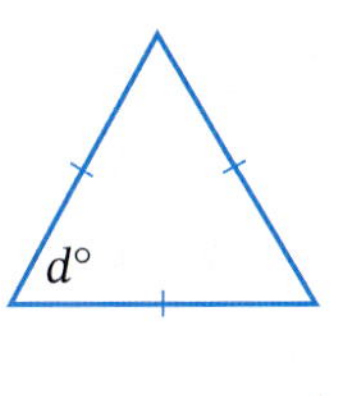

e

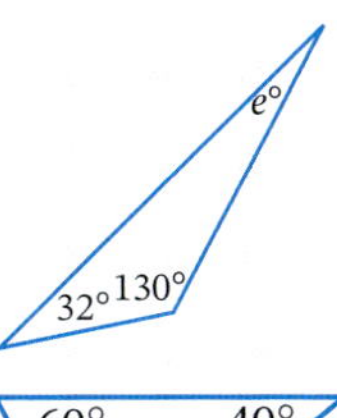

f

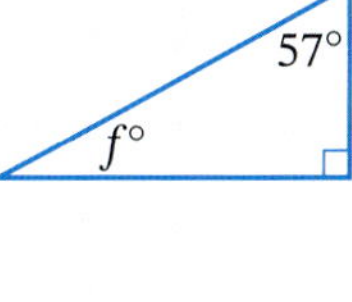

g

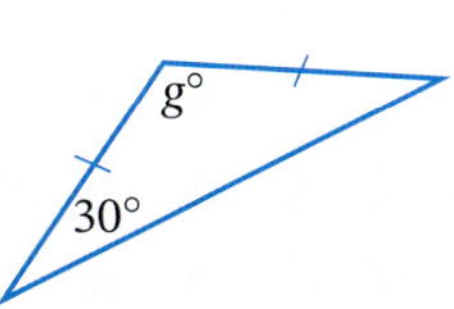

h

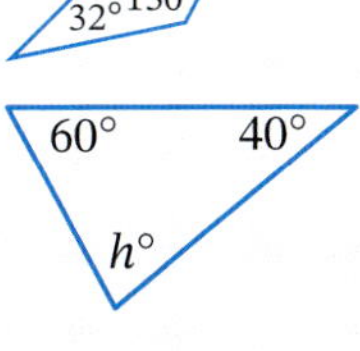

i

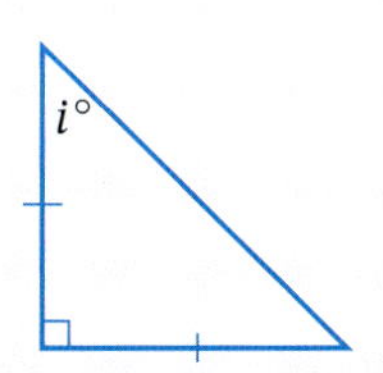

 Foundation  Standard Complex

EXAMPLE 5

2 Find the value of each variable.

Ⓡ **a** 85°, 42°, $a°$

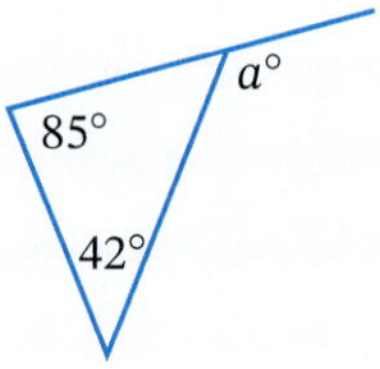

b 35°, 105°, $b°$

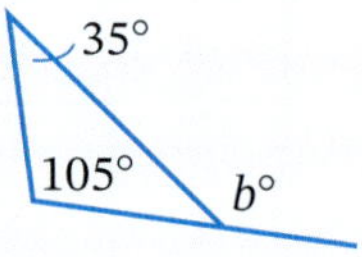

c $c°$, 70°, 130°

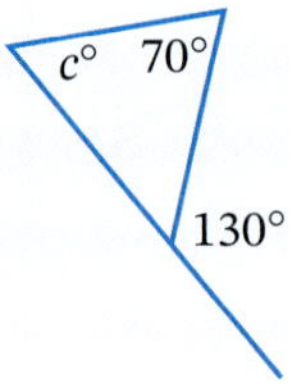

d $d°$

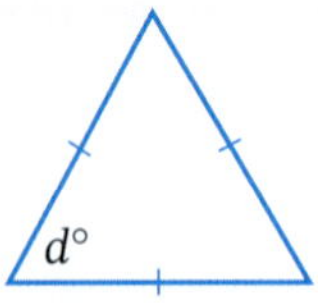

e $e°$, 32°, 130°

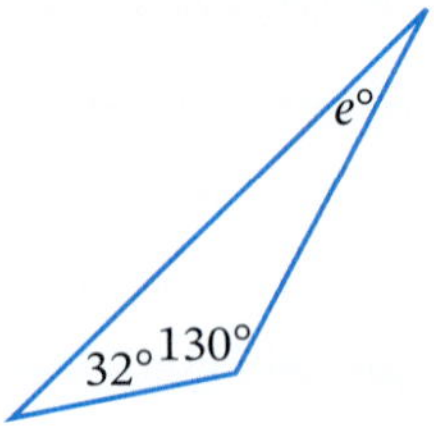

f 57°, $f°$

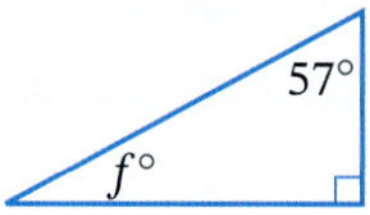

g 120°, 110°, $g°$

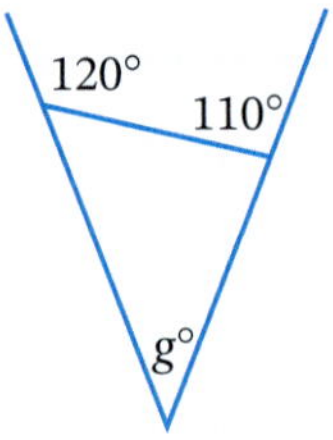

h $h°$, 68°

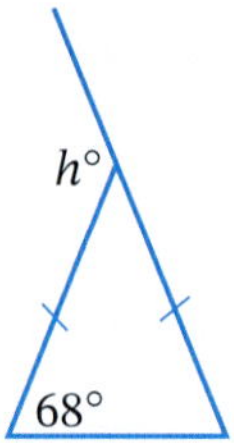

i 150°, $i°$

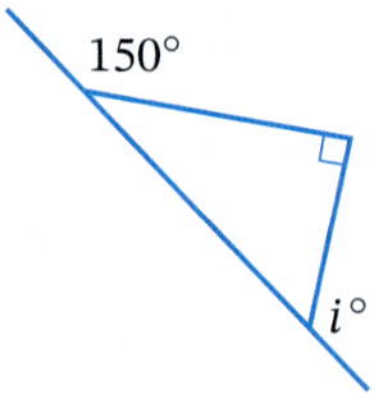

EXAMPLE 6

3 Find the value of each variable.

Ⓡ **a** 110°, 62°, 99°, $a°$

b 118°, 62°, $b°$, 118°

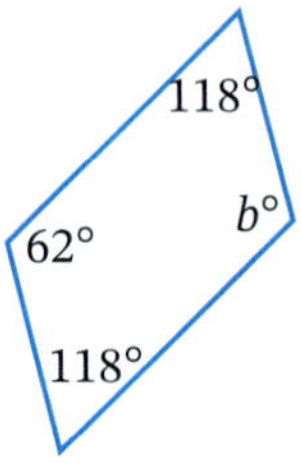

c $c°$, 98°, 71°, 82°

d $d°$

e 48°, 220°, 59°, $e°$

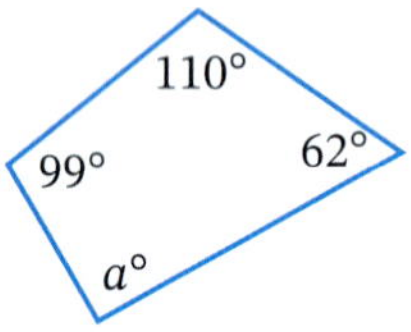

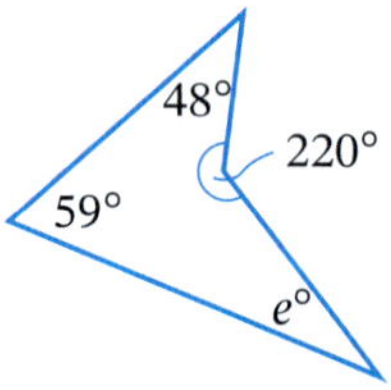

f 46°, 54°, $f°$, 15°

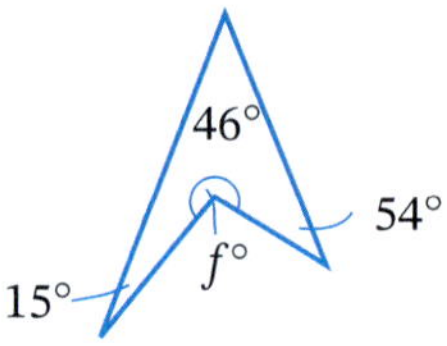

g 115°, 72°, 83°, $g°$

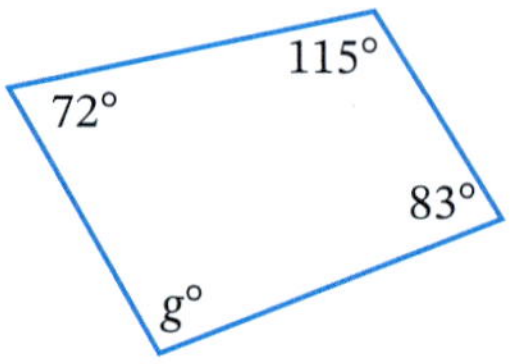

h $h°$

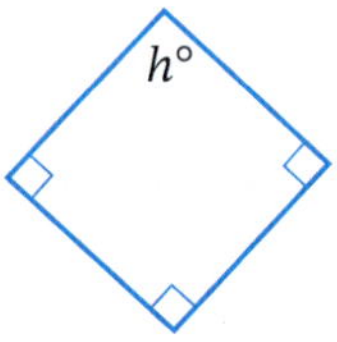

i $i°$, 40°, 40°, 140°

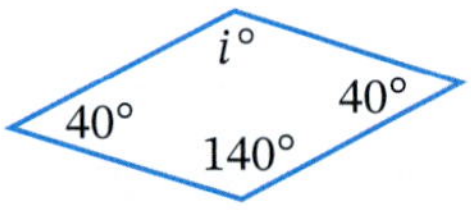

□ Foundation ○ Standard ⬡ Complex

4 Find the value of u. Select the correct answer **A**, **B**, **C** or **D**.

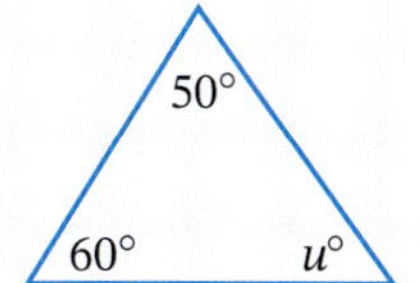

A 50 **B** 60

C 70 **D** 80

5 Find the value of r. Select the correct answer **A**, **B**, **C** or **D**.

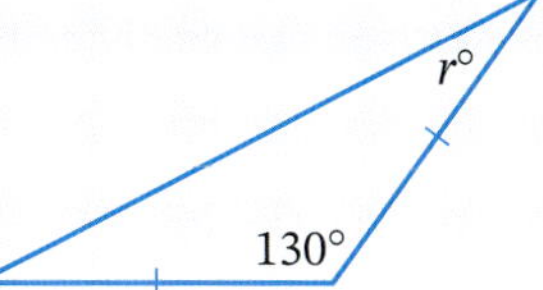

A 25 **B** 35

C 50 **D** 100

6 Which equation is correct for this triangle? Select the correct answer **A**, **B** or **C**.

R

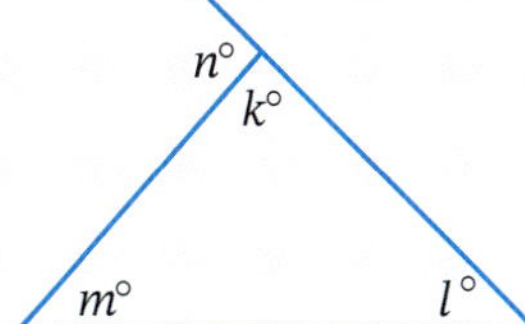

A $n = l + m$ **B** $n = k + m$

C $n = l + k$

7 Find the value of x. Select the correct answer **A**, **B**, **C** or **D**.

A 115 **B** 65

C 127 **D** 122

8 Find the size of angles x and y, giving reasons.

R

C

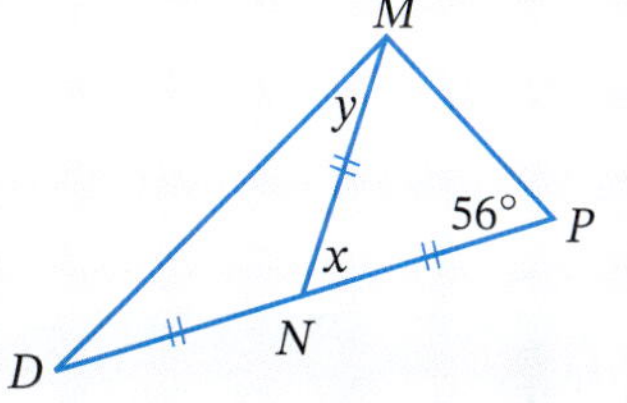

INVESTIGATION

Angle sum of a polygon

We know that the angle sum of a triangle is 180° and that the angle sum of a quadrilateral is 360°, but how do we find the angle sum of other convex polygons?

A hexagon can be divided into four triangles by drawing the diagonals from one vertex.

The sum of the angles in a hexagon = (angle sum of a triangle) × 4

$= 180° \times 4$

$= 720°$

An octagon can be divided into six triangles by drawing the diagonals from one vertex.

The total angles in an octagon = $180° \times 6 = 1080°$.

1 How is the number of sides related to the number of triangles formed in a shape?

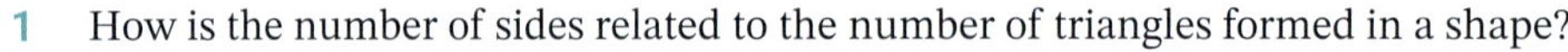

2 Copy and complete these sentences:

The angle sum of a polygon with n sides is

$A = 180 \times (\text{number of sides} - \_\_\_\_\_\_\_)°$

or: $A = 180 \times (n - \_\_\_\_\_\_\_)°$

3 Find the angle sum of each polygon.

a 11-sided polygon b 20-sided polygon c 14-sided polygon

Shutterstock.com/Gary Blakeley

Extension: Angle sum of a polygon 4.07

EXTENSION

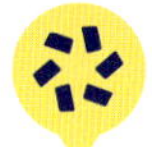

Worksheets
Angle sum of a polygon
Angles in regular polygons

A **convex polygon** has all vertices pointing outwards.

A **regular polygon** has all sides the same length and all angles the same size.

Example 7

These shapes are all hexagons (six sided).

a

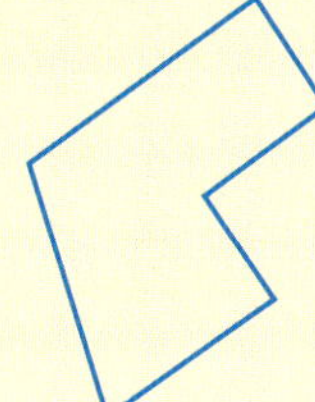

b

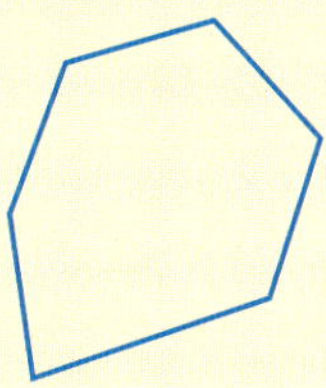

c

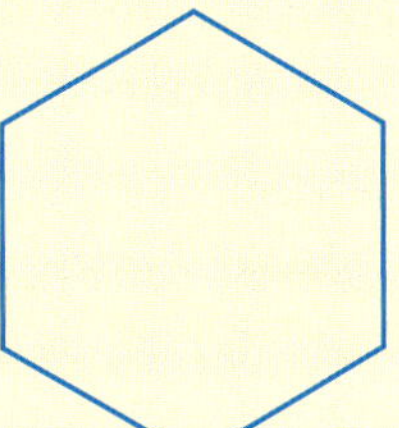

a Which hexagons are convex?

b Which hexagon is regular?

SOLUTION

a Hexagons **B** and **C** are convex because all of their vertices point outwards.

b Hexagon **C** is regular, because all its sides are equal and all its angles are equal.

ⓘ Angle sum of a polygon

The **angle sum of a polygon** with n sides is given by the formula:

$$A = 180(n - 2)°$$

This property applies to both convex and non-convex polygons.

EXTENSION

Example 8

Find the size of one angle in a regular hexagon.

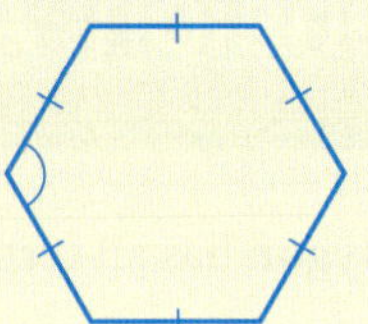

SOLUTION

A hexagon has six sides ($n = 6$).

Angle sum of a hexagon $= 180(6 - 2)°$

$= (180 \times 4)°$

$= 720°$

For a regular hexagon:

One angle $= 720° \div 6$

$= 120°$

$\therefore$ Each angle in a regular hexagon is 120°.

EXERCISE 4.07 ANSWERS ON P. 627

Angle sum of a polygon

EXAMPLE 7

1 Copy each polygon and write its correct name from this list.

C

hexagon	nonagon	heptagon	decagon
octagon	triangle	quadrilateral	pentagon

Also, state if each polygon is **regular** or **irregular**, **convex** or **non-convex**.

a 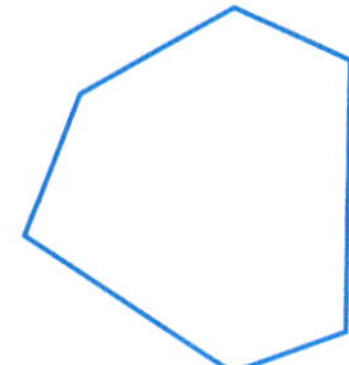

b 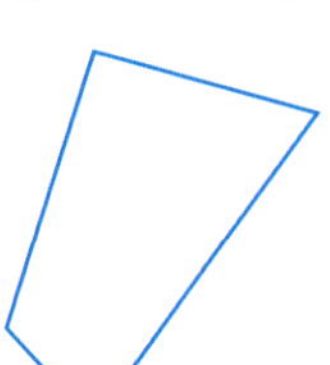

c 

d 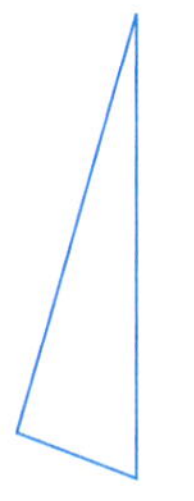

e 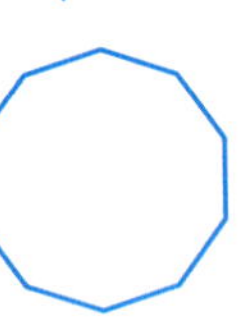

f 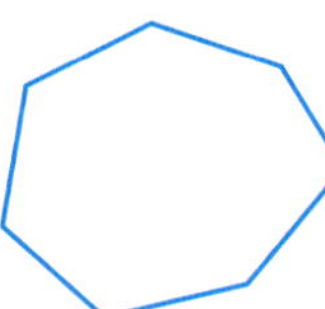

g 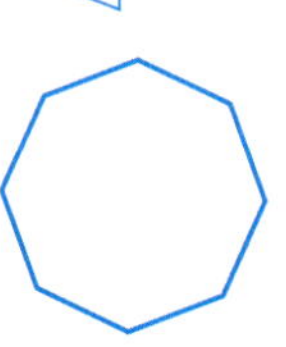

h 

Foundation Standard Complex

2 Copy and complete this table.

Polygon	Number of sides	Sum of angles inside polygon
hexagon		
heptagon		
octagon		
nonagon		
decagon		

3 Find the sum of the interior angles in:

a a 15 agon **b** a 20 agon **c** a 25 agon **d** a 100 agon

4 Find the value of each variable.

a

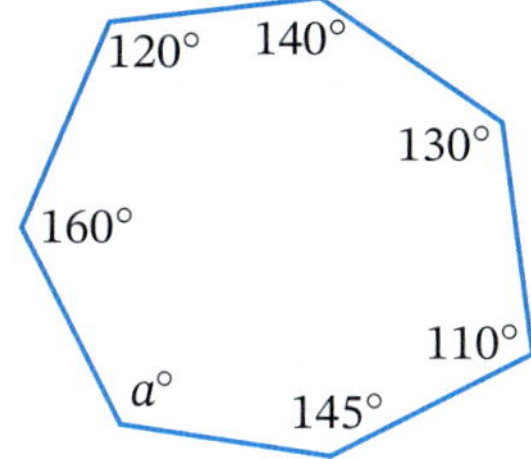

b

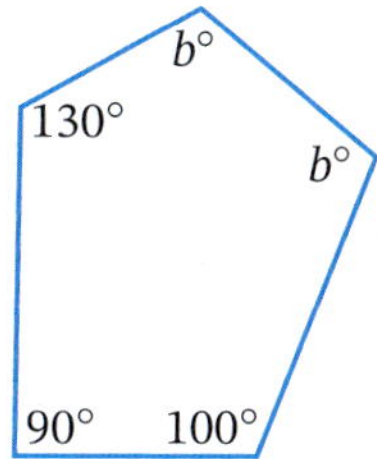

c

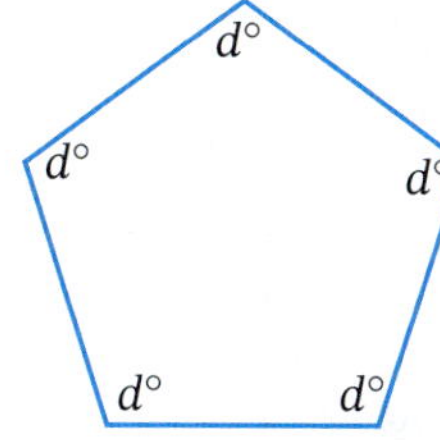

d

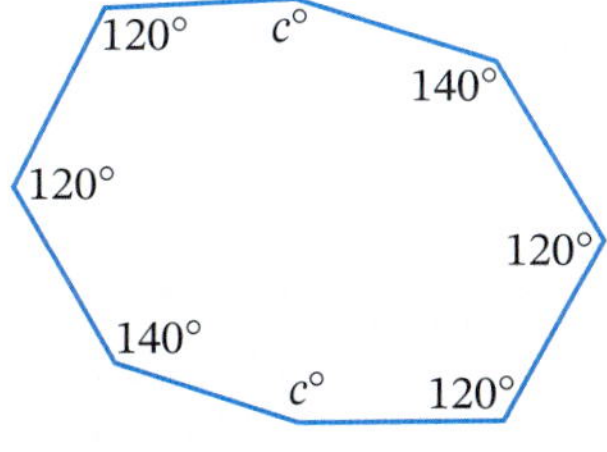

e

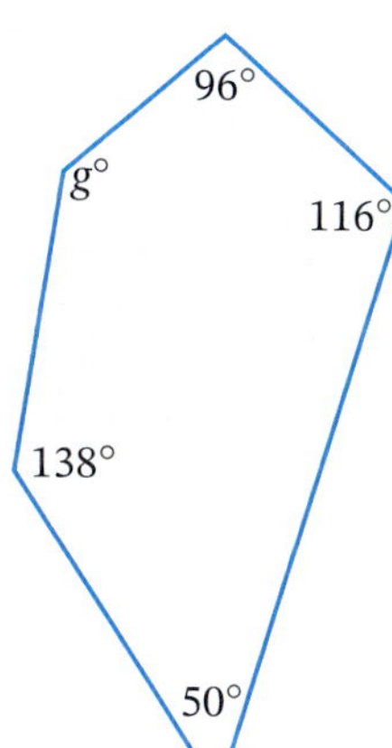

f

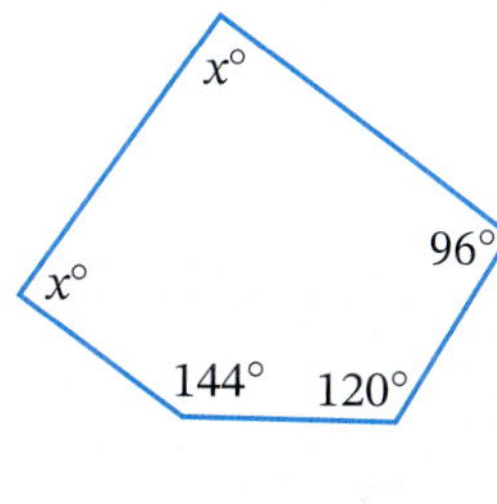

g

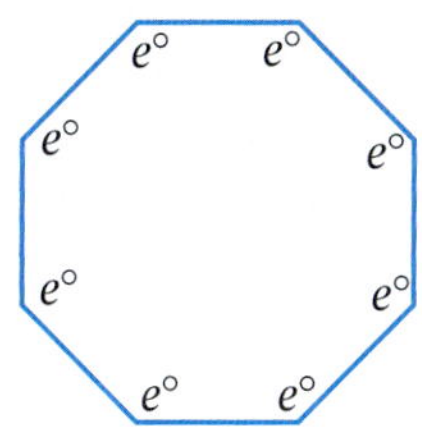

5 Find the size of one interior angle in each regular polygon.

EXAMPLE 8

a square **b** equilateral triangle **c** regular hexagon

d regular octagon **e** regular decagon **f** regular pentagon

g regular dodecagon

EXTENSION

4.07

Foundation Standard Complex

EXTENSION

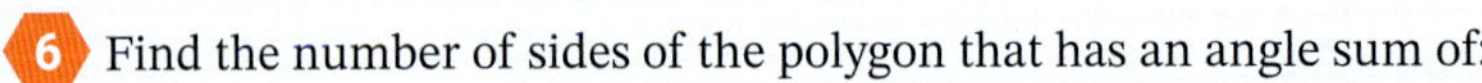

6 Find the number of sides of the polygon that has an angle sum of:

a 2160° **b** 5760° **c** 4320° **d** 9180° **e** 22 140°

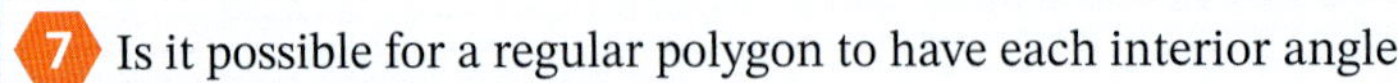

7 Is it possible for a regular polygon to have each interior angle:

Ⓡ **a** 144°? **b** 130°?

If so, find the number of sides.

Quiz
Mental skills 4B

☆ **MENTAL SKILLS** (4B) ANSWERS ON P. 627 **Maths without calculators**

Converting decimals and percentages to fractions

1 Consider each example.

a $0.35 = \frac{\cancel{35}^{7}}{_{20}\cancel{100}} = \frac{7}{20}$

(2 decimal places, two 0s in the denominator)

b $0.8 = \frac{\cancel{8}^{4}}{_{5}\cancel{10}} = \frac{4}{5}$

(one decimal place, one 0 in the denominator)

c $0.64 = \frac{\cancel{64}^{16}}{_{25}\cancel{100}} = \frac{16}{25}$

d $0.22 = \frac{\cancel{22}^{11}}{_{50}\cancel{100}} = \frac{11}{50}$

2 Now convert each decimal to a fraction.

a 0.75	**b** 0.28	**c** 0.3	**d** 0.14
e 0.06	**f** 0.85	**g** 0.32	**h** 0.49
i 0.56	**j** 0.9	**k** 0.72	**l** 0.65
m 0.2	**n** 0.24	**o** 0.53	

3 Consider each example.

a $26\% = \frac{\cancel{26}^{13}}{_{50}\cancel{100}} = \frac{13}{50}$

b $40\% = \frac{\cancel{40}^{2}}{_{5}\cancel{100}} = \frac{2}{5}$

c $8\% = \frac{\cancel{8}^{2}}{_{25}\cancel{100}} = \frac{2}{25}$

d $95\% = \frac{\cancel{95}^{19}}{_{20}\cancel{100}} = \frac{19}{20}$

4 Now convert each percentage to a fraction.

a 76%	**b** 10%	**c** 80%	**d** 45%
e 88%	**f** 56%	**g** 75%	**h** 31%
i 68%	**j** 5%	**k** 60%	**l** 54%
m 6%	**n** 49%	**o** 82%	

☐ Foundation ○ Standard ⬡ Complex

POWER PLUS ANSWERS ON P. 627

As questions become more complex, it may not be possible to find the answer in one step. It may be necessary to find another angle first.

Find the value of x in each diagram. (Give reasons for all steps.)

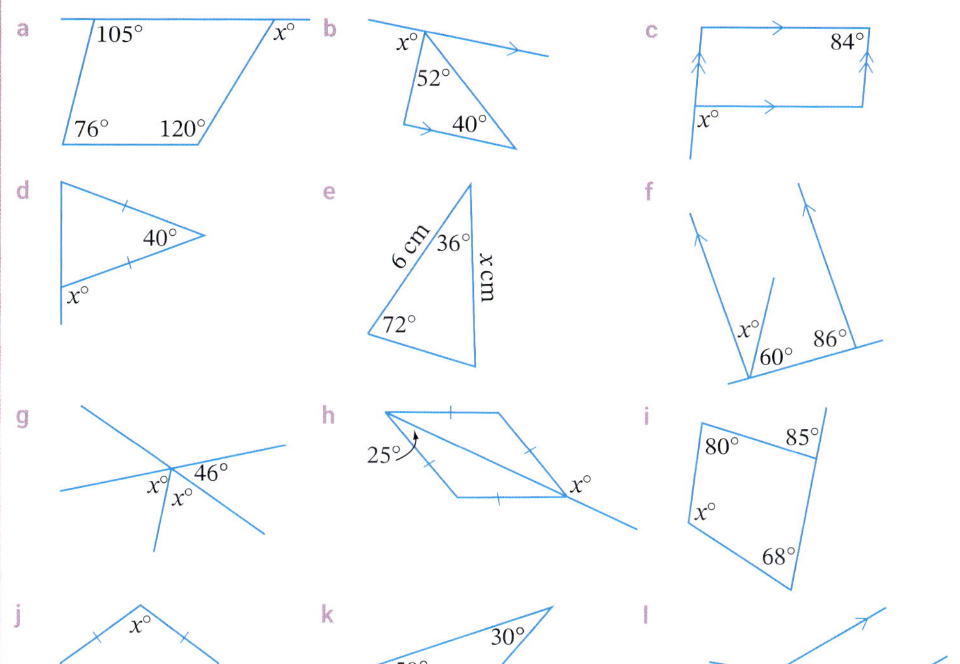

Puzzle
Shape puzzle 1

Worksheet
Equal angles

4.07

4 CHAPTER REVIEW

Quiz
Language of maths 4

Language of maths

acute	alternate	angle sum	co-interior
complementary	convex	corresponding	diagonal
equilateral	exterior angle	interior angle	isosceles
kite	obtuse	parallelogram	quadrilateral
reflex	rhombus	scalene	supplementary
trapezium	vertically opposite		

1 What are **supplementary angles**?

2 Name the types of angles associated with parallel lines cut by a transversal.

3 What type of triangle has one angle that is greater than 90°?

4 What is the **angle sum** of a quadrilateral?

5 Illustrate the difference between an **interior angle** and an **exterior angle** of a triangle.

6 What does **equilateral** mean? What is the common name for an 'equilateral parallelogram'?

Worksheet
Mind map: Geometry

Topic summary

- Do you think this chapter is useful? Why? What did you learn in this chapter?
- How confident do you feel with geometry?
- List anything in this chapter that you did not understand. Show your teacher.

Print (or copy) and complete this mind map of the topic, adding detail to its branches and using pictures, symbols and colour where needed. Ask your teacher to check your work.

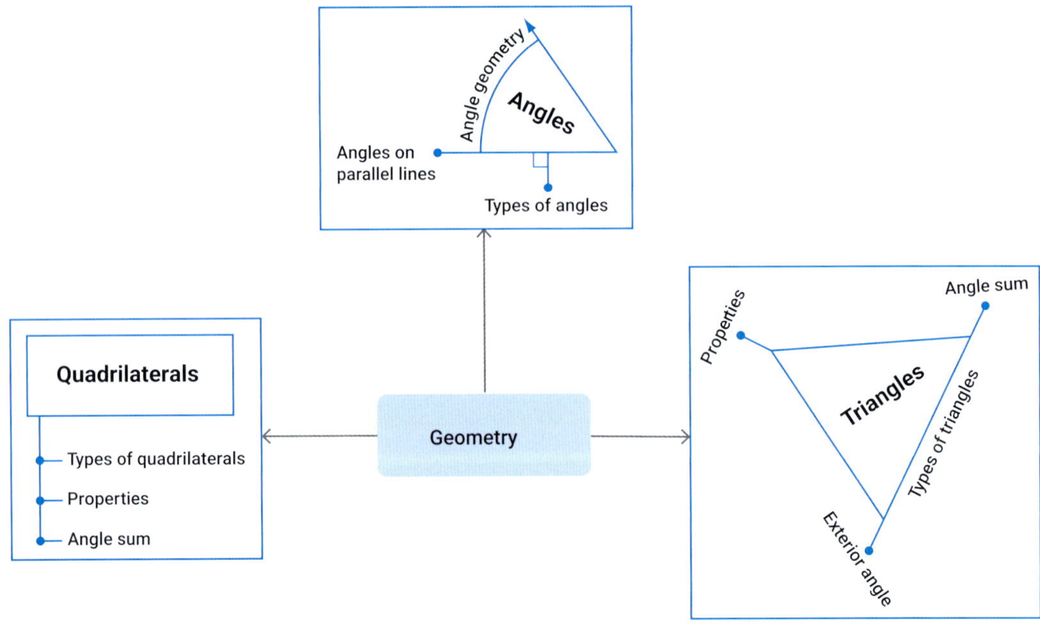

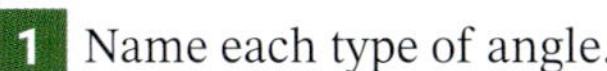

TEST YOURSELF

ANSWERS ON P. 627

1 Name each type of angle. 4.01

a

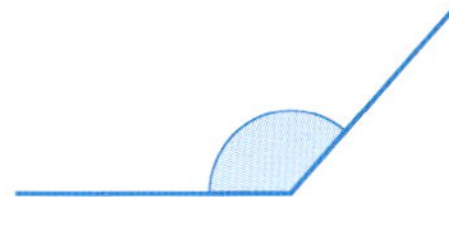

b

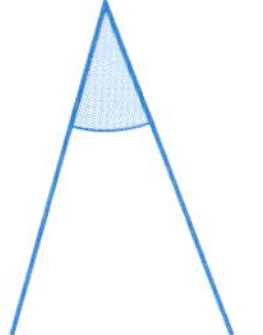

c

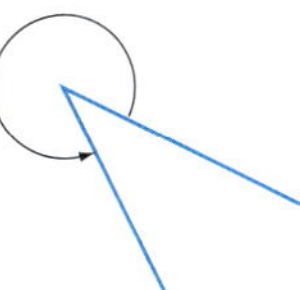

Quiz
Test yourself 4

2 Find the value of each variable, giving reasons. 4.01

a

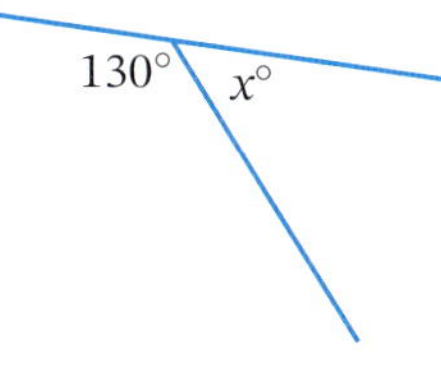

b

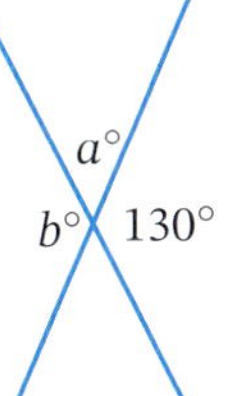

c

d

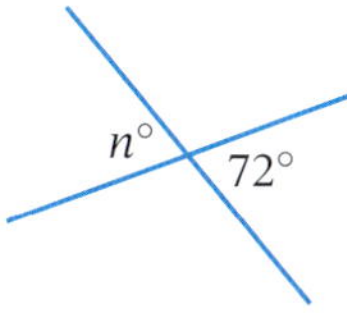

e

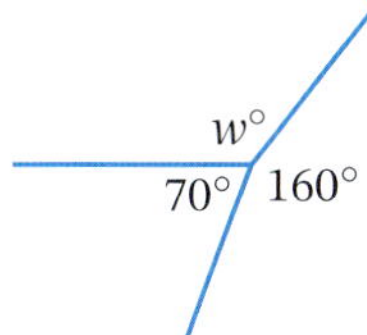

3 Name each type of angle pair. 4.02

a

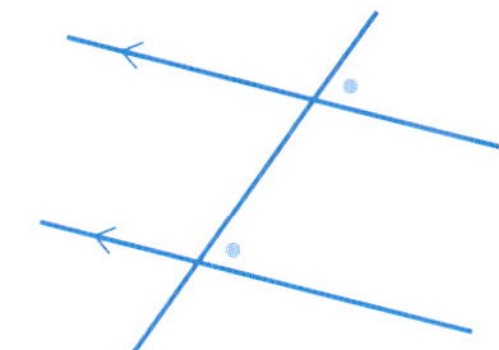

b

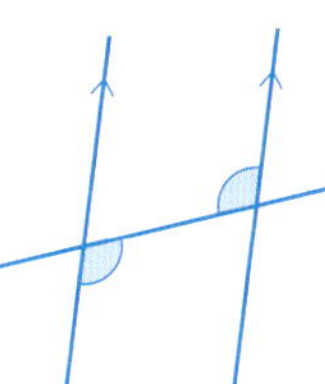

c

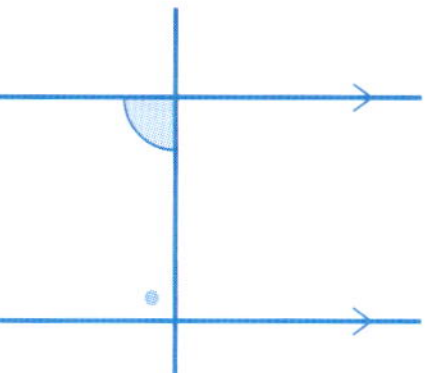

4 Find the value of each variable, giving reasons. 4.02

a

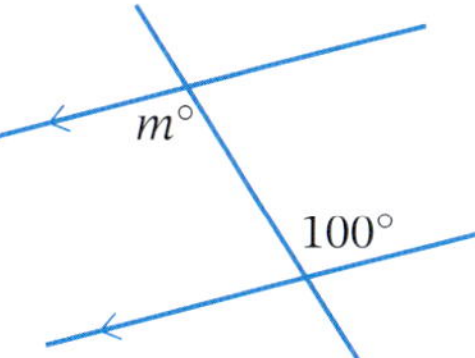

b

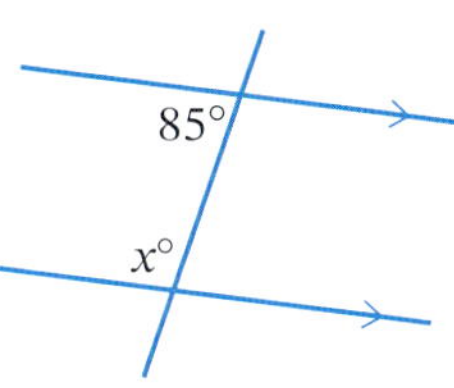

c

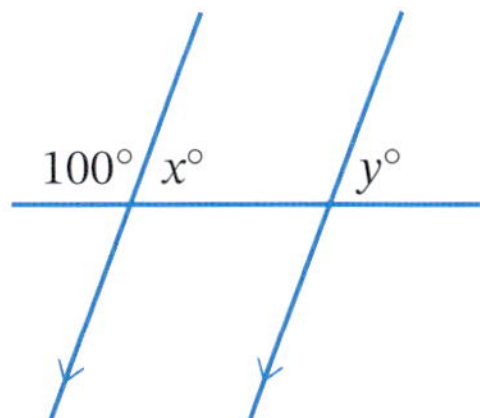

d

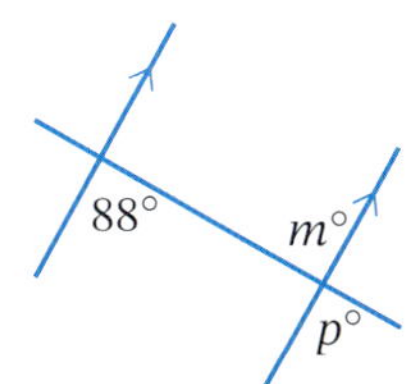

☐ Foundation ○ Standard ⬡ Complex

5 Classify each triangle by its sides and angles.

a

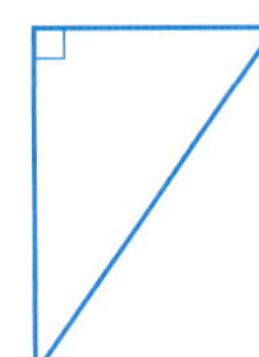

b

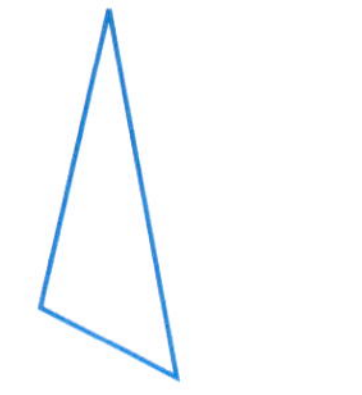

c

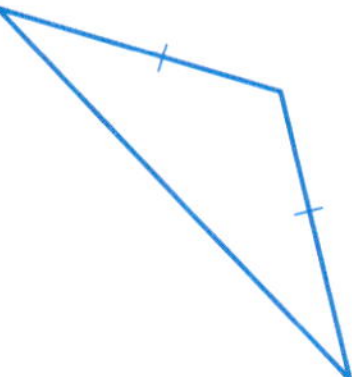

d

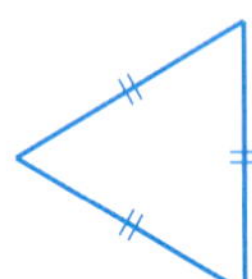

e

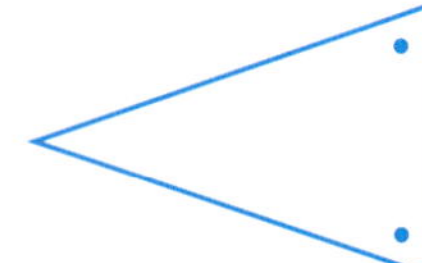

f

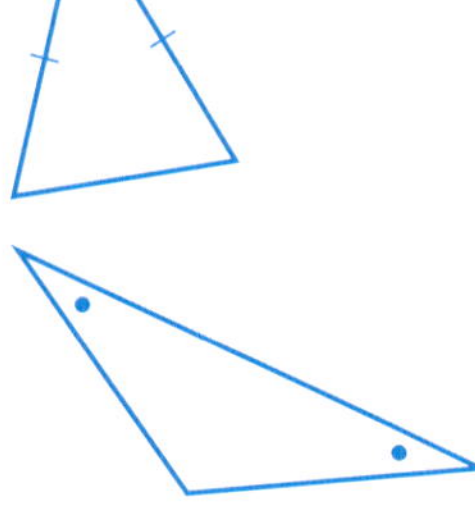

g

h

i

6 Sketch each quadrilateral.

a parallelogram **b** kite **c** trapezium

4.05

7 Name all the quadrilaterals that have each property.

a All sides equal in length.

b No parallel sides.

c All angles right angles.

d Two pairs of opposite angles equal.

e Diagonals bisect each other.

8 Find the value of each variable.

a

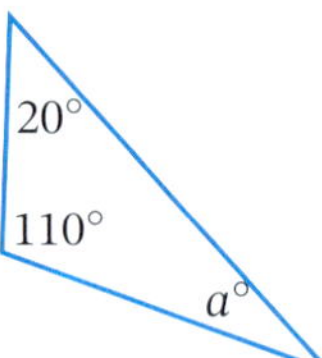

b

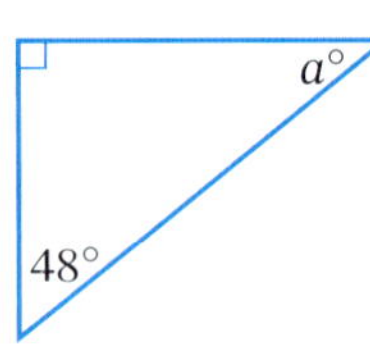

c

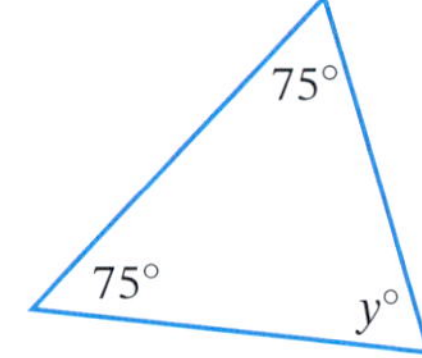

d

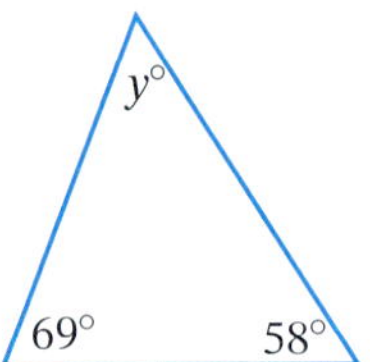

e

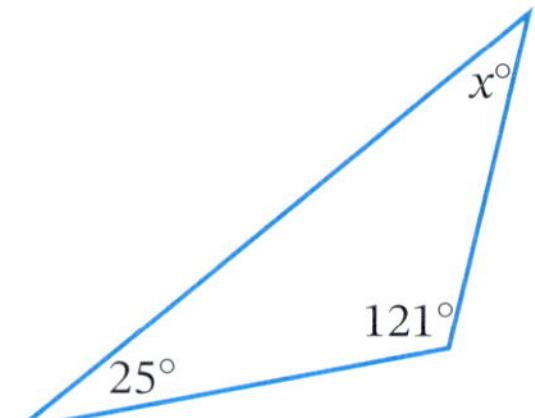

f

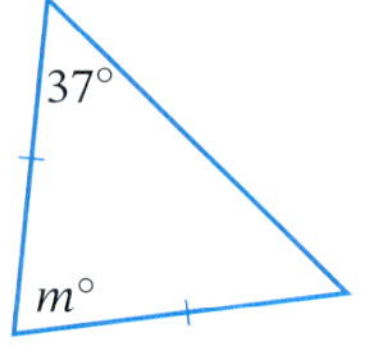

☐ Foundation ○ Standard ⬡ Complex

9 Find the value of each variable. 4.06

a
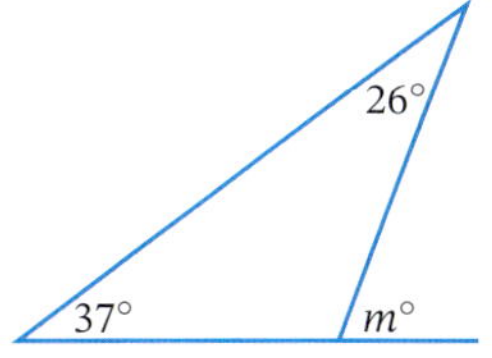

b
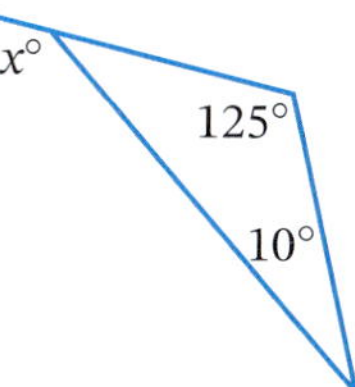

c
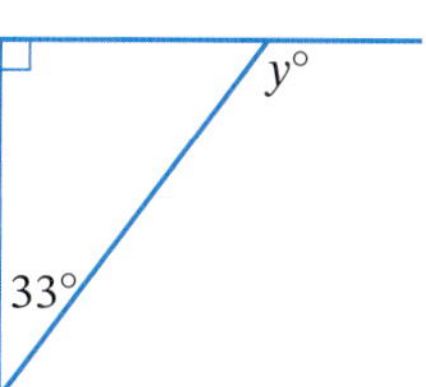

d
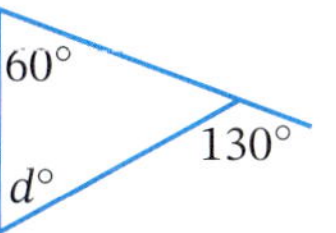

e
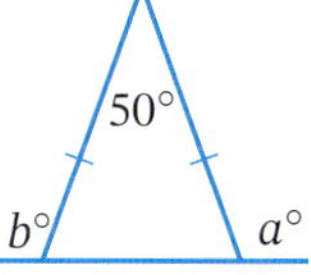

f
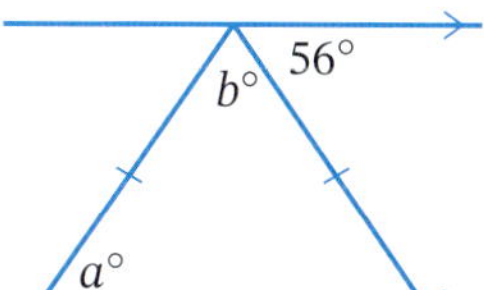

g
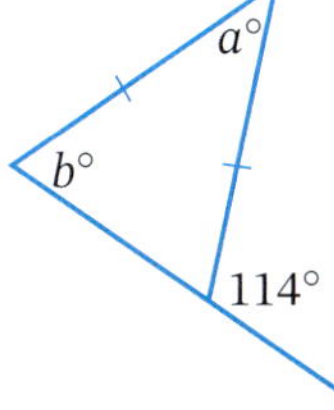

10 Find the value of each variable. 4.06

a
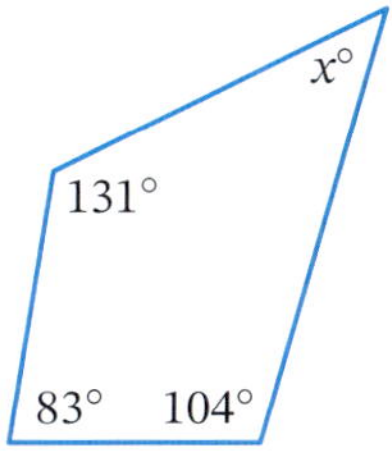

b
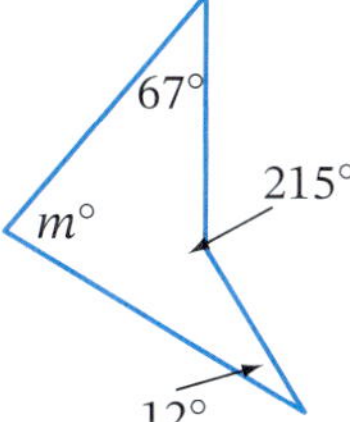

c
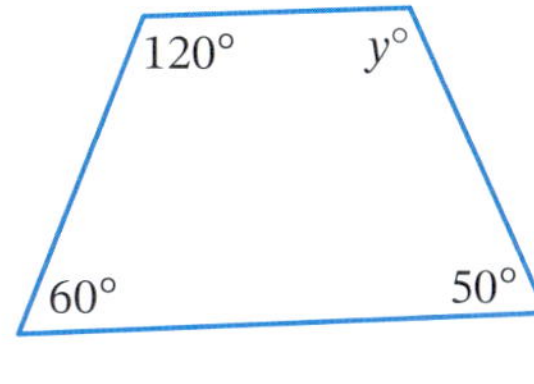

11 Find the value of x. Select the correct answer **A**, **B**, **C** or **D**. 4.06

A 65 **B** 25

C 55 **D** 80

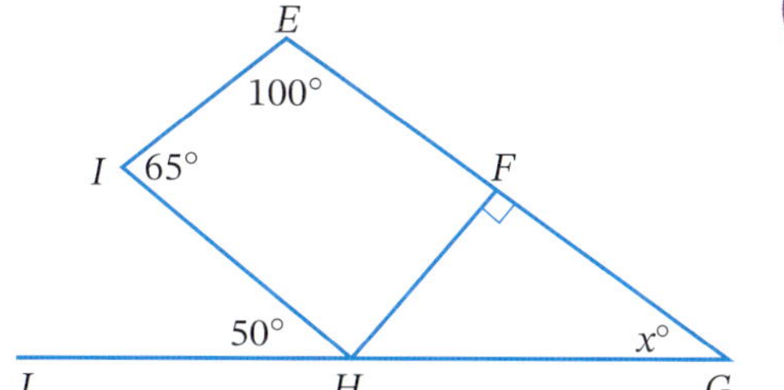

Foundation Standard Complex

5

MEASUREMENT

Area and volume

The International Space Station (ISS) is the largest man-made object in space. The construction of the ISS began in 1998. It circles the Earth every 90 minutes at a height of approximately 400 km above the Earth. ISS has a pressurised volume of 915 m^3 and it is 109 m long, the same length as a Boeing 707. The solar panels of the station cover an area of 2500 m^2 and produce enough electricity to power 10 average-sized homes.

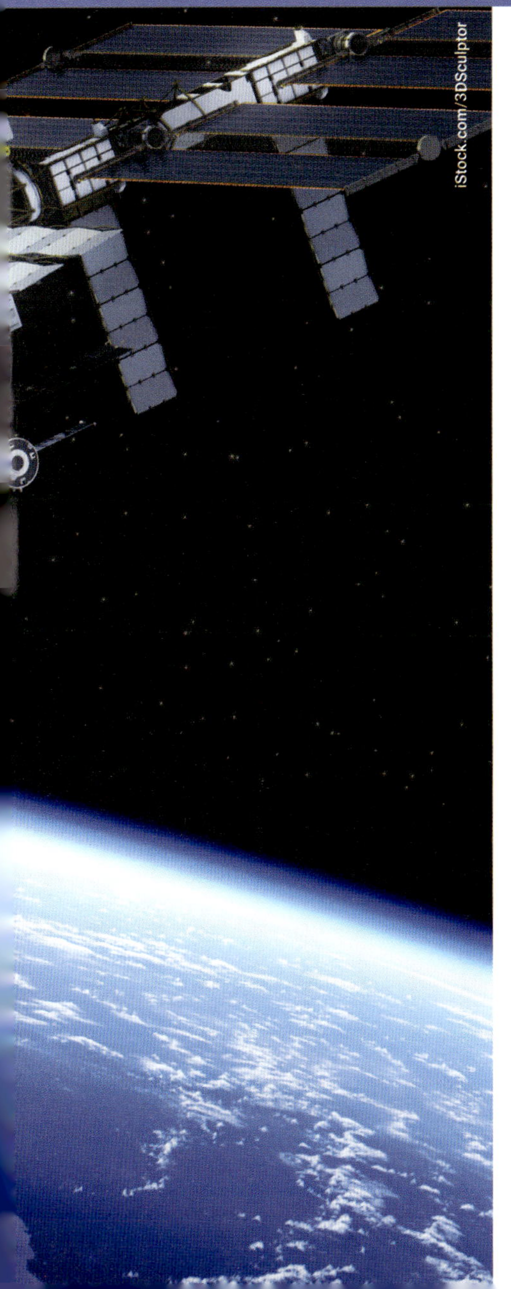

Chapter outline

*Extension

U = Understanding
F = Fluency
PS = Problem solving
R = Reasoning
C = Communication

Wordbank

capacity The amount of fluid (liquid or gas) in a container

composite shape A shape that is made up of two or more shapes

circumference The perimeter of a circle; a circle's outer boundary

cross-section A 'slice' of a solid, taken across the solid rather than along it

cubic metre The volume of a cube that measures 1 m by 1 m by 1 m

perpendicular height The height of a shape taken at right angles to its base

pi (π) A special irrational number, approximately 3.1416, used in calculating circular measurements

radius The distance from the centre of a circle to the circle's edge

Quiz
Wordbank 5

Videos (12):

5.01 Perimeter of composite shapes
5.02 Circumference of a circle
5.03 Metric units of area
5.04 Area formulas for triangles and quadrilaterals
5.05 Areas of composite shapes
5.06, 5.07 Area formulas for triangles and quadrilaterals • Area of a trapezium
5.07 Area formulas for triangles and quadrilaterals
5.08 Area of a circle
5.09 Perimeter and area of a sector
5.10 Metric units of volume
5.11 Volume of a prism
5.12 Volumes of prisms and cylinders
5.13 Volume and capacity

Twig videos (2):

5.02 Calculating pi: Archimedes
5.08 Beating the U-boats

PhET interactive (1):

5.03 Area builder

Quizzes (6):

- Wordbank 5
- SkillCheck 5
- Mental skills 5A
- Mental skills 5B
- Language of Maths 5
- Test Yourself 5

Skillsheets (4):

SkillCheck Multiplying by 10, 100, 1000
5.03 What is area?
5.04 What is area?
5.10 What is volume?
5.11 Solid shapes • What is volume?

Worksheets (20):

5.01 A page of composite shapes
5.02 Discovering pi • A page of circles • Circumference and area
5.03 Australian areas
5.04 Rectangle and triangle areas
5.05 A page of composite shapes • Odd areas
5.07 Quadrilateral and triangle areas • Approximating the area of a circle
5.08 Circumference and area • Area ID • Area and perimeter investigations
5.09 A page of circular shapes • Applications of area • Back-to-front problems • Area and perimeter investigations
5.10 Volume
5.11 Volume • Measuring shapes review
5.12 Volume and capacity • Volume and capacity match • Capacity

Mind map: Area and volume

Puzzles (8):

5.02 Circle crossword
5.04 Area group clues
5.05 Areas 1 • Area puzzles
5.07 Areas of quadrilaterals and triangles
5.08 Circle areas
5.09 Carpet talk
5.13 Capacity puzzle

Presentation (1):

5.01 Calculating the perimeter

Nelson MindTap

To access resources above, visit **cengage.com.au/nelsonmindtap**

5 **In this chapter, you will:**

✓ calculate perimeters and areas of triangles, quadrilaterals, circles and composite shapes.
✓ convert between units of area.
✓ (EXTENSION) calculate perimeters and areas of sectors.
✓ calculate volumes of prisms.
✓ (YEAR 9 EXTENSION) calculate volumes of cylinders.
✓ convert between units of volume.
✓ convert between units of volume and capacity.

SkillCheck

ANSWERS ON P. 628

Quiz SkillCheck 5

Skillsheet Multiplying by 10, 100, 1000

1 Copy and complete each conversion.

a 20 cm = _____ mm **b** 350 cm = _____ m **c** 2500 m = _____ km

d 46 mm = _____ cm **e** 4 km = _____ m **f** 5200 m = _____ km

g 600 m = _____ km **h** 8200 mm = _____ m **i** 0.4 m = _____ mm

2 Find the perimeter of each shape.

a

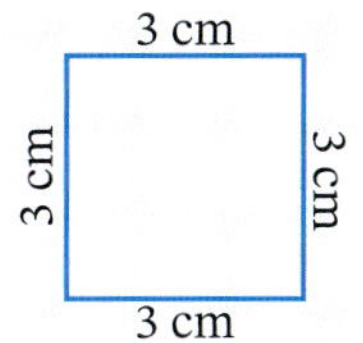

b

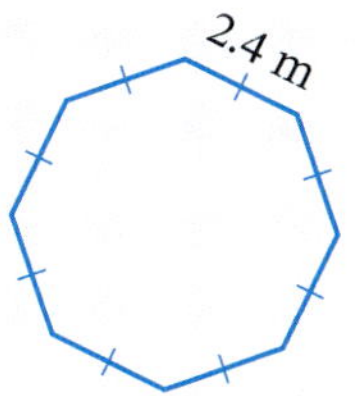

c

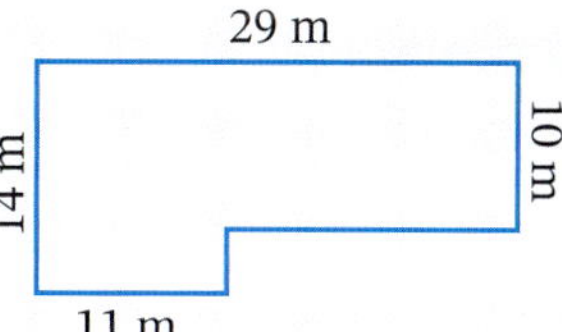

d

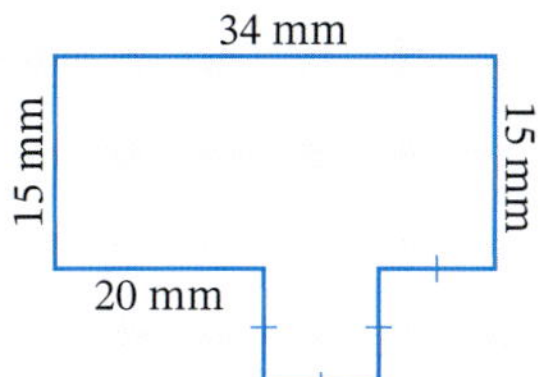

e

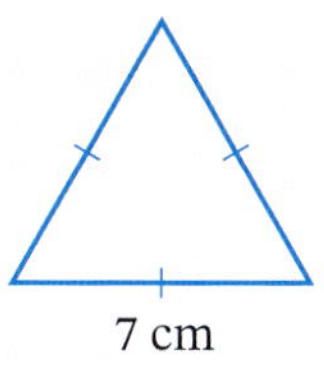

f

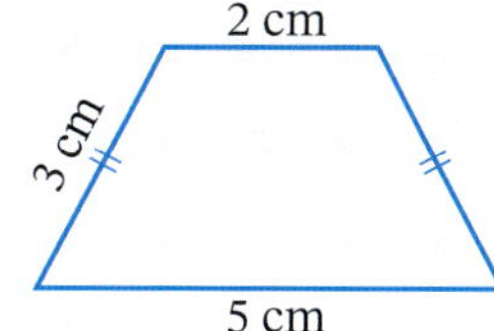

3 Find the area of each shape.

a

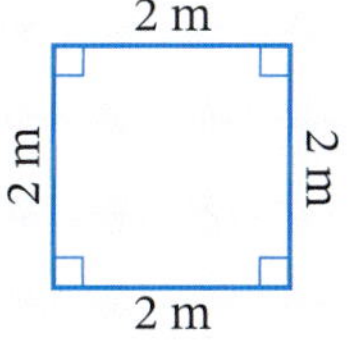

b

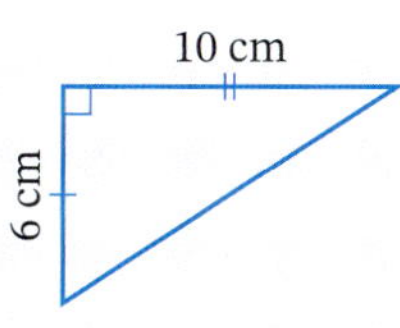

c

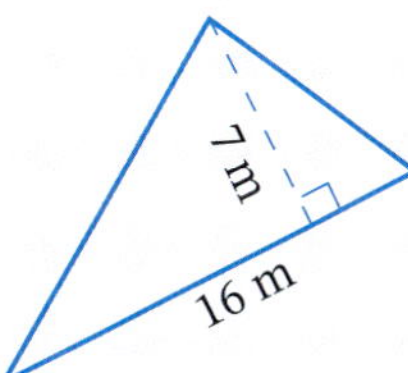

d

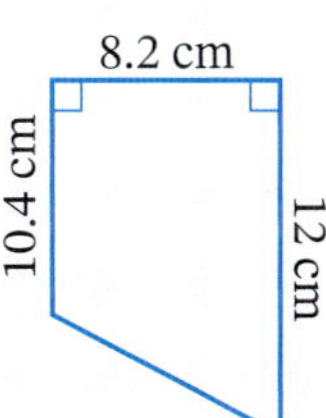

e

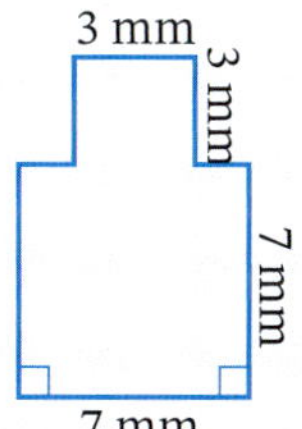

f

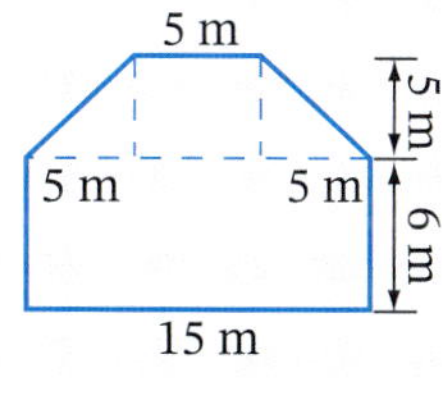

4 Write the name of each part of the circle shown choosing from the word list below.

arc chord circumference diameter

radius sector segment

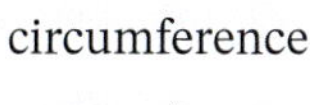

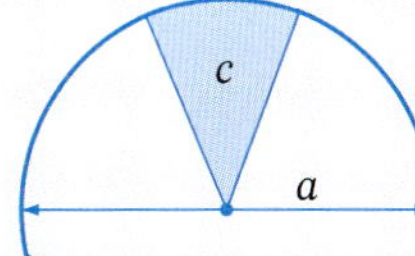

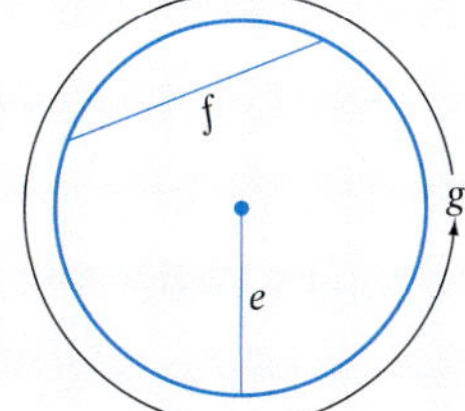

5.01 Perimeter

Worksheet
A page of composite shapes

Presentation
Calculating the perimeter

ⓘ Perimeter

The **perimeter** of a shape is the distance around the shape.

It is the sum of the lengths of the sides of the shape.

Example 1

Find the perimeter of each shape.

a

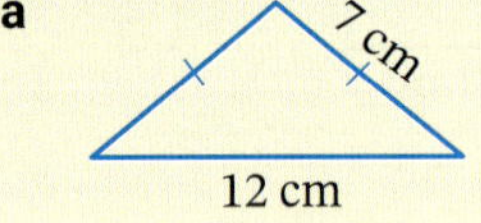

b

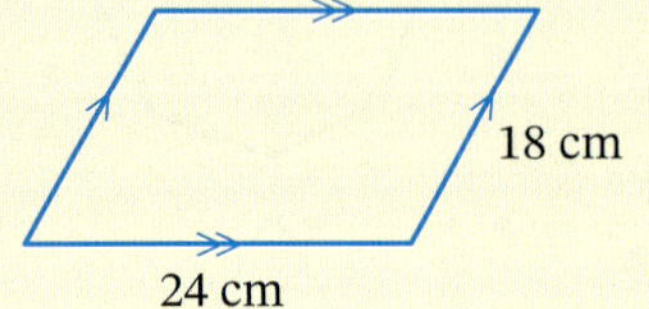

c

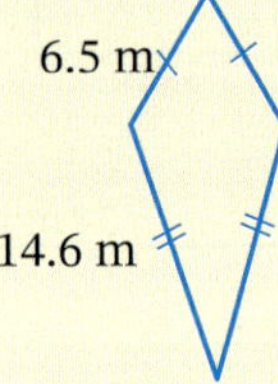

SOLUTION

a Perimeter $= 7 + 7 + 12$
$= 26$ cm

Adding the three sides of the isosceles triangle.

b Perimeter $= 24 + 18 + 24 + 18$
$= 84$ cm

In a parallelogram, opposite sides are equal.

c Perimeter $= 6.5 + 6.5 + 14.6 + 14.6$
$= 42.2$ m

In a kite, adjacent sides are equal.

Video
Perimeter of composite shapes

Example 2

Find the perimeter of each composite shape.

a

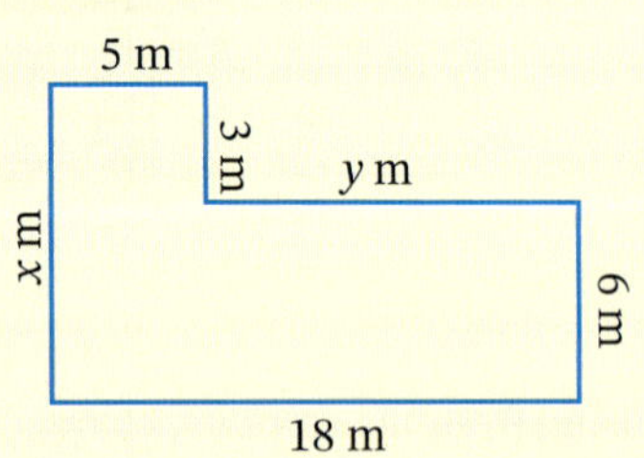

b

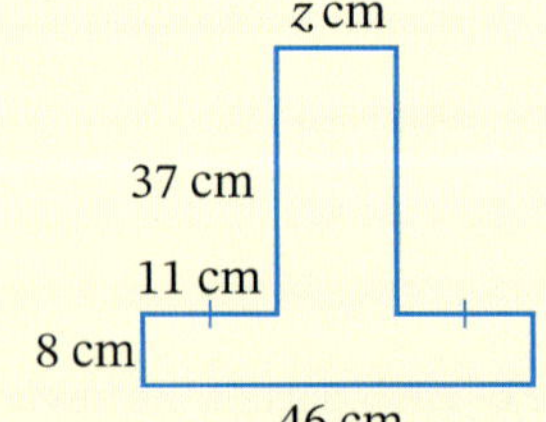

SOLUTION

a Find the unknown sides first.

$x = 3 + 6 = 9$

$y = 18 - 5 = 13$

Perimeter $= 9 + 5 + 3 + 13 + 6 + 18$
$= 54$ m

b Find the unknown sides first.

$z = 46 - 11 - 11 = 24$

Perimeter $= 8 + 11 + 37 + 24 + 37 + 11 + 8 + 46$
$= 182$ cm

EXERCISE 5.01 ANSWERS ON P. 628

Perimeter

U F PS R

1 Find the perimeter of each shape. EXAMPLE 1

a
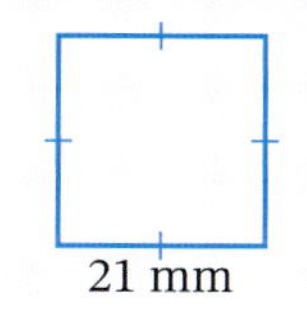

b
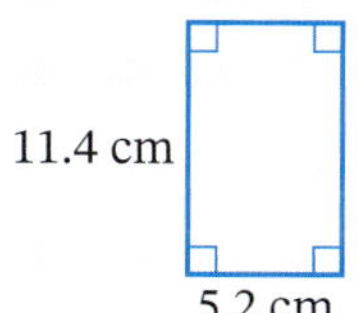

c
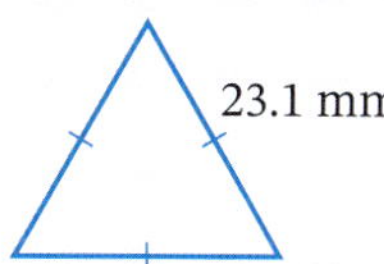

d
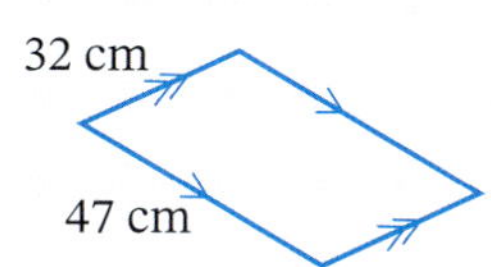

e
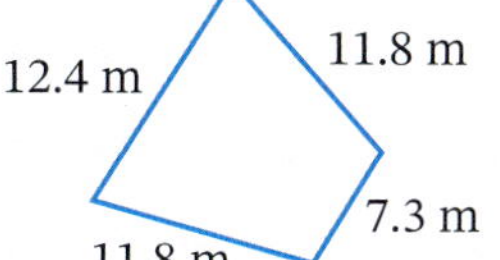

f
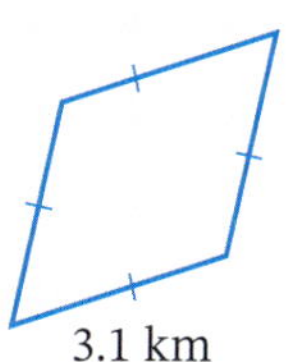

2 Find the perimeter of each shape. EXAMPLE 2

a
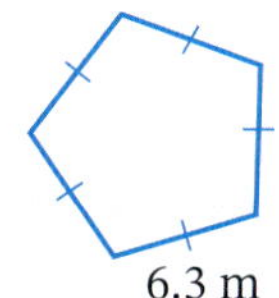

b
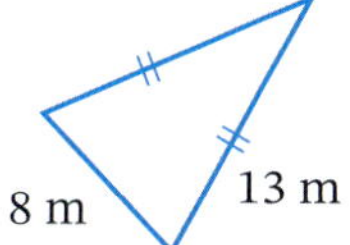

c
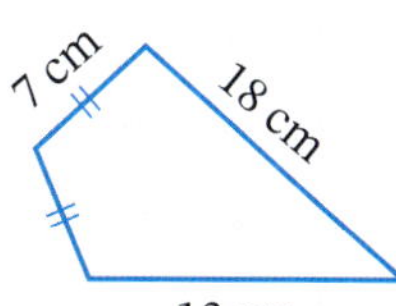

d
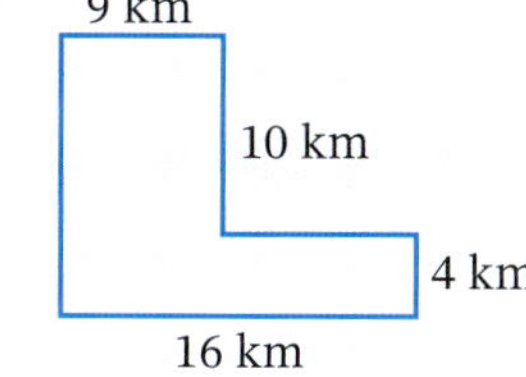

e
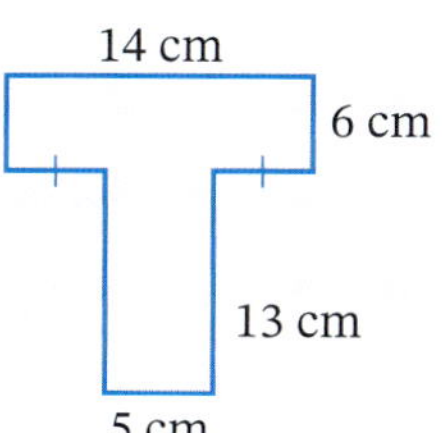

f
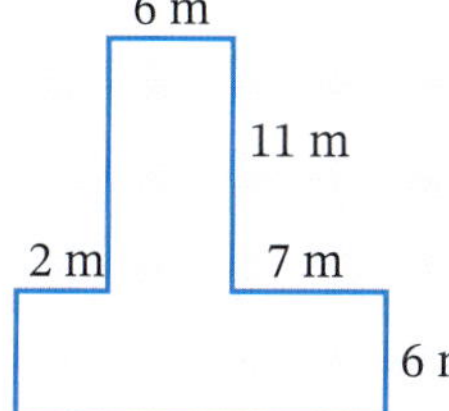

g
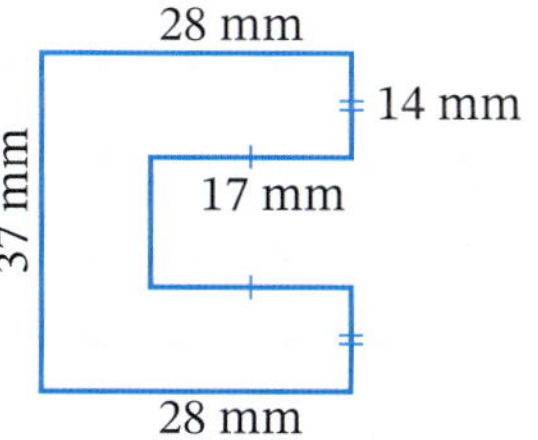

h
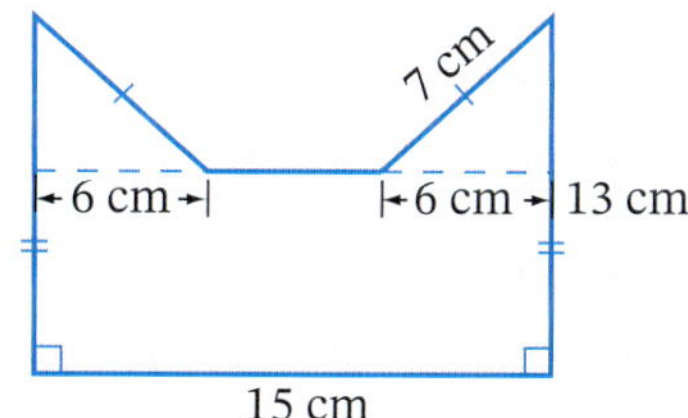

i
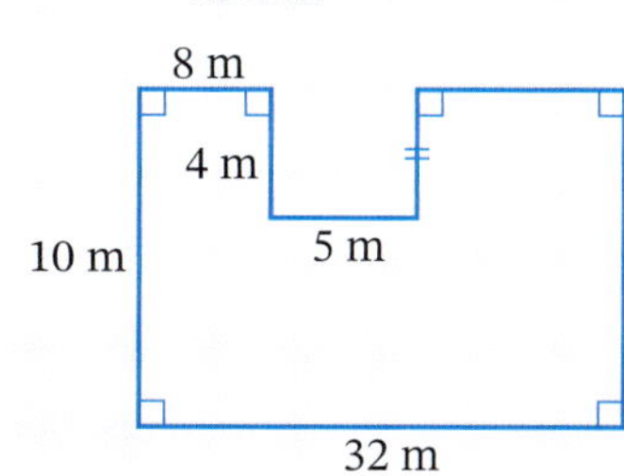

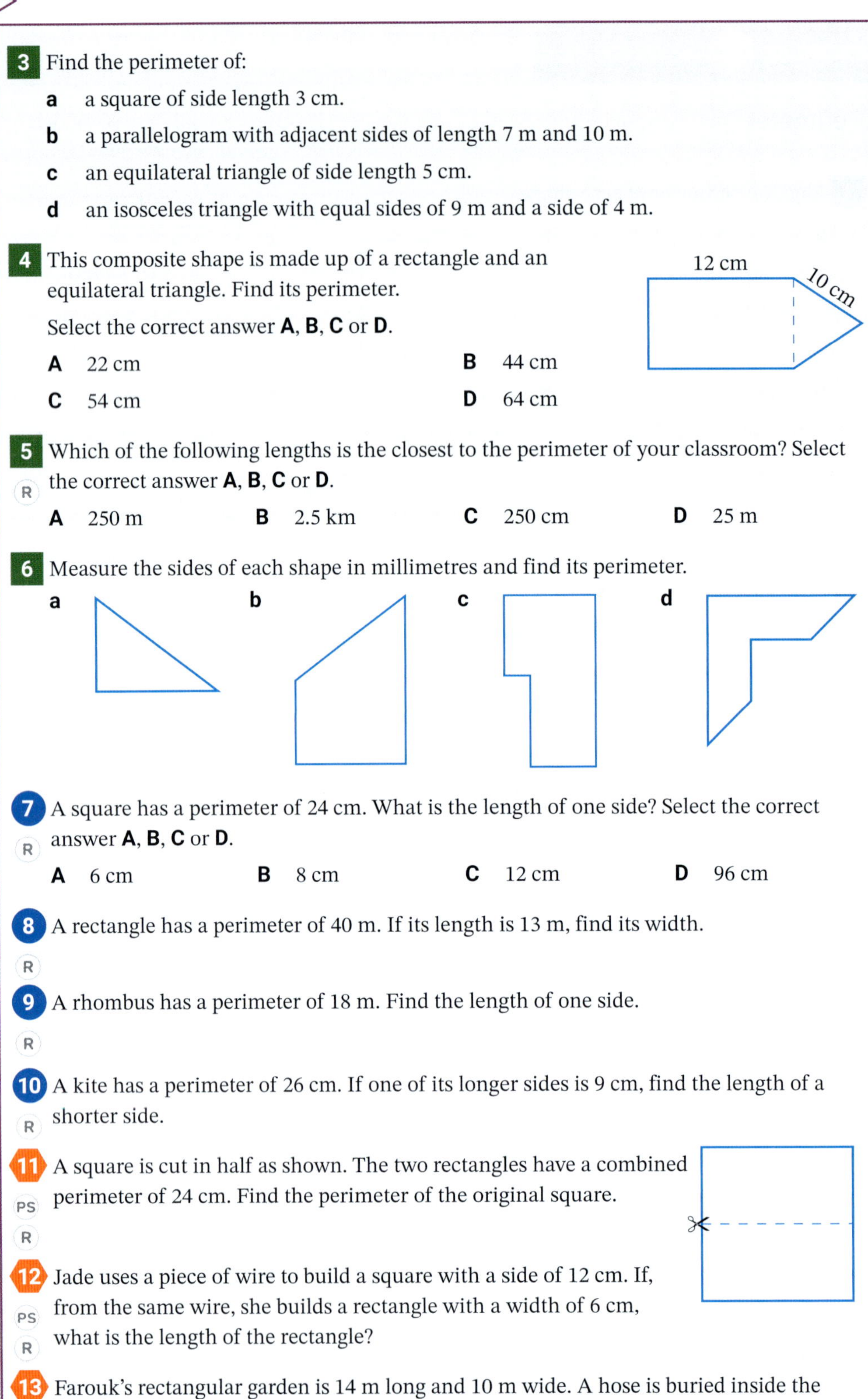

3 Find the perimeter of:

a a square of side length 3 cm.

b a parallelogram with adjacent sides of length 7 m and 10 m.

c an equilateral triangle of side length 5 cm.

d an isosceles triangle with equal sides of 9 m and a side of 4 m.

4 This composite shape is made up of a rectangle and an equilateral triangle. Find its perimeter.

Select the correct answer **A**, **B**, **C** or **D**.

A 22 cm **B** 44 cm

C 54 cm **D** 64 cm

5 R Which of the following lengths is the closest to the perimeter of your classroom? Select the correct answer **A**, **B**, **C** or **D**.

A 250 m **B** 2.5 km **C** 250 cm **D** 25 m

6 Measure the sides of each shape in millimetres and find its perimeter.

a **b** **c** **d**

7 R A square has a perimeter of 24 cm. What is the length of one side? Select the correct answer **A**, **B**, **C** or **D**.

A 6 cm **B** 8 cm **C** 12 cm **D** 96 cm

8 R A rectangle has a perimeter of 40 m. If its length is 13 m, find its width.

9 R A rhombus has a perimeter of 18 m. Find the length of one side.

10 R A kite has a perimeter of 26 cm. If one of its longer sides is 9 cm, find the length of a shorter side.

11 PS R A square is cut in half as shown. The two rectangles have a combined perimeter of 24 cm. Find the perimeter of the original square.

12 PS R Jade uses a piece of wire to build a square with a side of 12 cm. If, from the same wire, she builds a rectangle with a width of 6 cm, what is the length of the rectangle?

13 PS R Farouk's rectangular garden is 14 m long and 10 m wide. A hose is buried inside the garden $1\frac{1}{2}$ m from each side. How long is the hose?

Foundation Standard Complex

Circumference of a circle

5.02

The **circumference** of a circle can be found by multiplying its **diameter** by a special number called **pi** (pronounced 'pie'), represented by the Greek letter π. For any circle,

$$\frac{\text{circumference}}{\text{diameter}} = \pi = 3.14\ldots$$

Pi is often estimated as 3.14, but a more accurate value can be found on your calculator when you press the π key (you may need to press SHIFT first). As a decimal, the digits of π continue endlessly without any repeats or patterns:

$\pi = 3.141\,592\,653\,589\,793\ldots$

so, like $\sqrt{2}$, it is called an **irrational number**.

This number was named 'pi' by the Swiss mathematician Leonhard Euler in 1737.

The formula for the circumference of a circle is $C = \pi \times \text{diameter} = \pi d$

Because the diameter of a circle is double its radius, the circumference, C, of a circle with radius r is $C = \pi \times 2 \times \text{radius} = 2\pi r$.

Worksheets
Discovering pi
A page of circles
Circumference and area

Puzzle
Circle crossword

Video
Calculating pi: Archimedes

Circumference of a circle

The **circumference (perimeter) of a circle** is:

$C = \pi \times \text{diameter}$ or $C = 2 \times \pi \times \text{radius}$

$C = \pi d$ $\qquad$ $C = 2\pi r$

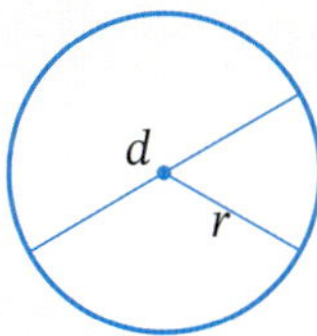

Example 3

a Estimate the circumference of a circle with a diameter of 5 cm.

b Calculate the circumference of the circle:

i correct to two decimal places.

ii in terms of π.

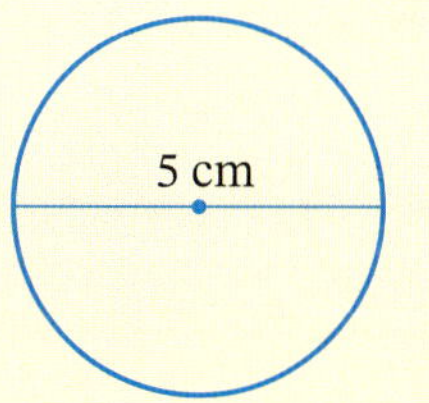

Video
Circumference of a circle

SOLUTION

a $C = \pi d$

$= \pi \times 5$

$\approx 3 \times 5$ $\quad$ π is approximately 3.

$= 15$ cm

b i $C = \pi d$ $\quad$ On your calculator enter π × 5 =.

$= \pi \times 5$

$= 15.707\,963\ldots$

≈ 15.71 cm

ii $C = \pi d$
$= \pi \times 5$
$= 5\pi$ cm

Like writing an answer as a surd in Pythagoras' theorem problems, writing an answer in terms of π is more exact as there is no rounding involved.

Example 4

Video Circumference of a circle

Calculate the circumference of the circle:

a correct to one decimal place

b in terms of π

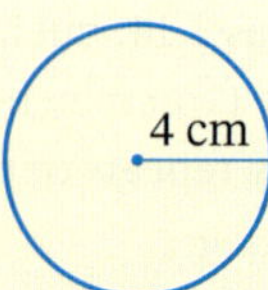

SOLUTION

a $C = 2\pi r$
$= 2 \times \pi \times 4$
$= 25.13274\ldots$
≈ 25.1 cm

b $C = 2\pi r$
$= 2 \times \pi \times 4$
$= 8\pi$ cm

EXERCISE 5.02 ANSWERS ON P. 628

Circumference of a circle

U F PS R

1 Estimate the circumference of each circle.

a

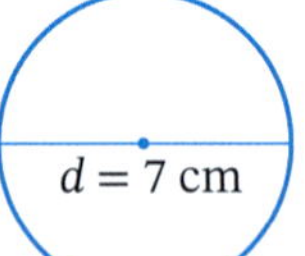

b

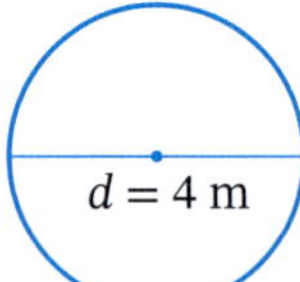

c

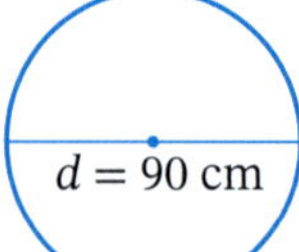

d

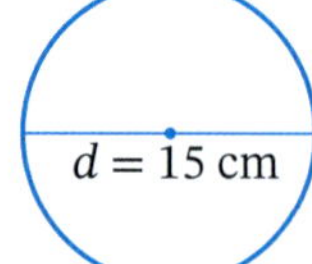

2 Calculate, correct to two decimal places, the circumference of each circle in question **1**.

3 Calculate the circumference of each circle in question **1** in terms of π.

EXAMPLE 4

4 Estimate the circumference of each circle.

a

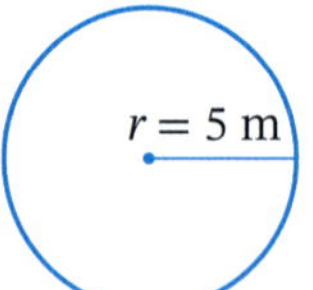

b $r = 13$ mm

c $r = 9$ cm

d $r = 6$ cm

5 Calculate, correct to one decimal place, the circumference of each circle in question **4**.

6 Calculate the circumference of each circle in question **4** in terms of π.

Foundation Standard Complex

7 A fish pond has a diameter of 2.4 m. Find its circumference correct to two decimal places.

8 Liam's Monster truck has wheels with a diameter of 168cm.

PS **a** How far does the monster truck move when the wheels turn one revolution? Answer to one decimal place.

b If Liam travels 792 m, how many complete turns does the monster truck wheel make?

9 Mars has a radius of approximately 3390 km. Determine the circumference of Mars.

10 A 20-cent coin has a radius of 16 mm. Calculate its circumference.

11 This tin of tomatoes has a diameter of 75 mm. If the label wraps around the tin completely, how long is the label? Answer correct to the nearest millimetre.

PS R

$d = 75$ mm

Shutterstock.com/Bahadir Yeniceri

12 Calculate, correct to two decimal places, the circumference of a circle with:

a a diameter of 5 cm **b** a diameter of 14 cm

c a radius of 9 cm **d** a radius of 17 cm

13 Which of the following intervals best shows the circumference of this circle? Select the correct answer **A**, **B**, **C** or **D**.

R

7.8 mm

A ____________________

B ____________________________

C _____________________________________

D ____

14 The ISS completes one circular orbit of the Earth at a height of 400 km. If the radius of the Earth is approximately 6400 km, find the distance travelled by the station. Select the closest answer **A**, **B**, **C** or **D**.

A 2513 km **B** 37 699 km **C** 40 212 km **D** 42 726 km

15 A circle has a circumference of 47.1m. Find its radius, correct to one decimal place.

R

16 The groundskeeper at a local soccer club needs to re-paint the line markings on a field. The measurements are given on the diagram below. If the groundskeeper takes 15 minutes to mark every 100 m of line, how long will it take him to mark this field. Answer to the nearest 15 minutes.

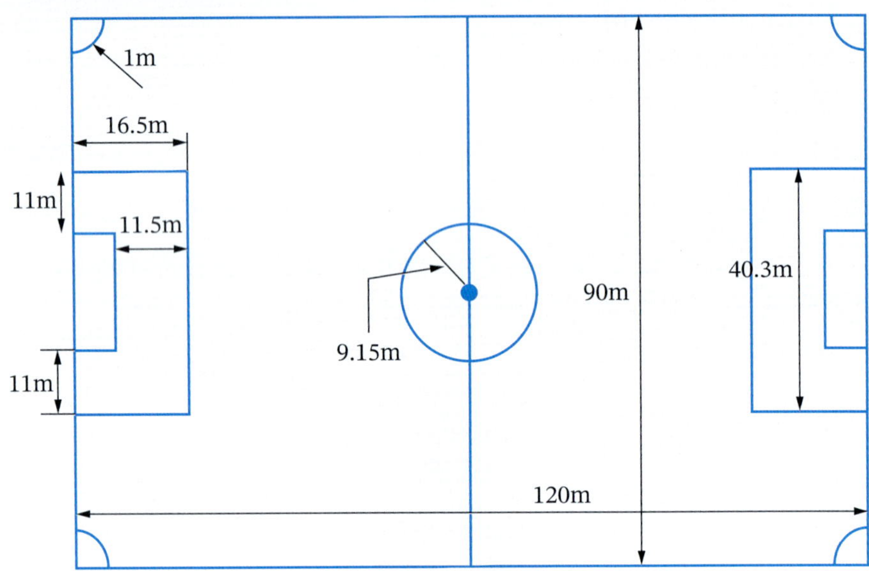

17 PS R Ali and Billy raced each other around this athletic track. Ali ran along the outside perimeter while Billy ran along the inside perimeter. After one lap of the track, who ran the longer distance and by how much? Answer correct to the nearest metre.

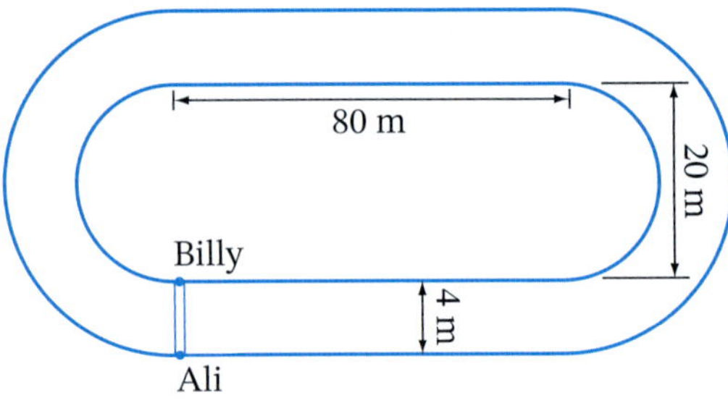

Metric units of area

5.03

The **area** of a shape is the amount of surface covered by the shape. Area is measured in **square units**.

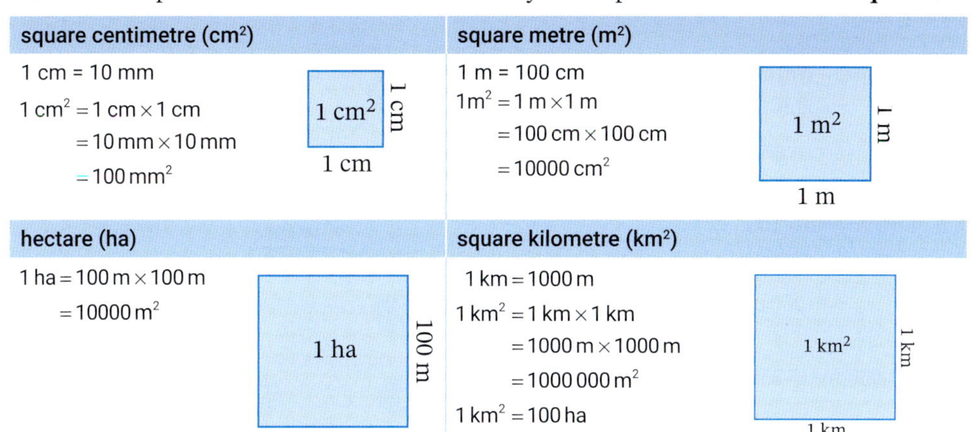

square centimetre (cm^2)	square metre (m^2)
$1\text{ cm} = 10\text{ mm}$ $1\text{ cm}^2 = 1\text{ cm} \times 1\text{ cm}$ $= 10\text{ mm} \times 10\text{ mm}$ $= 100\text{ mm}^2$	$1\text{ m} = 100\text{ cm}$ $1\text{m}^2 = 1\text{ m} \times 1\text{ m}$ $= 100\text{ cm} \times 100\text{ cm}$ $= 10000\text{ cm}^2$
hectare (ha)	**square kilometre (km^2)**
$1\text{ ha} = 100\text{ m} \times 100\text{ m}$ $= 10000\text{ m}^2$	$1\text{ km} = 1000\text{ m}$ $1\text{ km}^2 = 1\text{ km} \times 1\text{ km}$ $= 1000\text{ m} \times 1000\text{ m}$ $= 1000\,000\text{ m}^2$ $1\text{ km}^2 = 100\text{ ha}$

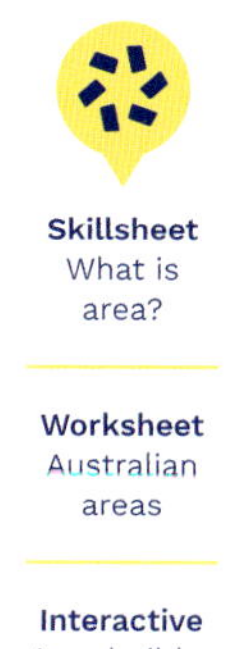

A **square centimetre** is about the area of a fingernail.

A **square metre** is approximately the area of the floor of a large shower recess.

A **hectare** is about the area of Sydney Football Stadium, or the grassed area inside an athletics (400 m) track.

A **square kilometre** is about the area of a theme park, like Dreamworld on the Gold Coast.

Note that, while $1\text{ cm} = 10\text{ mm}$, $1\text{ cm}^2 = 100\text{ mm}^2$ (double the number of 0s), and while $1\text{ m} = 100\text{ cm}$, $1\text{ m}^2 = 10\,000\text{ cm}^2$ (double the number of 0s).

Because a side length is multiplied by a side length to calculate an area, length conversion factors can also be squared to get an area conversion factor. This results in doubling the number of 0s in the length conversion factor.

ⓘ Metric units of area

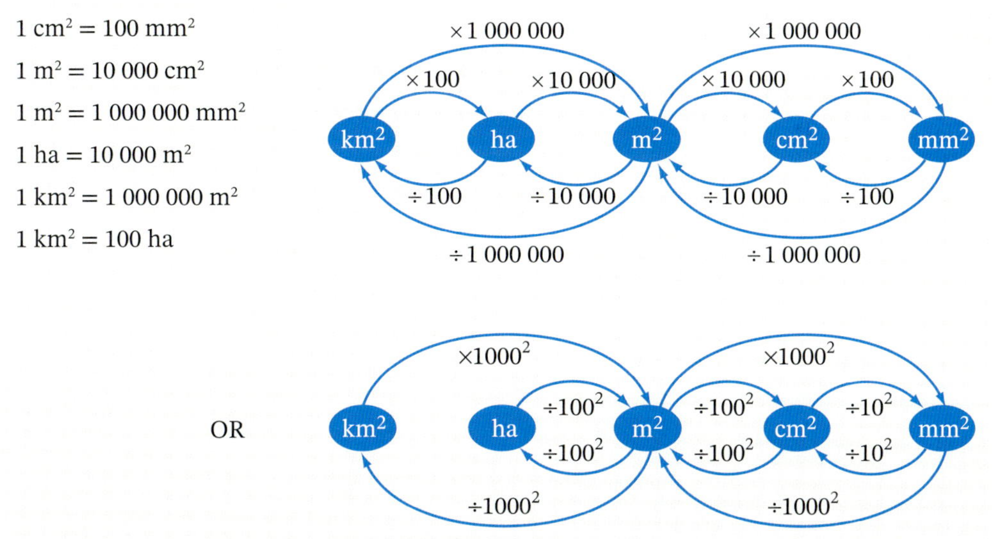

$1\text{ cm}^2 = 100\text{ mm}^2$

$1\text{ m}^2 = 10\,000\text{ cm}^2$

$1\text{ m}^2 = 1\,000\,000\text{ mm}^2$

$1\text{ ha} = 10\,000\text{ m}^2$

$1\text{ km}^2 = 1\,000\,000\text{ m}^2$

$1\text{ km}^2 = 100\text{ ha}$

Example 5

Convert:

a 800 mm² to cm² **b** 4.6 m² to cm² **c** 25 000 000 m² to ha

SOLUTION

a $800\ \text{mm}^2 = 800 \div 10^2$
$= 800 \div 100$
$= 8\ \text{cm}^2$

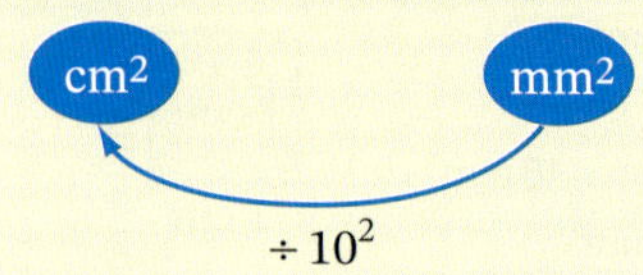

b $4.6\ \text{m}^2 = 4.6 \times 100^2$
$= 4.6 \times 10\ 000$
$= 46\ 000\ \text{cm}^2$

× 10 000
m²
cm²

c $250\ 000\ \text{m}^2 = 250\ 000 \div 100^2$
$= 250\ 000 \div 10\ 000$
$= 25\ \text{ha}$

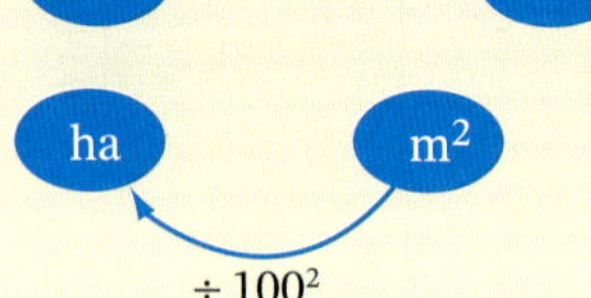

EXERCISE 5.03 ANSWERS ON P. 628

Metric units of area

1 What units of area would you use when measuring the area of:

(R) **a** your home? **b** your pencil case? **c** a football field?

(C) **d** the state of Queensland? **e** your body? **f** a desk top?

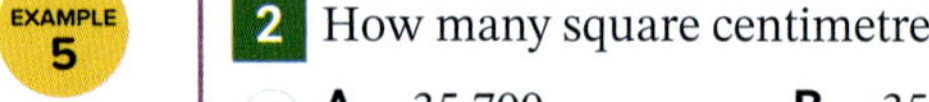

2 How many square centimetres are there in 3.57 m²? Select the correct answer **A**, **B**, **C** or **D**.

(R) **A** 35 700 **B** 357 **C** 0.357 **D** 0.0357

3 Copy and complete each conversion.

(R) **a** 100 mm² = ______ cm² **b** 750 mm² = ______ cm²

c 4500 cm² = ______ m² **d** 34 000 m² = ______ km²

e 60 000 mm² = ______ m² **f** 432 000 cm² = ______ m²

g 950 mm² = ______ m² **h** 7600 m² = ______ ha

i 4.8 mm² = ______ cm² **j** 30 m² = ______ cm²

k 850 cm² = ______ mm² **l** 6.9 km² = ______ m²

m 54 cm² = ______ mm² **n** 700 ha = ______ m²

o 0.48 m² = ______ cm² **p** 0.45 km² = ______ m²

▷

☐ Foundation ○ Standard ○ Complex

4 How many square metres are there in 1 km^2? Select the correct answer **A**, **B**, **C** or **D**.

R **A** 1000 **B** 10 000 **C** 100 000 **D** 1 000 000

5 The solar panels of the ISS cover an area of 2500 m^2. What is this area in km^2?

R C

6 A large cattle station in South Australia has an area of 30 028 km^2. What is this in hectares?

R C

7 The Cradle Mountain–St Clair National Park in Tasmania has an area of 134 805 ha. What is this in km^2?

R C

8 The centre of Sydney is approximately a rectangle bounded by George Street, Circular Quay, Macquarie and College Streets and Liverpool Street. This rectangle is about 1.75 km by 0.5 km. What is its area in:

R C

a km^2?

b hectares?

9 Copy and complete each conversion.

R

a 7600 mm^2 = ______ m^2

b 0.7 ha = ______ cm^2

c 8 m^2 = ______ mm^2

d 850 mm^2 = ______ cm^2

e 6.90 ha = ______ m^2

f 15.2 m^2 = ______ cm^2

g 8500 m^2 = ______ ha

h 49 000 000 m^2 = ______ km^2

10 The area of Australia is 7 682 300 km^2. Western Australia has an area of 2 526 000 km^2. Calculate, correct to one decimal place, the area of Western Australia as a percentage of the area of Australia.

11 Sydney Airport has an area of 881 ha. Sydney Harbour has an area of about 5500 ha.

R **a** What is the area of Sydney Airport in square kilometres?

C **b** How many 'Sydney Airports' would fit onto Sydney Harbour? Answer correct to one decimal place.

5.04 Areas of rectangles, triangles and parallelograms

Skillsheet What is area?

Worksheet Rectangle and triangle areas

Puzzle Area group clues

Video Area formulas for triangles and quadrilaterals

ⓘ Area formulas

Square

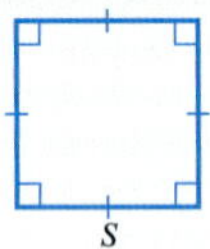

$A = (\text{side})^2$

$A = s^2$

Rectangle

$A = \text{length} \times \text{width}$

$A = lw$

Triangle

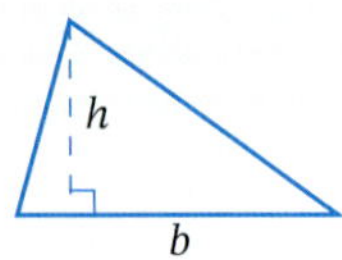

$A = \frac{1}{2} \times \text{base} \times \text{perpendicular height}$

$A = \frac{1}{2}bh$

Parallelogram

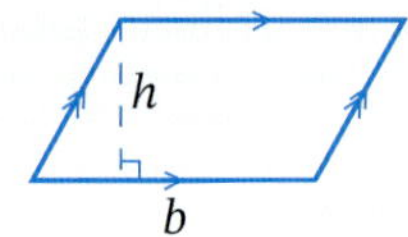

$A = \text{base} \times \text{perpendicular height}$

$A = bh$

Example 6

Find the area of each shape.

a

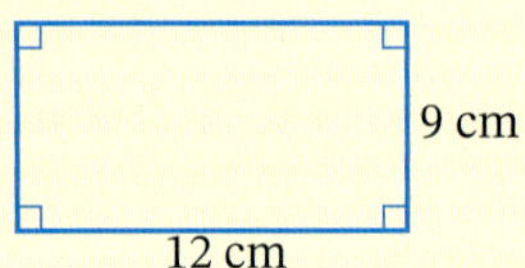

b

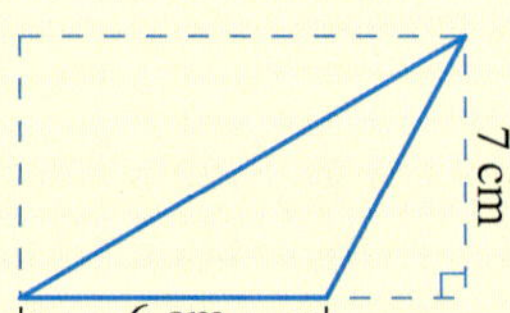

c

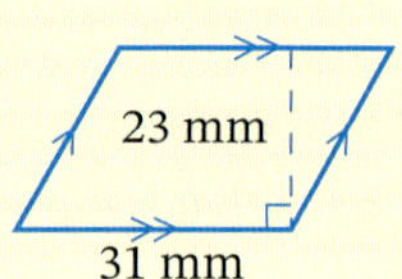

SOLUTION

a $A = lw$ — Area of a rectangle = length × width

$A = 12 \times 9$

$= 108 \text{ cm}^2$

b $A = \frac{1}{2}bh$ — Area of a triangle = $\frac{1}{2}$ × base × height

$= \frac{1}{2} \times 6 \times 7$

$= 21 \text{ cm}^2$

c $A = bh$ — Area of a parallelogram = base × height

$= 31 \times 23$

$= 713 \text{ m}^2$

EXERCISE 5.04 ANSWERS ON P. 628

5.04

Areas of rectangles, triangles and parallelograms

U F PS R

EXAMPLE 6

1 Find the area of each shape.

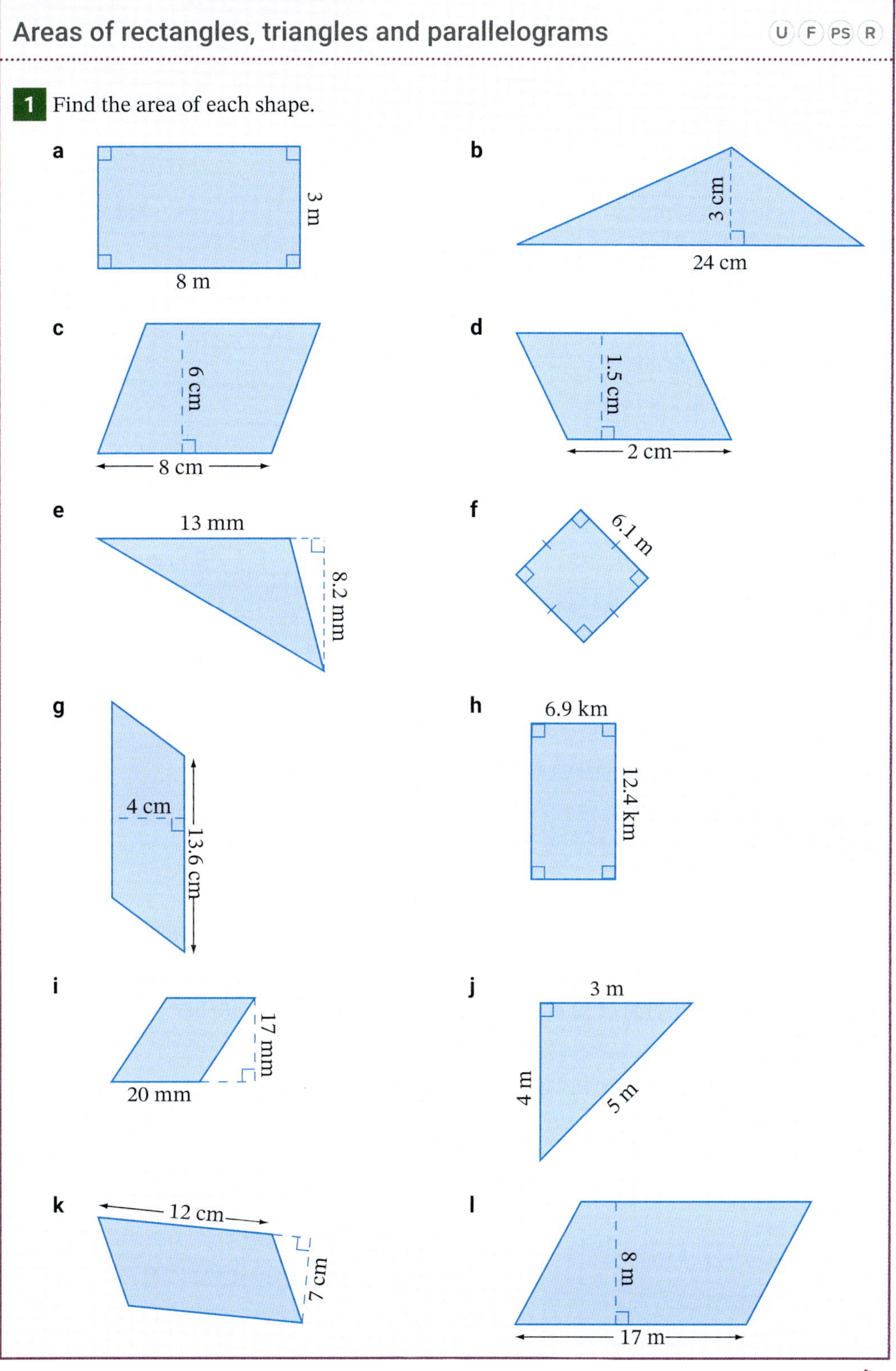

Foundation Standard Complex

2 Find the area of each shape. Make sure that you change all lengths to the same units first.

a

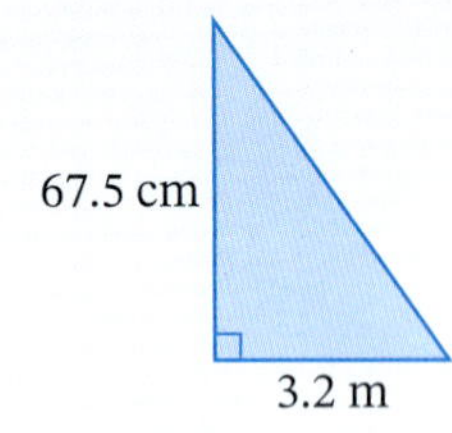

b

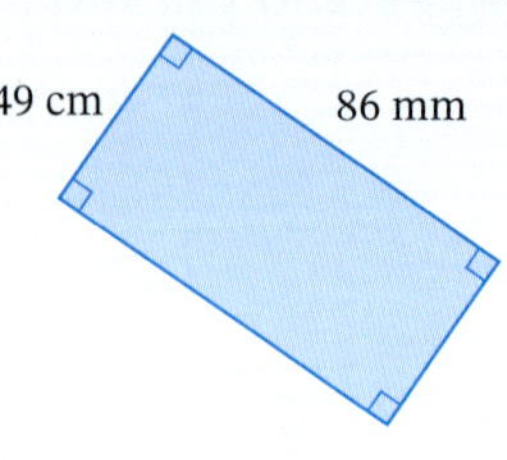

c

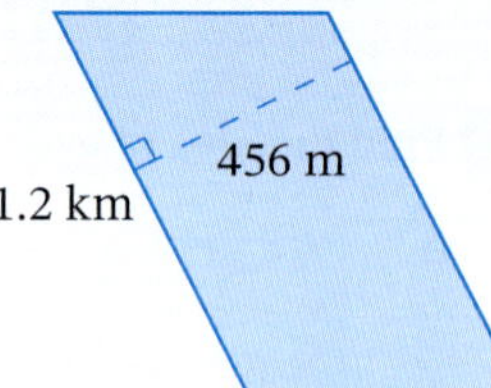

d

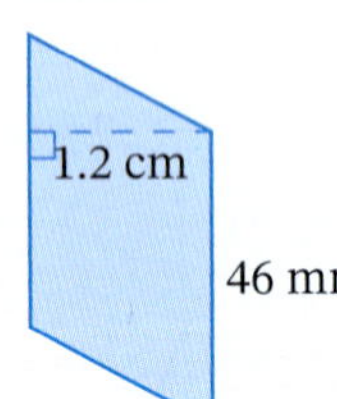

e

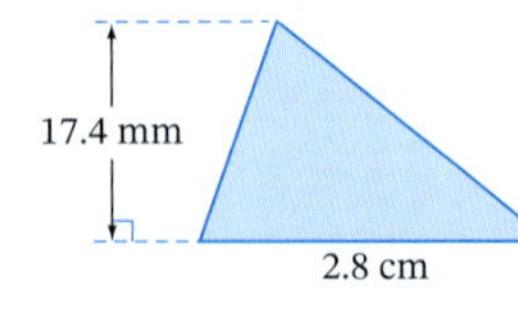

f

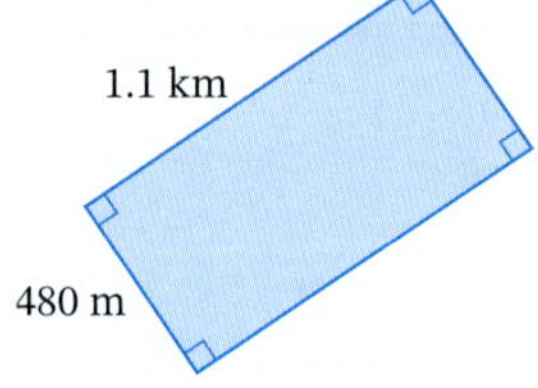

3 One hectare of land is subdivided into blocks. If the average block measures 20 m by 25 m, how many blocks would there be? Select the correct answer **A**, **B**, **C** or **D**.

PS R

A 50 **B** 25 **C** 40 **D** 20

4 Find the area of this shape. Select the correct answer **A**, **B**, **C** or **D**.

A 12.8 cm² **B** 24 cm^{2v}

C 25.6 cm² **D** 32 cm²

5 A square has an area of 25 cm². Find the length of one side.

R

6 A rectangle has an area of 27 m². If its length is 9 m, find its width.

R

7 A triangle has an area of 35 cm². If its base length is 10 cm, what is its perpendicular height?

R

8 A rectangle has an area of 72 mm². What could its length and width be?

R

9 The area of a square is 900 m². If the area of the square quadrupled, what would be the length of the new square?

PS R

☐ Foundation ○ Standard ⬡ Complex

10 The area of a triangle can be found using **Heron's formula** if you know the lengths of all three sides a, b and c:

$$\text{Area} = \sqrt{s(s-a)(s-b)(s-c)} \qquad \text{where } s = \text{half the perimeter} = \frac{a+b+c}{2}$$

Find the area of each triangle using Heron's formula, correct to one decimal place.

a

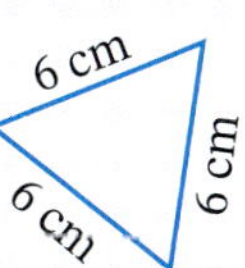

b

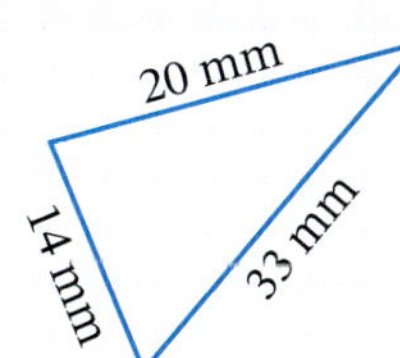

c

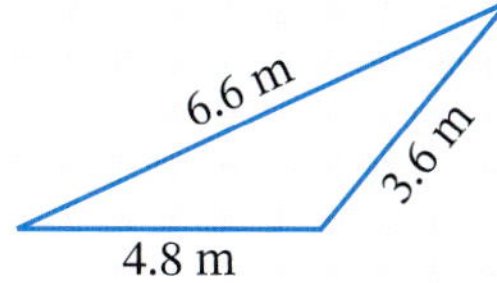

d

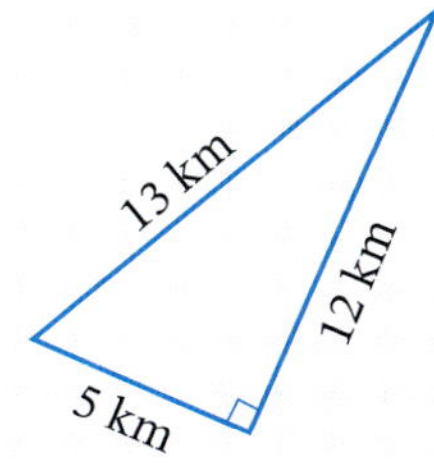

Areas of composite shapes 5.05

Shapes made by combining other simpler shapes are called **composite shapes.**

Video Areas of composite shapes

Worksheets A page of composite shapes

Odd areas

Puzzles Areas 1

Area puzzles

Example 7

Find the area of each composite shape.

a

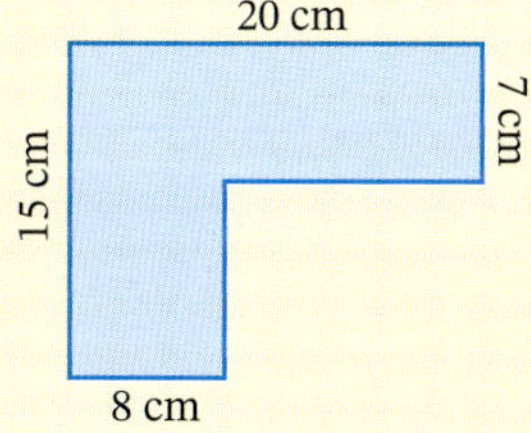

b

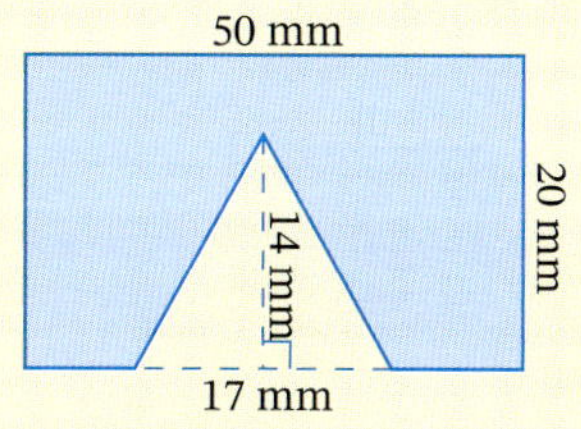

SOLUTION

a **Method 1**

Divide the shape into two rectangles, A and B, and find any unknown lengths.

$x = 20 - 8 = 12$

Area of shape = area of A + area of B

$= 8 \times 15 + 12 \times 7$

$= 204 \text{ cm}^2$

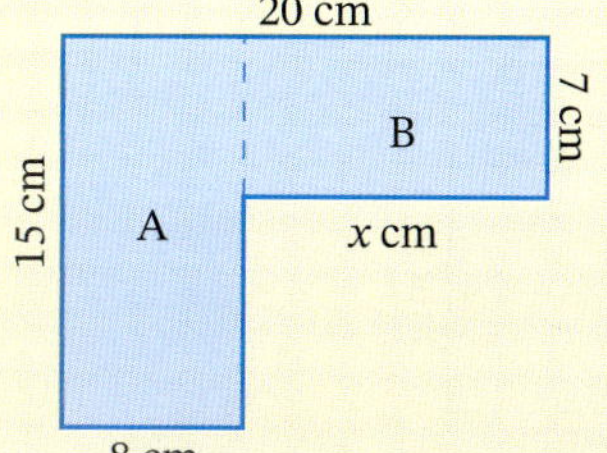

□ Foundation ○ Standard ⬡ Complex

Method 2

The area can also be found by subtracting areas.

$x = 20 - 8 = 12$

$y = 15 - 7 = 8$

Area = large rectangle − small rectangle

$= 20 \times 15 - 12 \times 8$

$= 204\ \text{cm}^2$

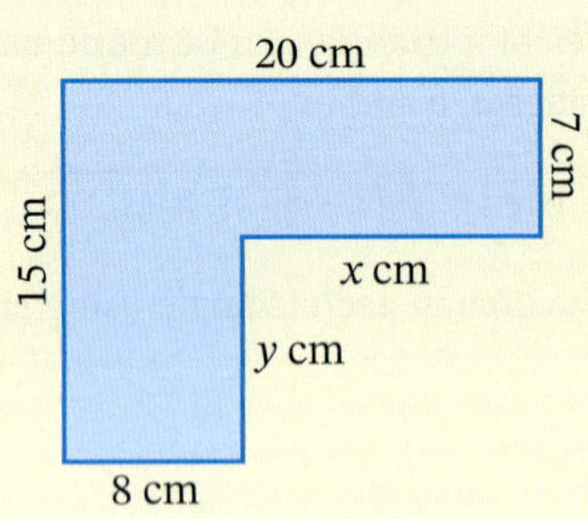

b Area = area of rectangle − area of triangle

$= 50 \times 20 - \frac{1}{2} \times 17 \times 14$

$= 1000 - 119$

$= 881\ \text{mm}^2$

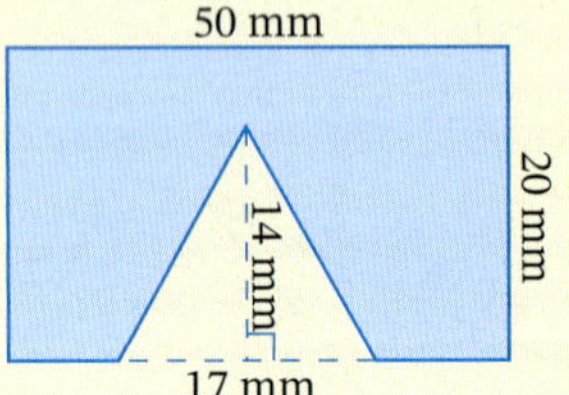

EXERCISE 5.05 ANSWERS ON P. 628

Areas of composite shapes

U F PS R

1 What is the area of this shape? Select the correct answer **A**, **B**, **C** or **D**.

R

A 21 m² **B** 27 m²

C 30 m² **D** 42 m²

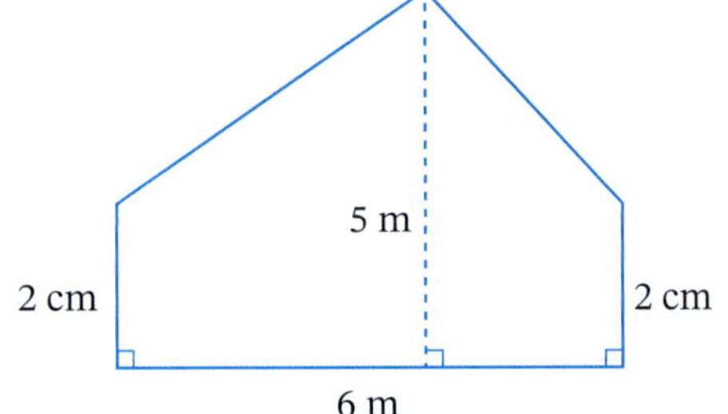

2 Find each shaded area.

PS R

a

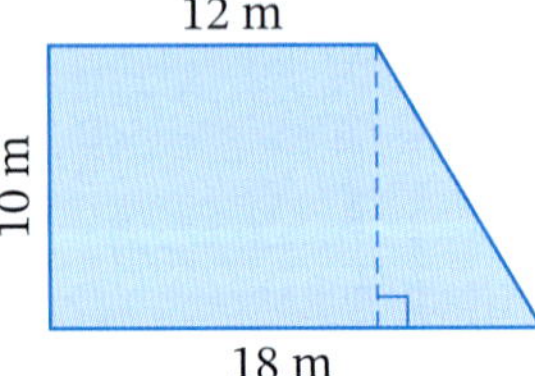

b

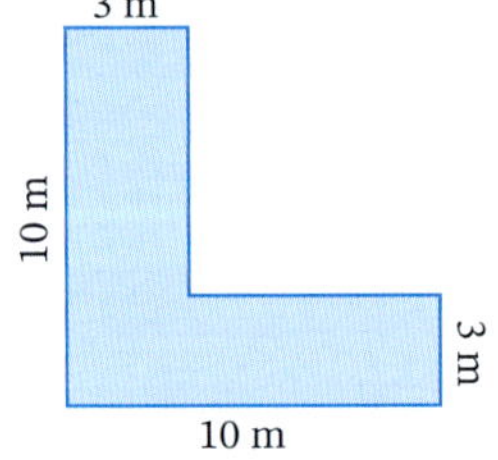

c

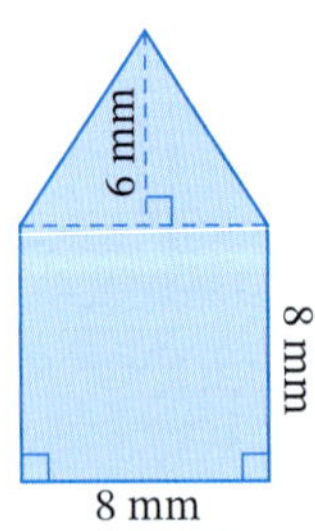

d

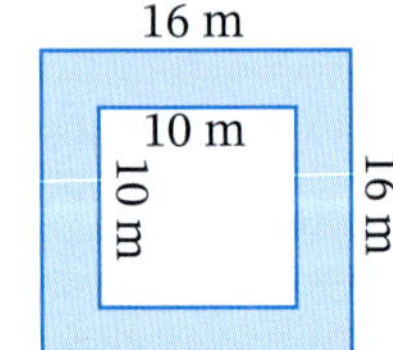

Foundation Standard Complex

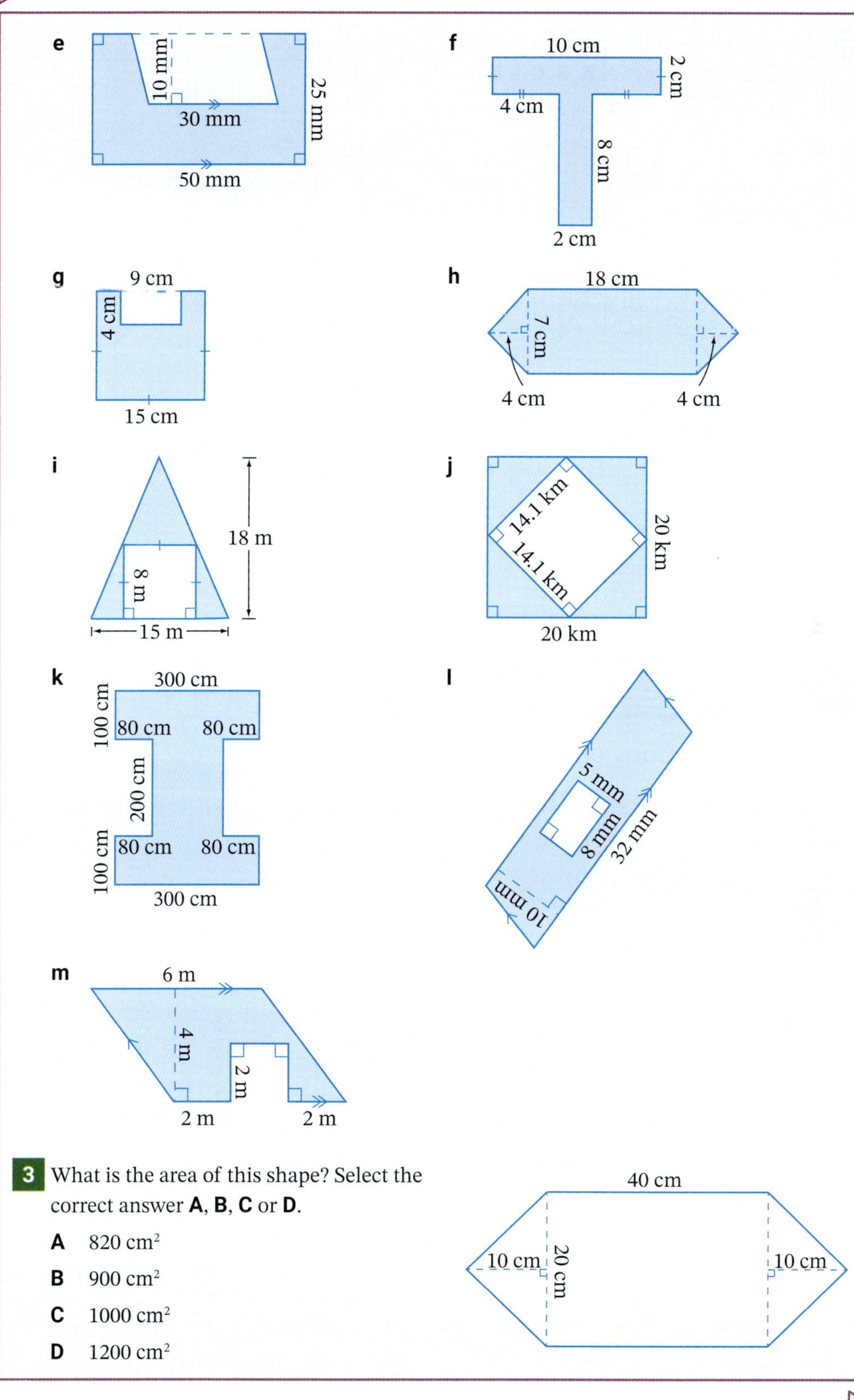

3 What is the area of this shape? Select the correct answer **A**, **B**, **C** or **D**.

A 820 cm^2

B 900 cm^2

C 1000 cm^2

D 1200 cm^2

Foundation Standard Complex

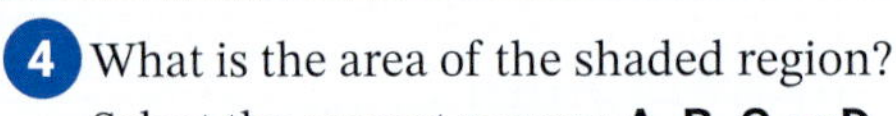

4 What is the area of the shaded region?

PS R Select the correct answer **A**, **B**, **C** or **D**.

A 14 cm² **B** 8 cm²

C 9 cm² **D** 12 cm²

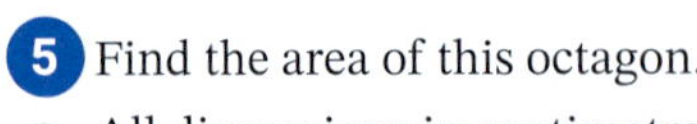

5 Find the area of this octagon.

PS R All dimensions in centimetres.

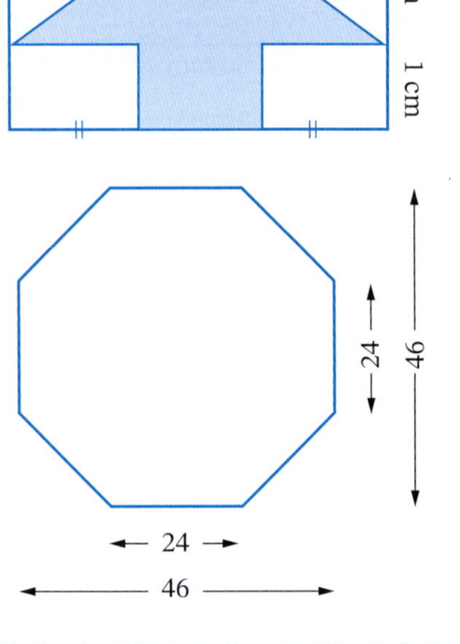

Quiz
Mental skills 5A

☆ **MENTAL SKILLS** 5A ANSWERS ON P. 628 **Maths without calculators**

Converting fractions and percentages to decimals

To convert a fraction into a decimal, change the denominator to a power of 10. This may require simplifying the fraction first.

1 Study each example.

a $\frac{31}{50} = \frac{31 \times 2}{50 \times 2} = \frac{62}{100} = 0.62$

b $\frac{18}{40} = \frac{18 \div 2}{40 \div 2} = \frac{9}{20} = \frac{9 \times 5}{20 \times 5} = \frac{45}{100} = 0.45.$

2 Now convert each fraction into a decimal.

a $\frac{7}{10}$ **b** $\frac{15}{25}$ **c** $\frac{14}{200}$

d $\frac{9}{20}$ **e** $\frac{49}{50}$ **f** $\frac{12}{30}$

To convert a percentage into a decimal, divide by 100 by moving the decimal point 2 places to the left.

3 Study each example.

a 18% = 0.18 **b** 70% = 0.70 = 0.7 **c** 11.4% = 0.114 **d** 2.9% = 0.029

4 Now convert each percentage to a decimal.

a 61% **b** 4% **c** 90%

d 38% **e** 27.1% **f** 0.7%

☐ Foundation ○ Standard ⬡ Complex

9780170465601

INVESTIGATION

Area of a trapezium

1 Draw two copies of this trapezium onto a sheet of paper. Draw a dotted line, parallel to the sides labelled a and b, halfway down the height of each trapezium and label the sides as shown.

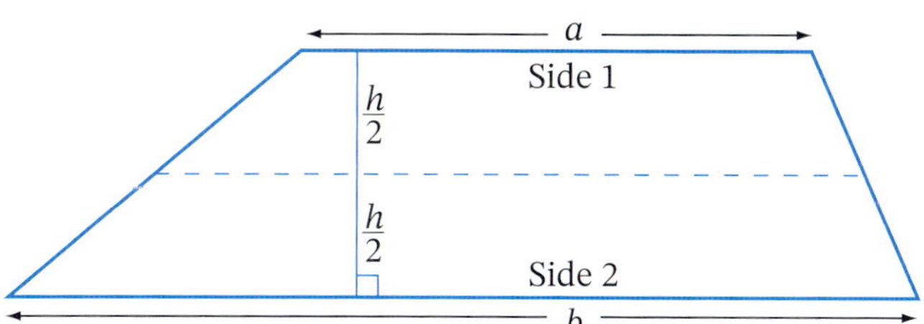

2 Cut out one trapezium and then cut along the dotted line to make two smaller trapeziums. Join the pieces to make a long parallelogram.

3 By measuring the height and base of the parallelogram to the nearest millimetre, find its area.

4 By comparing the measurements of the parallelogram to the measurements of the original trapezium, suggest a general formula for finding the area of any trapezium. Check your answer with your teacher.

5 Cut out the original trapezium and paste it and the parallelogram into your book.

Area of a trapezium 5.06

Video Area formulas for triangles and quadrilaterals

ⓘ Area of a trapezium

$$A = \frac{1}{2} \times \text{sum of lengths of parallel sides} \times \text{perpendicular height}$$

$$A = \frac{1}{2}(a+b)h$$

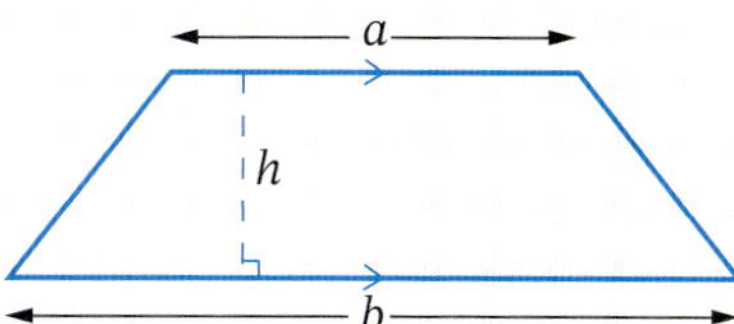

Video
Area of a trapezium

Example 8

Find the area of each trapezium.

a

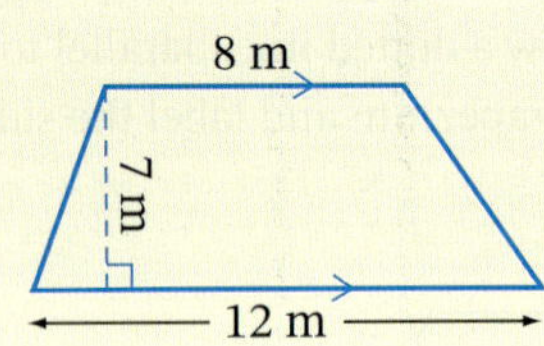

b

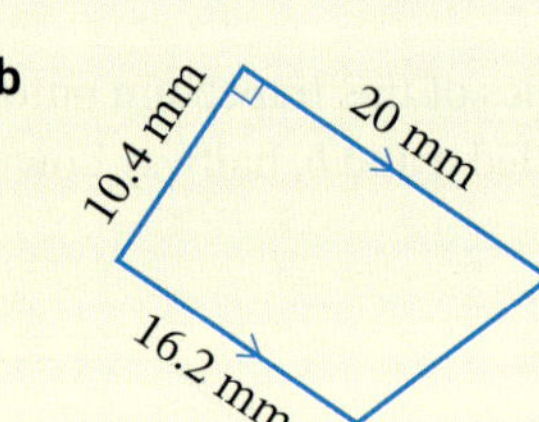

SOLUTION

a

$$A = \frac{1}{2}(a+b)h$$
$$= \frac{1}{2}(8+12) \times 7$$
$$= 70 \text{ m}^2$$

b

$$A = \frac{1}{2}(a+b)h$$
$$= \frac{1}{2}(20+16.2) \times 10.4$$
$$= 188.24 \text{ mm}^2$$

EXERCISE 5.06 ANSWERS ON P. 628

Area of a trapezium

U F PS R

EXAMPLE 8

1 Find the area of each trapezium.

a

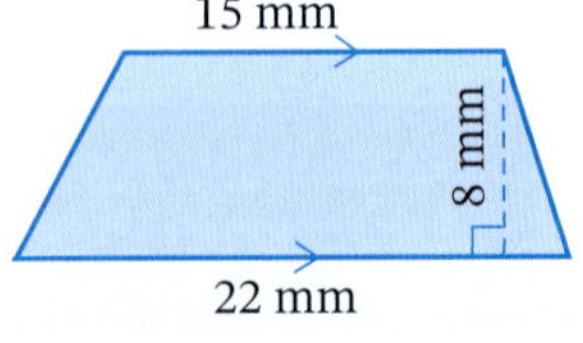

b

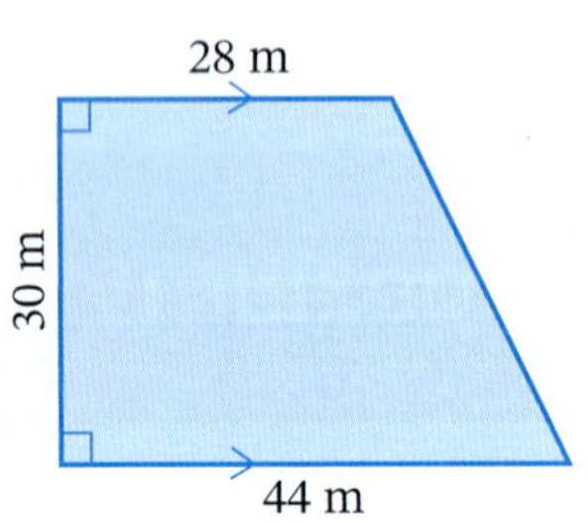

c

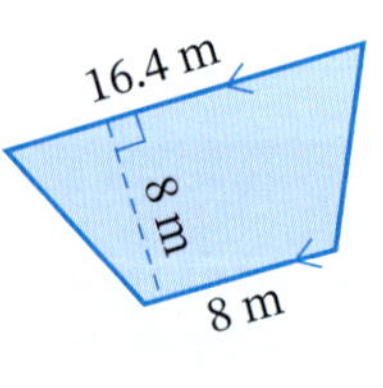

d

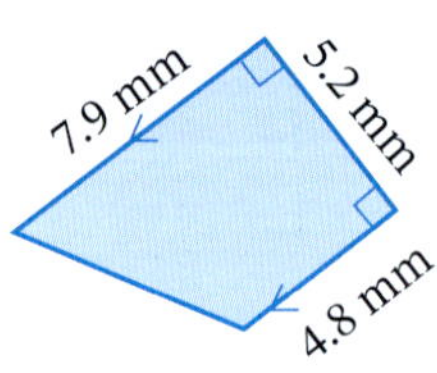

e

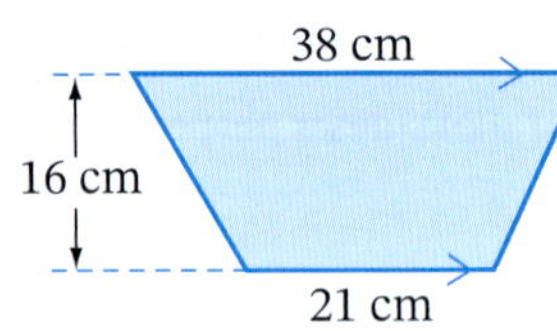

f

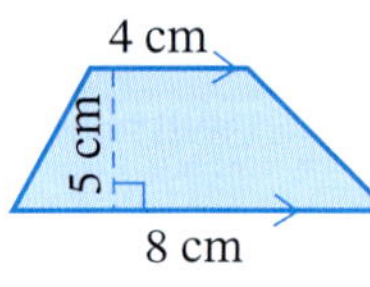

g

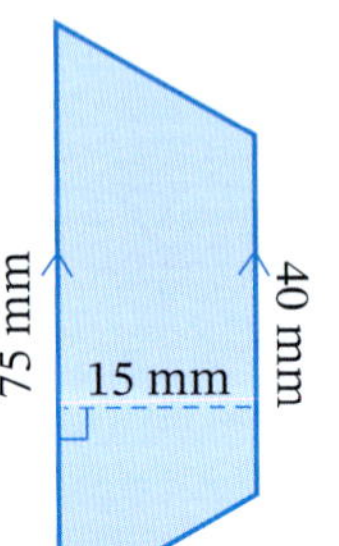

h

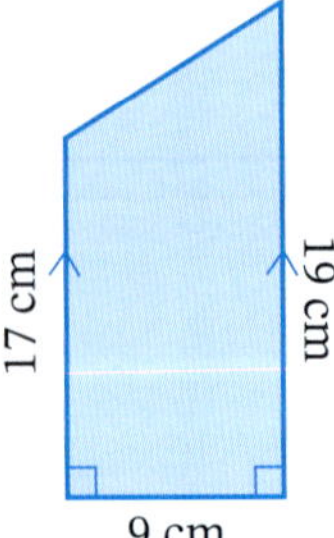

i

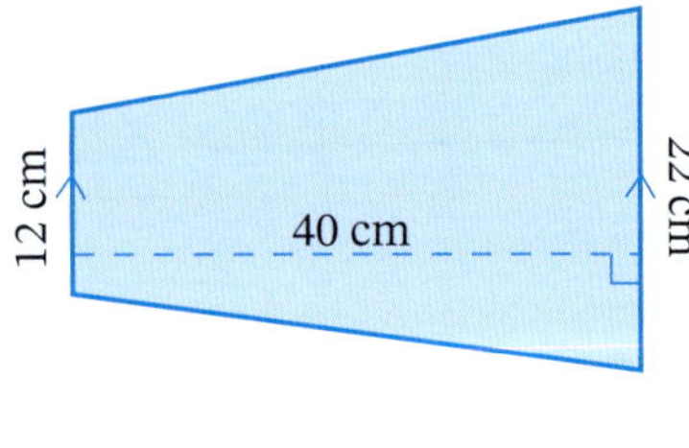

☐ Foundation ○ Standard ⬡ Complex

2 **a** Find the area of the block of land shown.

b Council regulations state that $\frac{1}{5}$ of the land must be reserved for gardens. How much land is available for building?

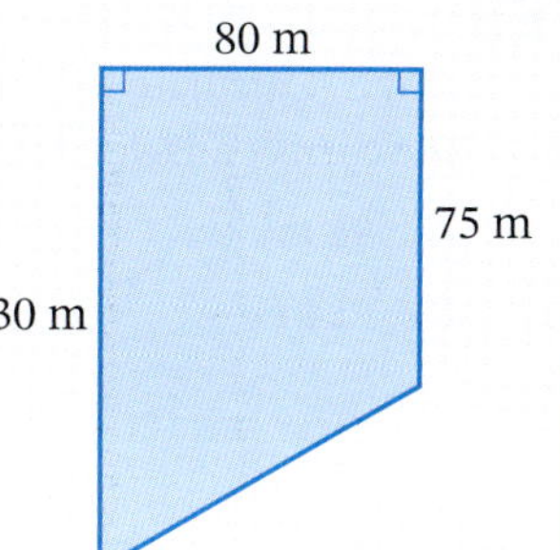

3 Find the area of the trapezoidal shape in the photo of the exterior of Federation Square, Melbourne.

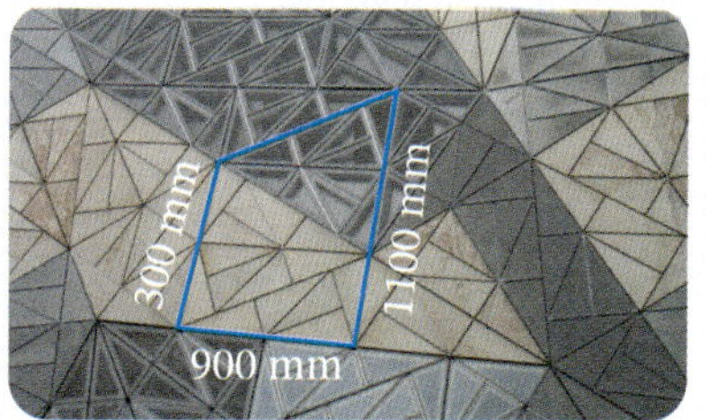

4 The travel graph shows a trip made by Katerina. What is the shaded area in the graph? Select the correct answer **A**, **B**, **C** or **D**.

A 44 units^2 **B** 48 units^2

C 42 units^2 **D** 40 units^2

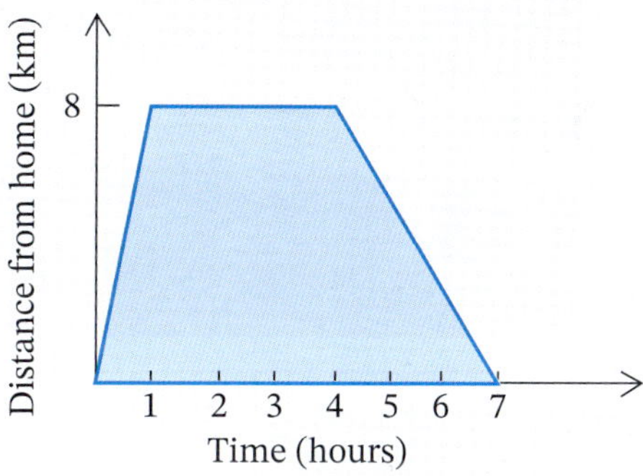

5 A magazine holder has trapezoidal sides as shown.

PS **a** Find the area of one of the trapezoidal sides.

R **b** Find the total area of cardboard needed to build this holder.

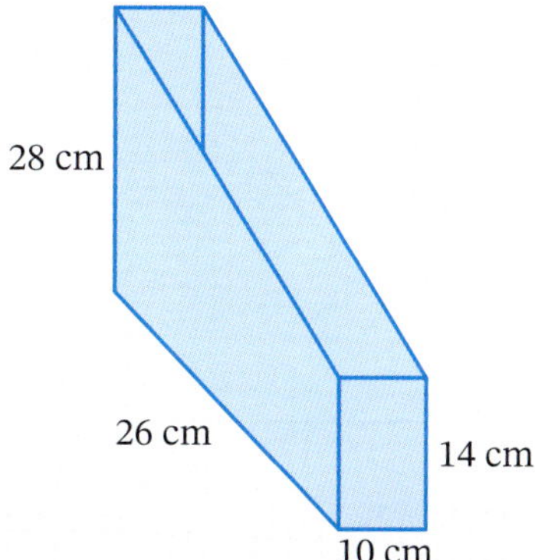

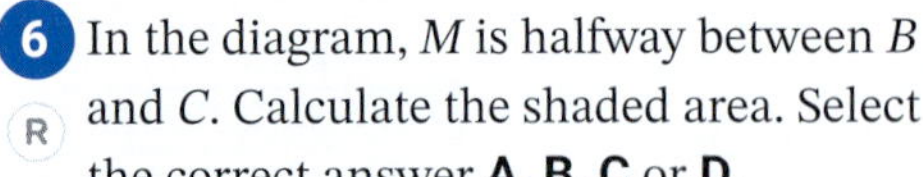

6 (R) In the diagram, M is halfway between B and C. Calculate the shaded area. Select the correct answer **A**, **B**, **C** or **D**.

A 96 m^2 **B** 132 m^2

C 224 m^2 **D** 144 m^2

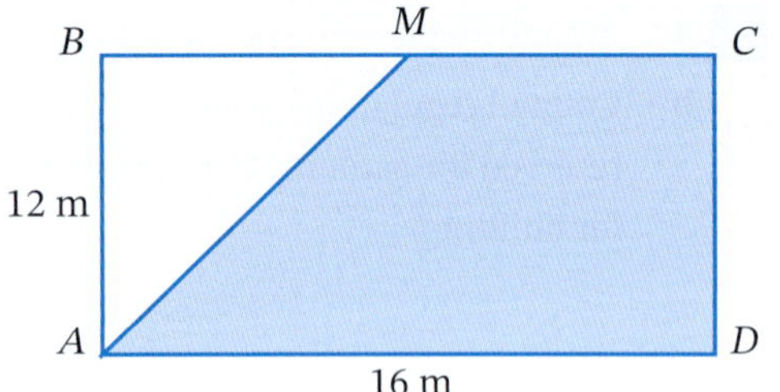

7 (PS) (R) One side of the stage where Grant's drama club performs is shaped like a trapezium with an area of 144 m^2. The trapezium's parallel sides are 30 m and 42 m. What is the height of the trapezium?

Alamy Stock Photo/Photononstop

INVESTIGATION

Area of a kite and a rhombus

1 Draw two copies of the kite and rhombus shown below on a sheet of paper. Draw the diagonals of length x and y and label them as shown. Notice that they cross at right angles. For the kite, one diagonal (x) is bisected by the other (y), while for the rhombus, *both* diagonals bisect each other.

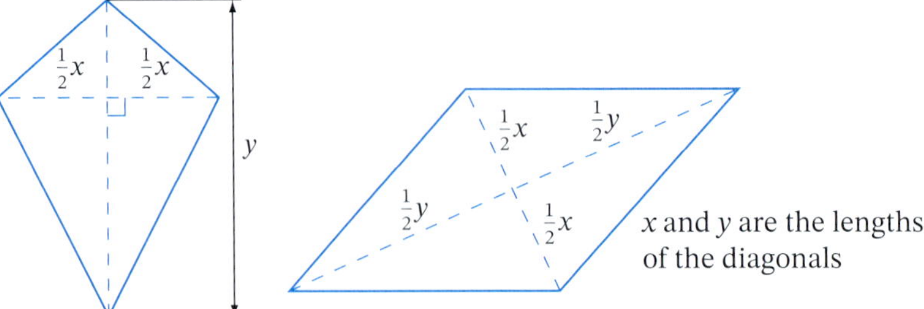

2 Cut out one of your kites and rhombuses and cut along their diagonals to make four triangles each. Join each set of triangles to make a rectangle.

Foundation | Standard | Complex

3 By comparing the measurements of the kite to the measurements of the rectangle formed from it, suggest a general formula for finding the area of any kite. Check your answer with your teacher.

4 a What is the definition of a kite?
   b Is a rhombus a special type of kite?

5 By comparing the measurements of the rhombus to the measurements of the rectangle formed from it, suggest a general formula for finding the area of any rhombus. Check your answer with your teacher.

6 Cut out the original kite and rhombus and paste all shapes into your book.

Areas of kites and rhombuses 5.07

Area of a kite or a rhombus

$$A = \frac{1}{2} \times (\text{diagonal}) \times (\text{other diagonal})$$

$$A = \frac{1}{2}xy$$

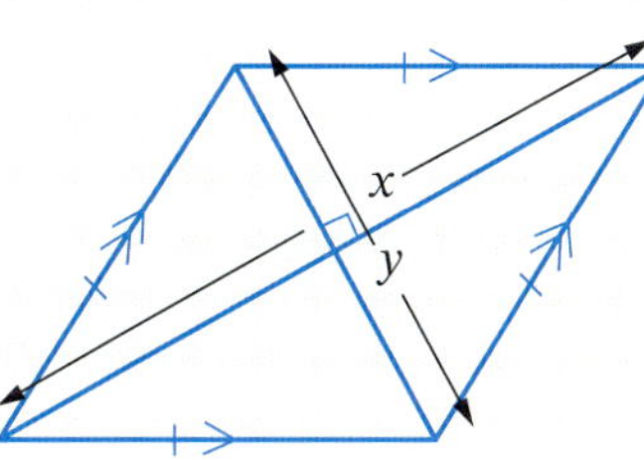

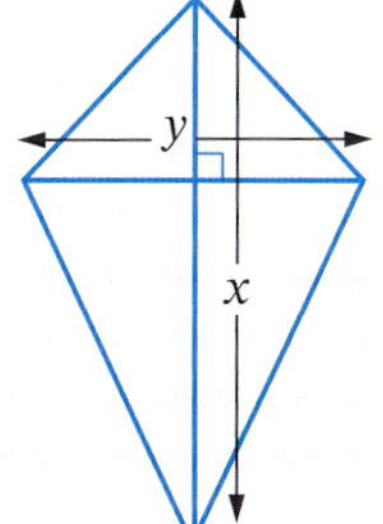

Video Area formulas for triangles and quadrilaterals

Worksheet Quadrilateral and triangle areas

Puzzle Areas of quadrilaterals and triangles

Example 9

Find the area of each shape.

a

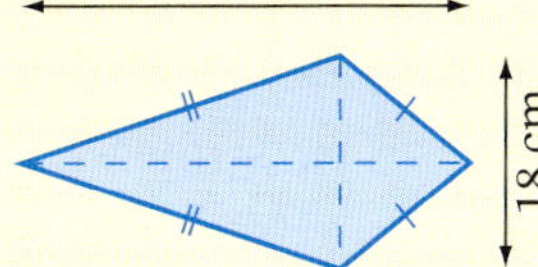

b

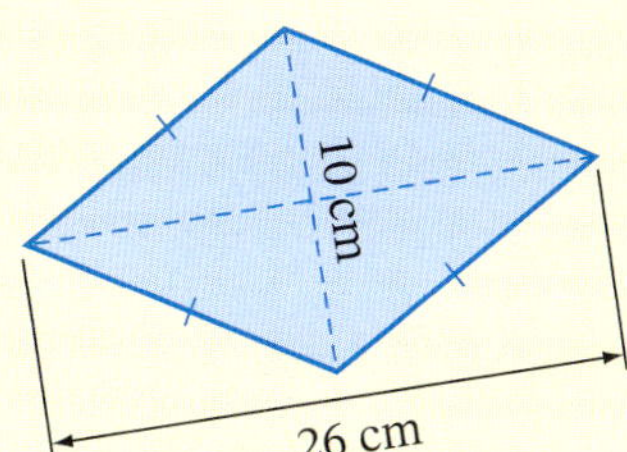

SOLUTION

a

$$A = \frac{1}{2}xy$$
$$= \frac{1}{2} \times 18 \times 52$$
$$= 468 \text{ cm}^2$$

b

$$A = \frac{1}{2}xy$$
$$= \frac{1}{2} \times 10 \times 26$$
$$= 130 \text{ cm}^2$$

EXERCISE 5.07 ANSWERS ON P. 628

Areas of kites and rhombuses

U F

1 Find the area of each kite.

a

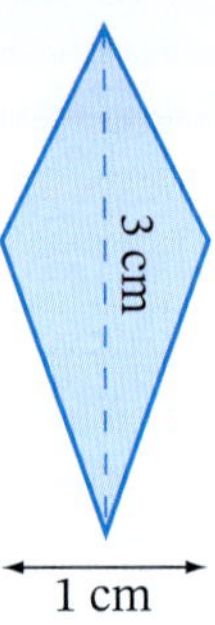

b

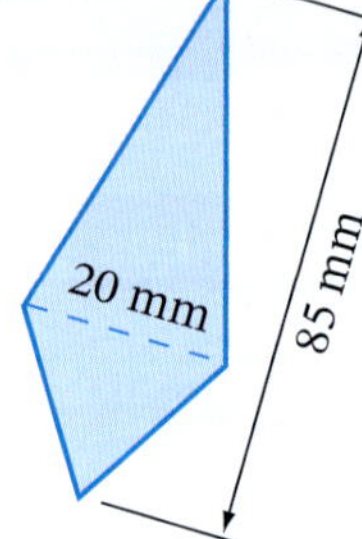

c

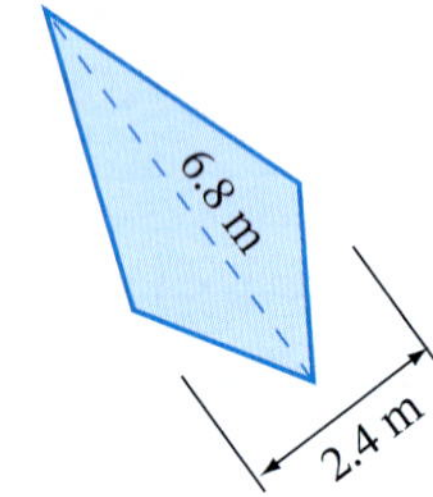

d

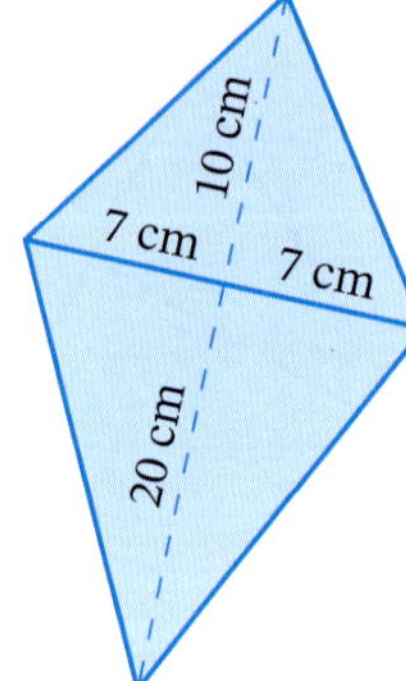

2 Find the area of each rhombus.

a

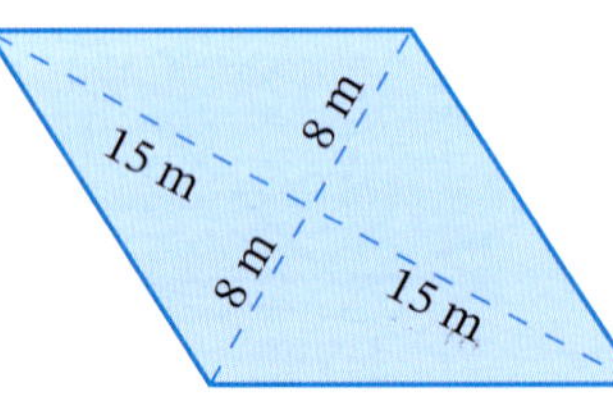

b

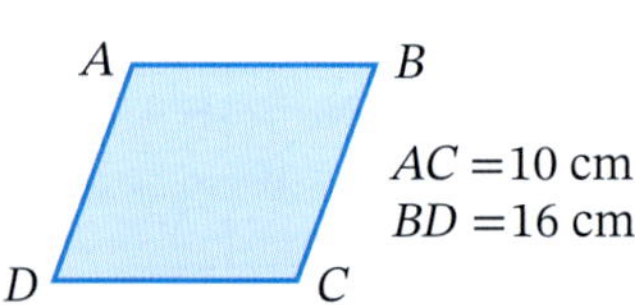

$AC = 10$ cm
$BD = 16$ cm

c

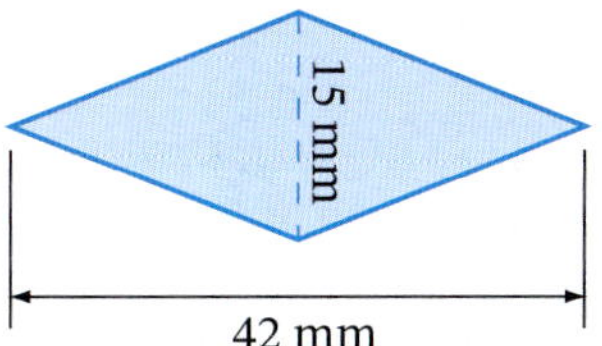

d

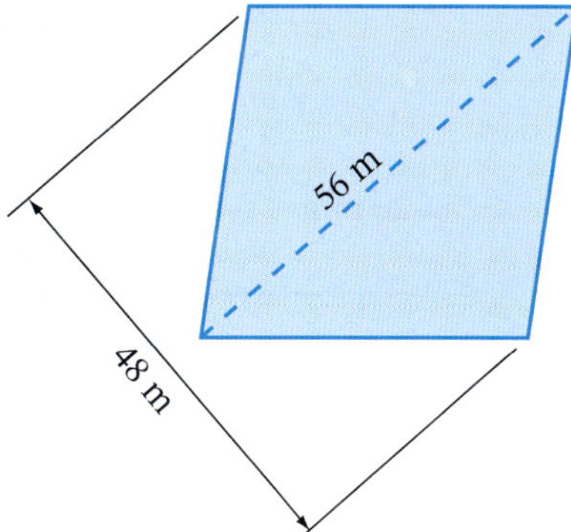

e

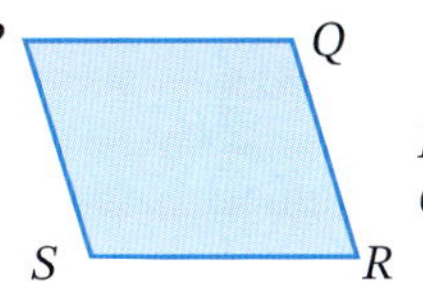

$PR = 2.4$ m
$QS = 1.5$ m

3 What area of material is needed to build this kite?

Select the correct answer **A**, **B**, **C** or **D**.

A 36.8 cm^2

B 3680 cm^2

C 73.6 cm^2

D 7360 cm^2

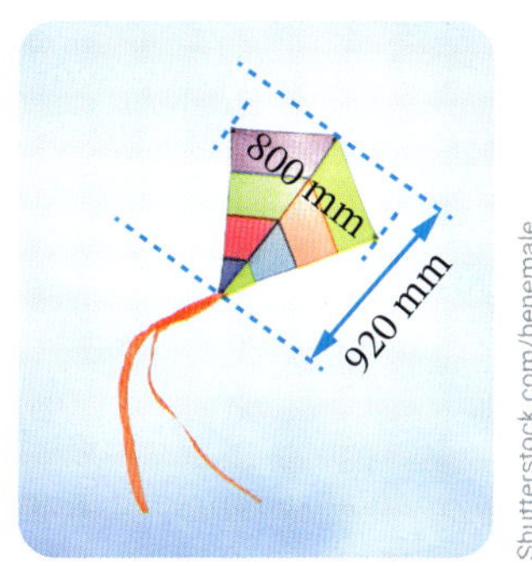

Shutterstock.com/benemale

4 Find the area of each shape.

a

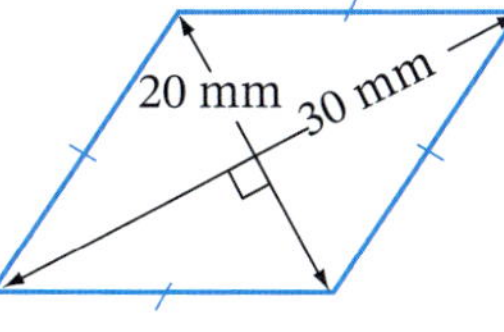

b

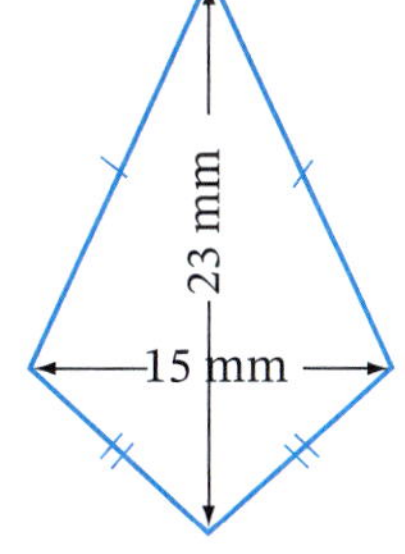

c

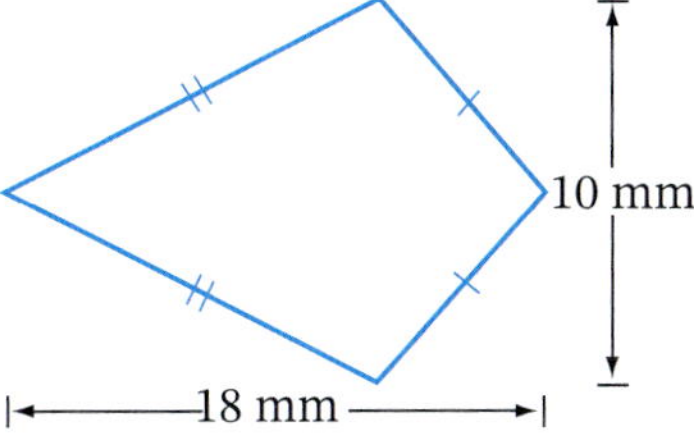

d

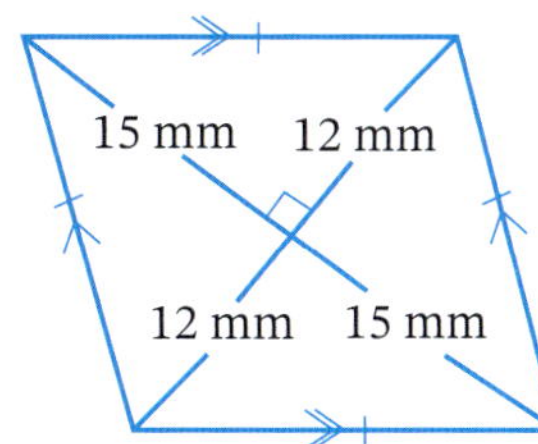

5.07

Foundation Standard Complex

e

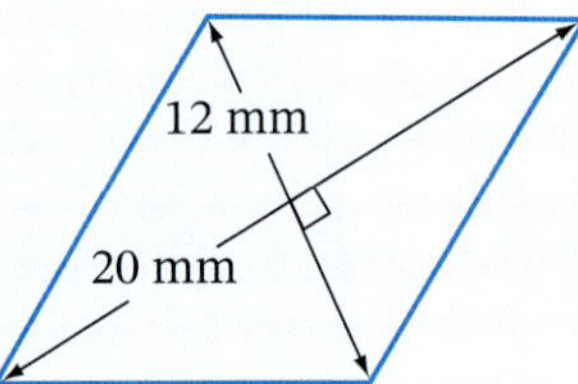

f

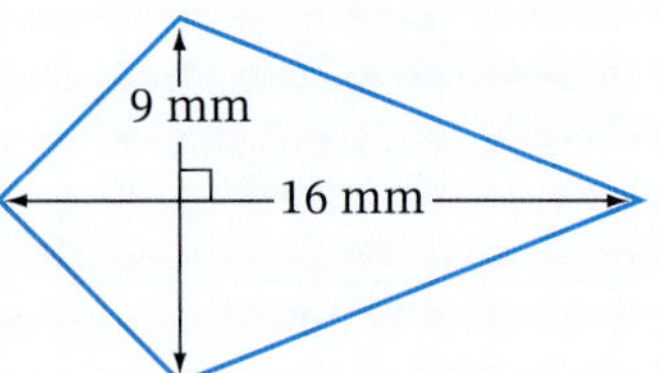

g

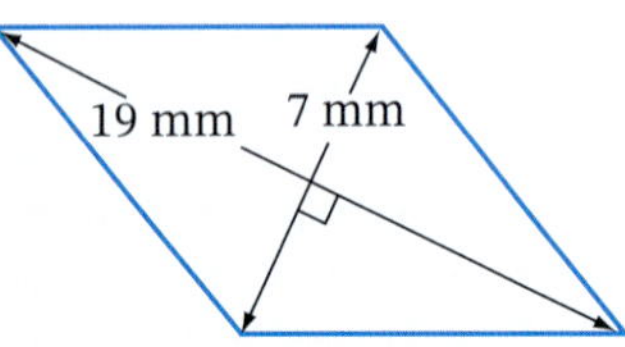

h

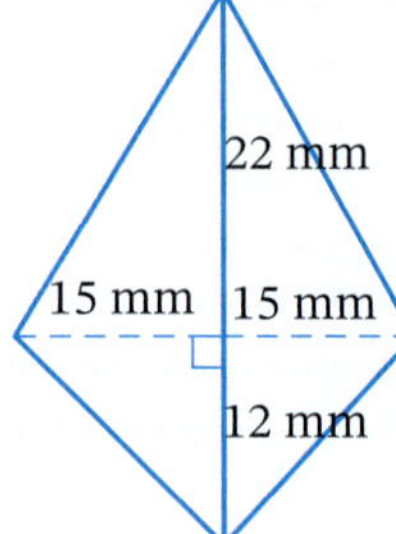

i

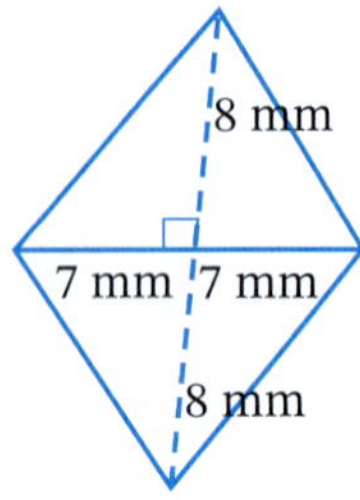

5 Anand used three identical kites to draw this logo.

a What are the lengths of the diagonals for each kite?

b Find the area of one kite.

c Find the area of the entire logo.

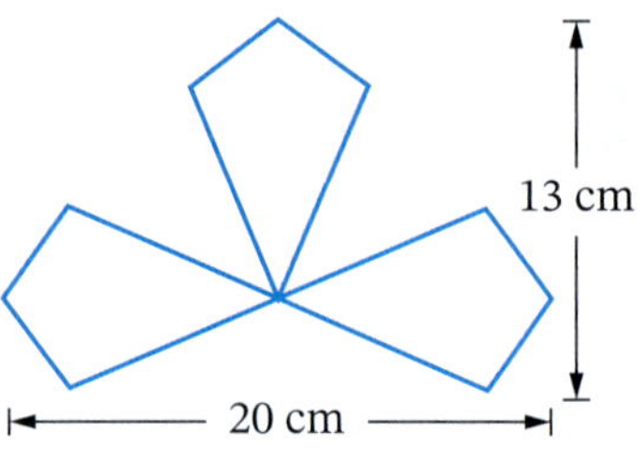

Quiz
Mental skills 5B

☆ **MENTAL SKILLS** 5B ANSWERS ON P.629 **Maths without calculators**

Comparing fractions, decimals and percentages

To compare or order fractions, express them with a common denominator first. To compare or order decimals, express them with the same number of decimal places first.

1 Study each example.

a Which fraction is smaller: $\frac{4}{10}$ or $\frac{3}{8}$?

Using a common denominator of 80 (8×10)

Using a common denominator of 40 (the LCM of 10 and 8)

☐ Foundation ○ Standard ⬡ Complex

$\frac{4}{10} = \frac{4 \times 8}{10 \times 8} = \frac{32}{80}$

$\frac{3}{8} = \frac{3 \times 10}{8 \times 10} = \frac{30}{80}$

By comparing numerators: $\frac{30}{80} < \frac{32}{80}$

$\therefore \frac{3}{8}$ is smaller.

$\frac{4}{10} = \frac{4 \times 4}{10 \times 4} = \frac{16}{40}$

$\frac{3}{8} = \frac{3 \times 5}{8 \times 5} = \frac{15}{40}$

By comparing numerators: $\frac{15}{40} < \frac{16}{40}$

$\therefore \frac{3}{8}$ is smaller.

b Arrange these decimals in ascending order: 0.407, 0.47, 0.047 and 0.4.

Express all decimals with three decimal places by inserting zeros at the end where necessary.

0.407 0.470 0.047 0.400

By comparing digits in the same decimal place, the decimals from smallest to largest are: 0.047 0.400 0.407 0.470

That is, 0.047, 0.4, 0.407, 0.47

2 Now find the smaller number in each pair.

a $\frac{2}{3}$ and $\frac{3}{5}$ **b** $\frac{1}{4}$ and $\frac{1}{3}$ **c** 0.15 and 0.105

d 3.826 and 3.68 **e** $\frac{3}{7}$ and $\frac{4}{10}$ **f** $\frac{5}{6}$ and $\frac{7}{8}$

g 2.87 and 2.817 **h** 0.5301 and 0.503 **i** $\frac{1}{5}$ and $\frac{2}{9}$

3 Arrange each set of numbers in ascending order.

a 0.81, 0.082, 0.821 **b** 3.5, 3.51, 3.55, 3.513 **c** $\frac{2}{3}, \frac{1}{6}, \frac{2}{5}$

d 0.007, 0.07, 0.7, 0.707 **e** $\frac{3}{8}, \frac{2}{10}, \frac{1}{2}$ **f** $\frac{1}{4}, \frac{4}{10}, \frac{3}{5}$

4 Study this example.

Arrange these numbers in descending order: 68%, $\frac{13}{20}$, $0.\dot{6}$, $\frac{3}{5}$.

To order fractions, decimals and percentages, express them all as percentages first.

As percentages: 68% = 68%, $\frac{13}{20}$ = 65%, $0.\dot{6}$ = 66.66...%, $\frac{3}{5}$ = 60%.

Largest to smallest, this is 68%, 66.66%, 65%, 60%, that is, 68%, $0.\dot{6}$, $\frac{13}{20}$, $\frac{3}{5}$

5 Now arrange each set of numbers in descending order.

a 0.25, $\frac{1}{6}$, 16%, $\frac{1}{5}$ **b** 27%, $\frac{1}{3}$, 0.4, 0.28 **c** 0.05, 50%, 6%, $\frac{1}{8}$

d $\frac{3}{4}$, 0.639, 55%, $\frac{2}{5}$ **e** 69%, 0.609, $\frac{2}{3}$, 0.6 **f** $\frac{2}{9}$, 0.105, 17%, 22.5%

INVESTIGATION

Belt around the Equator

Imagine that we wrapped a giant belt tightly around the Earth, along the Equator. This belt would touch the Earth at all points on its circumference, assuming the Earth was a perfect sphere or ball (with no mountains or valleys).

Now suppose we added one extra metre to the length of the belt. Then it would become loose and not touch the Earth any more. There would be a small gap between the Earth and the belt. How wide is this gap?

1 In a group of two to four people, guess whether you could:
   a slip your hand between the belt and the ground.
   b crawl under the belt.
   c sit under the belt.
   d stand under the belt.
2 If the diameter of the Earth is 12 755 m, calculate to two decimal places, the length of the tight belt.
3 Add 1 m to the length of the tight belt.
4 Calculate the diameter of the longer, looser belt.
5 So what is the length of the gap between the Earth and the loose belt?
6 Check whether your guesses in question 1 were correct.

INVESTIGATION

Area by cutting out sectors

Worksheet Approximating the area of a circle

You will need: compasses, a ruler, a protractor and a pair of scissors.

1 Draw a circle of radius 10 cm. Use your protractor to divide the circle into 12 sectors, each with angle size 30°.

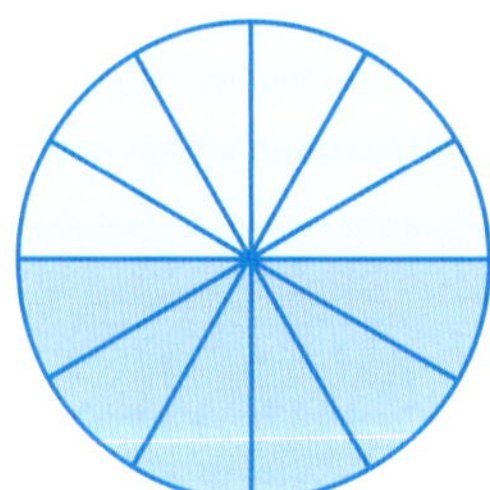

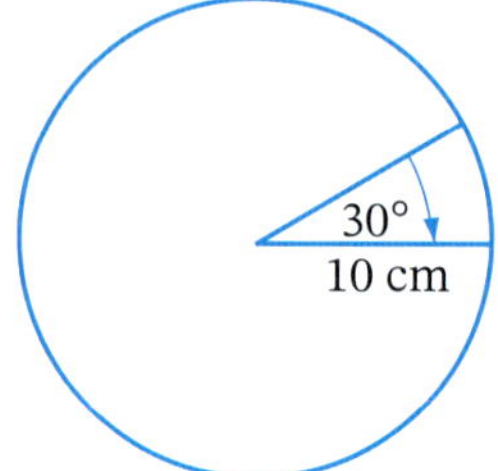

2 Cut out all the sectors and arrange them into a shape like a parallelogram.

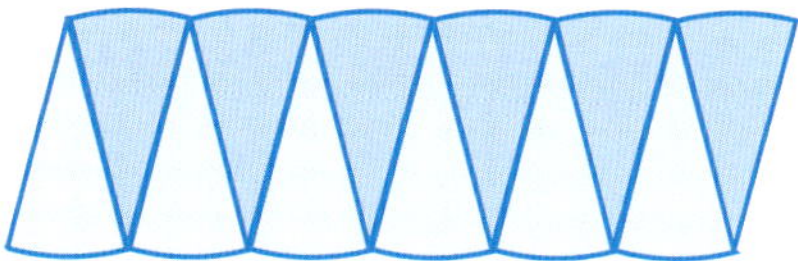

3 Use a ruler to measure the *base* and the *height* of your 'parallelogram'.

4 Now calculate the area of the 'parallelogram'.

5 Which formula gives the better approximation for the area of this circle: $A = 3 \times r^2$ or $A = 4 \times r^2$?

6 We will now find the actual formula for the area of a circle. A circle of radius r is cut into many sectors and rearranged into a 'parallelogram'.

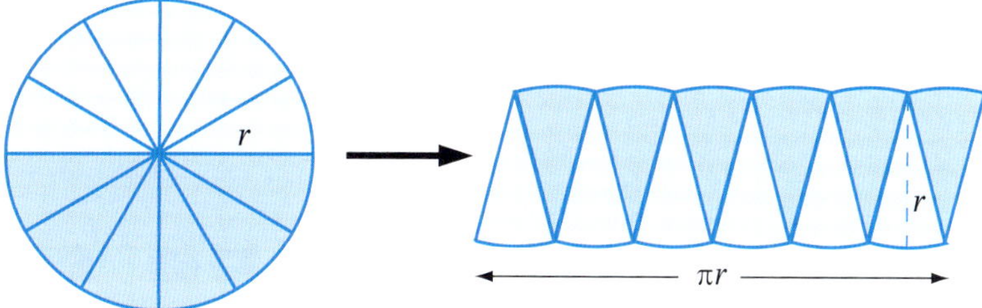

What is the formula for the circumference of this circle?

7 Explain why the length of the 'parallelogram' is πr.

8 What is the formula for the area of this 'parallelogram'?

9 Explain why the area of the circle is πr^2.

Area of a circle 5.08

The area of a square is calculated by squaring its length. $A = s \times s = s^2$

In a similar way, the area of a circle is calculated by squaring its radius and multiplying by π.

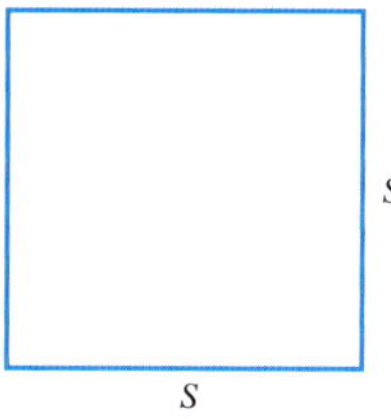

Worksheets
Circumference and area

Area ID

Area and perimeter investigations

Puzzle
Circle areas

Video
Beating the U-boats

ⓘ Area of a circle

$A = \pi \times (\text{radius})^2$

$A = \pi r^2$

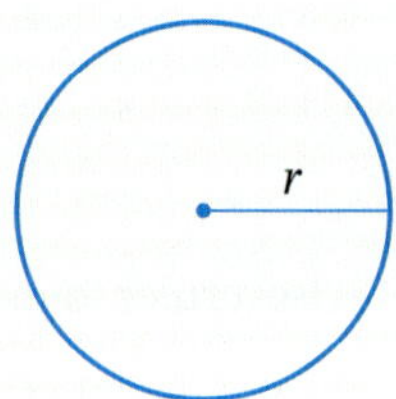

Example 10

Calculate the area of this circle:

a correct to two decimal places.

b in terms of π.

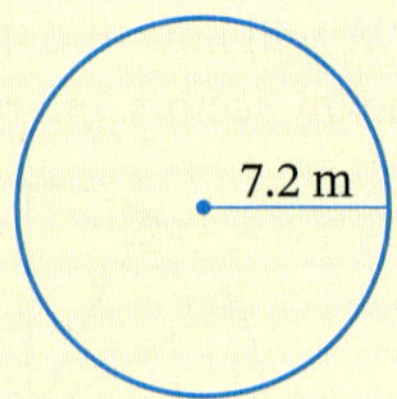

SOLUTION

a $A = \pi r^2$

$= \pi \times 7.2^2$

$= 162.8601\ldots$

$\approx 162.86 \text{ m}^2$

Area is measured in square units, such as m².

On a calculator enter:  × 7.2  =.

b $A = \pi r^2$

$= \pi \times 7.2^2$

$= 51.84\pi \text{ m}^2$

Example 11

Video
Area of a circle

Find the area of this circle:

a correct to one decimal place.

b in terms of π.

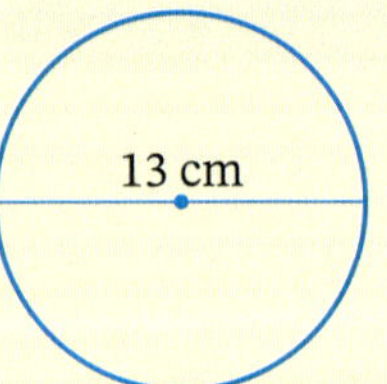

SOLUTION

a $r = \frac{1}{2} \times 13 = 6.5 \text{ cm}$

$A = \pi r^2$

$= \pi \times 6.5^2$

$= 132.7322\ldots$

$\approx 132.7 \text{cm}^2$

The radius is half of the diameter

b From **a**,

$$A = \pi \times 6.5^2$$
$$= 42.25\pi \text{ cm}^2$$

Example 12

Find, correct to two decimal places, the area of each shaded region.

a

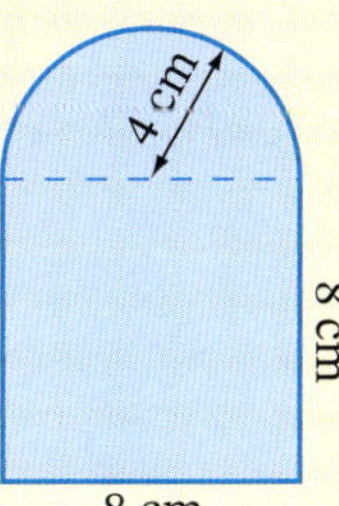

b

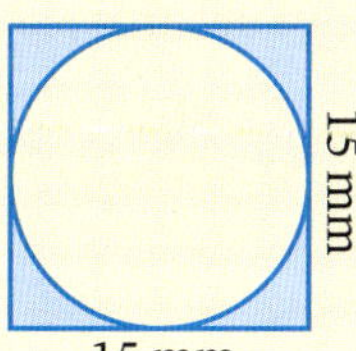

SOLUTION

a Area = area of square + area of semicircle

$$= s^2 + \frac{1}{2} \times \pi r^2$$
$$= 8^2 + \frac{1}{2} \times \pi \times 4^2$$
$$= 89.132741\ldots$$
$$\approx 89.13 \text{ cm}^2$$

b Shaded area = area of square − area of circle

$$= s^2 - \pi r^2$$
$$= 15^2 - \pi \times 7.5^2$$
$$= 48.285413\ldots$$
$$\approx 48.29 \text{ mm}^2$$

Radius: $r = \frac{1}{2} 15 = 7.5\text{mm}$

EXERCISE 5.08 ANSWERS ON P. 629

Area of a circle

U F PS R

1 Calculate, correct to one decimal place, the area of each circle.

EXAMPLE 10

a

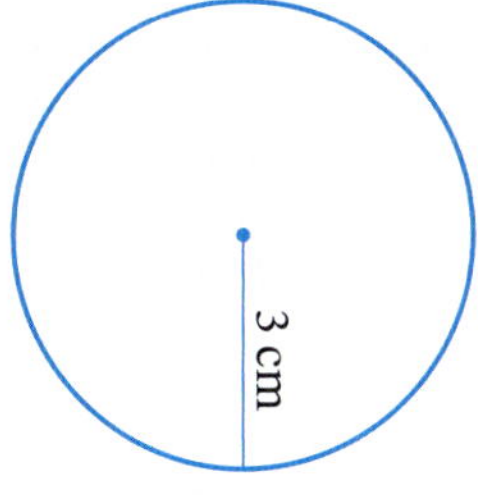

b

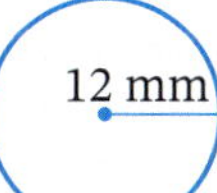

c

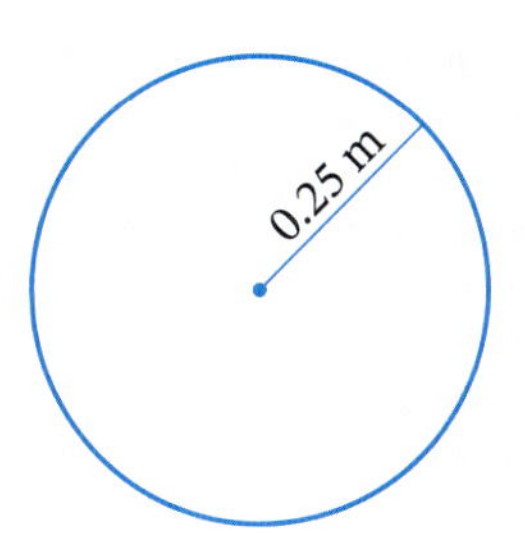

Foundation Standard Complex

d

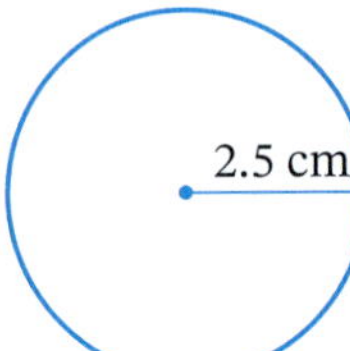

e

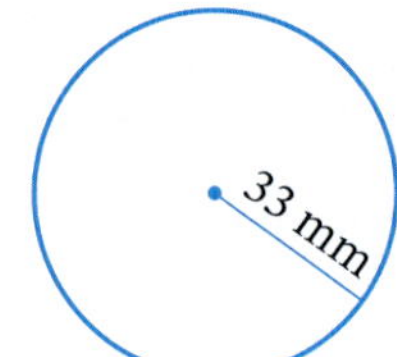

2 Calculate the area of each circle in question **1** in terms of π.

EXAMPLE 11

3 Calculate, correct to two decimal places, the area of each circle.

a

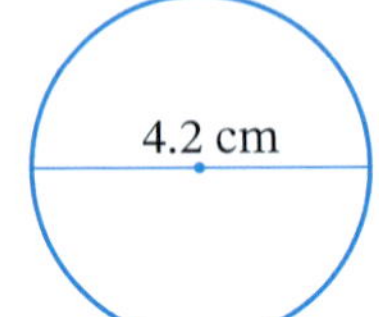

b

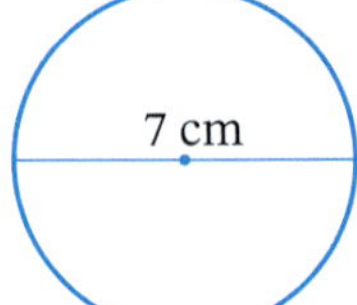

c

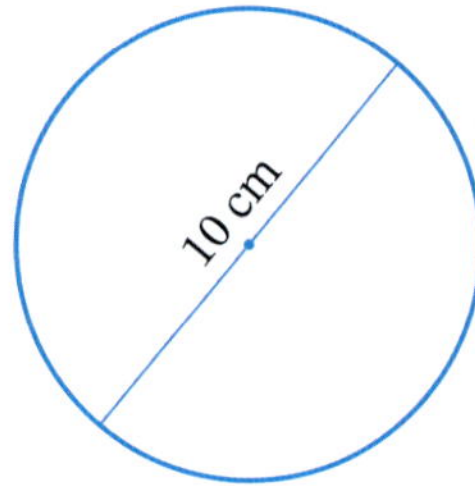

4 Calculate the area of each circle in question **3** in terms of π.

5 A sprinkler on the school playing field sprays water in a circular pattern of radius 13.1 m. Calculate the area being sprayed, to the nearest square metre.

6 Kevlar is a very strong light plastic. What area of it is needed to make a solid bicycle wheel of diameter 685.5 mm? Give your answer in square centimetres, correct to two decimal places.

7 A dinner plate has a radius of 14 cm. Calculate its area in terms of π.

8 What is the area of this circle, correct to one decimal place?
Select the correct answer **A**, **B**, **C** or **D**.

A 14.5 m^2 **B** 66.5 m^2
C 7.2 m^2 **D** 16.6 m^2

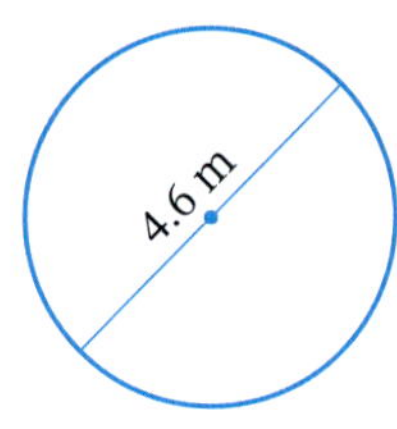

9 The circular floor of a fishpond is to be covered in plastic. Find, correct to one decimal place, the area of plastic needed if the diameter of the pond is 2.8 m.

10 Find the approximate radius of a circular tree trunk whose cross-sectional area is 454.40 cm^2. Give your answer correct to two decimal places.
R

EXAMPLE 12

11 The area of a DVD is 113 cm^2. Which of the following is closest to its diameter?
Select the correct answer **A**, **B**, **C** or **D**.
R

A 12 cm **B** 13 cm **C** 12.5 cm **D** 13.1 cm

Foundation Standard Complex

12 What is the area of this shape closest to? Select the correct answer **A**, **B**, **C** or **D**.

PS R

A 28 110 cm^2 **B** 22 455 cm^2

C 62 039 cm^2 **D** 11 570 cm^2

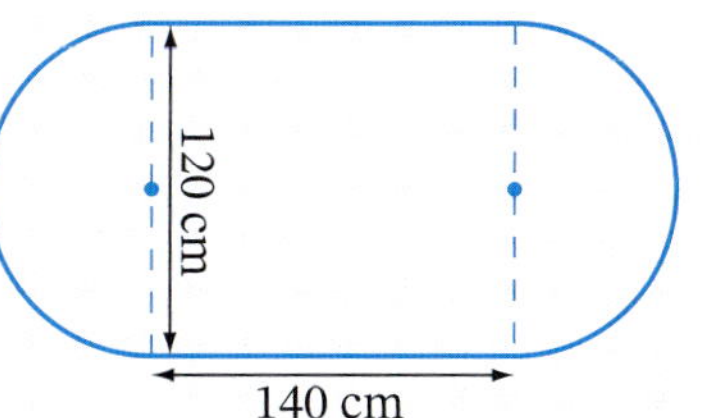

13 Find the area of each shaded region, correct to two decimal places.

a

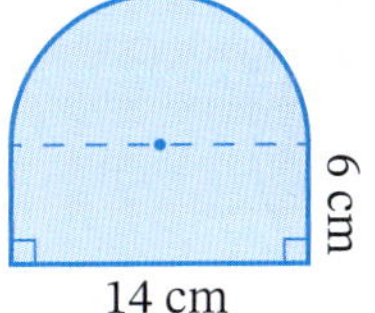

b

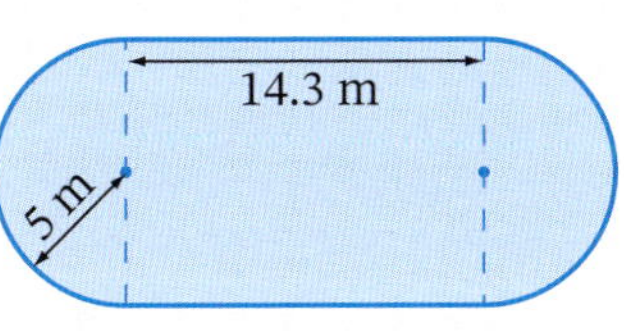

Hint: Combine two semicircles to make a whole circle.

c

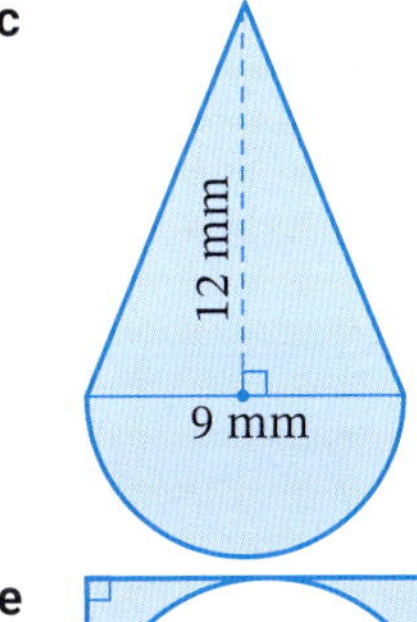

d

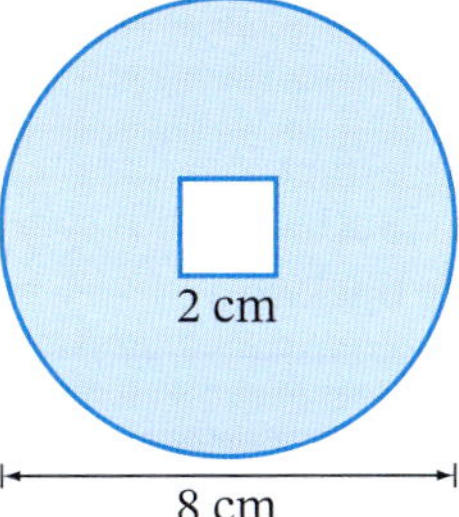

e

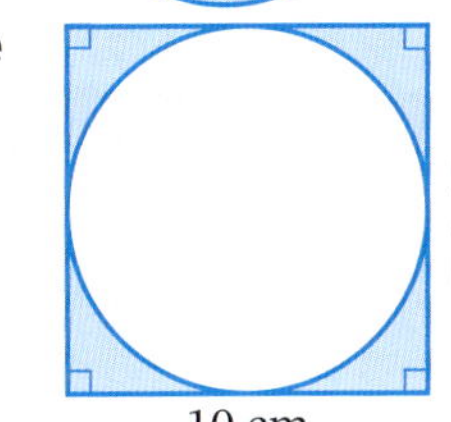

f

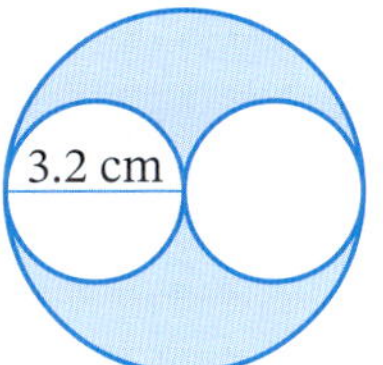

14 Each ring-shape or donut-shape shown below is called an **annulus**, made from two circles that are concentric (with the same centre) but of different size.

R

a

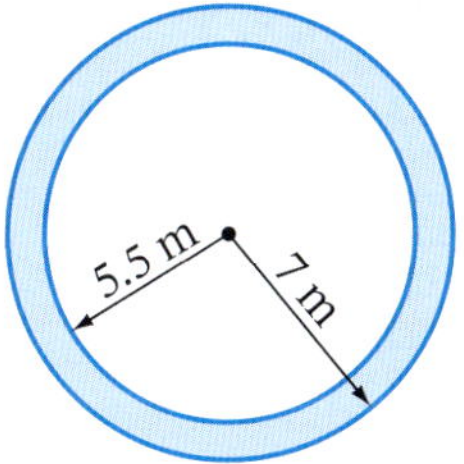

b

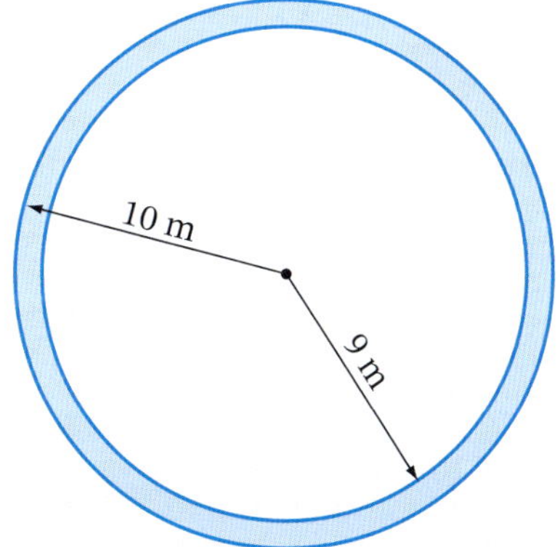

Foundation Standard Complex

For each annulus, find, correct to one decimal place, the area of:

i the larger circle.

ii the smaller circle.

iii the (shaded) annulus.

15 The dimensions of a running track are shown. The ends are circular.

PS

R

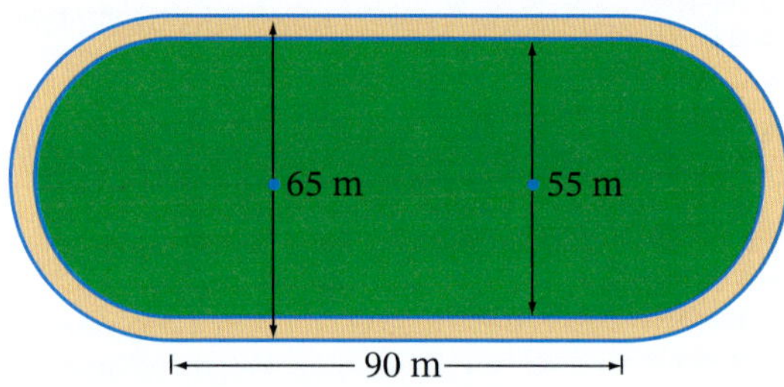

a Find the combined area of the (orange) track and the (green) grassed central area.

b Find the area of the grass.

c Find the area of the track.

d The running track is to be resurfaced with synthetic grass that costs $24.25/m². How much will the new running track cost?

5.09 Extension: Perimeter and area of a sector

EXTENSION

Video Perimeter and area of a sector

Worksheets A page of circular shapes

Applications of area

A sector is a fraction of a circle.

An arc is a fraction of a circle's circumference.

As there are 360° in a circle, the fraction is $\frac{\text{angle at the centre}}{360°}$.

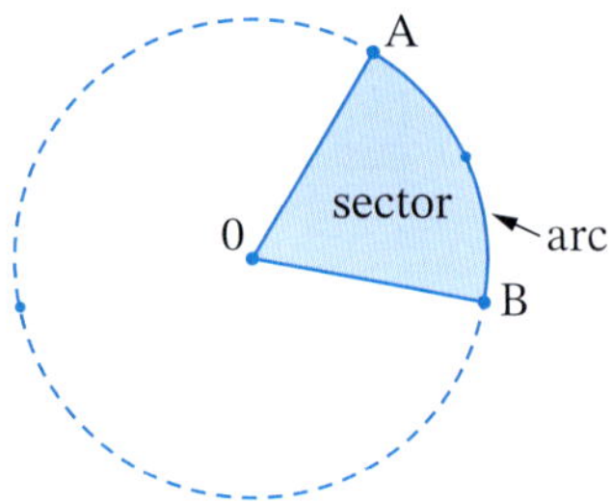

Example 13

Find, correct to two decimal places, the perimeter and area of each shape.

a

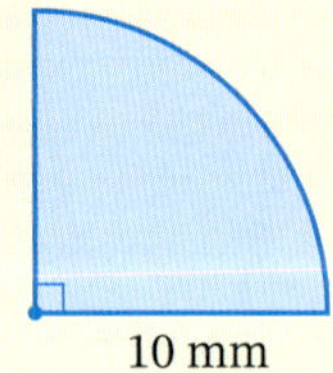

b

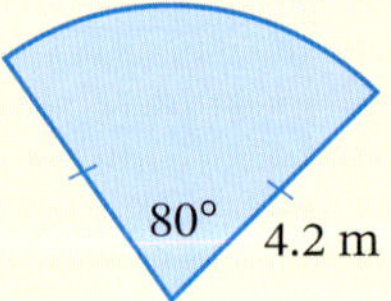

Foundation Standard Complex

SOLUTION

a This is a quadrant as the angle at the centre is 90°.

Fraction of circle is $\frac{90°}{360°} = \frac{1}{4}$

Perimeter = length of arc + radius + radius

Length of arc $= \frac{1}{4} \times$ circumference

$= \frac{1}{4} \times 2\pi r$

$= \frac{1}{4} \times 2 \times \pi \times 10$

$= 15.70796\ldots$

Perimeter of sector = 10 + 10 + 15.70796...

= 35.70796...

≈ 35.71 mm

Area of sector $= \frac{1}{4} \times$ area of circle

$= \frac{1}{4} \times \pi \times 10^2$

$= 78.5398\ldots$

$\approx 78.54\ \text{m}^2$

b Angle at the centre is 80°.

Fraction of circle is $\frac{80°}{360°} = \frac{2}{9}$

Perimeter = length of arc + radius + radius

Length of arc $= \frac{2}{9} \times$ circumference

$= \frac{2}{9} \times 2\pi r$

$= \frac{2}{9} \times 2 \times \pi \times 4.2$

$= 5.8643\ldots$

Perimeter of sector = 4.2 + 4.2 + 5.8643...

= 14.2643...

≈ 14.26 m

Area of sector $= \frac{2}{9} \times$ area of circle

$= \frac{2}{9} \times \pi \times 4.2^2$

$= 12.3150\ldots$

$\approx 12.32\ \text{m}^2$

EXTENSION

Worksheets
Back-to-front problems

Area and perimeter investigations

Puzzle
Carpet talk

5.09

EXERCISE 5.09 ANSWERS ON P. 629

Perimeter and area of a sector

U F PS R

1 What fraction of a whole circle is each sector?

a 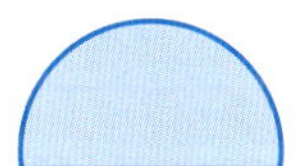

b 

c 

d 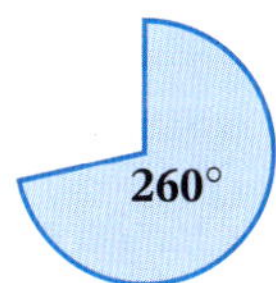

2 Find the perimeter of each sector, correct to one decimal place.

a 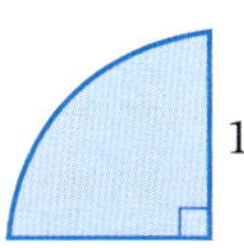

b 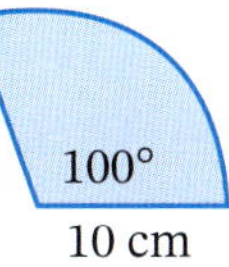

c 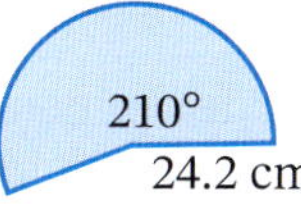

d 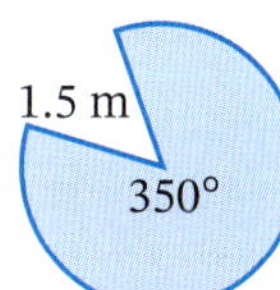

Foundation Standard Complex

EXTENSION

3 Find the area of this shape, correct to the nearest square centimetre. Select the correct answer **A**, **B**, **C** or **D**.

A 40 cm^2 **B** 63 cm^2

C 1200 cm^2 **D** 1885 cm^2

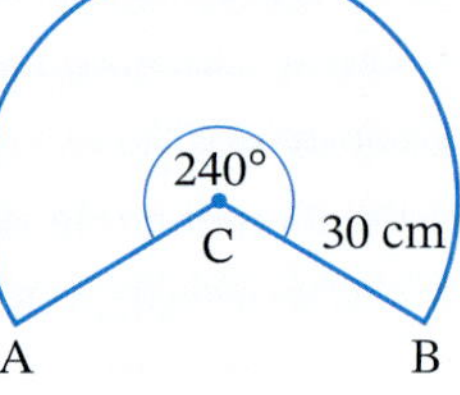

4 Find the area of each sector in question **2** correct to two decimal places.

5 Find, correct to one decimal place, the perimeter and area of each sector.

a 4 cm

b 25 mm

c

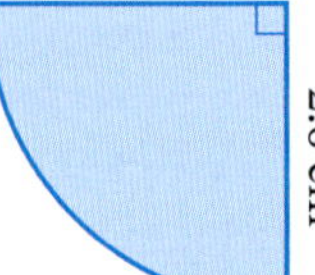

d 7 cm

e 120° 15 mm

f

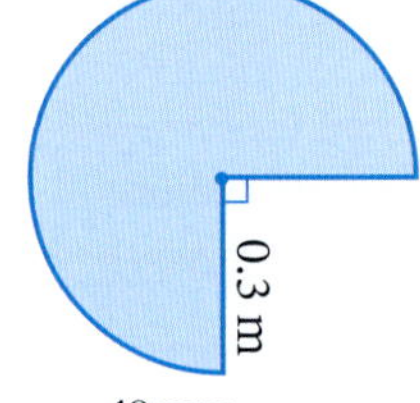

g 35 mm 60°

h

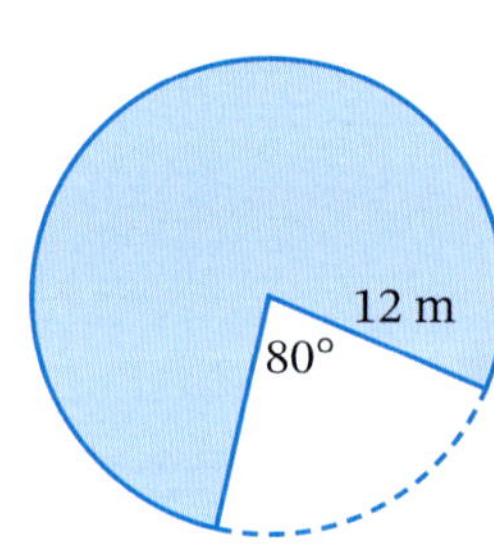

i

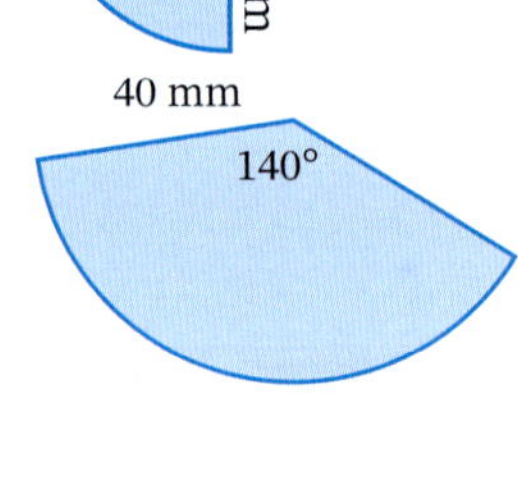

6 Find the perimeter of this shape, correct to the nearest whole square unit. Select the correct answer **A**, **B**, **C** or **D**.

A 64 **B** 63

C 61 **D** 73

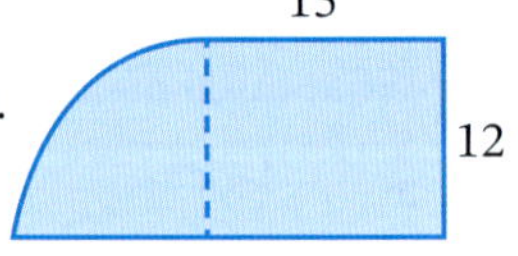

7 The length of the minute hand on a clock is 14 cm.

PS **a** How many degrees does the minute hand rotate through in one minute?

R **b** Find, correct to two decimal places, the area swept by the minute hand in one minute.

8 Find the area of the shaded section, correct to two decimal places.

R

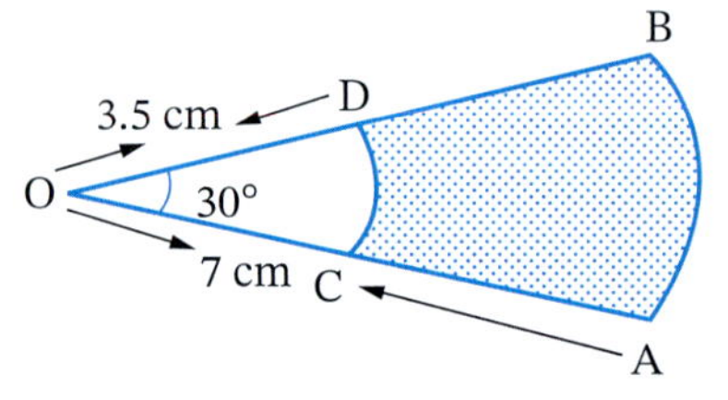

☐ Foundation ○ Standard ⬡ Complex

Metric units of volume

5.10

The **volume** of a solid is the amount of space it takes up. It is measured in **cubic units**.

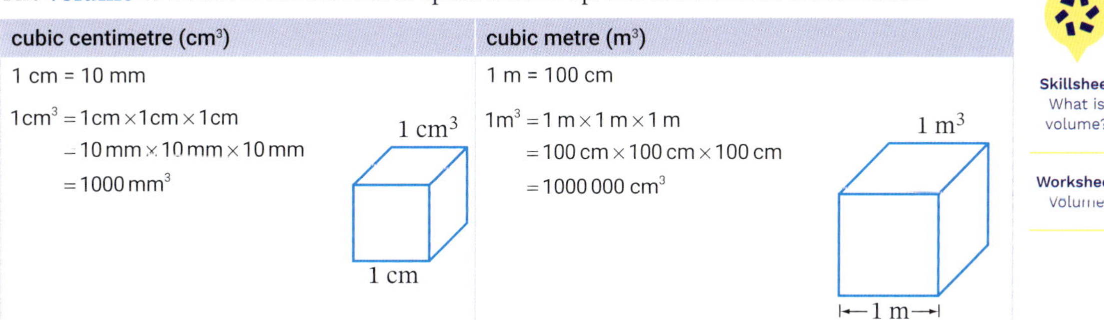

cubic centimetre (cm^3)	cubic metre (m^3)
1 cm = 10 mm	1 m = 100 cm
$1\text{cm}^3 = 1\text{cm} \times 1\text{cm} \times 1\text{cm}$ $= 10\text{mm} \times 10\text{mm} \times 10\text{mm}$ $= 1000\text{mm}^3$	$1\text{m}^3 = 1\text{m} \times 1\text{m} \times 1\text{m}$ $= 100\text{cm} \times 100\text{cm} \times 100\text{cm}$ $= 1000\,000\text{cm}^3$

Skillsheet
What is volume?

Worksheet
Volume

A **cubic centimetre** is about the volume of a person's tooth or a medical pill.

A **cubic metre** is about the volume of two washing machines.

Note that, while 1 cm = 10 mm, 1 cm^3 = 1000 mm^3 (triple the number of 0s), and while 1 m = 100 cm, 1 m^3 = 1 000 000 cm^3 (triple the number of 0s).

We can cube length conversion factors to find volume conversion factors.

ⓘ Metric units of volume

1 cm^3 = 1000 mm^3

1 m^3 = 1 000 000 cm^3

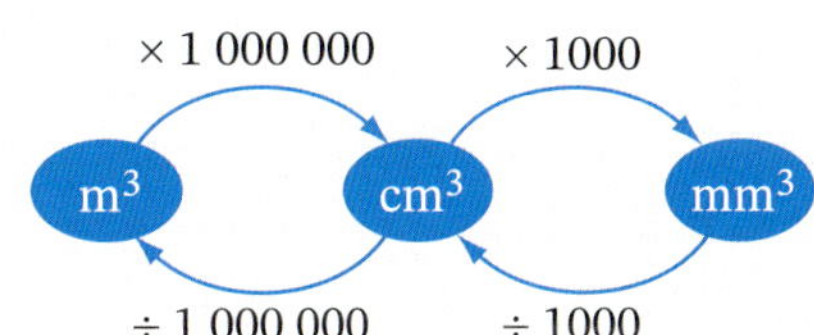

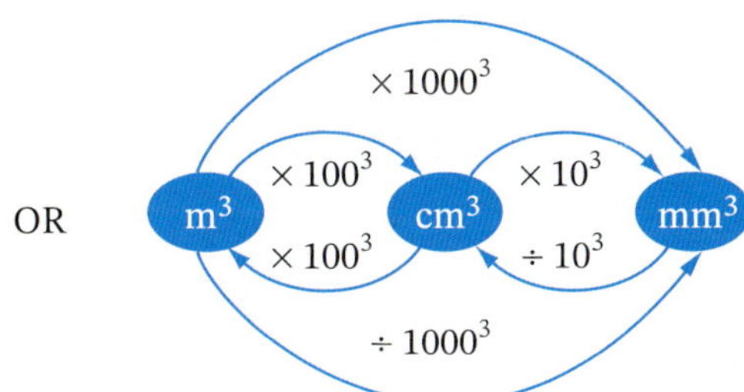

Video
Metric units of volume

Example 14

Convert:

a 36 cm^3 to mm^3 **b** 84 000 000 cm^3 to m^3

SOLUTION

a $36\text{ cm}^3 = 36 \times 1000\text{ mm}^3$

$= 36\,000\text{ mm}^3$

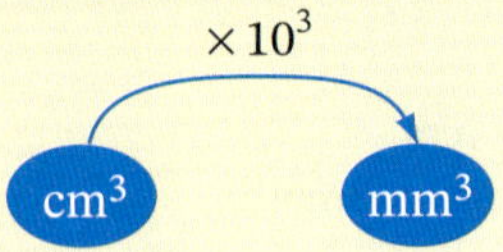

b $84\,000\,000\text{ cm}^3 = 84\,000\,000 \div 1\,000\,000\text{ m}^3$

$= 84\text{ m}^3$

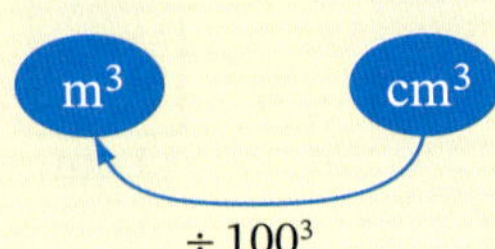

EXERCISE 5.10 ANSWERS ON P. 629

Metric units of volume

U F R C

1 What unit of volume should you use to measure the volume of:

R C

a a bedroom? **b** a backpack?

c a mobile phone? **d** a concert hall?

e your body? **f** an aeroplane?

2 A swimming pool has a volume of 38 m^3. How many cubic centimetres is this pool? Select the correct answer **A**, **B**, **C** or **D**.

R C

A 380 000 **B** 3 800 000

C 38 000 000 **D** 380 000 000

3 Copy and complete each conversion.

R C

a 5000 cm^3 = ________ mm^3 **b** 1.6 m^3 = ________ cm^3

c 2 cm^3 = ________ mm^3 **d** 4 m^3 = ________ mm^3

e 6000 cm^3 = ________ m^3 **f** 8.2 m^3 = ________ cm^3

g 0.007 m^3 = ________ mm^3 **h** 9 600 000 mm^3 = ________ m^3

i 4000 mm^3 = ________ cm^3 **j** 160 000 cm^3 = ________ m^3

k 250 mm^3 = ________ cm^3 **l** 12 cm^3 =________ mm^3

m 0.18 m^3 = ________ cm^3 **n** 200 000 cm^3 =________ m^3

4 The volume of a chest of drawers is 272 000 cm^3. Convert this to cubic metres.

R C

☐ Foundation ○ Standard ○ Complex

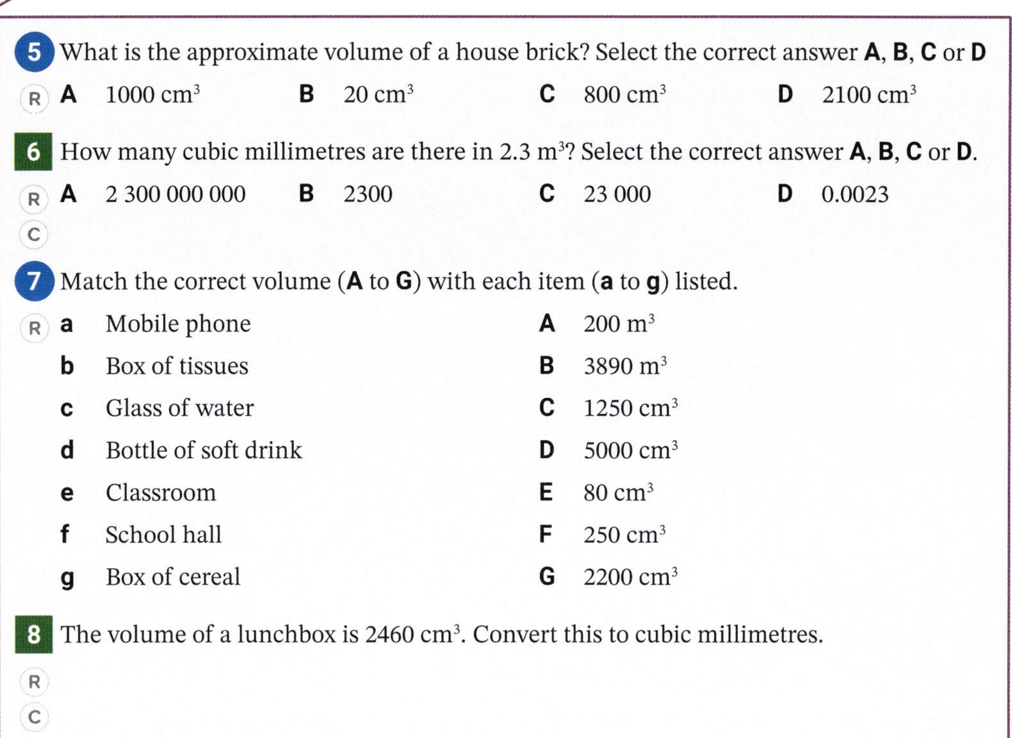

5 What is the approximate volume of a house brick? Select the correct answer **A**, **B**, **C** or **D**

R **A** 1000 cm^3 **B** 20 cm^3 **C** 800 cm^3 **D** 2100 cm^3

6 How many cubic millimetres are there in 2.3 m^3? Select the correct answer **A**, **B**, **C** or **D**.

R C **A** 2 300 000 000 **B** 2300 **C** 23 000 **D** 0.0023

7 Match the correct volume (**A** to **G**) with each item (**a** to **g**) listed.

R

a	Mobile phone	**A**	200 m^3
b	Box of tissues	**B**	3890 m^3
c	Glass of water	**C**	1250 cm^3
d	Bottle of soft drink	**D**	5000 cm^3
e	Classroom	**E**	80 cm^3
f	School hall	**F**	250 cm^3
g	Box of cereal	**G**	2200 cm^3

8 The volume of a lunchbox is 2460 cm^3. Convert this to cubic millimetres.

R C

Volume of a prism 5.11

A **cross-section** of a solid is a 'slice' of the solid, cut *across* it, parallel to its end faces, rather than along it.

If a solid has the same (uniform) cross-section along its length, and each cross-section is a **polygon** (with straight sides, not rounded), then the solid is called a **prism**. Here are some examples of prisms:

Skillsheets
Solid shapes

What is volume?

Worksheets
Volume

Measuring shapes review

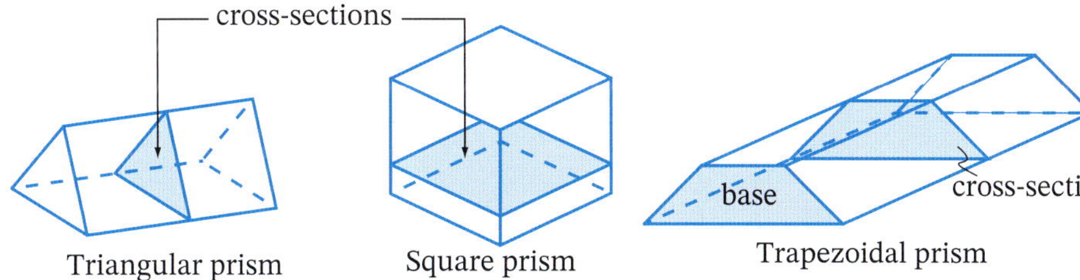

Prisms take their names from their cross-section. For example, the prism shown above is called a **trapezoidal prism** because its cross-sections are all trapeziums.

Because a prism is made up of identical cross-section 'slices' along its length, its volume can be calculated by multiplying the **area of the cross-section** by its **perpendicular height** (the length or depth of the prism).

ⓘ Volume of a prism

Volume = area of cross-section × perpendicular height

$V = Ah$

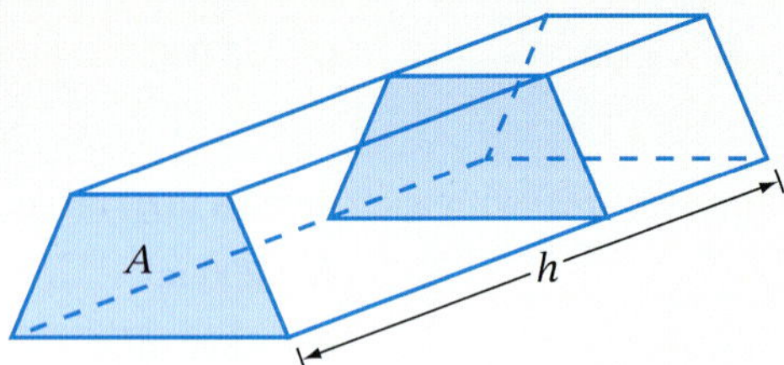

Example 15

a What is the name of this polygon?

b Draw a prism with this polygon as a cross-section.

c What shape are the side faces?

d What is the name of the prism?

SOLUTION

a Pentagon

b

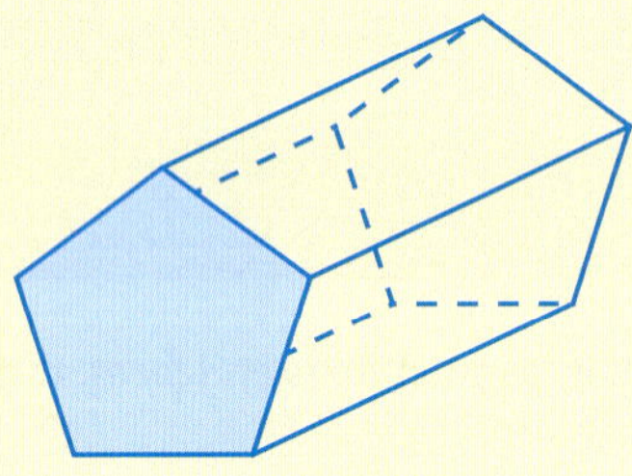

c The side faces are all rectangles.

d This is a pentagonal prism.

Video
Volume of a prism

Example 16

Find the volume of each prism.

a

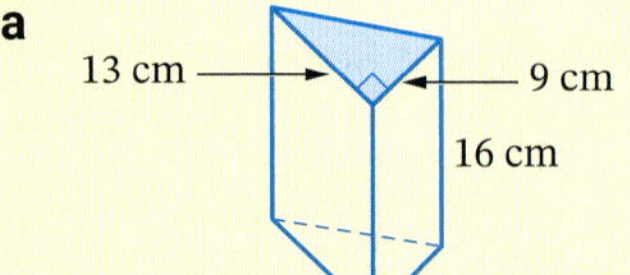

b

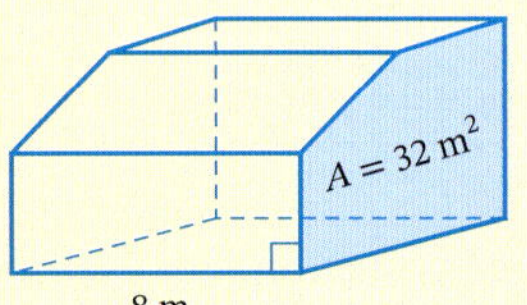

c

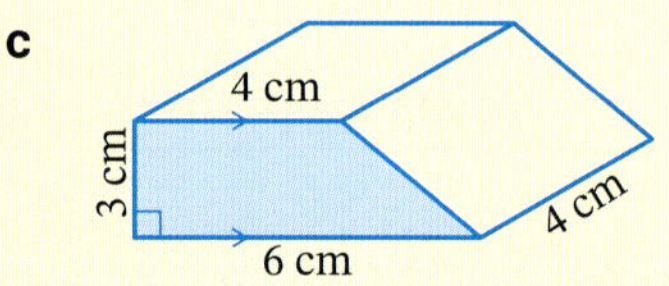

SOLUTION

a $A = \frac{1}{2} \times 9 \times 13$ — Area of shaded triangular cross-section

$= 58.5\text{ cm}^2$

$V = 58.5 \times 16$ — $V = Ah$, where height $h = 16$ cm.

$= 963\text{ cm}^3$

b $A = 32\text{ m}^2$ — Area of shaded cross-section.

$V = 32 \times 8$ — $V = Ah$, where height $h = 8$.

$= 256\text{ m}^3$

c $A = \frac{1}{2} \times (4+6) \times 3$ — Cross-section is a trapezium.

$= 15\text{ cm}^2$

$V = 15 \times 4$ — $V = Ah$, where height $h = 4$.

$= 60\text{ cm}^3$

EXERCISE 5.11 ANSWERS ON P. 629

Volume of a prism

1 State whether each solid is a prism or not.

C **a**

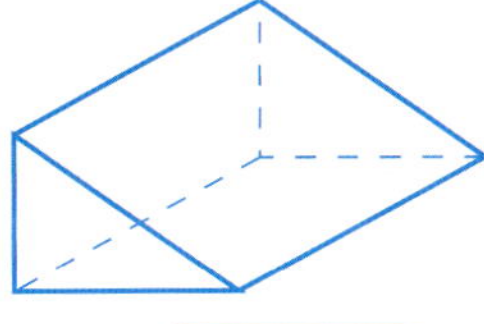

b

c

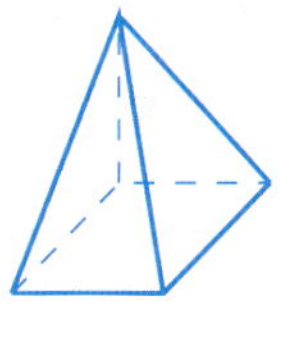

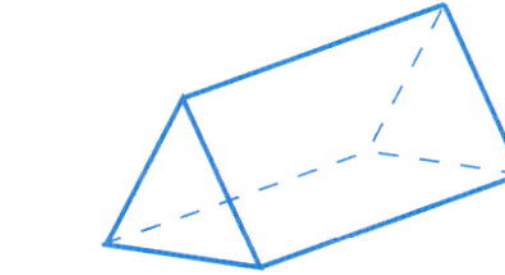

2 For each prism:

i draw its cross-section.

ii write the name of the prism.

a

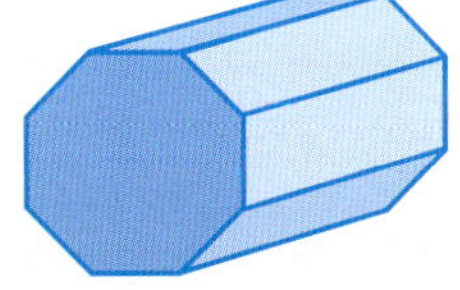

b

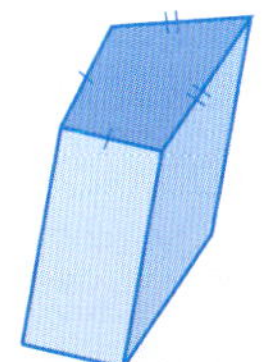

c

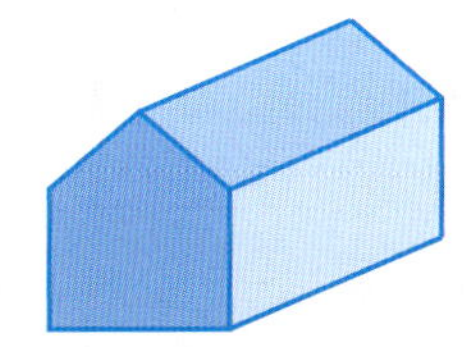

Foundation Standard Complex

3 Draw each prism and shade its base.

- **a** Square prism
- **b** Isosceles triangular prism
- **c** Trapezoidal prism
- **d** Hexagonal prism

EXAMPLE 16

4 Find the volume of each prism.

a

$A = 62$ m^2

3.4 m

b

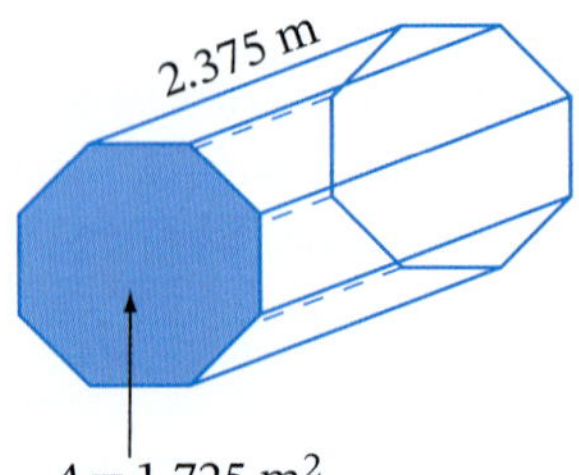

c

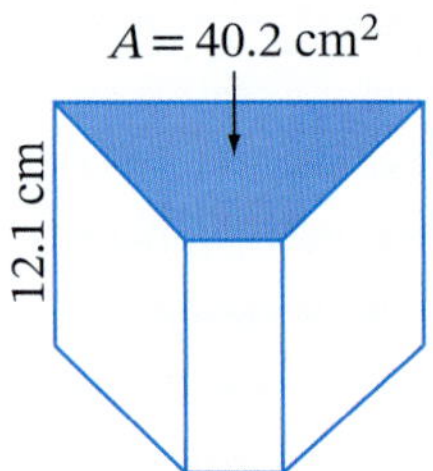

5 The volume of a rectangular prism can be found using the formula $V = lwh$. Use this formula to find the volume of each prism below.

a

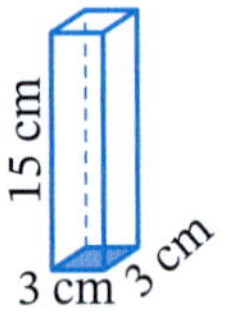

b

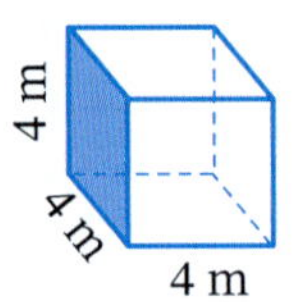

c

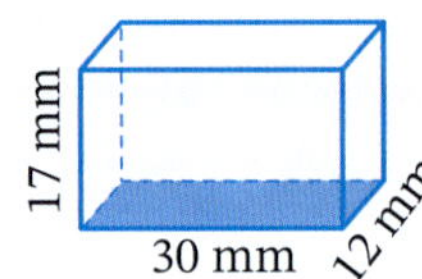

d

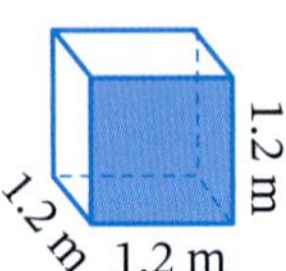

e

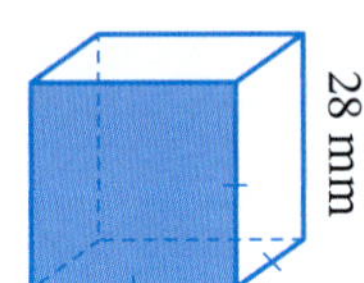

f

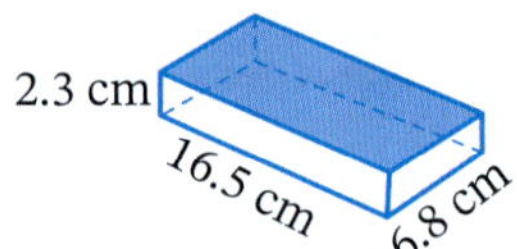

6 Find the volume of each triangular prism.

a

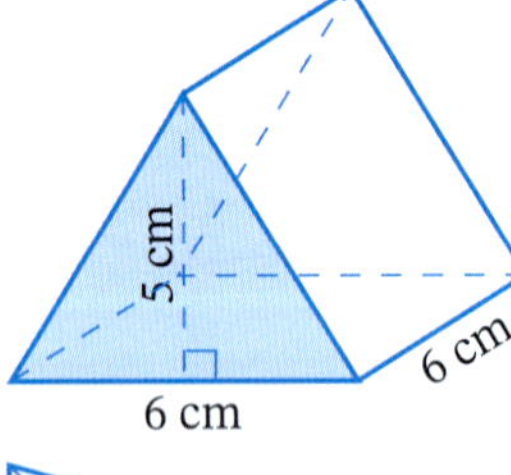

b

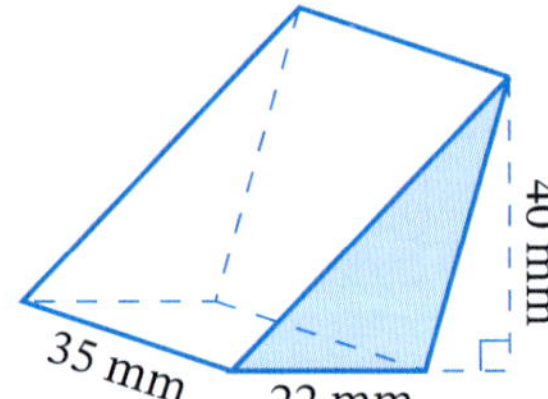

c

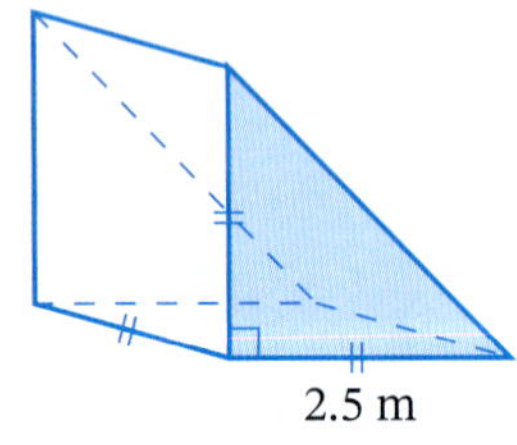

7 This shape is folded along the dotted lines to form a rectangular prism. Find the volume of the box in cubic units. Select the correct answer **A**, **B**, **C** or **D**.

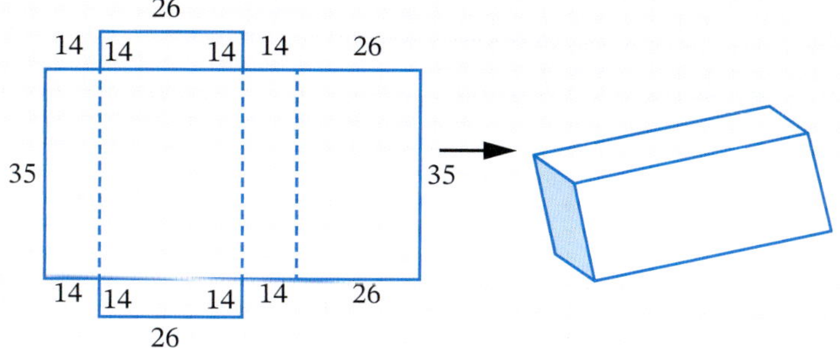

A 23 660 **B** 12 740 **C** 5096 **D** 258

8 Find the volume of each prism.

a

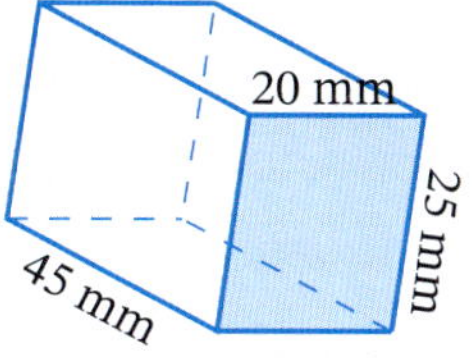

b

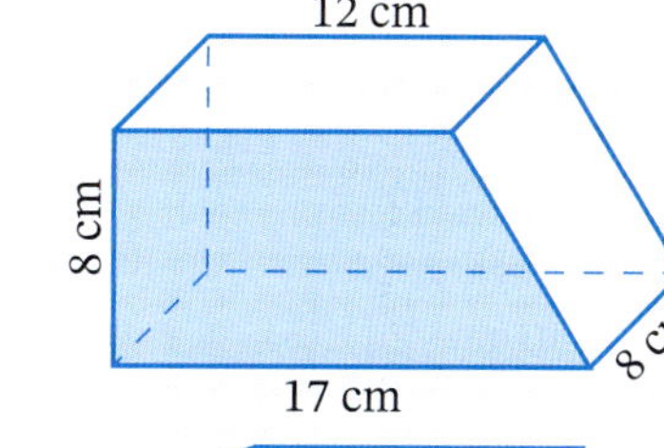

c

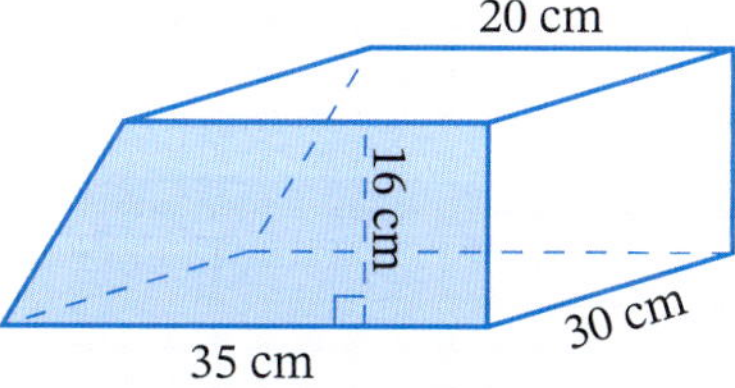

d

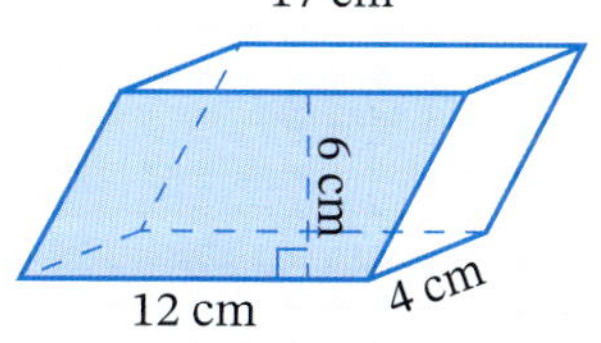

9 Find the volume of this chest of drawers.

Give the answer in m^3:

a by calculating the volume in cm^3, then converting to m^3.

b by converting each length to metres first, then calculating the volume.

Which method is easier?

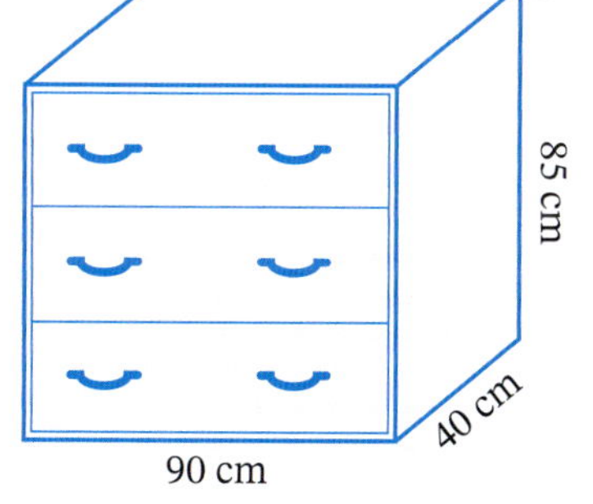

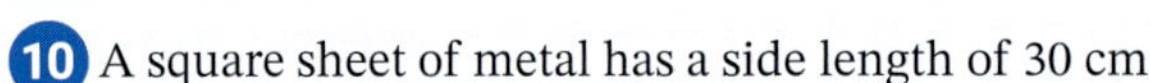

10 A square sheet of metal has a side length of 30 cm. Four identical squares of length 4 cm are cut away from the corners, as shown. Find the volume of the container formed when the remaining shape is folded along the dotted lines. Select the correct answer **A**, **B**, **C** or **D**.

PS R

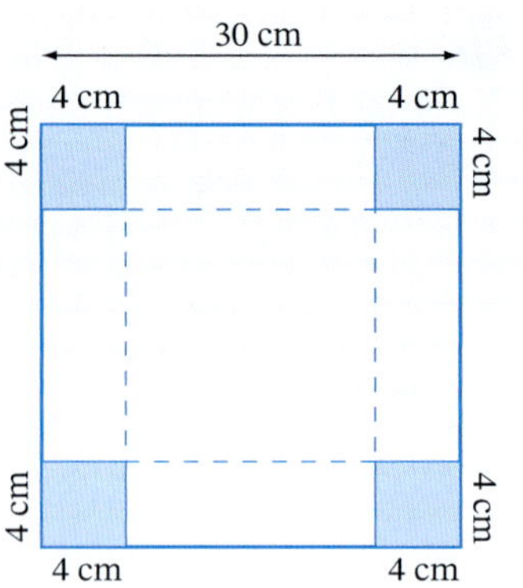

A 1936 cm² **B** 2704 cm²

C 64 cm² **D** 3600 cm²

11 A triangular prism has base length 6 m, height 10 m and volume 150 m³. What is its length?

R

12 A cube has a volume of 343 m³. What is its side length?

R

13 Calculate the volume of each prism.

PS R

a

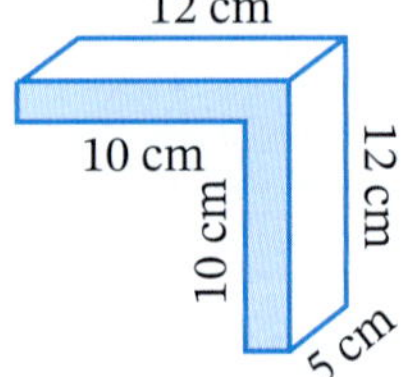

b

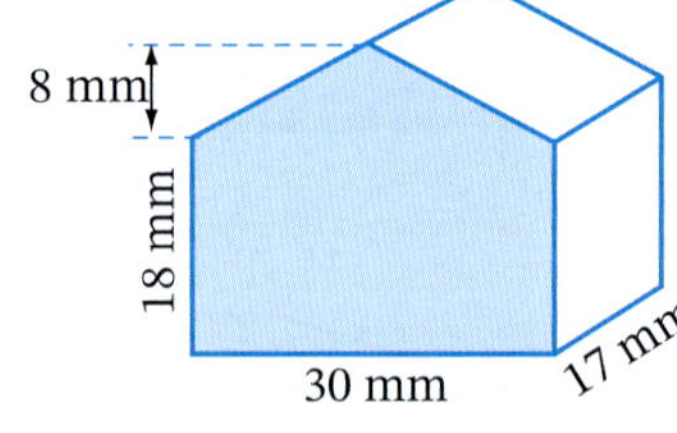

c

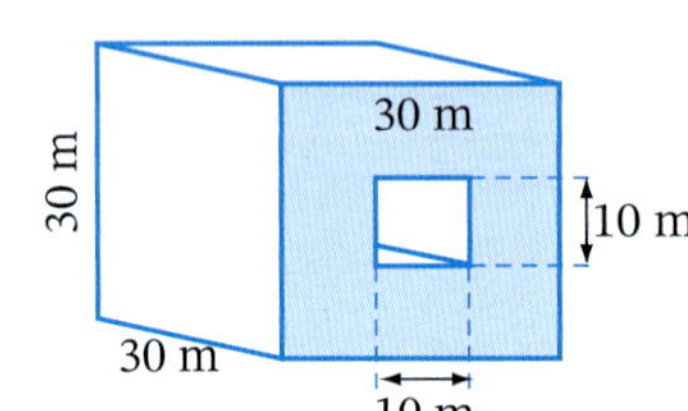

d

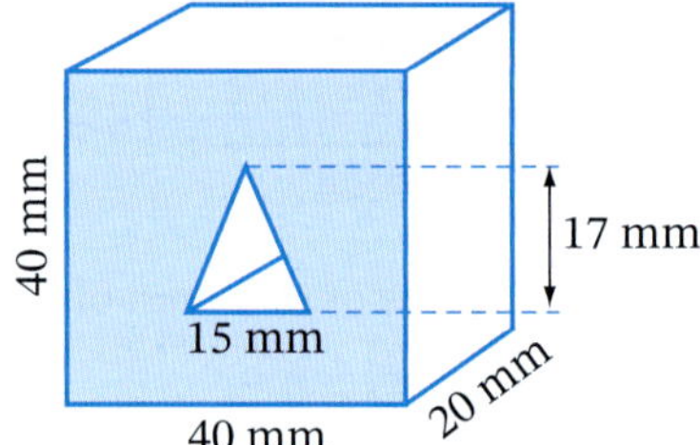

e

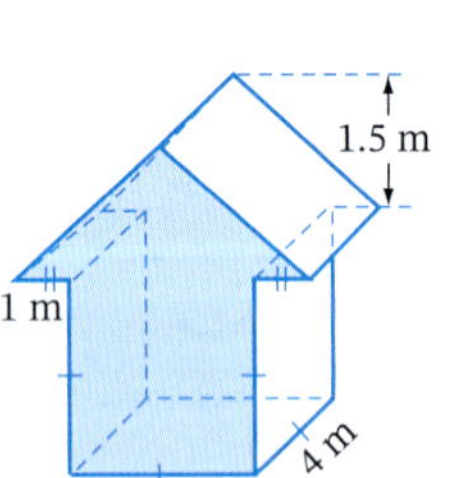

14 A rectangular garden is 12 m long and 4 m wide. It is filled with soil to a depth of 15 cm. Calculate the volume of the soil. Select the correct answer **A**, **B**, **C** or **D**.

PS R

A 0.72 m³ **B** 7.2 m³ **C** 72 m³ **D** 720 m³

15 A triangular prism has a volume of 36 cm³. What could its length, perpendicular height and base length be?

R

☐ Foundation ○ Standard ⬡ Complex

Extension: Volume of a cylinder 5.12

5.12

YEAR 9 EXTENSION

A **cylinder** is like a 'circular prism' because its cross-sections are all identical circles. Because of this, we can also apply the prism formula $V = Ah$ to calculate the volume of a cylinder.

For a circle, $A = \pi r^2$, so:

Volume $= Ah = \pi r^2 \times h = \pi r^2 h$

Video Volumes of prisms and cylinders

ⓘ Volume of a cylinder

$V = \pi \times (\text{radius})^2 \times \text{perpendicular height}$

$V = \pi r^2 h$

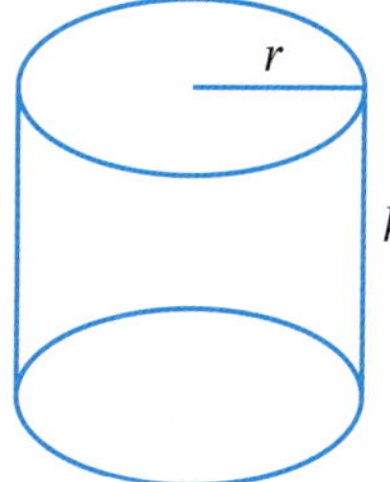

Example 17

Find the volume of this cylinder, correct to one decimal place.

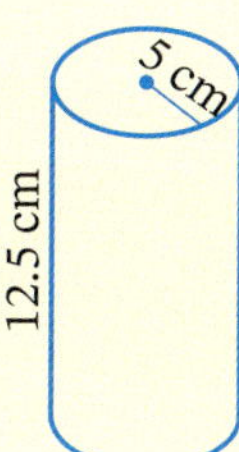

SOLUTION

$V = \pi \times 5^2 \times 12.5$ $V = \pi r^2 h$

$= 981.7477\ldots$

$\approx 981.7 \text{ cm}^3$

YEAR 9 EXTENSION

Example 18

Calculate the volume of concrete needed to make this pipe, correct to two decimal places.

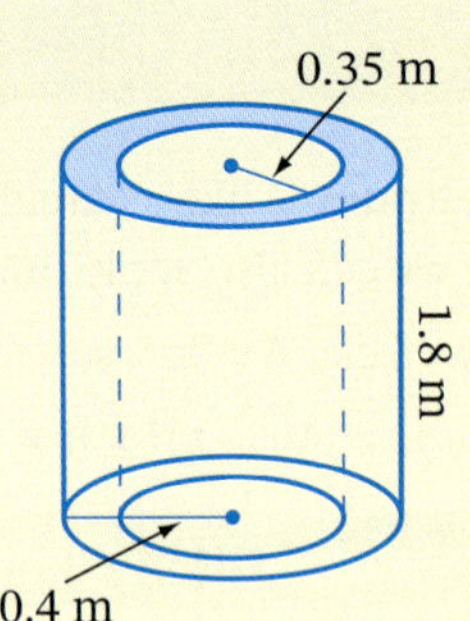

SOLUTION

Volume of concrete = volume of big cylinder − volume of small cylinder (hole)

$$= \pi \times 0.4^2 \times 1.8 - \pi \times 0.35^2 \times 1.8$$
$$= 0.212057\ldots$$
$$\approx 0.21\ \text{m}^3$$

EXERCISE 5.12 ANSWERS ON P. 630

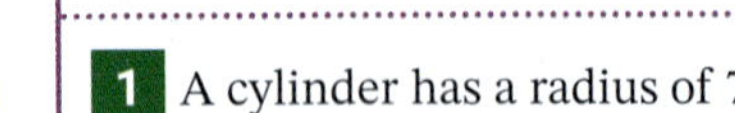

Volume of a cylinder

U F PS R

EXAMPLE 17

1 A cylinder has a radius of 7 m and a height of 12 m. Find correct to two decimal places:

a the area of its circular base.

b the volume of the cylinder.

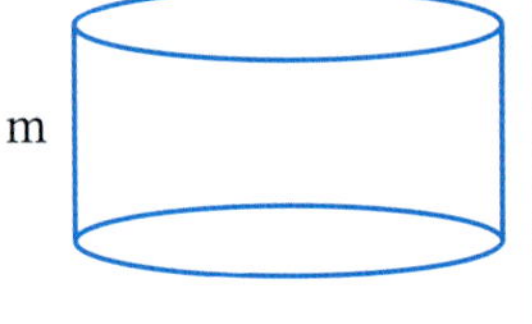

2 What is the volume of this cylinder? Select the closest answer **A**, **B**, **C** or **D**.

A $2\ \text{m}^3$

B $12\ \text{m}^3$

C $3\ \text{m}^3$

D $6\ \text{m}^3$

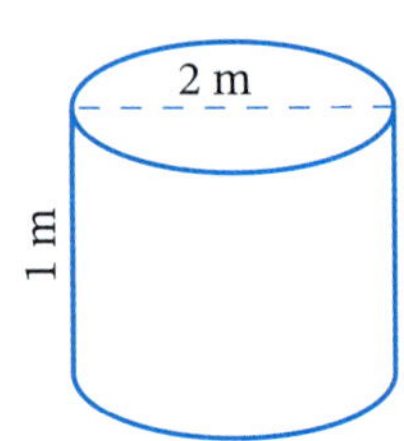

3 Find, correct to one decimal place, the volume of each cylinder.

a

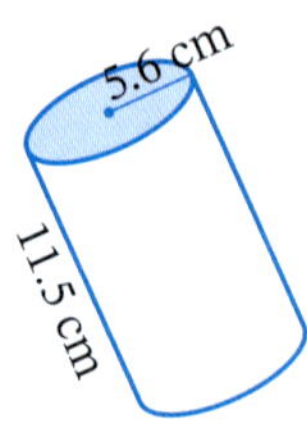

b

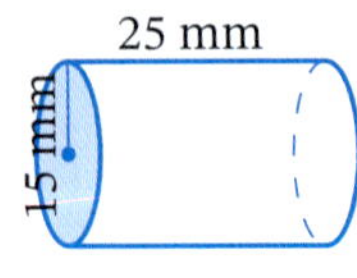

c

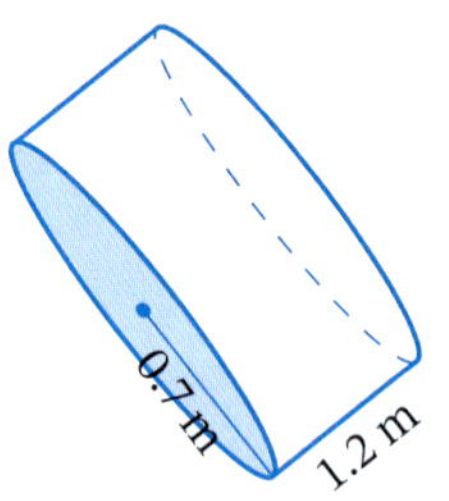

Foundation Standard Complex

9780170465601

d

e

24.8 cm
36.4 cm

f

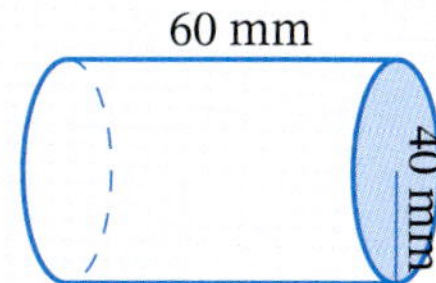

g

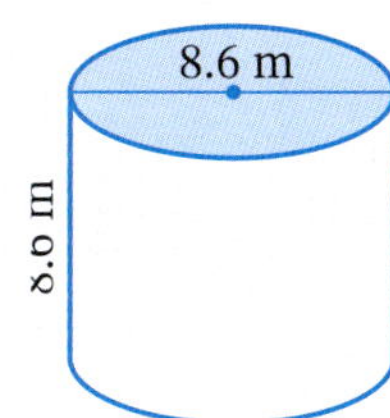

EXAMPLE 18

4 This metal pipe has an inner diameter of 8.5 cm and an outer diameter of 9.5 cm.

PS R

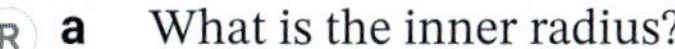

a What is the inner radius?

b What is the outer radius?

c Calculate, correct to two decimal places, the volume of metal needed to make the pipe.

5 A manufacturer is experimenting with a new size of can. Which of these options provides the larger volume?

PS R

A A can of radius 3.95 cm and height 11.8 cm.

B A can of diameter 8.1 cm and height 15 cm.

6 A cylinder has a radius of 5 cm and a volume of 628.32 cm^3. Find its height, correct to the nearest centimetre.

R

7 Find the volume of each solid, correct to one decimal place.

PS R

a

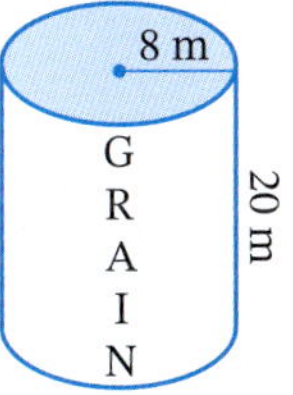

b

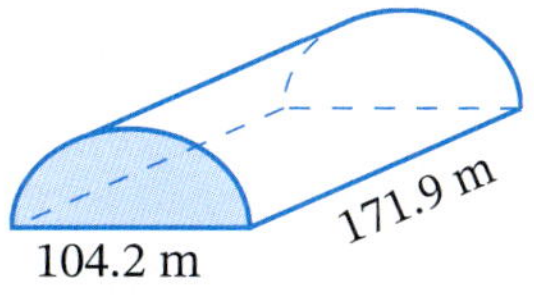

c

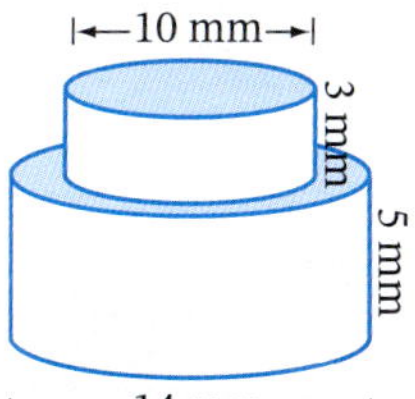

d

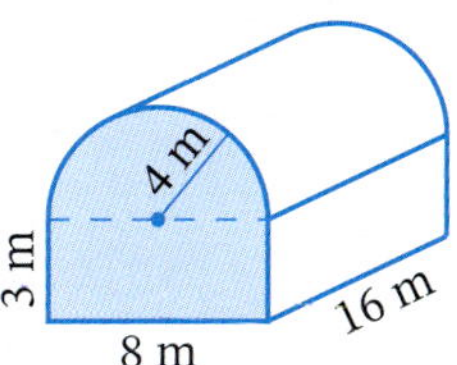

e

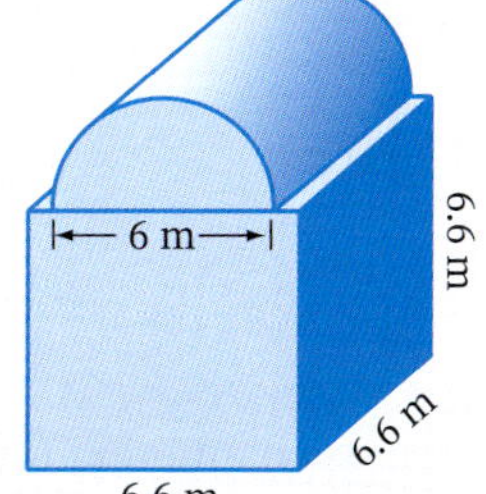

YEAR 9 EXTENSION

f

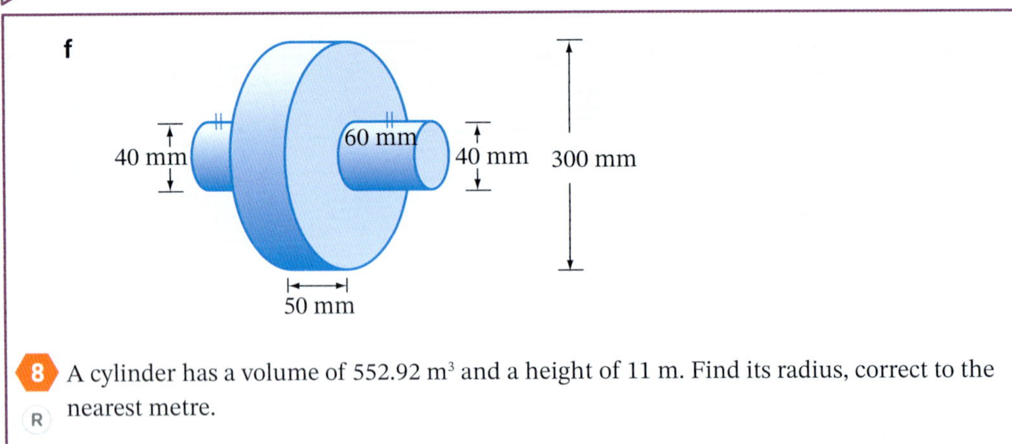

8 A cylinder has a volume of 552.92 m^3 and a height of 11 m. Find its radius, correct to the nearest metre.

R

5.13 Volume and capacity

Worksheets
Volume and capacity
Volume and capacity match
Capacity

Puzzle
Capacity puzzle

Capacity is the amount of fluid (liquid or gas) in a container.

The metric units of capacity are the litre (L), the millilitre (mL), the kilolitre (kL) and the megalitre (ML).

- A large drop of water is about 1 mL.
- A teaspoon holds about 5 mL.
- A tall standard carton of milk holds 1 L.
- A small rainwater tank holds about 1 kL.
- It takes 40 minutes to pump 1 kL of water out of a garden hose.
- Half an Olympic-sized swimming pool holds about 1 ML.

A can of drink holds 375 mL.

A can of paint holds 4 L.

Shutterstock.com/Oleksiy Mark
Shutterstock.com/Hintau Aliaksei

Metric units of capacity

Video
Volume and capacity

1 L = 1000 mL

1 kL = 1000 L

1 ML = 1000 kL = 1 000 000 L

This diagram shows how to convert between different units of capacity.

Foundation Standard Complex

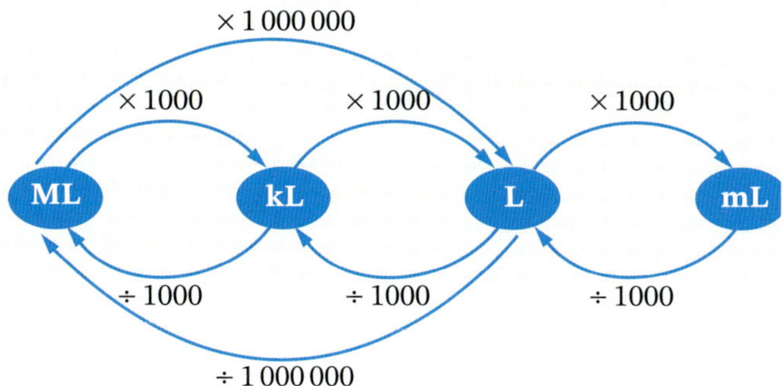

It is also useful to know the relationship between **volume and capacity**.

ⓘ Volume and capacity

1 cm^3 contains 1 mL

1 m^3 contains 1000 L or 1 kL

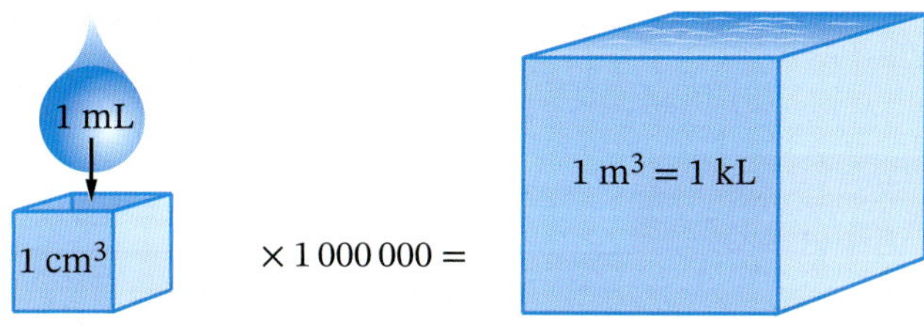

This means that a cubic centimetre can hold 1 mL of liquid, while a cubic metre can hold 1000 L of liquid.

Example 19

Convert:

a 3400 mL to L

b 2.9 m^3 to L

SOLUTION

a $3400\text{ mL} = 3400 \div 1000\text{ L}$

$= 3.4\text{ L}$

b $2.9\text{ m}^3 = 2.9\text{ kL}$

$= 2.9 \times 1000\text{ L}$

$= 2900\text{ L}$

Example 20

Find the capacity of this container in litres.

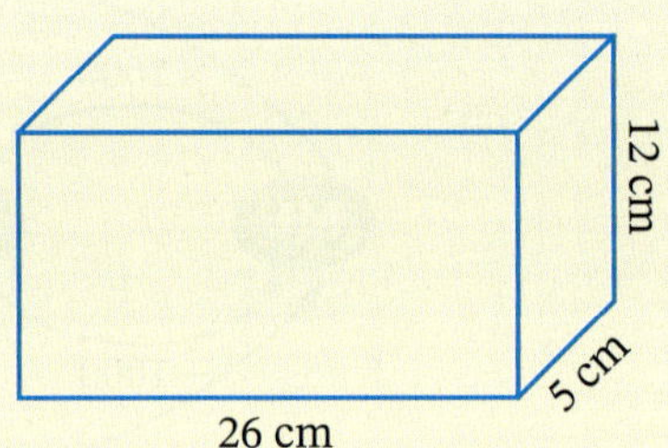

SOLUTION

$V = 32 \times 8 \times 15$ *$V = lwh$*

$\quad = 3840 \text{ cm}^3$ $1 \text{ cm}^3 = 1 \text{ mL}$

Capacity = 3840 mL

$\quad = 3840 \div 1000 \text{ L}$

$\quad = 3.84 \text{ L}$

Capacity = 3840 mL or 3.84 L

EXERCISE 5.13 ANSWERS ON P. 630

Volume and capacity

1 State what unit of capacity you would use when measuring:

R C

- **a** a glass of water
- **b** a dam
- **c** a car's petrol tank
- **d** a bottle of medicine
- **e** an office water cooler
- **f** an Olympic swimming pool

EXAMPLE 19

2 Copy and complete each conversion.

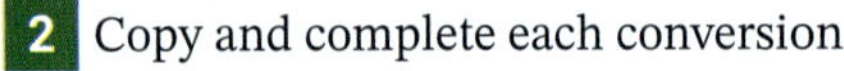

R C

- **a** 5000 mL = ________ L
- **b** 3.4 kL = ________ L
- **c** 1.6 L = _____ mL
- **d** 4000 cm³ = _____ L
- **e** 2980 kL = _____ ML
- **f** 7.1 ML = _____ L
- **g** 875 L = _____ kL
- **h** 8.2 m³ = _____ L
- **i** 0.8 ML = _____ kL
- **j** 1850 mL = _____ L
- **k** 5.4 kL = _____ L
- **l** 900 000 L = ___ m³
- **m** 6 kL = _____ L
- **n** 3500 mm³ = _____ mL
- **o** 1.2 m³ = _____ mL

3 What is the capacity of a regular bottle of cough medicine? Select the closest answer **A**, **B**, **C** or **D**.

R

A 200 mL   **B** 500 mL   **C** 1500 mL   **D** 2000 mL

4 The internal dimensions of a refrigerator are 150 cm (height), 60 cm (width) and 40 cm (depth). What is the capacity of the refrigerator? Select the closest answer **A**, **B**, **C** or **D**.

R C

A 350 L   **B** 400 L   **C** 250 L   **D** 300L

▷

☐ Foundation ○ Standard ○ Complex

9780170465601

5 PS R C A rectangular fish tank is 90 cm long and 30 cm wide, and is filled to a depth of 45 cm. Find the volume of water in:

a cubic centimetre **b** litres

6 PS R C A swimming pool is the shape of a rectangular prism 50 m × 18 m × 2 m. Find its volume in litres.

7 PS R C A tank in the shape of a rectangular prism has dimensions as shown.

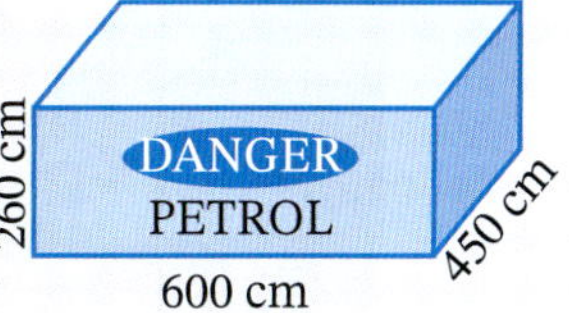

a Find the volume of the tank in cm^3.

b How many litres of petrol would fit in the tank?

8 PS R Samantha is inviting 18 friends to a party. She calculates that each person will drink 1300 mL of soft drink.

a How many litres of soft drink must she buy?

b If Samantha intends to buy large 2.5 L bottles of drink, how many bottles must she buy?

9 PS R C A gardener orders a cubic metre of topsoil. She wants to spread it in her garden to a depth of 10 cm. If her garden is rectangular, what might be its dimensions?

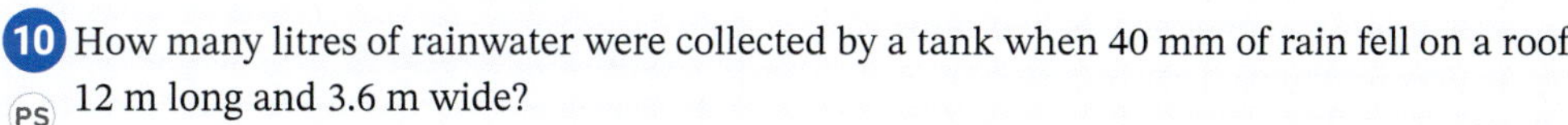

10 PS R C How many litres of rainwater were collected by a tank when 40 mm of rain fell on a roof 12 m long and 3.6 m wide?

11 PS R C A plastic road block barrier is shown. It is filled with water to weigh it down.

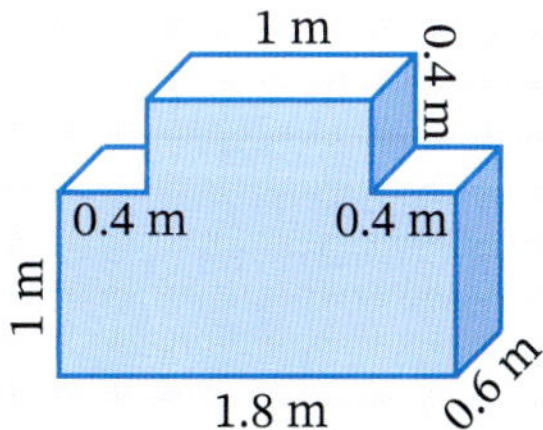

a Find the volume of the barrier.

b How many litres of water would it take to fill the barrier?

12 PS R C A can of drink has a radius of 3 cm and a height of 15 cm. Find its volume, correct to the nearest millilitre.

YEAR 9 EXTENSION

Foundation Standard Complex

13 The tank on a fuel tanker is in the shape of a cylinder 6 m long, with a diameter of 2 m.

PS R C

a Find, correct to three decimal places, the volume of the tank in cubic metres.

b How many litres of fuel can this tank hold?

Shutterstock.com/Krivosheev Vitaly

14 A cylindrical can holds 2 L of pesticide. What could its radius and height be?

PS R

15 A tap leaks 10 mL of water every 50 seconds. How many litres of water will be lost in:

PS R

a 1 hour? **b** 1 day?

16 The local school has installed a new six lane swimming pool in the grounds. The concrete has had time to set and the pool can now be filled. The school will be ordering 6000 L water trucks to fill the pool. How many trucks will they need to order if the pool has the dimensions shown in the diagram?

PS R

15 m
25 m
1 m
1.4 m

Foundation Standard Complex

DID YOU KNOW?

Water, water, everywhere

To help you better understand the size of a litre and a kilolitre, here are some examples of water use in and around the home:

- Washing your hands/face 5 L
- Brushing your teeth (tap running) 5 L
- Brushing your teeth (tap not running) 1 L
- Cooking and making coffee/tea 8 L/day
- Flushing the toilet 9–13 L
- Flushing the toilet (half flush) 4.5–6 L
- Household tap 18 L/minute
- Washing the dishes (hand) 18 L
- Washing the dishes (dishwasher) 25 L/cycle
- Bath 85–150 L
- Shower (8 minutes) 80–120 L
- Washing machine (front loading) 120 L/cycle
- Washing machine (top loading) 180 L/cycle
- Washing the car (with hose) 100–300 L
- Garden sprinkler 1–1.5 kL/hour
- Garden hose 1.8 kL/hour
- Swimming pool (backyard) 20–55 kL
- 50 m swimming pool 1870 kL

On average, a four-person house (with garden) uses 936 L of water per day. Half of it is used by outside taps or is flushed down a toilet.

How much water does your household use each day? Find out by asking your parents to show you the water bill.

POWER PLUS ANSWERS ON P. 630

1 Find, correct to one decimal place, the perimeter of each shape.

a

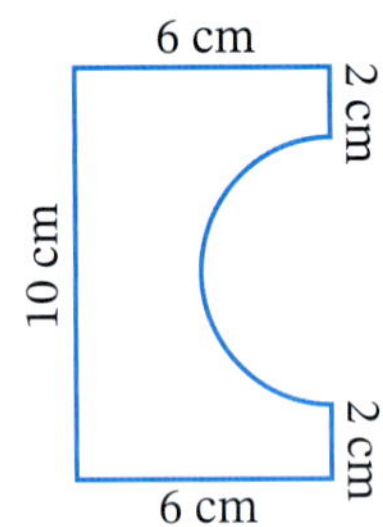

b

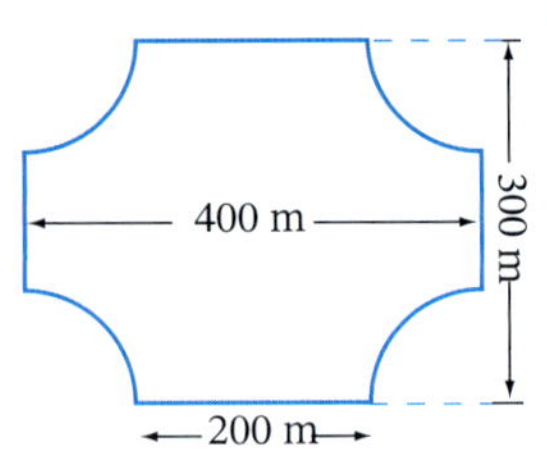

c

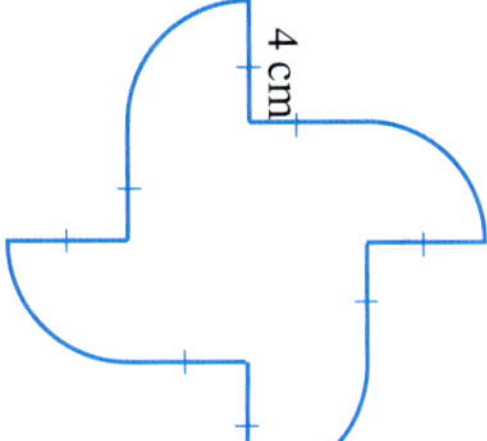

2 This diagram shows how a shotput field is marked out for an athletics event. Calculate, correct to one decimal place:

a the area of the field.

b the total length of the lines used.

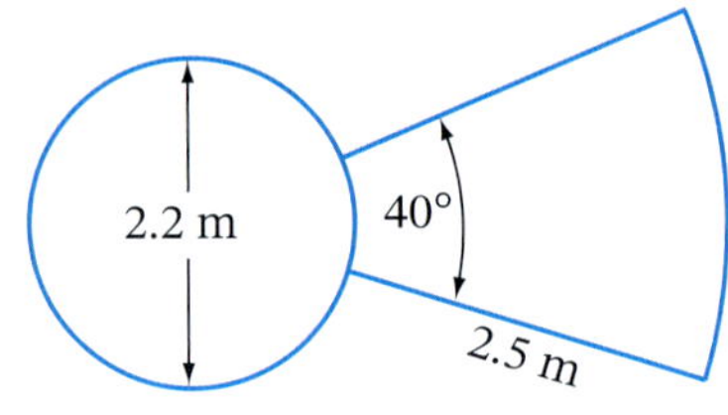

3 Mathsland has introduced a new 30-cent coin, as shown in the first diagram. △*ABC* is an equilateral triangle of length 2 cm.

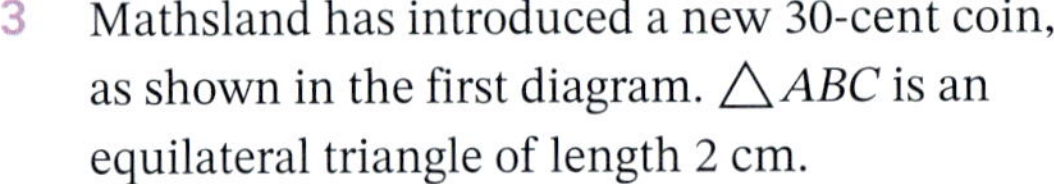

a Find, correct to three decimal places, the area of the sector in the second diagram.

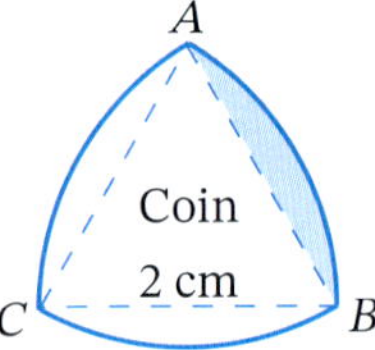

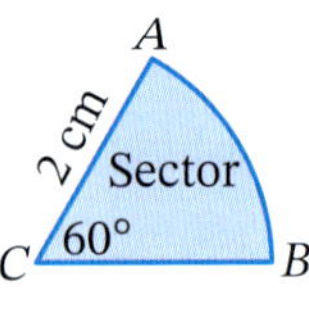

b Use Pythagoras' theorem to prove that the perpendicular height of △*ABC* is cm.

c Calculate the area of △*ABC*, correct to three decimal places.

d Calculate the area of the shaded segment in the first diagram.

e Hence calculate the area of the 30-cent coin, correct to two decimal places.

4 Find, correct to one decimal place, the area of each shape.

a

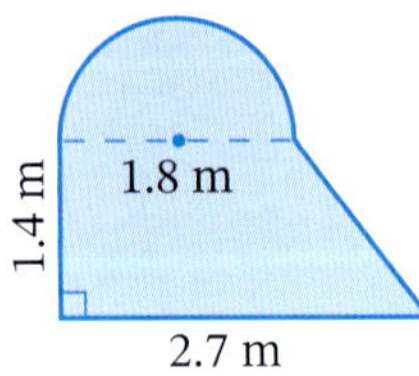

b

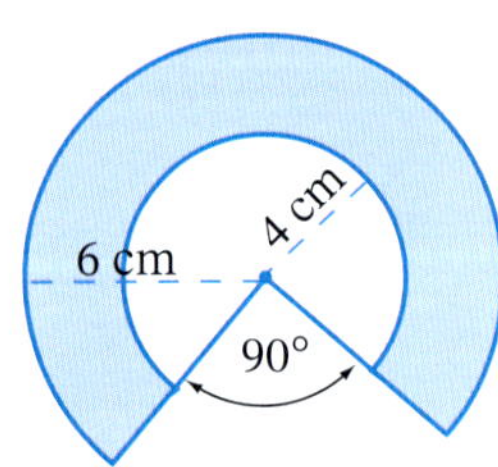

c

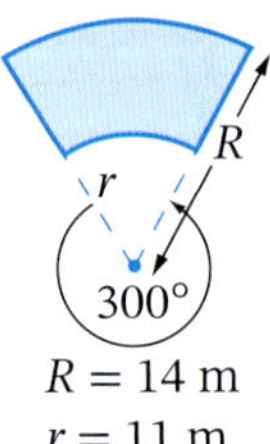

 9780170465601

5 CHAPTER REVIEW

Language of maths

Quiz
Language of maths 5

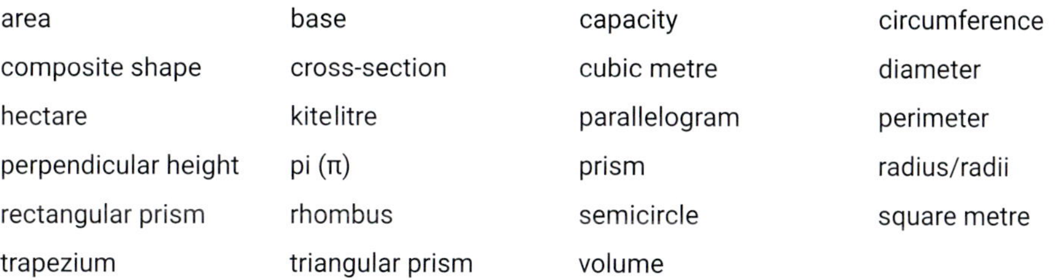

area	base	capacity	circumference
composite shape	cross-section	cubic metre	diameter
hectare	kitelitre	parallelogram	perimeter
perpendicular height	pi (π)	prism	radius/radii
rectangular prism	rhombus	semicircle	square metre
trapezium	triangular prism	volume	

1 Why are there two formulas for the circumference of a circle?

2 What is meant by the **perpendicular height** of a shape?

3 What is 1 million litres called?

4 In the formula $A = \frac{1}{2}xy$ for the area of a rhombus, what do x and y stand for?

5 What name is given to a 'slice' of a solid taken across the solid, rather than along it?

6 What measurement is double the radius of a circle?

Topic summary

Worksheet
Mind map: Area and volume

- What new things did you learn in this chapter?
- How often do you measure items? Give examples of the types of objects measured in your home.
- List four occupations that rely on being able to measure accurately.
- Did you have problems with any of the questions in this topic? If you did, discuss them with your teacher or a friend.

Print (or copy) and complete this mind map of the topic, adding detail to its branches and using pictures, symbols and colour where needed. Ask your teacher to check your work.

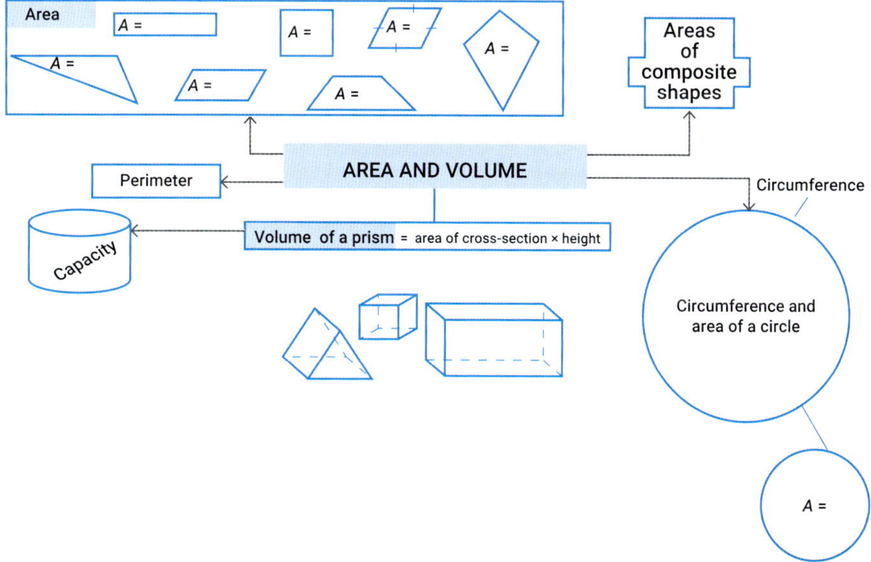

5 TEST YOURSELF

ANSWERS ON P. 630

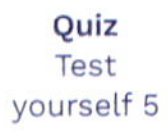

Quiz Test yourself 5

1 What is the perimeter of a rectangle with length 8 m and width 4 m? Select the correct answer **A**, **B**, **C** or **D**.

A 12 m **B** 16 m **C** 24 m **D** 48 m

2 Find the perimeter of each shape.

a 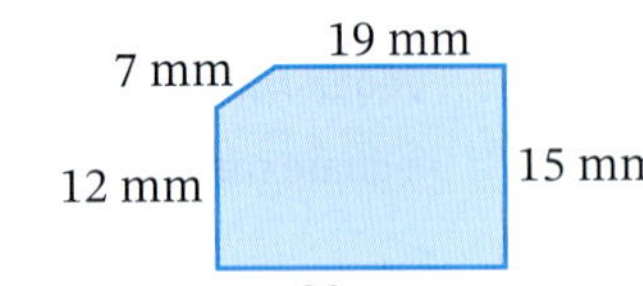

b 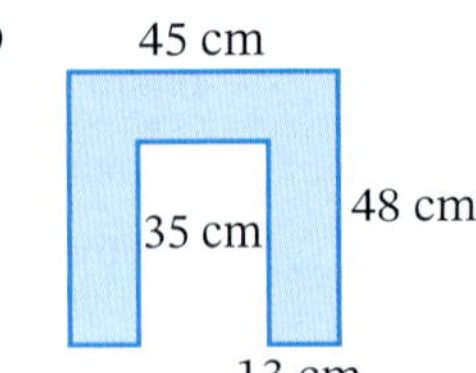

c 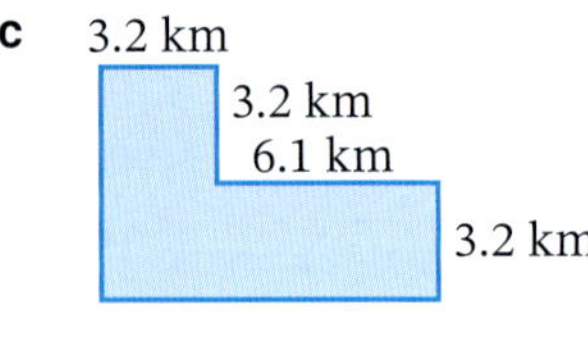

3 Calculate, correct to one decimal place, the circumference of each circle.

a 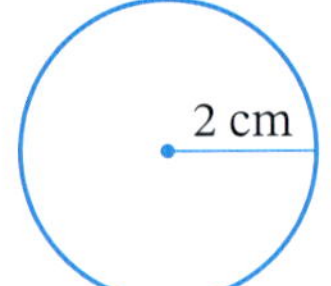

b 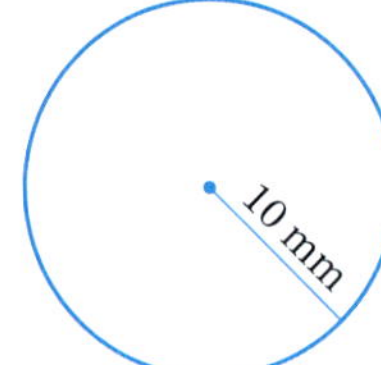

c 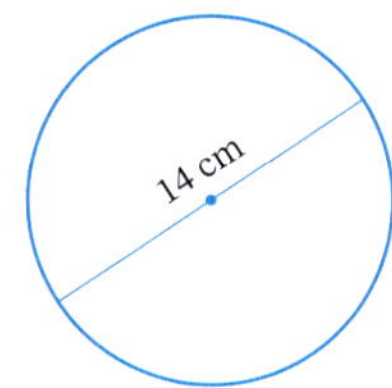

d 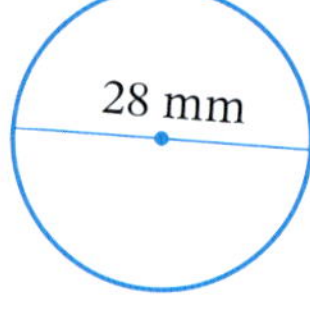

5.03

4 Copy and complete each conversion.

a $200\text{ mm}^2 = \_\_\_\_\_\text{ cm}^2$ **b** $7\text{ m}^2 = \_\_\_\_\_\text{ cm}^2$ **c** $8000\text{ cm}^2 = \_\_\_\_\_\text{ m}^2$

d $30\text{ km}^2 = \_\_\_\_\_\text{ m}^2$ **e** $70\,000\text{ cm}^2 = \_\_\_\_\_\text{ m}^2$ **f** $770\text{ mm}^2 = \_\_\_\_\_\text{ cm}^2$

g $700\text{ mm}^2 = \_\_\_\_\_\text{ m}^2$ **h** $7.2\text{ ha} = \_\_\_\_\_\text{ m}^2$ **i** $5.8\text{ mm}^2 = \_\_\_\_\_\text{ cm}^2$

5 Which triangle has an area of 64 m^2?

a 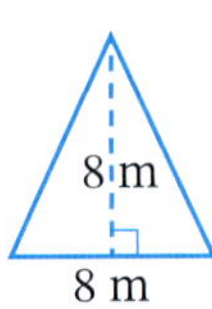

b 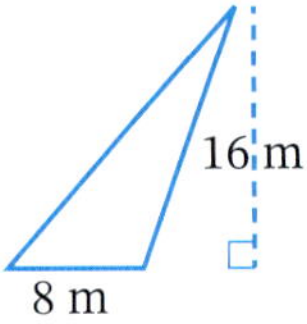

c 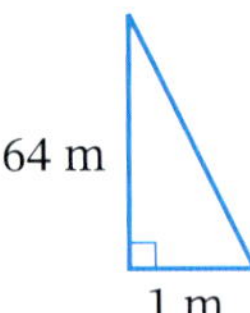

d 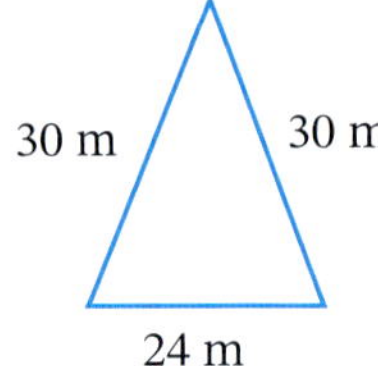

5.04

6 Find the area of each shape.

a 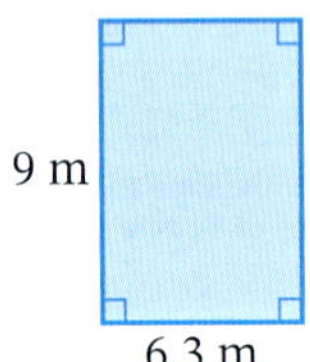

b 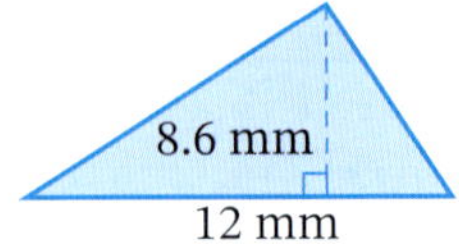

c 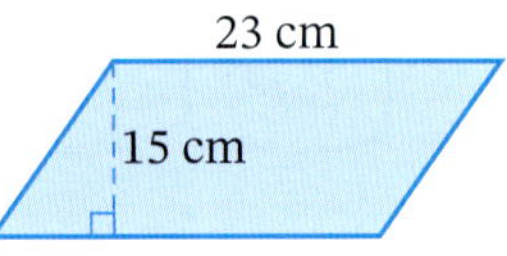

Foundation Standard Complex

9780170465601

d 7 km

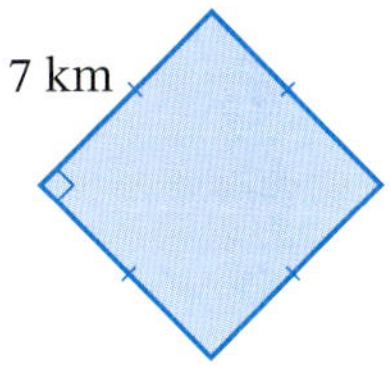

e 5.5 m, 12.6 m

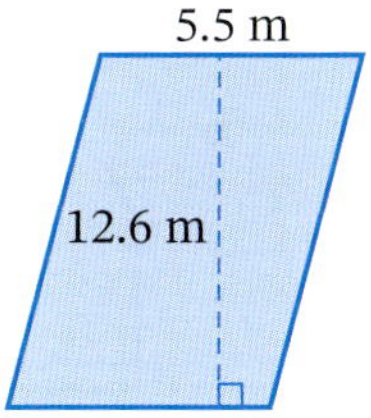

f 33 cm, 47 cm

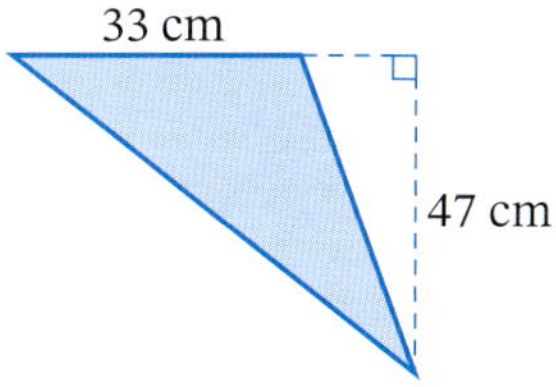

7 Find the area of each shape. 5.05

a 10 mm, 15 mm, 15 mm, 20 mm

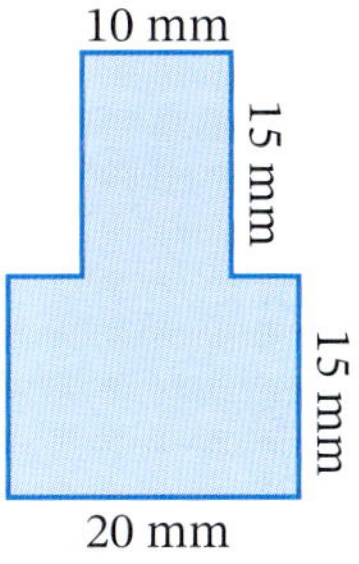

b 6 cm, 13 cm

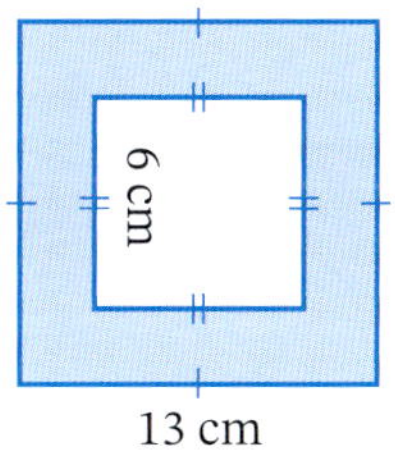

c 1.2 m, 1.5 m, 3 m

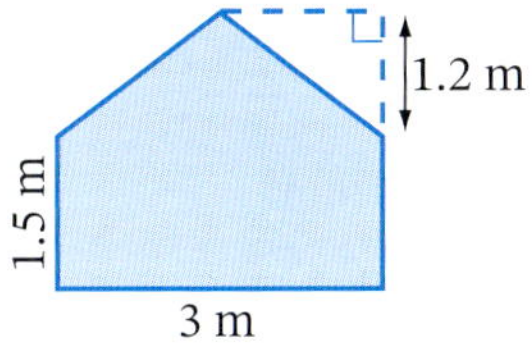

d 4 cm, 4 cm, 10 cm, 8 cm, 3 cm

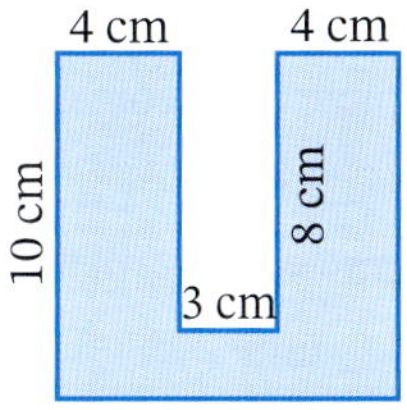

8 Find the area of each shape. 5.06

a 24 mm, 16.3 cm, 11.4 mm

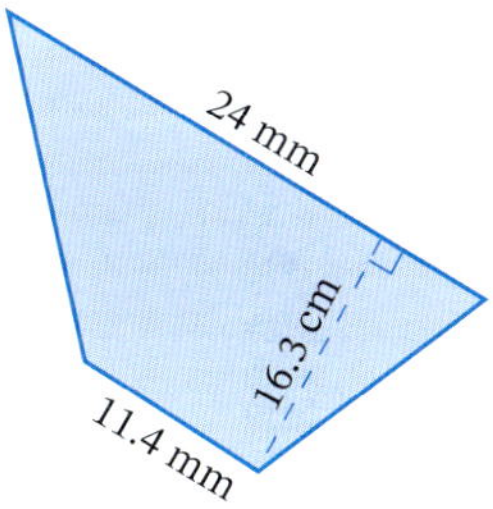

b 2.6 m, 5.2 m, 4 m

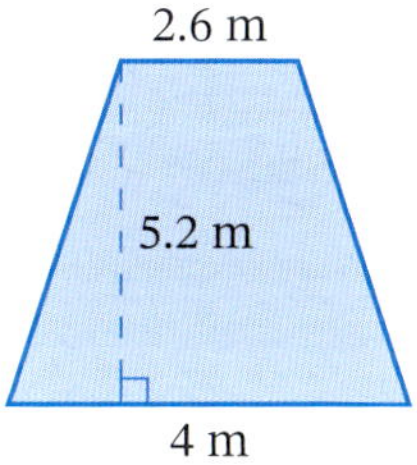

c 63 cm, 77 cm, 54 cm

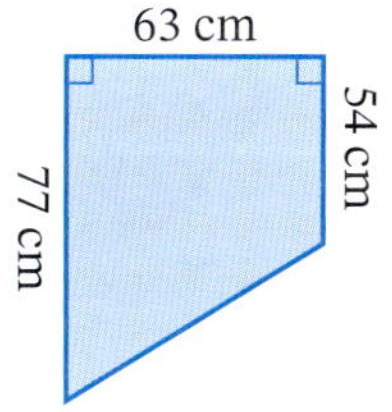

9 Find the area of each shape. 5.07

a 27 m, 7 m, 7 m

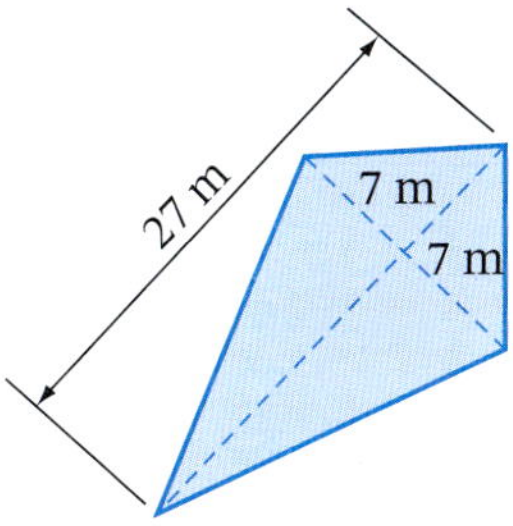

b 65 mm, 48 mm

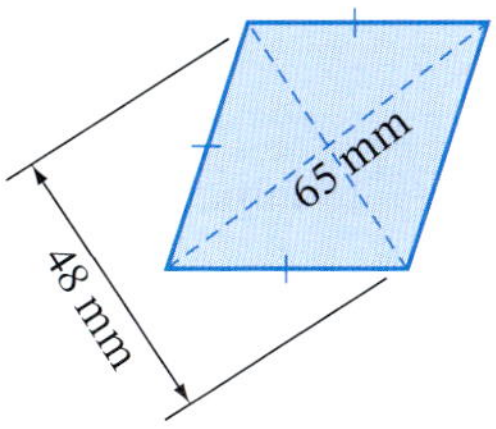

c 22 cm, 38 cm

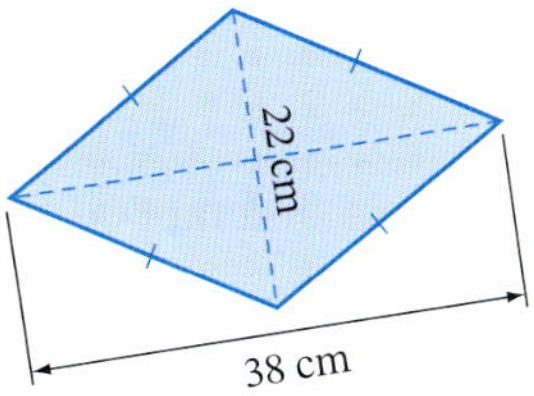

10 Calculate the area of each circle in Question 3:

i in terms of π **ii** correct to one decimal place

Foundation Standard Complex

EXTENSION

5.09

11 Calculate the perimeter of each shape, correct to two decimal places.

a
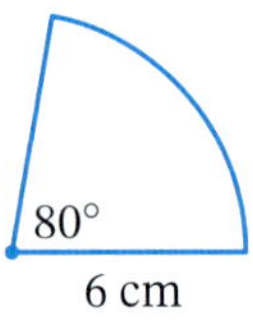

b
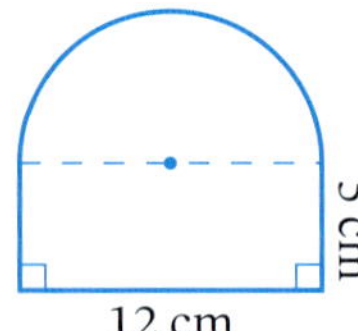

c
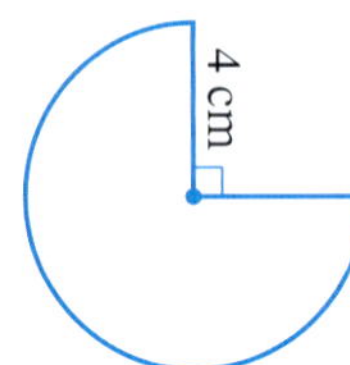

d
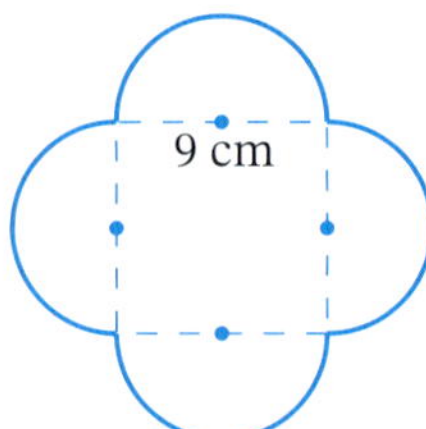

5.09

12 Calculate, correct to two decimal places, the area of each shape above.

5.10

13 Copy and complete each conversion.

a $8.2\text{ m}^3 =$ ______ cm^3

b $3.4\text{ cm}^3 =$ ______ mm^3

c $2\,000\,000\text{ cm}^3 =$ ______ m^3

d $1\text{ km}^3 =$ ______ m^3

e $8\text{ m}^3 =$ ______ mm^3

f $45\,000\text{ mm}^3 =$ ______ cm^3

5.11

14 Find the volume of each prism.

a
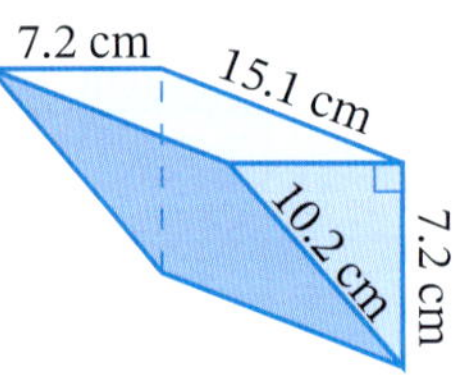

b
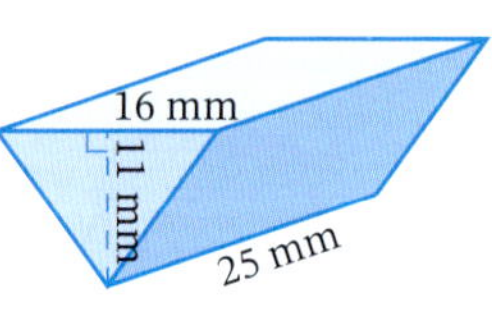

c
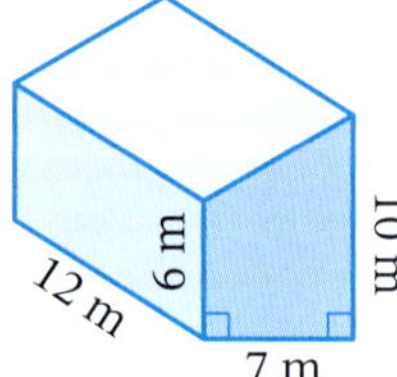

d
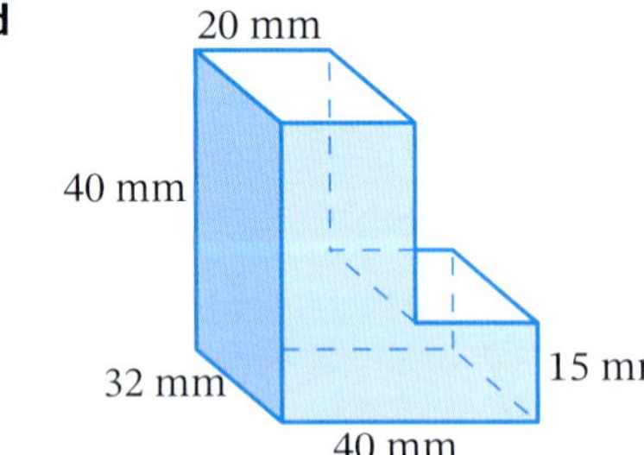

e
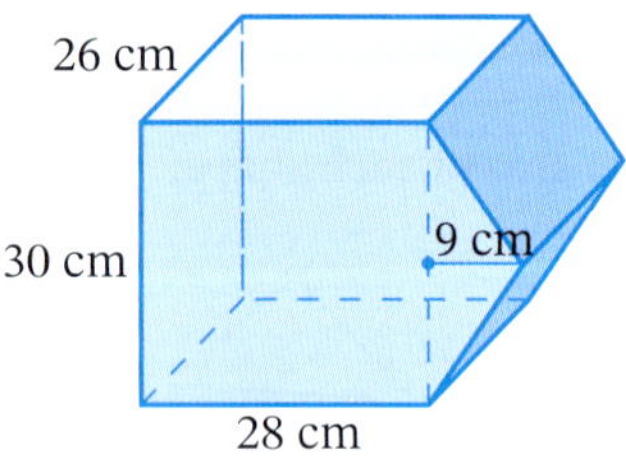

15 Find the volume of each solid, correct to two decimal places. YEAR 9 EXTENSION 5.12

a

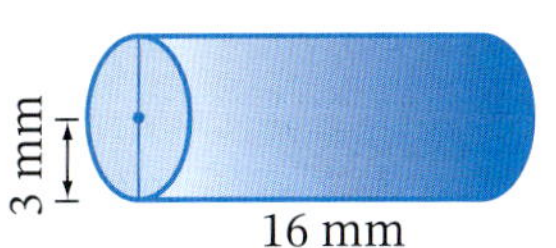

b

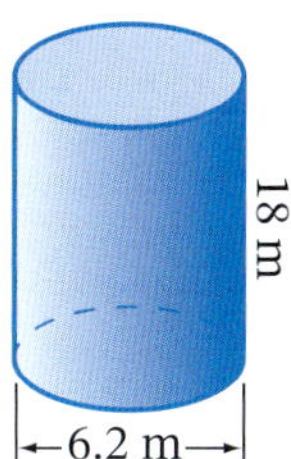

c

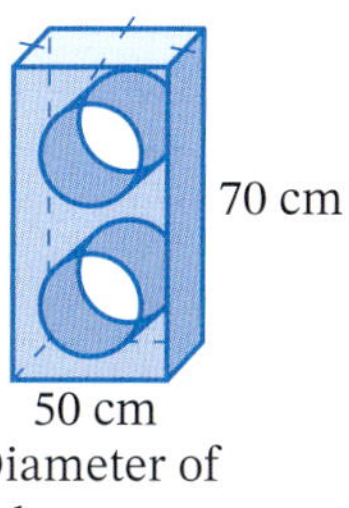

Diameter of circles = 16 cm

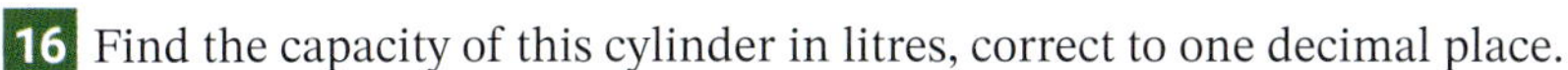

16 Find the capacity of this cylinder in litres, correct to one decimal place. 5.13

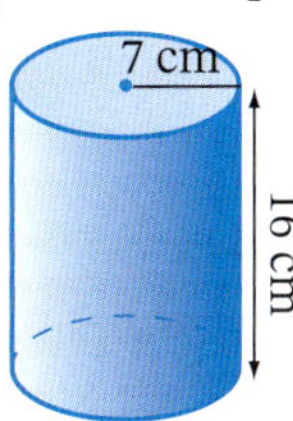

17 Copy and complete each conversion. 5.13

a 3 L = _________ mL

b 2500 mL = _________ L

c 6.5 kL = _________ L

d 7.2 mm^3 = _________ mL

e 120 mL = _________ L

f 35 cm^3 = _________ L

18 Find the capacity of each solid in litres. 5.13

a

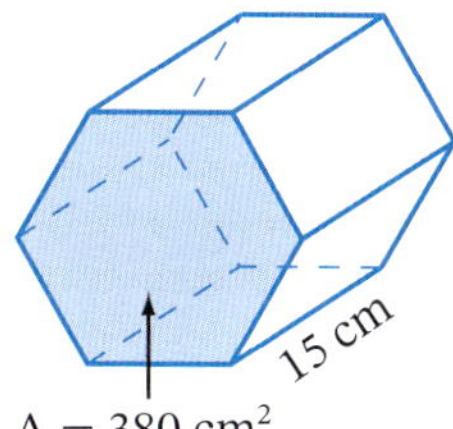

b

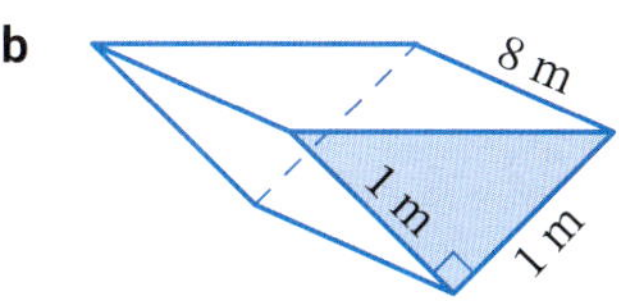

19 Shreya's pool has the shape of a trapezoidal prism. The shaded area is a trapezium. 5.13

a Find the shaded area.

b Find the volume of the pool.

c Find the capacity of the pool in litres.

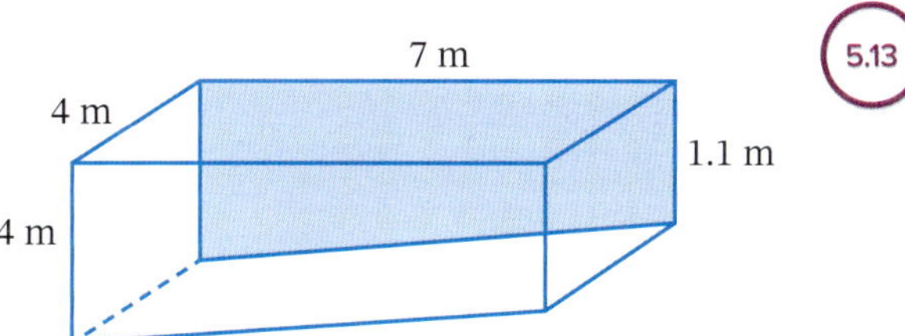

☐ Foundation ○ Standard ⬡ Complex

6

NUMBER

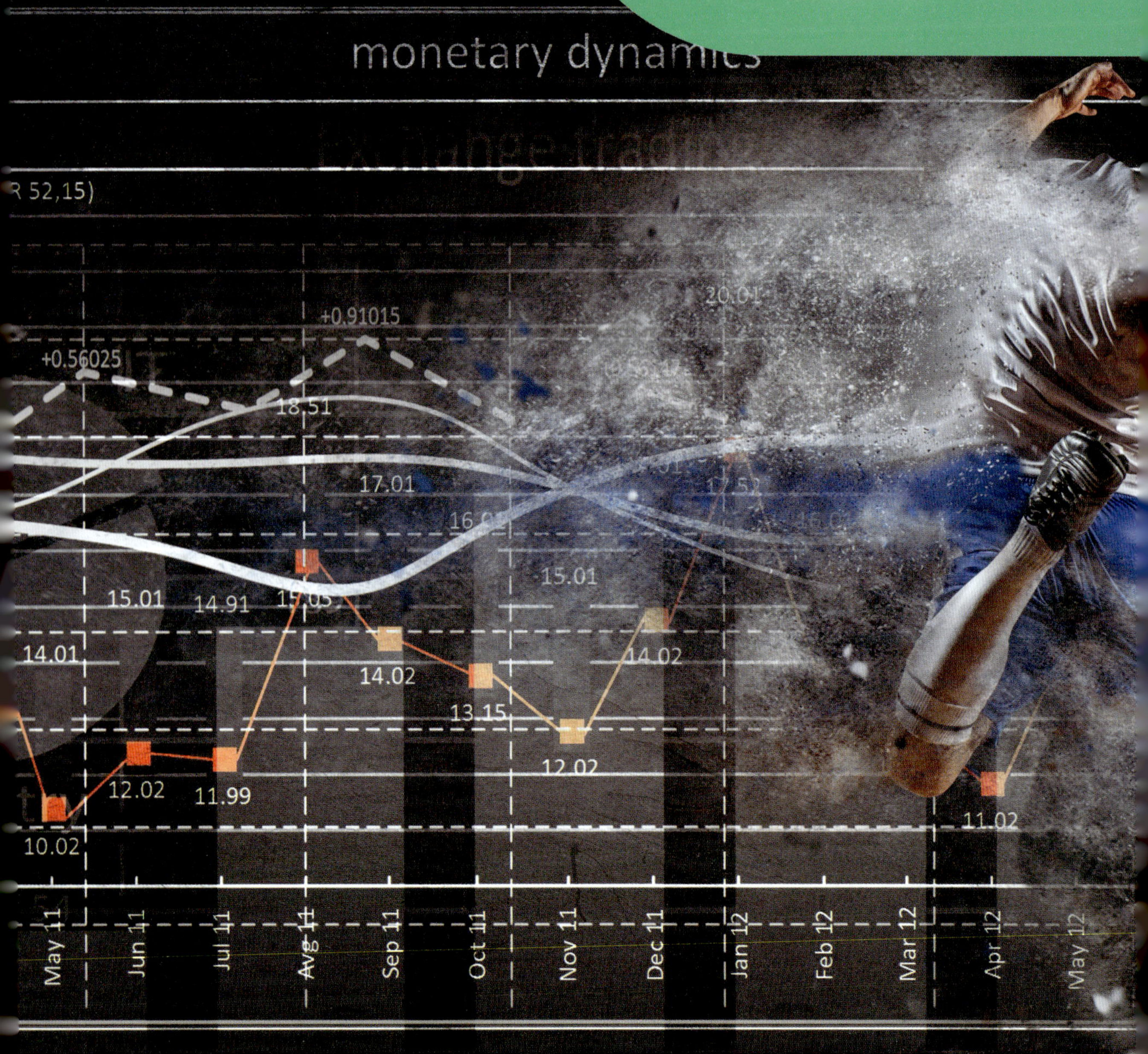

Fractions and percentages

Football players, tennis players and elite athletes often say 'I gave it 110%' after a match or event. What do they mean? Giving 100% is giving everything or the whole effort. They mean they have given extra effort or tried even harder than usual. When someone says 'I support them 110%', it is a similar thing. In both situations, the importance of the event or action is being emphasised.

Shutterstock.com/Sergey Nivens

Chapter outline

		Proficiencies
6.01	Fractions	U F R C
6.02	Adding and subtracting fractions	U F PS R
6.03	Multiplying and dividing fractions	U F PS R C
6.04	Percentages, fractions and decimals	U F R C
6.05	Fraction and percentage of a quantity	U F R
6.06	Expressing amounts as fractions and percentages	U F PS R
6.07	Percentage increase and decrease	U F PS R C
6.08	Percentages without calculators	U F PS R
6.09	The unitary method	U F PS R
6.10	Profit, loss and GST	U F PS R
6.11	Simple interest	U F PS
6.12	Percentage problems	U F PS R

U = Understanding
F = Fluency
PS = Problem solving
R = Reasoning
C = Communication

Wordbank

cost price The price of an item costs the retailer

discount The saving made between the original price of an item and the reduced price

improper fraction A fraction, such as $\frac{7}{3}$, in which the numerator is larger than or equal to the denominator

loss The amount lost when selling an item at a lower price

GST Goods and services tax, a 10% tax added to the original price of an item or service

profit The amount made when selling an item at a higher price

selling price The price at which an item is sold by the retailer

unitary method A method for finding the whole amount when a percentage of that amount is known, by first finding the size of 1%

Quiz
Wordbank 6

Videos (18):

SkillCheck Equivalent fractions • Using fractions
6.01 Improper fractions and mixed numerals • Simplifying fractions
6.02 Adding and subtracting fractions • Working out fractions
6.03 Multiplying and dividing fractions • Dividing fractions • Working out fractions
6.04 Converting percentages to fractions and decimals • Converting fractions and decimals to percentages
6.05 Fraction and percentage of a quantity Fraction of a quantity
6.06 Expressing quantities as fractions and percentages
6.07 Percentage increase and decrease
6.09 The unitary method
6.10 Business maths • Profit and loss • GST
6.11 Simple interest

Twig videos (3):

6.05 Percentages: Tax breaks
6.07 Hyperinflation: 1920s Germany • Inequalities: The most populous country

PhET interactives (5):

6.01 Fractions: Intro • Fraction matcher • Build a fraction • Fractions: Mixed numbers • Fractions: Equality

Quizzes (6):

- Wordbank 6
- SkillCheck 6
- Mental skills 6A
- Mental skills 6B
- Language of maths 6
- Test yourself 6

Skillsheets (7):

SkillCheck Fractions
6.01 Fractions • Equivalent fractions • Improper fractions and mixed numerals • Simplifying fractions
6.02 Adding and subtracting fractions
6.04 Multiplying and dividing by 10, 100, 1000
6.08 Mental percentages

Worksheets (26):

SkillCheck Fraction diagrams • Classifying fractions
6.01 Fractions wall • Pop stick calculator • Fraction strips
6.02 Fraction arithmagons • Pop stick calculator • Magic squares
6.03 Multiplying fractions using rectangles • Multiplying and dividing fractions number grids • What's the fraction question? • Fractions and decimals review
6.04 Fractions wall • Decimals wall • Percentages wall • Percentage shapes
6.05 Percentages of an amount
6.06 Calculating percentages • Money and percentages review
6.07 Discounts and special offers • Discounts
6.09 The unitary method • Working with percentages
6.10 Profit and loss
6.11 Simple interest
6.12 Percentage problems
Mind map: Fractions and percentages

Puzzles (14):

6.01 Mixed numerals to improper fractions
6.02 Adding and subtracting fractions
6.04 Converting percentages to fractions • Converting fractions to percentages • Converting decimals to percentages • Percentages to decimals • Percentominoes
6.05 Percentage crossnumber • Percentages • Fractions of words • Percentages of an amount
6.06 Calculating percentages
6.07 Discounts
6.10 Profit and loss

Technology (3):

6.06 Online discounts • Coffee shop sales
6.10 Profit and loss

Presentations (3):

6.01 Rational numbers • Improper fractions
6.04 Percentages, fractions and decimals

To access resources above, visit **cengage.com.au/nelsonmindtap**

6 In this chapter you will:

✓ order, add, subtract, multiply and divide fractions.
✓ convert between fractions, decimals and percentages.
✓ find a fraction or percentage of a quantity.
✓ express one quantity as a fraction or percentage of another.
✓ increase or decrease a quantity by a percentage.
✓ given a percentage of an amount, use the unitary method to find the whole amount.
✓ solve problems involving percentages, including profit, loss and GST.
✓ solve problems involving simple interest.

Quiz
SkillCheck
6

Videos
Equivalent fractions

Using fractions

Skillsheet
Fractions

Worksheets
Fraction diagrams

Classifying fractions

SkillCheck

ANSWERS ON P. 630

1 Copy and complete each pair of **equivalent** fractions.

a $\frac{1}{2}=\frac{}{6}$ **b** $\frac{2}{3}=\frac{}{12}$ **c** $\frac{4}{5}=\frac{16}{}$ **d** $\frac{7}{28}=\frac{1}{}$

e $\frac{3}{4}=\frac{}{8}$ **f** $\frac{15}{20}=\frac{}{4}$ **g** $\frac{6}{10}=\frac{24}{}$ **h** $\frac{3}{5}=\frac{9}{}$

2 Write each percentage as a fraction.

a 17% **b** 3% **c** 51% **d** 63% **e** 89%

3 Evaluate each difference.

a 100% − 30% **b** 100% − 95% **c** 100% − 39%

d 100% − 71% **e** 100% − 12% **f** 100% − 7%

4 Write the percentage that is shaded in each diagram.

a
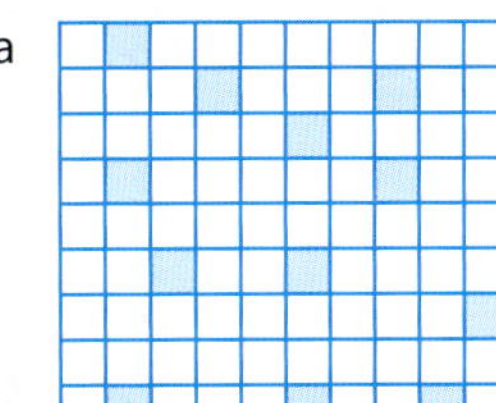

b
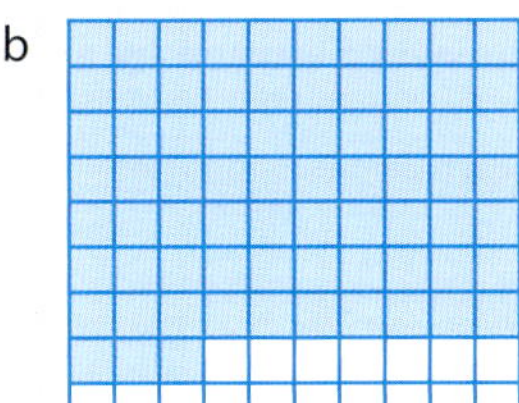

c
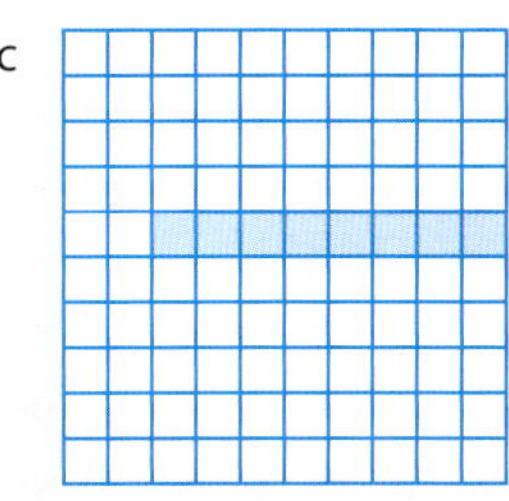

5 Write the percentage that is unshaded in each diagram in question **4**.

6 On a plane, 10% of the seats are first class and 25% are business class. The rest are economy class. What percentage of the seating is economy class?

7 Write all the factors of 100.

8 Evaluate each expression.

a 50% of \$80 **b** 50% of 12 kg **c** 25% of \$20

d 25% of 40 kg **e** 10% of \$30 **f** 10% of 20 kg

Fractions

6.01

Videos
Improper fractions and mixed numerals

Simplifying fractions

Scientific calculators have a fraction key for entering fractions: ▭/▭ or a b/c. Casio calculators have two modes for entering and displaying fractions. MATH mode shows fractions more realistically as it allows you to enter the numerator and denominator into two blank spaces on the calculator's screen, while LINE mode shows the fraction in a line (e.g. 1 ⌟ 3) and makes the fraction key act like a vinculum (fraction bar). Press SHIFT MODE to change between MATH and LINE modes. When MATH mode is selected, 'Math' appears on the calculator display. To ensure that mixed numeral answers are not converted to improper fractions, press SHIFT MODE and the down arrow to choose 'ab/c'.

Skillsheets
Fractions
Equivalent fractions
Improper fractions and mixed numerals
Simplifying fractions

Worksheets
Fractions wall
Pop stick calculator
Fraction strips

Puzzle
Mixed numbers to improper fractions

Presentations
Rational numbers
Improper fractions

Interactives
Fractions: Intro
Fraction matcher
Build a fraction
Fractions: Mixed numbers
Fractions: Equality

Example 1

Convert $\frac{27}{4}$ into a mixed numeral.

SOLUTION

$\frac{27}{4} = 27 \div 4$ — To find the number of 'wholes' in 27 quarters, divide 27 by 4.

$= 6$ remainder 3

$= 6\frac{3}{4}$ — Write the remainder in the numerator of a fraction.

Or on a calculator, enter: 27 [▭/▭] 4 [=] or 27 [a b/c] 4 [=].

For Casio MATH mode, press the [▭/▭] key first and use arrow keys to move the cursor to the two blank spaces to enter the values. Then press [=].

If the display shows $\frac{27}{4}$, press [SHIFT] [S⇔D] or [2nd F] [a b/c] to convert it to a mixed numeral.

[SHIFT] [S⇔D] or [2nd F] [a b/c] convert an answer between **mixed numeral** and **improper fraction** forms.

Example 2

Convert $4\frac{2}{5}$ to an improper fraction.

SOLUTION

$4\frac{2}{5} = \frac{5 \times 4 + 2}{5} = \frac{22}{5}$

This works because $4\frac{2}{5} = 4 + \frac{2}{5} = \frac{5 \times 4}{5} + \frac{2}{5}$.

Or on a calculator, enter: 4 [▭/▭] 2 [▭/▭] 5 [=] [SHIFT] [S⇔D] or 4 [a b/c] 2 [a b/c] 5 [=] [2nd F] [a b/c].

For entering mixed numerals in Casio MATH mode, press [SHIFT] [▭/▭] and move the cursor to the three blank spaces to enter the values of the mixed numeral. Then press [=].

Example 3

Which fraction is larger, $\frac{4}{10}$ or $\frac{3}{8}$?

SOLUTION

Method 1

Convert both fractions so that they share a common denominator of $10 \times 8 = 80$.

$\frac{4}{10} = \frac{4 \times 8}{10 \times 8} = \frac{32}{80}$ and $\frac{3}{8} = \frac{3 \times 10}{8 \times 10} = \frac{30}{80}$

Note that we multiply the numerator and denominator of each fraction by the denominator of the **other** fraction.

By comparing numerators, $\frac{32}{80} > \frac{30}{80}$

$\therefore \frac{4}{10}$ is larger.

6.01

Method 2

Another suitable denominator is the lowest common denominator (LCD) of both denominators.

The LCD of 10 and 8 is 40.

Convert both fractions.

$\frac{4}{10}=\frac{4\times4}{10\times4}=\frac{16}{40}$ and $\frac{3}{8}=\frac{3\times5}{8\times5}=\frac{15}{40}$

The LCD is the lowest common multiple of both denominators.

By comparing numerators, $\frac{16}{40}>\frac{15}{40}$.

$\therefore \frac{4}{10}$ is larger.

Example 4

Simplify each fraction.

a $\frac{10}{15}$ **b** $\frac{36}{60}$

SOLUTION

a $\frac{10}{15}=\frac{10\div5}{15\div5}=\frac{2}{3}$

Or on a calculator, enter: 10 [fraction key] 15 [=] or 10 [a b/c] 15 [=].

b $\frac{36}{60}=\frac{36\div6}{60\div6}=\frac{6}{10}=\frac{6\div2}{10\div2}=\frac{3}{5}$

or $\frac{36}{60}=\frac{36\div12}{60\div12}=\frac{3}{5}$

We can simplify in one step if we divide by 12, the HCF of 36 and 60.

Or on a calculator, enter: 36 [fraction key] 60 [=] or 36 [a b/c] 60 [=].

EXERCISE 6.01 ANSWERS ON P. 631

Fractions

U F R C

EXAMPLE 1

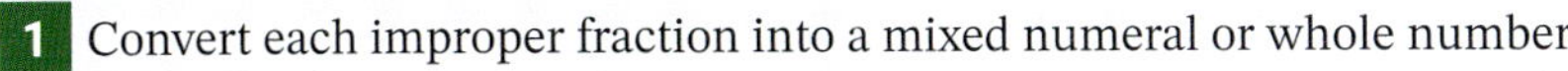

1 Convert each improper fraction into a mixed numeral or whole number.

C **a** $\frac{3}{2}$ **b** $\frac{11}{3}$ **c** $\frac{9}{4}$

d $\frac{12}{5}$ **e** $\frac{20}{4}$ **f** $\frac{47}{11}$

EXAMPLE 2

2 Convert each mixed numeral into an improper fraction.

C

a $3\frac{1}{2}$ **b** $4\frac{1}{3}$ **c** $5\frac{3}{4}$ **d** $7\frac{2}{3}$

e $7\frac{4}{5}$ **f** $7\frac{2}{9}$ **g** $10\frac{1}{7}$ **h** $9\frac{3}{10}$

Foundation Standard Complex

3 Write an improper fraction that can be converted into a whole number.

R C

4 Which is larger: a proper fraction or a mixed numeral? Explain your answer.

R C

EXAMPLE 3

5 For each pair of fractions, find the larger fraction.

a $\frac{1}{2}, \frac{1}{3}$ **b** $\frac{2}{3}, \frac{3}{4}$ **c** $\frac{3}{8}, \frac{1}{3}$

d $2\frac{3}{5}, 2\frac{7}{10}$ **e** $\frac{11}{15}, \frac{3}{5}$ **f** $\frac{5}{12}, \frac{2}{5}$

EXAMPLE 4

6 Simplify each fraction.

a $\frac{5}{10}$ **b** $\frac{4}{12}$ **c** $\frac{12}{26}$ **d** $\frac{18}{24}$

e $\frac{15}{25}$ **f** $\frac{16}{28}$ **g** $\frac{32}{48}$ **h** $\frac{60}{100}$

i $\frac{44}{77}$ **j** $\frac{20}{35}$ **k** $\frac{21}{35}$ **l** $\frac{72}{80}$

7 Which one of these fractions is **NOT** equivalent to $\frac{10}{25}$? Select the correct answer **A**, **B**, **C** or **D**.

A $\frac{1}{5}$ **B** $\frac{2}{5}$ **C** $\frac{20}{50}$ **D** $\frac{40}{100}$

8 Convert each improper fraction into a simplified mixed numeral.

a $\frac{9}{6}$ **b** $\frac{18}{6}$ **c** $\frac{45}{10}$ **d** $\frac{36}{20}$

DID YOU KNOW

The Torres Strait Islander Flag

The Torres Strait Islander flag was designed by Bernard Namok from Thursday Island who won a design competition for the flag in 1992. It was created as a symbol of unity and identity for Torres Strait Islander peoples.

The central blue panel of the flag represents the sea, the outer green panels represent the land and the black lines which separate the green and blue panels represent the Torres Strait Islander people. The centre of the flag shows a white dhari (dancer's headdress) with a white five-pointed star. The white colour is used to symbolise peace and the dhari represents Torres Strait Islander culture. The five points of the white star placed under the dhari represent the five island groups in the Torres Strait.

Foundation Standard Complex

The cultures and traditions of Torres Strait Islanders are strongly connected to the land, sea and sky – elements which are all represented in this flag.

Approximate what fraction of the flag is:

a **blue** **b** **green** **c** **black** **d** **white**

Ask your teacher to print a copy of this flag for you and then use a transparent grid placed over the flag to approximate.

Adding and subtracting fractions 6.02

Skillsheet
Adding and subtracting fractions

Videos
Adding and subtracting fractions

Working out fractions

Worksheets
Fraction arithmagons

Pop stick calculator

Magic squares

Puzzle
Adding and subtracting fractions

Adding and subtracting fractions

- To **add or subtract fractions**, convert them (if needed) so that they have the same denominator, then simply add or subtract just the numerators.
- To **add or subtract mixed numerals**, add or subtract the whole numbers and fractions separately.

Example 5

Evaluate each expression.

a $\frac{1}{3}+\frac{5}{6}$ **b** $\frac{5}{7}-\frac{2}{3}$ **c** $3-\frac{3}{4}$ **d** $4\frac{1}{5}-1\frac{2}{3}$

SOLUTION

a Common denominator $= 3 \times 6 = 18$.

Note that we multiply the numerator and denominator of each fraction by the denominator of the *other* fraction.

$$\frac{1}{3}=\frac{1\times6}{3\times6}=\frac{6}{18}$$
$$\frac{5}{6}=\frac{5\times3}{6\times3}=\frac{15}{18}$$
$$\frac{1}{3}+\frac{5}{6}=\frac{6}{18}+\frac{15}{18}$$
$$=\frac{21}{18}$$
$$=\frac{7}{6}$$
$$=1\frac{1}{6}$$

OR on a calculator, enter: 1 [fraction key] 3 [+] 5 [fraction key] 6 [=].

OR Using the LCD of 3 and 6 = 6:

$$\frac{1}{3}=\frac{1\times2}{3\times2}=\frac{2}{6}$$
$$\frac{5}{6}=\frac{5}{6}$$
$$\frac{1}{3}+\frac{5}{6}=\frac{2}{6}+\frac{5}{6}$$
$$=\frac{7}{6}$$
$$=1\frac{1}{6}$$

Note: [fraction key] or [a b/c].

b Common denominator = 7 × 3 = 21:

$\frac{5}{7}=\frac{5\times 3}{7\times 3}=\frac{15}{21}$

$\frac{2}{3}=\frac{2\times 7}{3\times 7}=\frac{14}{21}$

$\frac{5}{7}-\frac{2}{3}=\frac{15}{21}-\frac{14}{21}$

$=\frac{1}{21}$

OR 5 [fraction] 7 [−] 2 [fraction] 3 [=].

c $3-\frac{3}{4}=2+1-\frac{3}{4}$

$=2+\frac{4}{4}-\frac{3}{4}$

$=2\frac{1}{4}$

OR 3 [−] 3 [fraction] 4 [=].

d $4\frac{1}{5}-1\frac{2}{3}=4-1+\frac{1}{5}-\frac{2}{3}$

$=3+\frac{3}{15}-\frac{10}{15}$

$=3-\frac{7}{15}$

$=2+\frac{15}{15}-\frac{7}{15}$

$=2\frac{8}{15}$

OR 4 [fraction] 1 [fraction] 5 [−] 1 [fraction] 2 [fraction] 3 [=].

For MATH mode, press [SHIFT] [fraction] first.

EXERCISE 6.02 ANSWERS ON P. 631

Adding and subtracting fractions

1 Evaluate each expression.

a $\frac{1}{5}+\frac{3}{5}$ **b** $\frac{3}{4}-\frac{1}{4}$ **c** $\frac{3}{8}+\frac{2}{8}$ **d** $\frac{7}{8}+\frac{4}{8}$

EXAMPLE 5

2 Evaluate each expression.

a $\frac{2}{5}+\frac{3}{10}$ **b** $\frac{2}{3}-\frac{3}{7}$ **c** $\frac{3}{5}+\frac{1}{4}$ **d** $\frac{1}{2}-\frac{1}{4}$

e $\frac{1}{6}+\frac{3}{8}$ **f** $\frac{7}{10}-\frac{2}{3}$ **g** $\frac{4}{9}+\frac{5}{6}$ **h** $\frac{7}{8}-\frac{5}{12}$

i $\frac{5}{8}+\frac{1}{2}$ **j** $1-\frac{6}{7}$ **k** $4-\frac{4}{9}$ **l** $3-\frac{2}{5}$

3 R Novak bought a length of timber to build fences for his garden beds. He used $\frac{1}{5}$ of the wood for the first garden and $\frac{3}{8}$ of the wood for the second garden. What fraction of the wood remains for the third garden?

4 PS R What fraction goes in the blank? $\frac{7}{8}-\_\_\_\_=\frac{1}{4}$. Select the correct answer **A**, **B**, **C** or **D**.

A $\frac{6}{4}$ **B** $\frac{8}{12}$ **C** $\frac{3}{8}$ **D** $\frac{5}{8}$

5 PS R Copy and complete each equation with the correct fraction.

a $\frac{1}{4}+\_\_\_\_=\frac{2}{3}$ **b** $\_\_\_\_-\frac{2}{5}=\frac{1}{6}$

Foundation Standard Complex

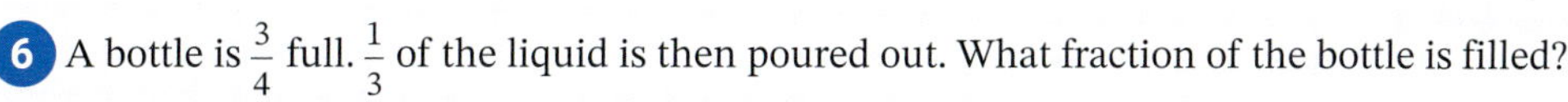

6 A bottle is $\frac{3}{4}$ full. $\frac{1}{3}$ of the liquid is then poured out. What fraction of the bottle is filled?

PS R

7 Danielle used half a sheet of adhesive plastic to cover her books, while Christina used $\frac{2}{5}$ of the same sheet. What fraction of the original sheet remains? Select the correct answer **A**, **B**, **C** or **D**.

PS R

A $\frac{3}{5}$ **B** $\frac{4}{5}$ **C** $\frac{1}{10}$ **D** $\frac{9}{10}$

8 Evaluate $2\frac{7}{8}+1\frac{3}{8}$. Select the correct answer **A**, **B**, **C** or **D**.

A $4\frac{1}{4}$ **B** $3\frac{4}{5}$ **C** $3\frac{1}{2}$ **D** $13\frac{1}{4}$

9 Evaluate each expression.

a $1\frac{1}{2}+\frac{2}{5}$ **b** $2\frac{1}{4}+1\frac{3}{8}$ **c** $7\frac{3}{8}+9\frac{1}{3}$ **d** $3\frac{4}{5}+1\frac{2}{9}$

e $6\frac{2}{5}-4\frac{7}{8}$ **f** $4\frac{3}{4}-2\frac{4}{5}$ **g** $2\frac{1}{3}-1\frac{1}{2}$ **h** $4\frac{3}{5}-2\frac{9}{10}$

10 In a magic square, each row, column and diagonal must add to the same total. Complete each magic square with appropriate fractions.

PS R

a

4	$1\frac{1}{2}$	
$\frac{1}{2}$		$4\frac{1}{2}$
3		

b

$\frac{1}{6}$		
	$\frac{2}{3}$	
	$\frac{1}{3}$	$1\frac{1}{6}$

Multiplying and dividing fractions 6.03

Multiplying and dividing fractions

- To **multiply fractions**, multiply the numerators and denominators separately, then simplify if possible (sometimes, it is easier to simplify the fractions first).
- To **divide by a fraction**, $\frac{a}{b}$, multiply by its reciprocal $\frac{b}{a}$.
- To **multiply or divide mixed numerals**, convert them to improper fractions first.

Videos
Multiplying and dividng fractions
Dividing fractions
Working out fractions

Worksheets
Multiplying fractions using rectangles
Multiplying and dividing fractions number grids
What's the fraction question?

Foundation Standard Complex

Worksheet Fractions and decimals review

Example 6

Evaluate each expression.

a $\frac{3}{5}\times\frac{2}{7}$ **b** $\frac{2}{3}\times\frac{3}{8}$ **c** $\frac{4}{5}\div\frac{2}{3}$ **d** $1\frac{1}{2}\times2\frac{2}{3}$

SOLUTION

a $\frac{3}{5}\times\frac{2}{7}=\frac{3\times2}{5\times7}$
$=\frac{6}{35}$

OR on a calculator, enter: 3 [fraction] 5 [×] 2 [fraction] 7 [=]

Note: [fraction] or [a b/c]

b $\frac{2}{3}\times\frac{3}{8}=\frac{2\times3}{3\times8}$
$=\frac{6}{24}$
$=\frac{1}{4}$

OR by simplifying first before multiplying:

$\frac{2}{3}\times\frac{3}{8}=\frac{\cancel{2}^1}{\cancel{3}_1}\times\frac{\cancel{3}^1}{\cancel{8}_4}$
$=\frac{1}{4}$

OR on a calculator, enter: 2 [fraction] 3 [×] 3 [fraction] 8 [=]

c $\frac{4}{5}\div\frac{2}{3}=\frac{4}{5}\times\frac{3}{2}$
$=\frac{12}{10}$
$=\frac{6}{5}$
$=1\frac{1}{5}$

OR by simplifying first before multiplying:

$\frac{4}{5}\div\frac{2}{3}=\frac{\cancel{4}^2}{5}\times\frac{3}{\cancel{2}_1}=\frac{6}{5}=1\frac{1}{5}$

OR on a calculator, enter: 4 [fraction] 5 [÷] 2 [fraction] 3 [=]

d $1\frac{1}{2}\times2\frac{2}{3}=\frac{3}{2}\times\frac{8}{3}$
$=\frac{\cancel{3}^1}{\cancel{2}_1}\times\frac{\cancel{8}^4}{\cancel{3}_1}$
$=\frac{4}{1}$
$=4$

OR on a calculator, enter: 1 [fraction] 1 [fraction] 2 [×] 2 [fraction] 2 [fraction] 3 [=]

For MATH mode, press [SHIFT] [fraction] first.

EXERCISE 6.03 ANSWERS ON P. 631

Multiplying and dividing fractions

U F PS R C

EXAMPLE 6

1 Evaluate each product.

a $\frac{1}{2}\times\frac{2}{3}$ **b** $\frac{2}{3}\times\frac{5}{7}$ **c** $\left(\frac{1}{8}\right)^2$ **d** $\frac{9}{10}\times\frac{5}{6}$

e $\frac{3}{5}\times\frac{3}{4}$ **f** $1\frac{1}{2}\times3$ **g** $3\frac{1}{4}\times3\frac{4}{5}$ **h** $1\frac{1}{3}\times2\frac{1}{3}$

2 What is the value of $\frac{7}{10}\div\frac{1}{5}$? Select the correct answer **A**, **B**, **C** or **D**.

A $\frac{7}{50}$ **B** $1\frac{1}{5}$ **C** $1\frac{2}{5}$ **D** $3\frac{1}{2}$

Foundation Standard Complex

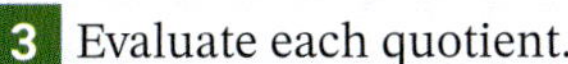

3 Evaluate each quotient.

a $\frac{1}{7}\div\frac{2}{3}$ **b** $\frac{2}{5}\div\frac{3}{4}$ **c** $\frac{5}{9}\div\frac{5}{6}$ **d** $\frac{9}{10}\div\frac{3}{5}$

e $4\div\frac{1}{4}$ **f** $6\div\frac{2}{3}$ **g** $\frac{5}{8}\div 5$ **h** $\frac{2}{5}\div 6$

i $\frac{7}{8}\div\frac{1}{4}$ **j** $10\div 3\frac{1}{7}$ **k** $3\frac{1}{2}\div 1\frac{1}{4}$ **l** $3\frac{3}{4}\div 1\frac{1}{10}$

3 From a box of chocolates, Lindy takes $\frac{3}{8}$ of the chocolates and shares them equally among her four children. What fraction of the box of lollies does each child receive?

PS R

5 **a** What number's reciprocal is itself?

R **b** What number does not have a reciprocal?

C **c** What is the product of a number and its reciprocal?

d What is the reciprocal of the reciprocal of $\frac{1}{3}$?

6 Copy and complete each equation.

R **a** $\frac{3}{4}\times\_\_\_\_=\frac{21}{32}$ **b** $\_\_\_\_\times\frac{1}{5}=\frac{2}{15}$

c $\_\_\_\_\div\frac{2}{3}=\frac{1}{7}$ **d** $\frac{4}{15}\div\_\_\_\_\ \frac{8}{21}$

7 **a** When a number is divided by a proper fraction, does the number increase or decrease? Give reasons for your answer.

R C **b** When a number is multiplied by an improper fraction, does the number increase or decrease? Give reasons for your answer.

Percentages, fractions and decimals 6.04

A **percentage** is a fraction whose denominator is 100.

Skillsheet Multiplying and dividing by 10, 100, 1000

Video Converting percentages to fractions and decimals

Example 7

Convert each percentage into a simplified fraction.

a 55% **b** 130% **c** $37\frac{1}{2}\%$

SOLUTION

a $55\%=\frac{55}{100}$
$=\frac{11}{20}$

b $130\%=\frac{130}{100}$
$=\frac{13}{10}$
$=1\frac{3}{10}$

c $37\frac{1}{2}\%=\frac{37\frac{1}{2}}{100}$
$=\frac{37\frac{1}{2}\times 2}{100\times 2}$
$=\frac{75}{200}$
$=\frac{3}{8}$

Foundation Standard Complex

Video
Converting percentages to fractions and decimals

Worksheets
Fractions wall
Decimals wall
Percentages wall
Percentage shapes

Presentation
Percentages, fractions and decimals

Video
Converting fractions and decimals to percentages

Puzzles
Converting percentages to fractions
Converting fractions to percentages
Converting decimals to percentages
Percentages to decimals
Percentominoes

Video
Converting fractions and decimals to percentages

Example 8

Convert each percentage into a decimal.

a 8% **b** 43.6% **c** $18\frac{1}{2}\%$

SOLUTION

a $8\% = \frac{8}{100}$
$= 8 \div 100$
$= 0.08$

b $43.6\% = \frac{43.6}{100}$
$= 43.6 \div 100$
$= 0.436$

c $18\frac{1}{2}\% = \frac{18\frac{1}{2}}{100}$
$= 18.5 \div 100$
$= 0.185$

To convert a percentage into a decimal mentally, move the decimal point two places to the left.

ⓘ Converting to a percentage

To convert a fraction or a decimal into a percentage, multiply it by 100%.

Example 9

Convert each fraction into a percentage.

a $\frac{1}{8}$ **b** $\frac{34}{40}$

SOLUTION

a $\frac{1}{8} = \frac{1}{8} \times 100\%$
$= 12\frac{1}{2}\%$

b $\frac{34}{40} = \frac{34}{40} \times 100\%$
$= 85\%$

Example 10

Convert each decimal into a percentage.

a 0.147 **b** 0.6

SOLUTION

a $0.147 = 0.147 \times 100\%$
$= 14.7\%$

b $0.6 = 0.6 \times 100\%$
$= 60\%$

To mentally convert a decimal into a percentage, just move the decimal point two places to the right.

ⓘ Ordering fractions, decimals and percentages

To order fractions, percentages and decimals, convert each into a percentage first.

Remember:

- Ascending order means from smallest to largest.
- Descending order means from largest to smallest.

Example 11

Arrange in ascending order: 0.667, 66%, $\frac{7}{11}$.

SOLUTION

Convert all numbers into percentages.

$0.667 \times 100\% = 66.7\%$

$66\% = 66\%$

$\frac{7}{11} \times 100\% = 63\frac{7}{11}\%$

Arrange the percentages from smallest to largest: $63\frac{7}{11}\%$, 66% and 66.7%.

So, in ascending order, the numbers are $63\frac{7}{11}$, 66% and 0.667.

EXERCISE  6.04 ANSWERS ON P. 631

Percentages, fractions and decimals

1 Convert each percentage into a simplified fraction. EXAMPLE 7

a 60%	**b** 75%	**c** 31%	**d** 8%
e 30%	**f** 85%	**g** 99%	**h** 3%
i 160%	**j** 135%	**k** 25%	**l** 250%

2 Which decimal is equal to $62\frac{1}{2}\%$? Select the correct answer **A**, **B**, **C** or **D**. EXAMPLE 8

A 62.12	**B** 62.5	**C** 0.625	**D** 0.0625

3 Convert each percentage to a decimal.

a 18%	**b** 82%	**c** 2%	**d** 50%
e 120%	**f** 51.1%	**g** 79%	**h** $12\frac{1}{2}\%$
i 16.3%	**j** 4%	**k** 18.7%	**l** $5\frac{1}{4}\%$

4 Convert each fraction into a percentage. EXAMPLE 9

a $\frac{17}{100}$	**b** $\frac{7}{10}$	**c** $\frac{13}{50}$	**d** $\frac{11}{20}$
e $\frac{5}{8}$	**f** $\frac{24}{25}$	**g** $\frac{2}{3}$	**h** $\frac{5}{4}$
i $1\frac{2}{5}$	**j** $\frac{27}{40}$	**k** $\frac{1}{16}$	**l** $\frac{4}{9}$

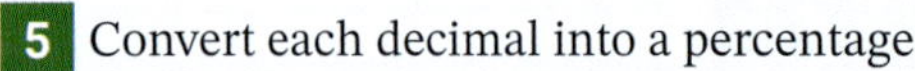

5 Convert each decimal into a percentage.

C

a 0.38	**b** 0.55	**c** 0.96	**d** 0.625
e 0.08	**f** 0.054	**g** 0.6	**h** 0.003
i 1.9	**j** 0.405	**k** 1.26	**l** 0.114

6 Copy and complete this table.

C

	Fraction	Decimal	Percentage
a		0.65	
b		0.6	
c			20
d			84
e	$\frac{1}{2}$		
f	$\frac{1}{8}$		
g			36
h	$\frac{5}{8}$		
i		0.73	
j	$\frac{1}{3}$		
k			$66\frac{2}{3}\%$

7 For each pair of numbers, determine which is larger.

R

a 75% and 0.73	**b** $\frac{4}{25}$ and 14%	**c** $\frac{2}{3}$ and 0.64
d 18% and $\frac{1}{7}$	**e** 0.22 and $\frac{1}{4}$	**f** 55% and $\frac{6}{11}$
g 0.93 and $\frac{19}{20}$	**h** 60% and $\frac{13}{25}$	**i** $\frac{1}{6}$ and 0.08

8 Arrange each set of numbers in ascending order.

R

a $\frac{4}{5}$, 78%, 0.75, $\frac{9}{11}$ **b** 22%, $\frac{1}{4}$, 0.29, $\frac{7}{20}$ **c** 0.62, $\frac{3}{5}$, 57%, 0.605

9 Which list of numbers is arranged in descending order? Select the correct answer **A**, **B**, **C** or **D**.

R

A $\frac{4}{11}$, $\frac{2}{5}$, 0.41, 43% **B** $\frac{2}{5}$, 0.41, $\frac{4}{11}$, 43%

C 43%, 0.41, $\frac{2}{5}$, $\frac{4}{11}$ **D** 0.41, $\frac{4}{11}$, 43%, $\frac{2}{5}$

10 Arrange each set of numbers in descending order.

R

a $\frac{9}{20}$, 0.47, $\frac{2}{5}$, 43% **b** 0.08, 86%, $\frac{21}{25}$, 0.88 **c** 0.905, $\frac{19}{20}$, 91%, $\frac{9}{10}$

Foundation Standard Complex

☆ MENTAL SKILLS 6A ANSWERS ON P. 632 Maths without calculators

Quiz Mental skills 6A

6.04

Finding 10%, 20% and 5%

To find 10% or $\frac{1}{10}$ of a number, simply divide the number by 10 by moving the decimal point one place to the left.

1 Study each example.

a 10% × 150 = 15 0. = 15 **b** 10% × $1256.80 = $125 6.8 = $125.68

c 10% × 4917 = 491 7. = 491.7 **d** 10% × $48.55 = $48.55 = $4.885

2 Now find 10% of each amount.

a 190 **b** $75 **c** 875 **d** $202

e $37.60 **f** 400 **g** $9.25 **h** 896

i $2700 **j** $3.80 **k** $1527.60 **l** $72.50

m 3154 **n** $10.70 **o** 426 **p** $24 317.60

20% is 10% doubled so to find 20% of a number, first find 10% then double it.

3 Study each example.

a 20% × 700
10% × 700 = 70
∴ 20% × 700 = 70 × 2 = 140

b 20% × $876
10% × $876 = $87.60
∴ 20% × $876 = $87.60 × 2 = $175.20

c 20% × 325
10% × 325 = 32.5
∴ 20% × 325 = 32.5 × 2 = 65

d 20% × $38.50
10% × $38.50 = $3.85
∴ 20% × $38.50 = $3.85 × 2 = $7.70

4 Now find 20% of each amount.

a 50 **b** 620 **c** $2450 **d** $8.60

e 72 **f** $12 700 **g** 390 **h** $5.80

i $45 **j** $84 **k** $4600 **l** 320

5% is half of 10%, so to find 5% of a number, first find 10% and then divide it by 2.

5 Study each example.

a 5% × 180
10% × 180 = 18
∴ 5% × 180 = 18 ÷ 2 = 9

b 5% × $76
10% × $76 = $7.60
∴ 5% × $76 = $7.60 ÷ 2 = $3.80

c 5% × 120
10% × 120 = 12
∴ 5% × 12 = 12 ÷ 2 = 6

d 5% × $142.20
10% × $142.20 = $14.22
∴ 5% × $142.20 = $14.22 ÷ 2 = $7.11

6 Now find 5% of each amount.

a 2000 **b** $12 **c** 50 **d** $27

e $36.80 **f** $84 **g** 800 **h** 130

i $9.60 **j** $138 **k** $72 **l** 840

6.05 Fraction and percentage of a quantity

Videos
Fraction and percentage of a quantity
Fraction of a quantity
Percentages: Tax breaks

Puzzles
Percentage crossnumber
Percentages
Fractions of words
Percentages of an amount

Worksheet
Percentages of an amount

Example 12

Find:

a $\frac{1}{5}\times 45$

b $\frac{7}{8}\times \$32$

c $\frac{1}{3}$ of one year

d $\frac{4}{5}$ of 2 kg (grams)

SOLUTION

a $\frac{1}{5}\times 45 = 45 \div 5$
$= 9$

b $\frac{7}{8}\times \$32 = \left(\frac{1}{8}\times \$32\right)\times 7$
$= \$4 \times 7$
$= \$28$

c $\frac{1}{3}$ of 1 year $= \frac{1}{3}\times 12$ months
$= 4$ months

Convert one year to 12 months.

d $\frac{4}{5}$ of 2 kg $= \frac{4}{5}\times 2000$ g
$= \left(\frac{1}{5}\times 2000 \text{ g}\right)\times 4$
$= 400 \text{ g} \times 4$
$= 1600$ g

Convert 2 kg to 2000 g.

ⓘ Percentage of a quantity

To find a **percentage of a quantity**, calculate:

$$\frac{\text{percentage}}{100}\times \text{quantity}$$

OR percentage ÷100 × quantity

Video
Fraction and percentage of a quantity

Example 13

Find:

a 8% of \$400

b 12.5% of 1 h (min)

c 20% of 3 m (cm).

SOLUTION

a 8% of \$400 $= \frac{8}{100}\times \$400$
$= \$32$

Or 8 ÷ 100 × \$400 or 0.08 × \$400

b 12.5% of 1 h = 12.5% × 60 min
$= \frac{12.5}{100}\times 60$ min
= 7.5 min

Convert 1 h to 60 min
or 12.5 ÷ 100 × 60 or 0.125 × 60

c 20% of 3 m $= 20\% \times 300$ cm — Convert 3 m to 300 cm

$= \frac{20}{100} \times 300$ cm — or $20 \div 100 \times 300$ or 0.2×300

$= 60$ cm

EXERCISE 6.05 ANSWERS ON P. 632

Fraction and percentage of a quantity

1 Find: (EXAMPLE 12)

- **a** $\frac{3}{5} \times 40$
- **b** $\frac{1}{4} \times 28$
- **c** $\frac{1}{6} \times 24$
- **d** $\frac{2}{3} \times 15$
- **e** $\frac{7}{10} \times 60$
- **f** $\frac{5}{8} \times 16$
- **g** $\frac{3}{4}$ of 1 km (metres)
- **h** $\frac{1}{3}$ of 1 day (h)
- **i** $\frac{2}{5}$ of 1 L (mL)
- **j** $\frac{1}{8}$ of 1 tonne (kg)
- **k** $\frac{5}{6}$ of 1 year (months)
- **l** $\frac{7}{12}$ of 1 h (min)

2 What is 35% of \$75? Select the correct answer **A**, **B**, **C** or **D**.

- **A** \$40
- **B** \$110
- **C** \$26.25
- **D** \$48.75

3 Find: (EXAMPLE 13)

- **a** 11% of \$500
- **b** 2% of 250 kg
- **c** 13% of 150 L
- **d** 21% of 600 cm
- **e** 15% of \$450
- **f** 5% of 5700 g
- **g** 42% of 1128 m
- **h** 112% of 256 km
- **i** 19.4% of 785 mL
- **j** 7.1% of \$220
- **k** 23.6% of \$380
- **l** 11.3% of 403 kg
- **m** 110% of 95
- **n** 150% of 302
- **o** 130% of \$2010
- **p** 105% of 120 m
- **q** 170% of 350 g
- **r** 115% of \$400

4 Taffy the cat had a mass of 2.7 kg when she was found. If her mass increased by $\frac{1}{4}$, how much did she gain?

5 A **discount** is a saving made between the original price of an item and the cheaper price.

- **a** How much do you *save* if you get a 25% discount on a \$420 games system?
- **b** What would be the discount price of the games system?

6 An Airbus can seat 351 passengers. If the plane flew with $\frac{2}{3}$ of the seats occupied, how many passengers were on the flight?

7 Calculate each amount (convert to a smaller unit first).

- **a** 12% of 8 m
- **b** 35% of 1 tonne
- **c** $\frac{1}{5}$ of 12 h
- **d** $2\frac{1}{2}\%$ of 3 L
- **e** $\frac{3}{4}$ of 15 kg
- **f** $7\frac{1}{2}\%$ of 12 km
- **g** 35% of 10 days
- **h** $\frac{1}{8}$ of 4 h
- **i** 67.5% of 40 g

☐ Foundation ○ Standard ○ Complex

8 12% of the 525 swimmers at the swimming carnival wore black caps. How many swimmers wore black caps?

9 Haroula earns $17.20 an hour in her part-time job. If she is given a 5% pay increase, by how much does her hourly rate increase?

10 If 42% of the 650 road deaths occurred on country roads, how many road deaths happened in the country?

11 In a city of 3 million people, 1% of the population are doctors. How many doctors are there? Select the correct answer **A**, **B**, **C** or **D**.

A 3 **B** 3000 **C** 300 **D** 30 000

12 $\frac{1}{20}$ of the cars produced in a factory were found to have steering defects. If 340 cars were produced, how many had steering defects?

13 How many children are there in a crowd of 40 530 if $\frac{3}{5}$ are children?

14 If 70% of the seats at a rock concert must be sold to make a profit, how many must be sold in a 2000-seat theatre?

15 Copy and complete each equation.

R **a** 20% × ______ = $18 **b** 15% × ______ = $75

TECHNOLOGY

Discounts

In this activity, you will calculate the discount and sale price of items in a department store given their original price.

1 Enter the following data into a spreadsheet. Make the headings **bold** in row 1.

	A	B	C	D	E
1	Item	Original Price	% Discount	Discount	Sale Price
2	Backpack	$69.95	10%	=B2*C2	
3	Headphones	$38.00	5%		
4	Beach towel	$20	30%		
5	Book	$29.95	15%		
6	Swimwear	$60	20%		
7	Media player	$145	10%		

2 Cell D2 shows the formula for calculating the discount on the backpack. Right click on D2 and **Fill Down** to copy the formula down to cell D7. This will calculate the discount that applies to each item.

3 In cell E2, enter the formula **=B2-D2** to calculate the sale price of the backpack. Use **Fill Down** to calculate the sale prices of the remaining items.

Foundation Standard Complex

Expressing amounts as fractions and percentages

Worksheets
Calculating percentages

Money and Percentages review

Percentages

Puzzle
Calculating percentages

Video
Expressing quantities as fractions and percentages

Technology
Online discounts

Coffee shop sales

ⓘ Expressing amounts as fractions and percentages

To write an amount as a fraction of a whole amount:

- write the amount in the numerator of the fraction.
- write the whole amount (total) in the denominator.

$$\text{Fraction} = \frac{\text{amount}}{\text{whole amount}}$$

To write an amount as a percentage of a whole amount:

- write the amount in the numerator of a fraction.
- write the whole amount (total) in the denominator.
- multiply by 100%.

$$\text{Percentage} = \frac{\text{amount}}{\text{whole amount}} \times 100\%$$

Example 14

There were 250 people at the school fete, and 160 of them were children.

a What fraction of the people at the fete were children?

b What percentage of the people at the fete were children?

SOLUTION

a $\text{Fraction} = \frac{\text{Number of children}}{\text{Total number of people}}$

$= \frac{160}{250}$

$= \frac{16}{25}$

b $\text{Percentage} = \frac{160}{250} \times 100\%$

$= 64\%$

Example 15

Express 36 minutes as:

a a fraction of an hour

b a percentage of an hour

SOLUTION

Quantities need to be expressed in the same units, so change 1 hour to 60 minutes.

a $\text{Fraction} = \frac{36}{60} = \frac{3}{5}$

b $\text{Percentage} = \frac{36}{60} \times 100\%$

$= 60\%$

EXERCISE 6.06 ANSWERS ON P. 632

Expressing amounts as fractions and percentages

U F PS R

1 Convert each test mark into a simplified fraction.

a 50 out of 100 **b** 38 out of 50 **c** 87 out of 100

d 8 out of 12 **e** 5 out of 20 **f** 45 out of 120

2 Convert each test mark in question **1** into a percentage.

3 A hockey team scored 8 goals. If one player scored 5 of them, what is this as a percentage of the team score?

4 In a class of 25 students, 6 ride bikes, 10 walk to school and the rest catch the bus.

a What fraction walk to school?

b What percentage catch the bus?

5 Wakeel answered 21 questions correctly out of 24 in his driving test. If the pass mark is 95%, did he pass?

PS R

6 The World Cup cricket final attracted 86 000 people to the game, but 94 500 tickets were sold before the match.

a What fraction of the sold tickets were used?

b What percentage (correct to the nearest whole number) of the sold tickets were not used?

7 Tania earns $1340 a week. She pays $428.80 in tax and saves $180 a week.

a What percentage of Tania's earnings is paid in tax?

b What fraction of her earnings does Tania save?

c What percentage (correct to one decimal place) of her earnings does Tania save?

8 Louise sells a house for $458 000. If the real estate agent is paid a commission of $22 900, what percentage of the sale price does the agent receive?

Foundation Standard Complex

9 The Great Gals are having a sale on microwave ovens. The sale price of each oven is listed below, along with the discount.

PS R

i

$149
Save $20

ii

$179
Save $26

iii

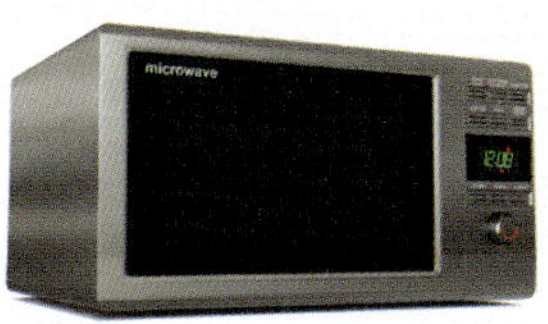

$219
Save $30

Shutterstock.com/Markus Gann
Shutterstock.com/Maxx-Studio
Dreamstime.com/Pioneer111

a Calculate the original price of each microwave oven.

b Calculate, to one decimal place, the percentage discount on each oven.

c Which oven has the greatest percentage discount?

10 Jamal bought a pair of roller skates for $180 and sold them at a profit of $45. What is the profit as a percentage? Select the correct answer **A**, **B**, **C** or **D**.

PS

A 25% **B** 4% **C** 45% **D** 20%

11 Express each measurement as a simplified fraction.

EXAMPLE 15

a 5 min of 1 h **b** 250 mL of 1 L **c** 700 m of 1 km
d 230 kg of 1 tonne **e** 75c of $6 **f** 40 min of 4 h
g 300 g of 2 kg **h** $3.80 of $14 **i** 12 h of 4 days
j 75 mm of 20 cm **k** 800 m of 1.5 km **l** 400 mL of 3.5 L

12 Express each measurement in question **11** as a percentage.

13 What is 40 minutes as a percentage of two hours? Select the correct answer **A**, **B**, **C** or **D**.

A 20% **B** 5% **C** $33\frac{1}{3}\%$ **D** 0.33%

14 A football team is scoring well in its matches if its points percentage is over 100, according to the formula:

$$\text{Points percentage} = \frac{\text{Points for}}{\text{Points against}} \times 100\%$$

where 'points for' are the total number of points the team has scored and 'points against' are the total number of points the other teams have scored when playing against that team.

Foundation Standard Complex

PS R

a Calculate, correct to one decimal place, the points percentage for each team listed in the table.

Team	Points for	Points against	Team	Points for	Points against
Broncos	391	313	Raiders	405	466
Bulldogs	410	366	Roosters	296	479
Cowboys	383	452	Sea Eagles	434	235
Dragons	321	395	Sharks	328	279
Eels	407	352	Storm	478	207
Knights	332	382	Tigers	498	642
Panthers	357	469	Titans	319	399
Rabbitohs	275	283	Warriors	688	454

b What are the top four teams based on these results?

DID YOU KNOW?

Where are the percentages?

Percentages are all around us.

- Financial institutions advertise interest rates they pay on investments and charge on loans as percentages.
- Companies report profits or losses as percentages. They also report dividends paid as percentages.
- When the value of a product increases or the cost to the consumer of different items increases, it is reported as a percentage.
- Many statistics about the population are given as percentages.

An understanding of percentages is vital to understanding the world we live in.

Find 5 other situations where percentages are used.

6.06

INVESTIGATION

Success rates in netball and basketball

Percentages are a good way of comparing sporting performances. Commentators often give statistics in the form of percentages so that we can compare the success of teams or players.

1 In netball, goal shooters rarely have the same number of shooting chances. To work out their success rates, we must use percentages to compare the number of shooting chances with the number of successful shots.

The following statistics were recorded in a match between Australia and New Zealand.

	Australia		New Zealand	
	Shooting chances	Goals	Shooting chances	Goals
1st Quarter	15	11	9	9
2nd Quarter	25	17	14	13
3rd Quarter	18	14	23	20
4th Quarter	27	20	18	17

a Calculate as a percentage correct to one decimal place:

i each team's success rate for each quarter.

ii the total success rate for each team.

b Comment on the relationship between the success rates and the final scores. What can you say about each team?

2 At any stage during the season in the National Basketball League (NBL), the teams will have played different numbers of games. The highest position on the ladder is awarded to the team with the highest **wins percentage**, which is calculated as follows:

$$\text{Wins percentage} = \frac{\text{Number of games won}}{\text{Number of games played}} \times 100\%$$

At one stage of a season the teams had won and lost the number of games shown below.

Team	Won	Lost
Adelaide 36ers	6	6
Cairns Taipans	8	7
Brisbane Bullets	7	4
Melbourne United	6	9
New Zealand Breakers	9	4
Perth Wildcats	3	11
Sydney Kings	11	3
SE Melbourne Phoenix	8	4
Illawarra Hawks	2	12

Calculate the wins percentage for each team and construct the 'NBL Ladder' showing the teams in the correct order.

6.07 Percentage increase and decrease

Video Percentage increase and decrease

Hyperinflation: 1920s Germany

The most populous country

Worksheets Discounts and special offers

Discounts

- Percentage increase means to increase (make bigger) a quantity by a percentage.
- Percentage decrease means to decrease (make smaller) a quantity by a percentage.

Example 16

Increase \$200 by 7%.

SOLUTION

Method 1

$7\%\text{ of }\$200 = \frac{7}{100} \times \$200 = \$14$ — Or 7 ÷ 100 × \$200 or 0.07 × \$200

$\$200 + \$14 = \$214$

Method 2

$(100\% + 7\%)\text{ of }\$200 = 107\%\text{ of }\$200 = \frac{107}{100} \times \$200 = \$214$

Increasing by 7% is the same as calculating 107%.

Or 107 ÷ 100 × \$200 or 1.07 × \$200

Example 17

Video Percentage increase and decrease

Puzzle Discounts

Decrease \$150 by 12%.

SOLUTION

Method 1

$12\%\text{ of }\$150 = \frac{12}{100} \times \$150 = \$18$ — Or 12 ÷ 100 × \$150 or 0.12 × \$150

$\$150 - \$18 = \$132$

Method 2

$(100\% - 12\%)\text{ of }\$150 = 88\%\text{ of }\$150 = \frac{88}{100} \times \$150 = \$132$

Decreasing by 12% is the same as calculating 88%.

Or 88 ÷ 100 × \$150 or 0.88 × \$150

Example 18

The price of a watch increases by 15%. If its original price was \$35, find its new price.

SOLUTION

Method 1

15% of \$35 = \$5.25

\$35 + \$5.25 = \$40.25

Method 2

$$(100\% + 15\%) \text{ of } \$35 = 115\% \text{ of } \$35$$
$$= \$40.25$$

Example 19

Find the price of a computer game system, originally priced at \$420, after a 9% discount.

Shutterstock.com/Saikorn

SOLUTION

Method 1

9% of \$420 = \$37.80

\$420 − \$37.80 = \$382.20

Method 2

$$(100\% - 9\%) \text{ of } \$420 = 91\% \text{ of } \$420$$
$$= \$382.20$$

EXERCISE 6.07 ANSWERS ON P. 632

Percentage increase and decrease

U F PS R C

1 Increase: EXAMPLE 16

a \$150 by 5%   b 400 by 20%   c 60 km by 22%

d \$2500 by 6%   e 95 kg by 60%   f 10 L by 33%

2 Decrease: EXAMPLE 17

a \$440 by 60%   b 120 by 15%   c 110 kg by 8%

d \$325 by 25%   e 2000 L by 38%   f \$1570 by 3%

3 Wooden chairs cost \$172 to make. Calculate the selling price if the chairs are marked up by 35% when sold. PS EXAMPLE 18

☐ Foundation ○ Standard ⬡ Complex

4 Julie buys a bike for $2700 and sells it a year later, making a 15% profit.

a How much profit did Julie make?

b What was the selling price of the bike?

5 Geeva's weekly pay of $980 increased by 4.5%. What is his new pay?

6 Yumi's height of 168 cm increased by 2%. What is her new height?

7 A department store has a mark-up of 200% on clothing. If the store buys a vest for $12, what will be its selling price after the mark-up? Select the correct answer **A**, **B**, **C** or **D**.

A $32	**B** $14.40	**C** $24	**D** $36

EXAMPLE 19

8 A car dealer offers a 15% discount on all new car purchases. What would you pay for a car marked at $21 990?

9 Calculate the sale price of a pair of sports shoes marked at $165 after a discount of 12%.

10 After speed cameras were installed, the road toll of 840 deaths decreased by 5%. What is the new road toll?

11 Aaron bought a house for $464 000 and sold it 10 years later, making a 150% profit. Calculate the selling price of the house.

12 What is the new price of a pair of jeans worth $75 if they are discounted by 25%? Select the correct answer **A**, **B**, **C** or **D**.

A $50	**B** $56.25	**C** $18.75	**D** $55

13 A netball sells for $45, but you receive 11% discount if you buy two. What is the discounted price of two netballs?
PS

14 The owner of a store buys a tablet device for $470. She adds a mark-up of $70 but, at sale time, offers a 30% discount.
PS

a Find the selling price after the mark-up.

b Find the discounted price at sale time.

c How much loss did the store owner makes on the tablet device?

d Calculate the percentage loss correct to one decimal place.

15 Find the number which, if increased by 18%, gives 767.
R

16 Winter coats priced at $830 were reduced by 15% at the end-of-season sale. On the last weekend of the sale, they were reduced by a further 40% off the discounted price. What was the final price of a coat?
PS R

17 Find the number which, if decreased by 40%, gives 306.
R

18 A portable air conditioner with an original price of $680 has 10% GST added to it. It is then sold at an end-of-year sale for '10% off'. Is the sale price of the air conditioner more than, less than or equal to its original price? Justify your answer by calculation.
PS R C

Foundation Standard Complex

Percentages without calculators

6.08

If we know the equivalent fractions for common percentages, then we can find percentages mentally.

Skillsheet Mental percentages

Fraction	$\frac{1}{4}$	$\frac{1}{2}$	$\frac{3}{4}$	$\frac{1}{3}$	$\frac{2}{3}$	$\frac{1}{5}$	$\frac{2}{5}$	$\frac{3}{5}$	$\frac{1}{10}$	$\frac{1}{8}$
Percentage	25%	50%	75%	$33\frac{1}{3}\%$	$66\frac{2}{3}\%$	20%	40%	60%	10%	$12\frac{1}{2}\%$

Example 20

Calculate each percentage mentally.

a $33\frac{1}{3}\%$ of \$1800 **b** $12\frac{1}{2}\%$ of 72 kg

SOLUTION

a $33\frac{1}{3}\%$ of $\$1800 = \frac{1}{3}$ of \$1800
$= \$600$

b $12\frac{1}{2}\%$ of 72 kg $= \frac{1}{8}$ of 72 kg
$= 9$ kg

Example 21

a Increase \$470 by 10%.

b Decrease \$75 by 20%.

SOLUTION

a 10% of $\$470 = \frac{1}{10} \times \$470$
$= \$47$
$\$470 + \$47 = \$517$

b 20% of $\$75 = \frac{1}{5} \times \$75$ (or $10\% \times \$75 \times 2$)
$= \$15$
$\$75 - \$15 = \$60$

Note: More examples and exercises on increasing and decreasing an amount by a percentage mentally can be found in Mental skills 6B: Percentage increase and decrease on p. 269.

EXERCISE 6.08 ANSWERS ON P. 632

Percentages without calculators

U F PS R

Complete this exercise without using a calculator.

EXAMPLE 20

1 Find 25% of each amount.

a 32 kg **b** 180 mm **c** 24 L **d** \$1000

2 Find 40% of \$60. Select the correct answer **A**, **B**, **C** or **D**.

A \$6 **B** \$12 **C** \$24 **D** \$48

Foundation Standard Complex

3. Find each amount.

a) 50% of 36 m
b) 75% of 12 g
c) $12\frac{1}{2}$% of 24 h
d) 20% of 5 L
e) $66\frac{2}{3}$% of $30
f) 60% of 20 km
g) 25% of 20 mL
h) $33\frac{1}{3}$% of 18 days
i) 75% of 100 kg

4. Find 10% of:

a) 60 tonnes
b) 40 h
c) 300 sheep
d) $90

5. Find each amount.

a) 10% of 70
b) 50% of $80
c) 40% of 120 min
d) 30% of $210
e) 60% of 250 cm
f) 20% of 400 mL

6. What is 5% of 80 days? Select the correct answer **A**, **B**, **C** or **D**.

A 8 days
B 4 days
C 40 days
D 80 days

7. Find $12\frac{1}{2}$% of each amount.

a) $72
b) 40 kg
c) 160 kL

EXAMPLE 21

8. Increase:

a) $80 by 25%
b) 140 by 10%
c) $150 by $33\frac{1}{3}$%
d) 28 by 50%
e) $240 by $12\frac{1}{2}$%
f) 45 by 20%

9. Decrease:

a) $310 by 10%
b) 120 by 20%
c) $80 by $12\frac{1}{2}$%
d) 60 by 5%
e) $900 by $66\frac{2}{3}$%
f) 2000 by 75%

10. A class of 28 students sat for a History test out of 40. After 25 minutes, Leonie had already answered 60% of the test correctly.

PS R

a) How many marks has Leonie already scored?
b) If Leonie's goal is to obtain 85% or more in the test, how many more marks will she need?

Foundation Standard Complex

☆ MENTAL SKILLS 6B ANSWERS ON P. 633 Maths without calculators

Quiz Mental skills 6B

6.08

Percentage increase and decrease

The fraction equivalents of commonly used percentages can help us when we need to increase or decrease a number by a percentage.

Percentage	10%	$12\frac{1}{2}\%$	20%	25%	$33\frac{1}{3}\%$	50%	$66\frac{2}{3}\%$	75%
Fraction	$\frac{1}{10}$	$\frac{1}{8}$	$\frac{1}{5}$	$\frac{1}{4}$	$\frac{1}{3}$	$\frac{1}{2}$	$\frac{2}{3}$	$\frac{3}{4}$

1 Consider each example.

a Increase 80 by 20%.

$$\begin{aligned}20\% \text{ of } 80 &= \tfrac{1}{5}\times 80 \\ &= 80\div 5 \\ &= 16\end{aligned} \quad \text{or} \quad \begin{aligned}10\% \text{ of } 80 &= 8 \\ 20\% \text{ of } 80 &= 8\times 2 \\ &= 16\end{aligned}$$

$$80\div 16 = 96$$

b Increase \$36 by $66\frac{2}{3}\%$.

$$\begin{aligned}66\tfrac{2}{3}\% \text{ of } 36 &= \tfrac{2}{3}\times 36 \\ &= \$36\div 3\times 2 \\ &= \$12\times 2 \\ &= \$24\end{aligned}$$

$$\$36+\$24=\$60$$

2 Now increase:

a \$280 by 10% **b** 45 by $33\frac{1}{3}\%$ **c** 25 by 20% **d** \$400 by $12\frac{1}{2}\%$

e 64 by 50% **f** \$72 by 25% **g** \$55 by 10% **h** 90 by $66\frac{2}{3}\%$

i 120 by 75% **j** \$80 by 5% **k** \$250 by 20% **l** 70 by 40%

3 Consider each example.

a Decrease 225 by $33\frac{1}{3}\%$.

$$\begin{aligned}33\tfrac{1}{3}\% \text{ of } 225 &= \tfrac{1}{3}\times 225 \\ &= 255\div 3 \\ &= 75\end{aligned}$$

$$225-75=150$$

b Decrease \$70 by 15%.

$$10\% \text{ of } \$70 = \tfrac{1}{10}\times \$70 = \$7$$

$$\therefore 5\% \text{ of } \$70 = \tfrac{1}{2}\times \$7 = \$3.50$$

$$\begin{aligned}\therefore 15\% \text{ of } \$70 &= (10\%\times \$70)+(5\%\times \$70) \\ &= \$7+\$3.50 \\ &= \$10.50\end{aligned}$$

$$\$70-\$10.50=\$59.50$$

4 Now decrease:

a \$480 by 25% **b** 60 by $33\frac{1}{3}\%$ **c** 110 by 20% **d** \$25 by 10%

e 900 by 50% **f** \$72 by $12\frac{1}{2}\%$ **g** \$320 by 75% **h** 150 by $66\frac{2}{3}\%$

i \$63 by 20% **j** \$100 by $12\frac{1}{2}\%$ **k** 250 by 10% **l** \$150 by 30%

6.09 The unitary method

Worksheets
The unitary method

Working with percentages

Video
The unitary method

If we know the percentage of an amount, but not the actual amount, we can use the **unitary method** to find the whole amount. The word **unit** means 'one', and with the unitary method we find 1% first.

Example 22

If 15% of an amount is $75, what is the amount?

SOLUTION

15% of amount = $75

1% of amount = $75 ÷ 15 = $5 — Find 1% first by dividing by 15.

100% of amount = $5 × 100 = $500 — Then find the whole amount by multiplying by 100.

The amount is $500. — Check: 15% × $500 = $75

ⓘ The unitary method

To find an amount given a **percentage** of the amount:

- first find 1% of the amount by dividing by the known percentage.
- then multiply by 100 to find the whole amount (100%).

Example 23

Farmer Kate lost 30% of her sheep in a flood. If she had 560 sheep left, how many sheep did she have before the flood?

SOLUTION

Because 30% of the total were lost,
100% − 30% = 70% of the total were left.
So: 70% of the total = 560 sheep

1% of the total = 560 ÷ 70
= 8 sheep

100% of the total = 8 × 100
= 800 sheep

Kate had 800 sheep before the flood. — Check: 70% × 800 = 560

EXERCISE 6.09 ANSWERS ON P. 633

6.09

The unitary method

U F PS R

EXAMPLE 22

1 What is the whole amount if:

a 10% of it is $70? **b** 25% of it is $140? **c** 17% of it is $782?

d 8.5% of it is $161.50? **e** 16% of it is 64 cm? **f** 20% of it is 56 kg?

g 42% of it is 1.26 m? **h** 30% of it is 45 min? **i** 70% of it is $280?

j 110% of it is $396? **k** 45% of it is 180 kg? **l** 120% of it is 132 min?

2 The deposit of 15% on a kitchen stove is $180. What is the price of the stove?

PS R

3 There were 51 870 men in the crowd at the football match. If this was 65% of the crowd, what was the total attendance?

PS R

4 Last year Joanna paid $23 009.28 in income tax. If this represents 28% of her income, calculate her income.

PS R

5 Jude sold his car for $12 200, making a loss of 20% on the original price. What was the original price?

6 Pip makes 18% profit on cakes she decorates. If she sells a cake for $330.40, what was the original cost of the cake? Select the closest answer **A**, **B**, **C** or **D**.

PS R

A $184 **B** $238

C $271 **D** $280

7 Binnsville's population increased by 5% or 1200 people. What was the old population?

PS R

8 The interest earned on an investment is $168. If the interest rate is 5.6%, how much was invested?

PS R

9 Venus buys a dress that has been discounted by 25%. If the sale price is $225, what was the original price of the dress?

PS R

10 Matthew buys a tent for $348 at a discount of 20%. What was the original price of the tent?

PS R

Foundation Standard Complex

6.10 Profit, loss and GST

Worksheet Profit and loss

Technology Profit and loss

Video Business maths

Puzzle Profit and loss

Video Profit and loss

Profit and loss

Retailers (store owners) buy goods and sell them to people.

- **Cost price** is how much they buy an item for.
- **Selling price** is how much they sell them for.
- If they sell an item for more than what they paid for it, they make a **profit**.
- If they sell an item for less than what they paid for it, they make a **loss**.

The **percentage profit** or **percentage loss** is usually calculated as a percentage of the cost price.

Example 24

The Book Bin buys books for \$18 each and sells them for \$22.50. For each book, calculate:

a the profit **b** the percentage profit

SOLUTION

a Cost price = \$18

Selling price = \$22.50

Profit = \$22.50 − \$18 (Selling price − cost price)

= \$4.50

b Percentage profit $= \frac{\$4.50}{\$18} \times 100\%$ ($\frac{\text{profit}}{\text{cost price}} \times 100\%$)

$= 25\%$

Video Profit and loss

Example 25

Renae buys a car for \$17 500 and sells it two years later for \$15 000. Find the percentage loss, correct to one decimal place.

SOLUTION

Cost price = \$17 500

Selling price = \$15 000

Loss = \$17 500 − \$15 000 (Cost price − selling price)

= \$2500

Percentage loss $= \frac{\$2500}{\$17\,500} \times 100\%$ ($\frac{\text{loss}}{\text{cost price}} \times 100\%$)

$= 14.2857\ldots\%$

$\approx 14.3\%$

Example 26

Cooper buys a mountain bike for \$850 and sells it a year later at 13% profit. Find the selling price.

SOLUTION

$$\text{Profit} = 13\% \times \$850$$
$$= \$110.50$$
$$\text{Selling price} = \$850 + \$110.50$$
$$= \$960.50$$

$$\text{OR Selling price} = (100\% + 13\%) \times \$850$$ Increasing \$850 by 13%
$$= 113\% \times \$850$$
$$= \$960.50$$

GST

Goods and Services Tax (GST) is a tax paid to the government on most goods (items) and services that we purchase. In Australia, GST is charged at 10% of the original price and is generally included in the marked price of the good or service.

Example 27

The selling price of a TV is \$722 + 10% GST. Calculate:

a the GST payable.

b the selling price.

Video
GST

SOLUTION

a $\text{GST} = 10\% \text{ of } \$722$
$= \$72.20$

b $\text{Selling price} = \$722 + \$72.20$
$= \$794.20$

OR $\text{Selling price} = 110\% \times \$722$
$= \$794.20$

Example 28

The selling price of a lounge suite is \$2695 with GST included. How much of this price is the GST?

Video
GST

SOLUTION

Selling price + GST = 100% + 10% = 110%

110% of the selling price = \$2695

$$10\% \text{ GST} = \$2695 \div 11$$
$$= \$245$$

EXERCISE 6.10 ANSWERS ON P. 633

Profit, loss and GST

U F PS R

1 In each situation, state:

- **i** the cost price.
- **ii** whether a profit or loss was made.

- **a** A car was bought for \$18 700 and sold for \$15 000.
- **b** A house was bought for \$543 000 and sold for \$674 000.
- **c** A video camera was sold for \$350 when it originally cost \$799.
- **d** An antique desk sold for \$8000 when it was purchased for \$6800.

EXAMPLE 24

2 For each situation, find:

- **i** the profit.
- **ii** the percentage profit, correct to one decimal place where necessary.

- **a** cost price \$85 selling price \$102
- **b** cost price \$415 selling price \$600
- **c** cost price \$2.75 selling price \$4.20

EXAMPLE 25

3 For each situation, find:

- **i** the loss.
- **ii** the percentage loss, correct to one decimal place where necessary.

- **a** cost price \$19 selling price \$14
- **b** cost price \$2700 selling price \$1450
- **c** cost price \$79 selling price \$58

4 Anthony bought a computer for \$519 and sold it three years later for \$220. Calculate his percentage loss, correct to one decimal place. Select the closest answer **A**, **B**, **C** or **D**.

A 26.2% **B** 42.4% **C** 57.6% **D** 172.8%

5 A 3D printer costing \$1250 is sold for \$1400. Find:

a the profit **b** the percentage profit

6 A car costing \$17 000 is sold for \$15 200. Find:

a the loss **b** the percentage loss, correct to one decimal place

EXAMPLE 26

7 Calculate the selling price for each situation.

- **a** cost price \$200, profit 17%
- **b** cost price \$42, profit 70%
- **c** cost price \$720, loss 35%
- **d** cost price \$1400, loss 9.5%
- **e** cost price \$2.50, profit 150%
- **f** cost price \$27, loss 40%

Foundation Standard Complex

9780170465601

8 A block of land is purchased by Erin and Mark for \$133 000. Six years later, they sell it for \$164 000. Calculate their percentage profit, correct to two decimal places.
PS

9 Rose buys a bed for \$2700. She later sells it for \$1500. Calculate her percentage loss, correct to one decimal place.
PS

10 For each item, calculate the GST payable and the final price.

- **a** a car priced at \$20 900
- **b** a home entertainment unit priced at \$1810
- **c** an ice cream priced at \$3.00
- **d** a theatre ticket priced at \$145
- **e** plumber's fees of \$180
- **f** a desk lamp priced at \$27

11 Given that the final price of each item has GST included, calculate the GST and the original price, correct to the nearest cent.
PS R

- **a** accountant's fees: \$792
- **b** piano lessons: \$198 per term
- **c** refrigerator: \$924
- **d** box of chocolates: \$18.70
- **e** travel guide: \$34.65
- **f** diary: \$16.70

12 A manufacturer sells TVs to stores for \$899 each. *Electric City* store adds on 60% profit for its selling price. It then must add on 10% GST
PS R

- **a** Calculate the store's selling price of one TV before GST is added. Round your answer to the nearest \$10.
- **b** Calculate the final selling price of each television after GST is added.
- **c** In the January sales all goods are sold at 40% off the marked price. Calculate the discounted price of a TV.
- **d** Find how much of this discounted price will be paid in GST by dividing by 11.
- **e** After paying GST to the government, does the store make a profit or a loss on each TV? How much profit or loss?
- **f** Calculate the percentage profit or loss. Round your answer to one decimal place.

INVESTIGATION

Uses of percentages

1. Prepare a one-page percentage collage for class display. Look through newspapers, magazines and brochures to cut out, for example. Find photographs or draw situations where percentages are used.
2. Select two examples of situations where percentages are used. Write a set of 10 questions for each example, using the rules about percentages you have learnt. Calculate the answers.
3. Swap the questions you wrote in question **2** with others in the class. Each person should answer at least three sets of questions. Ask the student who prepared the questions to mark your answers. If there are any disagreements, check with your teacher.

6.11 Simple interest

Worksheet
Simple interest

Banks, credit unions and other financial institutions reward you by paying you interest on your savings or investments. However, they also charge you interest if you borrow money from them.

- The original amount of money invested or borrowed is called the **principal**.
- **Interest** is calculated as a percentage of the principal.
- This percentage is called the **interest rate**, usually written as a rate **per annum** ('per year' or 'p.a.').
- **Simple interest** (or **flat rate interest**) is interest calculated simply on the original principal.

Example 29

Find the simple interest paid on \$7500 borrowed at 8% p.a. for:

a one year **b** three years

SOLUTION

a Interest = 8% of \$7500
= \$600

b Interest = 8% of \$7500 × 3
= \$1800

ⓘ The simple interest formula

$I = Prn$, where

I is the simple interest.

P is the principal.

r is the interest rate per year, expressed as a decimal.

n is the number of years.

Applying this to the above example,

$P = \$7500$, $r = 8\% = 0.08$, $n = 3$ years

$I = Prn$

$= \$7500 \times 0.08 \times 3$

$= \$1800$

The simple interest formula can also be written as $I = Pin$ or $I = Prt$.

Example 30

Find the simple interest earned on $8620 invested at 2.4% p.a. for 7 months.

SOLUTION

$P = \$8620$, $r = 2.4\% = 0.024$,

$n = 7 \text{ months} = \frac{7}{12} \text{ years}$

Divide by 12 to convert 7 months into years.

$I = Prn$

$= \$8620 \times 0.024 \times \frac{7}{12}$

$= \$120.68$

Video
Simple interest

6.11

EXERCISE 6.11 ANSWERS ON P. 633

Simple interest

U F PS

Round answers to the nearest cent if needed.

1 Which of the following is the simple interest earned on $600 at 5% p.a. for 4 years? Select the correct answer **A**, **B**, **C** or **D**.

A $20 **B** $30 **C** $80 **D** $120

2 Find the simple interest paid on a loan of:

a $2400 at 5% p.a. for 1 year
b $2400 at 5% p.a. for 6 years
c $820 at 8% p.a. for 4 years
d $75 at 4% p.a. for 2 years
e $16 250 at 3.5% p.a. for 3 years
f $6570 at 11.5% p.a. for 5 years
g $532 at 9.25% p.a. for 8 years
h $4200 at 7% p.a. for 2 years
i $985 at 4.3% p.a. for 1 year
j $14 990 at 5.4% p.a. for 3 years

3 Find the interest earned if you invest:

a $5000 at 9% p.a. for 3 years
b $7500 at 6% p.a. for 5 years
c $20 000 at 8% p.a. for 7 years
d $45 000 at 12% p.a. for 20 years
e $875 at 5.8% p.a. for 2 years
f $9000 at 6.95% p.a. for 10 years
g $11 452 at 7.25% p.a. for 6 years
h $4500 at 4% p.a. for 3 years
i $737 at 5% p.a. for 1 year
j $2340 at 8% p.a. for 2 years

4 Find the simple interest earned on:

a $1080 at 5% p.a. for 6 months
b $700 at 6.5% p.a. for 3 months
c $2500 at 9% p.a. for 1 month
d $40 000 at 3.4% p.a. for 9 months
e $475 at 2% p.a. for 5 months
f $8250 at 4.1% p.a. for 8 months
g $15 400 at 7% p.a. for 11 months
h $4700 at 2.8% p.a. for 17 months
i $350 at 3% p.a. for 14 months
j $12 900 at 4.7% p.a. for 27 months

☐ Foundation ○ Standard ⬡ Complex

5 Xavier borrowed $7000 from Nelson Bank for 3 years at 3.5% p.a. interest.

PS
- **a** How much interest was he charged?
- **b** How much, including interest, did Xander have to repay the bank?

6 Goran borrowed $30 000 from Nelson Finance to set up a catering business. He is charged 7% p.a. simple interest over 5 years.

PS
- **a** How much interest will Goran have to pay?
- **b** Calculate the total amount he must repay.
- **c** Goran agrees to repay the loan in monthly payments over five years. How much must he repay each month?

7 When Jesinta borrowed $18 000 from a credit union, she was charged 9% p.a. simple interest.

PS
- **a** How much interest would she pay in one year?
- **b** Her total interest bill was $6480. For how long did Jesinta borrow the money?

6.12 Percentage problems

Worksheet Percentage problems

EXERCISE 6.12 ANSWERS ON P. 633

Percentage problems

U F PS R

1 In a box of 180 oranges, 15% are not ripe. How many oranges are not ripe?

2 A test result is 65 out of 80. What percentage is this?

3 Southside High's hockey team wins 12 games out of 15. What percentage is this?

4 A shirt costing $55 is reduced by 25%. What is the new selling price?

5 In some mixed paint, 25% of the mixture is blue, 15% is yellow and the remainder is white. What percentage is white?

6 Westvale High's basketball team scored 1065 points in a season. Claire shot 22% of the total points. How many points did Claire score?

7 At Upper Darling High, 60% of students are girls. How many boys are there in this school of 870 students?

PS R

8 What percentage of 9 carat gold is pure gold, if pure gold is 24 carat?

PS R

Foundation Standard Complex

9 When a floor was being tiled, 830 green tiles were used out of a total of 2075 tiles. What percentage of the floor is made of green tiles?

PS R

10 In a Year 1 class of 24 students, there were three absent students. Find the percentage of students absent from the class.

PS R

11 A real estate agent earns $2\frac{1}{2}\%$ commission on a sale of \$345 000. How much does the agent earn?

12 A worker's pay rate of \$27.30/hour is increased to \$30.40/hour. Find the increase as a percentage of the original pay rate, correct to one decimal place.

PS R

13 Find how much rent is paid if it is 30% of a weekly income of \$875.

PS R

14 Anton receives a 7% pay rise. If his old salary was \$67 000 per year, what is his new salary? Select the correct answer **A**, **B**, **C** or **D**.

PS R

A \$67 007 **B** \$67 469 **C** \$71 690 **D** \$113 900

15 Maxine is a car salesperson. She is paid a retainer of \$620 per week to cover her living expenses, plus 5% of all sales she makes. Maxine's sales in the first week of December total \$74 999. Calculate her pay for that week.

PS R

16 Robert writes novels with a recommended retail price of \$34.95. He is paid 10% royalty on the first 4000 copies sold and 12.5% royalty on the remaining sales. Calculate the royalties Robert will receive if his latest novel sells 11 000 copies.

PS R

TECHNOLOGY

Weekly budget

In this activity, we will use a spreadsheet to calculate a weekly budget for Abdul, a full-time student with a part-time job in a café.

1 Enter Abdul's weekly expenses into a spreadsheet. Highlight cells B2 to B8, right click on **Format Cells** and **Currency** with **two decimal places**.

	A	B	C
1	Expenses	Weekly budget	% of total
2	Rent	157.40	
3	Food	80.00	
4	Fares	34.50	
5	Entertainment	37.00	

☐ Foundation ○ Standard ⬡ Complex

6	Mobile phone	11.75	
7	Savings	25.00	
8	Total expenses		

2 In cell B8, write a formula to calculate the sum of Abdul's weekly expenses.

3 In cell C2, write a formula to calculate the percentage of total expenses spent on rent. Right click on **Format Cells** and **Number** with **one decimal place**.

4 Select cell C2 and **Fill Down** to cell C8 to copy the formula and calculate the percentage of total expenses belonging to each item.

5 Highlight cells A1 to B7. Click **Insert**, **Pie Chart** to create a pie chart (sector graph) and give your pie chart an appropriate title.

6 Point your mouse (do **not** click) in each sector of the pie chart to read off the percentage breakdown (brackets) of Abdul's expenses.

POWER PLUS ANSWERS ON P. 633

1 A retailer pays \$55 for a chair, then marks it up 60% for the selling price, then adds 10% of this price for GST. Calculate the final selling price.

2 Some items increase in value over time, such as jewellery, antiques and real estate. This is called **appreciation**. A gold chain costs \$1500 and appreciates by 8.5% per year. Find:

a the value of the chain after one year.

b the value of the chain after two years.

3 The Blewes family buys a home for \$480 000, and it appreciates at 4% per year. Calculate the home's value:

a after one year b after two years c after five years

4 a A number is increased by 25%, then the result is decreased by 25%. Is the final answer more than or less than the original number?

b If a number is increased by 25%, then by what percentage must the result be decreased so that the answer is equal to the original number?

5 Most items lose value over time. This is called **depreciation**. A Ferrari car costs \$420 000 and depreciates at 15% per year. Calculate how much the Ferrari is worth (correct to the nearest \$100) after:

a 1 year b 2 years c 5 years.

6 a A number is decreased by $33\frac{1}{3}\%$, then the result is increased by $33\frac{1}{3}\%$. Is the final answer more than or less than the original number?

b If a number is decreased by $33\frac{1}{3}\%$, then by what percentage must the result be increased so that the answer is equal to the original number?

6 CHAPTER REVIEW

Language of maths

Quiz
Language of maths 6

commission	cost price	decimal	decrease
denominator	discount	fraction	goods and services tax (GST)
improper fraction	increase	loss	lowest common multiple
mark-up	mixed numeral	numerator	percentage
profit	proper fraction	quantity	reciprocal
reduction	selling price	simplify	unitary method

1 Which words in the list above refer to something getting:

a bigger? **b** smaller?

2 What is the product of a number and its reciprocal?

3 Name one good and one service that would have a **GST**.

4 The 'cent' in percentage means 100. Find other 'cent' words that are related to 100.

5 What do we mean when we say that someone gives '110%' effort or support to something?

6 Copy and complete: 'A **loss** is made by a retailer if the **cost price** is __________ than the **selling price**'.

Topic summary

Worksheet
Mind map: Fractions and percentages

- List any rules you remember from your work on fractions and percentages.
- Name at least three careers where percentages would be used.
- Is there anything you still do not understand about fractions or percentages? See your teacher.
- Give examples of where you may use fractions or percentages in the future.

Print (or copy) and complete this mind map of the topic, adding detail to its branches and using pictures, symbols and colour where needed. Ask your teacher to check your work.

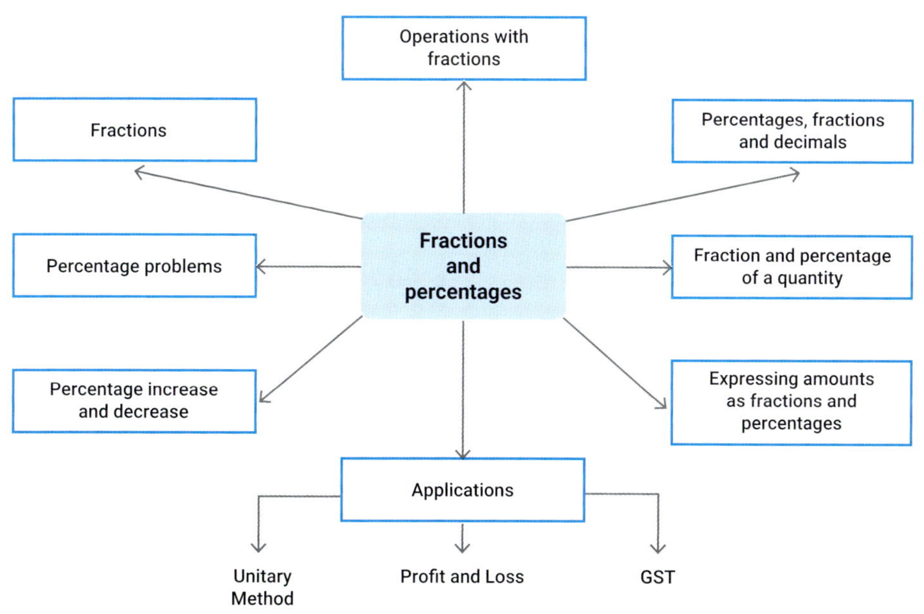

6 TEST YOURSELF

ANSWERS ON P. 633

1 Simplify each fraction.

a $\frac{6}{8}$ **b** $\frac{28}{48}$ **c** $\frac{35}{45}$

Quiz
Test yourself 6

2 Convert each improper fraction into a mixed numeral.

a $\frac{15}{4}$ **b** $\frac{22}{5}$ **c** $\frac{26}{6}$

3 Convert each mixed numeral into an improper fraction.

a $4\frac{1}{2}$ **b** $3\frac{2}{3}$ **c** $6\frac{4}{5}$

4 Arrange these fractions in descending order: $\frac{5}{6}, \frac{7}{12}, \frac{1}{2}, \frac{3}{4}$

5 Evaluate each expression.

a $\frac{1}{3} + \frac{2}{5}$ **b** $\frac{4}{7} - \frac{1}{2}$ **c** $\frac{3}{4} + \frac{2}{3}$

d $2\frac{1}{5} + 4\frac{1}{2}$ **e** $7 - \frac{5}{6}$ **f** $5\frac{3}{8} - 2\frac{1}{2}$

6 Santi buys 16 m of timber to fix her back deck. She uses $\frac{5}{8}$ of it to fix the deck. What fraction of the timber is left?

7 Evaluate each expression.

a $\frac{7}{8} \times \frac{4}{9}$ **b** $\frac{4}{5} \times \frac{2}{3}$ **c** $\frac{6}{7} \times \frac{5}{8}$

d $2\frac{1}{4} \times \frac{3}{10}$ **e** $\frac{1}{6} \div \frac{5}{12}$ **f** $5\frac{3}{8} \div 1\frac{1}{2}$

8 In Carrozza Park's population of 27 500, $\frac{3}{5}$ of the people are adults. How many people in Carrozza Park are adults?

9 Convert each percentage into a simplified fraction.

a 74% **b** 29% **c** 65% **d** 157%

10 Convert each percentage into a decimal.

a 12% **b** 16.2% **c** 2% **d** 5.4%

11 Convert each fraction into a percentage.

a $\frac{7}{8}$ **b** $\frac{12}{50}$ **c** $\frac{2}{9}$ **d** $\frac{33}{40}$

12 Convert each decimal into a percentage.

a 0.56 **b** 0.613 **c** 0.7 **d** 0.048

13 Arrange these numbers in ascending order: $\frac{3}{4}$, 0.725, 77%, 0.7

Foundation Standard Complex

14 Evaluate each expression. 6.05

a 15% of 40 b $\frac{1}{4}$ of 64 c 27.5% of \$880

d $\frac{2}{3}$ of 1 year (months) e 120% of 90 L f 35% of 6 m (cm)

15 Asam decided to go on a weight loss program. His starting weight was 187 kg. In the first six weeks he lost 13% of his body weight.

a How many kilograms did Asam lose in the first six weeks?

b What is his body weight now?

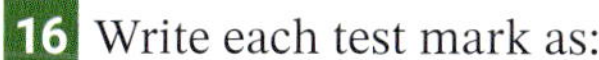

16 Write each test mark as:

i a simplified fraction ii a percentage

a 32 out of 40 b 45 out of 60 c 11 out of 20

17 Gianna scored 68 out of 80 in a test for her driving licence. Write her score as:

a a simplified fraction b a percentage

18 Westvale Secondary College has 375 boys and 405 girls. What percentage (correct to one decimal place) of the students are boys?

19 A refrigerator with a marked price of \$545 is discounted by 40%. Calculate its sale price.

20 Bella is to receive a salary increase of 3.5%. If she currently earns \$65 000, what will be her new salary?

21 Evaluate each expression without using a calculator.

a 20% of 55 tonnes b $12\frac{1}{2}$% of 96 cm c 75% of 12 min

22 A deposit of 18% paid on a car is \$26 480. What is the cost of the car?

23 James bought a guitar for \$750, then sold it for \$600. Calculate his percentage loss on the cost price.

24 The price of a mobile phone is \$95.70, including GST. Calculate:

a the amount of GST included in this price.

b the price of the phone before GST was added.

25 Zak sold \$850 worth of books last week. If he receives 5% commission on all sales, calculate his commission.

26 The cost of Evelyn's new glasses is \$425. Her health fund pays her a benefit of \$245. What percentage is her benefit of the cost of the glasses? Answer correct to one decimal place.

☐ Foundation ○ Standard ⬡ Complex

Practice set 2 ANSWERS ON P. 634

1 Simplify each fraction.

a $\frac{12}{24}$ **b** $\frac{7}{21}$ **c** $\frac{15}{25}$

2 Find the value of each variable, giving reasons.

a

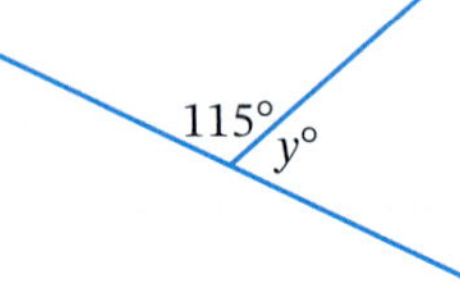

b

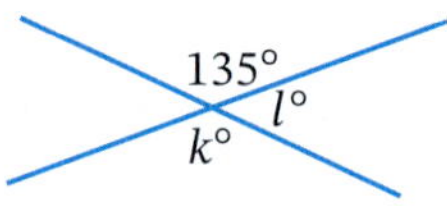

c

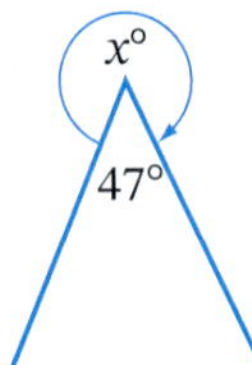

d

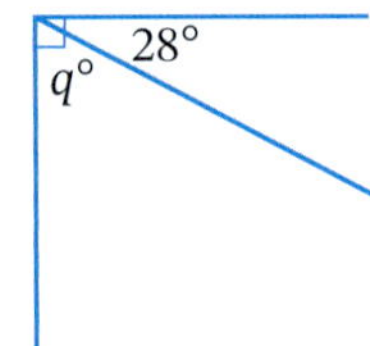

5.01

3 Find the perimeter of each shape.

a

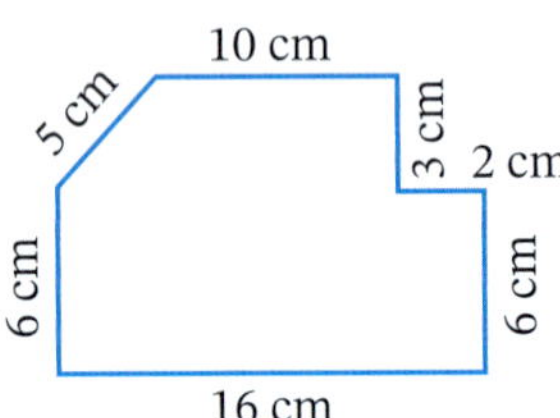

b

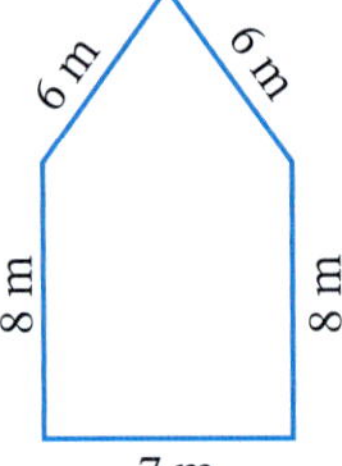

c

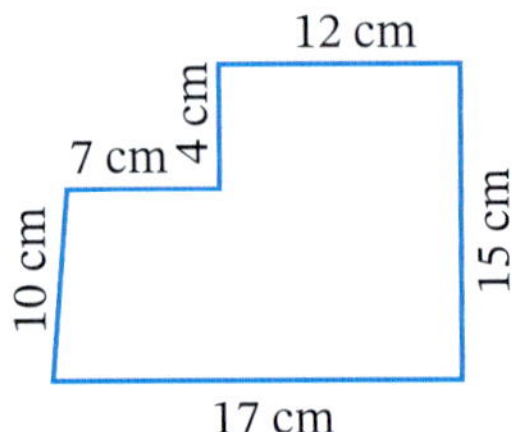

□ Foundation ○ Standard ⬡ Complex

4 Find the area of each shape. 5.04

a

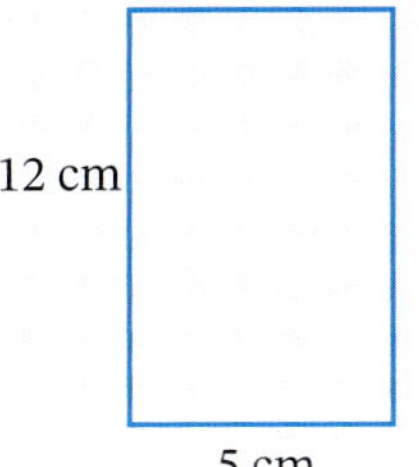

b

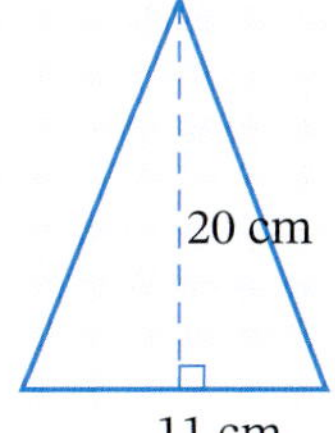

c

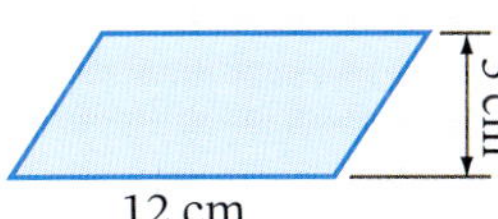

d

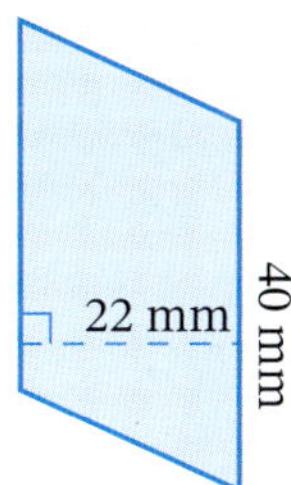

5 Classify each triangle: 4.03

i by sides **ii** by angles

a

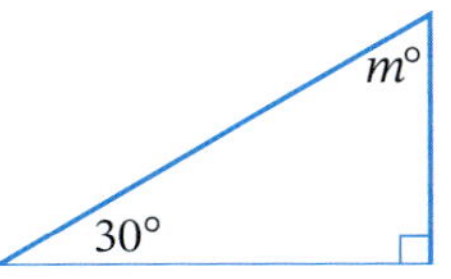

b

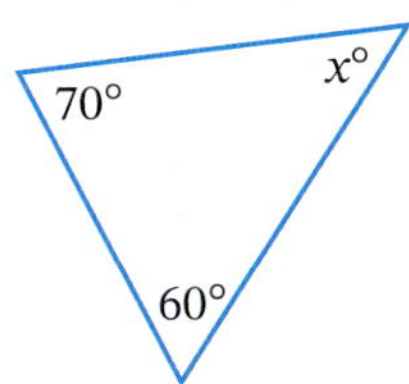

c

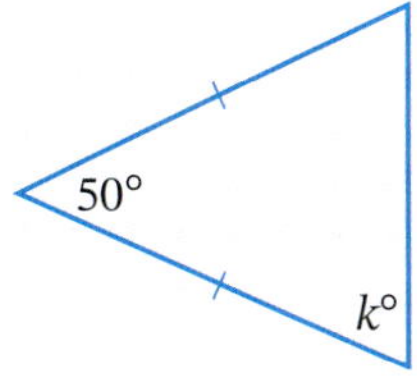

d

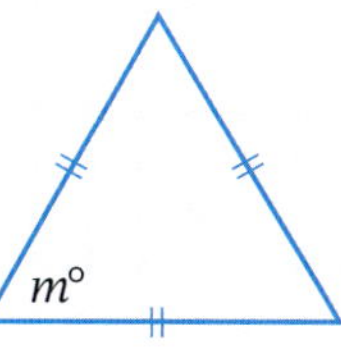

6 Find the value of each variable in question **5** above.

7 Convert each mixed numeral into an improper fraction.

a $2\frac{1}{2}$ **b** $1\frac{3}{4}$ **c** $4\frac{2}{5}$

8 Convert each percentage into: 6.04

i a simple fraction **ii** a decimal

a 25% **b** 40% **c** 85% **d** $12\frac{1}{4}\%$

9 Find the value of each variable, giving reasons. 4.02

a

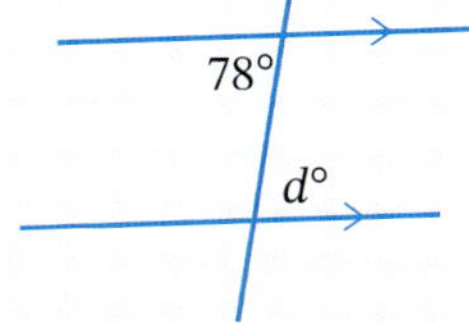

b

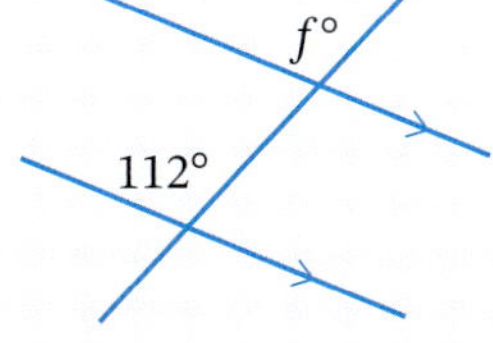

c

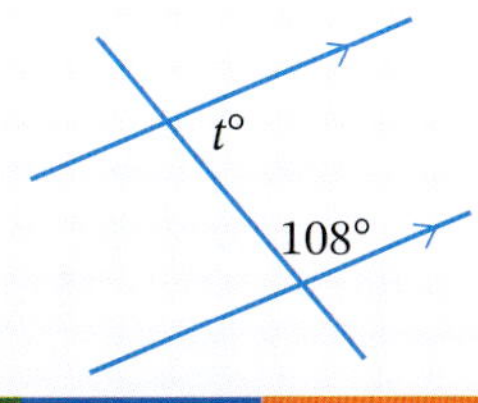

□ Foundation ○ Standard ⬡ Complex

PRACTICE SET 2

10 Convert each number into a percentage.

a 0.34 **b** $\frac{3}{4}$ **c** 0.215 **d** $\frac{5}{8}$

11 Name the (most general) quadrilateral with:

a opposite sides parallel.

b four right angles.

c four equal sides and four equal angles.

d two pairs of equal adjacent sides.

12 Find the area of each shape.

a

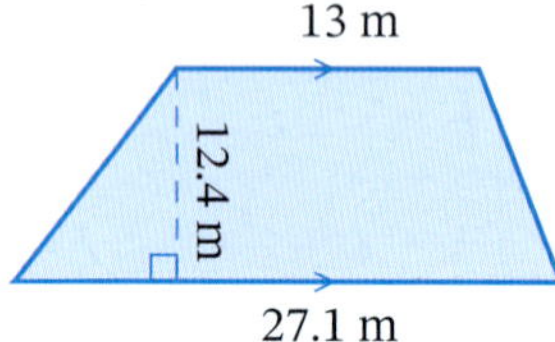

b

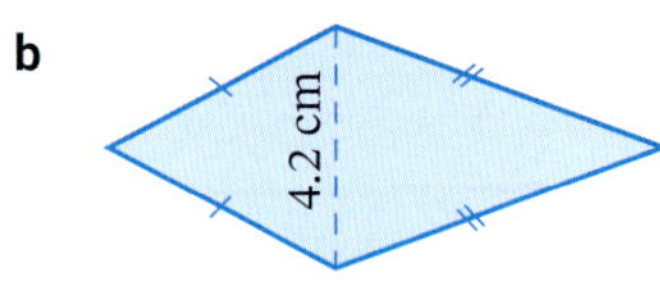

c

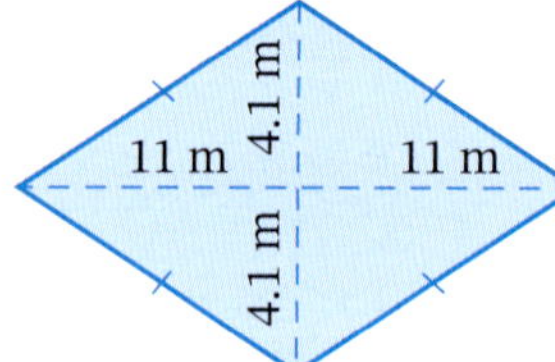

13 Convert each improper fraction into a mixed numeral.

a $\frac{11}{2}$ **b** $\frac{19}{3}$ **c** $\frac{14}{4}$

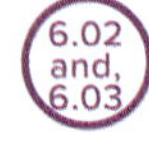

14 Evaluate each expression.

a $\frac{2}{5}+\frac{4}{5}$ **b** $\frac{1}{3}+\frac{2}{5}$ **c** $\frac{3}{5}-\frac{1}{4}$ **d** $1\frac{3}{5}+2\frac{1}{2}$

e $2-\frac{2}{3}$ **f** $\frac{1}{2}\times\frac{3}{5}$ **g** $\frac{3}{4}\times\frac{4}{5}$ **h** $\frac{4}{5}\div\frac{1}{2}$

15 Find the area of each shape.

a

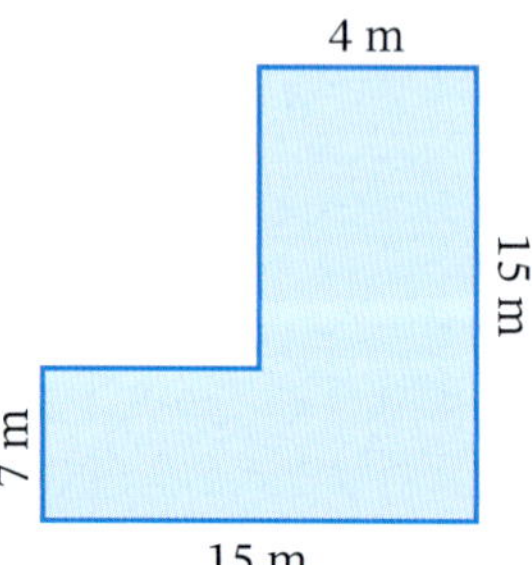

b

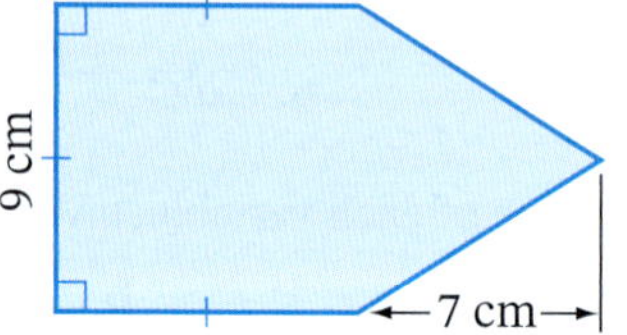

☐ Foundation ○ Standard ⬡ Complex

9780170465601

16 Find, correct to two decimal places, the circumference of a circle with: 5.02

a radius 5 cm **b** diameter 7 cm

17 Find, correct to two decimal places, the area of a circle with: 5.08

a radius 5 cm **b** diameter 7 cm

18 Find the value of each variable, giving reasons. 4.06

a

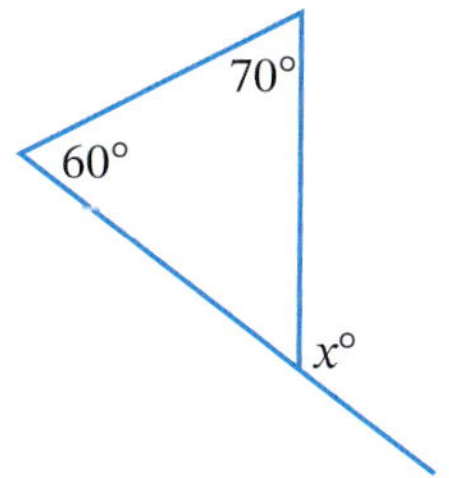

b

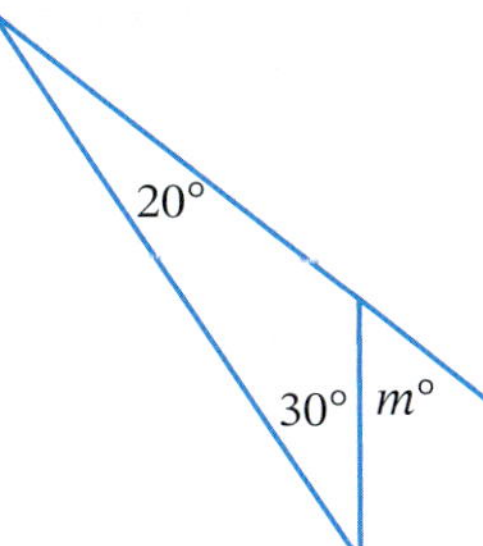

c

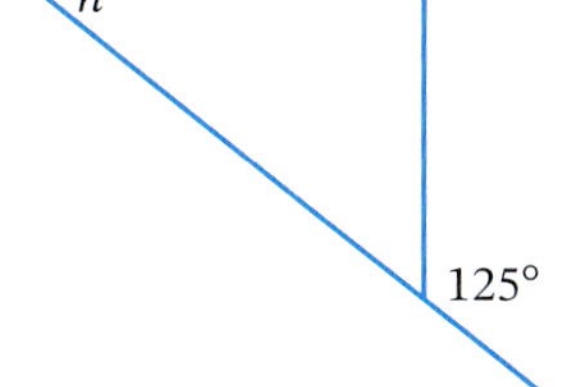

d

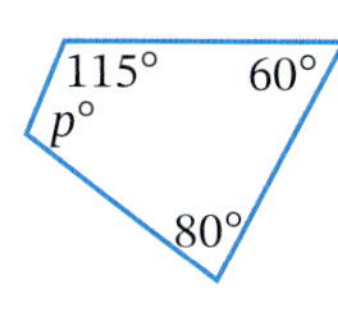

e

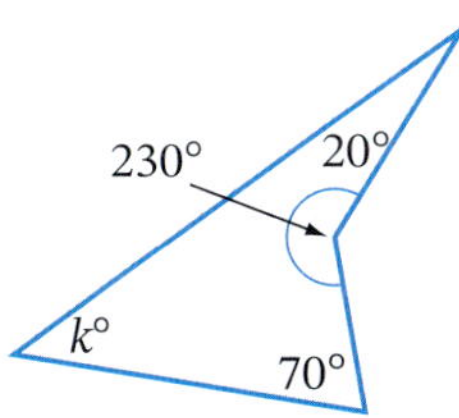

f

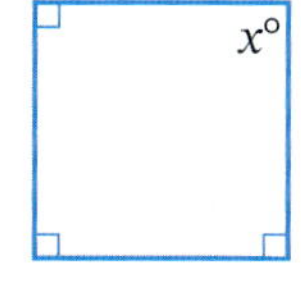

19 School canteen prices are to increase by 15%. Find the new price for a \$2 sandwich. 6.07

20 What is the capacity, in litres, of a box measuring 48 cm by 33 cm by 17 cm? 5.13

21 Find the volume of each prism. 5.11

a

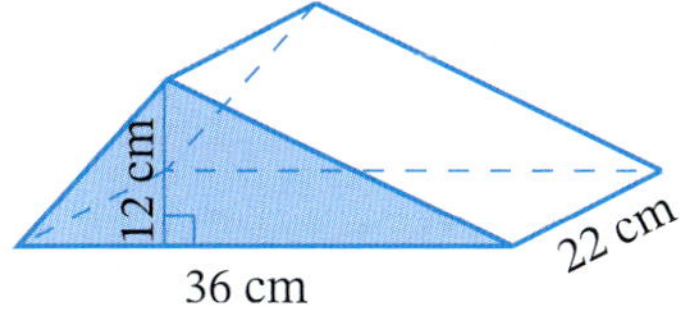

b

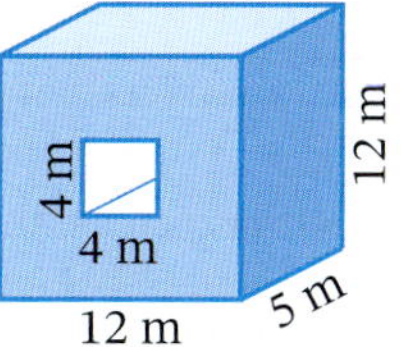

22 Convert:

a 3.6 m² to cm².

b 875 mm² to cm².

23 Arrange these numbers in ascending order: 63%, $\frac{3}{5}$, 0.67, $\frac{6}{7}$.

☐ Foundation ○ Standard ⬡ Complex

24 True or false?

a The sides of a rhombus are equal.

b The diagonals of a rectangle bisect each other at right angles.

c The opposite sides of a parallelogram are equal.

d The opposite sides of a trapezium are parallel.

6.05

25 Find:

a 2% of \$15.

b $\frac{2}{5}$ of 1320 kg.

c $\frac{3}{4}$ of \$6.40.

d 25% of 1 hour (in minutes).

26 What percentage is:

a 40 cm of 1 m?

b 350 mL of 1 litre?

c 45 minutes of $1\frac{1}{2}$ hours?

27 The population of Banksia decreased by 3.5%, from 3400 people. Find the new population.

EXTENSION

28 Find, correct to two decimal places, the perimeter of each shape.

a

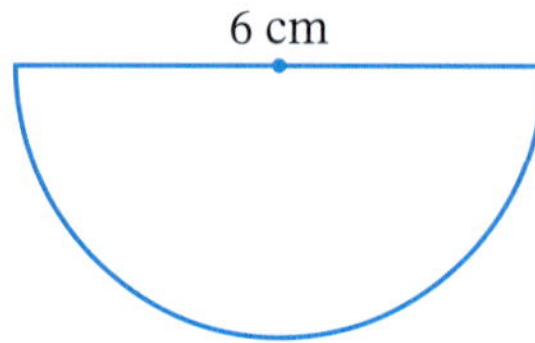

b

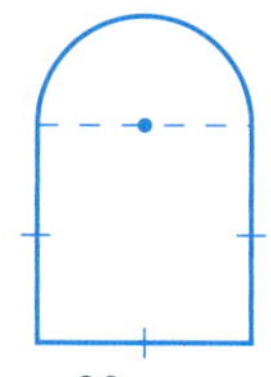

c

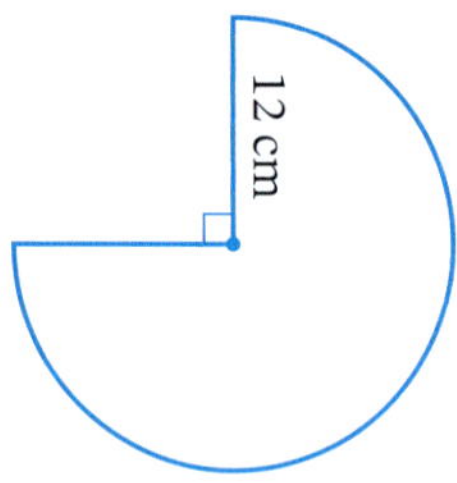

5.09

29 Find, correct to two decimal places, the area of each shape above.

5.09

30 Find the perimeter of this shape. Select the closest answer **A**, **B**, **C** or **D**.

A 109 cm B 77 cm C 737 cm D 129 cm

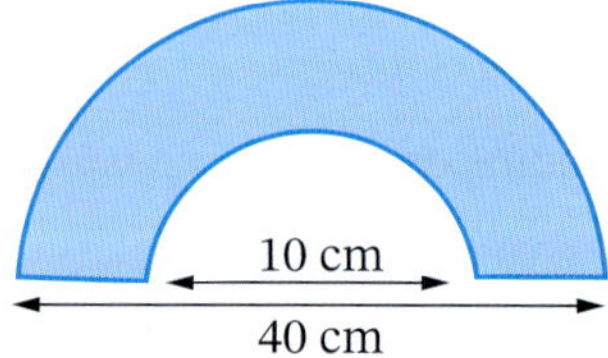

31 Find the area of the shape above, correct to one decimal place. 5.09

32 Mark saves 15% of his weekly income. If he saves $204, what is his weekly income? 6.09

33 A computer has a cost price of $1100 and a selling price of $1599.

a Calculate the profit made.

b Calculate, correct to one decimal place, the percentage profit on the cost price.

EXTENSION 5.09

34 Mr Moneybags wants to put a new garden with a path around it in the grounds of his country estate. The plan for this garden is shown. Find, correct to one decimal place:

a the area of the garden.

b the area of the path.

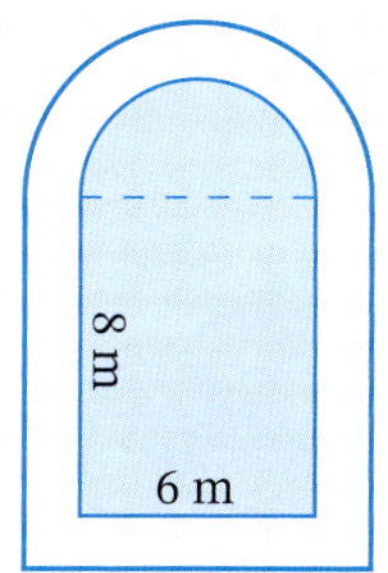

Width of path = 1.5 m

YEAR 9 EXTENSION

35 Find, correct to one decimal place, the volume of each cylinder.

a

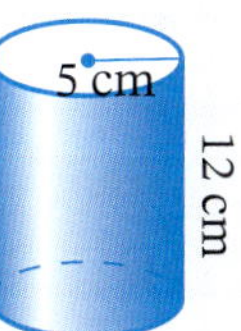

b

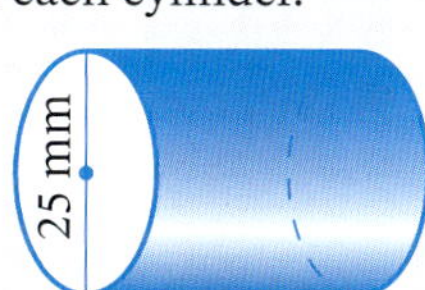

□ Foundation ○ Standard ⬡ Complex

7

STATISTICS

Investigating data

The Bureau of Meteorology collects weather data across Australia, including rainfall and temperatures. Polling organisations like Roy Morgan and Galaxy Research collect data on people's opinions on a wide range of issues. In NSW, the Bureau of Crime Statistics and Research (BOCSAR) collects data on crimes in different areas of the state. Health departments collect data on medical issues and health care. In all cases the data are analysed to evaluate decisions already made and to inform decision making in the future.

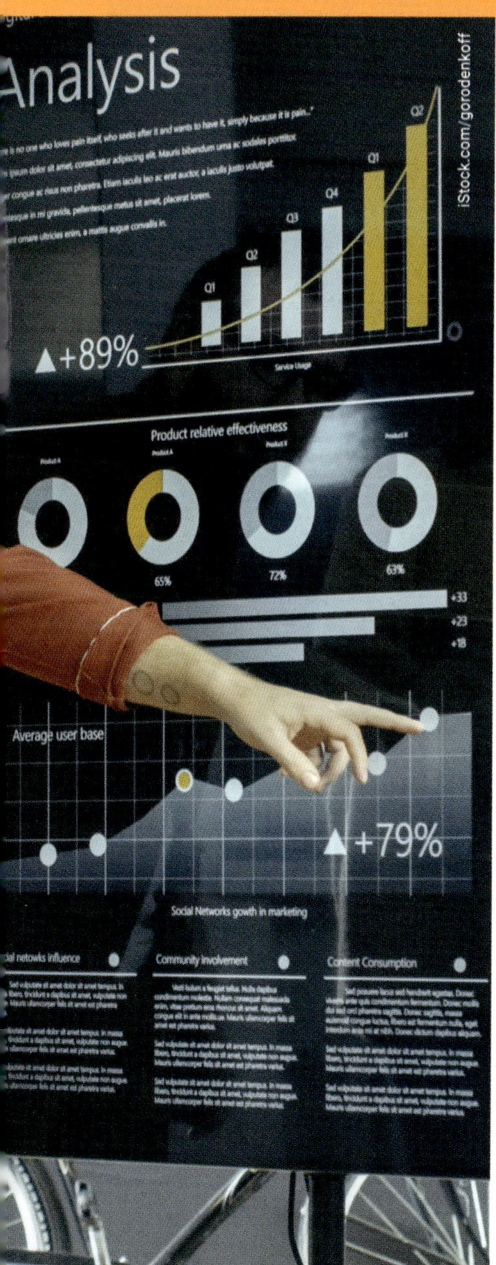

iStock.com/gorodenkoff

Chapter outline

		Proficiencies				
7.01	Organising and displaying data	U	F		R	C
7.02	The mean and mode	U	F		R	C
7.03	The median and range	U	F	PS	R	C
7.04	Analysing frequency tables	U	F			
7.05	Dot plots and stem-and-leaf plots	U	F	PS	R	C
7.06	Frequency histograms and polygons	U	F	PS	R	
7.07	Comparing data	U	F	PS	R	C
7.08	Census and samples	U	F		R	C
7.09	Designing survey questions	U	F	PS	R	C
7.10	Comparing samples and populations	U	F		R	C

U = Understanding
F = Fluency
PS = Problem solving
R = Reasoning
C = Communication

Wordbank

bias Something unwanted that causes a sample to not truly represent the population

categorical data Non-numerical data that can be classified into categories (e.g. people's hair colour or their favourite football team)

frequency How often a value appears in a data set

frequency histogram A special column graph that shows the frequencies of numerical data

frequency polygon A special line graph that shows the frequencies of numerical data

measure of central tendency An average, middle or typical value of a data set; a general name for the mean, median or mode

median The middle value of a data set when the values are arranged in order, or the average of the 2 middle values

numerical data Data that can be measured and counted using values (e.g. people's height or the number of devices they own)

Quiz
Wordbank 7

Videos (9):

SkillCheck Statistics

7.03 The mean, mode, median and range

7.04 Analysing frequency tables • Statistics from a frequency table

7.05 Analysing dot plots • Dot plots • Stem-and-leaf plots 1 • Stem-and-leaf plots 2 • Back-to-back stem-and-leaf plots

Twig videos (8):

7.01 Distorted graphs: Heat wave • Nightingale's diagram

7.02 Most popular pet

7.03 Average Joe

7.08 Counting crowds • Can fish oil make you smarter? • Can you trust your IQ? • The wrong guy won

PhET interactive (1):

7.03 Centre and variability

Quizzes (5):

- Wordbank 7
- SkillCheck 7
- Mental skills 7
- Language of Maths 7
- Test Yourself 7

Skillsheets (3):

7.01 Displaying data

7.02 Statistical measures

7.03 Statistical measures • Reading linear scales

Worksheets (12):

7.01 Every picture tells a story • Frequency tables

7.03 Mean, median, mode

7.04 The mean and fx tables

7.05 Stem-and-leaf plots • What data?

7.06 Analysing data

7.07 Analysing data • Comparing word lengths • Investigating young drivers

7.09 Canteen survey • Student survey form

Mind map: Investigating data

Puzzles (7):

7.01 Where all the cars are red

7.02 Looking for gold

7.03 Ranges and averages • Data puzzles • Mode, median and mean

7.06 Statistical match-up • Highway elephants

Technology (6):

7.01 Creating a graph

7.03 Students' marks • Finding the mean, median, mode and range • Statistical measures calculator

7.07 Population polygons • Daily rainfall

Presentations (1):

7.05 Stem-and-leaf plots

Spreadsheets (1):

7.03 Statistical measures calculator

To access resources above, visit **cengage.com.au/nelsonmindtap**

7 In this chapter you will:

- ✓ interpret and draw divided bar graphs, sector graphs, line graphs, stem-and-leaf plots and dot plots.
- ✓ organise data into a frequency table.
- ✓ calculate and interpret the mean, median, mode and range of a set of data, including those represented on a graph or plot.
- ✓ interpret and draw frequency histograms and polygons.
- ✓ investigate the effect of individual data values, including outliers, on the mean and median.
- ✓ identify ways of collecting data, including census, surveys, sampling and observation.
- ✓ design surveys and survey questions that are clear and unbiased.
- ✓ compare samples and populations, especially their means and proportions.

Quiz SkillCheck 7

Video Statistics

SkillCheck

ANSWERS ON P. 634

1 A group of school students visiting Japan have the following ages.

14 13 12 15 13 14 12 14 15 14 14 13 13

Show this information on a dot plot.

2 The masses (in kilograms) of some hospital patients are:

68 59 63 68 54 48 49 64 47 48 59 68 30

a Show this data on an ordered stem-and-leaf plot.

b What is the outlier?

3 A group of people was surveyed about the number of phone calls they made on a Saturday night. The results are shown below.

6	4	0	3	1	2	2	3	5	4	3	3
4	1	0	6	6	3	4	3	1	5	4	3

a Draw a column graph for this data.

b How many people were surveyed?

c What was the most common number of phone calls?

d What was the least common number of phone calls?

e What was the highest number of phone calls?

4 This back-to-back stem-and-leaf plot shows the points scored for and against the Koala Rugby team in each match, at one stage in the season.

For		Against
4	0	6
8 4	1	4 6 6
2 2 0 0	2	2 9 9
	3	0 1
8 1	4	7
3	5	

a What is the highest number of points scored by the Koalas in one game?

b How many matches have the Koalas played so far?

c How many points have been scored against the Koalas?

Organising and displaying data 7.01

Data is a collection of facts or raw information. Data can be organised and displayed in a variety of ways so that we can **analyse** (study) it in more detail. This table shows 3 examples of statistical graphs.

Skillsheet Displaying data

Worksheet Every picture tells a story

Puzzle Where all the cars are red

Videos Distorted graphs: Heat wave

Nightingale's diagram

Graph	Description
Column graph (or bar chart) 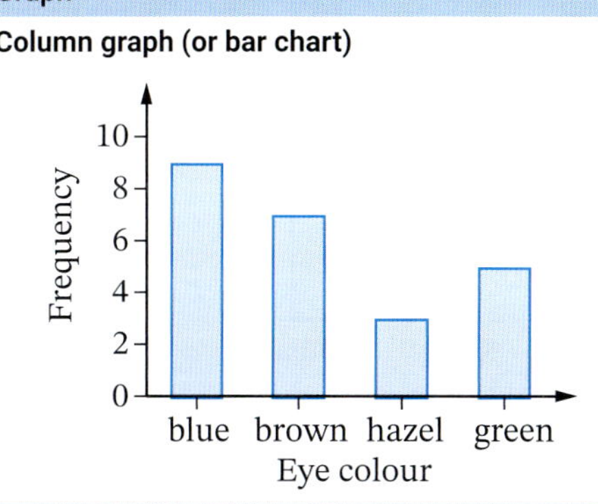 	• Mostly used for data that are in categories. • Uses columns to show the quantity or frequency of each category. • The horizontal axis (going across) shows the categories. • The vertical axis (going up) shows the frequency.

Sector graph (or pie chart) 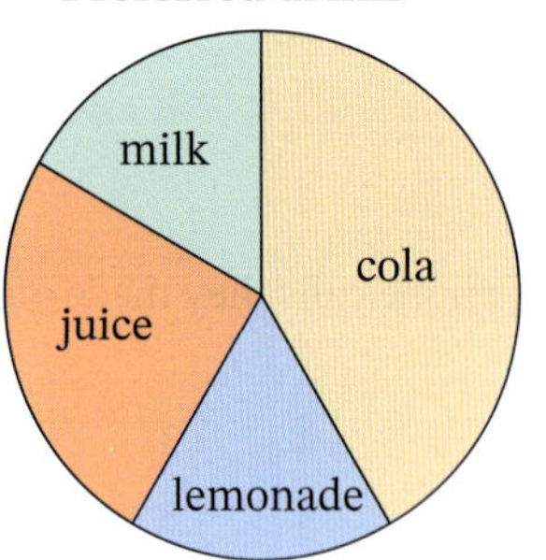	• Shows parts of a whole in circle form. • Uses sectors of a circle to show sizes of parts.
Divided bar graph 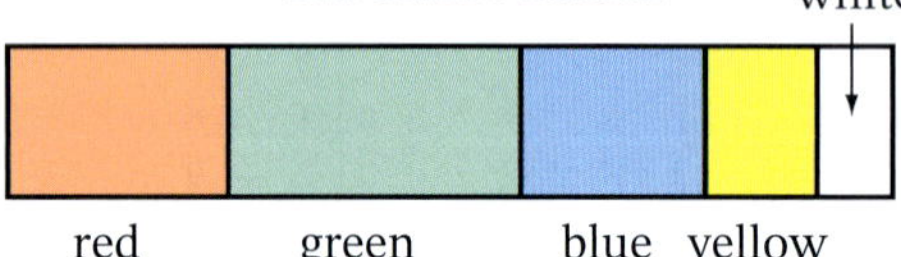	• Shows parts of a whole in rectangular form. • Uses sections of a rectangle to show sizes of parts.

Example 1

A sample of 80 people was surveyed on their favourite TV sport, and the results are shown.

TV sport	Australian football	Cricket	Motor racing	Rugby league	Rugby union	Football
Frequency	13	16	8	19	10	14

Display this data on a:

a column graph. **b** sector graph. **c** divided bar graph.

SOLUTION

a The highest frequency is 19 (for rugby league), so use a scale of 1 unit = 2 on the 'Frequency' axis, both axes, draw a column for each category and then the graph.

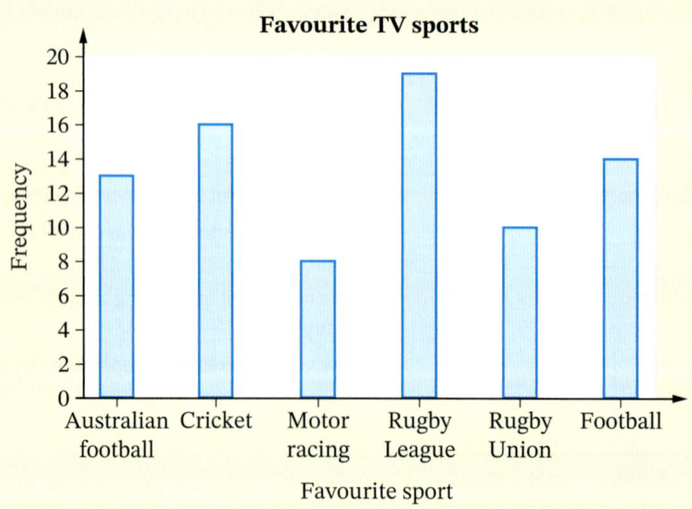

7.01

b Draw a large circle, then calculate the angle sizes for each sector using fractions of 360° (because there are 360° in a revolution).

Total number of people in sample = 80.

To find sector angle size, use the rule: $\dfrac{\text{frequency}}{\text{total}} \times 360^\circ$

Sector angles:

Australian football: $\dfrac{13}{80} \times 360^\circ = 58.5^\circ$

Cricket: $\dfrac{16}{80} \times 360^\circ = 72^\circ$

Motor racing: $\dfrac{8}{80} \times 360^\circ = 36^\circ$

Rugby League: $\dfrac{19}{80} \times 360^\circ = 85.5^\circ$

Rugby Union: $\dfrac{10}{80} \times 360^\circ = 45^\circ$

Football: $\dfrac{14}{80} \times 360^\circ = 63^\circ$

Construct the sector graph with the calculated angles, then the sectors and the graph.

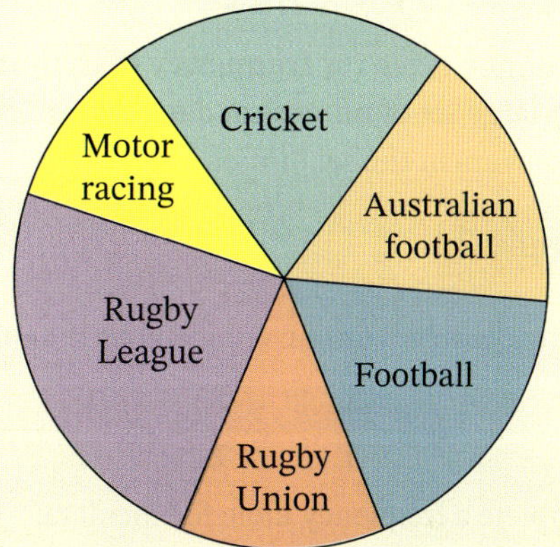

c Draw a rectangle whose length will represent 80, the total frequency (the whole), for example, is 12 cm. Then calculate the length of each section using fractions of 12 cm.

Section sizes:

Australian football: $\dfrac{13}{80} \times 12\text{ cm} = 1.95\text{ cm}$

Cricket: $\dfrac{16}{80} \times 12\text{ cm} = 2.4\text{ cm}$

Motor racing: $\dfrac{8}{80} \times 12\text{ cm} = 1.2\text{ cm}$

Rugby League: $\dfrac{19}{80} \times 12\text{ cm} = 2.85\text{ cm}$

Rugby Union: $\dfrac{10}{80} \times 12\text{ cm} = 1.5\text{ cm}$

Football: $\dfrac{14}{80} \times 12\text{ cm} = 2.1\text{ cm}$

Construct the divided bar graph with the calculated section lengths, then the sections and the graph.

Favourite TV sports

Australian football	Cricket	Motor racing	Rugby League	Rugby Union	Football

Shutterstock.com/Jeff Schultes

Frequency tables

Worksheet Frequency tables

A **frequency table** (or **frequency distribution table**) can be used to efficiently organise and count a large set of numerical data values. The **frequency** of a value is the number of times that value appears in the set of data.

Example 2

Twenty people were surveyed about the number of cars in their household.

2	2	1	2	0	3	2	1	1	4
1	1	1	2	2	0	3	2	1	2

Complete a frequency table for the data.

SOLUTION

Draw a table with columns for Score, Tally and Frequency, as shown in the next page. The lowest score is 0 and the highest is 4, so list the numbers 0–4 in the Score column.

For each value in the data set, draw a tally mark in the tally column. Tally means 'to count', and we are counting how often each value occurs.

The first value is 2, so mark '|' in the 'Tally' column beside 2. The second value is another 2, so mark another '|' there. Continue until you have counted all of the values.

Score	Tally	Frequency
0	II	
1	卌 II	卌 means a tally of 5.
2	卌 III	
3	II	
4	I	
Total		

For each value, count up the tally marks and write the answer in the 'Frequency' column. Then write the total in the final row and check that it is equal to the total number of values.

Score	Tally	Frequency
0	II	2
1	IIII II	7
2	~~IIII~~ III	8
3	II	2
4	I	1
Total		20

EXERCISE 7.01 ANSWERS ON P. 635

Organising and displaying data

1 A class of 30 PE students was surveyed about the day of the week that each student trained for their chosen sport.

C

Day	Monday	Tuesday	Wednesday	Thursday	Friday
Number of students	3	13	4	7	3

Display this data in a sector graph – the graph.

2 The colours of the jelly beans in a jar were counted.

C

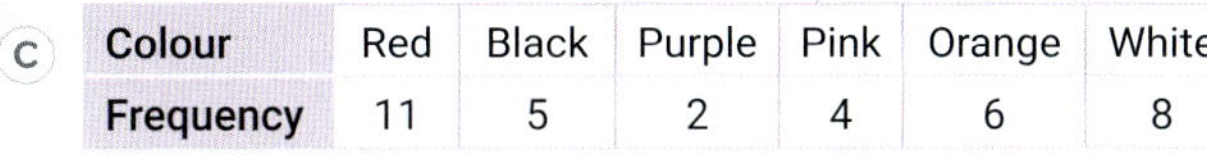

Colour	Red	Black	Purple	Pink	Orange	White
Frequency	11	5	2	4	6	8

Display this data as a column graph, using an appropriate scale – the graph and both axes.

Shutterstock.com/Hayati Kayhan

3 A sample of men were surveyed on the number of hours they slept the previous night.

C

Number of hours	6	7	8	9	10	11
Frequency	3	9	11	6	5	2

Display this data as a divided bar graph, 9 cm long – the graph.

4 This table shows the number of kilolitres of water used by a household in different areas over a year.

R

Area	Garden	Shower	Toilet	Kitchen
Water usage (in kilolitre)	90	520	140	50

If this information is shown in a divided bar graph of length 10 cm, which of the following is the length of the section representing 'Shower'? Select the correct answer **A**, **B**, **C** or **D**.

A 1.5 cm **B** 1.9 cm **C** 5.2 cm **D** 6.5 cm

EXAMPLE 2

5 The results of a spelling quiz (out of 10) for a class of students are as shown.

7	7	8	9	7	5	4	8	9
9	8	10	8	7	5	5	4	1
8	8	9	7	7	4	8	3	1

a Complete a frequency table for this data.

b What was the highest value?

c How many students scored 4 marks in the quiz?

d What was the most frequent value?

e Draw a column graph of this data.

6 Thirty customers were asked to name their favourite car colour. B = black, S = silver, W = white, U = blue, R = red.

B	B	W	R	W	R	B	S	U
W	R	B	S	B	U	W	B	W
S	B	B	R	R	U	W	B	B

a Complete a frequency table for this data.

b Draw a divided bar graph for this data.

c Which is the most popular colour for this group of customers?

7 A sample of women were asked how many siblings (brothers and sisters) they had.

4	0	2	3	2	4	3	5	4	3
0	0	1	2	3	2	2	1	4	3
1	1	1	5	2	5	0	1	2	1
0	1	3	2	3	4	1	1	3	3

a Complete a frequency table for this data.

b What was the highest number of siblings?

c What was the mode (the most frequent value)?

d Illustrate this data on a sector graph.

Foundation Standard Complex

9780170465601

TECHNOLOGY

Graphing life expectancy

Life expectancy describes the age to which a person is expected to live, based on statistical data covering different factors.

Technology
Creating a graph

7.01

1 Enter the data below into a new spreadsheet. It compares the life expectancy at birth for males and females in 10 countries.

	A	B	C	D
1	**Life expectancy at birth**			
2			**Years**	
3	**Country**	**Male**	**Female**	**Total**
4	Australia	80.5	85.0	82.7
5	Bangladesh	72.0	76.5	74.2
6	Brazil	71.2	78.4	74.7
7	Canada	81.1	85.9	83.4
8	Chad	56.5	60.1	79.4
9	China	74.0	78.4	76.1
10	India	68.4	71.2	69.7
11	Spain	79.0	85.2	82.0
12	Sweden	80.4	84.5	82.4
13	USA	78.0	82.5	80.3

Source: CIA

2 Highlight all 4 columns, including headings. Click **Insert** and **2-D Column Graph**. Give the graph an appropriate title and the axes.

3 Which country has:

a the lowest total life expectancy?
b the highest total life expectancy?
c the lowest male life expectancy?
d the lowest female life expectancy?
e the highest male life expectancy?
f the highest female life expectancy?

4 Suggest reasons why some countries would have higher life expectancies than others.

5 In column E, calculate the difference between the male and female life expectancies for each country.

6 Generally, do males or females live longer?

7 Which country has the greatest difference between its male and female life expectancies?

8 Which country has the least difference?

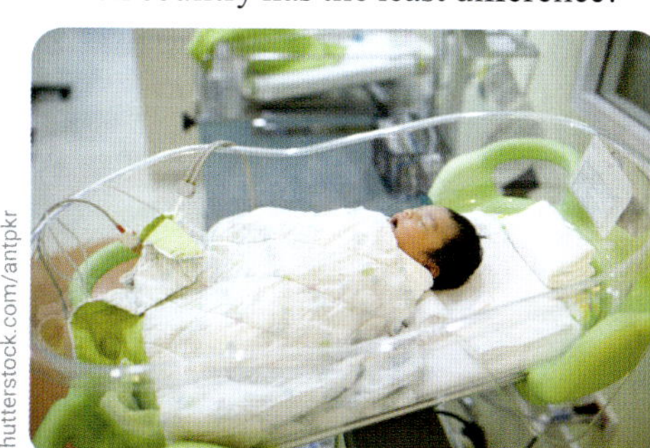
Shutterstock.com/antpkr

Shutterstock.com/Pressmaster

7.02 The mean and mode

Skillsheet Statistical measures

Puzzle Looking for gold

Video Most popular pet

When we analyse data, we summarise the information and try to find patterns and trends to draw some conclusions from them. We can use the **mean, mode, median** and **range** to make comparisons between sets of data.

The mean, mode and median are known as **measures of central tendency**: they are typical or central values that represent all data in the set.

The range is a **measure of spread**: it describes how the values are spread.

The **mean** is found by adding all the data values and dividing by the number of values. The mean is the most appropriate measure of centre to use when working with continuous data: data that sits on a smooth scale and the 'in-between' decimal values make a difference. Examples include height, weight and time.

ⓘ The mean

The symbol for the mean is $\bar{x}$.

$$\bar{x} = \frac{\text{sum of values}}{\text{number of values}}$$

- The mean is a single value that represents all of the data.
- *All* of the data values are used in the calculation of the mean.
- The mean is *not* necessarily one of the data values.

The **mode** is the most common or frequent value (or values). A set of data may have more than one mode or no mode at all. The mode is the most appropriate measure of centre to use when working with data that is described using words called **categorical data**.

ⓘ The mode

The **mode** is the data value (or values) that occurs the most often: the value with highest frequency. *Think:* mode = '**m**ost **o**ften'.

- The mode can be one or more values.
- The mode does not depend on all of the values in the data set.
- The mode is one (or more) of the data values.

Example 3

The number of goals scored by Courtney's soccer team in each match over a season were:

3 2 1 4 0 2 3 1 3 1 1 0

a Find the mean. **b** Find the mode.

SOLUTION

a Mean $\bar{x} = \dfrac{\text{sum of data values}}{\text{number of values}}$

$= \dfrac{3+2+1+4+0+2+3+1+3+1+1+0}{12}$

$= \dfrac{21}{12}$

$= 1.75$

Note that the value of the mean is at the centre of the data.

b The mode is 1.

It is the most common value (occurs 4 times).

The mean from a calculator's statistics mode

Scientific and graphics calculators have a statistics mode (SD or STAT). Follow the instructions in the table below to calculate the mean of the values from Example **3** using your calculator's statistics mode.

Operation	Casio scientific	Sharp scientific
Start statistics mode.	MODE STAT 1-VAR	MODE STAT =
Clear the statistical memory.	SHIFT 1 Edit, Del-A	2nd F DEL
Enter data.	SHIFT 1 Data to get table 3 = 2 =, etc. to enter in column. Press AC to leave table.	3 M+ 2 M+, etc.
Calculate the mean ($\bar{x} = 1.75$).	SHIFT 1 Var $\bar{x}$ =	RCL $\bar{x}$
Check the number of values (n = 12).	SHIFT 1 Var n =	RCL $\bar{x}$
Return to normal (COMP) mode.	MODE COMP	MODE 0

EXERCISE 7.02 ANSWERS ON P. 635

The mean and mode

1 The residents in a street were surveyed about the number of children living in each household. The results were:

2 2 1 2 0 3 2 1 1 4 1 0

a Find the mean, correct to one decimal place.

b Find the mode.

Foundation Standard Complex

2 For each set of data, find:

i the mean (rounded to 2 decimal places, if needed).

ii the mode(s).

a	6	8	7	6					
b	7	9	5	2	2	6	2		
c	11	7	7	6	1	2	4	1	
d	96	53	96	45	64	76	54	54	96
e	47.31	47.20	47.24	47.21	47.28	47.26			

3 Use your calculator's statistics mode to calculate the mean of each data set in question **2**.

4 Find the mode of each set of categorical data.

a Holden, Ford, Nissan, Holden, Hyundai, Ford, Hyundai, Nissan, Hyundai, Toyota and Toyota

b Heart, spade, spade, diamond, spade, club, heart, heart, diamond and spade

c Chinese, Korean, Japanese, Korean, Chinese, Japanese, Japanese, Korean and Korean

5 R C Why it is not possible to find the **mean** of each data set in question **4**?

6 R C In a diving competition, the judges awarded the following scores.

6.2 7.0 8.2 8.8 7.0 9.1 7.0 8.1 8.0

a Find the mean of these scores, correct to 2 decimal places.

b Find the mode of these scores.

c Which measure of central tendency represents this set of scores better? Give a reason for your answer.

7 What is the difference between the mean and the mode of the following set of scores?

20 60 20 50 40

Select the correct answer **A**, **B**, **C** or **D**.

A 20 **B** 12 **C** 18 **D** 2

8 R C The ages of the members of the Carrozza family are:

1 3 11 25 26 30 31 52 58

The ages of the members of the Binns family are:

5 8 16 21 25 29 40 63 93

a Calculate the mean age, correct to one decimal place, of each family.

b What is the main difference between these two sets of ages?

c What effect does this difference have on the means?

9 R C Two Year 8 classes scored the following marks for their PE assignments.

8Y: 5 5 5 5 6 6 6 7 7 7 8

Foundation Standard Complex

8Z: 3 3 4 5 5 6 6 7 9 9 10

a Calculate the mean mark, correct to one decimal place, for each class.

b Describe the differences between the way each class has its marks spread out.

c Class 8X scored the following marks, but Hannah was away.

8X: 4 5 5 6 6 6 7 7 7 10

What mark would Hannah have to score so that her class has the same mean as the other two classes?

10 The mean points score per game by a basketball team over a season of 30 games was 98. What was the total number of points scored over the season?

R

11 The takings at the school canteen were:

R

Monday $1600 Tuesday $2400 Wednesday $1500 Thursday $1750

a Find the average takings for these 4 days.

b How much should the canteen earn on Friday to make the average daily takings $2000?

12 **a** Find 4 values that have a mean of 10 and a mode of 11.

R **b** Find 7 values that have a mean of 13 and a mode of 5.

The median and range 7.03

Skillsheet Statistical measures

Worksheet Mean, median, mode

Puzzles Ranges and averages

Data puzzles

Mode, median and mean

Video Average Joe

Interactive Centre and variability

The **median** is the middle value of a set of data or the average of the two middle values. The median is the most appropriate measure of centre to use when working with **discrete data**: counting data that can only take on separate, distinct values with 'gaps' or 'jumps' ('in-between' values are not allowed). Examples include number of children in a family, shoe size, number of cars in a carpark.

ⓘ The median

When the values of a data set are ordered, the **median** is:

- the middle value if there is an odd number of values.
- the average of the two middle values if there is an even number of values.

Think: 'Median' sounds like 'medium', which is halfway between small and large.

- The median is a single value that represents all of the data.
- The median does not depend on all of the data values, so it is not affected by outliers (extremely high or low values).
- The median is often one of the data values.

A 'median' strip is the strip of grass or concrete down the ***middle*** of a highway.

☐ Foundation ◯ Standard ⬡ Complex

The **range** is a measure of the spread of a data set, and is simply the difference between the highest and lowest values. A high range indicates a large spread and low consistency in data scores, while a low range indicates a small spread and a high consistency in the data scores.

ⓘ The range

Range = highest value − lowest value

Example 4

Video The mean, mode, median and range

The ages of the members of the Carrozza family are:

3 1 25 58 26 52 30 31 11

Find:

a the median **b** the range

SOLUTION

a First, rewrite the ages in ascending (increasing) order.

1 3 11 25 (26) 30 31 52 58

4 values — Median — 4 values

Note that the value of the median is at the centre of the data.

There are 9 values, an odd number, so there is one middle value.

$\frac{1}{2} \times 9 = 4\frac{1}{2}$, so round up to 5 and count to the 5th value (circled above) to find the median.

Median = 26

b Range = 58 − 1 Highest value − lowest value

= 57

Example 5

The lap times (in seconds) of 6 cyclists were:

12.5 21.4 10.5 18.7 12.9 10.3

Find:

a the median **b** the range

SOLUTION

a Rewrite the values in ascending order:

10.3 10.5 (12.5) (12.9) 18.7 21.4

3 values — 2 middle values — 3 values

There are 6 values, an even number, so there are two middle values.

$\frac{1}{2} \times 6 = 3$, so count to the 3rd and 4th values (circled above) and find their average.

$$\text{Median} = \frac{12.5 + 12.9}{2}$$
$$= 12.7$$

12.7 is halfway between 12.5 and 12.9

b $\text{Range} = 21.4 - 10.3$
$= 11.1$

Highest value – lowest value

EXERCISE 7.03 ANSWERS ON P. 636

The median and range

U F PS R C

1 What is the median of this set of data?

EXAMPLE 4

10 3 5 9 4 3 10

Select the correct answer **A**, **B**, **C** or **D**.

A 3 **B** 5 **C** 7 **D** 10

2 What is the range of this set of data?

17.1 18.3 14.5 11.8 21.9 14.5

Select the correct answer **A**, **B**, **C** or **D**.

A 16.4 **B** 15.8 **C** 14.5 **D** 10.1

3 For each set of data, find:

EXAMPLE 5

i the median **ii** the range

a	11	9	7	4					
b	46	50	48	47	50	40			
c	12	13	11	14	10				
d	84	50	0	28	14	99	51	66	51
e	57	57	57	57	57	57	57		
f	64.2	99	99.2	71.2	93.5	61.8			

Foundation Standard Complex

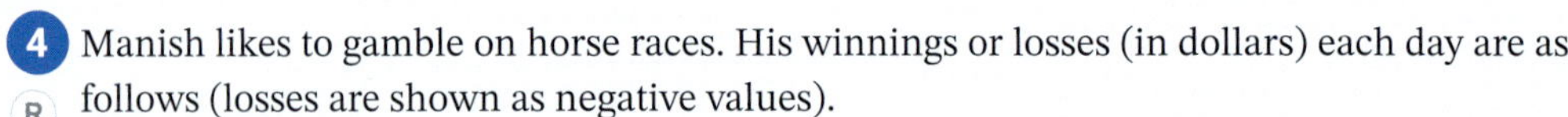

4 Manish likes to gamble on horse races. His winnings or losses (in dollars) each day are as follows (losses are shown as negative values).

R C

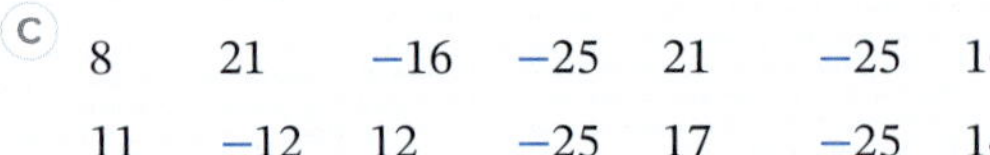

8	21	−16	−25	21	−25	16
11	−12	12	−25	17	−25	14

a Calculate the mean, correct to the nearest cent.

b Find the median and the mode.

c As Manish wins more often than he loses, should he continue to gamble on horse races? Give a reason for your answer.

5 Eleven houses were sold in Jacaranda Avenue. The selling prices are listed here.

R C

$580 000	$485 000	$660 000	$758 000	$644 000	$1 670 000
$815 000	$686 000	$666 500	$720 800	$805 000	

a Find the mode.

b Find the median.

c Find the mean.

d Which measure of central tendency best describes this set of house prices? Give a reason for your answer.

6 Two 10-pin bowlers recorded the following scores over 10 games.

PS R C

Aldo:	175	175	160	165	180
	145	150	177	177	177
Peter:	173	173	174	175	175
	176	176	177	179	179

a Calculate the mean for each bowler.

b Calculate the range for each bowler.

c Who would you choose, and why, if you were asked to present an award for:

i the more talented bowler?

ii the more consistent bowler?

d A third player, Rob, recorded the following scores over nine games.

Rob:	134	175	175	144	176
	150	187	166	180	

What score would Rob need to achieve in his 10th game to have the same mean score as Aldo?

7 The favourite party food for a group of 3-year-old children was recorded.

R

popcorn	fruit	chocolate	fruit	popcorn
chocolate	chocolate	fruit	fruit	fruit

Which is the only statistical measure that can be found for this data? Select the correct answer **A**, **B**, **C** or **D**.

A mean **B** median **C** mode **D** range

☐ Foundation ○ Standard ○ Complex

8 (R, C) Samir is the coach of the Panthers athletics team. He needs to pick his best runner for the 200 m race. He has recorded the times (in seconds) for the last few races of Daniel and Cristian.

Daniel:	29.4	31	27.5	32.5	28	
Cristian:	28.5	27	30.5	28.5	31.5	28.5

a Calculate the mean for each runner.

b Calculate the range for each runner.

c Who do you think Samir should choose as the runner for the next event? Give a reason for your answer.

9 (PS, R) A group of eight friends counted the number of letters in their surnames. The results were:

4 6 7 5 4 6 9 □

where □ represents a missing value.

Find a possible value of □ if the median is 6 and the range is 5.

10 (PS, R) **a** Find eight values that have a median of 8 and a range of 8.

b Find seven values from 3 to 9 inclusive that have a mode of 4 and a median of 6.

TECHNOLOGY

Mean, mode, median and range

1 Enter the data below into a new spreadsheet, listing the daily maximum temperatures in Camden for one week in January.

	A	B	C	D	E
1	Day	Temperature (deg C)			
2	Saturday	31.5		Mean	
3	Sunday	32.2		Mode	
4	Monday	33		Median	
5	Tuesday	36		Maximum	
6	Wednesday	36.8		Minimum	
7	Thursday	35.9		Range	
8	Friday	26.7			

2 Enter each formula into the given cell to calculate each statistic.

Cell E2: =**average(B2:B8)** Cell E3:=**mode(B2:B8)**

Cell E4:=**median(B2:B8)** Cell E5:=**max(B2:B8)**

Cell E6: =**min(B2:B8)** Cell E7: =**E5-E6**

Technology Students' marks

Finding the mean, median, mode and range

Statistical measures calculator

Spreadsheet Statistical measures calculator

Skillsheet Reading linear scales

Quiz Mental skills 7

☆ MENTAL SKILLS 7 ANSWERS ON P. 636 **Maths without calculators**

Reading linear scales

Understanding and reading the scale on a measuring instrument, on a number line or on the axis of a graph is an important mathematical skill.

1 Study each example.

a Complete the missing values on this scale.

- First, choose 2 values on the scale, say 100 and 120.
- Count the number of intervals ('spaces') between the 2 values.
  There are 4 intervals between 100 and 120.
- To find the size of each interval, divide the difference between the 2 values by the number of intervals.
- Difference = 120 − 100 = 20 km.
- Number of intervals = 4.
- Size of an interval = 20 ÷ 4 = 5 km.
- Use the calculated size of an interval to complete the missing values.

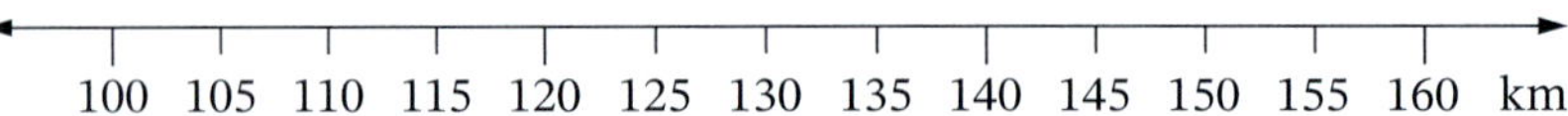

b Complete the values on this scale.

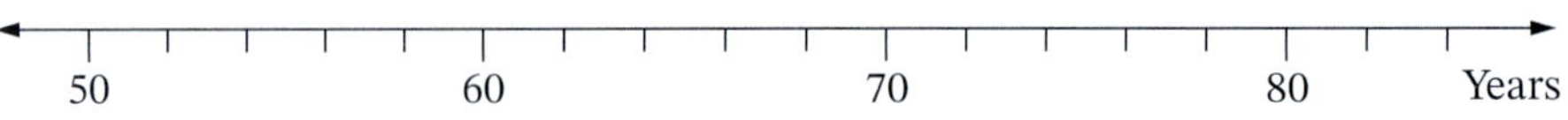

- Choose 50 and 60 on the scale.
- Number of intervals (between 50 and 60) = 5.
- Difference (between 50 and 60) = 60 − 50 = 10 years.
- Size of an interval = 10 ÷ 5 = 2 years.

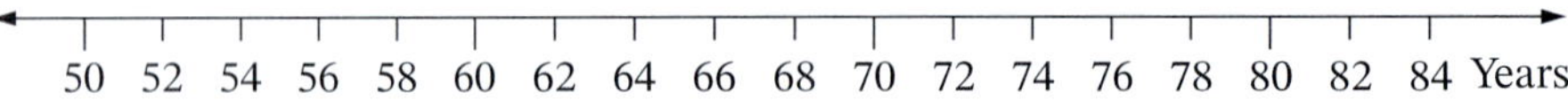

2 Now copy and complete the following scales.

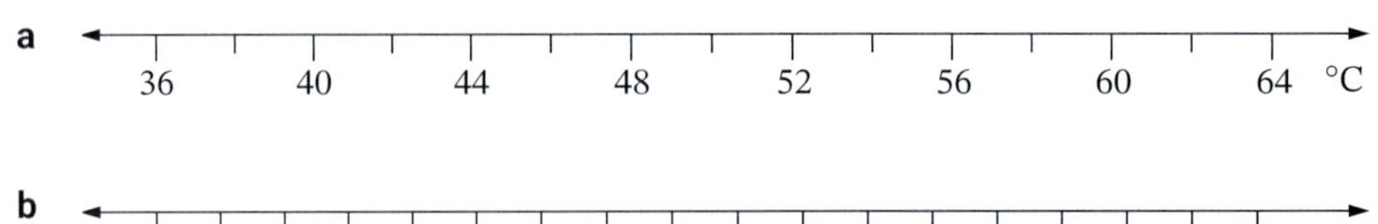

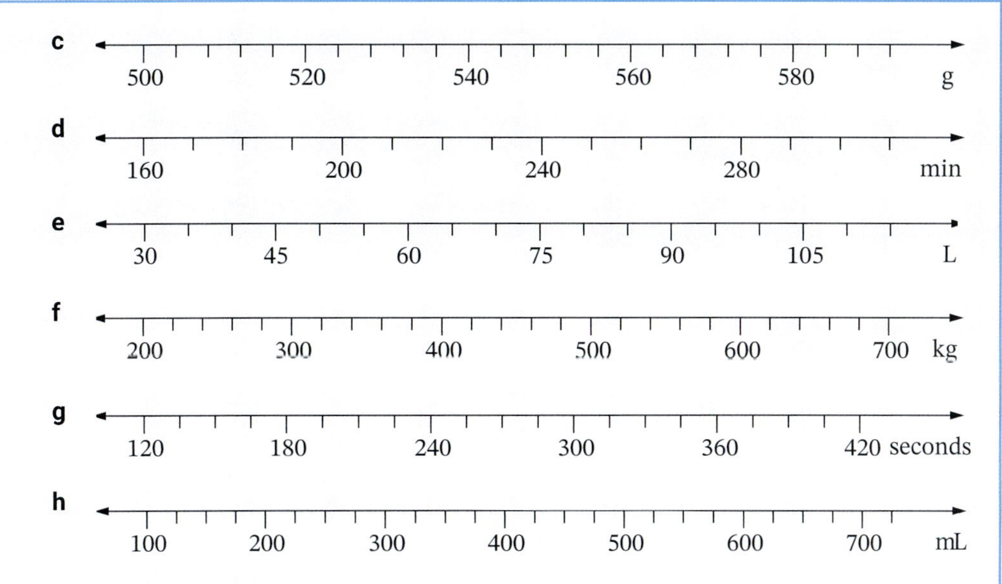

INVESTIGATION

Gym friends

Seven friends joined a gym to lose weight.

Name	Sex	Age	Weight loss
Brenda	F	36	4 kg
Peter	M	34	28 kg
Mark	M	28	21 kg
Fiona	F	33	2 kg
Manjeet	F	30	6 kg
Mia	F	34	3 kg
Natisa	F	29	6 kg

In the following questions, round all mean answers to two decimal places.

1 What is the mean age?

2 Find the mean age of the males.

3 What is the median age of the group? Who has this age?

4 What is the median age of the males? Who has this age?

5 If the men were not counted, would the mean age or the median age be affected more?

6 Predict what would happen to the mean age if Imraan (aged 21) joined the group. Test your prediction.

7 What was the mean age two years ago? How does this compare to the group's mean age now?

8 Compare the mean weight loss with the median weight loss of the whole group. Who has the median weight loss?

9 Compare the mean weight loss and the median weight loss of the females only. Who has the median weight loss now?

10 When Diego joined the group, the mean weight loss of the group became 20 kg. How much weight did Diego lose?

INVESTIGATION

How does an outlier affect a data set?

Outliers are extreme values that are very different from the other values in a data set. Outliers tend to be either very large values or very small values, when compared to the rest of the data set. It is important to understand how outliers may affect a data set when conducting a statistical investigation.

The heights, in centimetres, for a sample of 10 students were recorded from a particular class (shown below).

148 150 153 155 155 155 158 159 161 163

1 Copy the following table into your exercise books:

Measure	Original sample	Sample 2	Sample 3
Mean			
Median			
Mode			
Range			

2 Calculate the mean, median, mode and range for the original sample provided and record your answers in the table.

One student was removed from this sample and replaced by a different student. The new sample, sample 2 is shown below:

118 150 153 155 155 155 158 159 161 163

3 What do you notice about the height of this new student in comparison to the rest of the sample?

4 Calculate the mean, median, mode and range of sample 2 and record your findings in the table above.

5 Identify which statistical measures this short student has changed and explain how.

The short student was replaced by another student in sample 3. The heights of students in sample 3 are shown below:

150 153 155 155 155 158 159 161 163 187

6 What do you notice about the height of the new student in the sample in comparison to the rest of the sample?

7 Calculate the mean, median, mode and range of sample 3 and record your findings in the table above.

8 Identify which statistical measures this tall student has changed and explain how.

The really short student and the really tall student can be classified as outliers in these samples.

9 Use your finding from this investigation to copy and complete this summary table.

A low outlier	A high outlier ....
____________ the mean ______________	____________ the mean ______________
May ___________ the median __________	May ___________ the median __________
Does not affect the __________	Does not affect the __________
______________ the range ____________	______________ the range ____________

INVESTIGATION

Finding the mean of a large set of data

How do we find the mean of a large amount of data? Is there a simpler way than adding all the values, then dividing the total by the number of values?

A group of students was surveyed about the number of hours they watched TV on Saturday:

6	4	0	3	1	2	2	3	5	4	3	3
4	1	0	6	6	3	4	3	1	5	4	3

We can calculate the mean of this set of data using 3 different methods.

Method 1: Adding all the values and dividing by the number of values

1 Find the sum of all the data values.

2 How many values are there in the data set?

3 Hence show that the mean is approximately 3.17 hours.

Method 2: Arranging the values in order first to simplify the addition

1 Sort all of the values in ascending order and group the equal values:

0 0   1 1 1   2 2   3 3 3 3 3 3 3   4 4 4 4 4   5 5   6 6 6

2 To find the sum of all the values, notice that we can multiply each value by its frequency and add those amounts together.

Evaluate $(0 \times 2) + (1 \times 3) + (2 \times 2) + (3 \times 7) + (4 \times 5) + (5 \times 2) + (6 \times 3)$.

3 Hence show that the mean is approximately 3.17 hours.

Method 3: Using a frequency table with an extra column

Score	Frequency	Frequency × Score
0	2	0
1	3	3
2	2	4
3		
4		
5		
6		
Total		

This is a more organised version of Method 2.

1 Copy and complete the frequency table.

2 What is the total number of values?

3 What is the sum of all the values?

4 Hence show that the mean is approximately 3.17 hours.

7.04 Analysing frequency tables

Worksheet
The mean and fx tables

Videos
Analysing frequency tables

Statistics from a frequency table

Example 6

The number of minutes each train was late at a railway station in one afternoon were:

2 4 1 0 0 3 2 3
3 2 0 3 2 1 3

Copy and complete this frequency table for the data and use it to find:

a the range.

b the mode.

c the median.

d the mean, correct to one decimal place.

Score, x	Frequency (f)	fx
0		
1		
2		
3		
4		
Total		

SOLUTION

fx means $f \times x$, so in the fx column, multiply each score (x) by its frequency (f):

$3 \times 0 = 0$

$2 \times 1 = 2$

$4 \times 2 = 8$

Score, x	Frequency (f)	fx
0	3	0
1	2	2
2	4	8
3	5	15
4	1	4
Total	15	29

a Range $= 4 - 0$
$= 4$

Highest value – lowest value

b Mode = 3

The score with the highest frequency (5).

c There are 15 data values (an odd number), so the median (middle) is the **8th** value. According to the frequency table, the values in order are:

Half of 15 is $7\frac{1}{2}$, so round up to 8 and count the 8th value.

0, 0, 0,   1, 1   2, 2, (2), 2   3, 3, 3, 3,   4

The first 3 values are 0, the next 2 values are 1, the next 4 values are 2 and so on.

∴ The 8th value (circled) is 2.

Median = 2.

d Mean $\bar{x} = \dfrac{\text{sum of values}}{\text{number of values}}$.

The fx column multiplies each score by its frequency, so its total, 29, is the sum of all the data values. If we write all the values, we can see that the fx column groups and adds the values.

$\underbrace{0,0,0}_{\text{Sum } 0}\ \ \underbrace{1,1}_{2}\ \ \underbrace{2,2,2,2}_{8}\ \ \underbrace{3,3,3,3,3}_{15}\ \ \underbrace{4}_{4}$

The f column shows the frequency of each value so its total, 15, is the number of values.

$$\text{Mean} = \frac{\text{sum of } fx}{\text{sum of } f}$$
$$= \frac{29}{15}$$
$$= 1.9333\ldots$$
$$= 1.9$$

ⓘ Statistics for data in a frequency table

- The **mean** is $\bar{x} = \dfrac{\text{sum of } fx}{\text{sum of } f}$.
- The **mode** is the data value(s) with the highest frequency, f.
- The **range** is the difference between the last and first values in the score (x) column.

Example 7

The ages of children at a play centre are shown in the frequency table. Copy and complete the table and use it to find:

Score, x	Frequency, f	fx
5	3	
6	4	
7	5	
8	2	
9	2	
Total		

a the mode.

b the mean.

c the median.

d the range.

SOLUTION

Score, x	Frequency, f	fx
5	3	15
6	4	24
7	5	35
8	2	16
9	2	8
Total	16	108

a Mode = 7 — Has the highest frequency, 5.

b Mean:

$$\bar{x} = \frac{\text{sum of } fx}{\text{sum of } f} = \frac{108}{16} = 6.75$$

c There are 16 values (an even number), so there are 2 middle values.

5 5 5 6 6 6 6 7 7 7 7 7 8 8 9 9

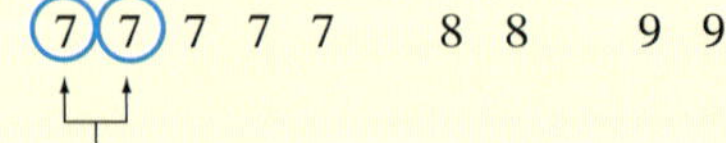

Median $= \frac{7+7}{2} = 7$	Half of 16 is 8, so count the 8th and 9th values (circled) and find the average of them.
d Range $= 9 - 5 = 4$	Note that the mean and median are close to each other and at the centre of the values.

The mean of data in a frequency table using a calculator's statistics mode

Data from a frequency table can be entered into a scientific or graphics calculator using the statistics mode. Follow the instructions in the table below to calculate the mean of the children's ages from the previous page using your calculator's statistics mode.

Operation	Casio scientific	Sharp scientific
Start statistics mode.	MODE STAT 1-VAR SHIFT MODE scroll down to STAT Frequency? ON	MODE STAT =
Clear the statistical memory.	SHIFT 1 Edit, Del-A	2nd F DEL
Enter data.	SHIFT 1 Data to get table 5 = 6 =, etc. to enter in x column 3 = 4 =, etc. to enter in FREQ column AC to leave table	5 2nd F STO 3 M+ 6 2nd F STO 4 M+, etc.
Calculate the mean. ($\bar{x} = 6.75$).	SHIFT 1 Var $\bar{x}$ =	RCL $\bar{x}$
Check the number of values. (n = 16)	SHIFT 1 Var n =	RCL n
Return to normal (COMP) mode.	MODE COMP	MODE 0

EXERCISE 7.04 ANSWERS ON P. 636

Analysing frequency tables

U F

In this exercise, round all means to one decimal place where necessary.

1 This table shows how many times a sample of children visited a doctor over 3 months. Copy and complete the table, and use it to find:

a the number of people surveyed.

b the median.

c the mean.

d the mode.

e the range.

No. of visits, x	Frequency, f	fx
0	5	
1	6	
2	14	
3	3	
4	5	
8	1	
Total		

2 A sample of hospital patients were surveyed on the number of hours they slept the previous night:

6	7	9	8	10	11	9	10	8	9	7	6
11	9	8	8	8	10	10	9	7	8	9	10

a How many people were surveyed?

b What was the highest number of hours people slept?

c Use a frequency table with an fx column to find the mean.

d Find the range.

e Find the median.

3 The judge's scores for students' speeches (out of 10) for an English class are as shown.

7	7	8	9	7	5	4	8	9
9	8	10	8	7	5	5	4	1
8	8	9	7	7	4	8	3	1

a Complete a frequency table for this data.

b What is the mode?

c What is the mean?

d What is the median?

e What is the range?

4 A class of students was asked how many children were there in their families.

2	4	5	1	2	3	1	3	2	4	2	4
2	4	2	3	1	5	6	2	2	1	3	3

Complete a frequency table for this data, including an fx column, and use it to find:

a the range **b** the mode **c** the median **d** the mean

5 This frequency table shows the heights in metres of a sample of students.

Find:

a the number of students in the sample.

b the mean, correct to 2 decimal places, using your calculator's statistics mode.

c the range.

d the median.

e the mode.

Height (m)	Frequency
1.42	4
1.44	5
1.46	8
1.48	11
1.5	6
1.52	3

☐ Foundation ○ Standard ⬡ Complex

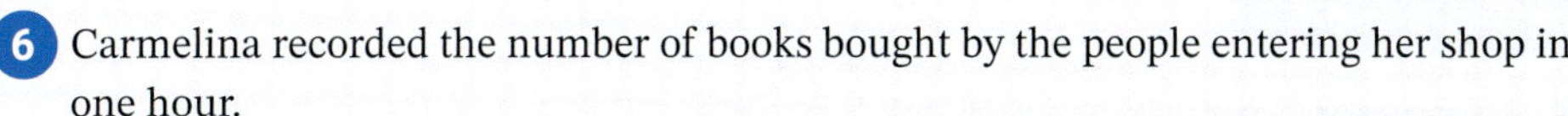

6 Carmelina recorded the number of books bought by the people entering her shop in one hour.

No. of books	Frequency
0	7
1	5
2	13
3	4
4	1
5	2
6	1

What is the median number of books bought? Select the correct answer **A**, **B**, **C** or **D**.

A 2 **B** 3

C 13 **D** 4

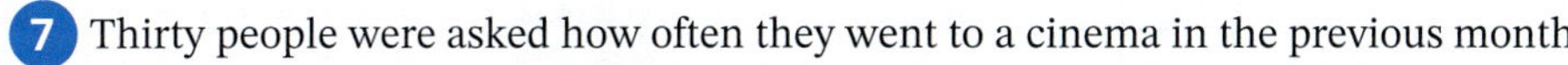

7 Thirty people were asked how often they went to a cinema in the previous month.

4	5	3	1	3	7	1	1	4	0
0	4	3	2	4	2	4	7	6	6
1	3	5	5	1	3	6	7	0	3

Construct a frequency table and use it to find:

a the range **b** the mode

c the median **d** the mean

9 A pigeon breeders' club has 25 members. This frequency table shows the number of pigeons each member has.

Score, x	Frequency, f
8	3
10	4
14	3
15	6
17	2
18	4
20	3

a What is the range?

b What is the median?

c Use your calculator's statistics mode to find the mean.

Foundation | Standard | Complex

TECHNOLOGY

Daily temperatures

1 Enter the following data into a spreadsheet: a frequency table showing the maximum daily temperatures in Fremantle in July.

	A	B	C
1	Temperatures (x)	Frequency (f)	*fx*
2	15	1	
3	16	3	
4	17	3	
5	18	7	
6	19	7	
7	20	2	
8	21	7	
9	22	0	
10	23	1	
11	Total:		
12	Mean:		
13	Range:		

2 How many temperatures are in this data set? In cell B11, enter an appropriate formula for the sum of the frequency column.

3 To calculate the mean, first complete the *fx* column. In cell C2, enter the formula for multiplying the values in the A and B columns, then **Fill Down** to copy the formula into cells C3–C10.

4 In cell C11, enter an appropriate formula for the sum of the *fx* column.

5 In cell B12, calculate the mean by entering the formula for dividing the sum of the *fx* column by the sum of the *f* column.

6 In cell B13, calculate the range by entering the formula for subtracting the lowest temperature from the highest temperature.

7.05 Dot plots and stem-and-leaf plots

Worksheets
Stem-and-leaf plots

What data?

Presentation
Stem-and-leaf plots

Dot plots

A **dot plot** is a graph that uses dots to show the frequency of each data value. A dot plot shows:

- any gaps in the data.
- any **clusters**, where values are grouped or bunched together.
- any **outliers**: extremely high or low values that are very different from the rest of the data.
- how the values are spread out.

An **outlier** 'lies outside' the other values in the data set. Outlier is pronounced 'out-ly-er'.

Example 8

An online music store recorded the daily sales figure of a particular album while it was popular, and displayed the results in a dot plot.

a On how many days were there exactly 13 sales?

b What is the range?

c Find the mode.

d Find the median.

e Calculate the mean, correct to one decimal place.

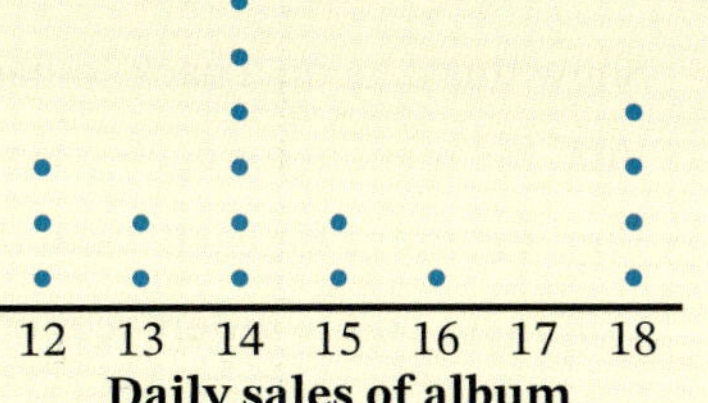

SOLUTION

a 13 sales were made on 2 days. — 2 dots above 13

b Range = 18 − 12 = 6 — Highest value − lowest value

c Mode = 14 — The value with the most dots.

d There are 18 values (18 dots). This is an even number, so there are 2 middle values (the 9th and 10th values). By counting the dots or by crossing out pairs of dots at each end, we can see that the middle values (circled) are 14 and 14.

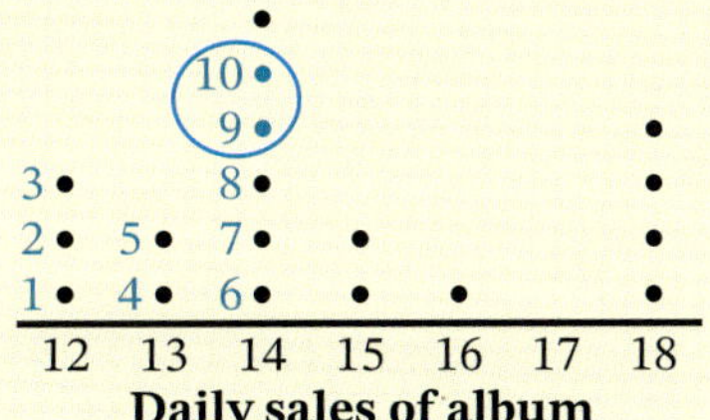

$$\text{Median} = \frac{14+14}{2} = 14$$

e $\text{Mean: } \bar{x} = \frac{\text{sum of values}}{\text{number of values}}$ $\left(\text{or } \frac{\text{sum of } fx}{\text{sum of } f}\right)$

Number of values = 18 — 18 dots

$$\bar{x} = \frac{3\times12+2\times13+6\times14+2\times15+16+4\times18}{18}$$

$$= 14.6666\ldots$$

$$= 14.7$$

Note that the mean and the median are close to each other and at the centre of the data.

Note: This answer can also be found using a frequency table and/or your calculator's statistics mode.

7.05

Videos
Analysing dot plots
Dot plots

Stem-and-leaf plots

Videos
Stem-and-leaf plots 1
Stem-and-leaf plots 2
Back-to-back stem-and-leaf plots

A **stem-and-leaf plot** (or **stemplot**) is like a sideways column graph, but one that lists the actual values on the horizontal columns. An **ordered** stem-and-leaf plot shows:

- all the data values, listed from smallest to largest (the smallest values are always closest to the stem).
- any modes (the most common values).
- any clusters or outliers.
- how the data are spread out.

We can use a **back-to-back stem-and-leaf plot** to make comparisons between 2 different data sets.

Example 9

The heights (in centimetres) of a sample group of Year 8 boys and girls were measured.

Girls:	148	155	153	137	155	137	146	165	157	168
Boys:	140	152	143	139	166	144	136	151	130	138

a Show this data on a back-to-back stem-and-leaf plot.

b Find the median for each set of data.

c Find the range for each set of data.

d Comment on the differences between the data for girls and boys.

SOLUTION

a

Heights of Year 8 students in centimetres

Girls		Boys
7 7	13	0 6 8 9
8 6	14	(0) (3) 4
7 5 (5) (3)	15	1 2
8 5	16	6

b Median: There are 10 values in each set.

The median is the average of the 5th and 6th values (circled in red on the plot).

Median: Girls $= \frac{153+155}{2} = 154$

Boys $= \frac{140+143}{2} = 141.5$

c Range: Girls = 168 – 137 = 31

Boys = 166 – 130 = 36

d The girls in the sample are generally taller (higher median), but the boys' heights are more spread out (higher range).

EXERCISE 7.05 ANSWERS ON P. 636

Dot plots and stem-and-leaf plots

U F PS R C

In this exercise, round all mean answers to one decimal place where necessary.

1 For each dot plot, find:

i the range **ii** the mode **iii** the median **iv** the mean

a

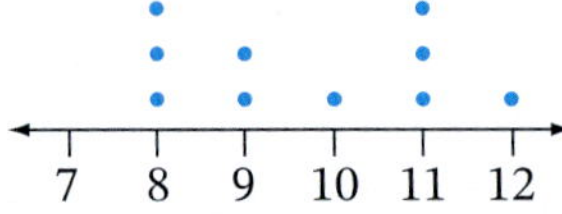

b

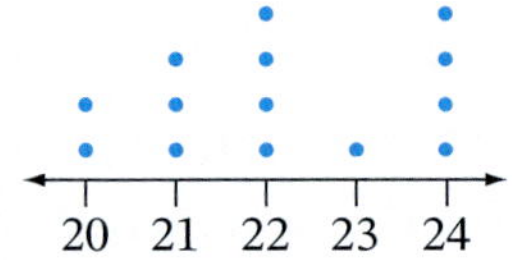

☐ Foundation ○ Standard ⬡ Complex

c

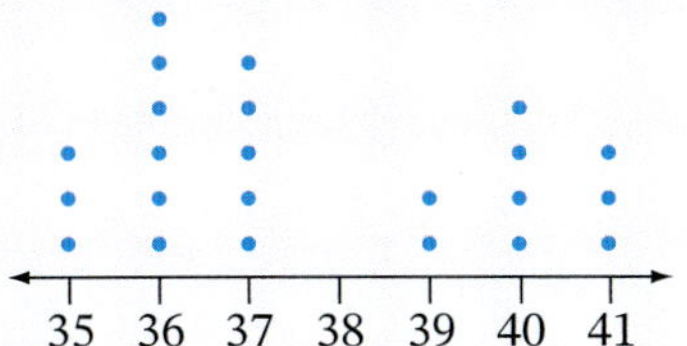

d

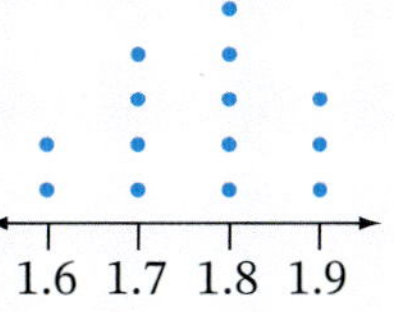

2 The number of goals a football player kicked for each match last season were:

2	7	3	2	4	0	3	2	4	3	5
3	1	1	2	4	2	5	3	0	1	0

a Draw a dot plot to represent this data.

b How many matches were there last season?

c Which value is an outlier?

d Calculate the median.

e Calculate the mean.

3 This dot plot shows the number of students in each class at Northvale Primary School.

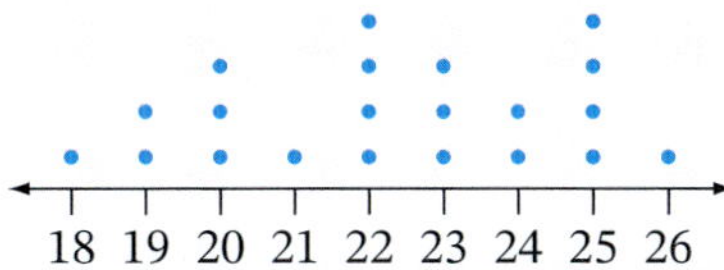

a How many students are there in the smallest class?

b Find the range.

c Find the mode.

d Calculate the mean.

e What percentage (to the nearest whole number) of classes have over 20 students?

f Which of the following values is the median? Select the correct answer **A**, **B**, **C** or **D**.

A 11 **B** 22 **C** 22.5 **D** 23

4 The number of songs Udani downloaded each day was:

R

1, 2, 0, 2, 3, 3, 3, 1, 1, 1, 6, 2, 3, 0

a Draw a dot plot for this data.

b How many days are represented by this dot plot?

c What is the outlier?

d Find the median.

e Calculate the mean.

f If this pattern continues in the following weeks, how many songs would she expect to download on average in a week?

Foundation Standard Complex

5 The quiz marks out of 10 for 2 Year 8 classes are shown.

R C

8 Hamper:	3	2	0	1	5	8	6	7	6	3	
	5	4	5	6	7	9	2	5	7		
8 Yen:	7	6	3	7	8	1	9	4	6	7	2
	7	2	8	10	9	9	5	7	8	9	10

a Draw a dot plot of the marks for each class.

b What is the mode of the marks for Class 8 Yen?

c What is the median of the marks for Class 8 Hamper?

d What is the range of the marks for Class 8 Yen?

e Draw a frequency table of the data for Class 8 Hamper and use it to calculate the mean.

f Which do you think is the 'better' class? Give a reason for your answer.

EXAMPLE 9

6 For each stem-and-leaf plot below, find:

i the range **ii** the mode **iii** the median **iv** the mean

a

Stem	Leaf
1	0 2 3
2	1 4 4 5 6
3	3 3 3 7
4	1 2 3 5 9

b

Stem	Leaf
7	3 4 5 7 7
8	1 2 2 3 8 9
9	0 4 4 4 4 6 7 9

c

Stem	Leaf
10	0 1 1 2
11	7 8 8 8 9
12	3 6 6 7
13	1 1 1 1 8
14	0

d

Stem	Leaf
0	5 5 6
1	1 4 4 7 7 7 7
2	0 3 8 8
3	
4	9 9 9

 Foundation  Standard  Complex

7 A survey shows that the number of students served at the school canteen daily over a 3-week period is:

105 76 97 108 114 106 124 101

112 98 95 105 117 101 92

a Show this information on a stem-and-leaf plot.

b What is the outlier?

c Find the mode.

d What was the highest value?

e On what stems were the values clustered?

f On what percentage (to one decimal place) of days were more than 100 students served?

8 The heights (in centimetres) of the students in a Science class are:

PS R C

156 154 158 167 164 163 155 166 159 158 137 163

171 160 164 169 157 161 160 165 145 175 167 167

a Make a stem-and-leaf plot for this data.

b On what stem are the data clustered?

c How many students were measured?

d How many students were shorter than 150 cm?

e Calculate the range.

f Find the median.

9 The time taken, in minutes, to drive between 2 towns along 2 different roads are shown on this back-to-back stem-and-leaf plot.

Fuller's Ridge Rd		Great Southern Drive
7 5 3 2	2	6 8
9 8 8 5 1 1	3	0 2 4 7 8
3 3 2 0 0	4	0 1 2 6 9 9
2	5	0 1 3 7
1 0	6	2
	7	1

a How many trips were made along the Great Southern Drive?

b How many trips made along Fuller's Ridge Road took over an hour?

c Which road is the faster one to take to drive between the 2 towns? Explain your answer.

d Which statement is true for the data sets above? Select the correct answer **A**, **B**, **C** or **D**.

A The range is the same for both data sets.

B There is a difference of 5 minutes in the median travel times.

C The mean travel time for Fuller's Ridge Road is 36 minutes.

D A travel time of 71 minutes along the Great Southern Drive is an outlier.

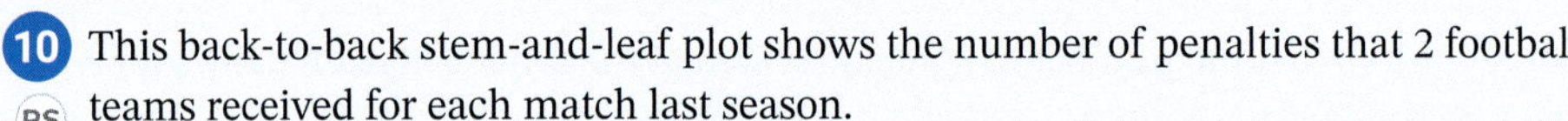

10 This back-to-back stem-and-leaf plot shows the number of penalties that 2 football teams received for each match last season.

PS R C

Wee Waa Reds		Narrabri Lions
9 8 4 4 2	0	7 9 9
8 6 5 5 2 2 2	1	2 3 3 6 8
3 1 1	2	5 6 8 9
0	3	0 0 1 3

a How many matches did these teams play in one season?

b How many times did the Narrabri Lions receive fewer than 20 penalties in a match?

c Find the highest number of penalties received by the Wee Waa Reds in a game.

d Find the mode for the Narrabri Lions.

e Which value was an outlier for the Wee Waa Reds?

f Which team generally had a higher number of penalties?

g Which team had a greater range in their number of penalties? Give a reason for your answer.

INVESTIGATION

Measures of average

1 Create 6 values in a data set where the mean equals the median.

2 Create 5 values in a data set where the median equals the mode.

3 Create 4 values where the mean, the median and mode all differ.

4 Create 5 values where the mean, median and mode are the same.

5 7 judges score a gymnast out of a possible 10. Their values are:

6 7 8 6 8 9 6

The scoring system requires that the lowest and highest scores are rejected and that the mean of the remaining 5 gives the final score. What is the final score?

6 In 4 games Dan scores 23, 11, 17 and 26 points. What score must Dan achieve in his next game so that he can average 20 on the 5 games? (Show the working or reasoning that you use.)

7 11 film students rate a movie out of a possible 10. Their ratings are:

7 9 9 4 6 4 8 9 7 8 5

The group cannot agree on a group rating for the movie. In one sentence, comment on each of these suggestions.

a Kim thinks that the rating should be 9 (the mode) because that was the rating given the most often.

b Vlado thinks that the rating should be 7 (the median) because it is not swayed by extremely high or extremely low ratings.

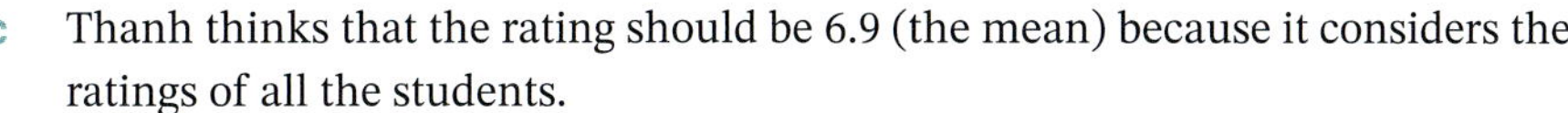

c Thanh thinks that the rating should be 6.9 (the mean) because it considers the ratings of all the students.

What overall rating would you give the film based on these opinions?

8 Find 3 numbers for which the median is 15, the mean is 13 and the range is 12.

9 Find 5 numbers for which the median is 3, the mean is 4 and the mode is 7.

Frequency histograms and polygons 7.06

Numerical data can be graphed on a **frequency histogram**, a special type of column graph, or a **frequency polygon**, a special type of line graph.

Worksheet
Analysing data

Puzzles
Statistical match-up

Highway elephants

Example 10

20 families were surveyed about how many phones they owned.

Graph this data as:

a a frequency histogram.

b a frequency polygon.

No. of phones	Frequency
0	2
1	7
2	8
3	2
4	1
Total	20

SOLUTION

a

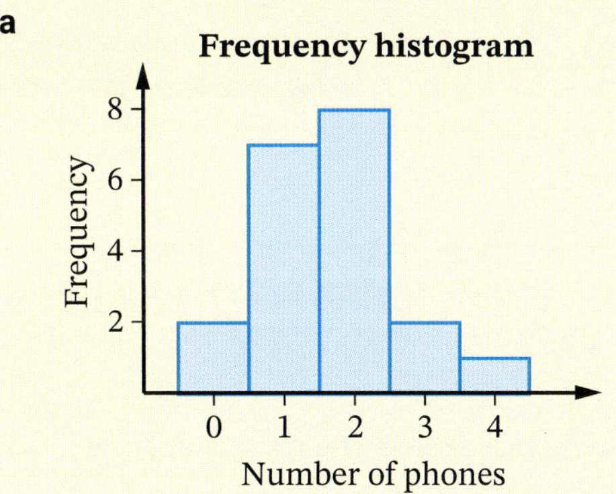

b

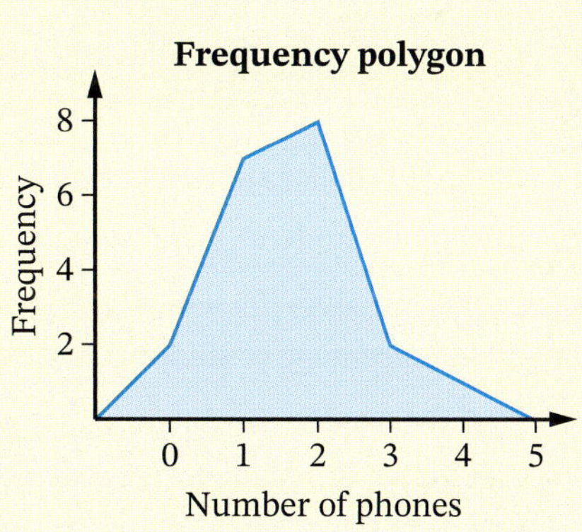

Note: A frequency polygon can be constructed from a frequency histogram by joining the midpoints of the tops of the columns.

ⓘ Frequency histogram

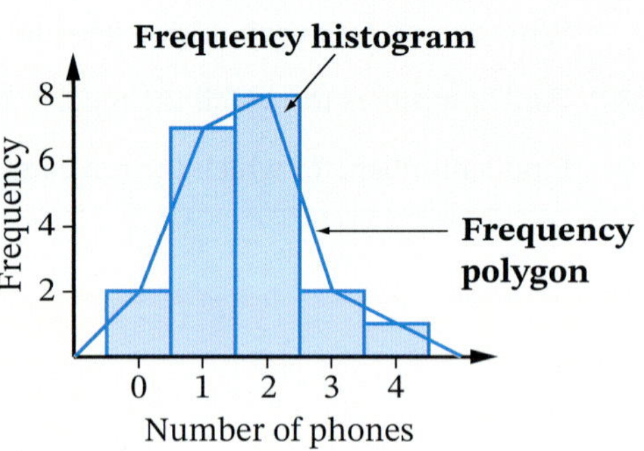

- A column graph that looks like a row of office buildings.
- Frequency is shown on the vertical axis.
- The values are shown on the horizontal axis.
- The columns stand together without gaps between them.
- The columns are centred on the values.
- There is a gap of half a column width on the left (between the frequency axis and the first column).

ⓘ Frequency polygon

- A line graph that looks like a mountain.
- Frequency is shown on the vertical axis.
- The values are shown on the horizontal axis.
- Begins and ends on the horizontal axis.
- Called a 'polygon' because the graph and the horizontal axis together make a shape with straight sides.

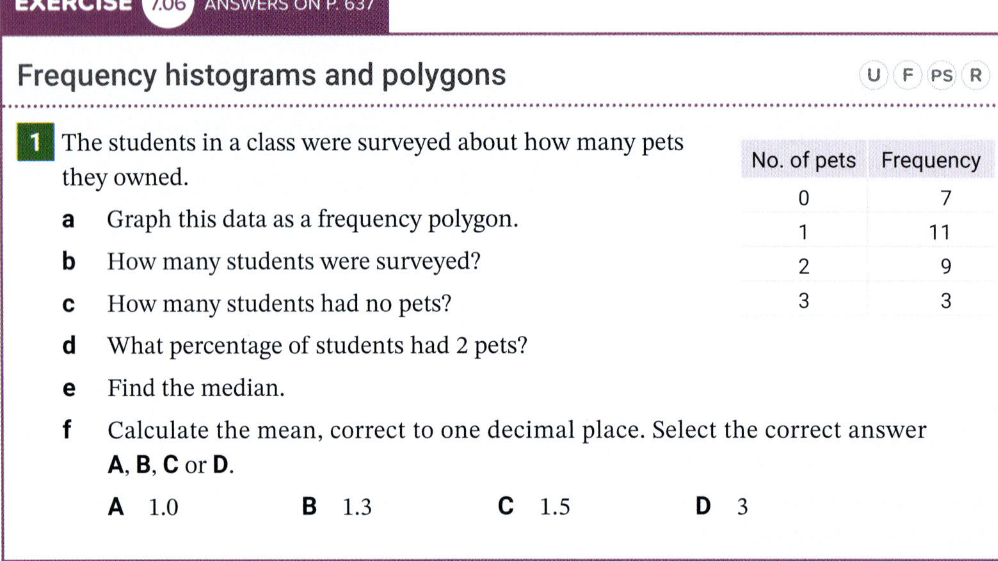

EXERCISE 7.06 ANSWERS ON P. 637

Frequency histograms and polygons

U F PS R

EXAMPLE 10

1 The students in a class were surveyed about how many pets they owned.

No. of pets	Frequency
0	7
1	11
2	9
3	3

a Graph this data as a frequency polygon.

b How many students were surveyed?

c How many students had no pets?

d What percentage of students had 2 pets?

e Find the median.

f Calculate the mean, correct to one decimal place. Select the correct answer **A**, **B**, **C** or **D**.

A 1.0 **B** 1.3 **C** 1.5 **D** 3

▷

☐ Foundation ○ Standard ○ Complex

2 Hadieya counted the number of phone calls she made each day for 30 days.

5	3	6	2	4	3	3	5	4	7
5	1	6	3	2	3	5	6	1	6
2	4	3	1	5	4	2	7	3	2

a Complete a frequency table for this data.

b Draw a combined frequency histogram and polygon of the data.

c On how many days did Hadieya makes 3 calls?

d How many phone calls were made over the 30 days?

e On how many days did Hadieya makes fewer than 3 calls?

f What was the mode?

g What was the mean, correct to one decimal place?

3 This histogram shows the number of computers found in a sample of homes.

a How many households were surveyed?

b Where are most of the values clustered?

c Find the mode.

d Find the range.

e Complete a frequency table, including an fx column, for the data and use it to calculate the mean.

f Find the median.

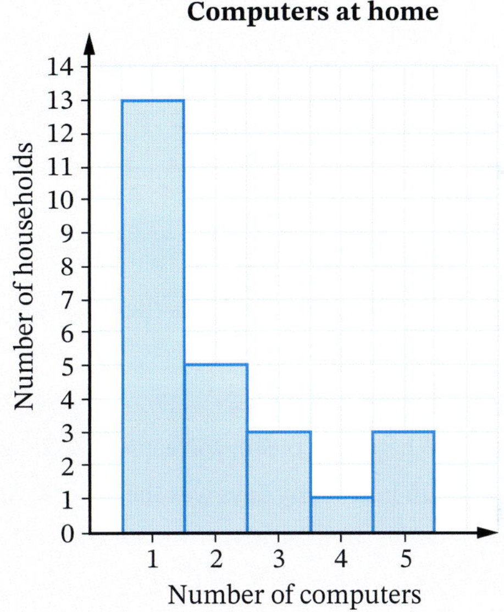

4 The numbers of goals scored by a soccer team in each match last season were:

2	0	1	2	3	4	3	1	2	0
2	1	4	3	2	5	1	2	0	3

a Complete a frequency table for this data.

b Draw a frequency histogram and frequency polygon of the data on the same axes.

c How many matches were played by the team?

d What was the highest number of goals scored?

e What was the mode?

f What was the mean?

g In how many matches were more than 2 goals scored?

h In what percentage of matches were no goals scored?

Foundation Standard Complex

5 The water level of a canal was recorded over 44 days and graphed on this histogram.

PS R **a** Use the histogram to copy and complete the frequency table.

Water level (cm)	Frequency
152	
153	
154	
155	
156	
157	
Total	

Water level of the canal

b On how many days was the water level 156 cm high?

c What is the range?

d What is the mode?

e What is the mean, correct to 2 decimal places?

f On how many days was the water level lower than 154 cm?

g What is the median?

h There needs to be at least 155 cm of water in the canal for a boat to travel on it safely. On how many days would the boat be able to use the canal?

☐ Foundation ○ Standard ⬡ Complex

Comparing data

7.07

This table describes the 4 statistical measures and when it is more appropriate to use each one.

Statistical measure	Features	When is it appropriate to use?
Mean		
$\bar{x} = \frac{\text{sum of values}}{\text{number of values}}$ For a frequency table: $\bar{x} = \frac{\text{sum of } fx}{\text{sum of } f}$	• For numerical data • Depends on all the values in the data set • Affected by outliers • Can be a decimal that is not one of the actual values	When the data set does not have many outliers. When working with numerical data.
Mode		
Most common value(s)	• For numerical and categorical data • There may be more than one mode, or no mode at all • Not affected by outliers	When the most common value or category is needed. When working with categorical data.
Median		
Middle value, or average of the 2 middle values, when values are arranged in order	• For numerical data • Can be one of the values • Not affected by outliers	When the data set has many outliers. When working with numerical data.
Range		
Highest value − lowest value	• For numerical data • Depends on the highest and lowest values only	When a measure of spread is needed.

Worksheets
Analysing data
Comparing word lengths
Investigating young drivers

Technology
Population polygons
Daily rainfall

Example 11

A group of 8 children was surveyed about the amount of pocket money (in dollars) they received each week.

20 32 32 40 18 32 18 48

a Find:

i the mean **ii** the mode **iii** the median **iv** the range

b If the last student's pocket money increased from \$48 to \$75, describe how this would affect:

i the mean **ii** the mode **iii** the median **iv** the range

c Which measure of central tendency best represents the new set of data?

SOLUTION

a **i**

$$\begin{aligned}\bar{x} &= \frac{\text{sum of values}}{\text{number of values}}\\ \text{Mean: } &= \frac{20+32+32+40+18+32+18+48}{8}\\ &= \frac{240}{8}\\ &= \$30\end{aligned}$$

ii Mode = \$32

iii Rewriting the 8 values in order:

18 18 20 (32) (32) 32 40 48

There are 2 middle values, 32 and 32 (circled above).

$$\text{Median} = \frac{32+32}{2}$$
$$= \$32$$

iv $\text{Range} = 48 - 18$
$$= \$30$$

b If the last student's pocket money increased to \$75, this value would be a high outlier compared to the others.

i The mean should increase.

$$\text{Mean: } \bar{x} = \frac{267}{8}$$
$$= \$33.375$$

ii The mode will not change because \$32 is still the most common value.

iii The median will not change because the order of the values has not been affected.

18 18 20 (32) (32) 32 40 75

The 2 middle values are still 32 and 32 (circled above).

iv The range will increase because the highest value is now even higher.

$\text{Range} = 75 - 18$
$$= \$57$$

c The median is the best measure of central tendency because it is not affected by the outlier of \$75.

Note:

- The mode (\$32) is a good measure, but it only represents 3 of the values.
- The mean (\$33.375) is the least appropriate because it has been affected by the outlier and is now higher than most of the values.

EXERCISE 7.07 ANSWERS ON P. 638

Comparing data

U F PS R C

In this exercise, round all mean answers to one decimal place where necessary.

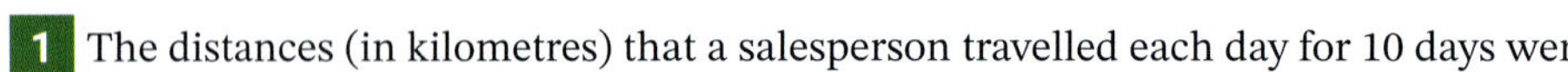

EXAMPLE 11

1 The distances (in kilometres) that a salesperson travelled each day for 10 days were:

49 77 63 56 49 40 56 40 74 49

a Find:

i the range **ii** the mode **iii** the median **iv** the mean

b Which measure of central tendency best represents the data?

Foundation Standard Complex

2 The ages of the mathematics teachers at Westvale Secondary College are:

R C

49 32 37 32 25 41 39

a Find:

i the range **ii** the mode **iii** the median **iv** the mean

b The 39-year-old teacher was replaced by a 22-year-old teacher. Without recalculating, describe how this will affect:

i the range **ii** the mode **iii** the median **iv** the mean

3 In a swimming competition, each swimmer was graded A, B, C, D or E. One squad achieved the results shown.

R C

Grade	A	B	C	D	E
No. of swimmers	4	9	8	6	3

a How many swimmers were in the squad?

b What is the mode?

c Why it isn't possible to find the mean or median of these grades?

4 The annual incomes (× \$1000) of 12 people are:

R C

84 57 121 75 81 134 66 484 210 93 125 90

a Find (in dollars):

i the range **ii** the median **iii** the mean

b Why isn't the mean or mode an appropriate measure of central tendency for the incomes?

c Remove the outlier from the set of data and recalculate the mean and median. Are they closer together now?

5 The heights (in centimetres) of a team of basketball players are:

R C

204 196 181 185 190 198 193

a Find:

i the mean **ii** the median **iii** the mode **iv** the range

b The 190 cm player is replaced by a 181 cm player. Without recalculating, describe how this will affect:

i the mean **ii** the median **iii** the mode **iv** the range

6 The cricket batting scores for Sunjeet and Chris over 6 innings are shown.

PS R C

Sunjeet: 29, 35, 40, 26, 28, 39 Chris: 2, 2, 3, 2, 188, 5

a What are the median scores for each player? Select the correct answer **A**, **B**, **C** or **D**.

A Sunjeet 32, Chris 2 **B** Sunjeet 40, Chris 2.5

C Sunjeet 33, Chris 2.5 **D** Sunjeet 32, Chris 2.5

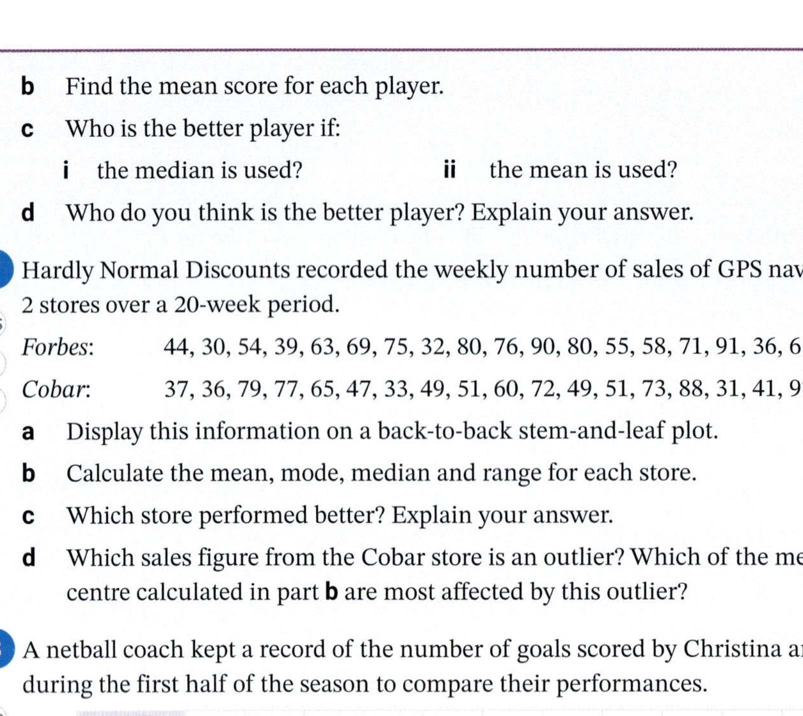

b Find the mean score for each player.

c Who is the better player if:

i the median is used? **ii** the mean is used?

d Who do you think is the better player? Explain your answer.

7 Hardly Normal Discounts recorded the weekly number of sales of GPS navigators at 2 stores over a 20-week period.

PS R C

Forbes: 44, 30, 54, 39, 63, 69, 75, 32, 80, 76, 90, 80, 55, 58, 71, 91, 36, 61, 65, 84

Cobar: 37, 36, 79, 77, 65, 47, 33, 49, 51, 60, 72, 49, 51, 73, 88, 31, 41, 97, 58, 30

a Display this information on a back-to-back stem-and-leaf plot.

b Calculate the mean, mode, median and range for each store.

c Which store performed better? Explain your answer.

d Which sales figure from the Cobar store is an outlier? Which of the measures of centre calculated in part **b** are most affected by this outlier?

8 A netball coach kept a record of the number of goals scored by Christina and Mary during the first half of the season to compare their performances.

PS R C

Christina	12	23	25	19	10	35	20	32	19	25	4	27
Mary	13	30	20	24	17	25	17	34	20	29	19	15

a Calculate the mean, mode, median and range for each player.

b Who do you think is the more consistent player? Explain.

c Who is the higher scorer?

d Who do you think is the better player? Why?

e What is the outlier value?

f If the outlier is removed, does it change:

i who the more consistent player is? **ii** who the higher scorer is?

9 Two judges rated 8 contestants in a singing contest.

R C

Delta	8	5	6	8	5	9	2	6
Guy	7	8	5	4	9	5	5	6

a Draw a dot plot for each judge's marks.

b Calculate the range and median for each judge.

c Describe differences in the way each judge allocates marks.

7.08 Census and samples

In statistics, the **population** means all of the items being studied, whether they can be people, cars, businesses or houses. Statistical data is collected in a variety of ways, such as surveys, Internet polls, interviews and questionnaires. To collect information, statisticians usually survey a large representative group called a **sample**. For example, to investigate opinions about school

☐ Foundation ○ Standard ⬡ Complex

uniform, we can survey a selection of students, teachers and parents. The bigger the sample, the more representative and accurate it will be.

To collect information about a whole population, **all** people or items must be surveyed. This is called taking a **census**. For example, to plan the locations of childcare centres, the government may need to count the number of babies in different areas.

Videos
Counting crowds

Can fish oil make you smarter?

Can you trust your IQ?

The wrong guy won

7.08

Sample	Census
Surveys a selected representative group of people or items.	Surveys **all** people or items (in the population).
Only gives approximate information about the population. Helps us make predictions about the population.	Gives **exact** information about the population.
Simple and inexpensive to survey.	Complex and expensive to survey.
Can be done quickly.	Takes a lot of time to collect and process the information.

In Australia, a national census is conducted every 5 years to count the Australian population and obtain important information about every person. It takes place in a year ending in 1 or 6, such as 2026 and 2031.

Example 12

Should a sample or a census be used to investigate each fact?

a The most popular car colour.

b Whether energy-saving light globes really use less electricity.

c The number of retired people living on the Gold Coast in Queensland.

d People's views on which nation will win the Rugby World Cup.

SOLUTION

a Sample, because it is easier and faster than asking all car-owners in Australia.

b Sample, because it is easier and faster than testing every light globe.

c Census, because the exact number is required.

d Sample, because it is very difficult to ask all people.

Random sampling and bias

Remember that in probability, a **random experiment** is a situation where every possible outcome has the same chance of occurring. When taking a sample, it is important that each person or item in the population has an equal chance of being chosen to be surveyed. This is called **random sampling**. To find a sample of students to survey about school uniform, we could place students' names in a box, mix them up and then choose 100 names without looking. Alternatively, we could use a computer or calculator to generate random numbers to select students.

A sample that is not random is called a **biased sample**, and is not truly representative of the population. For example, if you survey the first 100 students to arrive at school today, or the first 100 students on the school roll, then not all students in the school have an equal chance of being selected. Your sample would not represent students who arrive at school later or whose names begin with letters at the end of the alphabet.

Example 13

A random sample of students is to be selected for their views on a new school uniform. Will each of the following sampling methods provide a sample that is representative or biased? If biased, explain why.

- **a** Randomly selecting one student from each class.
- **b** Selecting 20 students from your mathematics class.
- **c** Selecting the students from the school's netball teams.
- **d** Randomly selecting students from the school roll by rolling a die and counting down the roll.

SOLUTION

- **a** Representative. Every student in the school has an equal chance.
- **b** Biased, because your mathematics class does not represent the other Year groups (or even the other Year 8 classes) at your school.
- **c** Biased, because the netball teams do not represent the whole school and will have mainly female students.
- **d** Representative. Every student in the school has an equal chance.

EXERCISE 7.08 ANSWERS ON P. 638

Census and samples

U F R C

1 Is a sample or a census more appropriate for each situation?

R **a** Testing coffee for taste.

C **b** Finding the number of migrants from New Zealand.

c Investigating people's views on road safety.

d Finding out people's opinions on whether Australia should become a republic.

e Testing a flu vaccine for effectiveness.

f Counting the number of Spanish-speaking children in Australian schools.

g Finding the population of Gundagai.

h Finding the number of people in Australia over 40, to plan for health care.

2 For each situation, state whether a sample or a census would be more appropriate and how the data could be collected.

R C **a** A car dealer wants to find the level of customer satisfaction with his business.

b To find the most popular TV program on Friday nights.

c To find the busiest airport in Australia.

d A principal wishes to find the least-liked subject in the school.

e The number of people who support whaling.

f The average height of students in your school.

g The amount of recyclable garbage each household produces in a week.

h The number of times the students in your class visit a restaurant in a month.

3 The parent association for Nelson Vale High School wants to know what people think of a new design for the school uniform. Which of the following samples are NOT biased? Select the correct answer **A**, **B**, **C** or **D**.

R

A Students visit the school office during the day.

B Parents as they are picking up their children after school.

C Randomly selected families from an alphabetical list of school families.

D Families watching the school sporting teams.

4 For each survey, describe the total population, give a suitable way of choosing a representative sample, write any problems that may be encountered and explain how bias could be minimised.

R C

a The attitudes of people who smoke in public.

b Whether girls are better than boys at mathematics.

c The choice of music at a school disco.

d The political parties people will vote at the next state election.

6 In each survey, state whether the sample is random. If not, explain why you think it is biased.

R C

a To find participants for a survey of students' study habits, a teacher wrote a message in the daily notices asking for volunteers.

b To ensure that items being made in a factory were of good quality, 2 items coming off the assembly line were tested at a randomly selected time each hour.

c In a poll to find voters' preferences in the next election, every 100th person listed in the White pages phone database was called. If there was no answer, the next person listed was called and so on.

d In a parent survey about school uniform, the school council posted letters to all parents of students at the school asking for their opinion on the issue.

e In a survey about the best song of the decade, listeners of a rock radio station were asked to ring and nominate a song.

7 The student council wants to survey students on their attitude to mobile phone usage at school. Fifteen students from each year group were selected to take part in the survey.

R C

a Do you believe the results of this survey would be reliable? Explain your answer.

b What factors should have been considered in selecting the sample?

c How would you have selected the students to be surveyed?

TECHNOLOGY

Random numbers

Random numbers can be used to select people or items for a sample. A calculator or spreadsheet can be set up to generate a random number from 1 to 100, for example, and that number can be matched to the name of a person or item from a list. For a scientific calculator:

Casio	Sharp
Enter the formula RanInt#(1,100) as shown: ALPHA . (1 SHIFT) 100) =	2nd F 7 (RANDOM), select 3 R-INT, + 1 =

This generates a random number from 1 to 100. Press = more times to generate more random numbers.

Alternatively, on a **spreadsheet**, type **=INT(RAND()*100+1)** into a cell and press the ENTER key, then the F9 key repeatedly for more random numbers.

1 Use your calculator or spreadsheet to generate 12 random numbers from 1 to 100.

2 Use your calculator or spreadsheet to generate 12 random numbers from 1 to the number of students in your class.

3 Use a class roll to number the students in your class from 1 onwards. Suppose you need to choose a random sample of nine students from your class for a survey on homework. Use your calculator or spreadsheet to generate nine random numbers to select the nine students for your survey.

4 Choose 4 students from each Year 8 maths class to answer some questions. Devise several different methods to ensure that your selection is random. Prepare a short class talk about your methods.

INVESTIGATION

Statistics in the media

We live in a world of 24-hour news: from TV, newspapers, websites, Facebook, Twitter, blogs, etc. To detect bias in news, we need to consider where the news item comes from and what types of samples the statistics are based on. Is the information supplied by a reporter, the police, a company executive, a government official or an opinionated blogger? Are the statistics based on a small sample, large sample, unrepresentative sample or a phone or Internet poll of volunteers? Each may have a particular bias that may influence how the story is reported.

Find examples of news items or surveys reported in a newspaper, magazine or the Internet and investigate the following questions.

a Where did it come from?

b Who wrote the story?

c Does it show any bias?

d Can it be supported by other news providers?

e What type of sampling was used? Was it representative?

f How many people were questioned?

g When was the survey conducted?

DID YOU KNOW?

The Australian Census

The first population counts in Australia were known as musters, where people assembled at a location to be counted. The first census when people were recorded at their dwellings was in NSW in 1828. The first simultaneous census across all states was in 1881 and the population was counted as 2.25 million.

Following Federation in 1901, the first national census was developed and undertaken by the Commonwealth in 1911. From 1961, the Census has been held every 5 years in the years ending in 1 and 6.

The current census asks questions relating to the family, where you live, income, religion, level of education and other aspects of modern life.

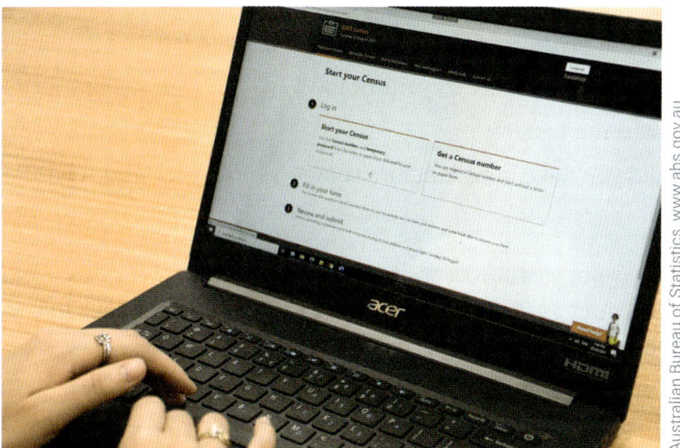

Australian Bureau of Statistics, www.abs.gov.au

What significant change was introduced to the Census in 2016?
Find 5 questions that were asked in the most recent census.

Designing survey questions 7.09

Surveys are often used to gather information when conducting a statistical investigation. When conducting a statistical investigation, you should consider using these steps:

1 Choose a context or specific aspect for your investigation.
2 Identify the variables being investigated and the subjects of the investigation.
3 Plan and carry out procedures for data collection to obtain random, fair and representative samples. This may involve conducting surveys, experiments or observations.
4 Explore, summarise and represent the data using tables and graphs and calculating statistics.
5 Analyse, discuss and interpret the results.
6 Communicate the results of your investigation by preparing a report.

Worksheets
Canteen survey

Student survey form

When designing a survey, it is important that the questions asked are clear, fair and not biased. A good survey question:

- uses plain English.
- is not vague and does not have a 'double meaning'.
- is not too open-ended, and does not allow too many different answers.
- asks for only one piece of information.
- does not try to influence a particular answer unfairly (this is also called bias).
- protects a person's privacy and does not ask for personal information, such as phone numbers.
- provides choices, such as check boxes or a scale of 1–5 to select from.

Example 14

Explain what is wrong with each survey question and suggest an alternative.

a 'How many phones are there in your home?'

b 'What is your e-mail address?'

c 'How long did you have to wait before someone answered your call?'

SOLUTION

a This question is not clear. Does it include mobile phones? Do 2 or more phones using the same line count as one phone? A better question would be: 'How many fixed-line phone numbers does your home have?'

b This question is too personal. This should be an optional question, perhaps adding 'May we contact you for further information if required?'

c This question has too many possible answers. It would be better to provide options:

☐ <3 minutes ☐ 3–5 minutes ☐ >5 minutes

EXERCISE 7.09 ANSWERS ON P. 638

Designing survey questions

U F PS R C

1 Explain what is wrong with each survey question and suggest an alternative.

R **a** 'Are you happy?'

C **b** 'How did you enjoy your meal at our restaurant?'

c 'What is your religion?'

d 'Soldiers have died under our Australian flag, so do you think it should be changed?'

e 'How much do you earn?'

f 'How often do you eat takeaway food?'

g 'What brand of computer do you own?'

A Apple **B** HP **C** Toshiba

☐ Foundation ○ Standard ○ Complex

2 Some Year 8 students designed this questionnaire to investigate the breakfast choices of students. They were particularly interested in whether there is a difference between what boys and girls have for breakfast.

R C

BREAKFAST QUESTIONNAIRE

Please complete each blank or tick the box or boxes that apply to you.

Age : ______________ Male ☐ Female ☐

1 What do you eat for breakfast?

Porridge	☐	Rice bubbles	☐	Eggs	☐
Toast	☐	Corn flakes	☐	Muesli	☐

2 What do you drink?

Orange juice	☐	Milk	☐	Tea	☐
Coffee	☐	Apple juice	☐		

3 Do you think you have a healthy breakfast? Yes ☐ No ☐

Give reasons for your answer:

....

....

In a small group, discuss these questions.

a Do you think there are enough questions on the questionnaire? If you think more questions need to be added, write them down?

b Do you think the wording of the questions can be improved? If yes, replace the questions?

c Do the questions take into account cultural differences? Is it possible to improve the questions to minimise this problem?

3 As a group activity, design a survey about:

- the types of cars owned by the families of students in the class; or
- the most popular types of food sold in the school canteen.

PS C

Determine the suitable wording of each question and provide check box choices where appropriate.

4 Select one topic from the list below and carry out a statistical investigation

PS R C

Does a person's age influence their reaction time?

Is a person's arm span related to their height?

Does the distance you live away from school affect how you travel to school?

Do students prefer school canteen food over packed lunches from home?

A topic of your choice approved by your teacher.

☐ Foundation ○ Standard ⬡ Complex

7.10 Comparing samples and populations

Statisticians do not study just one sample. They generally take a number of samples from a population and calculate the statistics of each sample. The results of each sample are meant to estimate the statistics of the whole population. Also, the larger the sample sizes, the better the estimates of the population.

Example 15

The ages of 100 people working at a theme park are:

18	19	28	27	20	20	24	40	24	19
30	35	26	24	22	19	27	27	23	29
28	40	21	17	20	22	23	21	24	23
22	26	36	25	16	24	36	25	19	25
34	39	45	20	35	21	33	27	19	33
18	27	34	30	25	34	37	29	25	27
24	24	16	18	25	21	26	31	26	25
18	19	26	22	25	22	45	38	34	43
57	24	48	18	20	30	28	22	26	35
29	21	21	28	16	22	35	24	18	23

a Randomly select 5 samples of 12 ages from this population of employees, and for each sample calculate the mean to one decimal place, the median, the number of employees aged under 20 and the number aged between 30 and 39.

b From your sample statistics, estimate the population mean, median, the total number of employees aged under 20 and the total number aged between 30 and 39.

SOLUTION

a Sample 1: 40, 19, 17, 26, 25, 27, 34, 18, 19, 43, 22 and 23

Sample 2: 19, 19, 25, 36, 23, 24, 16, 18, 17, 21, 24 and 18

Sample 3: 24, 22, 43, 25, 16, 34, 35, 25, 24, 22, 20 and 40

Sample 4: 35, 24, 20, 35, 48, 26, 28, 18, 17, 25, 22 and 22

Sample 5: 27, 22, 19, 25, 25, 25, 35, 34, 35, 39, 34 and 24

Each sample of 12 ages is selected randomly from the above population.

	Mean	Median	No. aged under 20	No. aged 30–39
Sample 1	26.1	24	4	1
Sample 2	21.7	20	6	1
Sample 3	27.5	24.5	1	2
Sample 4	26.7	24.5	2	3
Sample 5	28.7	26	1	5

The values are similar for different samples but not close, because each sample is small and does not represent the whole population accurately.

b We can estimate the population statistics by taking averages of the samples.

$$\text{Estimate of mean} = \frac{26.1+21.7+27.5+26.7+28.7}{5}$$
$$= 26.14$$

$$\text{Estimate of median} = \frac{24+20+24.5+24.5+26}{5}$$
$$= 23.8$$

$$\text{Average number of employees aged under 20 per sample} = \frac{4+6+1+2+1}{5}$$
$$= 2.8$$

$$\therefore \text{Estimated number of employees aged under 20 in population} = \frac{2.8}{12} \times 100$$
$$= 23.333\ldots$$
$$\approx 23$$

$$\text{Average number of employees aged between 30 and 39 per sample} = \frac{1+1+2+3+5}{5}$$
$$= 2.4$$

$$\therefore \text{Estimated number of employees aged between 30 and 39 in population} = \frac{2.4}{12} \times 100$$
$$= 20$$

EXCERCISE 7.10 ANSWERS ON P. 639

Comparing samples and populations

1 R C **a** Copy this table.

	Mean	Median	No. aged under 20	No. aged 30–39
Sample 1				
Sample 2				
Sample 3				
Sample 4				
Sample 5				

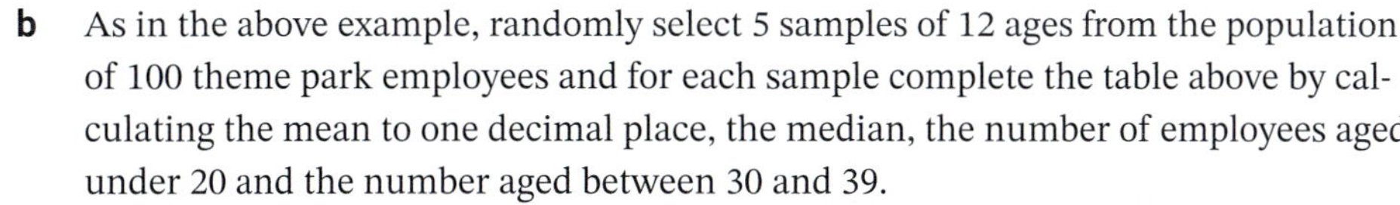

b As in the above example, randomly select 5 samples of 12 ages from the population of 100 theme park employees and for each sample complete the table above by calculating the mean to one decimal place, the median, the number of employees aged under 20 and the number aged between 30 and 39.

c Are the values similar or different across the samples? Why?

d From your sample statistics, estimate the population mean, median, the total number of employees aged under 20 and the total number aged between 30 and 39.

e Add a row of the table for 'Population' and calculate the population mean, median, the total number of employees aged under 20 and the total number aged between 30 and 39 using all 100 numbers.

Foundation Standard Complex

▷

f How do the statistics of each sample compare to the population statistics?

g How do your estimated statistics from part **d** compare to the population statistics?

2 **a** Copy this table.

R C

Sample size	Mean	Median	% aged under 20	% aged 30–39
12				
24				
36				
48				
60				
Population				

b Select another random sample of 12 ages from the population given in Example **14** and calculate and record, in the table, correct to one decimal place (where appropriate), the mean, the median, the *percentage* of employees aged under 20 and the *percentage* aged between 30 and 39.

c Repeat for random samples of 24, 36, 48 and 60 ages, recording your results in the table.

d Are the values similar or different across the samples? Why?

e Copy your results from question **1 e** for the population of 100 ages into the last row of the table.

f How do your sample statistics compare to the population statistics?

g Do larger samples represent the population better? Are their statistics closer to the population statistics?

Foundation Standard Complex

POWER PLUS ANSWERS ON P. 639

7.10

1 A small class achieved these marks in an exam:

66 68 74 76 82 79

a Find the mean of these marks, correct to one decimal place.

b The teacher realised there was an error in the marking and added 3 to each mark. Find the mean of the new marks.

c What effect did the extra 3 marks have on the mean?

d What effect does adding or subtracting the same number for all of the data have on the mean?

2 Yelena sat 5 exams. Her average mark was 74%. What mark should Yelena obtain in her next exam if she wishes her average mark for the 6 exams to be 77%?

3 If the mean of a set of values is 20, must 20 be one of the values? Give an example to illustrate your answer.

4 Five values of data were collected, but the figures were lost. The mean of the data was known to be 8 and the median was 9.

a What was the total of the data?

b What could the data have been?

c If you are told that the range is 7, what could the set of data have been?

5 Four sisters work at the same bank. One earns \$500 per week, another earns \$800 per week. The mean weekly wage of the 4 sisters is \$2000. Is this possible? How?

7 CHAPTER REVIEW

Quiz
Language of maths 7

Language of maths

bias	categorical	census	continuous
cluster	discrete	dot plot	frequency histogram
frequency polygon	frequency table	mean	measure of central tendency
measure of spread	median	mode	numerical
outlier	population	random	range
sample	stem-and-leaf plot	survey	

1 Name the **measures of central tendency.**

2 What word means how often a value is listed in a set of data?

3 What type of data is described by words rather than numbers?

4 Is a **frequency polygon** a column graph or a line graph?

5 What is a survey of the whole population called?

6 What word describes a sample that misrepresents the population or a survey question that tries to lead you to answer in a particular way?

Worksheet
Mind map: Investigating data

Topic summary

- What did you find easy in this chapter? What don't you understand? Talk to your teacher.
- Give some examples of situations where statistics and measures of central tendency are used. Explain, where possible, which measure of central tendency is being used.

Print (or copy) and complete this mind map of the topic, adding detail to its branches and using pictures, symbols and colour where needed. Ask your teacher to check your work.

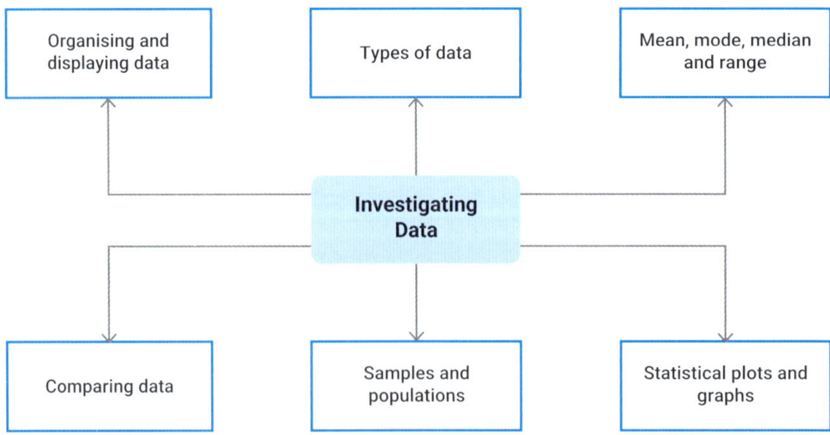

7 TEST YOURSELF

ANSWERS ON P. 639

Quiz
Test yourself 7

In this exercise, round all mean values to 2 decimal places where necessary.

1 Twenty families were asked how many computers they owned.

3	1	2	3	4	3	2	1	4	3
3	2	4	6	3	5	4	2	3	4

a Complete a frequency table for this data.

b What was the most frequent value?

c How many people owned 4 computers?

d What percentage of people owned one computer?

e Draw a sector graph of this data.

2 For each set of data, find:

i the mean **ii** the mode **iii** the median **iv** the range

a 6 12 11 12 10 6 6 10 6

b 8 4 1 1 4 1 3 6

c 20 21 23 24 25

3 The number of goals scored in each soccer match by the Southwest Strikers is shown in this frequency table.

a Copy and complete the frequency table.

b How many matches did the Strikers play?

c Find the range.

d Find the mode.

e Find the median.

f Find the mean.

Score, x	Frequency, x	fx
0	3	
1	7	
2	6	
3	4	
4	2	
5	1	
6	1	
Total		

☐ Foundation ○ Standard ⬡ Complex

4 For each plot below, find:

i the range **ii** the median **iii** the mode **iv** the mean

a

33	34	35	36	37
•	•	•	•	•
•	•	•	•	
•		•	•	
•			•	
•				

b

Steam	Leaf
12	3 5 6
13	1 1
14	0 8 8 8 8
15	7 8

7.06

5 The daily number of people visiting a house that was for sale was:

1	3	5	6	5	4	5	4	3	5
3	5	6	6	5	4	3	3	5	4
1	2	2	5	4	3	6	6	2	5

a Arrange this information into a frequency table.

b Draw a frequency histogram and frequency polygon.

c On how many days did the house receive more than 4 visitors?

d Find the mode.

e Find the median.

6 21 Year 8 students were asked to find out the ages of their mothers.
The results were:

30	35	35	35	36	36	36	36	36	37	37
37	37	38	39	39	41	43	43	45	45	

a Find the median.

b Find the mode.

c Find the mean.

d Which measure of central tendency is the most appropriate for representing this data?

e Predict how many 39-year-old mothers there would be in a Year 8 group of 120 students.

f Suppose one of the ages is incorrect: one of the 39s should actually be 40. Without recalculating, describe how this will affect:

i the mean **ii** the median **iii** the mode **iv** the range

7 Determine whether a sample or a census would be more appropriate for each survey.

a Electing a Year 8 representative for student council.

b Investigating audience reaction to a new film.

c Testing the safety of a new car.

d Finding how many people live in Australia.

☐ Foundation ○ Standard ○ Complex

8 A school of 1200 students wants to survey a sample of students to determine how often they use mobile phones. 

a State a suitable way of choosing a representative sample of 50 students.

b Write one problem that may be encountered in conducting the survey.

c Explain how bias could be minimised.

9 Write 4 questions that could be used in the survey described in question **8**. 

10 Explain what is wrong with each survey question and suggest an alternative. 

a 'How big is your house?'

b 'Is it time that Australia grew up and became a republic?'

8

SPACE

Congruent and similar figures

A **tessellation** is a 'tiling pattern' of repeated congruent shapes covering a flat surface so that there is no overlapping or gaps. Regular tessellations use only one shape and use the three regular polygons, an equilateral triangle, the square and the regular hexagon.

M. C. Escher (1898–1972) was a Dutch graphic artist who used congruent figures to create many interesting tessellations using birds, fish and other creatures. These are called non-regular tessellations.

Alamy Stock Photo/Ant Smith

Chapter outline

		Proficiencies
8.01	Transformations	U F R C
8.02	Congruent figures	U F R C
8.03	Constructing triangles	U F PS R C
8.04	Congruent triangles	U F PS R C
8.05	Proving properties of quadrilaterals	U F PS R C
8.06	Similar figures	U F R C
8.07	Similar triangles	U F PS R C

U = Understanding
F = Fluency
PS = Problem solving
R = Reasoning
C = Communication

Wordbank

Quiz
Wordbank 8

congruence test One of four tests for proving that triangles are congruent: SSS, SAS, AAS and RHS

congruent Identical; exactly the same (symbol: ≡)

image A transformed shape after it has been translated, reflected or rotated

included angle The angle between two given sides of a shape

scale factor The amount by which a figure is enlarged or reduced, equal to $\frac{\text{image length}}{\text{original length}}$

similar To have the same shape but not necessarily the same size (symbol '|||')

superimpose To place one figure on top of another congruent figure so that sides and angles match

transformations Translations (slides), reflections (flips) or rotations (spins or turns)

Videos (2):

8.01 Transformations

8.07 Finding an unknown side in similar figures

***Twig* videos (4):**

8.01 Transformations: Skateboarding

8.02 Tessellated designs

8.06 Jai Singh (sundial) • Fractals: The Koch snowflake

Quizzes (6):

- Wordbank 8
- SkillCheck 8
- Mental skills 8A
- Mental skills 8B
- Language of maths 8
- Test yourself 8

Skillsheet (1):

8.07 Finding sides in similar figures

Worksheets (9):

8.01 Transformations • 5 mm grid paper

8.03 Drawing different triangles

8.05 Proving properties of quadrilaterals

8.06 A page of similar figures • Enlargements and reductions • Enlarging a logo

8.07 A page of similar figures • Enlargements and reductions • Congruent triangle proofs

Mind map: Congruent and similar figures

Puzzles (3):

8.03 Triangle constructions group clues

8.06 Cartoon enlargement

8.07 Similar figures

Nelson MindTap

To access resources above, visit **cengage.com.au/nelsonmindtap**

8 In this chapter you will:

- ✓ translate, reflect and rotate shapes, including a number plane using coordinates, where the reflections are across the *x*- or *y*-axis and the rotations are of multiples of 90°.
- ✓ define congruent figures using transformations.
- ✓ use the symbol '≡' in a congruence statement, naming the vertices in matching order.
- ✓ investigate minimum conditions for constructing triangles.
- ✓ identify the congruent triangles tests (SSS, SAS, AAS and RHS).
- ✓ investigate and prove the properties of the special quadrilaterals using congruent triangles.
- ✓ define congruent figures using pairs of matching sides and angles.
- ✓ calculate the scale factor for corresponding sides in shapes to confirm similarity.
- ✓ use the symbol '|||' in a similarity statement, naming the vertices in matching order.
- ✓ identify the similar triangles tests (SSS, SAS, AA and RHS).
- ✓ determine whether a pair of triangles are similar.

Quiz
SkillCheck 8

SkillCheck

ANSWERS ON P. 639

1 Draw each special quadrilateral and mark all axes of symmetry.

a rectangle **b** square **c** parallelogram

d rhombus **e** trapezium **f** kite

2 From the list in question **1**, name **all** quadrilaterals that have:

a all angles 90° **b** opposite sides parallel

c opposite sides equal **d** one pair of parallel sides

e four equal sides **f** two pairs of adjacent sides equal

3 State the transformation (translation, rotation or reflection) used on each original figure to form the image.

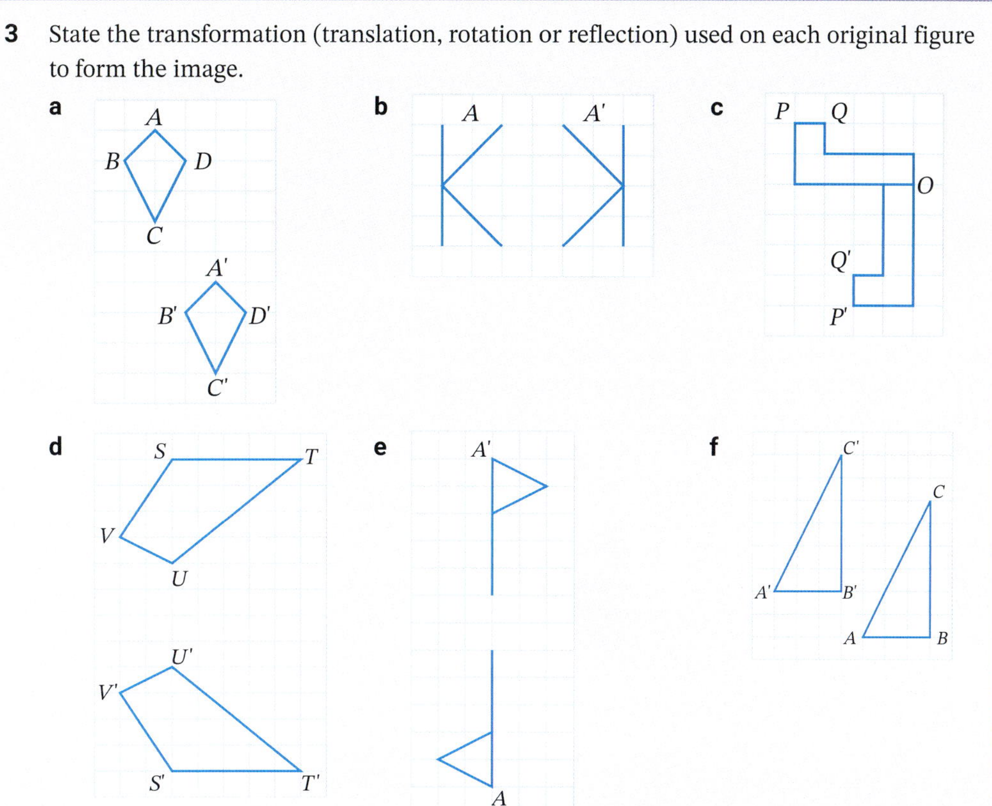

Transformations 8.01

There are three congruence **transformations**.

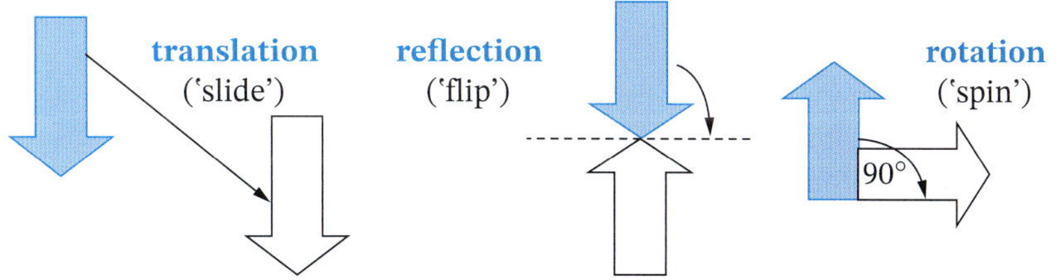

Worksheet Transformations

Videos Transformations

Transformations: Skateboarding

'Congruence' means identical and 'transformation' means change. Even though a shape has changed position (transformed), it still has the same shape and size (congruence).

- The original shape is called the **original.**
- The transformed shape is called the **image.**

Composite transformations

Combinations of translations, reflections and rotations can be applied one after the other. These are called **composite transformations**.

Example 1

Reflect the triangle below across the dotted line and then rotate the image 180° about X.

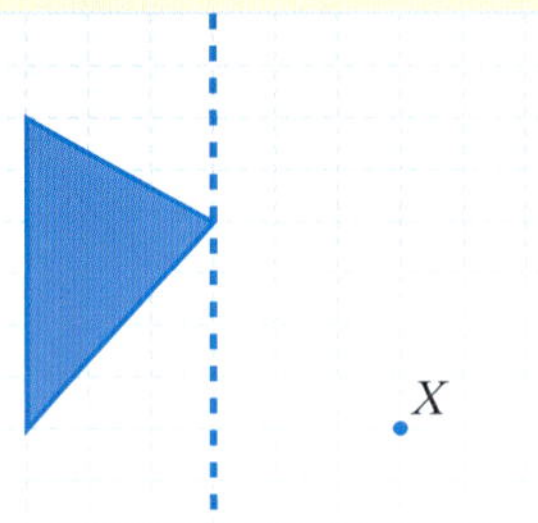

SOLUTION

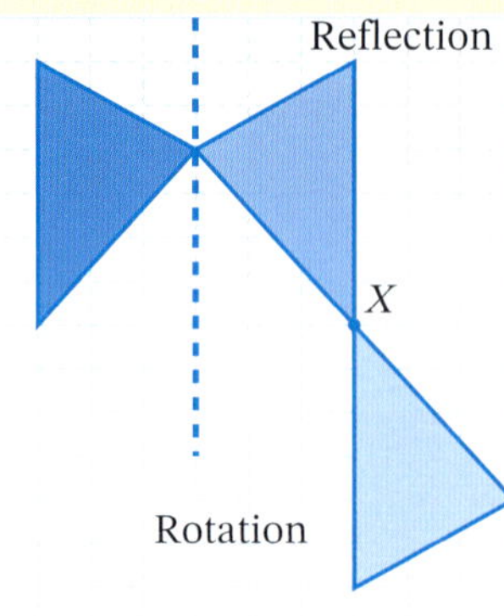

Transformations on a number plane

When a point or vertex P of an original shape is transformed, the corresponding point or vertex on the image is labelled P', pronounced 'P-dash' or 'P-prime'.

Example 2

a Reflect the pentagon $ABCDE$ across the y-axis to create the image $A'B'C'D'E'$.

b Compare the coordinates of each vertex of $ABCDE$ to its matching vertex in $A'B'C'D'E'$.

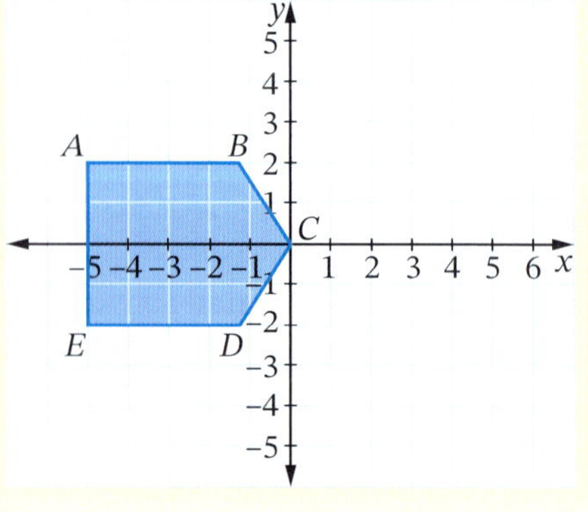

SOLUTION

a

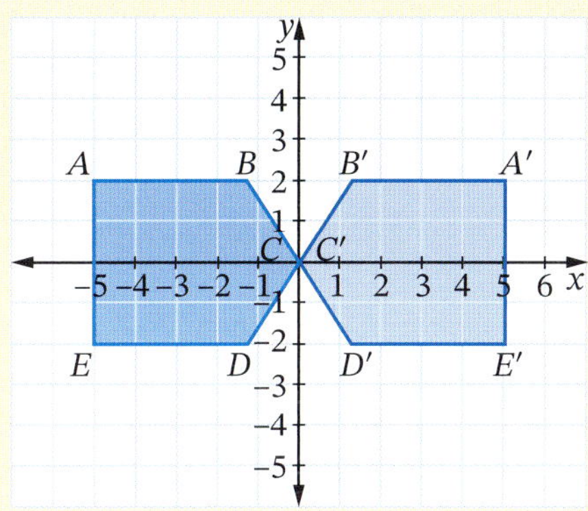

b $A(-5, 2) \rightarrow A'(5, 2)$

$B(-1, 2) \rightarrow B'(1, 2)$

$C(0, 0) \rightarrow C'(0, 0)$

$D(-1, -2) \rightarrow D'(1, -2)$

$E(-5, -2) \rightarrow E'(5, -2)$

When reflected in the y-axis, the x-coordinate of each vertex changes sign while the y-coordinate stays the same.

EXERCISE 8.01 ANSWERS ON P. 640

Transformations

U F R C

1 For each diagram, state the combination of two transformations used on the original figure to form the image.

R

a

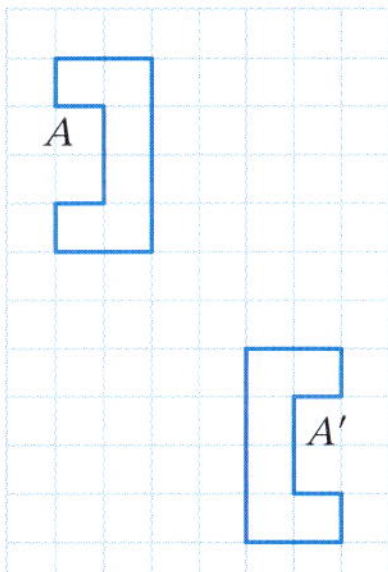

b

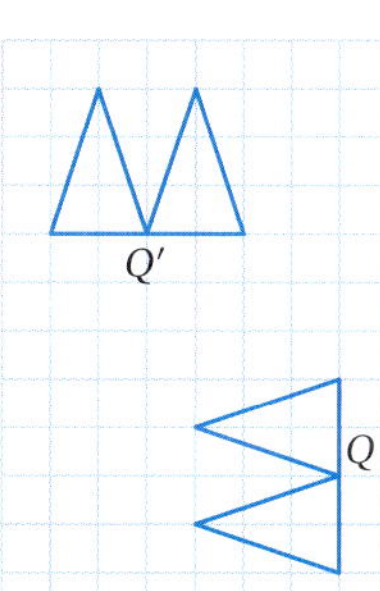

c

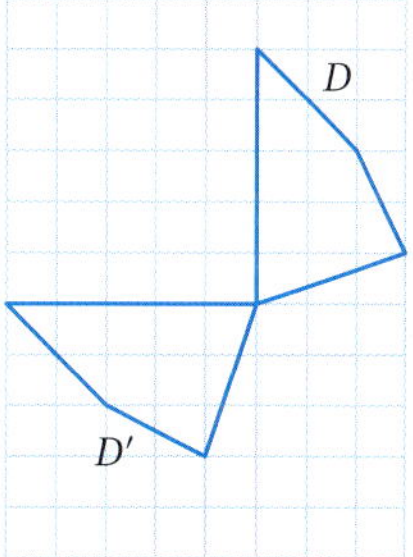

2 Copy each shape onto grid paper and draw the image after performing the stated composite transformations.

Worksheet
5 mm grid paper

a Reflect across the line, rotate 90° clockwise about X.

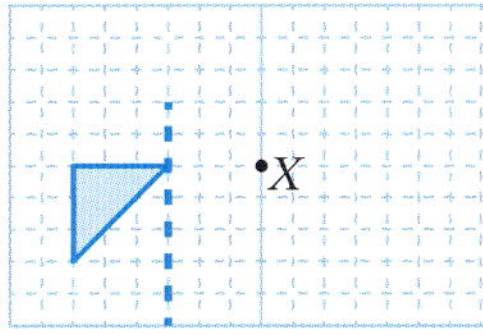

b Translate 5 units right, then reflect across the line.

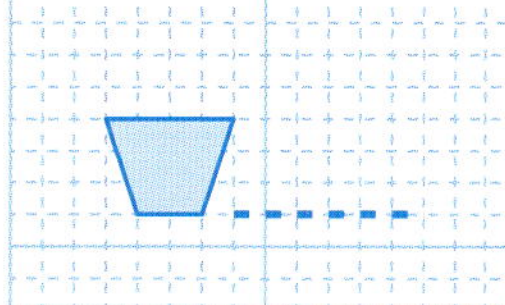

c Rotate 90° clockwise about P, then reflect across the line, then translate 3 units left.

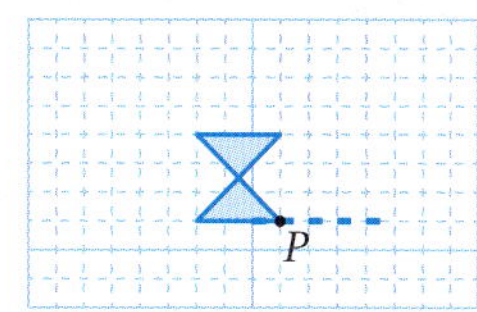

Foundation Standard Complex

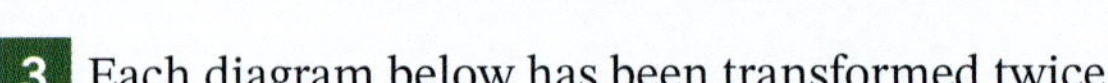

3 Each diagram below has been transformed twice.

R **i** Name the two transformations that have been performed.

ii Name one transformation that would give the same result as the two transformations.

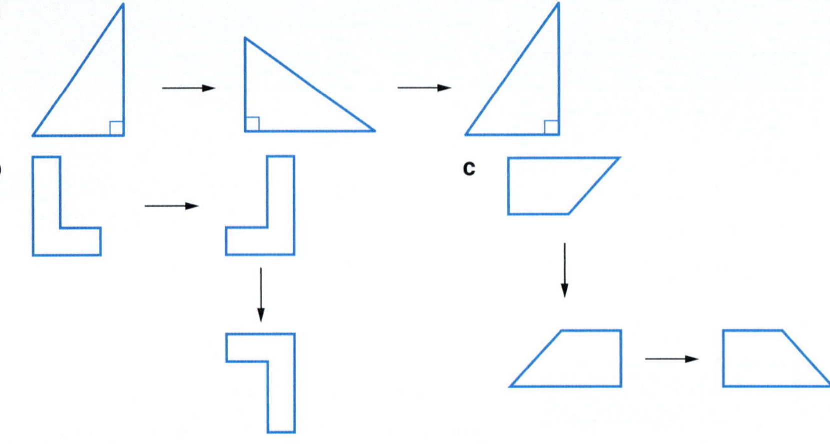

4 **a** Reflect the house shape across line *l* and then across line *m*.

R **b** Which single transformation would give the same result as the two reflections in part **a**?

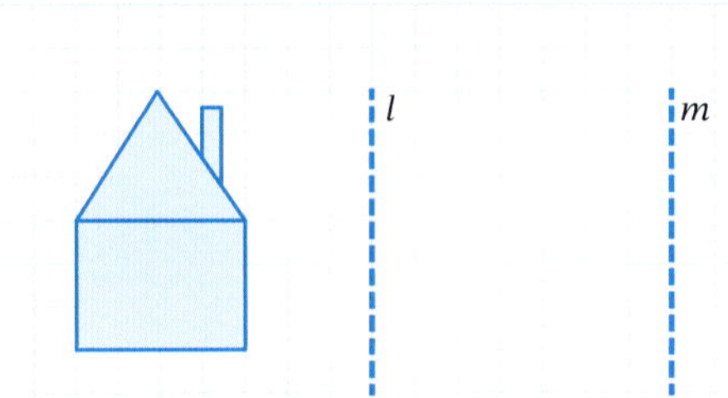

EXAMPLE 2

5 **a** Copy and reflect the *L*-shape across the *x*-axis to create the image $S'T'U'V'W'X'$.

R C **b** Compare the coordinates of each vertex in the original figure to its matching vertex in the image figure.

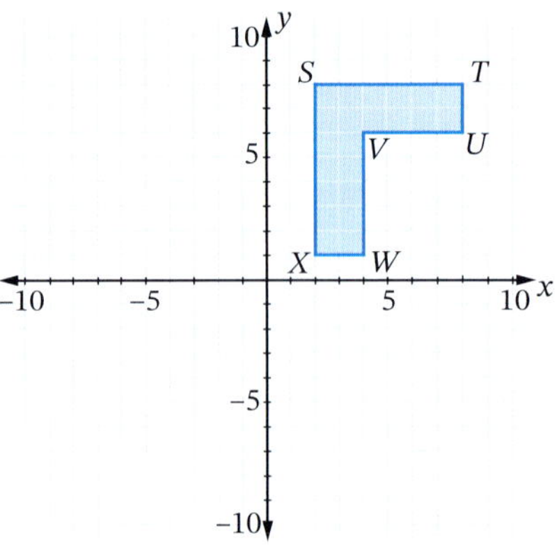

6 **a** Accurately describe the transformation that has been performed on *STUVWXYZ*.

R C **b** Compare the coordinates of each vertex in the original shape to its matching vertex in the image.

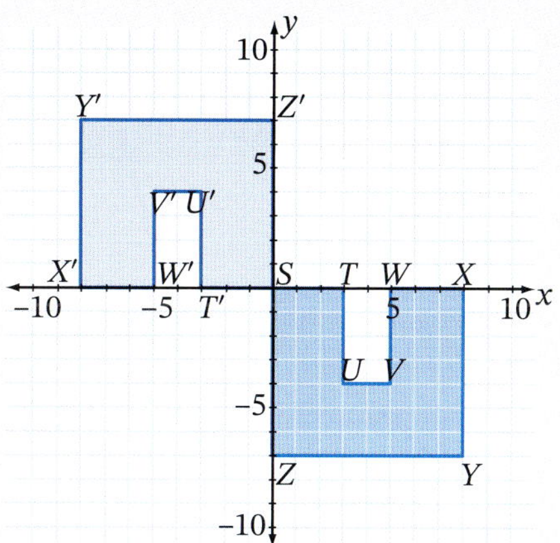

7 **a** Copy and rotate *OPQRST* about *O* through an angle of 180°.

R C **b** Write the coordinates of the new position of *Q* and compare them to its original coordinates.

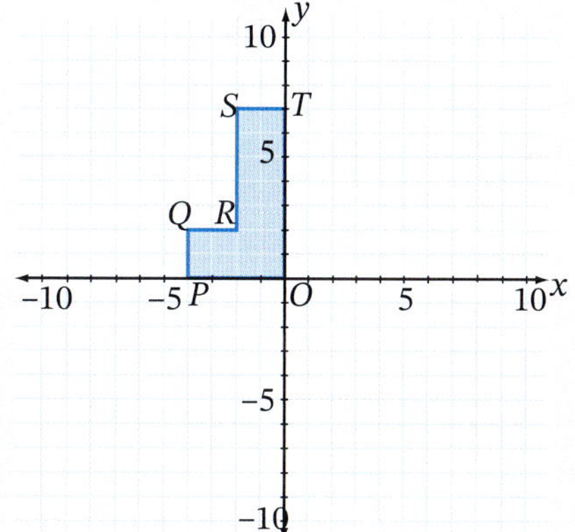

8 **a** Accurately describe the transformation that has been performed on $\triangle WXY$.

R C **b** Compare the coordinates of the vertices of $\triangle WXY$ to those of the image $\triangle W'X'Y'$.

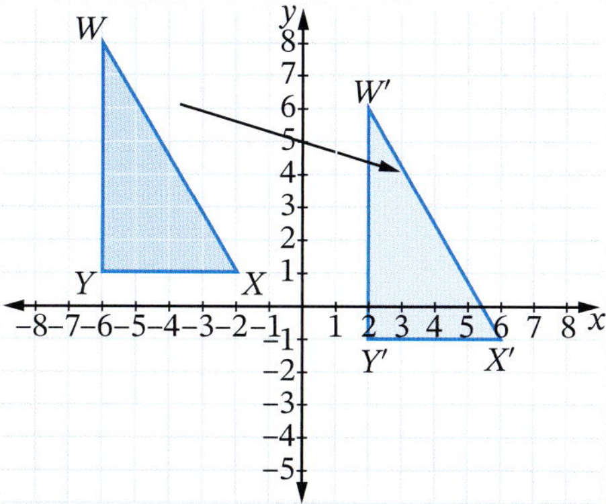

8.02 Congruent figures

Two shapes are **congruent** if they have exactly the same shape and size. Congruent means 'identical'.

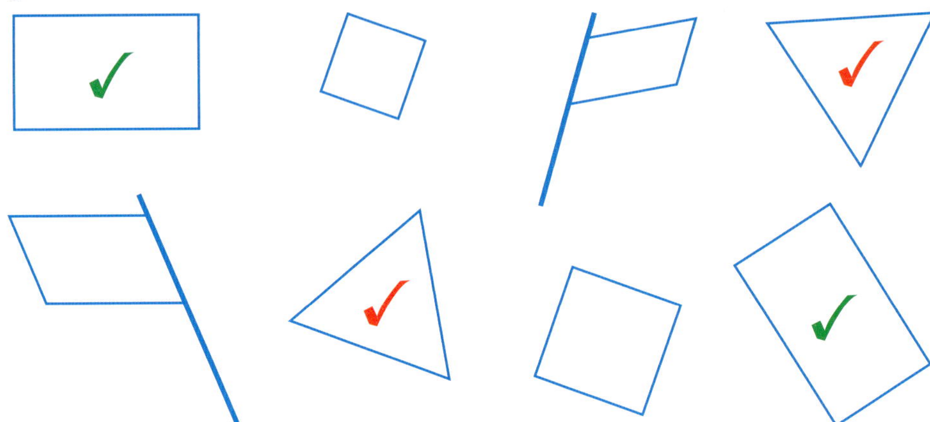

- The 2 rectangles are congruent (marked by green ticks).
- The 2 triangles are congruent (marked by red ticks).
- The 2 squares are not congruent, because they are not the same size.
- The 2 flags are not congruent, because they are not the same size.

One way of testing whether 2 shapes are congruent is to **superimpose** one shape onto the other, that is, to move it to a position on top of the other so that the sides and angles match.

Congruent figures may be identified by superimposing them through a combination of translations, rotations and reflections.

These 2 figures are congruent because $ABCDEF$ can be superimposed onto $A'B'C'D'E'F'$ by translation.

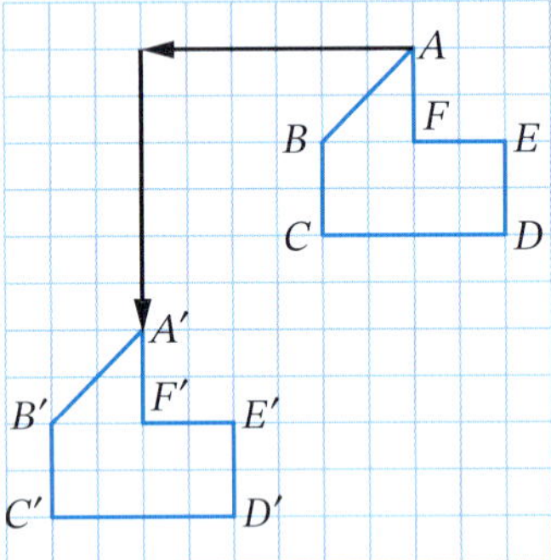

These 2 figures are congruent because arrow P can be superimposed onto arrow P' by rotation.

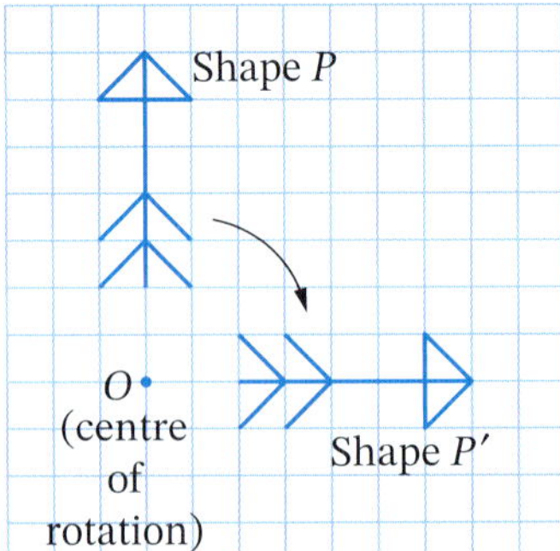

These two quadrilaterals are congruent because $ABCD$ can be superimposed onto $A'B'C'D'$ by reflection.

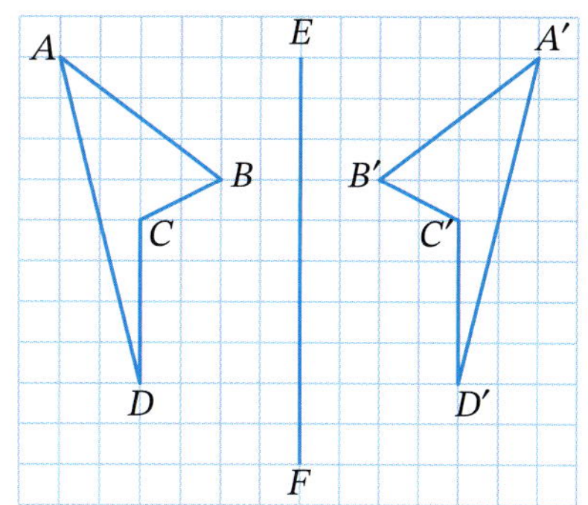

The congruence symbol ≡

The symbol for 'is congruent to' is a special equals sign, written as '≡' (which also means 'is identical to'). The two triangles at right are congruent, so we can write $\triangle ABC \equiv \triangle XYZ$, which is read 'triangle ABC is congruent to triangle XYZ'.

When using this notation, we must make sure that the vertices (angles) of the congruent figures are written in matching order: $\angle A = \angle X$, $\angle B = \angle Y$, $\angle C = \angle Z$.

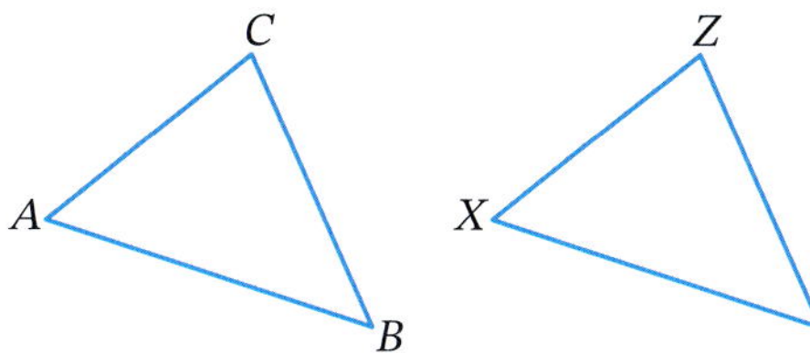

Example 3

For each pair of congruent figures, identify the required matching side and angle, then write the statement relating the figures using the congruency symbol.

a

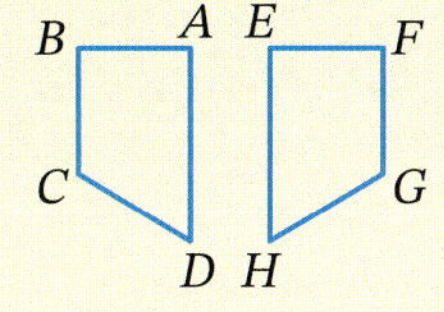

b

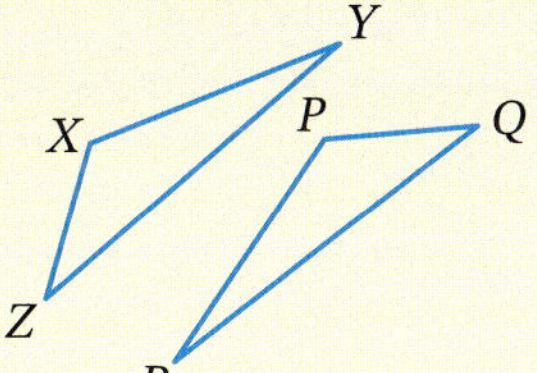

c

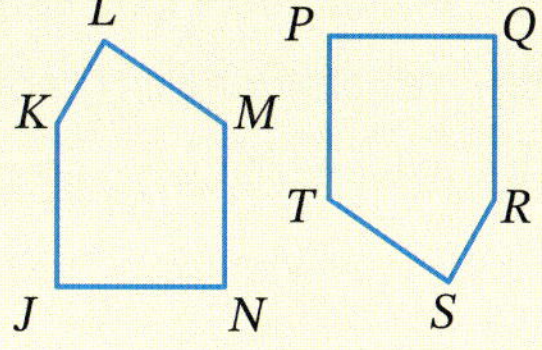

SOLUTION

a $\angle B = \angle F$

$CD = GH$

$ABCD \equiv EFGH$

b $PQ = XZ$

$\angle R = \angle Y$

$\triangle XYZ \equiv \triangle PRQ$

c $\angle J = \angle Q$

$TS = ML$

$JKLMN \equiv QRSTP$

Note that the vertices of congruent figures are named in matching order: equal angles are paired.

ⓘ Congruent figures

- Matching sides are equal.
- Matching angles are equal.
- The symbol for 'is congruent to' is ≡.
- Vertices are named in matching order.

EXERCISE 8.02 ANSWERS ON P. 640

Congruent figures

U F R C

1 By measuring or examining, find the four matching pairs of congruent shapes from the diagrams below.

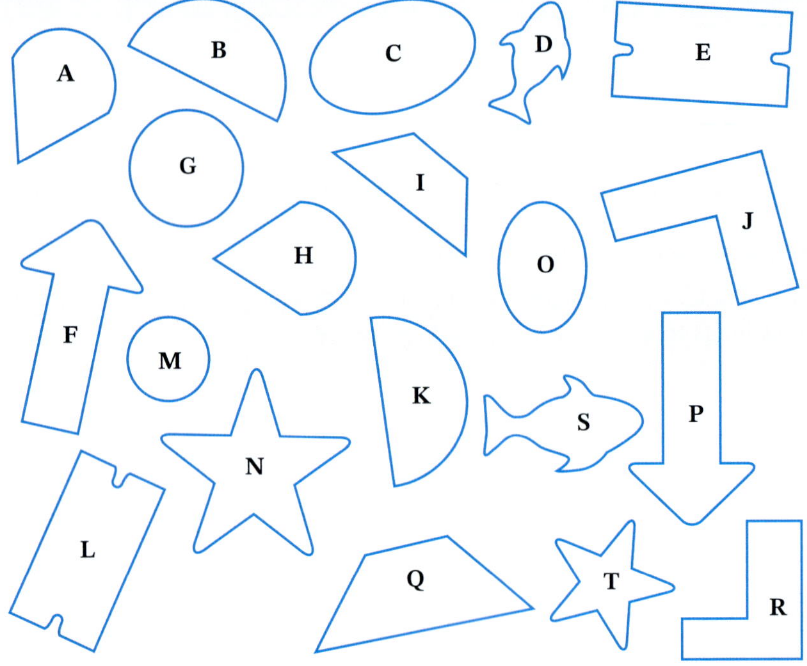

2 Of the triangles below:

a which two triangles are congruent to triangle *A*?

b which triangle is congruent to triangle *B*?

c which triangle is congruent to triangle *K*?

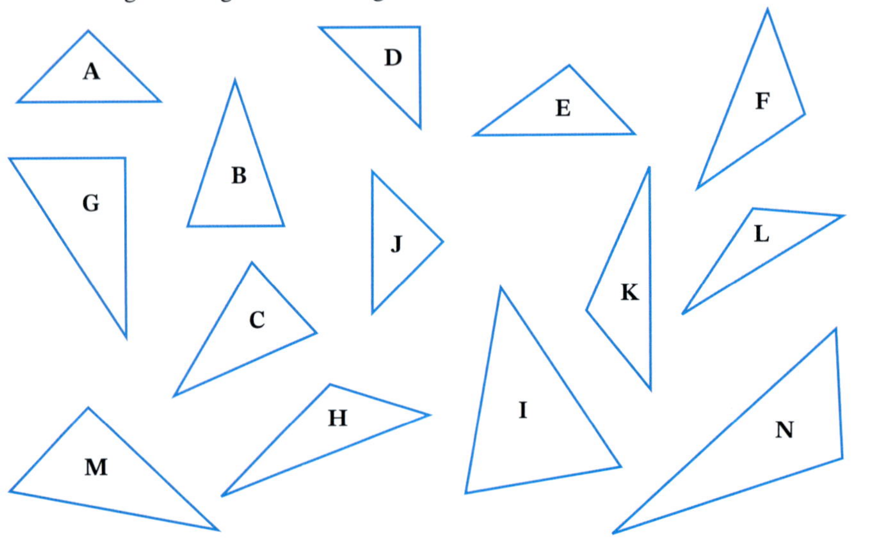

□ Foundation ○ Standard ○ Complex

3 For each set of shapes, identify the two congruent figures, using the correct notation and matching order of vertices.

R C

a

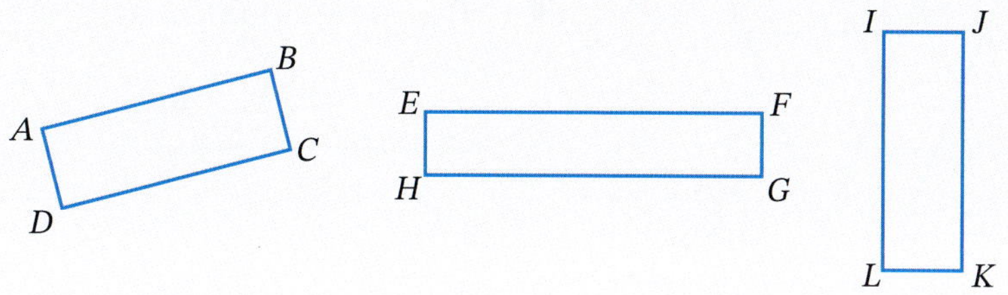

b

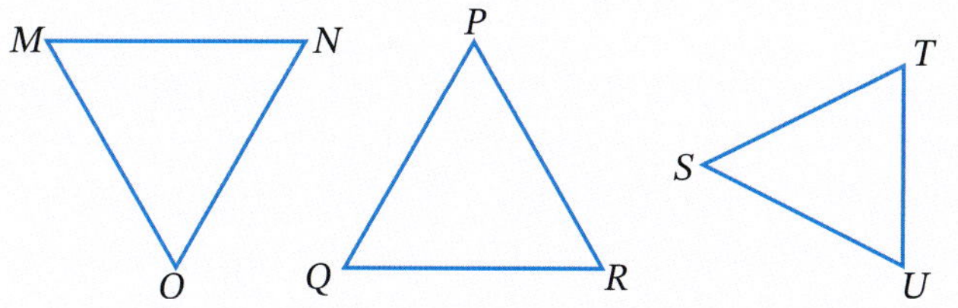

c

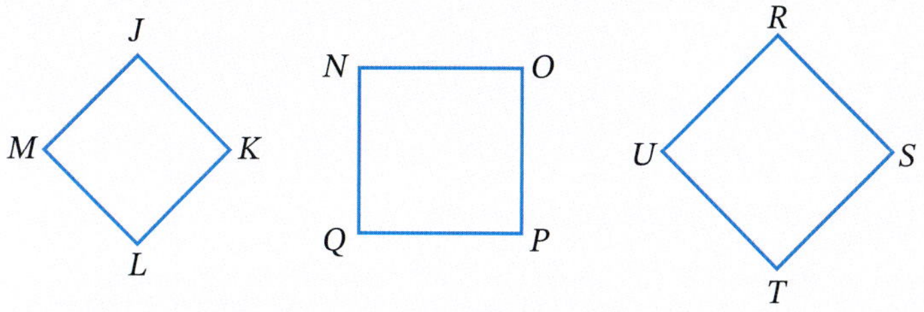

d

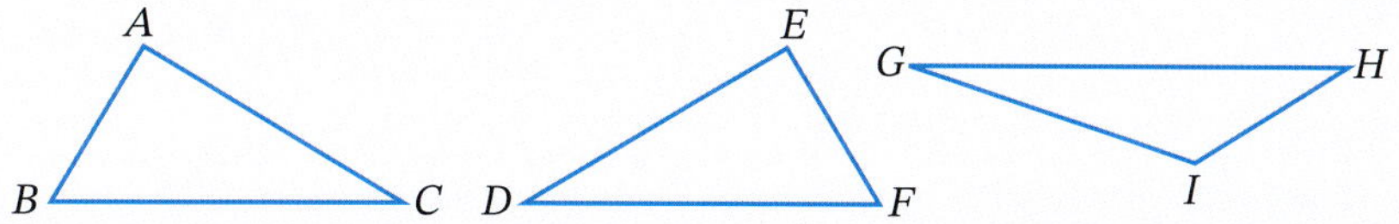

e

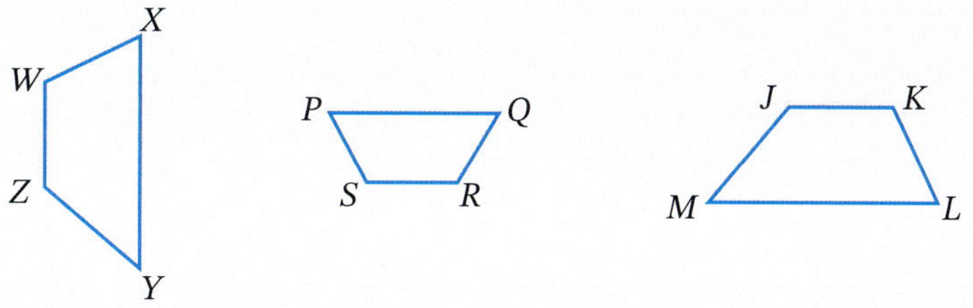

f

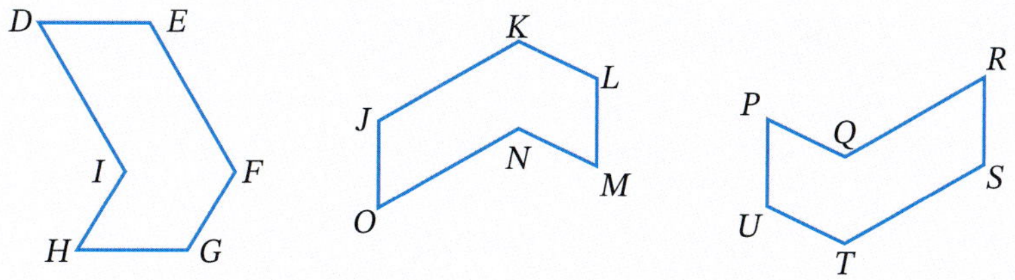

Foundation Standard Complex

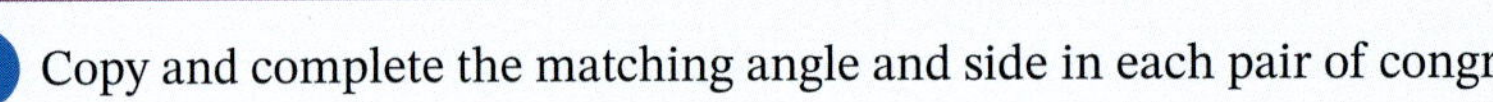

4 Copy and complete the matching angle and side in each pair of congruent figures.

R C

a

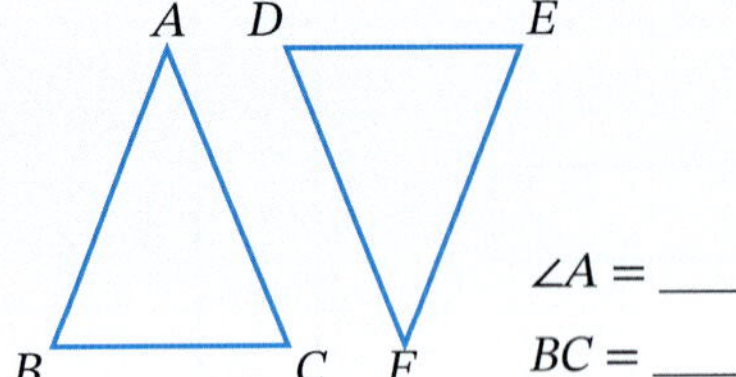

$\angle A =$ ____

$BC =$ ____

b

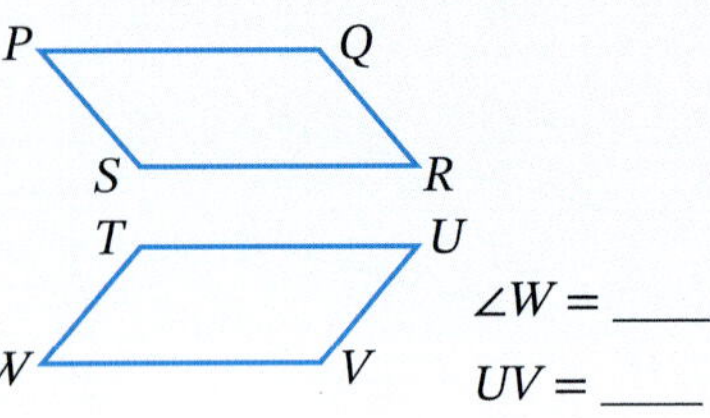

$\angle W =$ ____

$UV =$ ____

c

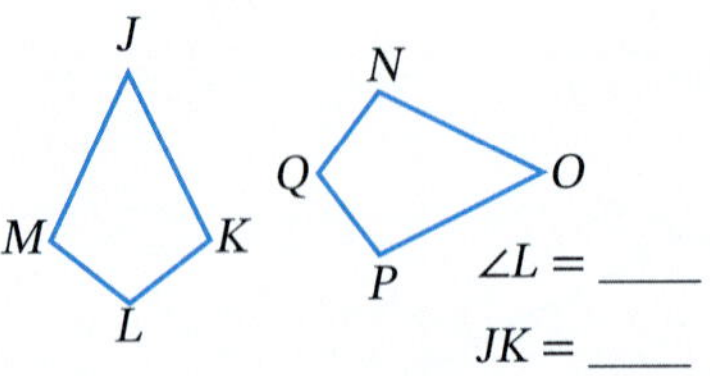

$\angle L =$ ____

$JK =$ ____

d

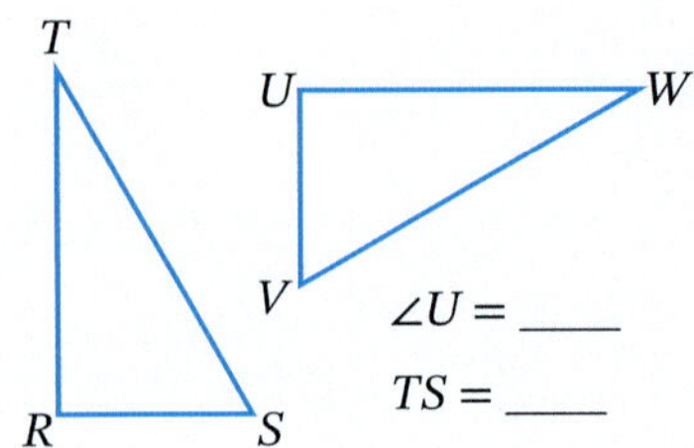

$\angle U =$ ____

$TS =$ ____

e

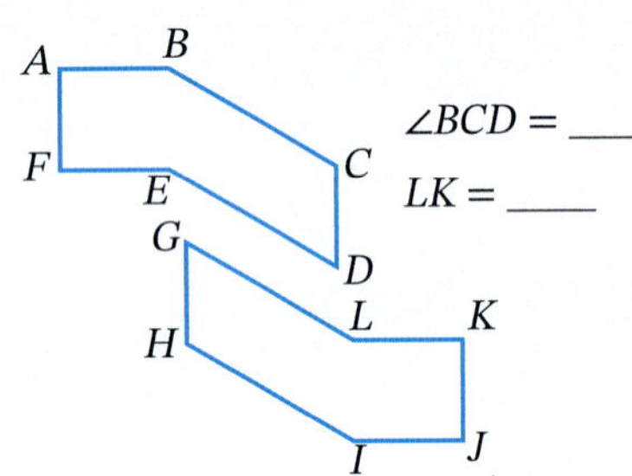

$\angle BCD =$ ____

$LK =$ ____

f

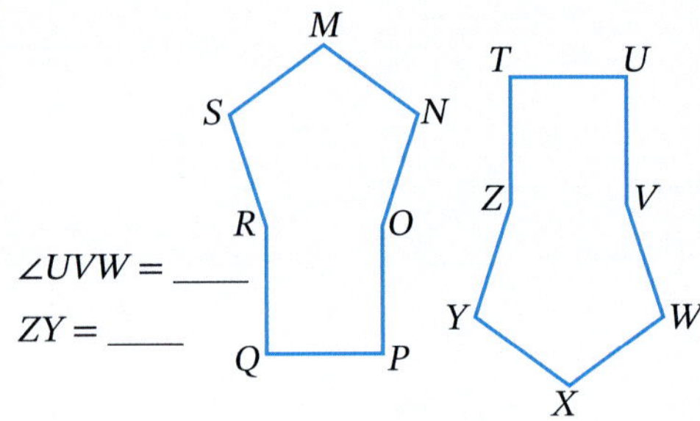

$\angle UVW =$ ____

$ZY =$ ____

g

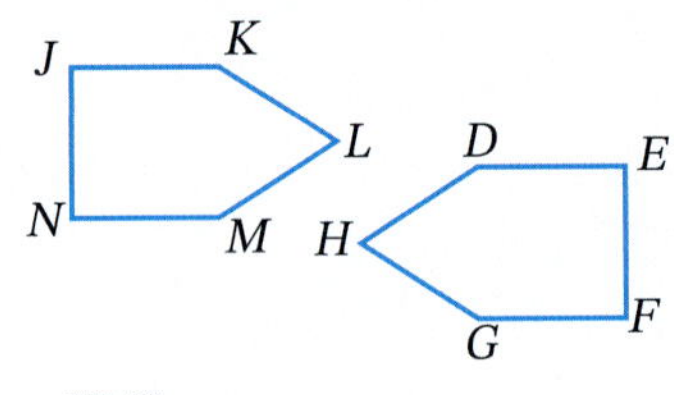

$\angle NML =$ ____

$EF =$ ____

h

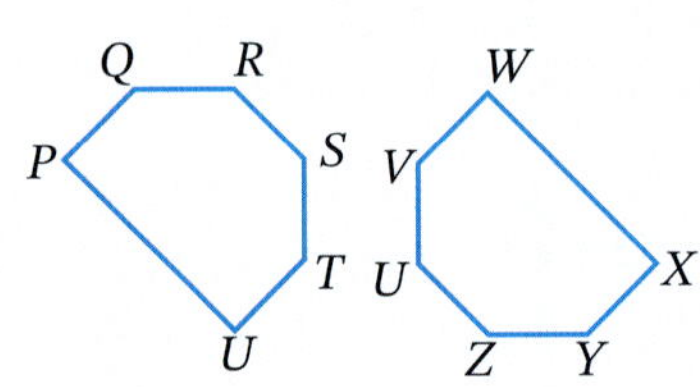

$\angle QPU =$ ____

$RS =$ ____

5 Which triangles below are congruent? Select the correct answer **A**, **B**, **C** or **D**.

R

A X and Y **B** X and Z **C** Y and Z **D** X, Y and Z

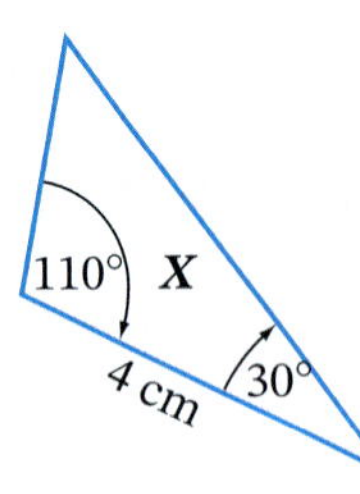

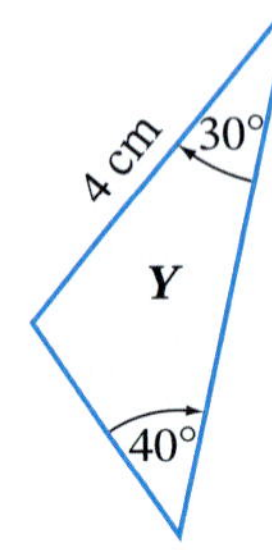

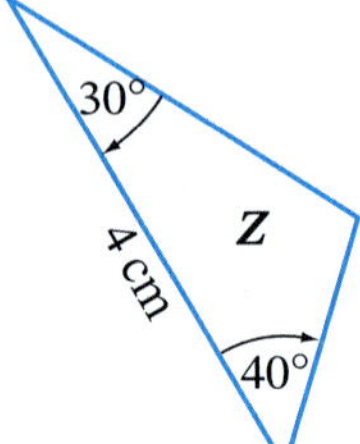

Foundation Standard Complex

6 For each pair of congruent triangles, list:

R C **i** the three pairs of matching sides **ii** the three pairs of matching angles

a

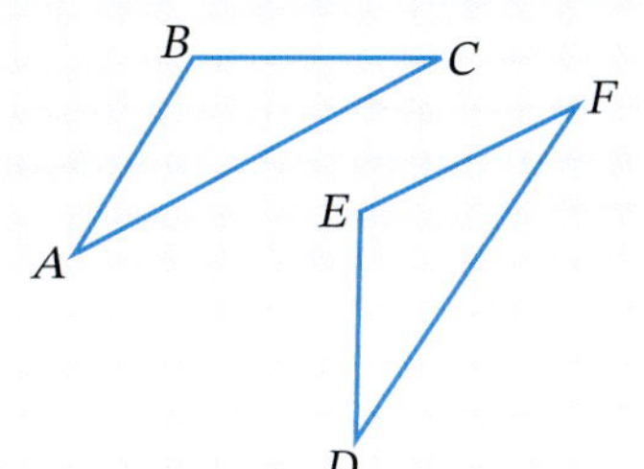

b

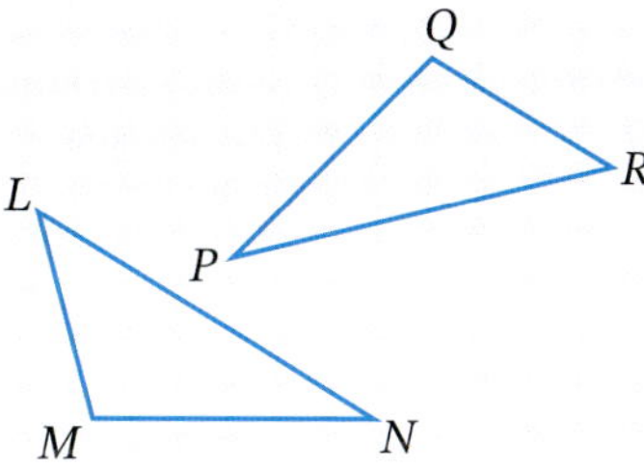

c

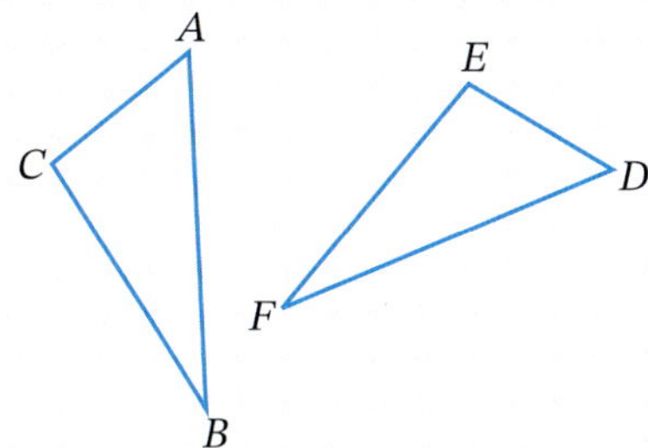

d

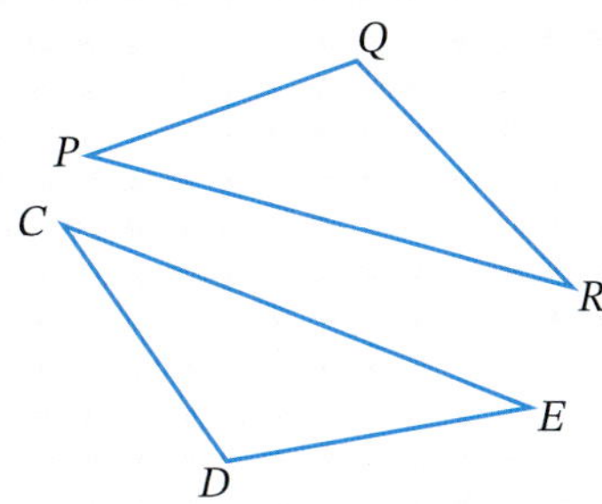

7 Draw a rectangle and its diagonals. Now, you have four triangles. Shade congruent triangles using the same colours.

8 Draw a parallelogram and its diagonals. Now, you have four triangles. Shade congruent triangles using the same colours.

Congruence in design

Congruent figures appear in a variety of art and design work. Some examples are shown below.

iStock.com/poweroforever

iStock.com/AlizadaStudios

Video
Tessellated designs

1 Find five examples of congruent figures in the real world. Look at logos and design; artwork, including the work of Escher; tessellations in paving and walls; and decoration in other cultures, including Aboriginal and Islamic design. Put together a presentation of your examples, showing clearly the congruent shapes that have been used.

2 Produce your own design or piece of art based on congruent shapes.

☐ Foundation ○ Standard ⬠ Complex

8.03 Constructing triangles

Worksheet Drawing different triangles

Puzzle Triangle constructions group clues

To construct (draw accurately) a triangle, we need to know the length of its sides and the size of its angles. We can construct triangles using geometrical instruments (ruler, protractor and compasses) or dynamic geometry software. *Hint*: Draw a rough sketch before beginning the construction.

Example 4

Construct an isosceles triangle with sides of length 3, 3 and 4 cm.

SOLUTION

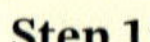

Step 1:

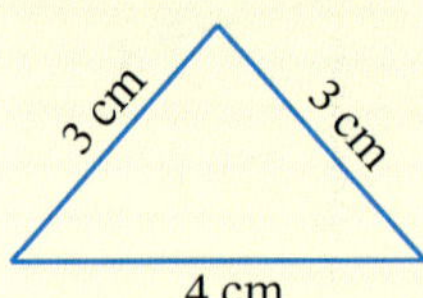

Do a rough sketch.

Step 2:

4 cm

Use a ruler.

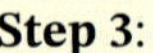

Step 3:

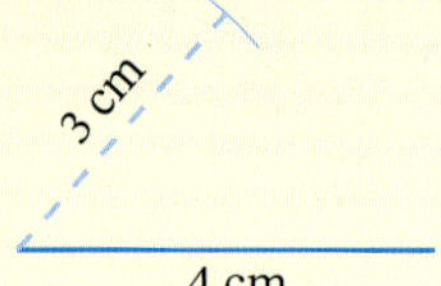

Use a ruler and compasses.

Step 4:

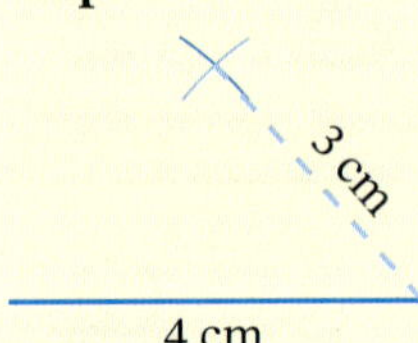

Use a ruler and compasses.

Step 5:

3 cm

3 cm

4 cm

Use a ruler to complete the triangle.

Example 5

Construct a triangle with two sides of length 3 and 4 cm and an **included angle** of 100°.

SOLUTION

Step 1:

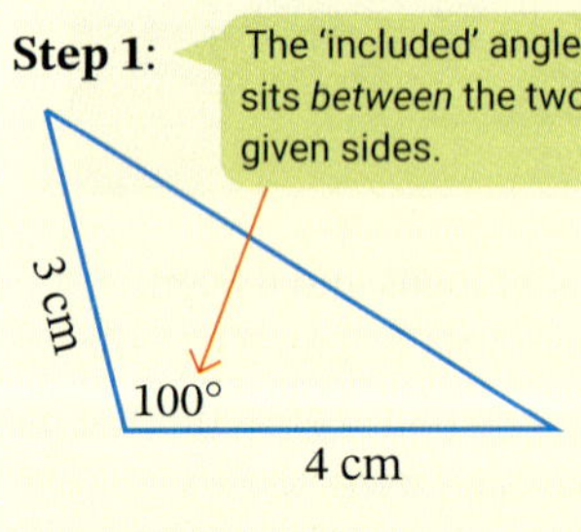

Do a rough sketch.

Step 2:

4 cm

Use a ruler to draw the 4 cm side.

Step 3:

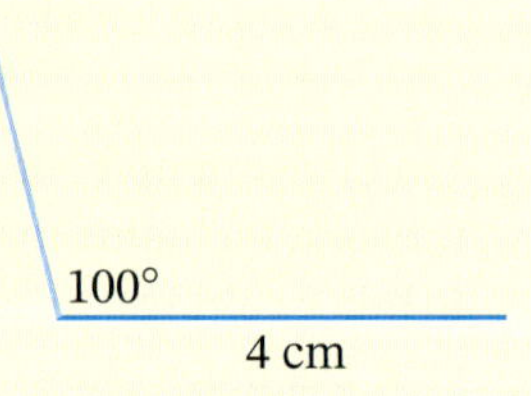

Use a protractor to measure 100°.

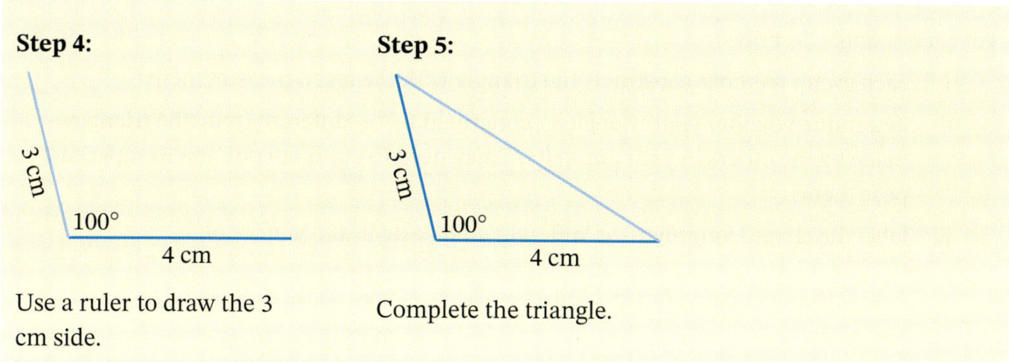

Step 4: Use a ruler to draw the 3 cm side.

Step 5: Complete the triangle.

EXERCISE 8.03 ANSWERS ON P. 641

Constructing triangles

U F PS R C

EXAMPLE 4,5

1 Construct each triangle accurately.

a Triangle TUV: TV = 4 cm, TU = 6 cm, VU = 7 cm

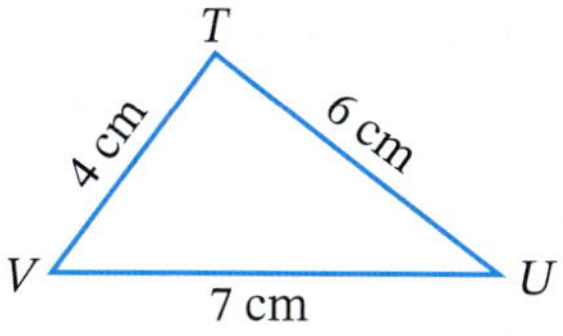

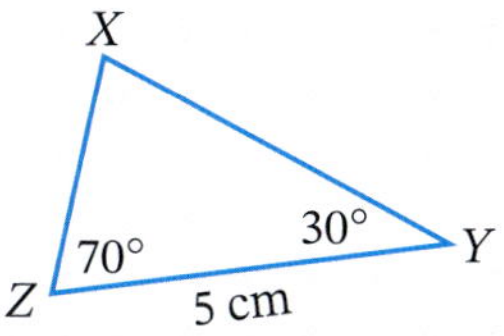

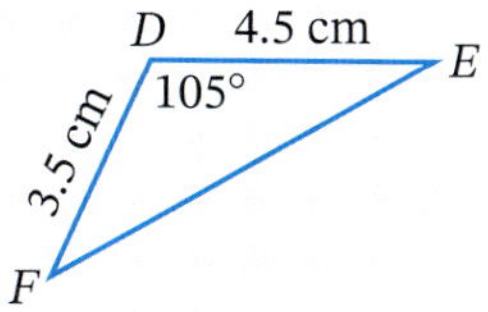

b Triangle XYZ: $\angle Z = 70°$, $\angle Y = 30°$, ZY = 5 cm

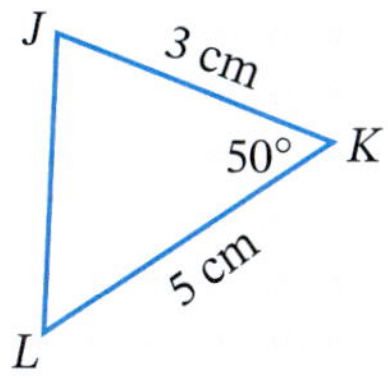

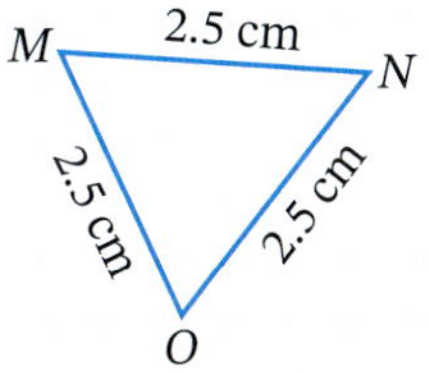

c Triangle DEF: DE = 4.5 cm, DF = 3.5 cm, $\angle D = 105°$

d Triangle JKL: JK = 3 cm, KL = 5 cm, $\angle K = 50°$

e Triangle MNO: MN = 2.5 cm, MO = 2.5 cm, NO = 2.5 cm

f Triangle HIJ: HI = 4 cm, $\angle H = 60°$, $\angle I = 30°$

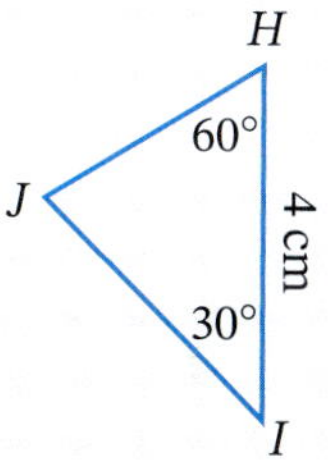

2 **a** Which triangle in question **1** is equilateral?

b What is the size of each angle?

3 **a** Which triangle in question **1** is right-angled?

b Measure the length of its shortest side.

4 PS R How many measurements about the sides and angles of a triangle are needed to construct a congruent triangle? As a group activity for three to four students, for each triangle described below, try to construct different triangles (if possible) with the same measurements. If you can't, and all the triangles drawn are **congruent**, then the measurements that were given are enough. But if one or more of the triangles drawn are different, then the measurements that were given are not enough.

A similar investigation using dynamic geometry software can be found in the Technology activity on page 367.

▷

For each triangle described:

i each group member constructs the triangle with the given measurements.

ii determine whether your triangle is 'congruent to' or 'different from' the triangles made by the others in your group (you may need to cut them out first to superimpose them).

iii label the shape 'congruent' or 'different' and paste it into your book.

a One side and one angle: 5 cm and any angle 35°.

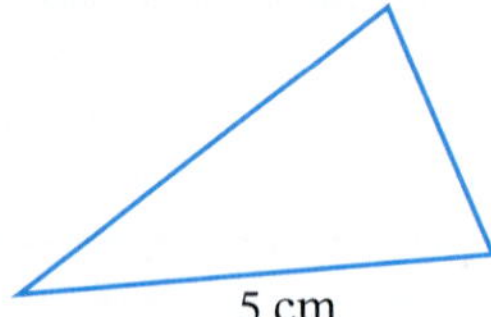

b 3 sides: 4 cm, 6 cm and 8 cm.

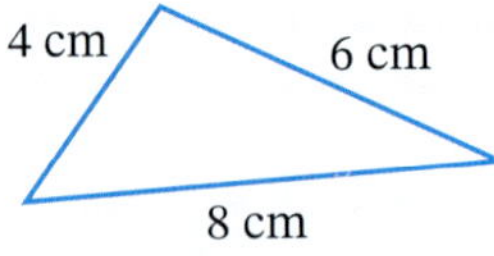

c 3 angles: 25°, 95° and 60°.

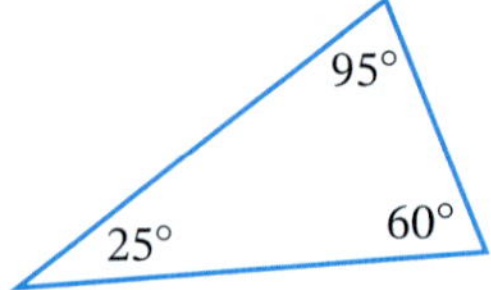

d 2 angles and one side: 55°, 75° and any side 8 cm.

e 2 sides and one angle: 4 cm, 7 cm and any angle 40°.

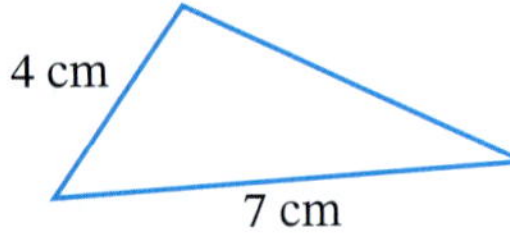

f 2 sides and included angle: 5 cm, 6 cm and included angle 80°.

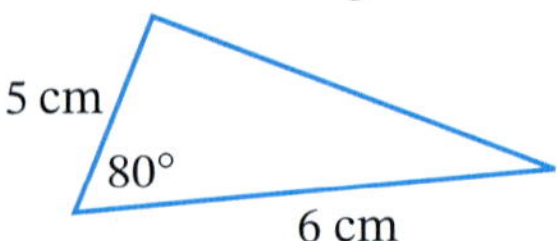

g 2 angles and particular side: 60°, 40° and the side opposite the 60° angle 5 cm.

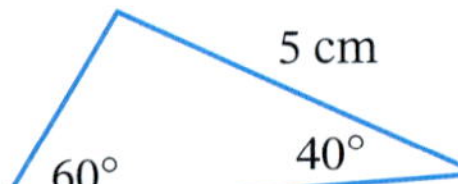

5 R C Consider the triangles you constructed in question **4**. Which set of measurements produced congruent triangles? Suggest a reason why.

9780170465601

 Maths without calculators

Quiz Mental skills 8A

Time before and time after

1 Study each example.

a What is the time 7 hours and 40 minutes after 11:45 a.m.?

11:45 a.m. + 7 hours = 6:45 p.m.

Count: '11:45, 12:45, 1:45, 2:45, 3:45, 4:45, 5:45, 6:45'

$$\begin{aligned} 6:45 \text{ p.m.} + 40 \text{ minutes} &= 6:45 \text{ p.m.} + 15 \text{ minutes} + 25 \text{ minutes} \\ &= 7:00 \text{ p.m.} + 25 \text{ minutes} \\ &= 7:25 \text{ p.m.} \end{aligned}$$

OR:

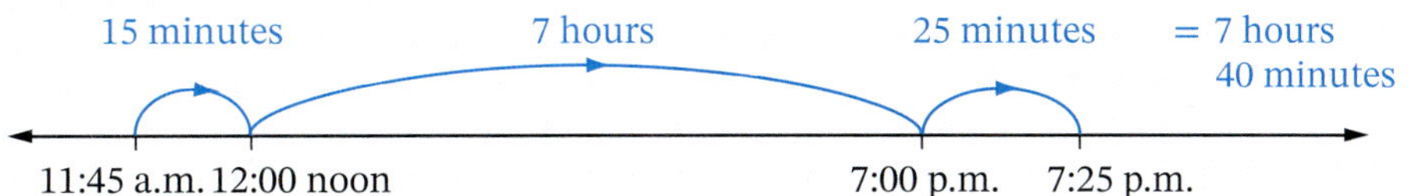

b What is the time 10 hours and 15 minutes after 18:50?

18:50 + 10 hours = 04:50 (next day).

Count: '18:50, 19:50, 20:50, 21:50, 22:50, 23:50, 00:50, 1:50, 2:50, 3:50, 4:50'

$$\begin{aligned} 04:50 + 15 \text{ min} &= 04:50 \text{ hours} + 10 \text{ min} + 5 \text{ min} \\ &= 05:00 + 5 \text{ min} \\ &= 05:05 \text{ (next day)} \end{aligned}$$

OR:

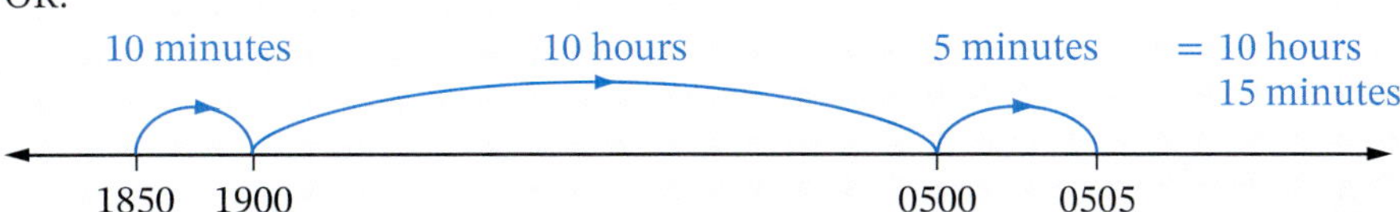

2 Now find the time of day:

a 3 hours 20 minutes after 9:05 a.m.

b 5 hours 40 minutes after 7:30 p.m.

c 4 hours 35 minutes after 6:15 p.m.

d 11 hours 10 minutes after 11:45 a.m.

e 2 hours 45 minutes after 03:25 a.m.

f 7 hours 5 minutes after 17:05 p.m.

g 8 hours 30 minutes after 12:40 a.m.

h 4 hours 55 minutes after 10:20 p.m.

i 6 hours 25 minutes after 04:35 a.m.

j 2 hours 15 minutes after 20:50 p.m.

k 9 hours 50 minutes after 2:30 p.m.

l 3 hours 10 minutes after 8:25 a.m.

3 Study each example.

a What is the time 2 hours and 40 minutes before 7:20 p.m.?

7:20 p.m. − 2 hours = 5:20 p.m. Count back: '7:20, 6:20, 5:20'

$$\begin{aligned} 5:20 \text{ p.m.} - 40 \text{ minutes} &= 5:20 \text{ p.m.} - 20 \text{ minutes} - 20 \text{ minutes} \\ &= 5:00 \text{ p.m.} - 20 \text{ minutes} \\ &= 4:40 \text{ p.m.} \end{aligned}$$

OR:

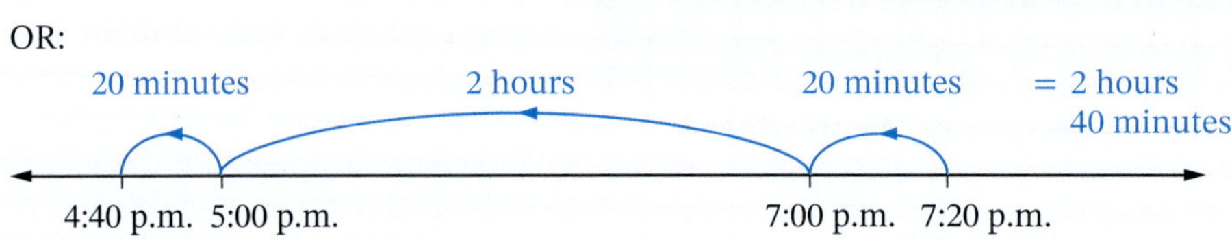

b What is the time 8 hours and 45 minutes before 11:15?

11:15 − 8 hours = 03:15

Count back: '11:15, 10:15, 09:15, 08:15, 07:15, 06:15, 05:15, 04:15, 03:15'
(or 11 − 8 = 3).

$$\begin{aligned} 03:15 - 45 \text{ min} &= 03:15 - 15 \text{ min} - 30 \text{ min} \\ &= 03:00 - 30 \text{ min} \\ &= 02:30 \end{aligned}$$

OR:

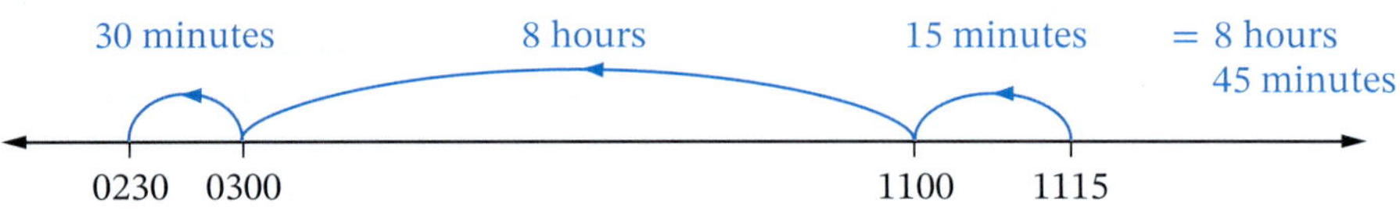

4 Now find the time of day:

a 1 hour 15 minutes before 7:20 p.m.

b 4 hours 40 minutes before 11:20 a.m.

c 3 hours 20 minutes before 3:30 p.m.

d 5 hours 35 minutes before 8:25 a.m.

e 2 hours 10 minutes before 14:55 p.m.

f 3 hours 45 minutes before 07:40 a.m.

g 5 hours 25 minutes before 4:15 a.m.

h 9 hours 30 minutes before 9:45 p.m.

i 4 hours 20 minutes before 20:05 p.m.

j 2 hours 15 minutes before 06:15 a.m.

TECHNOLOGY

Tests for congruent triangles

We will use dynamic geometry software to test whether two triangles are congruent. It is best to do this activity in small groups.

Given three sides: SSS

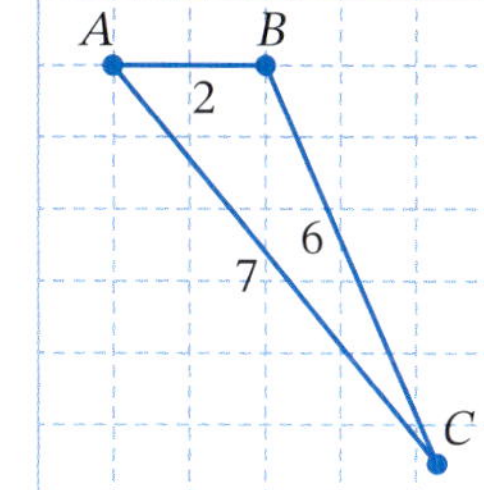

1 Construct this triangle.

2 Can you construct a **different** triangle with the same three side lengths? What do you notice about the second triangle? Is it congruent (exactly the same shape and size) to the first triangle?

Given two sides and the included angle: SAS

1 Construct this triangle.

2 Can you construct a different triangle with the same two sides and an included angle of 80°? Or is it congruent to the first triangle?

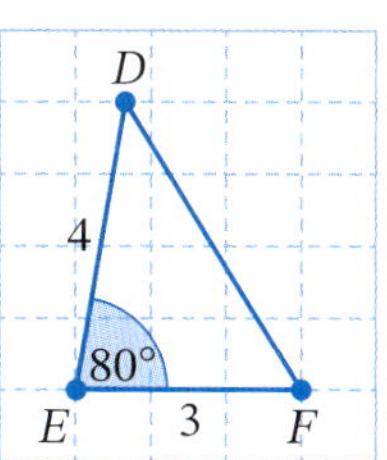

Given two angles and a side: AAS

1 Construct this triangle with a side length of 5 units between two angles of size 30° and 70°.

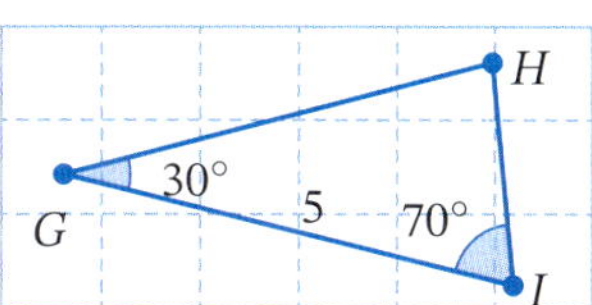

2 Can you construct a different triangle with the same two angles and a matching side of 5 cm? Or is it congruent to the first triangle?

Given a right angle, hypotenuse and side: RHS

1 Construct this right-angled triangle with one of the shorter sides being 3 cm and the hypotenuse 5 cm.

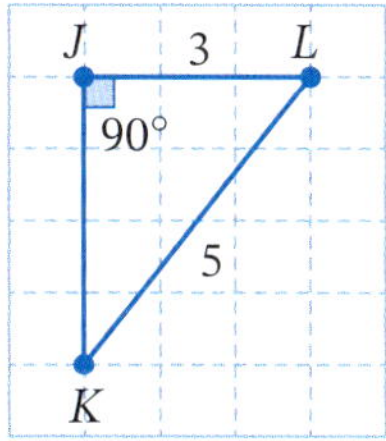

2 Can you construct a different triangle with the same two sides and right angle? Or is it congruent to the first triangle?

Given three angles: AAA

1 Construct this triangle with angles 77°, 62° and 41°.

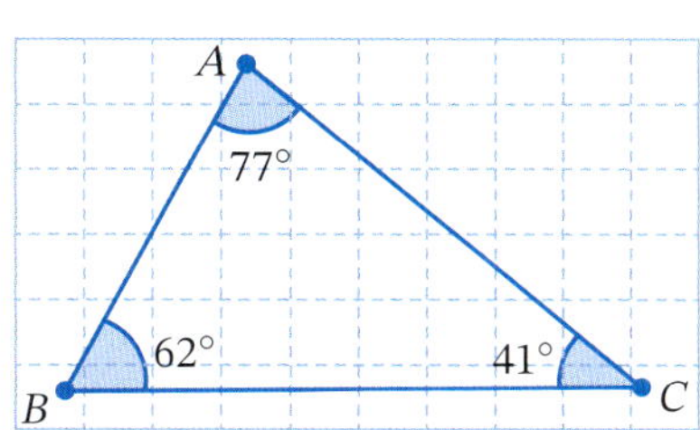

2 Can you construct a different triangle with the three angles? Or is it congruent to the first triangle?

3 Are all triangles with angles 77°, 62° and 41° congruent?

INVESTIGATION

Tests for congruent triangles

You will need: paper, scissors and geometrical instruments.

How much information about a triangle is necessary for us to be able to draw a congruent triangle? This group investigation will look at the conditions required to show that two triangles are congruent, based on your answers to questions from the previous exercise. It is best to complete this investigation in small groups, with every person constructing their own triangle.

For each triangle described below:

- a every person in the group constructs the triangle accurately.
- b cut the triangle out and compare it with others in the group.
- c decide whether all of the triangles in your group are congruent.

1 Given three sides: **SSS**.

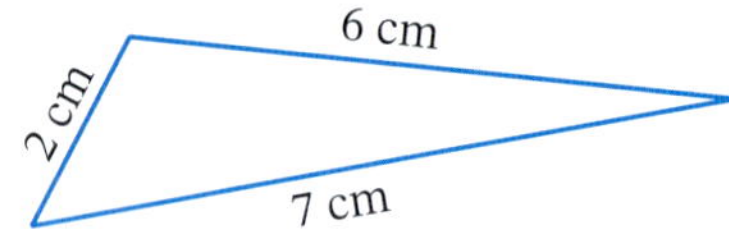

2 Given two sides and the included angle: **SAS**.

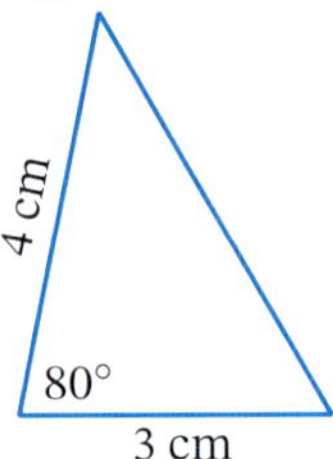

3 Given two angles and a side: **AAS**.

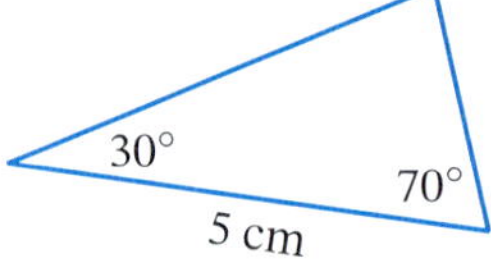

4 Given a right angle, hypotenuse and side: **RHS**.

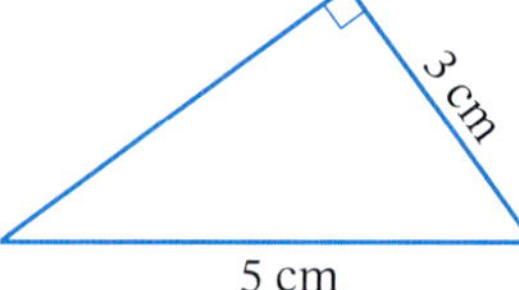

From this investigation, you can see that you don't need to know all the measurements (sides and angles) of two triangles to determine whether they are congruent.

Only **three** measurements are needed, which gives us a method of testing for congruent triangles (described in the next section).

 9780170465601

Congruent triangles

ⓘ Tests for congruent triangles

There are 4 **congruence tests** for proving congruent triangles: **SSS**, **SAS**, **AAS** or **RHS**.
Two triangles are congruent if:

- the 3 sides of one triangle are equal to the 3 sides of the other triangle **(SSS rule).**

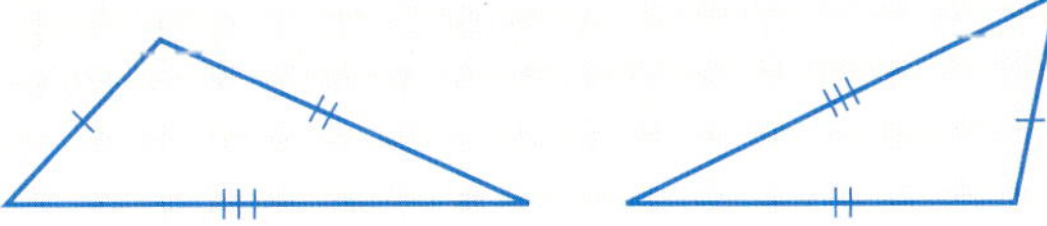

- 2 sides and the **included** angle of one triangle are equal to 2 sides and the **included** angle of the other triangle **(SAS rule).**

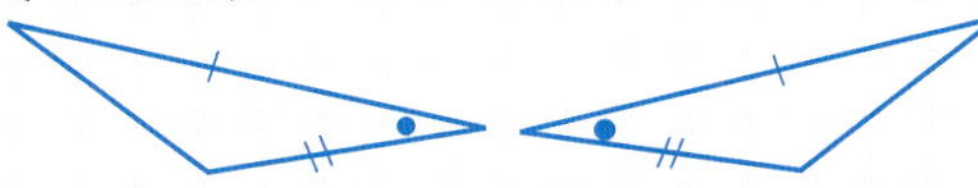

- 2 angles and one side of one triangle are, respectively, equal to 2 angles and the matching side of the other triangle **(AAS rule).**

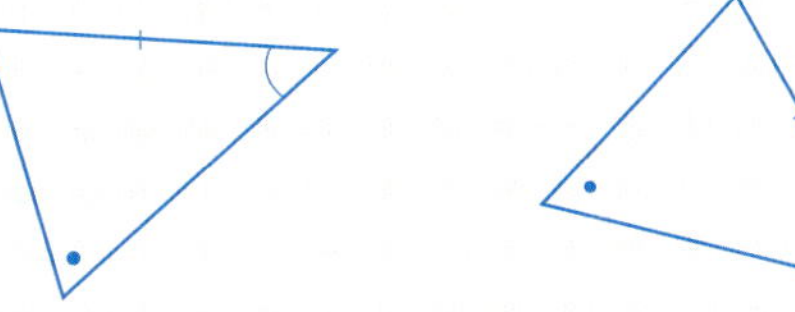

- they are right-angled and the hypotenuse and another side of one triangle are, respectively, equal to the hypotenuse and another side of the other triangle **(RHS rule)**.

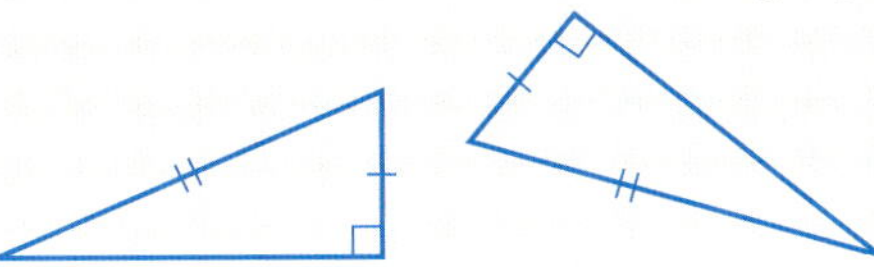

Example 6

a Which test shows that these 2 triangles are congruent?

b Use the correct notation to write a congruency statement relating these 2 triangles.

c Find x.

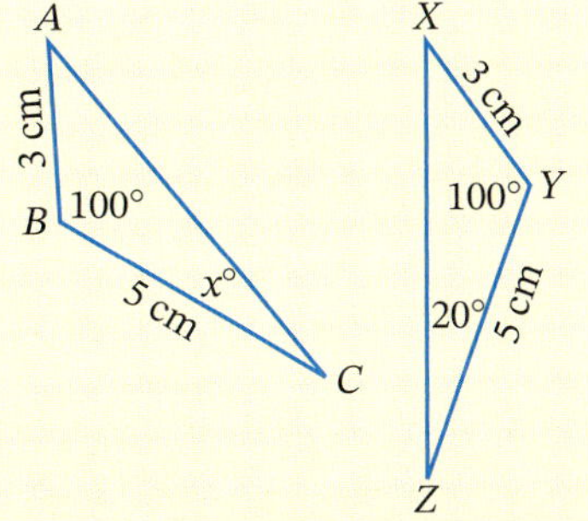

SOLUTION

a $AB = XY = 3$ cm S

$\angle B = \angle Y = 100°$ A

$BC = YZ = 5$ cm S

The congruence test is SAS.

b $\triangle ABC \equiv \triangle XYZ$ Matching order of vertices.

c $\angle C$ matches $\angle Z$.

$\therefore x = 20$

Congruent triangles

U F PS R C

1 Which test can be used to show that these 2 triangles are congruent? Select the correct answer **A**, **B**, **C** or **D**.

R C

A SSS

B SAS

C AAS

D RHS

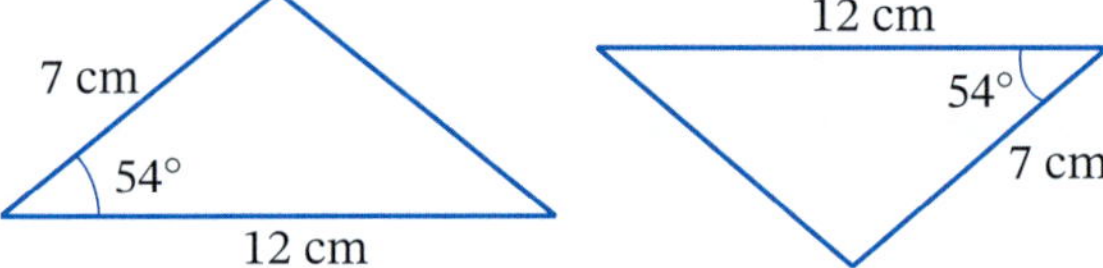

2 For each pair of triangles, state which rule (SSS, SAS, AAS or RHS) shows that they are congruent.

R C

a

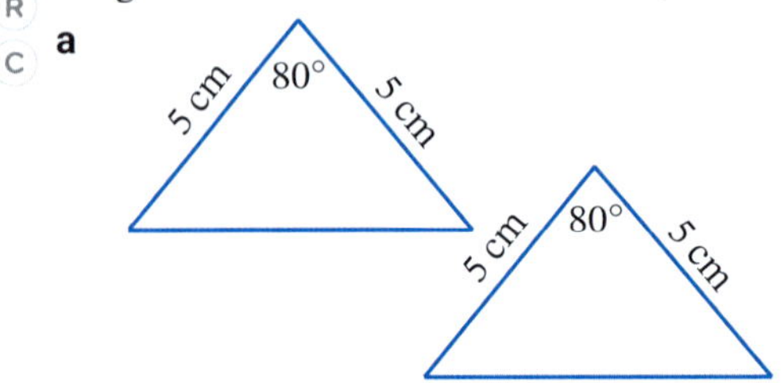

b

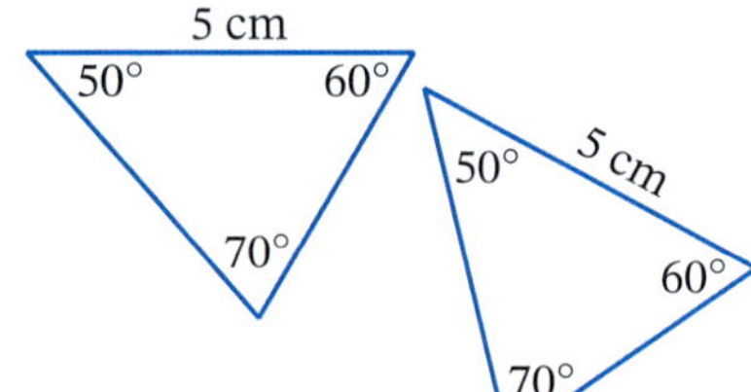

c

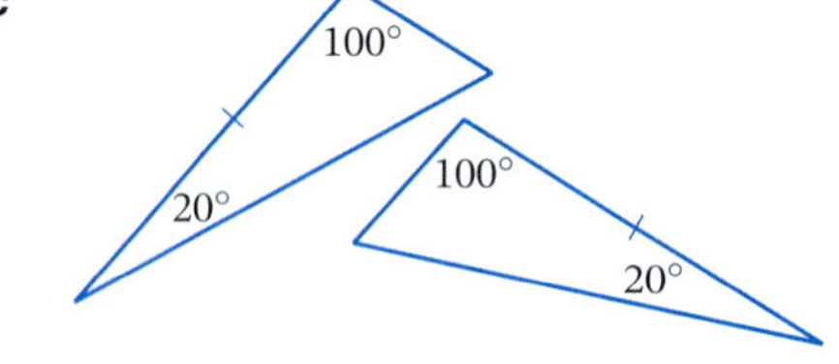

d

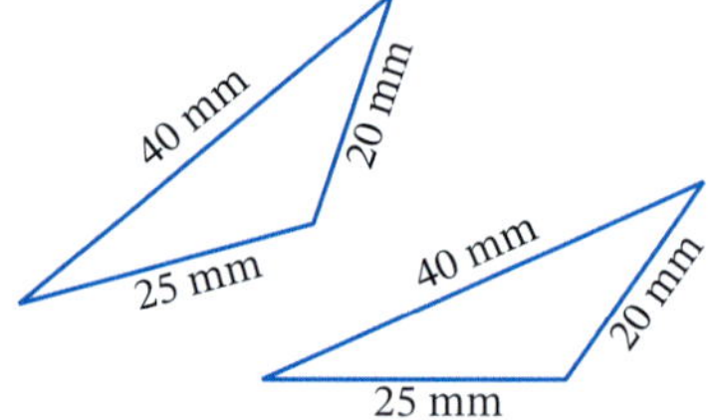

e

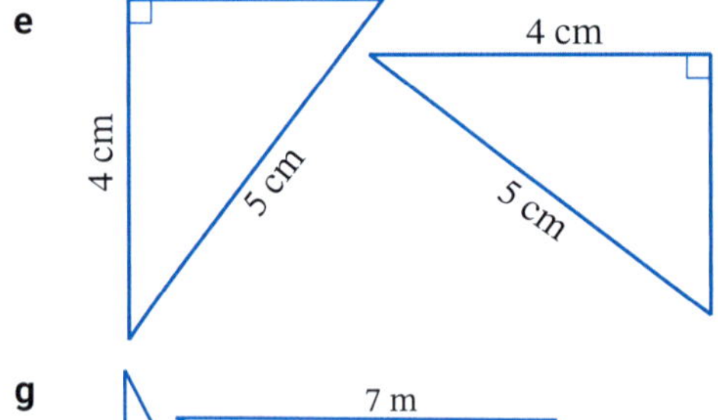

f

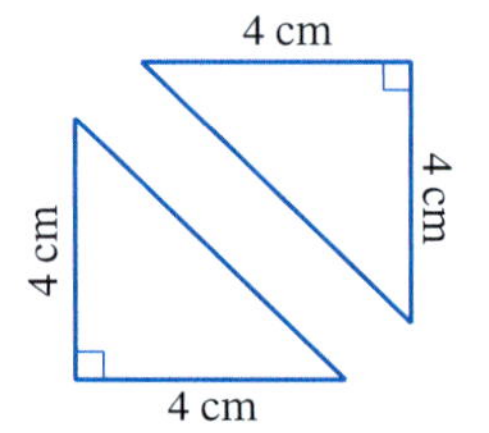

g

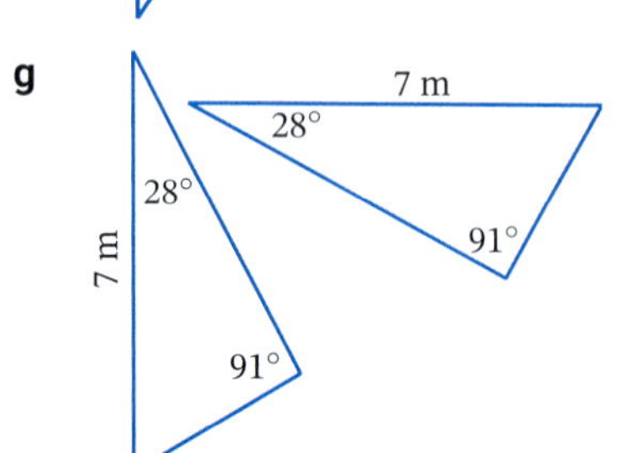

h

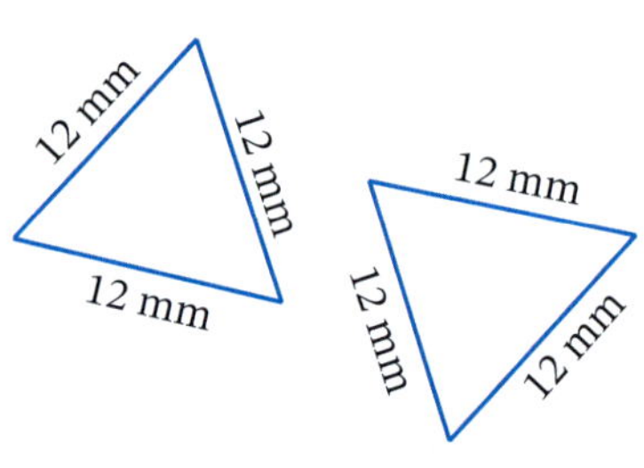

☐ Foundation ○ Standard ⬡ Complex

i

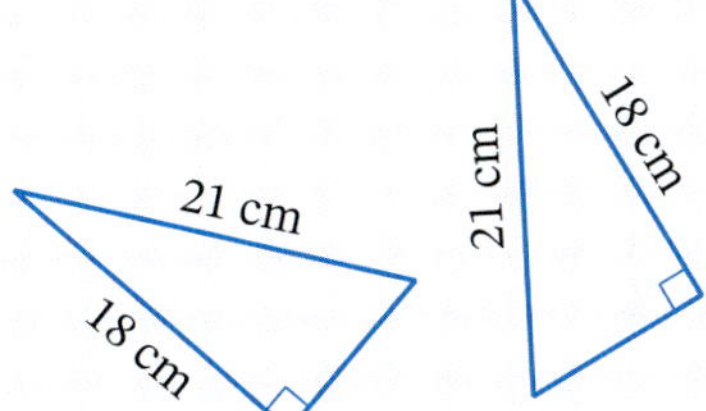

j

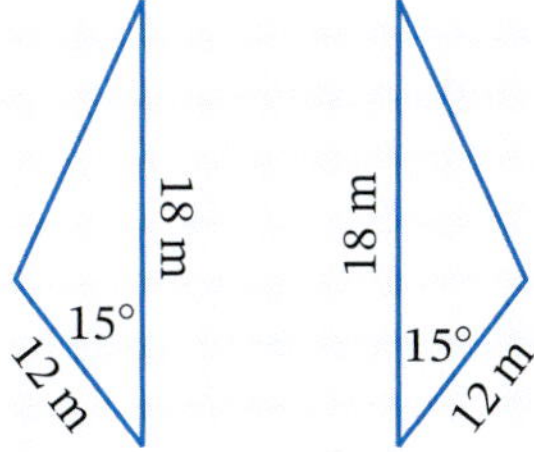

3 R C **a** What additional information is needed to prove that these 2 triangles are congruent? Select the correct answer **A**, **B**, **C** or **D**.

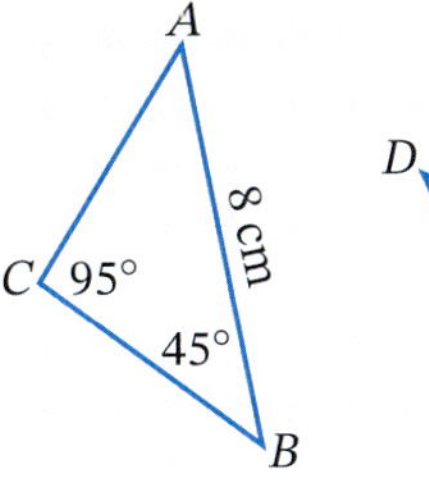

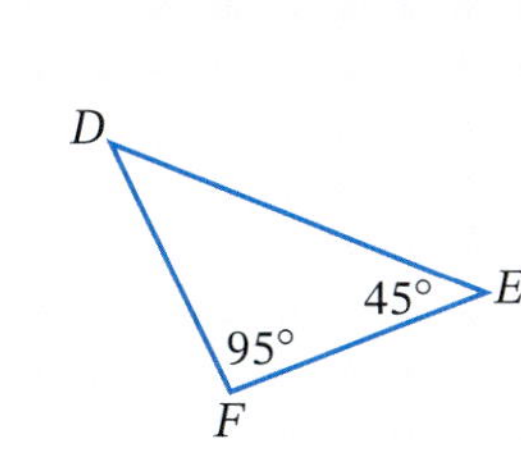

A $\angle EDF = 40°$ **B** $DE = 8$ cm

C $DF = 8$ cm **D** $EF = 8$ cm

C **b** Use the correct notation to write a congruency statement relating these two triangles.

4 Find the value of each variable for each pair of congruent triangles.

R **a**

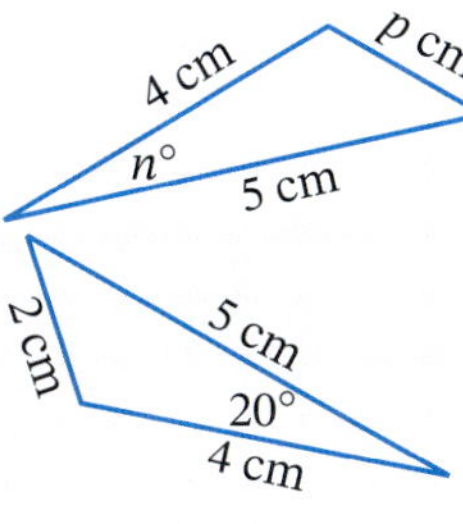

b

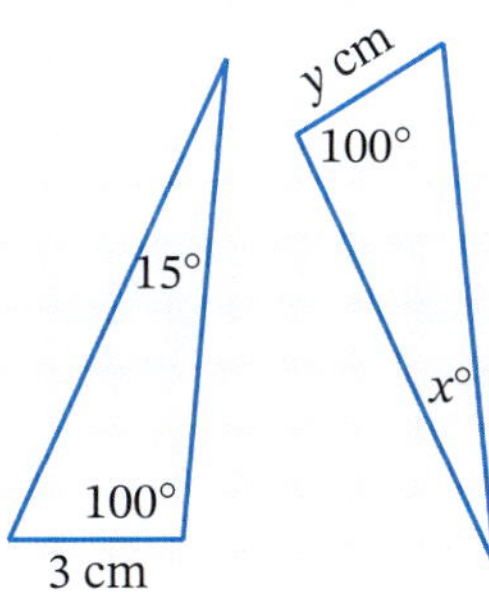

c

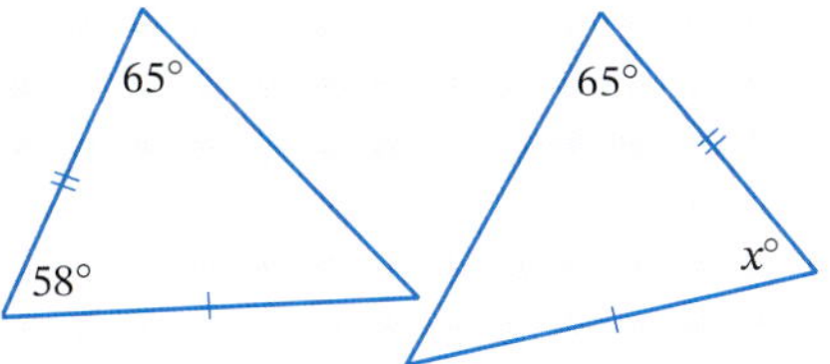

d

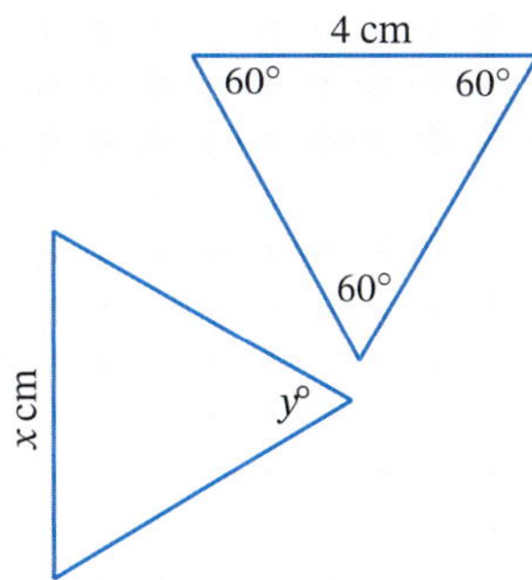

5 For each pair of triangles, state which test can be used to prove that they are congruent and then find the value of the variable.

R C

a

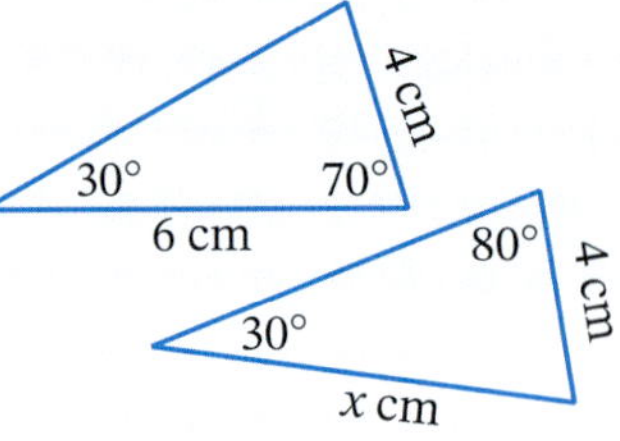

b

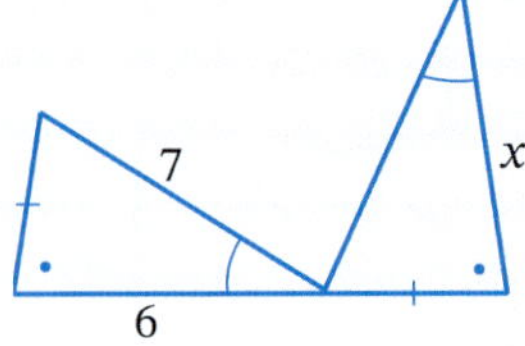

6 PS R C

a What geometrical rule shows that $x = 46$ in the diagram?

b Which test can be used to prove that the two triangles are congruent?

c Use the correct notation to write a congruency statement relating these two triangles.

d Find y.

e Which angle in RST is equal in size to $\angle P$?

f Hence why are sides PQ and TS parallel?

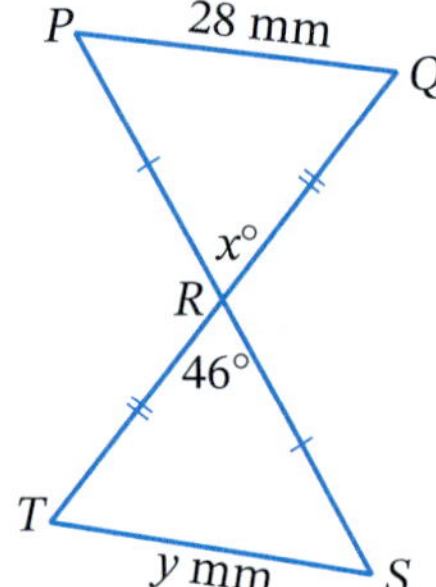

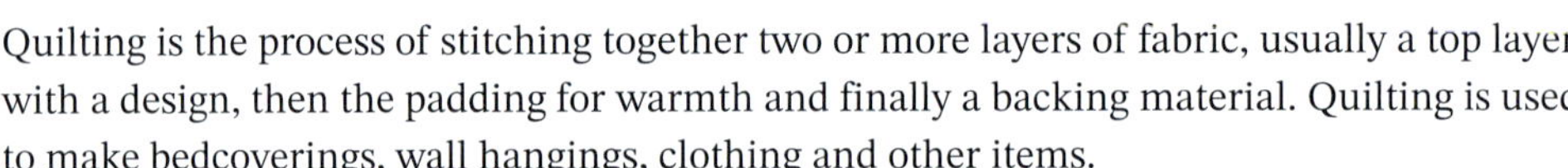

Quilting

Quilting is the process of stitching together two or more layers of fabric, usually a top layer with a design, then the padding for warmth and finally a backing material. Quilting is used to make bedcoverings, wall hangings, clothing and other items.

Shutterstock.com/MaxCab

iStock.com/MaxCab

Often blocks of design are sewn together into rows and columns to create a quilt. The design blocks mostly use congruent shapes to create a pattern for the quilt.

Find examples of quilting patterns and highlight the congruent figures in them.

Proving properties of quadrilaterals

Worksheet
Proving properties of quadrilaterals

In Chapter 4, *Geometry*, we examined many properties of the special quadrilaterals.

Quadrilateral	Properties
Trapezium 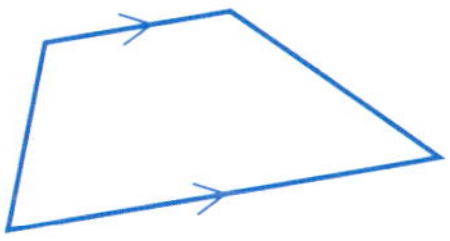	• One pair of parallel sides. • No axes of symmetry.
Kite 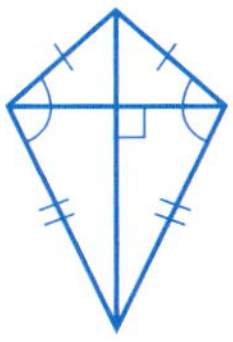	• 2 pairs of equal adjacent sides. • One pair of opposite angles equal. • One axis of symmetry. • Diagonals intersect at right angles.
Parallelogram 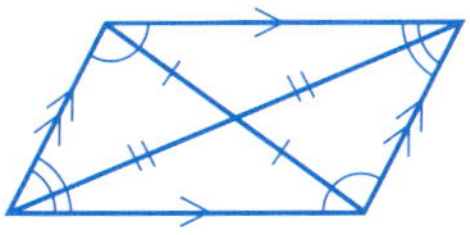	• Opposite sides are parallel and equal. • Opposite angles are equal. • No axes of symmetry. • Diagonals bisect each other.
Rhombus 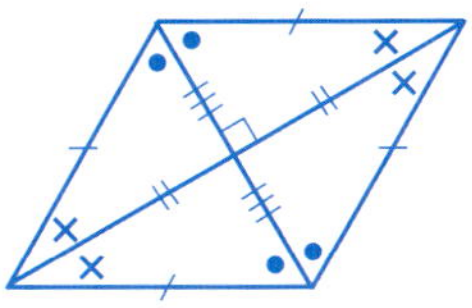	• 4 equal sides. • 2 axes of symmetry. • A special type of parallelogram. • Diagonals bisect each other at right angles. • Diagonals bisect the angles of the rhombus.
Rectangle 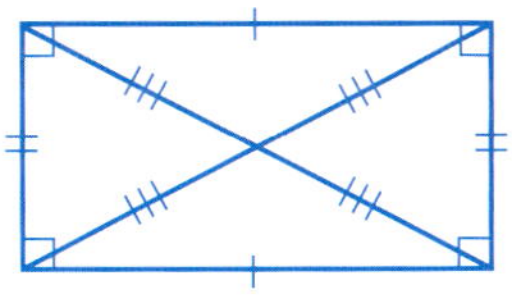	• All 4 angles measure 90°. • 2 axes of symmetry. • A special type of parallelogram. • Diagonals are equal and bisect each other.
Square 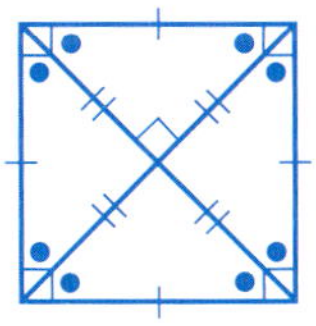	• 4 equal sides, four angles of 90°. • 4 axes of symmetry. • A special type of rhombus and rectangle. • Diagonals are equal and bisect each other at right angles. • Diagonals bisect the angles of the square.

We can now use congruent triangles to prove some of these properties.

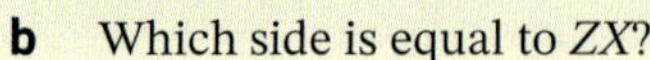

Example 7

ZYXW is a rectangle, so its opposite sides are equal and all angles are 90°.

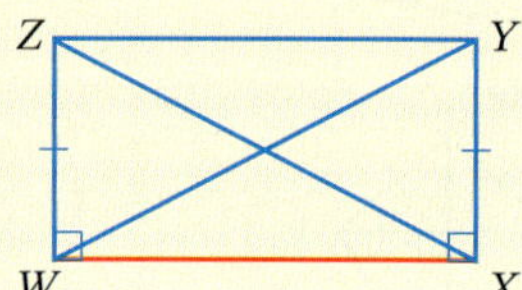

a $\triangle ZXW$ and $\triangle YWX$ are both right-angled triangles that have *WX* as their base. Which test proves that they are congruent?

b Which side is equal to *ZX*?

c What does this prove about the diagonals of a rectangle?

SOLUTION

a $ZW = YX$, $\angle ZWX = \angle YXW = 90°$, *WX* is shared (the common side in red). Therefore, SAS proves that they are congruent.

b *YW* is equal to *ZX*.

c The diagonals of a rectangle are equal in length.

EXERCISE 8.05 ANSWERS ON P. 641

Proving properties of quadrilaterals

U F PS R C

1 *ABCD* is a kite, so it has two pairs of equal adjacent sides. The diagonal *DB* is its axis of symmetry that divides it into two congruent triangles.

PS R C

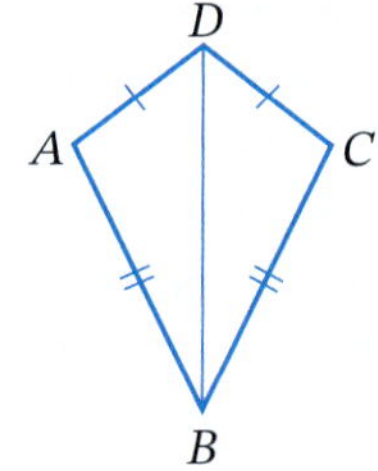

a Which test proves that the two triangles are congruent?

b Which angle is equal to $\angle A$?

c Which angle is equal to $\angle ADB$? Copy the diagram and mark both angles with a dot.

d Draw the other diagonal *AC*, intersecting *DB* at point *X*.

e This creates four triangles. Looking at your diagram, which congruence test proves that $\triangle DAX \equiv \triangle DCX$?

f Which side is equal to *AX*? Mark both sides with three dashes.

g Which angle is equal to $\angle DXA$?

h What is the size of $\angle DXA$? Mark this on your diagram.

i What does this prove about the diagonals of a kite?

2 *SPQR* is a parallelogram, so its opposite sides are parallel.

PS R C

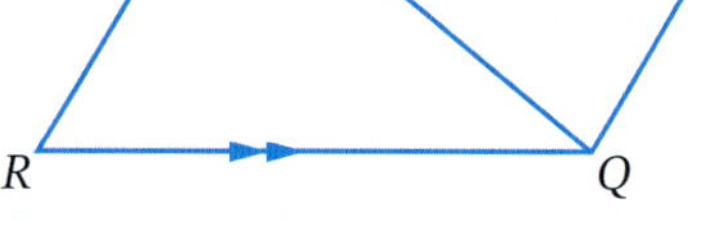

a Why is $\angle PSQ = \angle RQS$? What type of angles are they? Copy the diagram and mark both angles with a dot.

b Which angle is equal to $\angle PQS$? Mark both angles with an arc.

c The diagonal *SQ* divides the parallelogram into two congruent triangles. Which test proves that they are congruent?

□ Foundation ○ Standard ○ Complex

d Which side is equal to SP? Mark both sides with a dash.

e Which side is equal to PQ? Mark both sides with two dashes.

f What does this prove about the opposite sides of a parallelogram?

g Which angle is equal to $\angle P$? Mark both angles with a cross.

h What does this prove about the opposite angles of a parallelogram?

3 Draw the other diagonal PR of the parallelogram from question **2**, crossing SQ at X.

PS R C

a Which angle is equal to $\angle SPX$? Mark both angles with little circles.

b Which test proves that the top and bottom triangles joined at X are congruent?

c Which side is equal to SX? Mark both sides with three dashes.

d Which side is equal to PX? Mark both sides with four dashes.

e What does this prove about the diagonals of a parallelogram?

4 $PQST$ is a rhombus, so all sides are equal. A rhombus is a special type of parallelogram so its diagonals should also bisect each other as shown.

PS R C

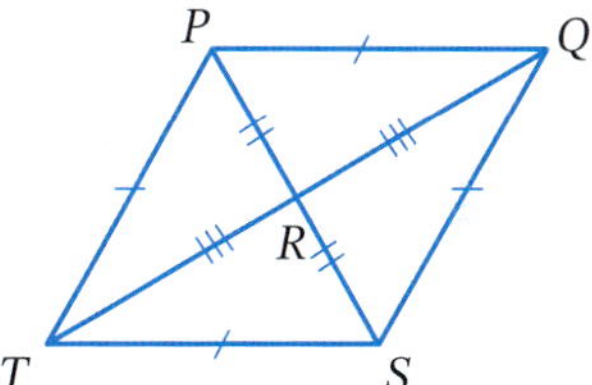

a The two diagonals divide the rhombus into four congruent triangles. Which test proves that the four triangles are congruent?

b What is the size of the four angles around point R? Why? Copy the diagram and mark those angles with the correct symbol.

c What does this prove about the diagonals of a rhombus?

d The bigger triangle, $\triangle PQT$, is isosceles so mark its two equal angles with a dot.

e Mark the other two angles that are equal to the angles marked with a dot.

f Mark the remaining angles in the congruent triangles with a cross since they are equal.

g What does this prove about the diagonals of a rhombus?

5 $WXYZ$ is a square, so all sides are equal and all angles are 90°.

PS R C

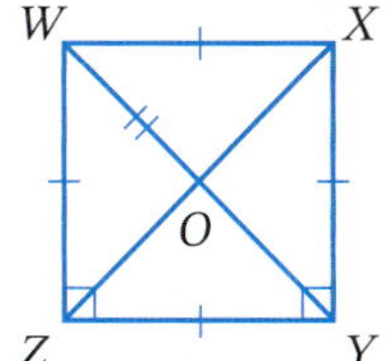

a Which congruence test proves that $\triangle WYZ \equiv \triangle XZY$?

b Which side is equal to WY?

c What does this prove about the diagonals of a square?

d A square is a special type of parallelogram, so its diagonals bisect each other. Copy the diagram and mark all sides that are equal to WO with double dashes.

e The two diagonals divide the square into four congruent triangles. Which test proves that they are congruent?

f What is the size of the four angles around point O? Mark those angles with the correct symbol.

g What does this prove about the diagonals of a square?

h As the four congruent triangles are also isosceles triangles, mark all equal angles with a dot. What is the size of these angles?

i What does this prove about the diagonals of a square?

□ Foundation ○ Standard ⬡ Complex

INVESTIGATION

Similar figures

1 Look at the following figures and complete the questions.

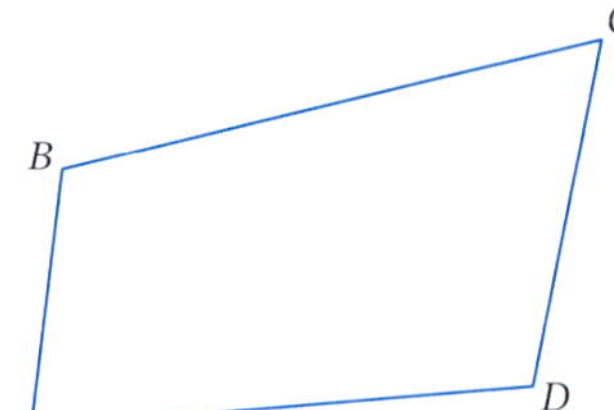

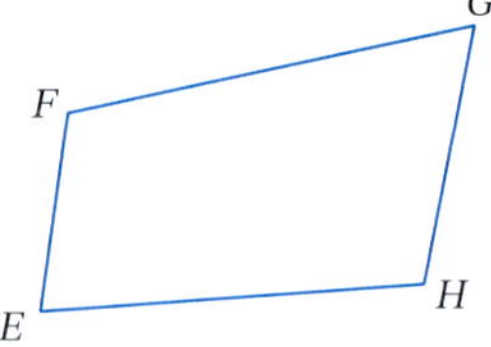

a Measure the angles in each figure as accurately as you can. Copy and complete the table.

$\angle A=$	$\angle E=$
$\angle B=$	$\angle F=$
$\angle C=$	$\angle G=$
$\angle D=$	$\angle H=$

b Copy and complete:

The matching pairs of angles are ________________.

c Measure the sides in each figure as accurately as you can, then copy and complete the table.

Length of sides (mm)		Ratio of matching sides
$AB=$	$EF=$	$\frac{AB}{EF}=$
$BC=$	$FG=$	$\frac{BC}{FG}=$
$CD=$	$GH=$	$\frac{CD}{GH}=$
$DA=$	$HE=$	$\frac{DA}{HE}=$

d Copy and complete:

The ratios of the matching sides are ________________.

2 Look at these figures:

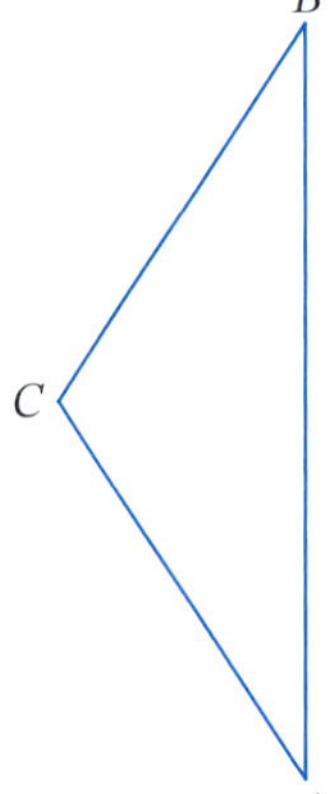

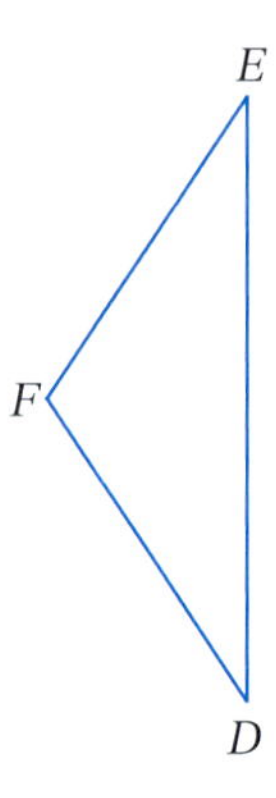

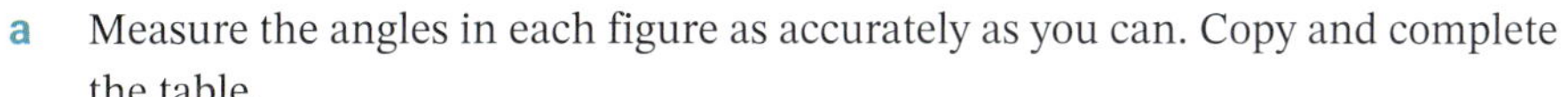

a Measure the angles in each figure as accurately as you can. Copy and complete the table.

∠A=	∠E=
∠B=	∠F=
∠C=	∠G=

b Copy and complete:

The matching pairs of angles are ________________.

c Copy and complete the table.

Length of sides (mm)		Ratio of matching sides
AB=	DE=	$\frac{AB}{DE}=$
BC=	EF=	$\frac{BC}{EF}=$
CA=	FD=	$\frac{CA}{FD}=$

Matching sides are in the same ratio.

Length of sides (mm) Ratio of matching sides

8.06

Similar figures 8.06

Figures that have the *same shape* but not necessarily the same size are said to be **similar**.

The teddy bears below are similar shapes because the height of the second bear is 1.5 the height of the first bear.

Worksheets
A page of similar figures

Enlargements and reductions

Enlarging a logo

Puzzle
Cartoon enlargement

Videos
Jai Singh (sundial)

Fractals: The Koch snowflake

The letters *N* below are similar shapes because the height of the second *N* is half the height of the first *N*.

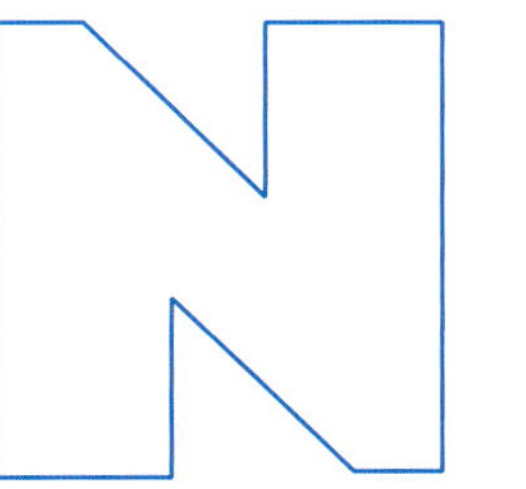

The two triangles below are also similar figures. A statement can be written to indicate these triangles are similar using the symbol '|||' which means 'is similar to'. For this pair of triangles the statement would be '$\triangle ABC \,|||\, \triangle DEF$', which is read 'triangle ABC is similar to triangle DEF'.

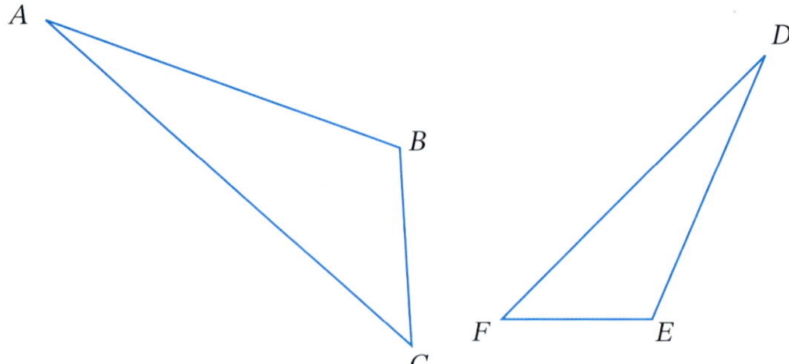

When using this notation, we must write the vertices (angles) of the shapes in matching pair order.

For the pair of triangles above:

- $\angle A$ matches $\angle D$.
- $\angle B$ matches $\angle E$.
- $\angle C$ matches $\angle F$.

ⓘ Similar figures

Similar figures have the same shape. Their matching angles are equal and their matching sides are in the same ratio.

Example 8

Which two of these parallelograms are similar?

A B C D
3 cm
2 cm
120°
E F G H
1 cm
1.5 cm
120°
J K L M
4 cm
3 cm
110°

SOLUTION

For similar figures, matching angles must be equal and matching sides must be in the same ratio.

$\angle A = 120°$ and $\angle E = 120°$ but $\angle J = 110°$.

Also, checking the ratios of matching sides in $ABCD$ and $EFGH$:

$\frac{EF}{AB} = \frac{1.5}{3} = \frac{1}{2}$ $\qquad$ $\frac{EH}{AD} = \frac{1}{2}$

$\therefore \frac{EF}{AB} = \frac{EH}{AD}$ $\qquad$ So matching sides are in the same ratio.

$\therefore$ The two similar parallelograms are $ABCD$ and $EFGH$.

EXERCISE  8.06 ANSWERS ON P. 641

Similar figures

1 Find the pair of similar figures in each set of shapes below. Then write a similarity statement of the form '$PQRS ||| WXYZ$' naming the vertices in the correct order.

R C

a

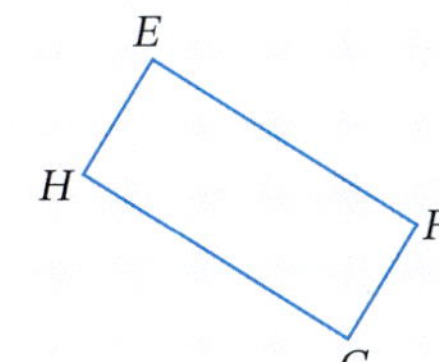

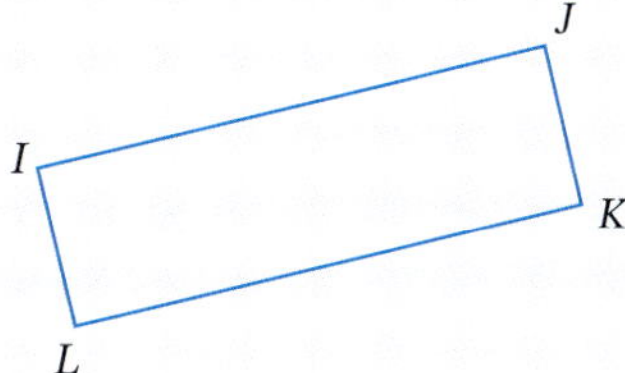

b

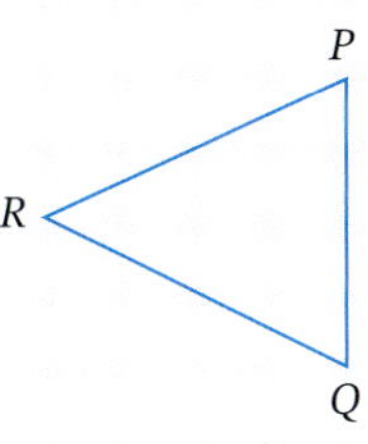

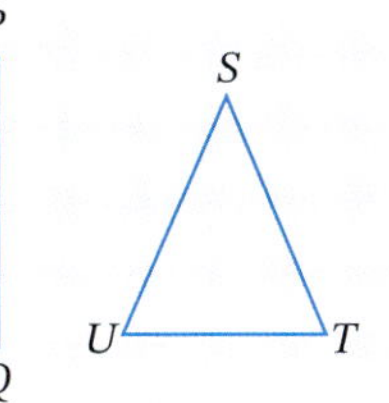

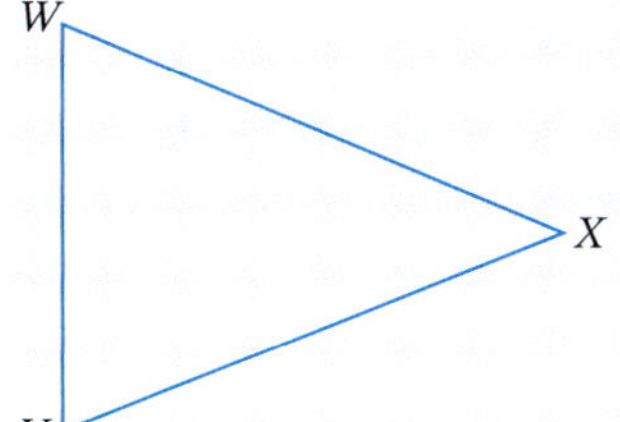

c

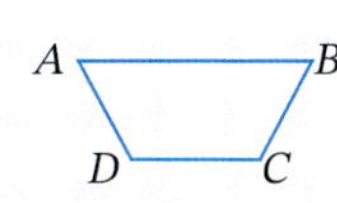

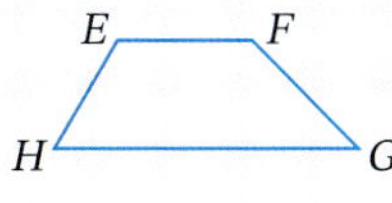

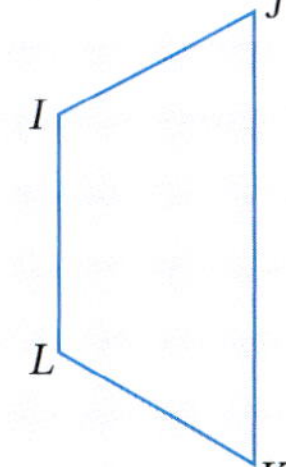

d

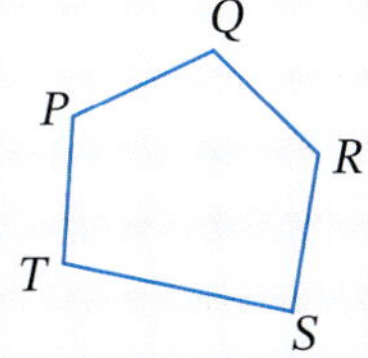

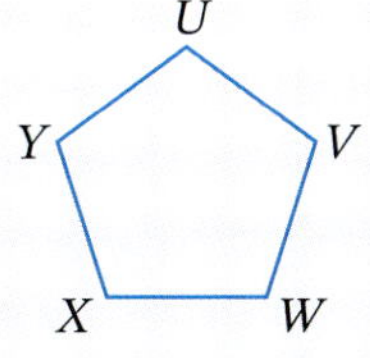

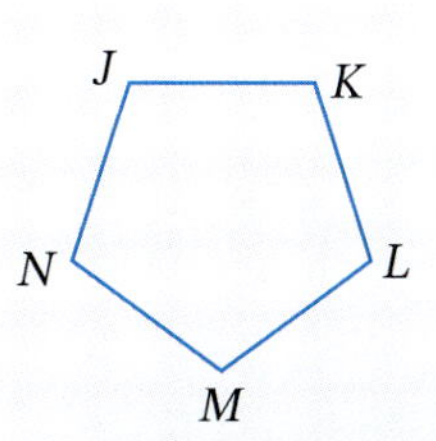

 Foundation  Standard Complex

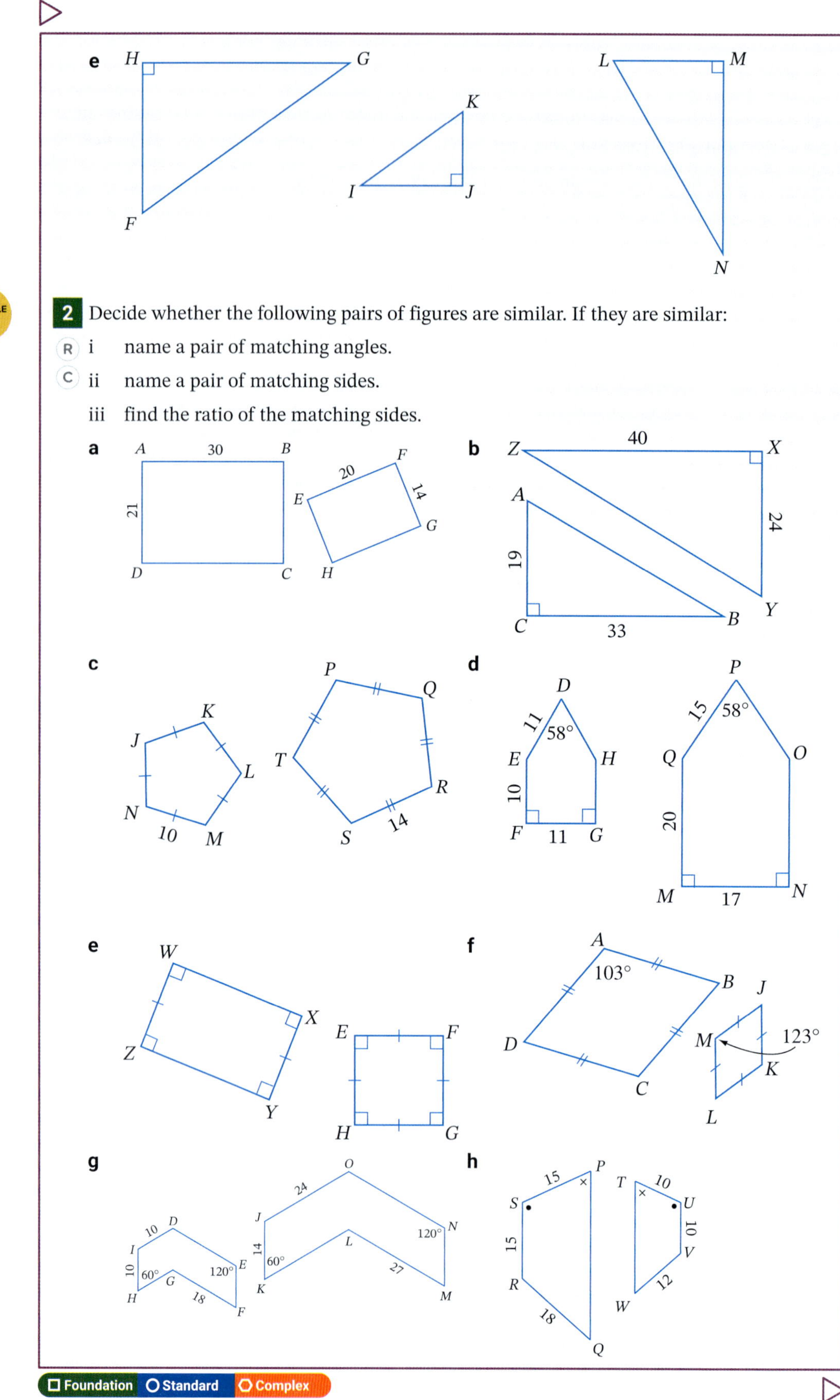
e
H
G
F
K
I
J
L
M
N
2 Decide whether the following pairs of figures are similar. If they are similar:
R i name a pair of matching angles.
C ii name a pair of matching sides.
iii find the ratio of the matching sides.
a
A 30 B
21
D C
F 20 E 14 G H
b
Z 40 X
24
Y
A 19 C 33 B
c
K J L N 10 M
P Q T R S 14
d
D 11 58° E H 10 F 11 G
P 15 58° Q O 20 M 17 N
e
W X Z Y
E F H G
f
A 103° B D C
J M 123° K L
g
10 D I 10 60° G 120° E H 18 F
O 24 J 14 60° L 120° N K 27 M
h
S 15 P 15 R 18 Q
T 10 U 10 V 12 W

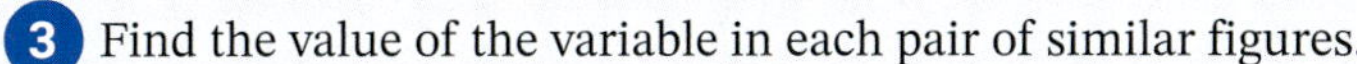

3 Find the value of the variable in each pair of similar figures.

R **a**

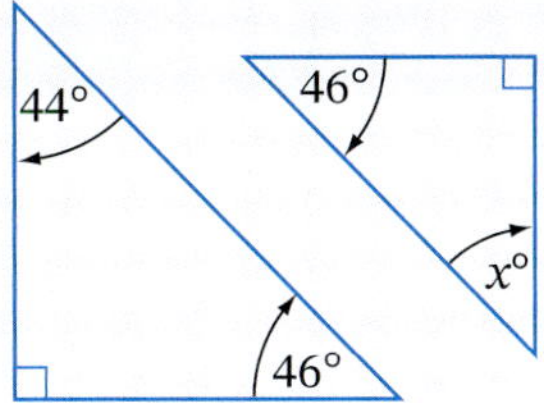

b

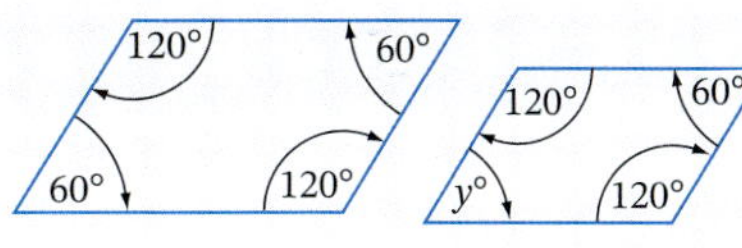

c

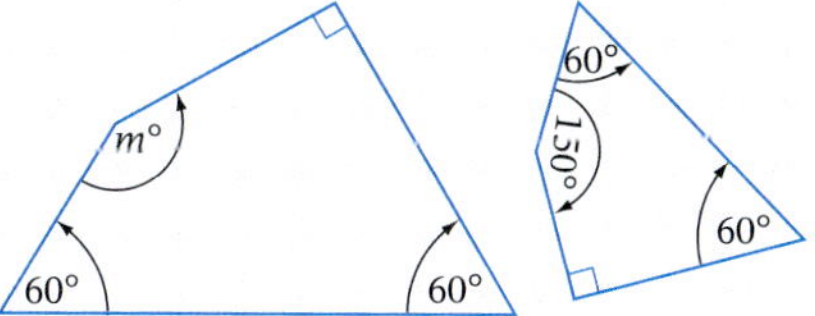

d

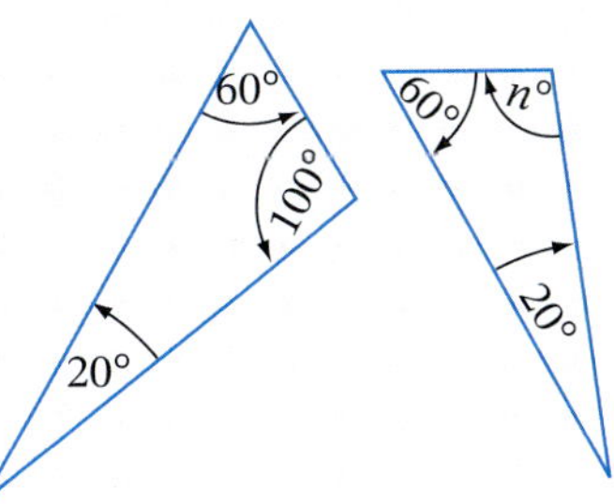

e

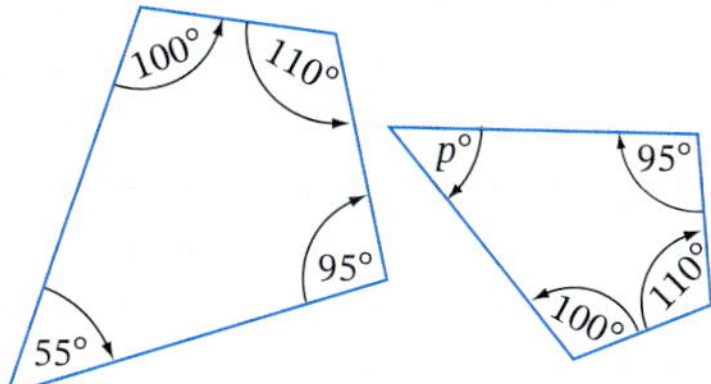

f

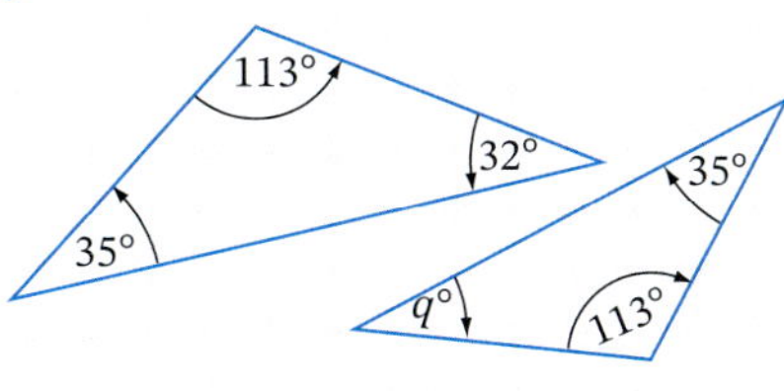

4 **a** State whether the figures in each pair below are similar.

R **i**

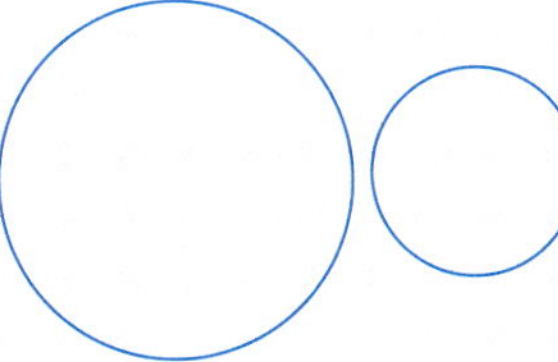

ii

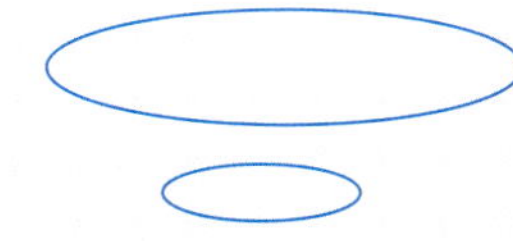

iii

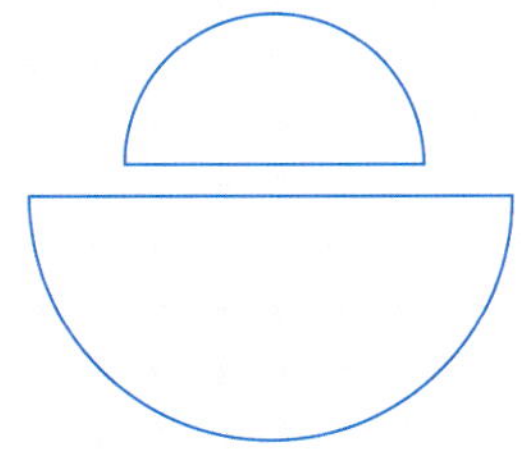

5 **a** Draw two squares that are similar. State the ratio of the matching sides.

R **b** Draw two circles that are similar.

C **c** Draw two rectangles that are similar. State the ratio of the matching sides.

d Draw two similar right-angled triangles. State the ratio of the matching sides.

e Make up your own shape and draw it. Draw another that is similar to it.

6 Discuss these statements and decide whether they are true or false.

R **a** All squares are similar.

b All rectangles are similar.

C **c** All triangles are similar.

d All circles are similar.

Foundation | Standard | Complex

8.07 Similar triangles

Scale factor

Skillsheet
Finding sides in similar figures

Worksheets
A page of similar figures

Enlargements and reductions

Puzzle
Similar triangles

Video
Finding an unknown side in similar figures

A similar figure is usually an enlargement or a reduction of another figure. The figure we start with is called the **original** and the enlarged or reduced figure is called the **image**.

The **scale factor** describes the change in size between the original and the image.

For example, if the triangle ABC is enlarged by doubling each side, the scale factor is 2.

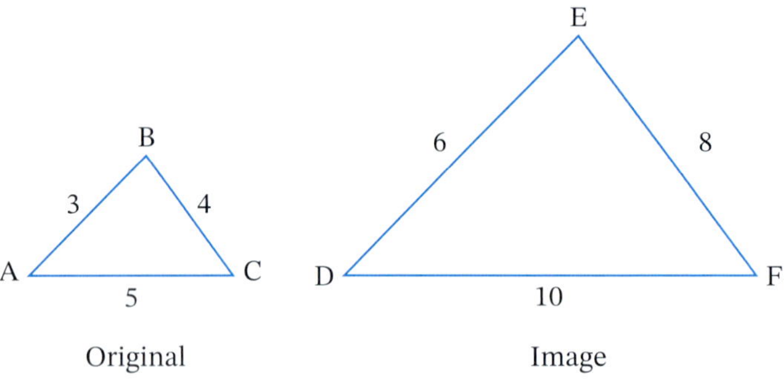

$AB = 2 \times DE$

$BC = 2 \times EF$

$CA = 2 \times FD$

$\therefore$ Scale factor $= 2$

Unless otherwise specified, the original should be on the left (or top) and the image on the right (or bottom).

To check if a pair of figures are similar figures, we must check that matching sides have the same scale factor.

Scale factor

The scale factor between a pair of similar shapes can be calculated using the rule:

$$\text{Scale factor} = \frac{\text{image length}}{\text{original length}}$$

- For an enlargement, the scale factor between the original and the image will be greater than 1.
- For a reduction, the scale factor between the original and the image will be smaller than 1.

9780170465601

Example 9

8.07

Determine whether this pair of triangles is similar.

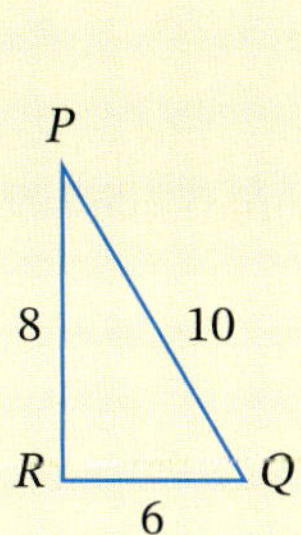

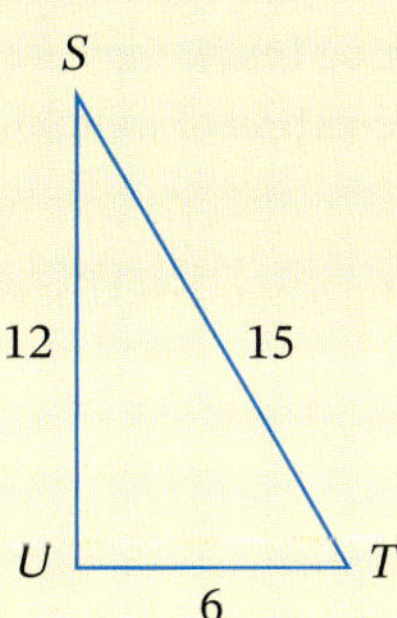

SOLUTION

Calculate scale factor for a pair of matching sides, say ST and PQ.

Image is an enlargement, therefore the scale factor is greater than 1.

$$\text{Scale factor} = \frac{\text{image length}}{\text{original length}} = \frac{ST}{PQ} = \frac{15}{10} = 1.5$$

Calculate scale factor for another pair of matching sides, TU and QR.

Scale factor also 1.5.

$$\text{Scale factor} = \frac{TU}{QR} = \frac{9}{6} = 1.5$$

Calculate scale factor for the third pair of matching sides, US and RP.

Scale factor also 1.5.

$$\text{Scale factor} = \frac{US}{RP} = \frac{12}{8} = 1.5$$

All pairs of matching sides have the same scale factor.

Note that the vertices of the triangles are written in matching order.

Therefore, $\triangle PQR \,|||\, \triangle STU$.

ⓘ Tests for similar triangles

There are 4 **similarity tests** for proving that two triangles are similar: **'SSS'**, **'SAS'**, **'AA'** or **'RHS'**. As with the **congruence tests,** they are based on checking pairs of matching sides and angles, but looking at the **ratios** of matching sides.

Two triangles are similar if:

- all matching sides are in the same ratio (have the same scale factor) **('SSS').**

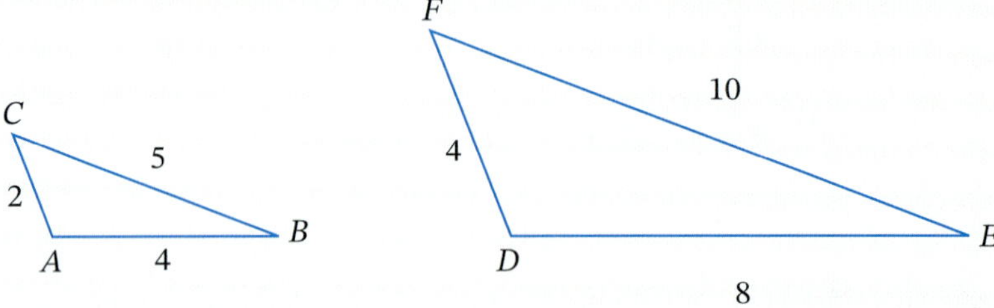

- 2 pairs of matching sides are in the same ratio (have the same scale factor) and the included angles are equal in size **('SAS').**

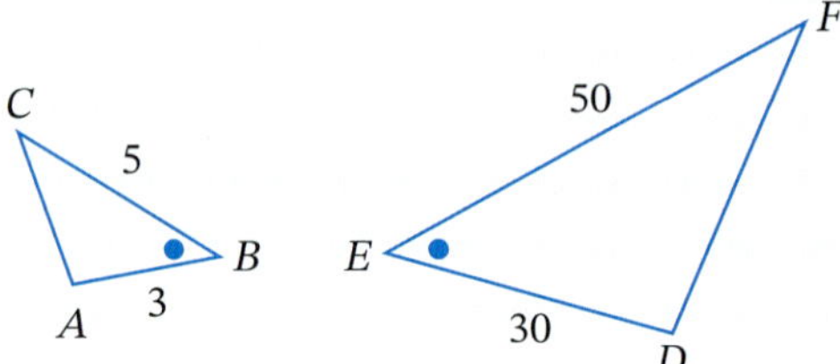

- 2 pairs of matching angles are equal in size **('AA'** or **'equiangular').**

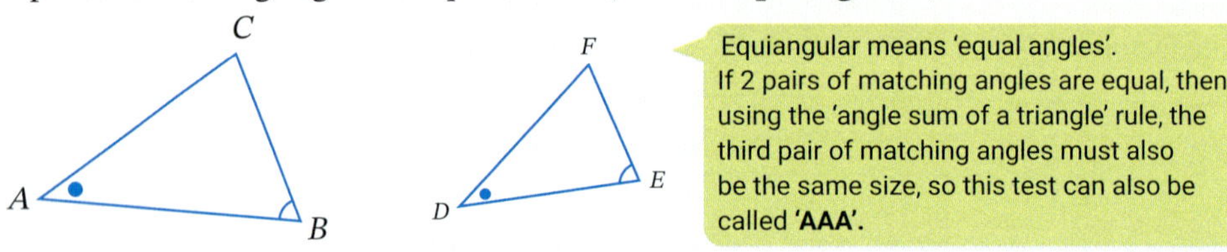

- they are right-angled, matching hypotenuses are in the same ratio, and one other pair of matching sides are in the same ratio **('RHS').**

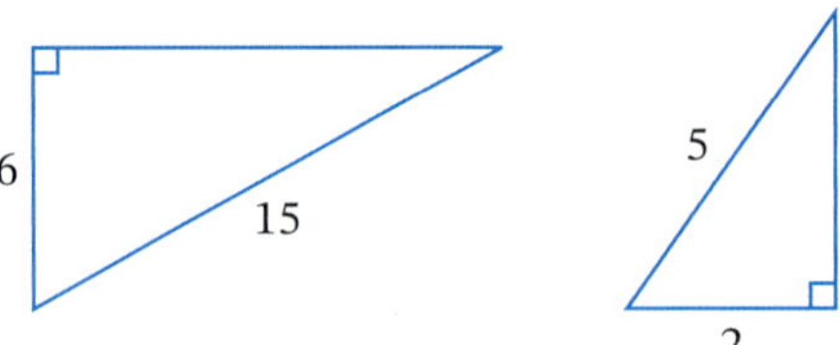

Video
Similar triangles

Example 10

Determine if each pair of triangles are similar.

a

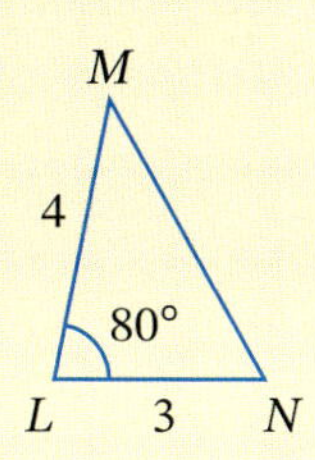

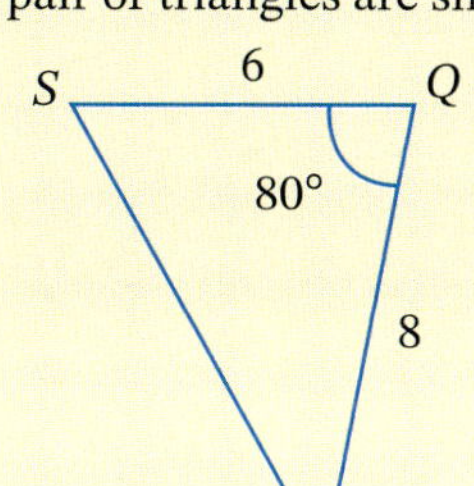

b

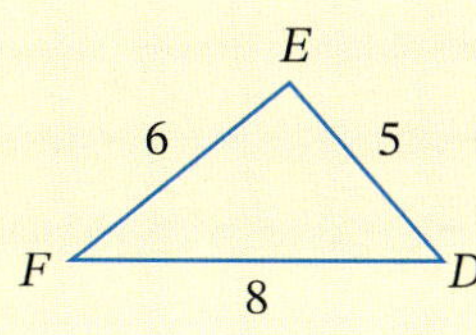

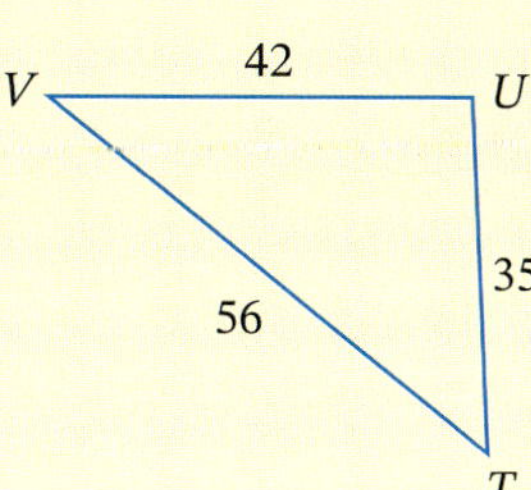

c

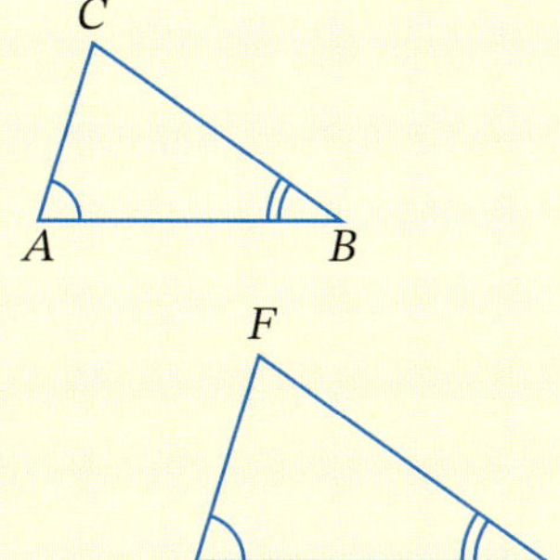

d

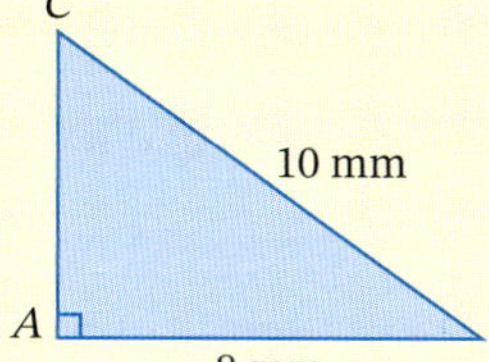

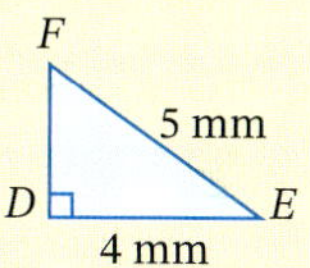

SOLUTION

a Both triangles have an 80° angle between two sides, so try the 'SAS' test.

Check the scale factors for matching pairs of sides.

For QR and LM:

Scale factor $= \frac{8}{4} = 2.$

For QS and LN:

Scale factor $= \frac{6}{3} = 2.$

Two pairs of matching sides have the same scale factor and their included angles are equal: $\angle L = \angle Q = 80°$.

$\therefore \triangle LMN ||| \triangle QRS$ using 'SAS'.

b Both triangles have all three sides given, so try the 'SSS' test.

For UT and ED:

Scale factor $= \frac{35}{5} = 7$.

For VU and FE:

Scale factor $= \frac{42}{6} = 7$.

For TV and DF:

Scale factor $= \frac{56}{8} = 7$.

Three pairs of matching sides have the same scale factor.

$\therefore \triangle EDF \parallel\!| \triangle UTV$ using 'SSS'.

c Both triangles have two pairs of matching angles equal, so use the 'AA' test.

$\angle A = \angle D$

$\angle B = \angle E$

$\therefore \triangle \text{ABC} \parallel\!| \triangle \text{DEF}$ using 'AA'.

d Both triangles are right-angled and have matching hypotenuses and another side given, so try the 'RHS' test.

$\angle A = \angle D = 90°$

For FE and CB (hypotenuses):

Scale factor $= \frac{5}{10} = \frac{1}{2}$

For DE and AB:

Scale factor $= \frac{4}{8} = \frac{1}{2}$

Both triangles are right-angled, and the matching hypotenuses and another pair of matching sides have the same scale factor.

$\therefore \triangle ABC \parallel\!| \triangle DEF$ using 'RHS'.

Finding an unknown side in similar triangles

Example 11

Find the value of the variable in each pair of similar triangles.

a

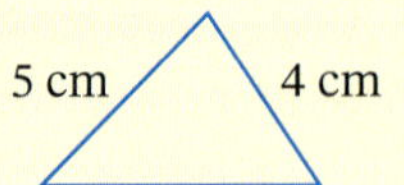

b

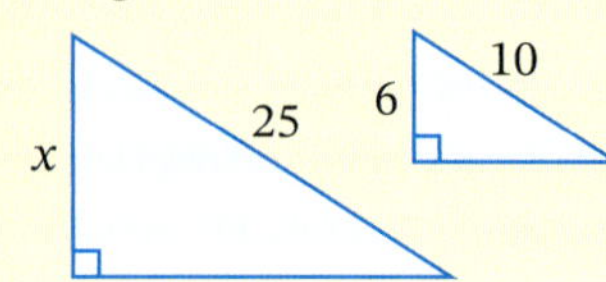

SOLUTION

a Scale factor $= \frac{8}{4} = 2$.

Triangle on right has been enlarged by a scale factor of 2.

So, 5 cm must also be multiplied by 2 to make x.

$x = 5 \times 2$

$= 10$

b Scale factor $= \frac{10}{25} = \frac{2}{5}$.

Triangle on right has been reduced, by a scale factor of $\frac{2}{5}$.

But x is on the original triangle (on the left).

So 6 m must be divided by $\frac{2}{5}$ to make x.

$x = 6 \div \frac{2}{5}$

$= 6 \times \frac{5}{2}$

$= 15$

EXERCISE 8.07 ANSWERS ON P. 642

Similar triangles

U F PS R C

1 Calculate the scale factor for each pair of similar figures.

a

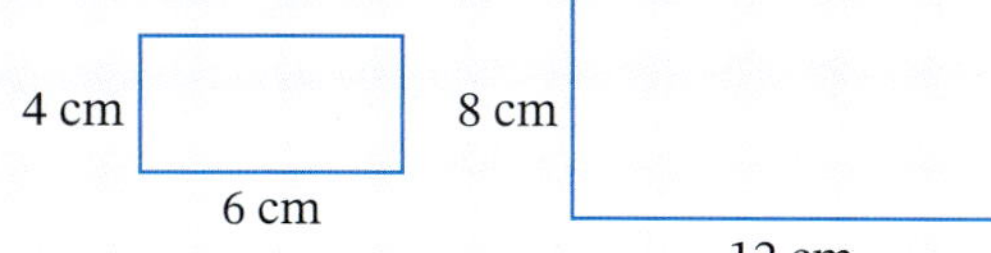

b

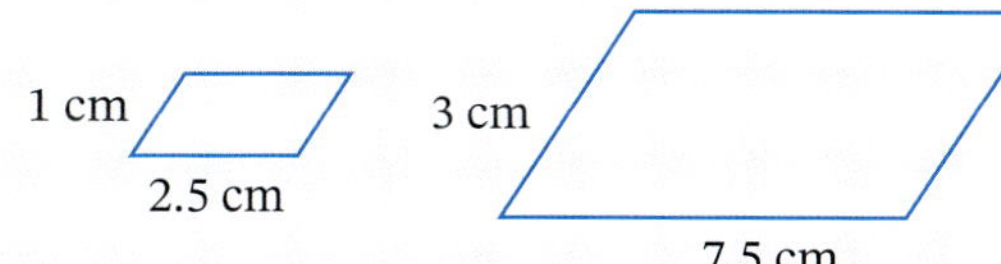

c

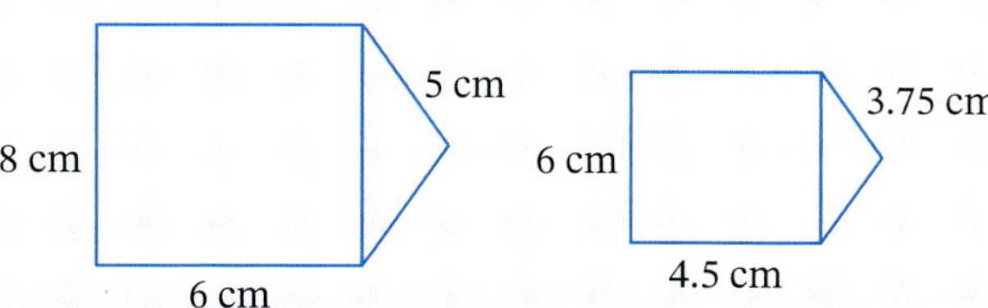

▷

☐ Foundation ○ Standard ⬡ Complex

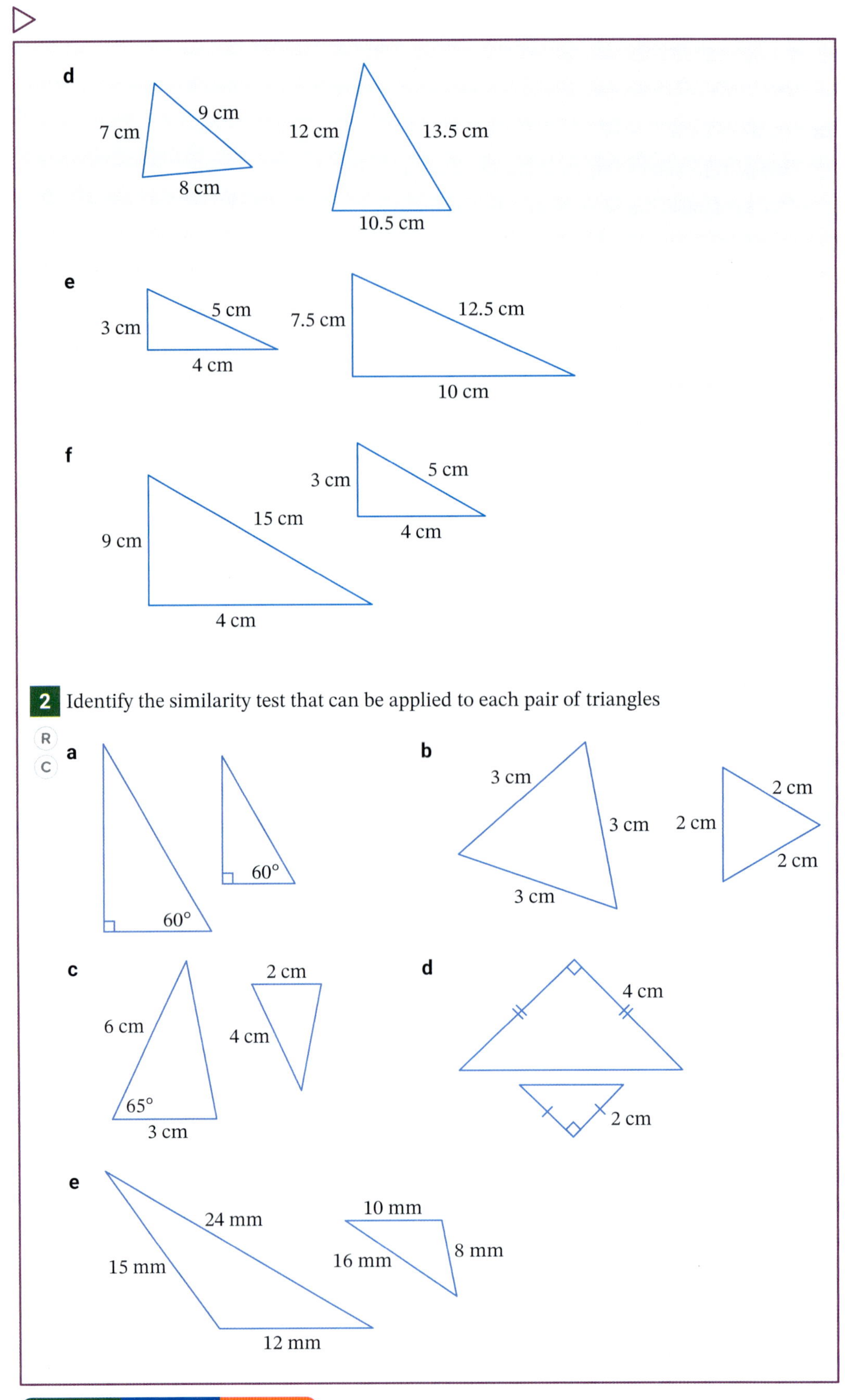

Foundation Standard Complex

f

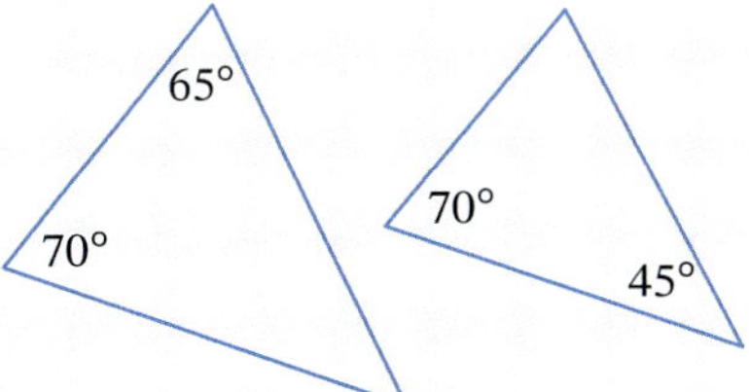

3 Explain why only two pairs of angles are needed to prove that two triangles are similar.

R C

4 Determine if the following pairs of triangles are similar.

EXAMPLE 10

R C

a

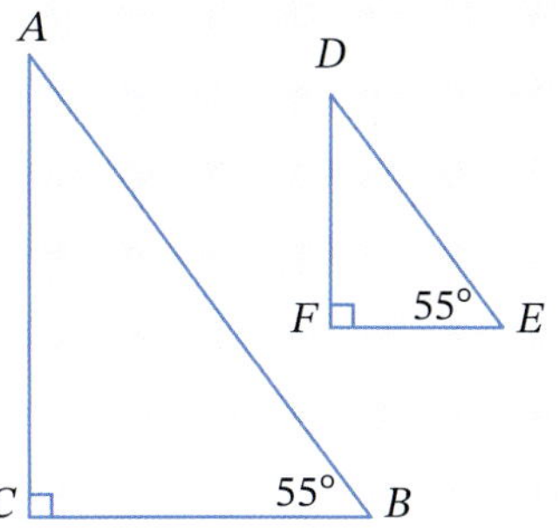

b

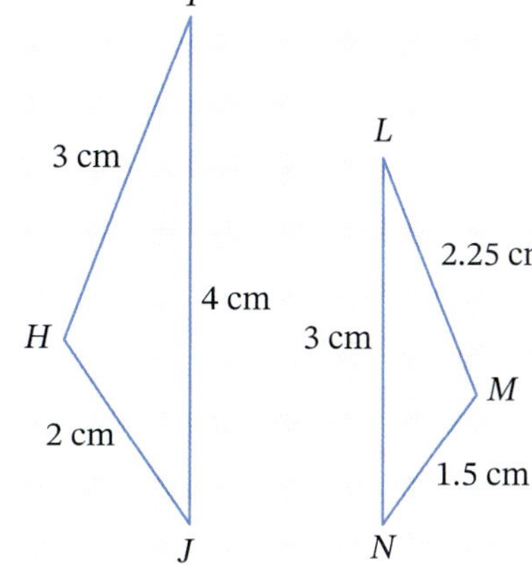

c

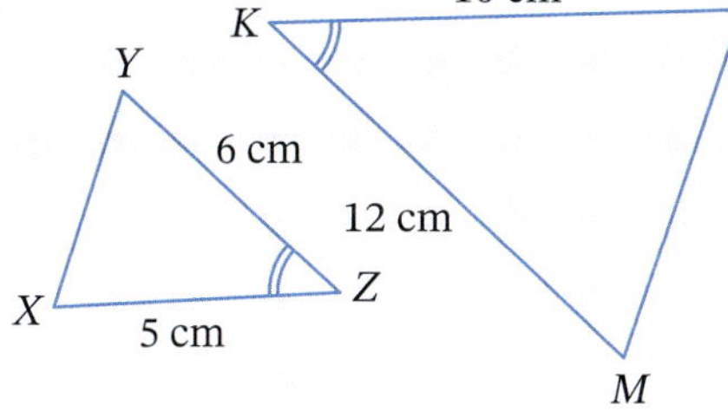

d

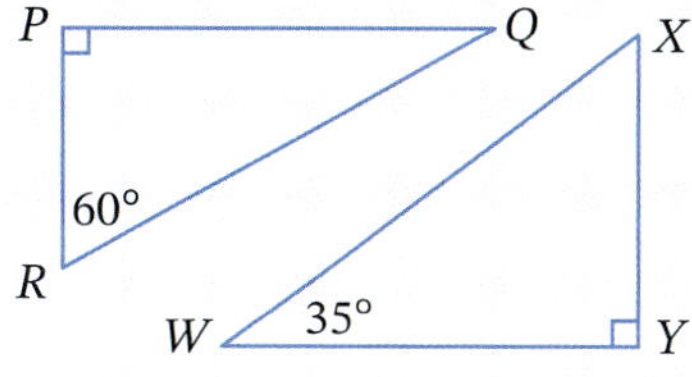

e

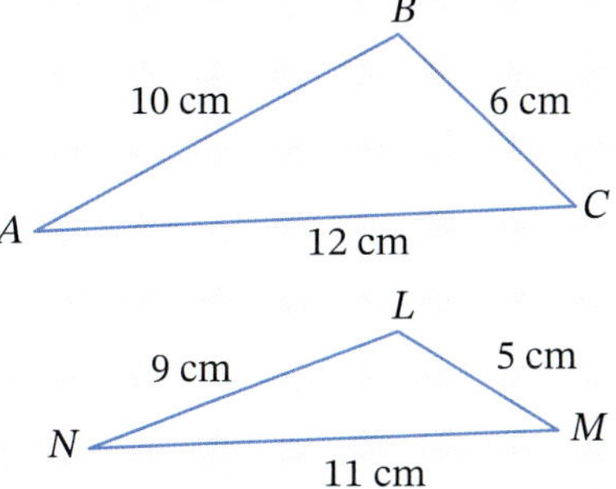

f

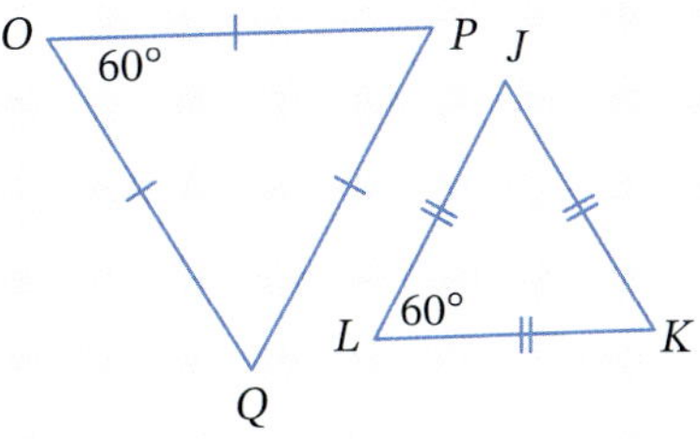

Foundation Standard Complex

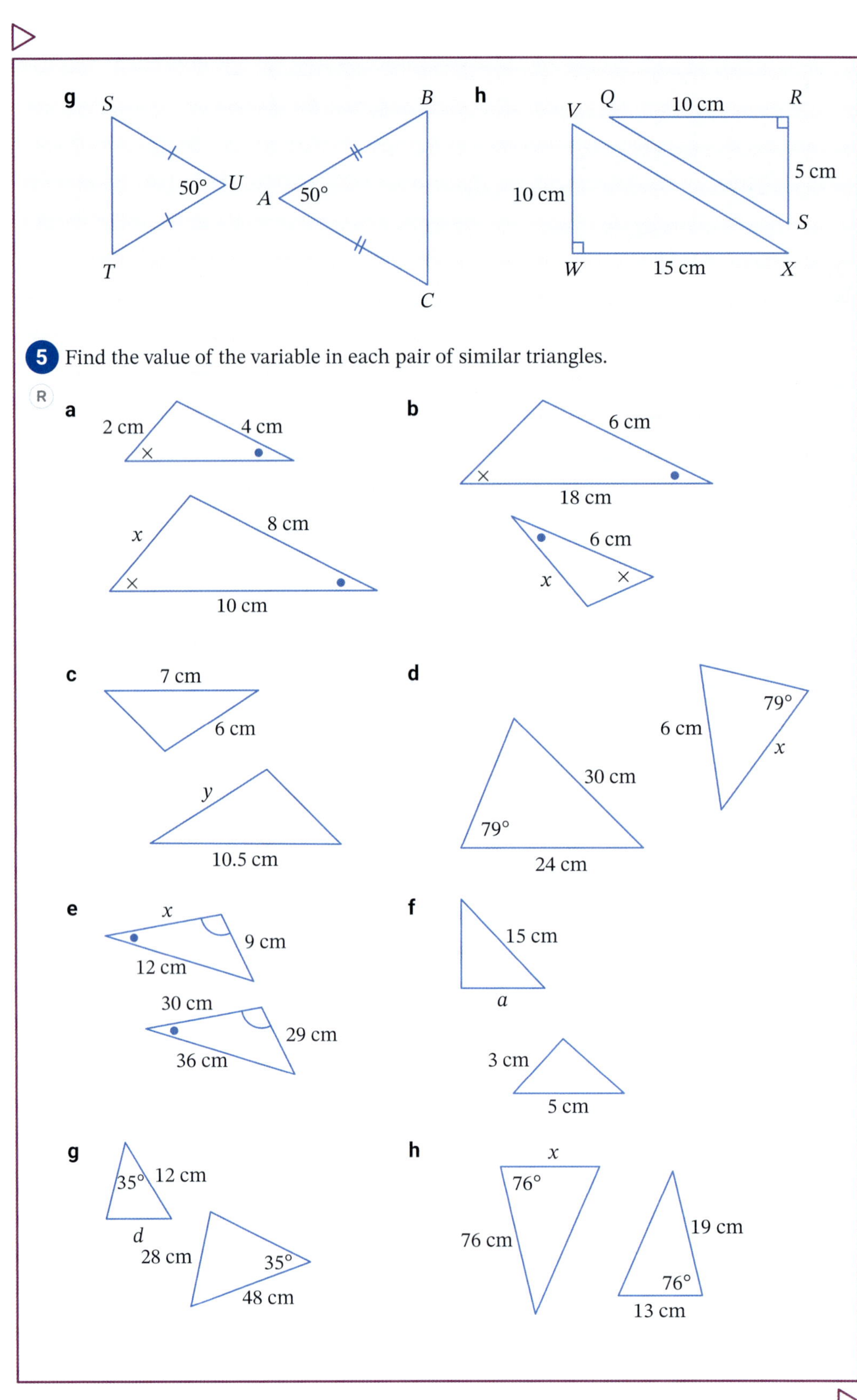

5 Find the value of the variable in each pair of similar triangles.

6 Find the unknown side length in each pair of similar triangles.

R

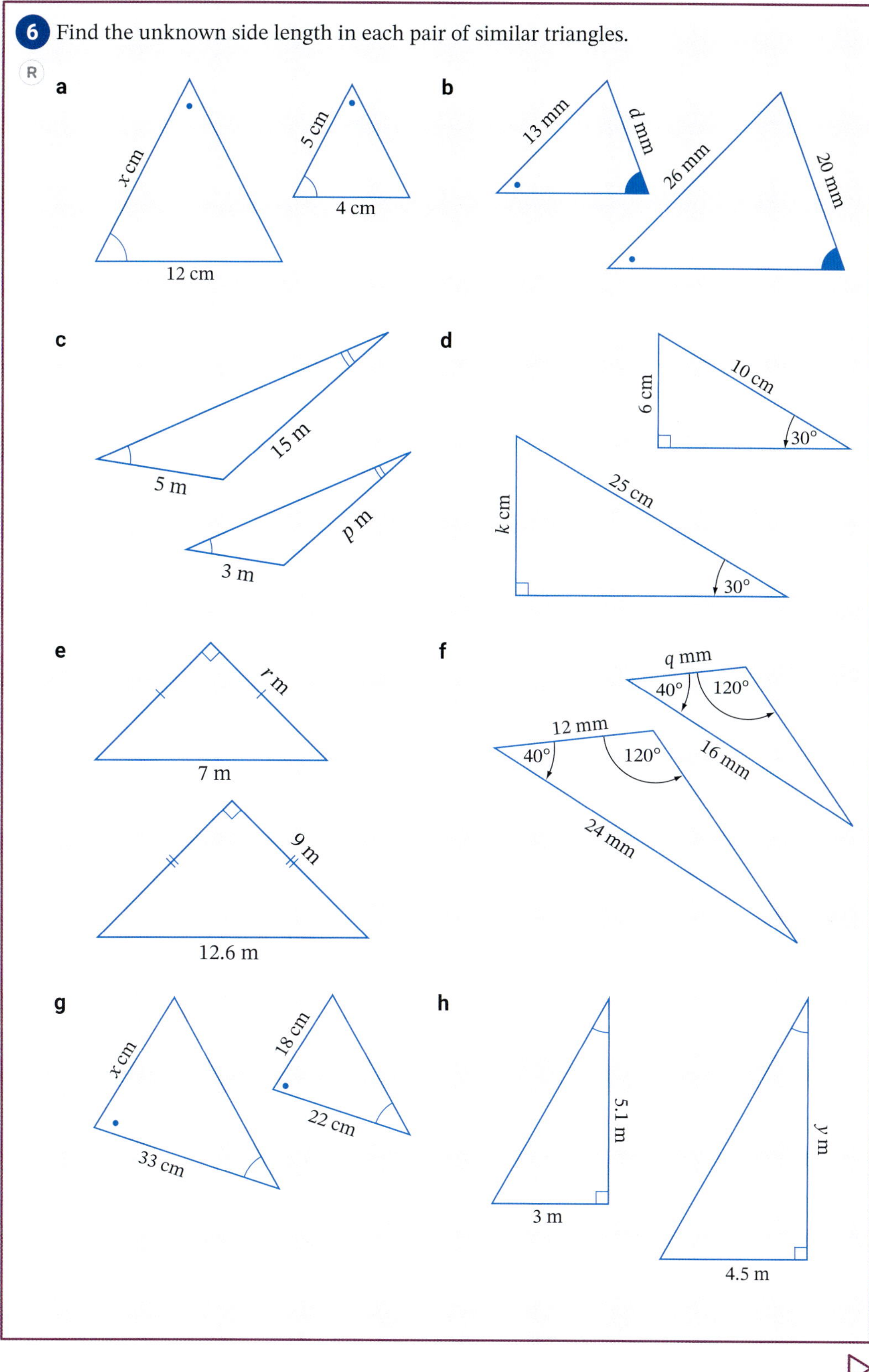

8.07

Foundation Standard Complex

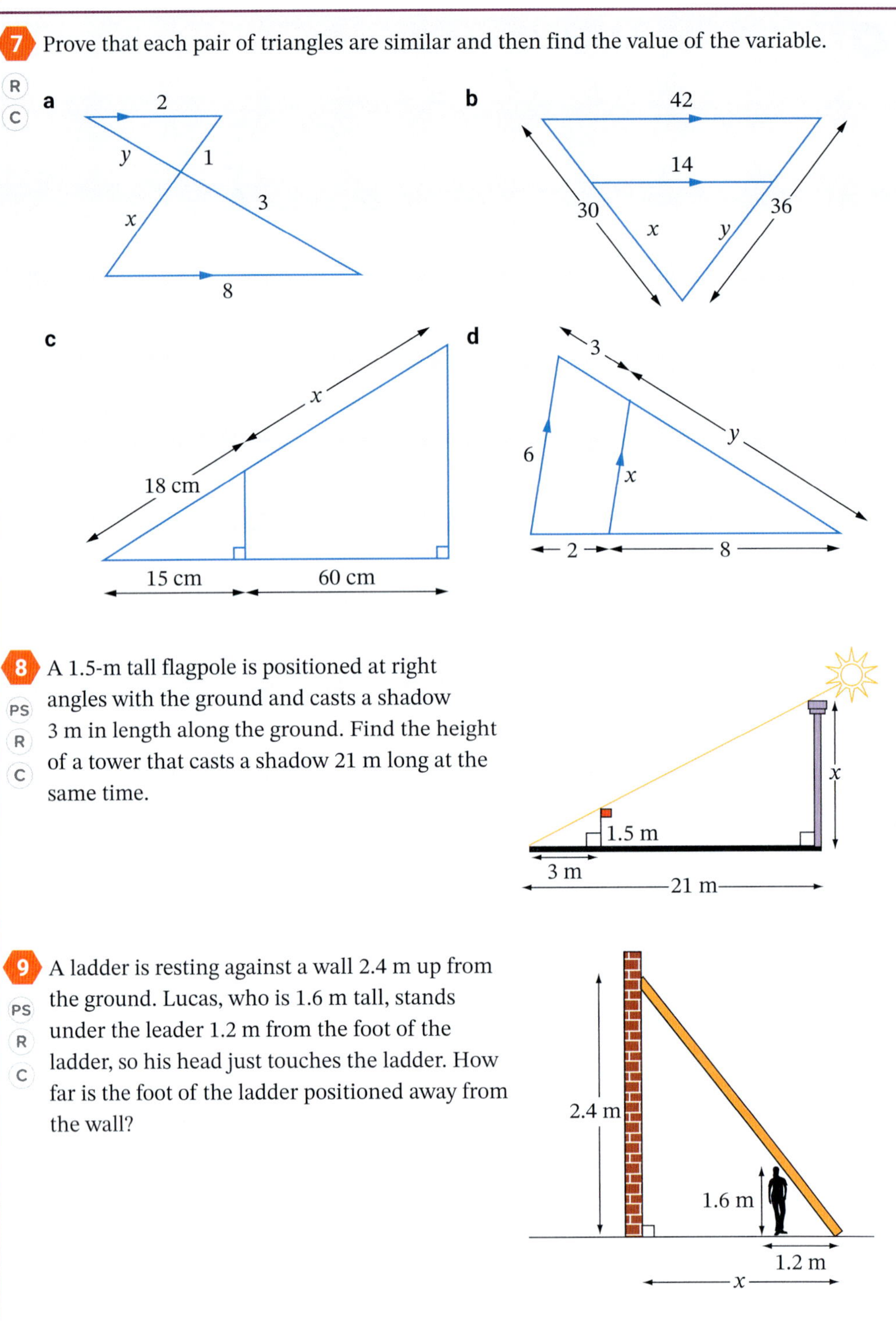

7 Prove that each pair of triangles are similar and then find the value of the variable.

R C

8 A 1.5-m tall flagpole is positioned at right angles with the ground and casts a shadow 3 m in length along the ground. Find the height of a tower that casts a shadow 21 m long at the same time.

PS R C

9 A ladder is resting against a wall 2.4 m up from the ground. Lucas, who is 1.6 m tall, stands under the leader 1.2 m from the foot of the ladder, so his head just touches the ladder. How far is the foot of the ladder positioned away from the wall?

PS R C

☐ Foundation ◯ Standard ⬡ Complex

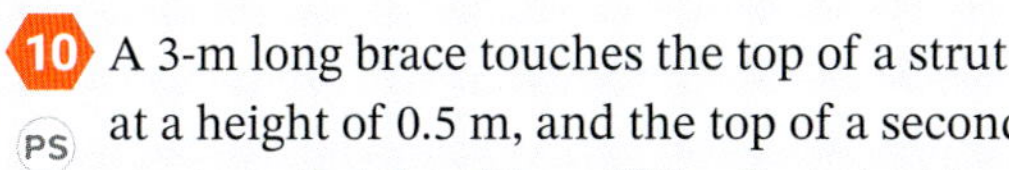

10 PS R C A 3-m long brace touches the top of a strut at a height of 0.5 m, and the top of a second strut at a height of 1 m. If the first strut is 1.4 m from the base of the brace, how far is the second strut from the base of the brace?

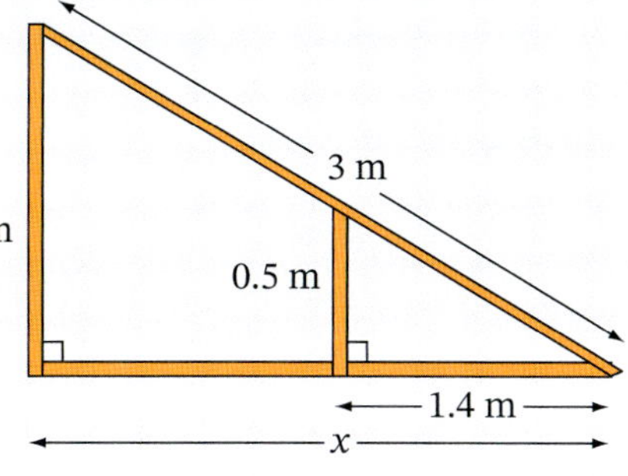

11 PS R C A tree that is 3.8 m back from one side of a river bank is sighted from the other side of the river. Other measurements are taken and are shown on this diagram. What is the width of the river?

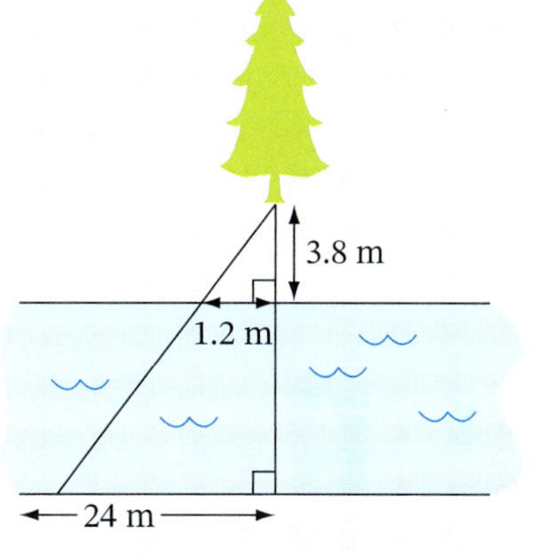

Quiz
Mental
skills 8B

☆ MENTAL SKILLS 8B ANSWERS ON P. 642

Maths without calculators

Time differences

1 Study each example.

a What is the time difference between 11:40 a.m. and 6:15 p.m.?

From 11:40 a.m. to 5:40 p.m. = 6 hours

Count: '11:40, 12:40, 1:40, 2:40, 3:40, 4:40, 5:40'

From 5:40 a.m. to 6:00 p.m. = 20 min

From 6:00 p.m. to 6:15 p.m. = 15 min

5 hours + 20 min + 15 min = 6 hours 35 min

OR:

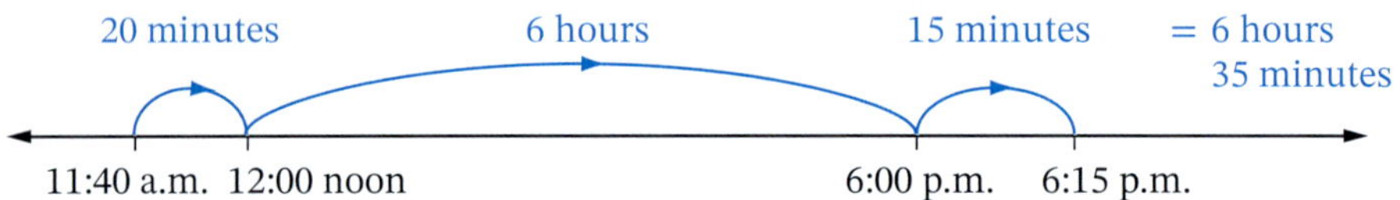

b What is the time difference between 16:45 and 23:20?

From 16:45 to 22:45 = 6 hours (22 − 16 = 6)

From 22:45 to 23:00 = 15 minutes

From 23:00 to 23:20 = 20 minutes

6 hours + 15 minutes + 20 minutes = 6 hours 35 minutes

OR:

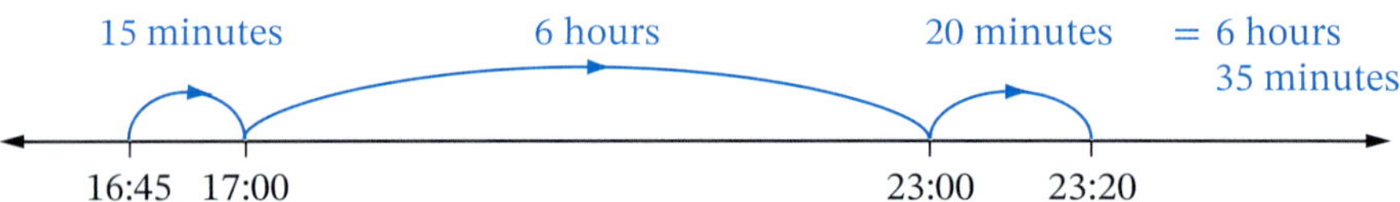

2 Now find the time difference between:

a 11:10 a.m. and 7:40 p.m.

b 6:20 p.m. and 12:00 midnight

c 4:45 p.m. and 8:10 p.m.

d 2:35 a.m. and 10:50 a.m.

e 1:05 p.m. and 12:30 a.m.

f 9:35 a.m. and 11:15 a.m.

g 04:25 and 09:35

h 14:40 and 20:25

i 7:55 a.m. and 3:50 p.m.

j 2:40 p.m. and 10:20 p.m.

9780170465601

POWER PLUS ANSWERS ON P. 642

Worksheet Congruent triangle proofs

1 Copy and complete each formal proof of congruent triangles.

a $WXYZ$ is a kite.

In $\triangle WYZ$ and $\triangle WYX$:

$WZ =$ ______ (equal sides of a kite)

$ZY =$ ______ (________________)

Side ______ is common to both triangles.

$\triangle WYZ \equiv \triangle WYX$ (SSS)

b ABC is an isosceles triangle, where $AB = AC$.

D is the midpoint of BC.

In $\triangle ABD$ and $\triangle ACD$:

$AB =$ ______ (________________)

$BD =$ ______ (D is ____________________)

AD is ____________ to both triangles.

$\triangle ABD \equiv \triangle ACD$ (______)

c $GHJK$ is a parallelogram.

In $\triangle GHJ$ and $\triangle JKG$:

$GH =$ ______ (________________________)

$HJ =$ ______ (________________________)

$\angle GHJ = \angle$______ (________________________)

$\triangle$______ $\equiv \triangle$ ______ (______)

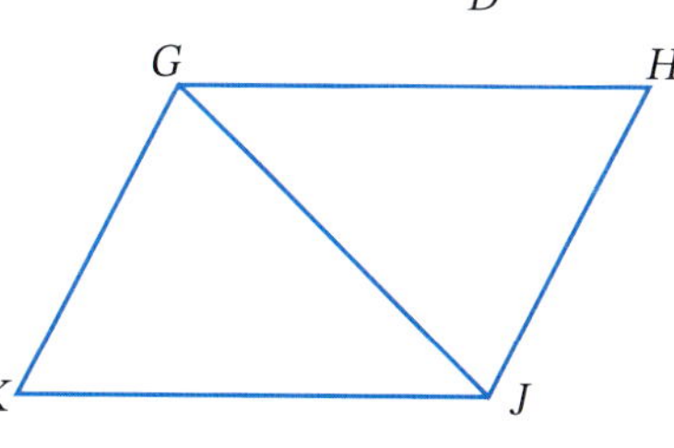

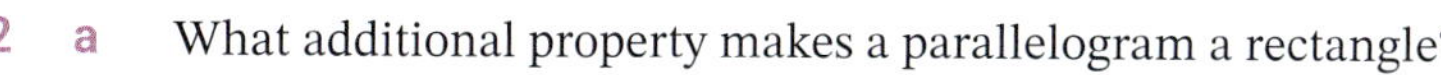

2 **a** What additional property makes a parallelogram a rectangle?

b What makes a kite a rhombus?

c What makes a rectangle a square?

3 Name all quadrilaterals whose:

a opposite angles are equal.

b diagonals intersect at 90°.

c diagonals are equal.

d angles are all 90°.

e opposite sides are parallel.

f diagonals bisect each other.

8 CHAPTER REVIEW

Language of maths

Quiz
Language of maths 8

AA	AAS	congruence test	congruent (≡)
construct	diagonal	equiangular	hypotenuse
image	included angle	matching	original
reflection	RHS	rotation	SAS
scale factor	similar	similarity test	SSS
superimpose	transformation	translation	vertex/vertices

1 What are the three **congruence transformations**?

2 What word means:

  a how much the image is larger or smaller than the original shape?

  b to draw accurately?

  c to place one shape on top of another shape so that it fits?

3 What does the phrase 'vertices must be named in matching order' mean?

4 What does **RHS** stand for:

  a in congruent triangles?

  b in solving equations?

5 What is an **included angle**?

6 What type of figures has the same shape and angles but not the same size?

Topic summary

- How useful do you think this chapter is to you?
- Which sections did you find the most interesting? Which section did you find the most difficult?

Print (or copy) and complete this mind map of the topic, adding detail to its branches and using pictures, symbols and colour where needed. Ask your teacher to check your work.

Worksheet
Mind map: Congruent and similar figures

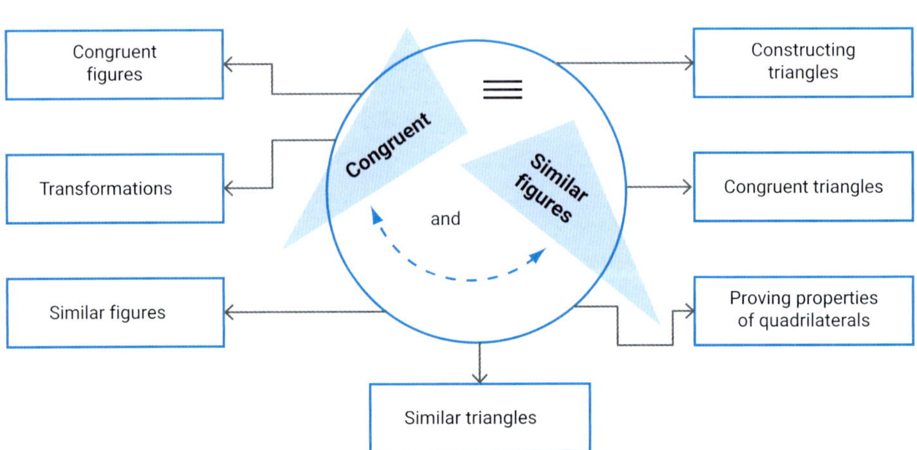

8 TEST YOURSELF

ANSWERS ON P. 642

Quiz
Test yourself 8

1 Copy this diagram, reflect the shape across the line, then rotate it 270° clockwise about the point X.

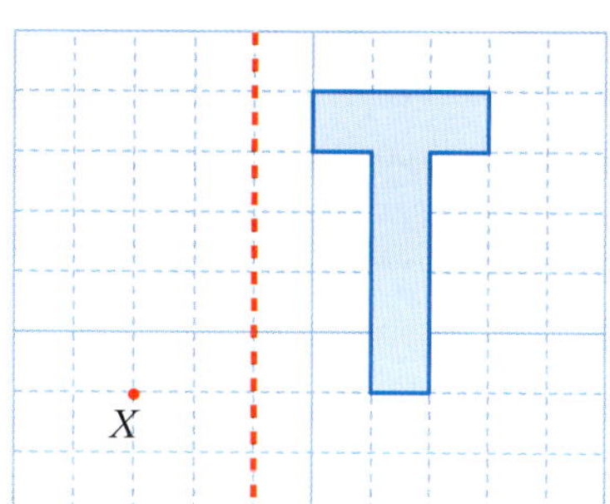

2 **a** Accurately describe the transformation that has been performed on the green L-shape $ABCDEF$.

b Write the coordinates of the new position of C and compare them to its original coordinates.

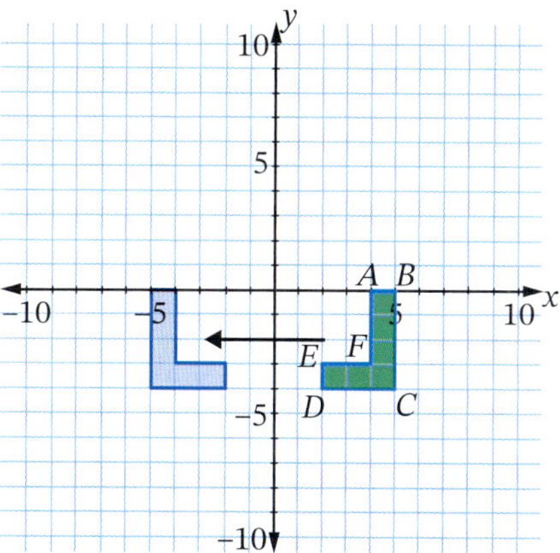

3 From the diagrams, match 4 pairs of congruent triangles and for each pair:

a name one pair of matching sides.

b name one pair of matching angles.

c write the congruence statement using the ≡ notation and match the order of the vertices.

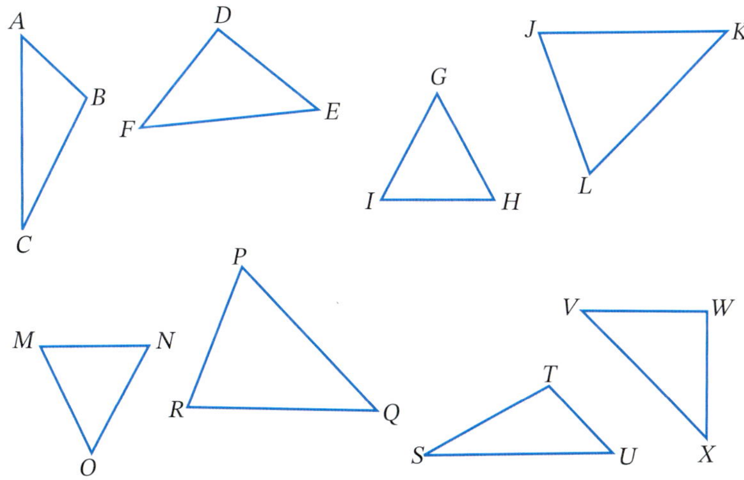

Foundation Standard Complex

4 This triangle has a base of length 6 cm that is bounded by angles of size 78° and 29°.

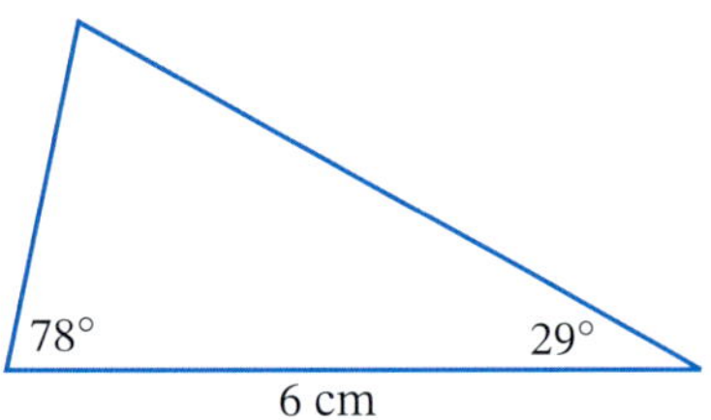

a Construct a triangle that is congruent to this triangle.

b Is it possible to construct another triangle with the same measurements that is not congruent?

8.04

5 Which congruence test (SSS, SAS, AAS or RHS) proves that each pair of triangles is congruent?

a

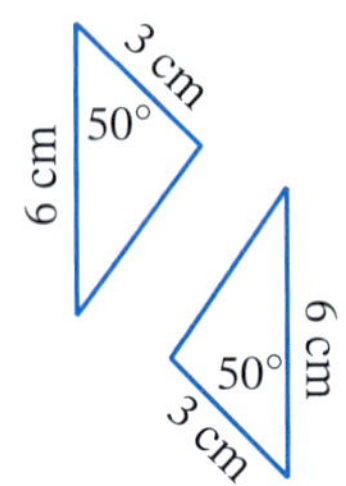

b

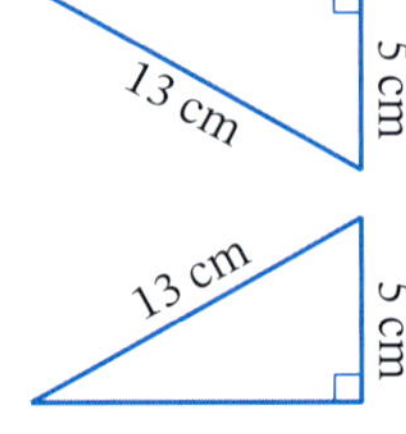

c

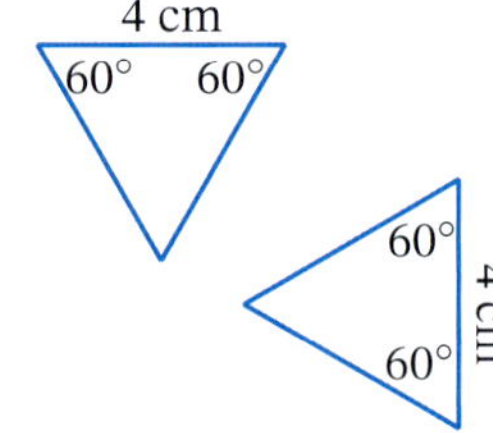

6 The diagonals of this rhombus divide the rhombus into 4 congruent triangles.

a Which test proves that the 4 triangles are congruent?

b Hence explain why the diagonals cross at right angles.

c Name the 3 angles that are equal to $\angle PQR$.

d True or false? The diagonals of a rhombus:

- **i** are equal.
- **ii** bisect each other.
- **iii** bisect the angles of the rhombus.

7 For this pair of similar figures:

a name a pair of corresponding sides.

b name a pair of corresponding angles.

c write a similarity statement of the form _____ ||| _____.

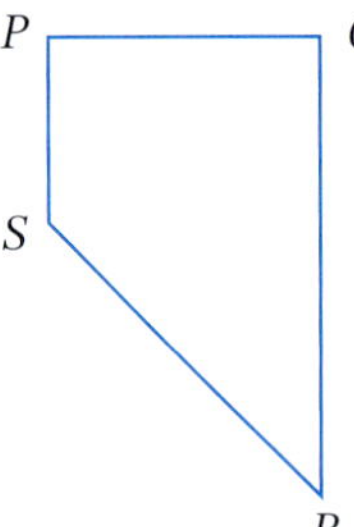

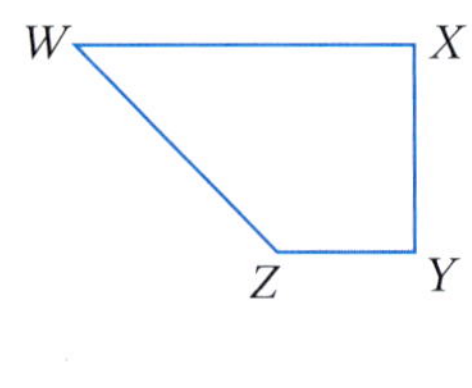

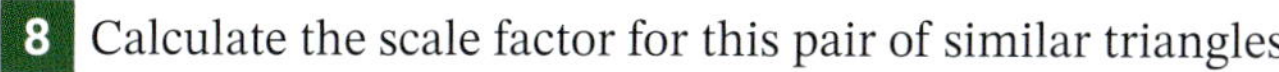

8 Calculate the scale factor for this pair of similar triangles. 8.07

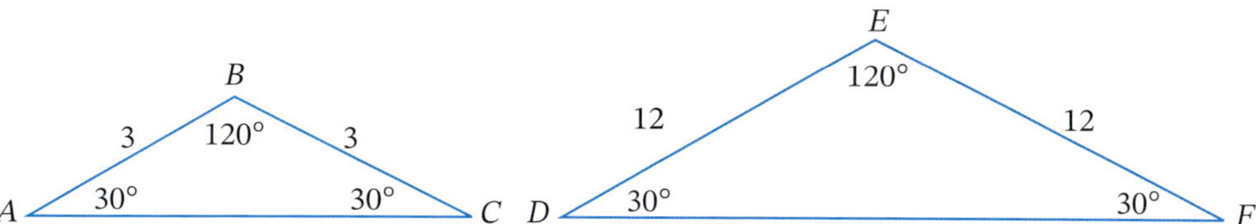

9 List the 4 similar triangles tests and explain how they are used.  8.07

10 Find the value of the variable in each pair of similar triangles.  8.07

a

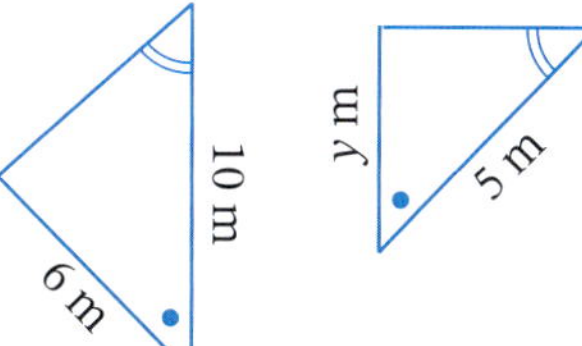

b

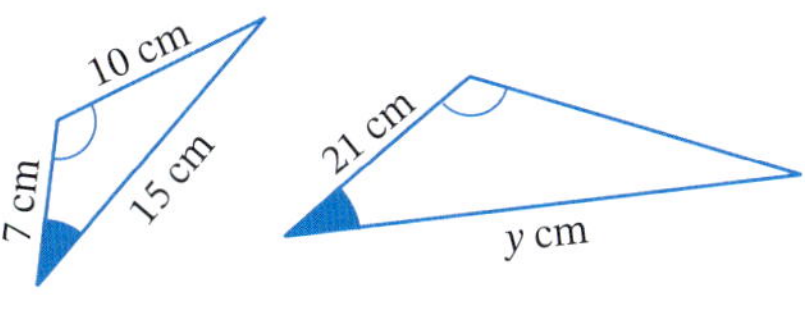

c

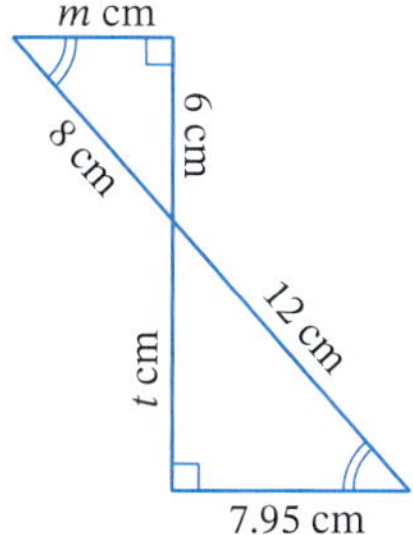

d

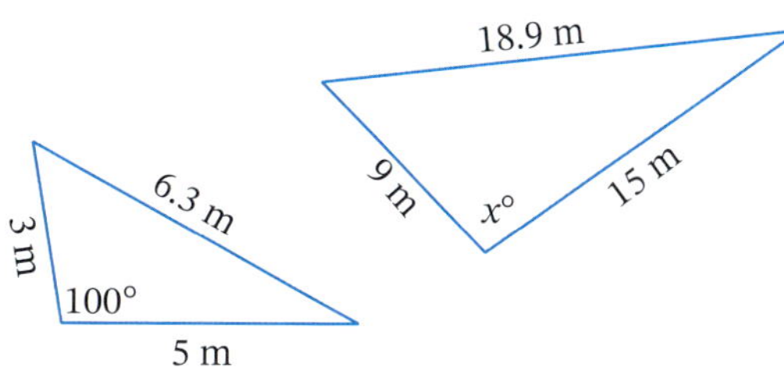

11 Determine whether the pair of triangles shown are similar, giving reasons.  8.07

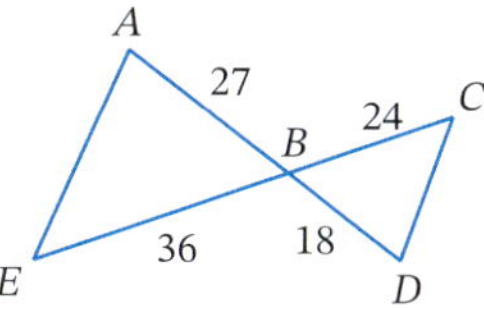

12 The top of a scoreboard is supported by a 32 m long wire which is bolted to the ground 20 m behind the base of the scoreboard. The wire runs through the top of a 10 m high pole, which is 12 m behind the scoreboard. How high is the scoreboard?  8.07

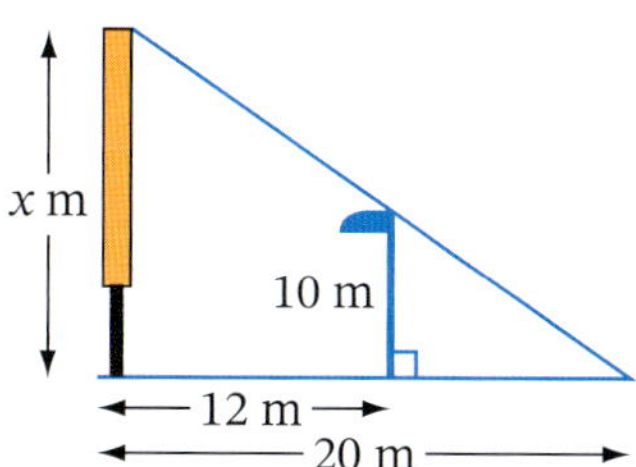

Foundation | Standard | Complex

9

PROBABILITY

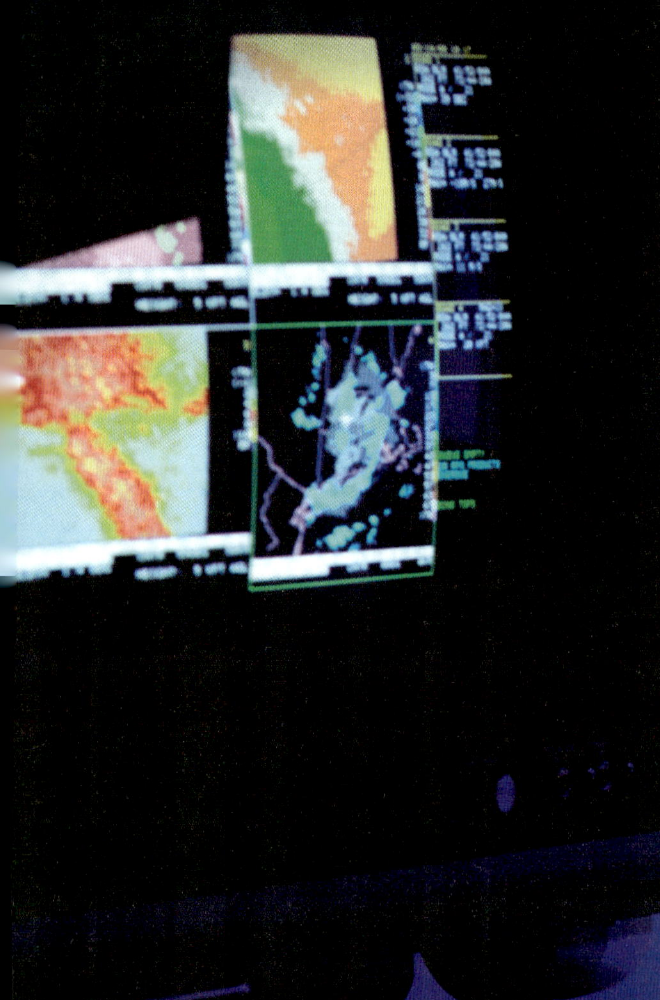

Probability

Probability plays an important part in our daily lives. The study of chance has applications in weather forecasting, predicting disease, political campaigns, insurance and gambling. Insurance companies use probability to assess the risk of an event occurring. Statistics show that the age and sex of a driver affects the chances of a car accident, and the more distance a person drives, the higher the risk of an accident. These statistics determine how much drivers pay to insure their cars.

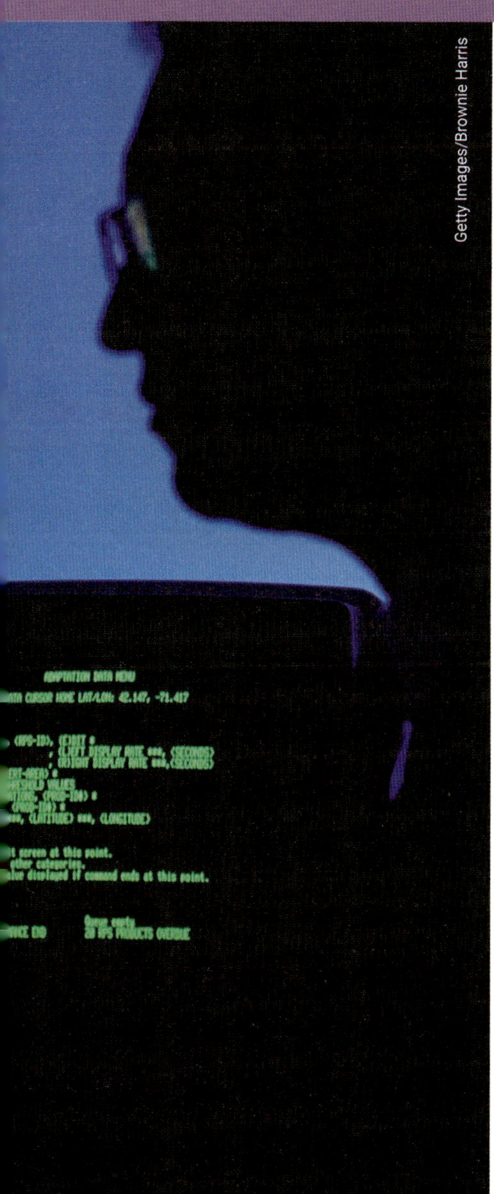

Getty Images/Brownie Harris

Chapter outline

		Proficiencies				
9.01	Probability	U	F		R	R
9.02	Complementary events	U	F	PS	R	C
9.03	Venn diagrams	U	F	PS	R	C
9.04	Two-way tables	U	F		R	C
9.05	Tree diagrams	U	F	PS	R	C
9.06	Probability problems	U	F	PS	R	C
9.07	Observed and expected frequencies	U	F		R	C
9.08	Probability simulations	U	F		R	C

U = Understanding
F = Fluency
PS = Problem solving
R = Reasoning
C = Communication

Wordbank

at least Referring to the smallest number, for example, 'at least 2' means 2, 3, 4, ..., that is, '2 or more'

complementary event The 'opposite' event (e.g. the complementary event to selecting an Ace from a deck of cards is ***not*** selecting an Ace)

expected frequency The expected number of times an event will occur over repeated trials

mutually exclusive events Events or categories that have no items in common

trial One go or run of a repeated probability experiment, for example, one roll of a die

two-way table A table that shows the number of items belonging to overlapping categories

Venn diagram A diagram that uses circles (usually overlapping) to group items into categories

Quiz
Wordbank 9

Videos (8):

SkillCheck Probability
9.01 Probability of an event
9.02 Complementary events
9.03 Venn diagrams 1 • Venn diagrams 2
9.04 Two-way tables
9.05 Tree diagrams 3
9.07 Experimental probability • Observed and expected frequencies

Twig videos (2):

SkillCheck Probability: Irrational fears
9.03 Venn diagrams: Global habitats

PhET interactive (1):

9.07, 9.08 Plinko probability

Quizzes (5):

- Wordbank 9
- SkillCheck 9
- Mental skills 9
- Language of Maths 9
- Test Yourself 9

Skillsheets (2):

SkillCheck The language of chance
9.01 Sample space

Worksheets (17):

SkillCheck Chance cards • Describing probability
9.01 Basic probability • Games of chance • Probability problems
9.02 Theoretical probabilities
9.04 Two-way tables • Two-way probability tables
9.05 Probability review • Tree diagrams
9.07 Relative frequencies • Dice probability • Spinning chance • A page of spinners • Experimental probabilities • Probability review • Coin toss experiment
Mind map: Probability

Puzzles (12):

9.01 Greedy Pig game • Spinner game • Basic probability • Theoretical probabilities
9.02 Complementary events
9.03 Venn diagrams group clues • Venn diagrams • Venn diagrams matching activity
9.04 Two-way probability tables
9.05 Combined events
9.06 Matching probabilities
9.07 Experimental probability • Duelling dice

Technology (1):

9.07 Simulating a spinner

Presentations (4):

9.03 Venn diagrams 1 • Venn diagrams 2
9.05 Combined events
9.07 Theoretical probability

Spreadsheets (3):

9.07 Coin-tossing simulator • Die-rolling simulator • Spinner simulator

Nelson MindTap

To access resources above, visit **cengage.com.au/nelsonmindtap**

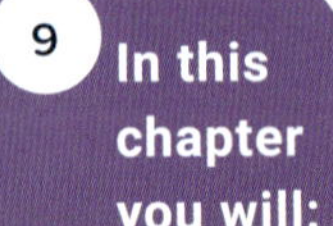

9 **In this chapter you will:**

- ✓ use sample space to calculate the probability of an event.
- ✓ solve problems involving complementary events.
- ✓ use Venn diagrams, two-way tables and tree diagrams to represent sample spaces and events to solve probability problems.
- ✓ identify two or more events that are mutually exclusive or overlapping.
- ✓ describe events using words such as 'and', exclusive 'or', inclusive 'or' and 'at least'.
- ✓ compare the observed frequency of an event with its expected frequency.
- ✓ conduct probability simulations to investigate observed frequencies of experiments.

SkillCheck

ANSWERS ON P. 643

Quiz SkillCheck 9

1 Convert each number into a decimal.

a $\frac{17}{20}$ **b** $\frac{3}{8}$ **c** 2% **d** 30%

2 List all the possible outcomes for each situation.

a tossing a coin
b rolling a die
c the gender of a baby
d the colour of traffic lights
e the result of a football game
f the result of a driving test

3 Convert each number into a percentage.

a 0.7
b 0.56
c $\frac{2}{3}$
d $\frac{6}{25}$

4 Evaluate each expression.

a $1-\frac{1}{3}$
b $1-\frac{2}{5}$
c $1-\frac{3}{8}$

5 Which term best describes the chance that the next baby born in Australia is a girl? Select the correct answer **A**, **B**, **C** or **D**.

A certain
B definite
C even chance
D probable

6 Convert each number into a simplified fraction.

a 0.28
b 0.02
c 64%
d 80%

7 Is the chance of each event greater than or less than $\frac{1}{2}$?

a You being home by 4 p.m. this afternoon.
b You winning Lotto one day.
c You listening to the radio today.
d You learning to drive next year.
e You going interstate this year.
f You sending a text message on your phone today.

Videos
Probability

Probability: Irrational fears

Skillsheet
The language of chance

Worksheets
Chance cards

Describing probability

Probability 9.01

In probability (chance) situations, the set of all possible outcomes is called the **sample space**. For example, if a coin is tossed, the sample space is {heads, tails}. If each outcome has an equal chance, then we say that each outcome is **equally likely**.

An **event** consists of one or more outcomes. For example, the event that a baby is born in a summer month consists of the outcomes {December, January, February}. An event has the symbol E, and we can calculate its probability as a fraction.

Skillsheet
Sample space

Worksheets
Basic probability

Games of chance

Probability problems

Video
Probability of an event

ⓘ Probability of an event

$P(E)$ means 'the probability of an event, E (occurring)'.

If all possible outcomes are **equally likely**, then:

$$P(E)=\frac{\text{number of favourable outcomes}}{\text{total number of outcomes}}$$

$$\text{or } P(E)=\frac{\text{number of outcomes matching } E}{\text{number of outcomes in the sample space}}$$

Puzzles Greedy Pig game

Spinner game

Basic probability

Theoretical probabilities

A **favourable outcome** is one of the outcomes in the event that you want, whose probability you are calculating.

The **probability** of an event can be written as a fraction, decimal or percentage. For example, the probability of tossing tails on a coin can be written as $\frac{1}{2}$, 0.5 or 50%.

Example 1

a Write the sample space for this spinner.

b Is each outcome equally likely?

c Find the probability that the spinner lands on red.

d Find the probability that the spinner lands on a 'traffic light' colour.

SOLUTION

a The sample space is {red, yellow, green, blue}.

b Each coloured region is equal in size $\left(\frac{1}{4}\text{ of the circle}\right)$, so each outcome is equally likely.

c $P(\text{red}) = \frac{1}{4}$ One chance in 4.

d $P(\text{traffic light colour}) = P(\text{red or yellow or green})$ 3 favourable outcomes out of 4.

$= \frac{3}{4}$

The language of probability

Probability term	In Example 1
An **experiment** (or **chance experiment**) is a situation involving chance that leads to results called outcomes.	Spinning a spinner.
A **trial** is one go or run of the experiment.	One spin of the spinner.
An **outcome** is the result of an experiment.	The arrow landing on red.
The **sample space** is the set of all possible outcomes.	{red, yellow, green, blue}
An **event** is one or more outcomes of an experiment.	The arrow landing on a 'traffic light' colour: red, yellow or green
In a **random** experiment, every possible outcome has the same chance of occurring.	All spins on this spinner are random because every colour has the same chance.

The range of probability

Because probability is a fraction, its value must range from 0 to 1, or as a percentage, from 0% to 100%.

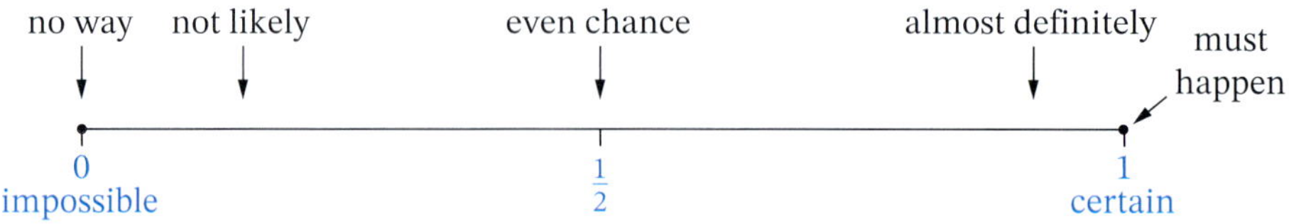

Example 2

A die is rolled. Find the probability that the number rolled is:

a divisible by 3.

b a factor of 6.

c less than 7 (answer as a percentage).

d **at least** 4 (answer as a decimal).

SOLUTION

There are 6 possible outcomes when a die is rolled: {1, 2, 3, 4, 5, 6}.

a $P(\text{divisible by } 3) = \frac{2}{6} = \frac{1}{3}$ 2 numbers {3 and 6}.

b $P(\text{a factor of } 6) = \frac{4}{6} = \frac{2}{3}$ 4 numbers {1, 2, 3 and 6}.

c $P(\text{less than } 7) = \frac{6}{6} = 100\%$ All outcomes are less than 7: a certain event.

d $P(\text{at least } 4) = \frac{3}{6} = \frac{1}{2} = 0.5$ 3 numbers {4, 5 and 6}.

'At least 4' means the smallest number is 4; that is, 4 or more.

EXERCISE 9.01 ANSWERS ON P. 643

ANSWERS ON P. 643

Probability

U F R C

1 For each spinner: EXAMPLE 1

C **i** write down the sample space.

ii for one spin, calculate the probability that the spinner lands on red.

a

b

c

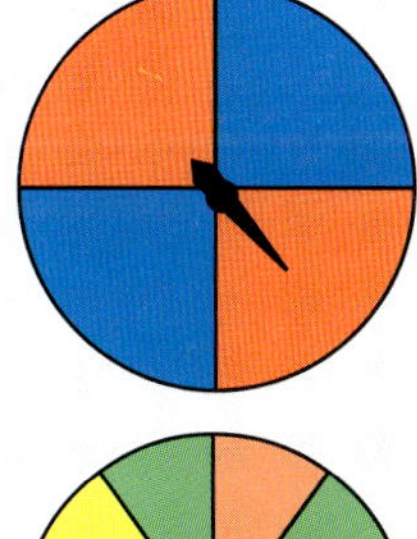

d

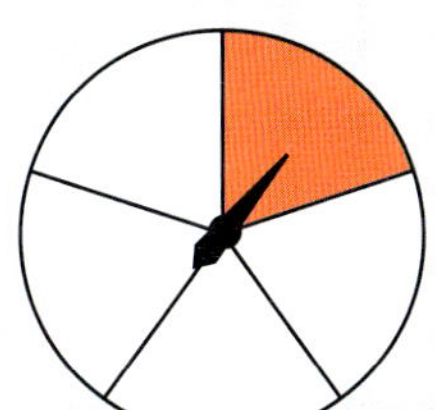

e

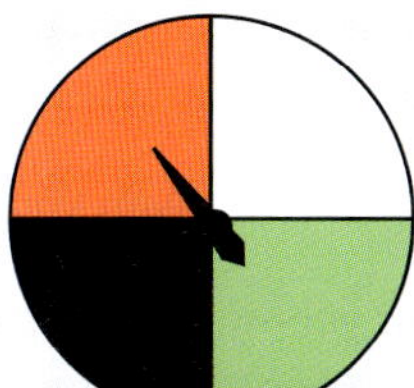

f

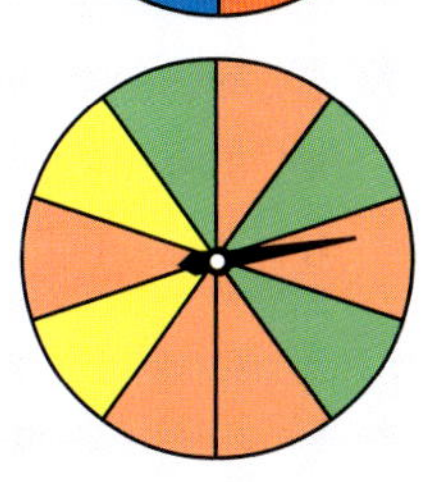

Foundation Standard Complex

2 For each experiment, count the number of possible outcomes and state whether each outcome is equally likely.

C

a Tossing a coin.

b The result of a rugby match when Australia plays New Zealand.

c The first letter of a person's name.

d The gender of a baby.

e The last digit of a car number plate.

f The result of a driving test.

3 List the outcomes in each event.

C

a Rolling an odd number on a die.

b Selecting a vowel from the letters of the alphabet.

c Having a house number greater than 4 but less than 10.

d Having a birthday in a month beginning with A.

e Living in a state of Australia.

f Being in a high school grade.

EXAMPLE 2

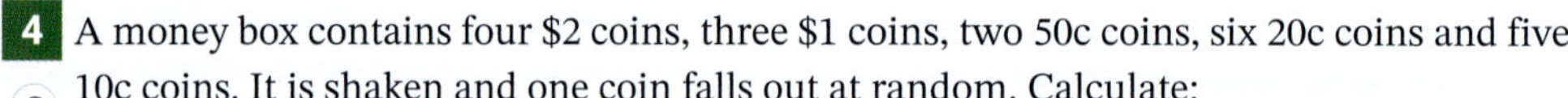

4 A money box contains four \$2 coins, three \$1 coins, two 50c coins, six 20c coins and five 10c coins. It is shaken and one coin falls out at random. Calculate:

C

a P(50c coin)

b P(\$1 coin) (answer as a decimal)

c P(10c or 20c coin), as a percentage

d P(not a 10c coin)

e P(gold coin), as a decimal

f P(a coin under \$1), as a percentage

5 A jar of lollies contains five red, three green, six yellow and two blue lollies. Taylor selects one lolly from the jar at random. Calculate each probability.

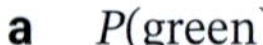

a P(green)

b P(yellow or blue)

c P(traffic light colour), as a decimal.

d P(not yellow), as a percentage.

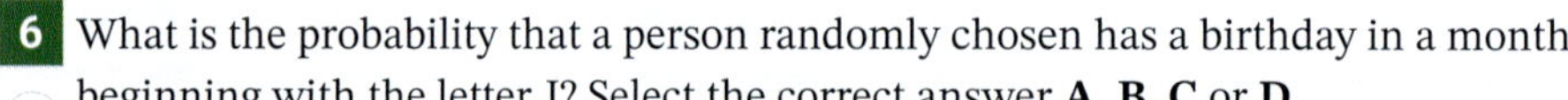

6 What is the probability that a person randomly chosen has a birthday in a month beginning with the letter J? Select the correct answer **A**, **B**, **C** or **D**.

C

A $\frac{1}{12}$ **B** $\frac{1}{6}$ **C** $\frac{1}{4}$ **D** $\frac{1}{3}$

7 Queensland is playing NSW in a rugby league match.

R **a** List the sample space for the outcomes of the match.

C **b** Are all outcomes equally likely? Explain your answer.

8 A packet of jelly beans has four yellow, three red, six green and three black jelly beans remaining. You tip the packet and one jelly bean rolls out at random. What is the probability that it is not a red jelly bean? Select the correct answer **A**, **B**, **C** or **D**.

A $\frac{4}{12}$ **B** $\frac{3}{16}$ **C** $\frac{3}{13}$ **D** $\frac{13}{16}$

Foundation Standard Complex

9780170465601

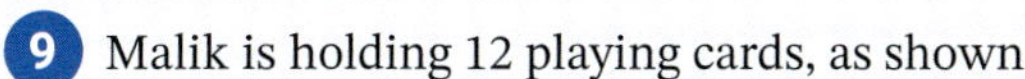

9 Malik is holding 12 playing cards, as shown. His friend Lara picks a card without looking and calls out the number on the card.

R C

a How many 5s are there in the 12 cards?

b List all the possible numbers that Lara tcould pick.

c Copy and complete this table.

Outcome	5	6	7	8
Probability as a fraction.	$\frac{1}{4}$			
Probability as a decimal.	0.25			
Probability as a percentage.	25%			

10 Elena has two \$5 notes, four \$10 notes and three \$20 notes in her wallet. If she selects one note at random, what is the probability that it is a \$20 note? Select the correct answer **A**, **B**, **C** or **D**.

A $\frac{3}{35}$ **B** $\frac{60}{110}$ **C** $\frac{1}{3}$ **D** $\frac{20}{35}$

11 Match each probability value to its correct description.

C

a $\frac{1}{2}$ **b** 0 **c** 90% **d** 1

e $\frac{3}{4}$ **f** 0.1 **g** 0.6 **h** 2%

A cannot happen **B** better than average chance **C** even chance

D good chance **E** very likely **F** almost impossible

G slim chance **H** must happen

12 The 52 cards in a standard deck of playing cards are shown below, divided evenly into four suits: hearts, diamonds, clubs and spades.

Hearts:

Diamonds:

Clubs:

Spades:

The cards are shuffled and one is taken out at random.

a How many outcomes are in the sample space?

b Is each outcome equally likely?

Foundation Standard Complex

c How many Aces are in the deck?

d How many hearts cards are there?

e How many red 7s are there?

f Find the probability of selecting:

i a red card ii a clubs card iii the King of spades

iv a Queen v a card with an even number vi a black picture card

13 The weather forecaster says that the probability of a rainy day in April is 30%.

a Is this a high probability or a low probability?

b About how many rainy days are expected in April?

14 Your maths teacher calls out a name randomly from your class roll. What is the probability that it is:

a your name? b a girl's name?

c someone aged 14? d someone with brown hair?

15 R An Esky contains eight cans of lemonade, five cans of orange drink and two cans of lime drink. How many cans of cola must be added so that the probability of randomly selecting:

a a lemonade is 50%? b an orange drink is 0.25?

c a lime drink is $\frac{1}{12}$? d a cola can is $\frac{2}{5}$?

INVESTIGATION

Complementary events

In everyday life, we use the word 'complementary' to describe things that go together and 'complete the picture' when they are together. For example, when dressing for an occasion:

- a shirt and a matching tie is complement to each other.
- a dress and a matching pair of shoes is complement to each other.

Remember also that 'complementary' angles add to 90°.

In probability, **complementary events** are events that together make up all the possible outcomes.

Event	Complementary event
Tossing a tail on a coin.	Tossing a head on a coin.
Rolling a 6 on a die.	Rolling any of the other numbers, from 1 to 5, on a die.
Raining	Not raining.
Being born on a Monday.	Being born on a day other than Monday.

1 Suppose that there is an equal chance of being born on any day of the week: Monday to Sunday.

a What is P(Tues), the probability of being born on a Tuesday?

b What is P(not Tues), the probability of being born on a day other than Tuesday?

c What do you notice about $P(\text{Tues}) + P(\text{not Tues})$?

2 A fruit bowl contains seven apples, four oranges and nine bananas. One piece of fruit is selected at random from the bowl.

a Find P(orange).

b Find P(not orange).

c What do you notice about $P(\text{orange}) + P(\text{not orange})$?

3 A baby is selected at random from the maternity section of a large hospital. There is an equal chance of the baby being a boy or a girl.

a Find P(boy).

b Find P(girl).

c What do you notice about $P(\text{boy}) + P(\text{girl})$?

4 Copy and complete the following sentence.

The probability of an event _____ the probability of its complementary event must always equal _____.

DID YOU KNOW?

Two-up

Two-up is an Australian coin-tossing game. The game was very popular with soldiers during World War I and it is only legal to play it on Anzac Day (25 April). Two-up involves the tossing of two coins and betting on 'heads' or 'tails' on both coins.

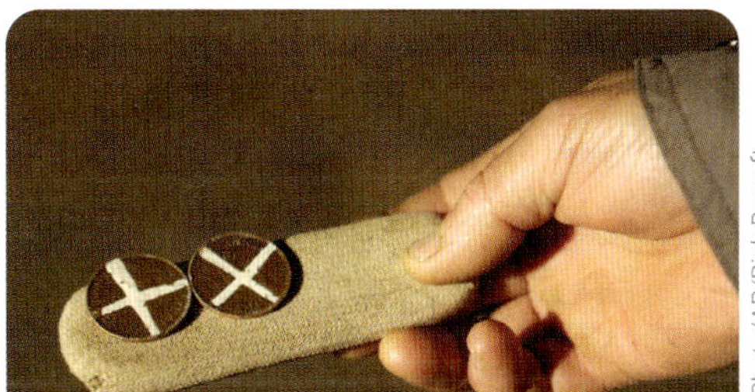

AAP Photo/AP/Rick Rycroft

Heads: both coins land heads up.

Tails: both coins land tails up.

Odds: one head and one tail.

There is no winner if 'odds' are thrown and the coins are tossed again.

Write down the sample space for this game.

What is the probability of throwing 'odds'?

9.02 Complementary events

Puzzle
Complementary events

Worksheet
Theoretical probabilities

In any situation, the probabilities of all possible outcomes must add to 1.

Complementary events are two events that together make up all the possible outcomes, such as a **head** and a **tail** when tossing a coin. The **complement** of an event E, are all those outcomes that are not E, or that are the 'opposite' of E.

Because an event and its **complement** cover all possible outcomes, the sum of their probabilities must be equal to 1.

ⓘ Complementary events

$P(E) + P(\text{not } E) = 1$

or $P(\text{not } E) = 1 - P(E)$

or $P(\text{complementary event}) = 1 - P(\text{event})$

or $P(\text{event not occurring}) = 1 - P(\text{event occurring})$

If the probability is written in percentage form, then $P(\text{not } E) = 100\% - P(E)$.

Example 3

Video
Complementary events

A pack of 20 cards contains 10 red, 6 yellow and 4 green cards. One card is drawn from the pack at random. Find the probability that this card is:

a yellow

b not yellow

c not green

SOLUTION

a $P(\text{yellow}) = \frac{6}{20} = \frac{3}{10}$

b $P(\text{not yellow}) = 1 - P(\text{yellow})$

$= 1 - \frac{3}{10}$

$= \frac{7}{10}$

The complement of 'yellow' is 'not yellow', which is 'red or green'.

Note that $P(\text{yellow}) + P(\text{not yellow}) = \frac{3}{10} + \frac{7}{10} = 1$, which covers all possible outcomes.

c $P(\text{green}) = \frac{4}{20} = \frac{1}{5}$

The complement of 'green' is 'not green', which is 'red or white'.

$P(\text{not green}) = 1 - P(\text{green})$

$= 1 - \frac{1}{5}$

$= \frac{4}{5}$

Note that $P(\text{green}) + P(\text{not green}) = \frac{1}{5} + \frac{4}{5} = 1$, which covers all possible outcomes.

Complementary events

U F PS R C

1 Write the complement of each event.

C **a** Tossing tails on a coin.

b Cloudy day tomorrow.

c Selecting a white chocolate from a box containing white and brown chocolates.

d Rolling a 6 on a die.

e Winning a game of hockey.

f Selecting a hearts card from a deck of cards.

g Being born in winter.

h Being under 15 years old.

2 A pile of 11 cards is placed on a desk. There are six red cards and the rest are blue. If one card is selected at random, what is the probability that it is not red? Select the correct answer **A**, **B**, **C** or **D**.

EXAMPLE 3

A $\frac{5}{11}$ **B** $\frac{6}{11}$ **C** $\frac{6}{5}$ **D** $\frac{5}{6}$

3 A die is rolled. What is the probability that the result is:

a a 3? **b** not a 3? **c** not even?

d not prime? **e** at least 2? **f** at most 4?

4 There are 6 books on a shelf: 3 Mathematics, 2 History and 1 Sport. If one book is chosen at random, what is the probability of selecting:

a a History book? **b** a book that is not History?

c a book that is not Sport? **d** a book that is not Science?

5 A car park has 450 cars, 100 motor bikes and 10 buses. One of them is selected at random.

a What is the probability that it is a car?

b What is the probability that it is not a car? Select the correct answer **A**, **B**, **C** or **D**.

A $\frac{45}{56}$ **B** $\frac{110}{450}$ **C** $\frac{10}{560}$ **D** $\frac{11}{56}$

6 What is the probability that your maths teacher was born in a month:

a beginning with the letter M?

b does not begin with the letter M?

7 What is the decimal probability that a mobile phone number selected at random does not end in 0 or 1?

8 Tahnee buys five tickets in a raffle in which 1000 tickets are sold and there is only one prize. What is the probability of Tahnee not winning the prize?

☐ Foundation ○ Standard ⬡ Complex

9 A jar contains 40 red, 25 blue, 50 black and 15 white jelly beans. One jelly bean is selected at random. What is the probability that it is:

a white? **b** not white? **c** not yellow?

d not blue or black? **e** not red? **f** not black or white?

10 Write the probability of the event that is complementary to each event.

C **a** The probability of choosing a Jack from a pack of cards is $\frac{1}{13}$.

b The chance of shooting a basketball hoop is 61%.

c The probability of winning a prize is 0.15.

11 In a bag of toy cars there are only three colours: red, blue and white. If you take out a car at random, the chance of it being red is 0.4 and the chance of it being white is 0.25.

R **a** What is the chance of selecting 'red or white'?

b What is the chance of the car you select being blue?

c If the bag holds 40 cars, how many of each colour would you expect to find?

d What is the chance of the car you select being pink?

12 Four students, Sue, Liam, Emily and Matt, write their names on cards and place the cards in a bag. A card is chosen, without looking, to select the class captain.

a Find the probability that Emily was not chosen.

b Find the probability that the captain is a boy.

c What is the chance that the captain is not a boy?

d What is the chance that the captain is the teacher?

13 The probability that it will rain this weekend is 85%. What is the probability that it won't rain?

14 Which of the following is the complementary event to 'winning a race'? Select the correct answer **A**, **B**, **C** or **D**.

R C **A** coming last **B** coming second or third

C not winning the race **D** coming second

15 A game involves throwing a coin into a grid of squares. For a player to win, the coin must land in a yellow square. If the coin lands outside the grid or between squares, the throw is not counted and the player has another try.

For each throw, what is the probability of:

a not winning?

b winning?

16 The letters of the word PROBABILITY are written on separate cards and one is selected randomly. What is the probability that a letter drawn out is:

a not P? **b** not a vowel? **c** not I? **d** neither A nor B?

17 In a family there are Mum and Dad, two daughters and a son. Each day, they take it in turns to wash the dishes after dinner. If you visit the family, what is the chance that the person washing up is:

a male? **b** not a parent? **c** not Mum? **d** not male?

18 PS R A random four-digit number is to be formed from the digits 1, 3, 5 and 8. What is the probability that the number will:

a start with the digit 5? **b** be an odd number? **c** be greater than 5000?

9.03

Venn diagrams 9.03

A **Venn diagram** is a diagram that uses circles (usually overlapping) to group items into categories. A rectangle represents the whole group while the circles represent categories. Items common to two or more categories are placed in the intersection (overlapping region) of the circles. The Venn diagram was invented in 1880 by English mathematician and priest, John Venn (1834–1923).

Videos
Venn diagrams 1
Venn diagrams: Global habitats

Puzzles
Venn diagrams group clues
Venn diagrams
Venn diagrams matching activity

Example 4

A group of people were surveyed on the type of vehicle they drove, and the results are shown on the Venn diagram.

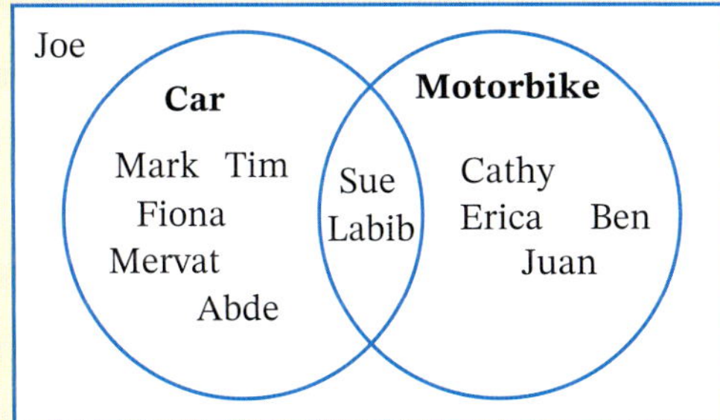

a How many people were surveyed?

b How many people drive a car?

c How many people drive a car or ride a motorbike?

d Who drives a car and rides a motorbike?

e Who only rides a motorbike?

f Why is Joe outside of the circles?

Foundation Standard Complex

SOLUTION

a Twelve people were surveyed.

b Seven people drive a car. The people within the car circle.

c Eleven people drive a car or ride a motorbike. All the people within the circles.

d Sue and Labib can drive a car and ride a bike. The people in the intersection of the circles.

e Cathy, Erica, Ben and Juan can only ride a bike.

f Joe does not drive a car or ride a bike.

Mutually exclusive vs overlapping events

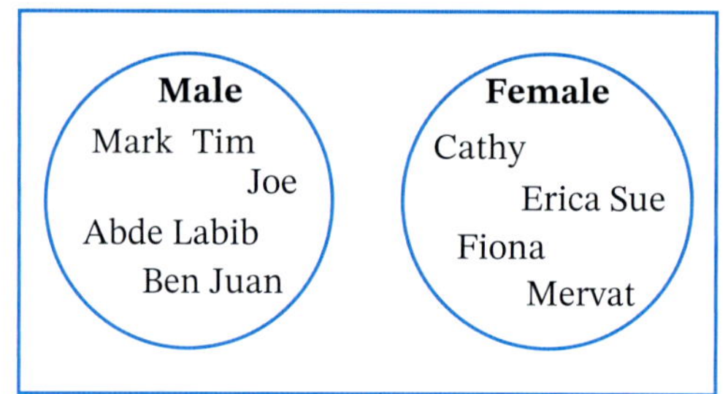

Sometimes, two categories must be represented on a Venn diagram as two separate circles because it is not possible for them to overlap. Here is an example:

In this case, it is not possible to be male *and* female. This is an example of a **mutually exclusive event**. Mutual means 'shared feature' and exclusive means 'belonging to one group only' (such as an exclusive party or an exclusive news story).

However, most Venn diagrams (as in Examples **4** and **5**) describe **overlapping events** or **non-mutually exclusive events**.

The meaning of 'A and B'

For two categories or events A and B, the phrase **'A and B'** means to have both of them occurring together. For example, 'to drive a car' **and** 'to ride a motorbike' in Example **4** means to do both things and is the overlapping section in the centre of the Venn diagram.

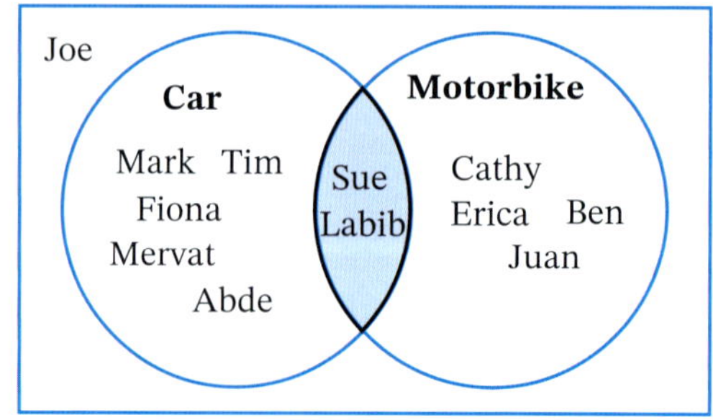

The meaning of 'A or B'

If A and B are **overlapping**, the phrase **'A or B'** means to have A or B or both. For example, 'to drive a car' **or** 'to ride a motorbike' in Example **4** means to drive a car only (but not a motorbike), or to ride a motorbike only (but not a car), or to do both. In this case, 'A or B' actually **includes** 'A and B' so this is an example of an **inclusive** 'or'.

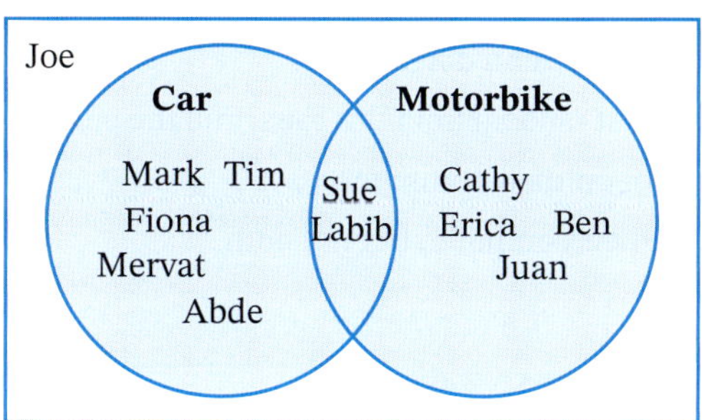

If A and B are **mutually exclusive**, the phrase **'A or B'** means to have A only or B only (but not both). For example, 'male' **or** 'female' means to be male or female, but not both. In this case, 'A or B' **excludes** 'A and B' so this is an example of an **exclusive** 'or'.

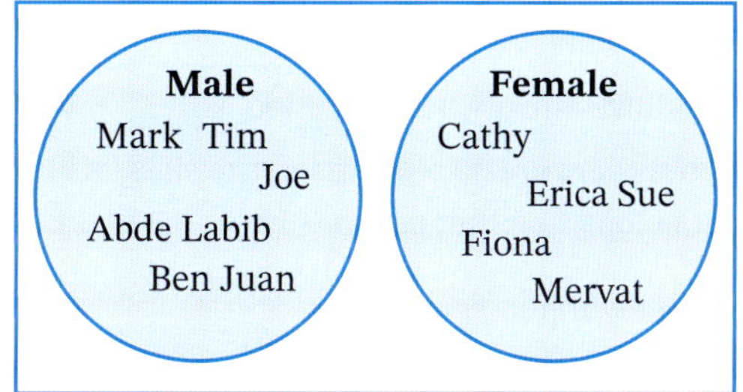

The meaning of 'neither A nor B'

The phrase **'neither A nor B'** means that none of the categories. In Example **4**, 'neither to drive a car nor ride a motorbike' means to not drive a car and to not ride a motorcycle, and is the region outside both circles of the Venn diagram.

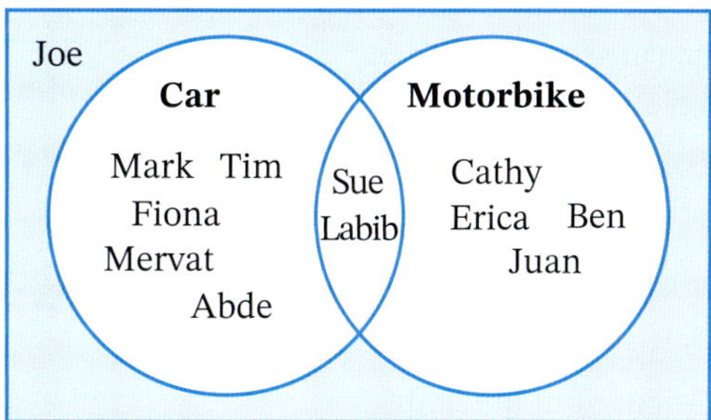

Example 5

Video
Venn diagrams 2

Presentations
Venn diagrams 1
Venn diagrams 2

Thirty students in a class were surveyed on how they relaxed after school.

- Eighteen play on their computer.
- Fifteen play sport.
- Six play sports and play on their computer.

a Show this information on a Venn diagram.

b If one student is selected at random from this class, find the probability that the student:

i plays sport, but not on their computer.

ii plays sport or on their computer, but not both.

SOLUTION

a When drawing Venn diagrams always begin with the intersection (the overlapping section).

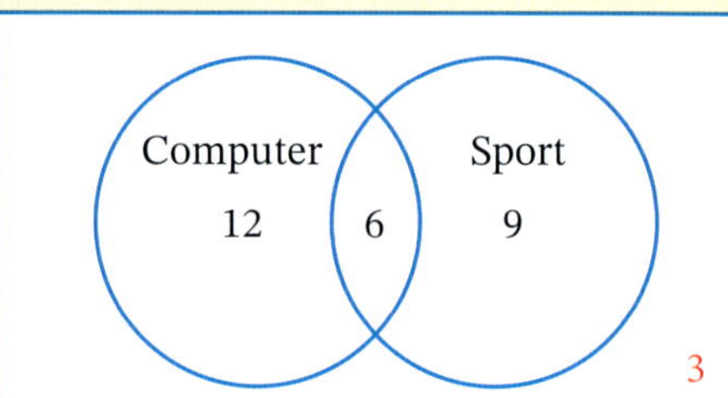

- 6 students belong to both groups.
- 18 play on the computer but 6 have already been counted, so 18 – 6 = 12 play on the computer only.
- 15 play sport but 6 have already been counted, so 15 – 6 = 9 play sport only.
- 12 + 6 + 9 = 27, but there are 30 students in the class, which means that 3 students do not belong to either group.

b **i** $P(\text{plays sport but not computer}) = \frac{9}{30} = \frac{3}{10}$

ii $P(\text{plays sport or computer but not both}) = \frac{12+9}{30} = \frac{21}{30} = \frac{7}{10}$

EXERCISE 9.03 ANSWERS ON P. 644

Venn diagrams

1 A group of students were asked whether they liked Maths or Science at school. The Venn diagram shows the results of this survey. How many students:

Maths
Science
21
9
12
8

a were surveyed?

b liked Maths but not Science?

c liked Maths or Science?

d liked Maths or Science but not both?

e did not like either subject?

f liked Maths and Science? Select the correct answer **A**, **B**, **C** or **D**.

A 9 **B** 21 **C** 33 **D** 42

2 This Venn diagram shows the sports played by a number of female students.

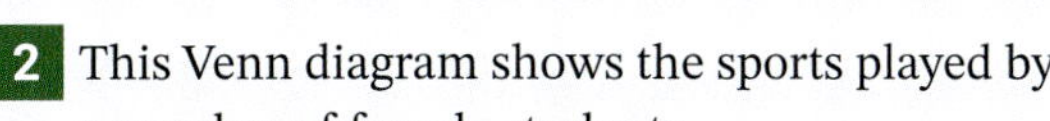

R C

6
Netball
21
Hockey
19

a Are netball and hockey mutually exclusive or not?

b How many students play neither netball nor hockey?

c How many students play netball or hockey?

d Find the total number of students surveyed.

e How many students play netball and hockey?

3 This Venn diagram compares the categories female and left-handed.

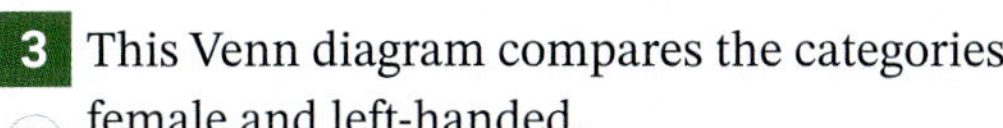

R C

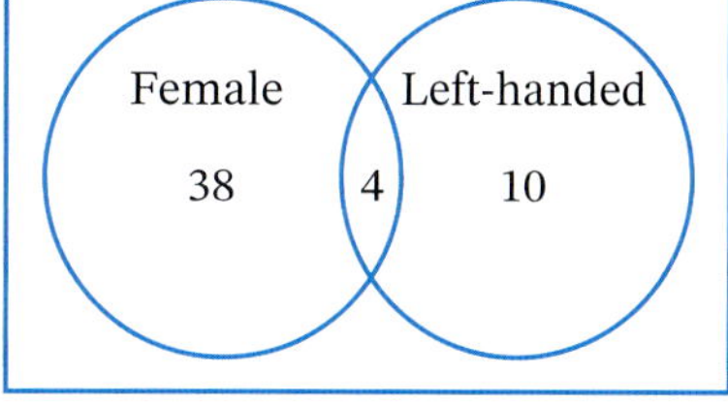

a Is female and left-handed mutually exclusive or not?

b Find the total number of people represented in the diagram.

c How many left-handed females are there?

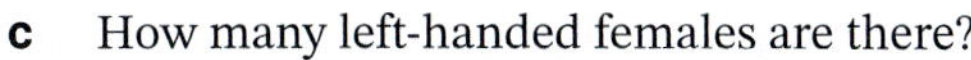

d What is the decimal probability (correct to three decimal places) that a person randomly chosen from this group is:

i female?

ii female but not left-handed?

iii female or left-handed?

iv a left-handed male?

e Describe the type of person that would be represented outside the two circles on the Venn diagram.

4 A survey was carried out by an ice cream shop to decide whether chocolate or strawberry was the more popular flavour.

R C

Strawberry
120
57
Chocolate
248
15

a How many people were surveyed?

b How many people liked strawberry or chocolate, but not both?

c How many people liked neither strawberry nor chocolate?

d If a person is randomly chosen from the survey, what is the probability that the person likes both chocolate and strawberry? Select the correct answer **A**, **B**, **C** or **D**.

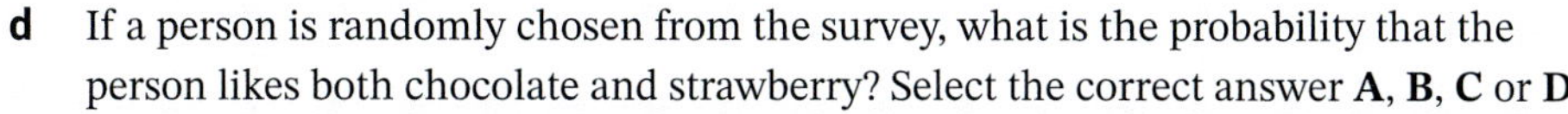

A $\frac{57}{440}$ **B** $\frac{368}{440}$ **C** $\frac{425}{440}$ **D** $\frac{57}{425}$

Foundation Standard Complex

5 The Tourism Council surveyed 130 people to find whether they prefer New South Wales or Queensland as a holiday destination.

R C

NSW 44 Queensland 66 NSW and Queensland 20

a Construct a Venn diagram to represent the results.

b How many people like NSW or Queensland?

c How many people like NSW or Queensland, but not both?

d What is the probability that a person likes NSW exclusively?

e What is the probability that a person does not like NSW or Queensland?

6 **a** Draw a Venn diagram representing the eye colours of this group of students.

R C

Blue eyes	32
Brown eyes	38
Neither blue nor brown eyes	10

b Are these groups mutually exclusive or not? Why?

c How many students are in the group?

d How many students have blue or brown eyes?

e How many students have blue and brown eyes?

f What is the percentage probability that a student chosen at random from this group does not have brown eyes?

7 A retailer surveyed the first 30 customers to see what they bought.

PS R C

20 bought milk 13 bought bread 1 bought neither milk nor bread

a Show this information on a Venn diagram.

b How many customers bought milk and bread?

c How many customers bought milk only?

d What is the probability that a customer randomly chosen:

i bought bread only? **ii** did not buy milk?

iii bought milk or bread? **iii** bought milk or bread, but not both?

8 There is a class of 30 students, all of whom study Information Technology or Chemistry. Seven of them study both subjects and 20 study Information Technology.

PS R C

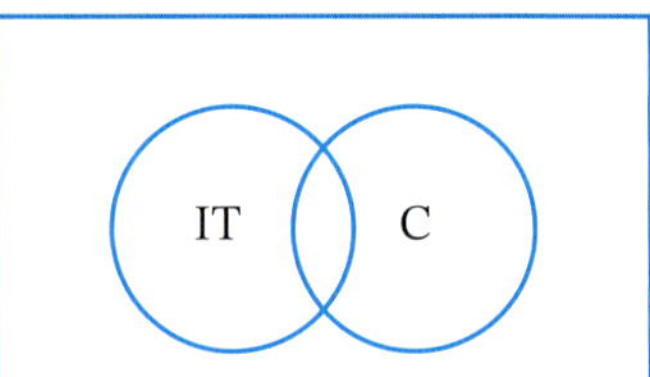

a Copy and complete the Venn diagram to find how many students study Chemistry.

b If one student is chosen at random, what is the probability that the student studies Chemistry?

Foundation Standard Complex

Two-way tables

9.04

A **two-way table** is another way of grouping items into overlapping categories, especially when there are many overlaps that cannot be represented by Venn diagrams easily.

Video
Two-way tables

Worksheets
Two-way tables

Two-way probability tables

Puzzle
Two-way probability tables

Example 6

Fifty students were surveyed on whether they preferred dogs or cats more as pets.
The results were sorted into this two-way table.

	Preferred pet	
	Dog	**Cat**
Boys	12	4
Girls	11	15

a How many students do not like dogs or cats?

b How many students like cats?

c How many boys like cats?

d How many students:

  i like cats or are boys?

  ii like cats or are boys, but not both?

e What is the probability that a student selected at random from the survey likes dogs?

SOLUTION

a $12 + 4 + 11 + 15 = 42$ — Adding the values in the table.

Students who do not like dogs or cats $= 50 - 42$ — Fifty students in survey.

$= 8$

b Students who like cats $= 4 + 15$

$= 19$

c Boys who like cats $= 4$

d **i** Students who like cats or who are boys $= 12 + 4 + 15$

$= 31$

ii Students who like cats or who are boys but not both $= 12 + 15$

$= 27$

e Students who like dogs $= 12 + 11$

$= 23$

Probability of selecting a student who likes dogs $= \frac{23}{50}$

The relationships between events A and B shown on a Venn diagram can also be represented on a two-way table:

		Category 1		
		A	Not A	
Category 2	B	'Both A and B'	'B not A' OR 'Only B'	Total of 'B'
	Not B	'A not B' OR 'Only A'	'Neither A nor B'	Total of 'Not B'
		Total of 'A'	Total of 'Not A'	Overall Total

Example 7

A group of 50 Year 9 students were grouped in a Venn diagram according to whether they took Chinese or Art as an elective subject.

Represent this information in a two-way table.

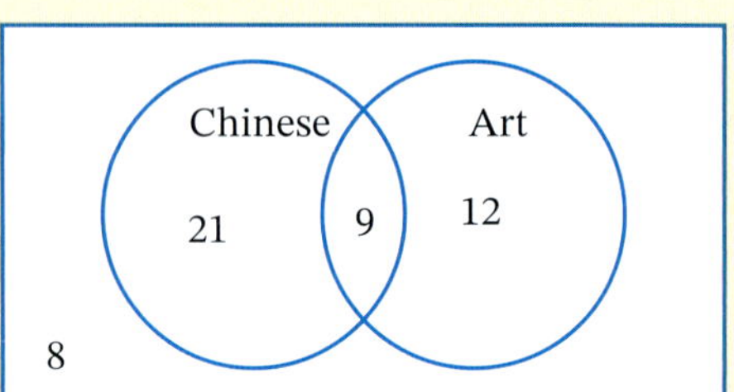

SOLUTION

From the Venn diagram:

- 9 students take Chinese and Art.
- 21 students take Chinese, but not Art.
- 12 students take Art, but not Chinese.
- 8 students do not take Chinese or Art.

	Chinese	Not Chinese	
Art	9	12	21
Not Art	21	8	29
	30	20	50

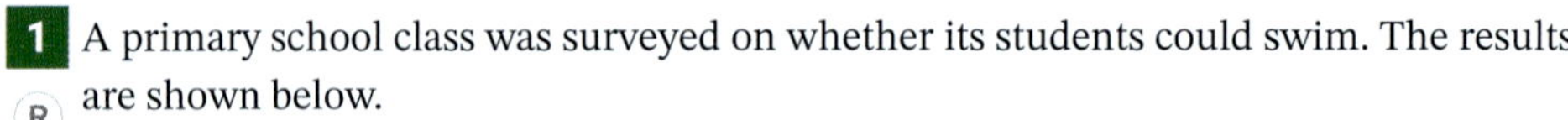
EXERCISE 9.04 ANSWERS ON P. 644

Two-way tables

U F R C

EXAMPLE 6

1 A primary school class was surveyed on whether its students could swim. The results are shown below.

R C

	Can swim	Cannot swim
Boys	13	2
Girls	9	3

a How many students are in the class?

b How many students are boys or cannot swim?

c How many students are boys and cannot swim?

d What is the probability that a student randomly selected from this class is a girl?

e What is the probability that a student selected at random is:

i a non-swimmer? **ii** a girl who can swim?

Foundation Standard Complex

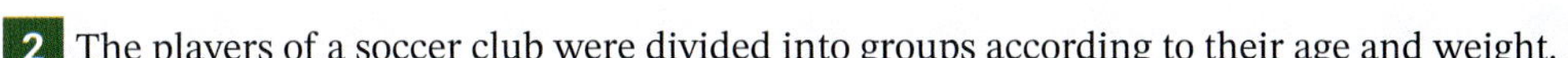

2 The players of a soccer club were divided into groups according to their age and weight.

R C

	Heavy	Light
Junior	64	96
Senior	144	32

a How many players does the club have?

b How many players are juniors or light?

c How many players are juniors or light, but not both?

d What is the probability that a player selected at random is:

i a senior? **ii** a junior and heavy? **iii** a senior or light?

3 This incomplete table describes the audience watching a movie at a cinema.

R C

	Under 18	Over 18	Total
Female		142	198
Male	45		
Total			344

a Copy and complete the table.

b How many males were in the audience?

c How many under 18 females were there?

d If a person is selected at random from the audience, what is the probability that the person:

i is male and over 18? **ii** is male or over 18?

iii is male or over 18, but not both? **iv** is over 18?

4 A group of children were asked whether they liked carrots.

R C

	Likes carrots	Dislikes carrots	Total
Boys	75		200
Girls		40	
Total	145		

a Copy and complete the table.

b How many children dislike carrots?

c What is the probability that a child randomly selected is:

i a girl? **ii** a boy or dislikes carrots?

iii a boy and dislikes carrots?

d What is the chance that a child randomly selected is a girl and likes carrots? Select the correct answer **A**, **B**, **C** or **D**.

A $\frac{29}{62}$ **B** $\frac{11}{31}$ **C** 1 **D** $\frac{7}{31}$

Foundation Standard Complex

5 The Venn diagram below shows the number of students who participate in swimming or bowling regularly. Copy and complete the two-way table for this data.

R C

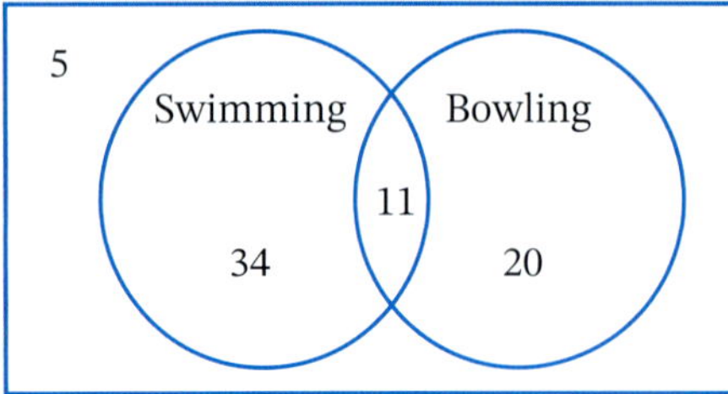

	Swimming	Not swimming	Total
Bowling			
Not bowling			
Total			

6 The Venn diagram compares two features of the cars on display at Rusty Motors.

R C

a Copy and complete the two-way table.

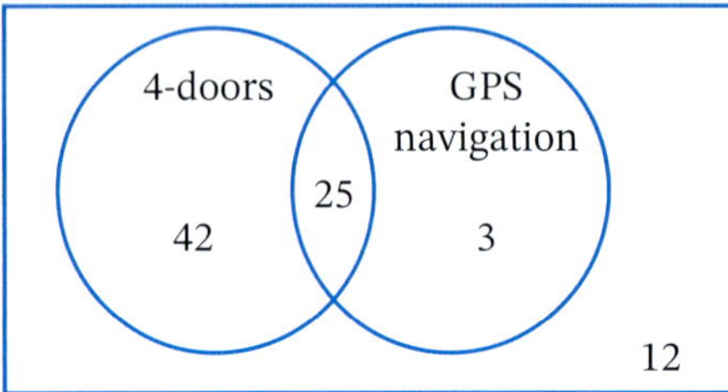

	GPS navigation	No GPS navigation	Total
4-doors			
Not 4-doors			
Total			

b A car is chosen at random from the car yard. What is the probability that it has:

i GPS navigation?
ii four doors?
iii four doors and GPS navigation?
iv no GPS navigation?

7 A group of university students were tested to see if they needed glasses. The table shows the results.

R C

	Needs glasses	Does not need glasses	Total
Male	22		82
Female		85	98
Total		145	

a Copy and complete the table.

b If a student is chosen randomly from this group, what is the probability that the student:

i is female?
ii needs glasses?
iii is female and needs glasses?
iv is male or does not need glasses?
v is female or needs glasses?
vi is male or does not need glasses, but not both?

9.04

TECHNOLOGY

Rolling a die

In this activity, you need a scientific calculator or spreadsheet to simulate the rolling of a die.

1 Copy this table.

Outcome	Frequency
1	
2	
3	
4	
5	
6	

2 Random numbers from 1 to 6 can be generated on a calculator or spreadsheet using the RanInt, RANDOM or =RAND() functions.

Casio scientific calculator	Sharp scientific calculator
Enter the formula RanInt#(1,6) as shown: ALPHA . 1 SHIFT) 6) =	2nd F 7 (RANDOM), select 1 R-DICE, =

Press = more times to generate more random integers from 1 to 6.

On a spreadsheet, type **=INT(RAND()*6+1)** into a cell and press the Enter key, then the F9 key repeatedly for more random numbers.

3 Repeat the simulation 20 times and record the results in your table.

4 Are certain numbers more likely to be rolled than others? For example, 2 more likely to be rolled than 5? Do your results reflect this?

5 Compare your results with the simulated results of other members in your class. Are they similar or different? Are they what you and your classmates expected? Discuss.

MENTAL SKILLS 9 ANSWERS ON P. 644 Maths without calculators

Quiz Mental skills 9

Multiplying or dividing by a multiple of 10.

1 Study each example.

- **a** $4 \times 700 = 4 \times 7 \times 100 = 28 \times 100 = 2800$
- **b** $5 \times 60 = 5 \times 6 \times 10 = 30 \times 10 = 300$
- **c** $12 \times 40 = 12 \times 4 \times 10 = 48 \times 10 = 480$
- **d** $3.2 \times 30 = 3.2 \times 3 \times 10 = 9.6 \times 10 = 96$ (by estimation, $3 \times 30 = 90 \approx 96$)
- **e** $4.5 \times 50 = 4.5 \times 5 \times 10 = 22.5 \times 10 = 225$ (by estimation, $5 \times 50 = 250 \approx 225$)
- **f** $9.4 \times 200 = 9.4 \times 2 \times 100 = 18.8 \times 100 = 1880$ (by estimation, $9 \times 200 = 1800 \approx 1880$)

2 Now evaluate each product.

a	8×2000	**b**	3×70	**c**	11×900
d	2×300	**e**	4×4000	**f**	5×80
g	7×70	**h**	1.3×40	**i**	2.5×600
j	6.8×200	**k**	3.9×50	**l**	4.4×4000

3 Study each example.

a $8000 \div 400 = 80\cancel{0}\cancel{0} \div 1\cancel{0}\cancel{0} \div 4 = 80 \div 4 = 20$

b $200 \div 50 = 20\cancel{0} \div 1\cancel{0} \div 5 = 20 \div 5 = 4$

c $6000 \div 20 = 600\cancel{0} \div 1\cancel{0} \div 2 = 600 \div 2 = 300$

d $282 \div 30 = 282 \div 10 \div 3 = 28.2 \div 3 = 9.4$

e $3520 \div 40 = 352\cancel{0} \div 1\cancel{0} \div 4 = 352 \div 4 = 88$

f $8940 \div 2\cancel{0}\cancel{0} = 8940 \div 1\cancel{0}\cancel{0} \div 2 = 89.4 \div 2 = 44.7$

4 Now evaluate each quotient.

a	$560 \div 70$	**b**	$2500 \div 50$	**c**	$3200 \div 400$
d	$440 \div 20$	**e**	$160 \div 40$	**f**	$1500 \div 30$
g	$450 \div 50$	**h**	$744 \div 80$	**i**	$2550 \div 300$
j	$846 \div 200$	**k**	$576 \div 60$	**l**	$2040 \div 50$

9.05 Tree diagrams

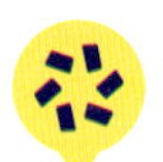

Worksheet Probability review

Tree diagrams

Presentation Combined events

Video Tree diagrams 3

A **two-step experiment** or **two-stage experiment** is a chance experiment that has 2 parts or stages, for example:

- rolling 2 dice
- tossing a coin and die together
- drawing 2 prizes in a raffle
- observing the weather each day over a weekend.

The sample space for a two-step experiment can be displayed using lists, tables or tree diagrams.

Example 8

A coin is tossed twice.

a Show all the outcomes in the sample space using:

i a list **ii** a table **iii** a tree diagram

b What is the probability of tossing:

i exactly one tail? **ii** 2 tails? **iii** at least one tail?

SOLUTION

a Let H = heads and T = tails.

i Using a list: — Finding the 4 outcomes by listing.

Possible outcomes are {HH, HT, TH, TT}.

ii Using a table:

		1st coin	
		H	T
2nd coin	H	HH	TH
	T	HT	TT

Using a table ensures that all outcomes are counted.

iii Using a tree diagram:

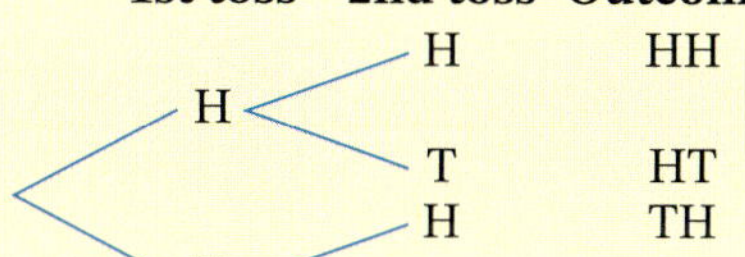

A tree diagram shows all the different alternative pathways for outcomes at each stage.

For the 1st toss, you could get heads or tails. If you get heads, then for the 2nd toss, you could get heads or tails. If you get tails for the first toss, then for the 2nd toss, you could get heads or tails.

b **i** $P(\text{exactly one tail}) = P(\text{HT or TH}) = \frac{2}{4} = \frac{1}{2}$ — 2 chances out of 4.

ii $P(\text{2 tails}) = P(TT) = \frac{1}{4}$ — 1 chance out of 4.

iii $P(\text{at least one tail}) = P(\text{1 or 2 tails})$ — 'At least one' means '1 or more'.

$= P(\text{HT or TH or HH})$ — 3 chances out of 4.

$= \frac{3}{4}$

A **tree diagram** lists all the possible outcomes of each stage. Branches stretch out to show the possible pathways of outcomes at each step or stage. The outcomes column at the end of the diagram lists the sample space.

When there are a large number of outcomes in each step of a two-step experiment, it is best to use a table or tree diagram to list all the possible outcomes. The table and tree diagram are systematic ways of ensuring all possible outcomes have been included in the sample space.

Foundation Standard Complex

Video Tables and tree diagrams

Example 9

There are 8 people in a dance class: 3 men (Grant, Rick, Stefan) and 5 women (Bethany, Tania, Elise, Chloe, Felicity).

a List all possible male-female dancing couples, using:

i a table **ii** a tree diagram

b How many dancing couples are possible?

c If one of these dancing couples is selected at random, what is the probability that:

i it is Rick and either Tania or Chloe?

ii it does not include Grant or Felicity?

iii it includes Stefan but not Bethany?

SOLUTION

a This experiment has 2 steps: the male dancer and the female dancer.

For the male dancer, there are 3 possible men.

For the female dancer, there are 5 possible women.

Let: G = Grant R = Rick S = Stefan B = Bethany

T = Tania E = Elise C = Chloe F = Felicity

i Using a table:

		Men		
		G	R	S
Women	B	GB	RB	SB
	T	GT	RT	ST
	E	GE	RE	SE
	C	GC	RC	SC
	F	GF	RF	SF

ii Using a tree diagram:

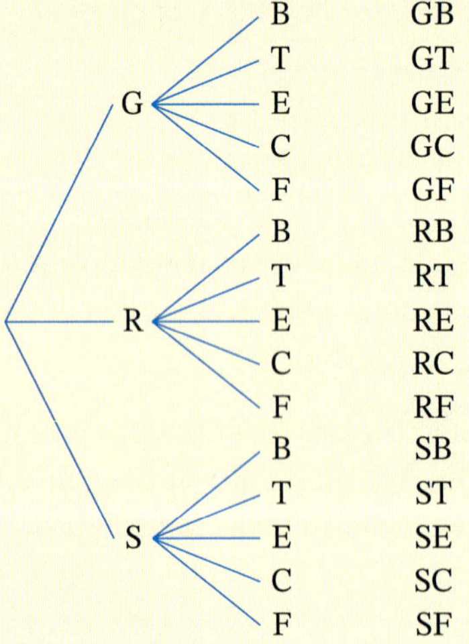

b There are $3 \times 5 = 15$ possible dancing couples.

c i $P(\text{Rick, Tania/Chloe}) = P(\text{RT or RC})$
$$= \frac{2}{15}$$

ii $P(\text{not Grant, not Felicity}) = P(\text{all outcomes without a G or F})$
$$= \frac{8}{15}$$

iii $P(\text{Stefan, not Bethany}) = P(\text{ST or SE or SC or SF})$
$$= \frac{4}{15}$$

EXERCISE 9.05 ANSWERS ON P. 645

Tree diagrams

U F PS R C

1 R C

a How many outcomes are possible when:

i tossing a coin? ii tossing a die?

EXAMPLE 8

b A coin and a die are tossed together. Show all outcomes in the sample space, using:

i a list

ii a copy of this table

iii a tree diagram.

		Die					
		1	2	3	4	5	6
Coin	H						
	T						

c How many outcomes in the sample space?

d Find the probability of tossing:

i heads and an even number

ii tails and a number less than 3

iii heads and not 1.

2 PS R C

Two dice are rolled together and their sum is calculated.

a Copy and complete this table of possible sums.

b What is the total number of possible outcomes?

c Find the probability of rolling a sum:

i of 10 ii of 6

iii less than 4 iv of 1

v that is odd vi of 7 or 11.

		1st die					
		1	2	3	4	5	6
2nd die	1						
	2					7	
	3						
	4		6				
	5						
	6						

d Find the percentage probability of rolling:

i a 'double 1' (1 and 1)

ii a 'double' anything (the same number twice).

e Which sum(s) is:

i most likely? **ii** least likely?

3 PS R C The digits 4, 5, 7 and 8 are written on separate cards and placed in a box. Two cards are drawn out at random to form a 2-digit number. Note that the same digit cannot be used twice, so '75' is allowed but '77' is not.

a Use a tree diagram to list all possible 2-digit numbers.

b How many 2-digit numbers are possible?

c Find, as a percentage, the probability that the number drawn:

i has a first digit of 7 **ii** is odd

iii is less than 60 **iv** is divisible by 5.

4 PS R C When staying at a hotel, for breakfast Simone can choose one item from each course of a set menu.

1st course	2nd course
Cereal (C)	Bacon and eggs (B)
Raisin toast (R)	Pancakes (P)
Watermelon (W)	Sausages and hash browns (S)
Yoghurt (Y)	

a Use a table to list all the different 2-course breakfasts available.

b How many 2-course breakfasts are possible?

c If one of the combinations is chosen at random, find the probability that it includes:

i cereal or raisin toast **ii** watermelon but not pancakes

iii yoghurt or pancakes **iv** yoghurt and pancakes.

5 PS R C Henryk, Ramy, Anastacia, Krystal and Tom are on the student council. They need to select a chairperson and a secretary.

a Use a table to determine all the possible pairings of chairperson and secretary.

b If each person is equally likely to be selected, find the decimal probability that:

i Krystal is chairperson and Tom is secretary

ii Ramy is either chairperson or secretary

iii Anastacia is secretary

iv Henryk and Tom fill the positions (in any order).

6 PS R C The weather this weekend will be either sunny or rainy each day, with each outcome being equally likely.

a Draw a tree diagram to show the possible outcomes for Saturday and Sunday.

b What is the probability that:

i it rains on both days?

ii the weather is the same on both days?

iii it rains on Sunday?

iv it is sunny on at least one of the days?

7 A coin is tossed 3 times.

PS **a** Show all the possible outcomes of heads (H) and tails (T) using a list.

R **b** Copy and complete this tree diagram.

C

1st coin **2nd coin** **3rd coin** **Outcomes**

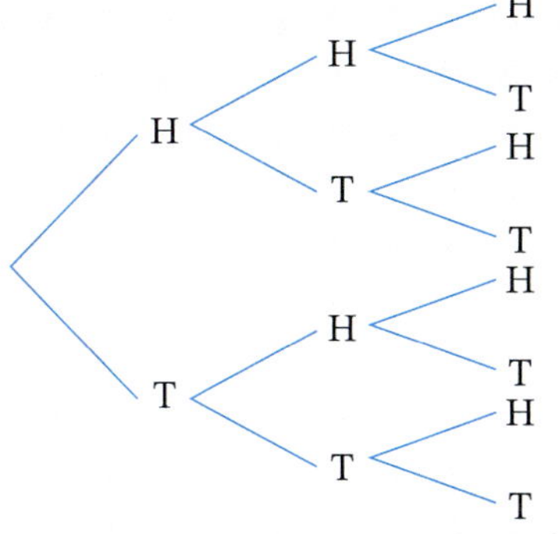

HHH

b How many outcomes are there in the sample space?

c Find, as a decimal, the probability of tossing:

i exactly one tail

ii at least 2 heads

iii 3 heads

iv more tails than heads.

Probability problems 9.06

The following exercise involves all the probability ideas that have been covered so far.

Puzzle
Matching probabilities

EXERCISE  ANSWERS ON P. 646

Probability problems

1 There were 28 students on the school trip to Japan: 17 were born in Australia, 3 in Italy, 4 in Vietnam, 1 in Japan and 3 in China. One student was chosen at random. Find the probability that the student was:

a born in Australia **b** born in Asia **c** not born in Vietnam.

2 R C The 7 letters of the word SUCCESS are written on separate cards. A card is selected at random and the letter noted.

a List the sample space.

b Is each letter equally likely to be selected? Explain.

c Find the chance that the letter chosen:

i is an S
ii is a vowel
iii is not a C
iv is also a letter of the word FAIL

3 R C Alex selects one sock at random from a bag containing 2 black, 2 blue and 2 red socks.

a List the sample space.

b Find the probability that Alex selects:

i a blue sock **ii** a black sock **iii** a pink sock

c What is the complementary event to choosing a red sock? What is the probability of this event?

4 A spinner is evenly divided into 5 sections numbered 1, 2, 3, 4 and 5. For one spin, find the probability, as a percentage, that it lands on:

a 2
b an even number
c a number less than 5
d a number at most 5

5 Sophia bought 4 tickets in a lottery in which there were 100 000 tickets. What is the probability that she won first prize? Select the correct answer **A**, **B**, **C** or **D**.

A 0.000 0025 **B** 0.000 25 **C** 0.04 **D** 0.000 04

6 One ball is selected at random from a barrel of balls numbered 1–100. Find the probability that the number shown on the ball is:

a 12 **b** greater than 40 **c** even **d** at most 85

7 R Stathis flipped a coin 7 times and a tail showed each time. What is the chance of a tail showing in the next toss?

8 A computer selected a random number from 1 to 15 inclusive. Find the probability that the number is:

a odd
b not prime
c a multiple of 3 or 5
d a factor of 12

9 R C In a tennis tournament there are 60 players. Of these, 34 are from NSW, 20 are from Victoria and 6 are from Queensland. Find the probability that a player selected at random from the tournament is:

a from NSW
b from Queensland
c not from Victoria
d from Queensland or Victoria

10 PS R C 40 students at a school camp can select kayaking or bushwalking as an activity, or both. A total of 35 students chose kayaking, 20 chose bushwalking, including some who chose both. All students chose an activity.

Kayaking	Both	Bushwalking
	15	

a Copy and complete the Venn diagram representing these students.

b What is the probability that a student, chosen at random:

i chose both activities? ii did not choose bushwalking?

11 The probability that a carton of eggs contains any broken eggs is 0.1. Find the probability that a carton contains no broken eggs.

12 A sample of students from a sports high school were surveyed on whether they participated in hockey or judo. The results are shown in this two-way table.

R C

	Judo	Not Judo	Total
Hockey	128		360
Not hockey	100	40	
Total			

a Copy and complete the table.

b What is the probability that a student chosen randomly:

i does judo? ii doesn't do hockey?

iii does judo, but not hockey? iv does hockey and judo?

v does hockey or judo? vi does hockey or judo, but not both?

13 In a bag there are lollies of three different colours: red, green and purple. The chance of taking out a red lolly at random is 0.5, the chance of it being green is 0.2.

PS R C

a What is the chance of the lolly being purple?

b If the bag holds 10 lollies, how many of each colour would you expect to find?

c If a second bag had 100 lollies, how many of each would you expect to find?

d Which bag would give you a better chance of picking a red lolly? Explain your answer.

Observed and expected frequencies 9.07

Probability calculated using the formula:

$$P(E) = \frac{\text{number of favourable outcomes}}{\text{total number of outcomes}}$$

is more specifically called **theoretical probability**.

We can also determine the probability of an event based on the results of an experiment or trial that has been repeated many times, such as the safety testing of different cars, or rely on past statistics, such as the number of rainy days in April. This type of probability is called **experimental probability** or **observed probability**, which is based on **relative frequency**, the number of times an event occurred as a fraction of the total frequency of outcomes.

Worksheets
Relative frequencies
Dice probability
Spinning chance
A page of spinners
Experimental probabilities

Interactive
Plinko probability

ⓘ Relative frequency

$$P(E) = \frac{\text{number of times the event happened}}{\text{total number of trials}}$$

$$\text{or } P(E) = \frac{\text{frequency of } E}{\text{total frequency}}$$

As in statistics, 'frequency' means the number of times something happens.

Expected frequency is the expected number of times an event will occur over repeated trials, based on the theoretical probability of the event taking place.

Worksheet Probability review

Puzzles Experimental probability

Duelling dice

Videos Experimental probability

Observed and expected frequencies

Presentation Theoretical probability

ⓘ Expected frequency

Expected frequency = theoretical probability × number of trials

Observed frequency is the actual number of times an event occurs over repeated trials in an experiment.

Example 10

Declan rolled a die 60 times and recorded the results in a table.

Outcome	1	2	3	4	5	6
Frequency	10	11	13	7	8	11

a What is the theoretical probability of rolling a 5 or 6 on a die?

b What is the observed frequency of rolling a 5 or 6 over 60 rolls?

c What is the expected frequency of rolling a 5 or 6 over 60 rolls? How does this compare with the observed frequency?

d What is the experimental probability of rolling a 5 or 6?

SOLUTION

a $P(5 \text{ or } 6) = \frac{2}{6} = \frac{1}{3}$

b From the table, the observed frequency of 5s and 6s = 8 + 11 = 19

c $\text{Expected frequency of 5s or 6s} = \frac{1}{3} \times 60$ $= 20$ — Probability × number of trials

which is close to the observed frequency, 19.

d $\text{Experimental } P(5 \text{ or } 6) = \frac{8+11}{60} = \frac{19}{60}$

EXERCISE 9.07 ANSWERS ON P. 646

Observed and expected frequencies

U F R C

1 Harper rolled a biased die 60 times and the number 2 came up 15 times.

a What is the experimental probability of rolling a 2?

b If the same die was rolled 500 times, what is the expected frequency of rolling a 2?

2 **a** Copy this table.

Outcome	Tally	Frequency
Head		
Tail		
	Total	

(C) **b** Flip a coin 50 times and record the results in the table.

c Find, as a decimal, the experimental probability of flipping:

i a head **ii** a tail

d Find, as a decimal, the theoretical probability of flipping:

i a head **ii** a tail

e How do the experimental probabilities compare with the theoretical probabilities?

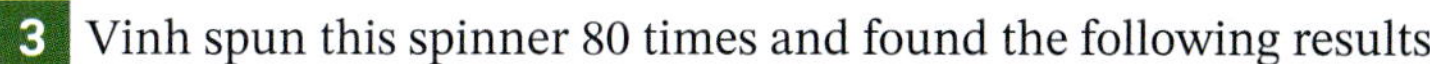

3 Vinh spun this spinner 80 times and found the following results:

(C) (C)

Outcome	Red	Green	Yellow
Frequency	44	25	11

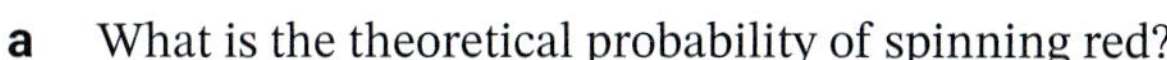

a What is the theoretical probability of spinning red?

b For 80 spins, what is the expected frequency of spinning red? How does this compare with the observed (actual) frequency?

c What is the experimental probability of spinning red?

d What is the theoretical probability of spinning yellow?

e Find the expected frequency of spinning yellow over 80 spins and compare this with the observed frequency.

f What is the experimental probability of spinning yellow?

4 The matches in a sample of matchboxes were counted. The number of matches in each box are recorded below.

(R)

Number of matches	48	49	50	51	52	53
Number of matchboxes	3	6	20	16	4	1

a How many matchboxes were tested?

b What is the experimental probability of finding 51 matches in a box?

c What is the experimental probability of finding fewer than 50 matches in a box?

d In 1000 matchboxes, how many matchboxes would you expect to contain:

i exactly 50 matches? **ii** at least 50 matches?

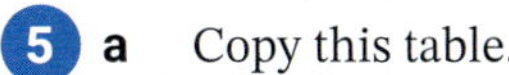

5 **a** Copy this table.

Outcome	Tally	Frequency
1		
2		
3		
4		
5		
6		
	Total	

(R) (C) **b** Roll a die 72 times and record the results in the table.

c Find the theoretical probability of rolling:

i 4.

ii a number greater than 2.

iii an even number.

d When rolling a die 72 times, what is the expected frequency of rolling:

i 4?

ii a number greater than 2?

iii an even number?

e How do your expected frequencies compare with the observed frequencies from the table?

f Find the experimental probability of rolling:

i 4.

ii a number greater than 2.

iii an even number.

g If a die was rolled 1000 times, how many times should 4 come up:

i based on the theoretical probability?

ii based on the experimental probability?

6 Sefina tossed a coin many times and got 135 heads and 115 tails. Calculate, as a percentage, the experimental probability of tails with this coin.

7 A die was rolled 80 times, with the results shown below.

R C

Outcome	1	2	3	4	5	6
Frequency	11	13	9	13	12	22

a Is each outcome equally likely?

b Do you think this die is biased (unfair)? Give a reason for your answer.

c Write, as a percentage, the experimental probability of rolling a 1 on this die.

d If this die was rolled 100 times, what is the expected frequency of 3 coming up?

8 A pair of dice was rolled 50 times and their sum calculated each time.

Sum	2	3	4	5	6	7	8	9	10	11	12
Frequency	0	2	4	6	5	5	9	6	8	3	2

a Find, as a decimal, the experimental probability of rolling a sum:

i of 11

ii of 5 or 6

iii that is odd

iv of 6 or more

v of at most 6

vi that is a composite number

b Which sum(s):

i was most likely?

ii had a probability of $\frac{3}{25}$?

iii was least likely?

iv had a probability of 16%?

v was second-most likely?

vi had a probability of $\frac{1}{10}$?

Foundation Standard Complex

TECHNOLOGY

Tossing a coin

In this activity, a spreadsheet is used to simulate the tossing of a coin. The computer can quickly generate either 1 (representing heads) or 2 (tails). We will use the command RAND to generate these random numbers.

1 In cell A1 enter the label 'Tossing a coin'.

2 In cell A2, enter the formula **=INT(RAND()*2+1)**. Press Enter and either 1 or 2 should appear randomly in the cell.

3 Select cell A2 and **Fill Down** to cell A11 to generate random numbers (1 or 2), to simulate the tossing of a coin 10 times.

4 Your results will probably be different from those of other students in the class. Compare.

5 Copy the table below and in the first blank row record your results for Trials 1–10 (the numbers of heads and tails in your first 10 'tosses').

Trial	Number of heads	Number of tails
1–10		
11–20		
21–30		
31–40		
41–50		
51–60		
61–70		
71–80		
81–90		
91–100		
Total		

6 Simulate another 10 tosses of the coin by making the spreadsheet generate another set of random numbers. Do this by placing the cursor between the tops of columns A and B (so that it turns into a 'double arrow') and clicking.

7 Enter the results for Trials 11–20 in your table.

8 Repeat 8 more times so that you have 100 tosses of the coin recorded in the table.

9 a For 100 tosses of a coin, what is the expected frequency of heads based on the theoretical probability?

b Compare this with the observed frequency.

10 Calculate, as a decimal, the theoretical and experimental probabilities of tossing heads.

11 Compare your results with those of other students in your class. Briefly explain any differences and why they may have occurred.

12 Combine the results of students in your class to calculate the experimental probability of tossing heads, as a decimal. How does this compare with the theoretical probability?

Worksheet
Coin toss experiment

Spreadsheets
Coin-tossing simulator

Die-rolling simulator

Spinner simulator

Technology
Simulating a spinner

9.08 Probability simulations

Interactive
Plinko
probability

Conducting probability experiments allow you to gather data that can be used to investigate and calculate the probabilities of events. If you need a large data set to work with, it is sometime easier to use an online **simulator** to gather the data and then calculate the probabilities needed. It is always interesting to see how these observed probabilities compare with theoretical probability for the events.

EXERCISE 9.08 ANSWERS ON P. 646

Probability simulations

U F R C

Luna is investigating the probabilities related to tossing a coin.

1 What is the theoretical probability of tossing a head?

2 Toss a coin 20 times, complete the table below and use it to calculate the experimental probability of tossing a head.

Result	Tally	Frequency
Heads		
Tails		

3 Now toss the coin 50 times, complete the table below and use it to calculate the experimental probability of tossing a head.

Result	Tally	Frequency
Heads		
Tails		

4 Access an online coin toss simulator such as www.statcrunch.com/applets (choose 'Probability of a head with a fair coin') and run a simulation that tosses a coin 1000 times. Determine the experimental probability of tossing a head in the simulation.

5 Compare your probability results from questions 1 to 4 and write a few sentences to explain any similarities and differences.

R C

Harpreet is investigating the probabilities related to rolling a die.

6 What is the theoretical probability of rolling a 3?

7 If you roll the die 30 times, how many times would you expect each number to be rolled, and what is the theoretical probability of rolling each number on the dice?

8 R C Roll a die 30 times, complete the table below and then compare the results you achieved with your expected results from question **2**. Explain any difference in the results.

Result	Tally	Frequency	Experimental probability
1			
2			
3			
4			
5			
6			

9 Access an online die roll simulator such as www.academo.org/demos (choose 'Statistics of rolling dice') and run a simulation that rolls a die 600 times. Determine the experimental probability of tossing a head in the simulation.

10 R C Compare your experimental probabilities from questions 8 and 9 to your theoretical probabilities from question **7** and write a few sentences to explain any similarities and differences.

Zain is investigating the probabilities related to selecting a specific suit (hearts ♥, diamonds ♦, clubs ♣ or spades ♠) from a standard deck of 52 cards.

11 What is the theoretical probability of selecting each different suit from a regular deck of cards?

12 Use a standard deck of c2 cards and select a card 20 times. Make sure you place the selected card back into the deck and shuffle the deck before each consecutive selection. Copy and complete the table below and use it to calculate the experimental probability of each result.

Result	Tally	Frequency	Experimental probability
Hearts ♥			
Diamonds ♦			
Clubs ♣			
Spades ♠			

13 R Did your theoretical probability from question **11** match the experimental probability of selecting a heart from question **12**?

14 Access the online card simulator at http://random-cards.com/, select the one shuffled deck option and run the simulation of selecting a card 100 times. Fill out a tally table like the one in question **12** above for this simulation. Use your results to determine the experimental probability of each possible result for this simulation.

☐ Foundation ○ Standard ⬡ Complex

15 (R) (C) Compare your experimental probabilities from questions 12 and 14 to your theoretical probabilities from question **11** and write a few sentences to explain any similarities and differences.

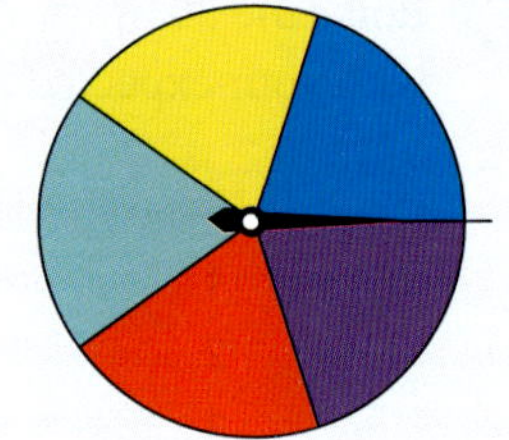

16 Trinh is investigating the probabilities related to spinning the spinner shown here.

What is the theoretical probability of spinning each different colour in this spinner above?

17 Access the online spinner at www.nctm.org/adjustablespinner/ and change the number of sectors in the spinner to 5. Set the spinner to spin 20 times and run the simulation. Copy and complete this results table:

Result	Frequency	Experimental probability
Blue		
Yellow		
Cyan		
Read		
Purple		

18 (R) Did your theoretical probability from question **16** match the experimental probability of spinning each colour in question **17**?

19 Start the simulation again using the same spinner but complete a total of 200 spins this time. Complete a table like the one in question **17** above for this simulation.

20 (R) (C) Compare your probability results from questions 16 to 19 and write a few sentences to explain any similarities and differences.

POWER PLUS ANSWERS ON P. 646

9.08

1 A market research survey of 50 people shows that 42 people own a car, 20 own a tablet device and 8 own a UHD TV. Of these:

- 4 own a car and a UHD TV.
- 3 own a UHD TV and a tablet.
- 14 own a car and a tablet.
- 1 person has a car, a tablet and a UHD TV.

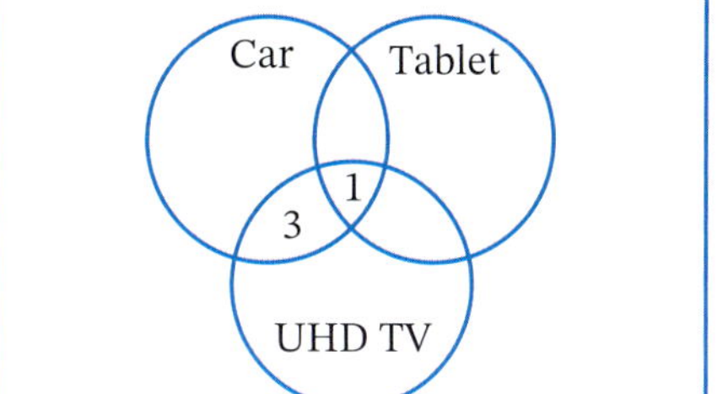

Copy and complete the following Venn diagram to represent the survey results.

2 Two dice are rolled and the numbers are **multiplied** together to arrive at a score.

a Copy and complete this table to show all possible products.

	1st die						
	×	1	2	3	4	5	6
2nd die	1	1	2	3			
	2	2	4				
	3	3	6				
	4			12			
	5						
	6						

b How many different products are possible?

c Why isn't each product equally likely?

d Which product is most likely?

e Which product is least likely?

f What is the probability of a product of:

i 6? ii 20? iii at least 20?

g Which product has a probability of $\frac{1}{12}$?

3 Three friends decide to play a game with 2 dice. Danielle wins if the sum of the numbers is 3 or 5, and Vanessa wins if the total is 6 or 8. Any other total means that Karla wins. Is the game fair? Explain your answer.

4 A committee of 2 people is to be selected from 2 boys (Paul and Sumeet) and 2 girls (Tash and Nadine).

a List all the committees you can form.

b If you were to choose a committee at random, what is the probability that it would include Tash?

9 CHAPTER REVIEW

Language of maths

Quiz
Language of maths 9

certain	complementary	event	exclusive
expected frequency	experimental probability	improbable	impossible
inclusive	likely	mutually exclusive	observed frequency
outcome	overlapping	probable	random
relative frequency	sample space	theoretical probability	tree diagram
trial	two-way table	unlikely	Venn diagram

1 What is the meaning of **sample space**?

2 What is the **complementary event** to winning a soccer match?

3 What does it mean when a person is selected '**at random**' for a survey?

4 Draw a Venn diagram with categories 'Year 7 students' and 'Year 8 students'. Are these categories **mutually exclusive** or not?

5 What term means the number of times an event should occur over repeated trials?

6 Explain what 'is right-handed or drives a car' means exactly, given that they are **overlapping** events.

Topic summary

Worksheet
Mind map: Probability

- Write in your own words what you have learnt in this chapter.
- Write which parts of the chapter were new to you.
- Copy and complete:

  The things I understand about probability are...

  The things I am still not confident in doing in this chapter are...

Print (or copy) and complete this mind map of the topic, adding detail to its branches and using pictures, symbols and colour where needed. Ask your teacher to check your work.

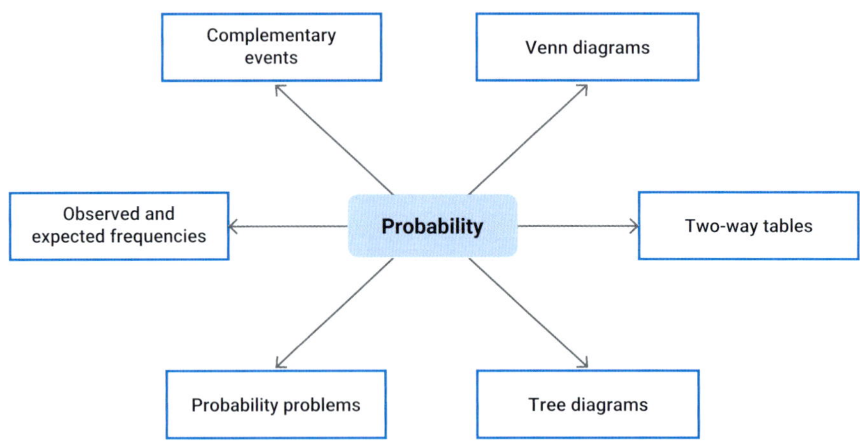

9 TEST YOURSELF
ANSWERS ON P. 646

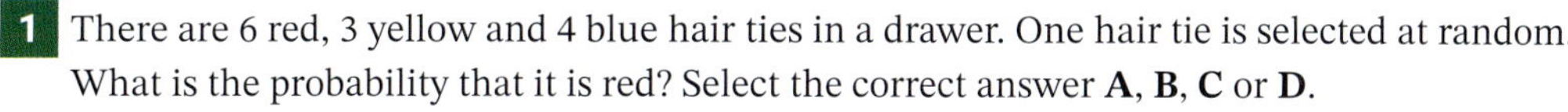

1 There are 6 red, 3 yellow and 4 blue hair ties in a drawer. One hair tie is selected at random. What is the probability that it is red? Select the correct answer **A**, **B**, **C** or **D**. 

A $\frac{1}{13}$ **B** $\frac{6}{13}$ **C** $\frac{6}{7}$ **D** $\frac{13}{6}$

2 A die is rolled. Find the probability that the number that comes up is:

a 1 **b** more than 2 **c** odd **d** composite

3 **a** Write the sample space for this spinner. 

b Find, as a percentage, the probability that the number spun is:

i 5 **ii** at least 5

iii 5 or less **iv** less than 5

4 Write the complement of each event and its probability. 9.02

a Choosing an Ace from a standard deck of cards.

b Rolling a factor of 6 on a die.

c Buying the winning ticket out of 1000 tickets sold.

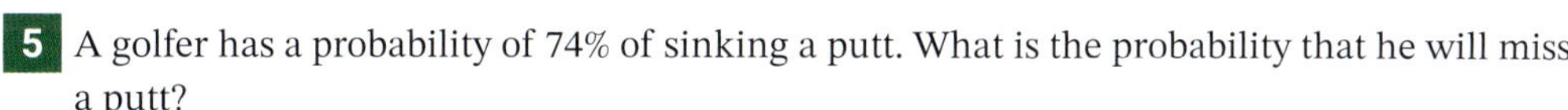

5 A golfer has a probability of 74% of sinking a putt. What is the probability that he will miss a putt? 

6 A truck carries 240 boxes of hand sanitiser, 305 boxes of soap, 335 boxes of paper and 120 boxes of pens. The driver chooses one box at random. What is the probability that it is: 

a a box of hand sanitiser? **b** a box of pens or soap?

c a box of paper? **d** not a box of paper?

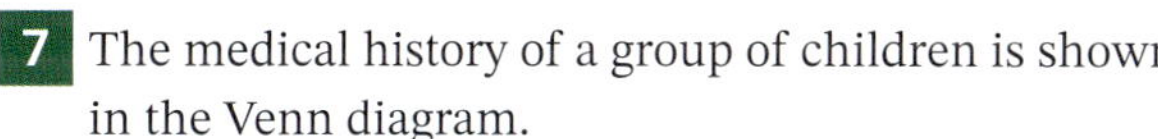

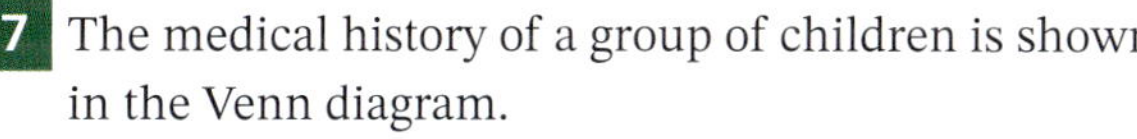

7 The medical history of a group of children is shown in the Venn diagram. 

a How many children are in the group?

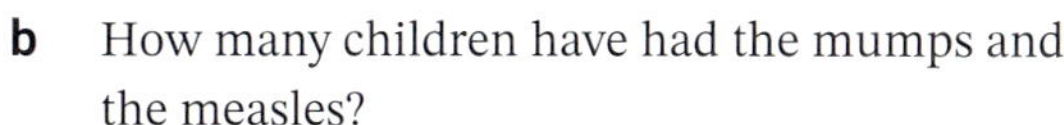

b How many children have had the mumps and the measles?

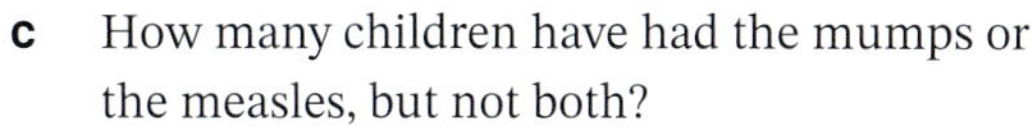

c How many children have had the mumps or the measles, but not both?

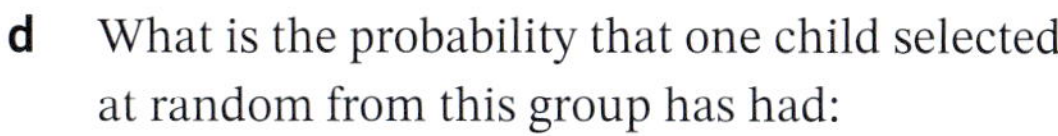

d What is the probability that one child selected at random from this group has had:

i the measles?

ii the mumps or the measles?

iii neither the mumps nor the measles?

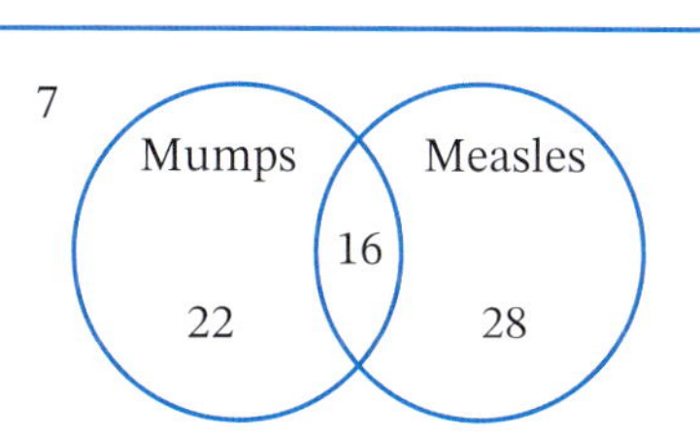

Foundation Standard Complex

8 A group of people were surveyed on whether they watched cricket or soccer on TV.

The results were sorted into a two-way table.

	TV sport	
	Cricket	**Not cricket**
Soccer	28	15
Not soccer	20	7

a How many people were surveyed?

b How many people watch soccer?

c How many people watch soccer or cricket?

d What is the probability that a person selected at random from the survey:

- **i** watches neither soccer nor cricket?
- **ii** watches soccer or cricket, but not both?
- **iii** does not watch cricket?
- **iv** does not watch soccer?

9 Represent the data in the Venn diagram on the two-way table.

	Tall	**Not tall**
Blond		
Not blond		

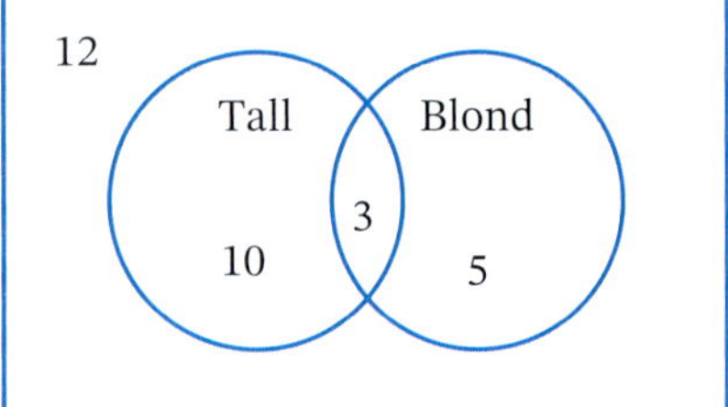

9.05

10 A tennis squad has 3 boys, Marcus, Wang and Piyush, and 3 girls, Souraya, Diane and Amia. The coach selects a boy and a girl at random from the squad to represent the school in a mixed doubles competition.

a Construct a tree diagram to determine all the possible mixed doubles pairs.

b What is the probability that Washi and Diane will be selected to represent the school?

c What is the probability that neither Washi nor Diane will be selected?

9.06

11 An 8-sided die has 2 red, 3 white, 1 blue and 2 yellow faces. If the die is rolled, find the decimal probability that the face that comes up is:

a blue **b** yellow **c** not red **d** white or yellow

9.07

12 A die was rolled 80 times and the numbers 1 or 6 came up 25 times.

a What is the experimental probability of rolling 1 or 6, as a percentage?

b What is the theoretical probability of rolling 1 or 6, as a percentage?

c Calculate the expected frequency of rolling 1 or 6 over 80 trials.

Practice set 3 ANSWERS ON P. 647

1 Identify the 3 pairs of congruent shapes from the shapes below.

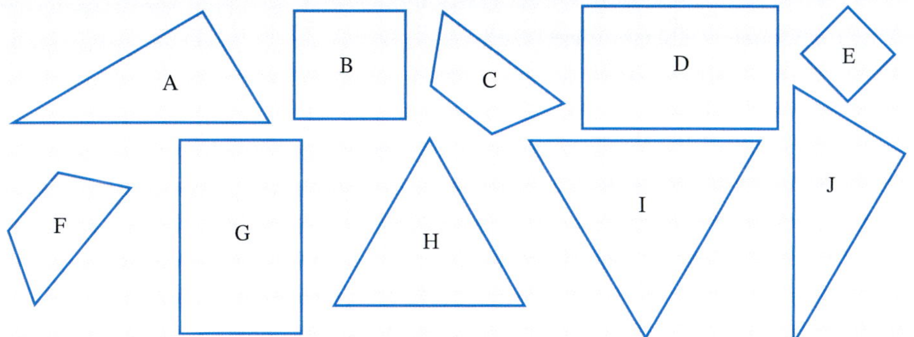

2 If this spinner is spun, find the probability, as a percentage, that the arrow lands on:

9.01

a green

b a traffic light colour

c a colour that is not red.

3 Mrs Smart gave her class a quick quiz. The results are shown in the table.

a Copy and complete the table.

b How many students are there in Mrs Smart's class?

c Construct a frequency histogram and polygon for these results.

Mark	Tally	Frequency
6	卌	
7	卌 ||	
8		5
9	卌 卌	
10		3

4 a Name the 2 congruent triangles below, using '△ ____ ≡ △ ____' notation.

8.04

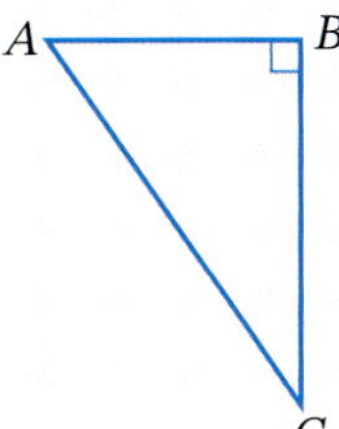

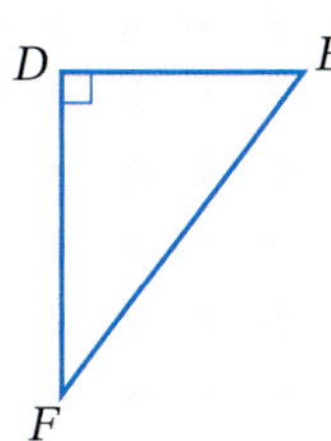

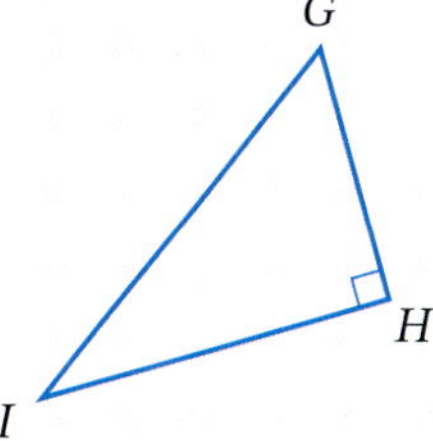

b For the congruent triangles, name one pair of matching sides and one pair of matching angles.

☐ Foundation ○ Standard ⬡ Complex

5 A die was rolled 40 times and the results are shown here.

5	1	5	2	3	2	6	5	2	5
4	2	5	2	6	2	6	4	3	1
5	5	3	6	4	2	6	5	6	5
5	2	3	3	3	3	3	1	2	1

a Draw a frequency table for these results.

b Draw a divided bar graph to represent this information.

6 For this pair of similar figures:

a name a pair of matching sides.

b name a pair of matching angles.

c write a similarity statement of the form '_____ ||| _____'.

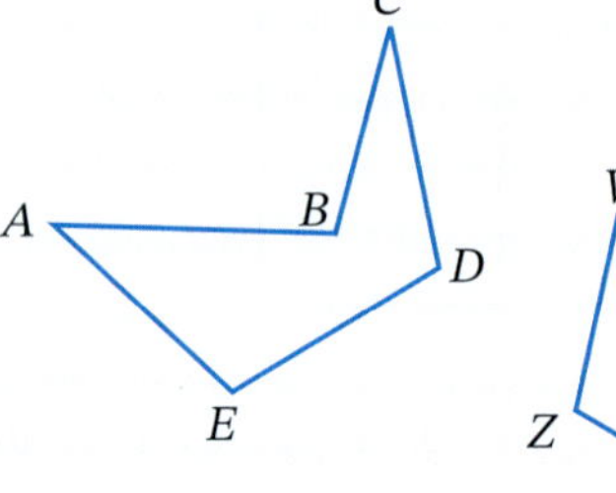

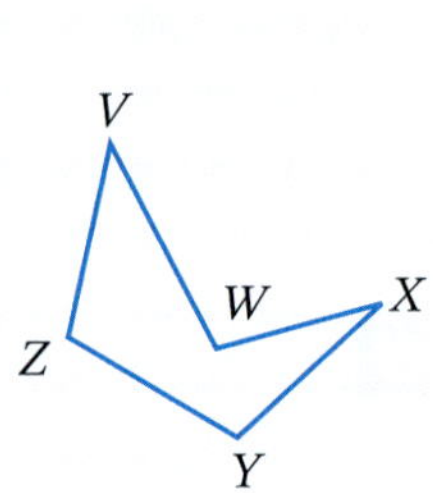

7 A die is rolled. Find the probability that the number that comes up is:

a even

b at least 2

c a composite number.

8 Construct a dot plot for these History assignment marks.

22	16	10	8	12	11	15	13	12	13	11	16
11	12	10	10	12	12	15	11	13	15	17	13

a Find the median.

b Find the mode.

c What is the outlier?

d Where are the marks clustered?

9 Construct a triangle with:

a sides of 7 cm, 5 cm and 9 cm.

b sides of 4 cm and 5 cm and an included angle of 65°.

☐ Foundation ○ Standard ⬡ Complex

10 Find the value of each variable in each pair of similar triangles. 8.07

a

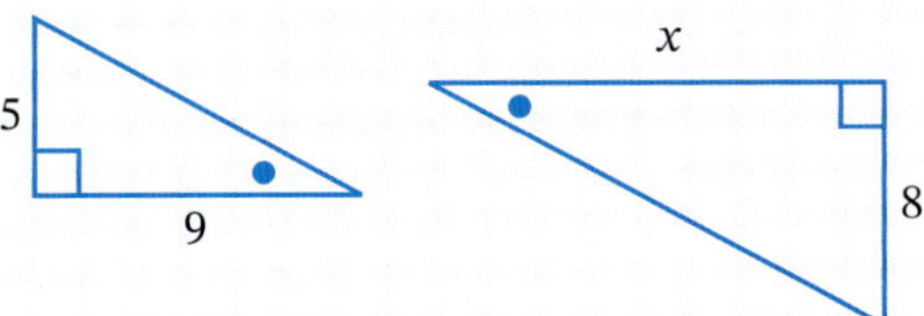

b

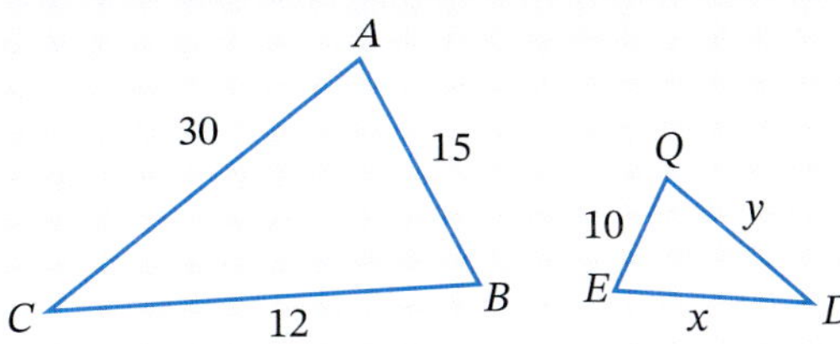

11 For these Science exam marks, find: 7.02, 7.03

90 89 95 92 90 92 97 93 90 97

- **a** the mean
- **b** the mode
- **c** the median
- **d** the range

12 A pack of cards contains 10 blue, 6 purple and 4 pink cards. One card is drawn from the pack at random. Find the probability that this card is not purple. 9.02

13 Calculate the scale factor for this pair of similar triangles. 8.07

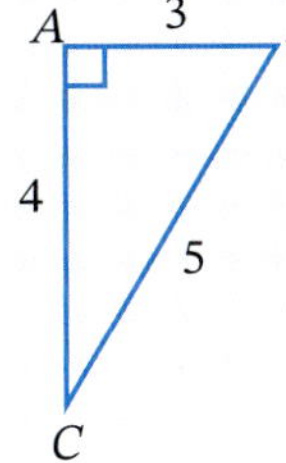

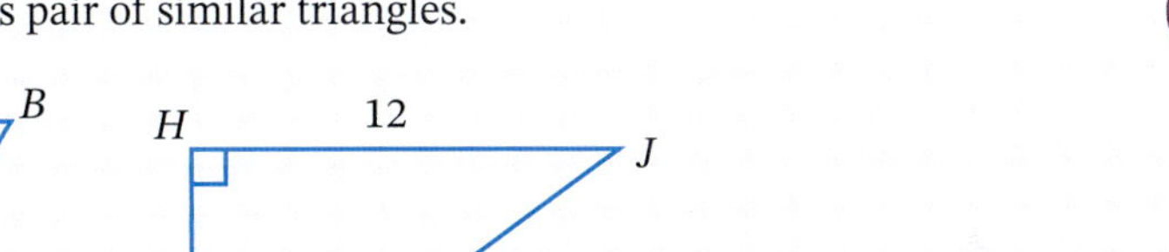

14 The ages of people taking out a policy with an insurance company are shown. 7.04

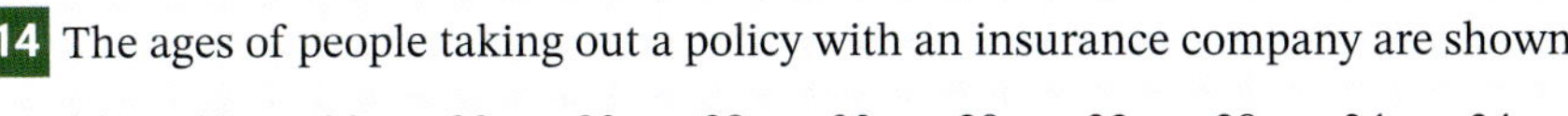

34	32	28	30	33	28	30	29	33	28	34	34
28	30	32	32	34	31	30	31	28	30	28	30

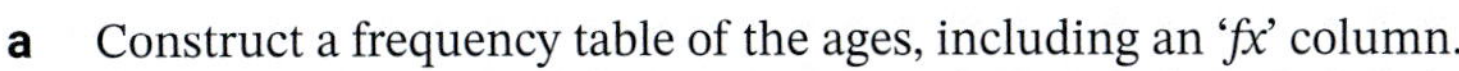

- **a** Construct a frequency table of the ages, including an 'fx' column.
- **b** What fraction of the people are under 30 years of age?
- **c** For the above data, find:
  - **i** the mean (to one decimal place)
  - **ii** the median
  - **iii** the mode
  - **iv** the range

☐ Foundation ○ Standard ○ Complex

9.03

15 In a group of 50 Year 9 students, 34 study History, 25 study Japanese and 11 study both.

a Represent this information in a Venn diagram.

b How many students study Japanese but not History?

c What is the probability that a student randomly chosen from this group studies neither History nor Japanese?

8.04

16 **a** Name the 4 tests for congruent triangles.

b Which test can be used to prove that these triangles are congruent?

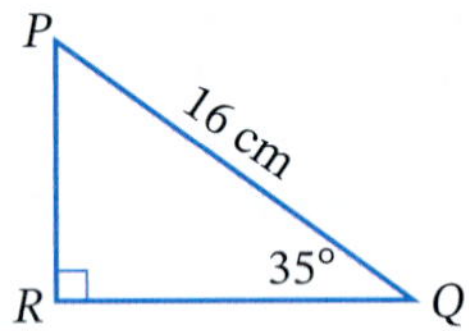

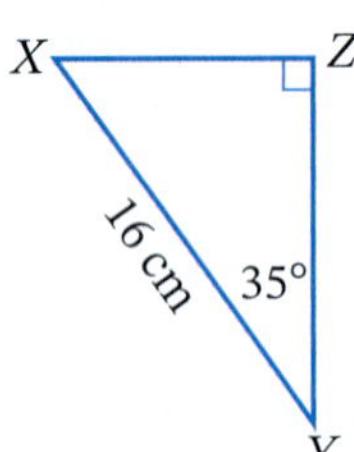

9.06

17 Fareeba spun this spinner 120 times and found the following results:

Outcome	Red	Green	Yellow	Blue
Frequency	23	52	24	21

a What is the theoretical probability of spinning red?

b For 120 spins, what is the expected frequency of spinning red? How does this compare with the observed frequency?

c What is the experimental probability of spinning red?

d What is the theoretical probability of spinning blue?

e Find the expected frequency of spinning blue over 120 spins and compare this with the observed frequency.

f What is the experimental probability of spinning blue?

Foundation Standard Complex

18 Year 8 students were surveyed on whether they participated in weekend sport. 9.04

a Copy and complete this two-way table.

	Plays weekend sport	Does not play weekend sport	Total
Male	49		
Female		45	84
Total		68	

b How many students are in Year 8?

c How many Year 8 students play weekend sport?

d If a student is chosen at random from Year 8, what is the probability that the student:

i is female?

ii is male or plays sport but not both?

19 State which similar triangle test can be applied to each pair of triangles. 8.07

a

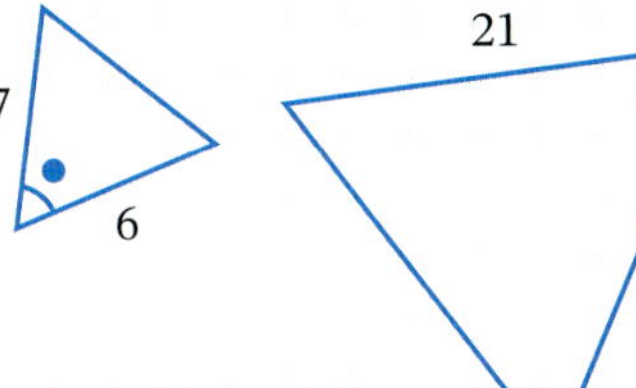

b

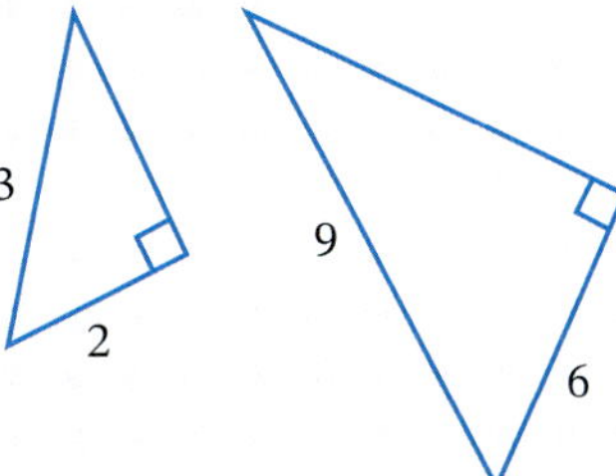

20 The marks scored by 10 students in Maths and Science exams are shown. 7.07

Maths	35	58	87	43	52	96	23	35	48	58
Science	43	69	73	62	54	88	24	40	40	60

a Show these marks on a back-to-back stem-and-leaf plot.

b Find the median and the range for each exam.

c Based on these results, which exam do you think was more difficult?
Justify your answer.

21 Determine whether the pair of triangles below are similar triangles by using an appropriate similar triangles test and then find the value of x. 8.07

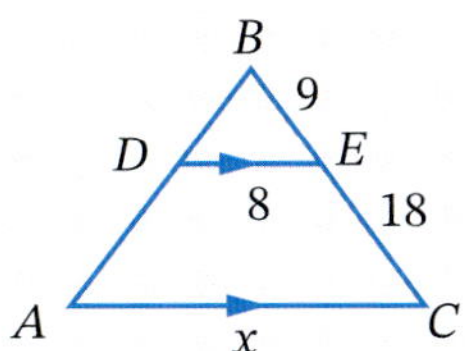

Foundation Standard Complex

22 The nationalities of players in a junior tennis tournament are shown below.

Australia	USA	UK	Asia	Europe
3	9	4	20	28

Find the probability that a player selected at random from this group:

a is Australian.

b comes from Asia.

c is not American.

d is from Europe or America.

23 Jess wants to find the most popular TV program in Australia.

a Should she use a sample or a census? Give a reason for your answer.

b Suppose she decides to use a sample. How could she choose a representative sample?

c Give an example of a biased question she could ask.

d Change your question from part **c** so that it is NOT biased.

24 A tree's shadow is 20 m long. At the same time of day, a 2 m stick held vertically on the ground will cast a shadow 1 m long. Use a diagram to determine the height of the tree.

25 The weather this weekend will be either sunny or rainy each day, with each outcome being equally likely.

a Draw a tree diagram to show the possible outcomes for Saturday and Sunday.

b What is the probability that:

i it is sunny on only one of the days?

ii it rains both days?

iii it rains on at least one day?

iv it rains on at most one day?

Complex

10
ALGEBRA

Equations and inequalities

Equations are extremely important for solving the same problem many times with different values. In animation and film-making, programmers use complex equations to make action figures move like real people. Most modern appliances are now smart devices with responses that rely on the use of equations.

Chapter outline

	Proficiencies
10.01 Equations	U F
10.02 One-step equations	U F R C
10.03 Two-step equations	U F R C
10.04 Equations with variables on both sides	U F R C
10.05 Equations with brackets	U F R C
10.06 Equation problems	U F PS R C
10.07 Extension: Simple quadratic equations $x^2 = c$*	U F R C
10.08 Extension: Equations and formulas*	U F PS R C
10.09 Inequalities	U F R C

*YEAR 9 EXTENSION

U = Understanding
F = Fluency
PS = Problem solving
R = Reasoning
C = Communication

Wordbank

backtracking method A method of solving equations by 'undoing' or performing inverse (opposite) operations in reverse order

balancing method A method of solving equations by using the same operations on both sides of the equation

equation A mathematical statement that two quantities are equal, involving algebraic expressions and an equals sign (=)

inequality A mathematical statement that two quantities are unequal, involving algebraic expressions and an inequality sign (>, <, ≥ or ≤)

inverse operation An opposite used in solving an equation, for example, the inverse operation of multiplying is adding

solve (an equation) To find the value of an unknown variable in an equation

solution The answer to an equation or problem, the correct value(s) of a variable that makes an equation true

substitute To replace a variable with a number

Quiz
Wordbank 10

Videos (7):

10.02 One-step equations

10.03 Two-step equations • Solving equations

10.04 Equations with variables on both sides • Solving equations

10.05 Equations with brackets • Solving equations

10.06 Equation problems • Solving equations

10.09 Solving inequalities

***Twig* videos (2):**

10.02 The Arabic science of balancing

10.09 Inequalities: The world's most populous country

***PhET* interactives (3):**

10.02 Equality explorer: basics • Equality explorer

10.09 Number line: Integers

Quizzes (5):

- Wordbank 10
- SkillCheck 10
- Mental skills 10
- Language of Maths 10
- Test Yourself 10

Skillsheets (4):

10.01 Guess-and-check

10.03 Solving equations by balancing • Solving equations by backtracking • Solving equations using diagrams

Worksheets (13):

10.02 Equations 1 • One-step equations • Guess-and-check

10.03 Backtracking • Equations writing activity

10.04 Equations with unknowns on both sides

10.05 Equations 2 • Equations 3 (Extension) • Checking solutions

10.06 Solving equations review • Angle problems with algebra

10.08 Working with formulas

Mind map: Equations and inequalities

Puzzles (8):

10.02 Equations dominoes • One-step equations

10.03 Solving equations

10.04 Equations with unknowns on both sides

10.05 Equations match • Equations • Equations order activity

10.06 Writing and solving equations

Presentations (3):

10.03 Solving linear equations

10.04 Formally solving equations

10.06 Writing equations

Spreadsheet (1):

10.03 Linear equation solver

Nelson MindTap

To access resources above, visit **cengage.com.au/nelsonmindtap**

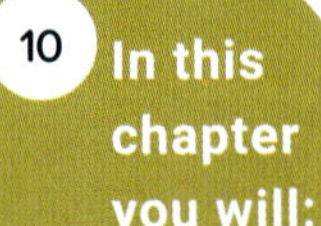

✓ solve a variety of linear equations, including those with variables on both sides, and brackets.

✓ solve problems by using equations.

✓ (EXTENSION) solve simple quadratic equations of the form $x^2 = c$

✓ (EXTENSION) solve equations involving formulas to solve problems

✓ solve simple inequalities, including those involving multiplying or dividing by a negative number

SkillCheck

ANSWERS ON P. 648

Quiz
SkillCheck 10

1 Find the number represented by □ to make each equation true.

a	$\square + 3 = 10$	**b**	$4 \times \square = 28$	**c**	$12 - \square = 4$
d	$\square \div 3 = 9$	**e**	$9 + \square = 15$	**f**	$\square - 7 = 16$
g	$\square \times 6 = 30$	**h**	$24 \div \square = 8$	**i**	$10 + \square = 20$
j	$\square - 5 = 0$	**k**	$\frac{1}{2} \times \square = 4$	**l**	$\square \div 5 = 4$

2 Evaluate each expression.

a	$5 - 8$	**b**	$-8 + 3$	**c**	$3 \times (-4)$
d	$-6 \div 3$	**e**	$-2 + 5$	**f**	$-2 - 9$
g	$-4 \times (-9)$	**h**	$-12 \div (-4)$	**i**	$-3 - 7$
j	$80 \div (-10)$	**k**	$6 - 15$	**l**	-8×9

3 If $d = 7$, evaluate:

a $2d + 3$ **b** $3d - 4$ **c** $\frac{d+1}{2}$ **d** $18 - 2d$

4 If $x = -3$, evaluate:

a $5x + 9$ **b** $4 - x$ **c** $\frac{x-3}{-1}$ **d** $2x - 8$

5 Simplify each expression.

a $3n - 2n$ **b** $8x + 3x$ **c** $5r - r$ **d** $6t + 2t$

6 Expand each expression.

a	$2(m + 3)$	**b**	$3(x - 2)$	**c**	$4(k + 7)$
d	$5(d - 1)$	**e**	$6(a - 3)$	**f**	$4(b + 8)$
g	$5(2q - 6)$	**h**	$9(5j + 1)$		

7 What is the opposite operation to:

a adding? **b** dividing?

c multiplying? **d** subtracting?

Equations 10.01

In Year 7, we learned that an **equation** is a statement involving a variable (such as x), numbers and an equals (=) sign, for example, $\mathbf{4x + 5 = 57}$. When we find the correct value that makes the equation true, we **solve** the equation. The value is called the **solution** to the equation.

There are different ways of solving equations. The simplest method is the 'guess, check and improve' method.

The 'guess, check and improve' method

The 'guess, check and improve' method involves:

- guessing a number for the solution.
- checking the guess by substituting the number into the equation.
- improving the guess by testing better numbers until the correct solution is found.

Skillsheet
Guess-and-check

Example 1

Use 'guess, check and improve' to solve the equation $4x + 5 = 57$ for x.

SOLUTION

Guess	Check	Comment
$x = 3$	$4 \times 3 + 5 = 17$	Smaller than 57. Try a bigger number.
$x = 10$	$4 \times 10 + 5 = 45$	Still smaller than 57. Try a bigger number.
$x = 20$	$4 \times 20 + 5 = 85$	Bigger than 57. Try a number between 10 and 20.
$x = 15$	$4 \times 15 + 5 = 65$	Still bigger than 57. Try a number between 10 and 15.
$x = 13$	$4 \times 13 + 5 = 57$	Correct.

The solution is $x = 13$.

EXERCISE 10.01 ANSWERS ON P. 648

Equations

U F

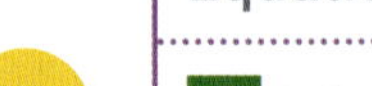

EXAMPLE 19

1 Solve each equation.

a $x + 3 = 6$ **b** $x - 3 = 6$ **c** $a + 12 = 17$

d $m - 5 = 15$ **e** $b + 1 = 11$ **f** $5c = 20$

g $4k = 24$ **h** $d - 3 = 10$ **i** $10 + m = 25$

j $14 - x = 9$ **k** $\frac{y}{3} = 5$ **l** $\frac{n}{5} = 5$

2 What is the solution to $2x - 6 = 24$? Select the correct answer **A**, **B**, **C** or **D**.

A $x = 12$ **B** $x = 15$ **C** $x = 14$ **D** $x = 9$

3 Solve each equation.

a $2x + 4 = 14$ **b** $3p - 5 = 16$ **c** $4k + 6 = 26$

d $5x - 9 = 11$ **e** $7x + 6 = 27$ **f** $8x - 30 = 34$

g $\frac{m}{4} - 3 = 4$ **h** $\frac{a}{2} + 12 = 18$ **i** $10 - 3d = 4$

j $12 - 5n = 7$ **k** $\frac{r-9}{2} = 3$ **l** $\frac{k+1}{3} = 4$

m $2p + 3 = -1$ **n** $\frac{r}{5} + 4 = 6$ **o** $\frac{2y}{3} = 10$

p $2 \times (x + 1) = 6$

4 Solve the equation $6\left(\frac{x}{2} - 1\right) = 12$. Select the correct answer **A**, **B**, **C** or **D**.

A $x = \frac{13}{3}$ **B** $x = 6$ **C** $x = 3$ **D** $x = 2$

Foundation Standard Complex

One-step equations

The 'guess, check and improve' method is simple but slow. There are two **algebraic** methods for solving equations that are faster and more efficient:

The **balancing method** involves representing both sides of an equation as balance scales, and solving the equation by performing the same operation on both sides to keep it 'balanced'.

Shutterstock.com/defpicture

The **backtracking method** involves using a flow chart to show what operations are performed on the variable (say, x) to create the equation, then using a reverse flow chart to undo each operation by performing the inverse (opposite) operation in reverse order.

Shutterstock.com/rnl

Worksheets
Equations 1
One-step equations
Guess-and-check
The Arabic science of balancing

Puzzles
Equations dominoes
One-step equations

Interactives
Equality explorer: Basics
Equality explorer

ⓘ Solving equations

To solve an equation, aim to have the variable (such as x) on one side of the equation and a number on the other side, in the form:

$x = \_\_\_\_$

Check your solution by substituting it back into the equation.

Example 2

Solve each equation.

a $m + 4 = 12$ **b** $3y = 18$ **c** $k - 3 = 5$ **d** $\frac{t}{8} = 6$

SOLUTION

a Method 1: The balancing method

Let m represent an unknown number in an envelope, and ◯ represent 1. We can represent the equation $m + 4 = 12$ on balance scales as shown.

$m + 4 = 12$

The 2 sides of the balance are equal. To find the value of $\boldsymbol{m}$, we can remove 4 balls from both sides.

$m + 4 - 4 = 12 - 4$

$m = 8$

The solution is $m = 8$.

We can check our answer by substituting it back into the equation to see if it is true.

Check: $8 + 4 = 12$.

Subtracting 4 from both sides.

Method 2: The backtracking method

First use a flowchart as shown to show how we get from $\boldsymbol{m}$ to $\boldsymbol{m} + 4$.

In this equation, $\boldsymbol{m} + 4 = 12$. To backtrack (get back) to $\boldsymbol{m}$, we need to undo the operation 'add 4', which is 'subtract 4', as shown in the reverse flowchart.

$m + 4 = 12$

$m = 12 - 4$

$\boldsymbol{m = 8}$.

The solution is $\boldsymbol{m = 8}$.

Check: $8 + 4 = 12$.

Undo '+ 4' by subtracting 4.

Starting at 12 on the flowchart and ending at 8.

Undoing what has been done to m.

Video One-step equations

b Method 1: The balancing method

$3y = 18$

$\frac{3y}{3} = \frac{18}{3}$

$y = 6$

Check: $3 \times 6 = 18$

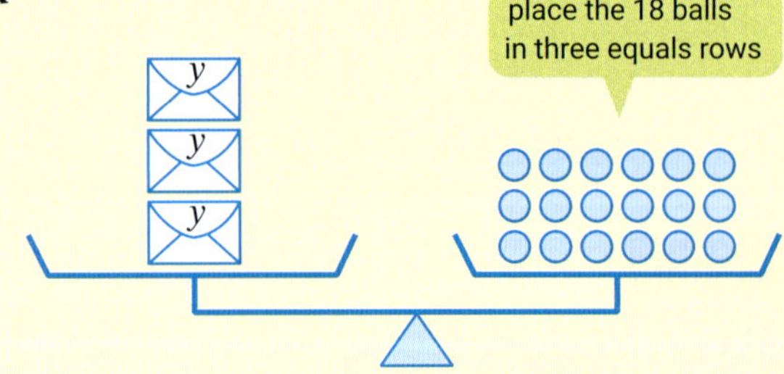

Dividing both sides by 3.

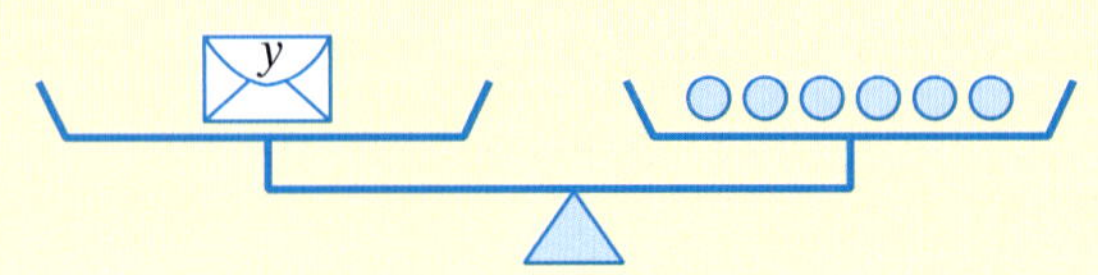

Method 2: The backtracking method

$3y = 18$

$y = \frac{18}{3}$

$y = 6$

y	$\xrightarrow{\times 3}$	$3y$
=		=
6	$\xleftarrow{\div 3}$	18

Undo '× 3' in 3***y*** by dividing by 3.

Check: 3 × 6 = 18.

c Method 1: The balancing method

$k - 3 = 5$

$k - 3 + 3 = 5 + 3$ Adding 3 to both sides.

$k = 8$

Check: $8 - 3 = 5$

Method 2: The backtracking method

$k - 3 = 5$

$k = 5 + 3$ Undo '– 3' by adding 3.

$k = 8$ Check: 8 – 3 = 5.

d Method 1: The balancing method

$\frac{t}{8} = 6$

$\frac{t}{8} \times 8 = 6 \times 8$ Multiplying both sides by 8.

$t = 48$ Check: $\frac{48}{8} = 6$.

Method 2: The backtracking method

$\frac{t}{8} = 6$

$t = 6 \times 8$ Undo '÷ 8' by multiplying by 8.

$t = 48$ Check: $\frac{48}{8} = 6$.

These are all 'one-step' equations because they require only one step or one inverse operation to solve.

ⓘ Inverse operations

Operation	Inverse operation
+	−
−	+
×	÷
÷	×

EXERCISE 10.02 ANSWERS ON P. 648

One-step equations

U F R C

1 Solve each equation, showing the working.

a $x + 5 = 12$ **b** $y + 13 = 36$ **c** $m + 9 = 21$ **d** $k + 27 = 54$

e $k + 6 = 2$ **f** $s + 3 = 1$ **g** $n + 9 = -4$ **h** $p + 17 = -3$

i $m + 8 = 0$ **j** $g + 4.5 = 20$ **k** $a + 1\frac{3}{4} = 12$ **l** $k + 2.7 = -5$

2 Solve each equation, showing the working.

a $t - 11 = 40$ **b** $w - 7 = 4$ **c** $a - 5 = 2$ **d** $j - 23 = 51$

e $q - 12 = 17$ **f** $f - 42 = 68$ **g** $x - 9 = 24$ **h** $g - 10 = 1$

i $j - 3 = -5$ **j** $w - 9 = -2$ **k** $m - 12 = -27$ **l** $d - 1 = -1$

m $y - 17 = 3.9$ **n** $n - 2.1 = 5.9$ **o** $b - \frac{4}{5} = \frac{1}{4}$ **p** $s - 2\frac{1}{3} = 4\frac{1}{2}$

3 Solve each equation, showing the working.

a $6x = 24$ **b** $4g = 16$ **c** $10y = 90$ **d** $5b = 45$

e $12d = 108$ **f** $9c = 81$ **g** $23p = 115$ **h** $2m = 26$

i $5x = -25$ **j** $7q = -42$ **k** $-3c = 54$ **l** $-15x = -120$

m $4p = 37$ **n** $8r = 18$ **o** $3q = -10$ **p** $9w = -12$

4 Solve each equation, showing the working.

a $\frac{s}{3} = 5$ **b** $\frac{m}{7} = 6$ **c** $\frac{d}{10} = 12$ **d** $\frac{w}{2} = 29$

e $\frac{a}{3} = -4$ **f** $\frac{f}{-2} = 9$ **g** $\frac{r}{5} = -3$ **h** $\frac{m}{-7} = 1$

i $\frac{x}{-8} = -6$ **j** $\frac{t}{2} = \frac{3}{4}$ **k** $\frac{n}{-4} = -4$ **l** $\frac{x}{3} = -\frac{1}{6}$

5 Solve $3x = -12$. Select the correct answer **A**, **B**, **C** or **D**.

A $x = -9$ **B** $x = -36$ **C** $x = -4$ **D** $x = -15$

6 Write a one-step equation whose solution is:

R C

a $m = 4$ **b** $x = 0$ **c** $r = 7$ **d** $d = -5$

☐ Foundation ◯ Standard ⬡ Complex

Two-step equations

10.03

Two-step equations require 2 steps or 2 inverse operations to solve.

Videos
Two-step equations
Solving equations

Skillsheets
Solving equations by balancing
Solving equations by backtracking
Solving equations using diagrams

Worksheets
Backtracking
Equations writing activity

Puzzle
Solving equations

Presentation
Solving linear equations

Spreadsheet
Linear equation solver

Example 3

Solve each equation.

a $3x + 2 = 17$ **b** $\frac{k}{5} + 4 = 7$

SOLUTION

a Method 1: The balancing method

$3x + 2 = 17$

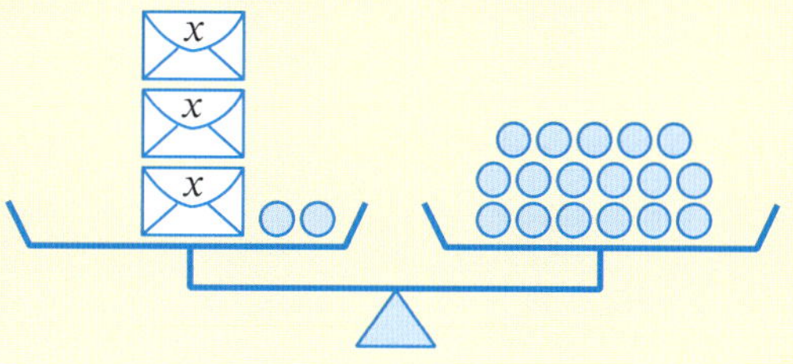

$3x + 2 - 2 = 17 - 2$

Step 1: Subtracting 2 from both sides.

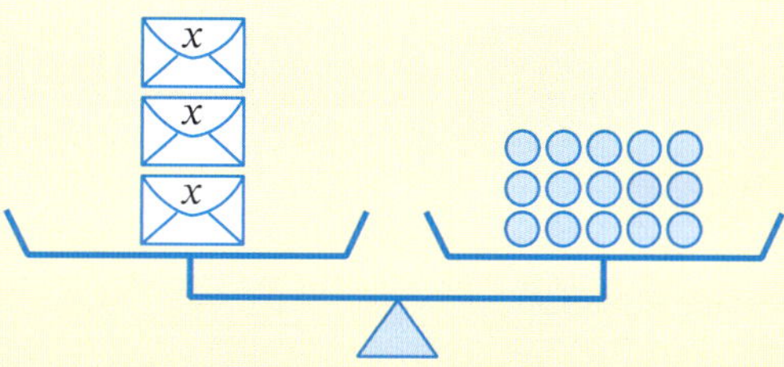

$3x = 15$

$\frac{3x}{3} = \frac{15}{3}$

Step 2: Dividing both sides by 3.

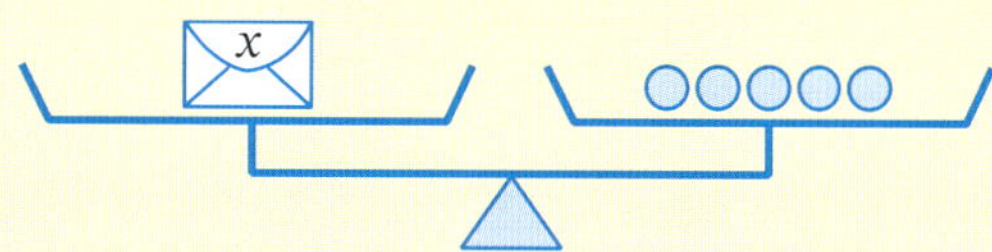

$x = 5$

Check: $3 \times 5 + 2 = 17$.

Method 2: The backtracking method

$3x + 2 = 17$

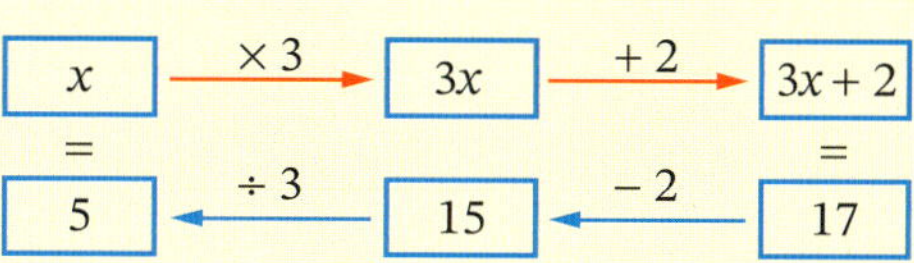

$3x = 17 - 2$

$x = \frac{15}{3}$

$x = 5$

Step 1: Undo '+ 2' by subtracting 2.

Step 2: Undo '× 3' in $3x$ by dividing by 3.

Check: $3 \times 5 + 2 = 17$.

b Method 1: The balancing method

$$\frac{k}{5} + 4 = 7$$

$$\frac{k}{5} + 4 - 4 = 7 - 4$$ ***Step 1***: Subtracting 4 from both sides.

$$\frac{k}{5} = 3$$

$$\frac{k}{5} \times 5 = 3 \times 5$$ ***Step 2***: Multiplying both sides by 5.

$$k = 15$$ Check: $\frac{15}{5} + 4 = 3 + 4 = 7$.

Method 2: The backtracking method

$$\frac{k}{5} + 4 = 7$$

$$\frac{k}{5} = 7 - 4$$ ***Step 1***: Undo '+ 4' by subtracting 4.

$$\frac{k}{5} = 3$$

$$k = 3 \times 5$$ ***Step 2***: Undo '÷ 5' by multiplying by 5.

$$k = 15$$ Check: $\frac{15}{5} + 4 = 3 + 4 = 7$.

Video Two-step equations

Example 4

Solve each equation.

a $\frac{2h}{3} = 4$ **b** $\frac{x+1}{2} = 7$

SOLUTION

a Method 1: The balancing method

$$\frac{2h}{3} = 4$$

$$\frac{2h}{3} \times 3 = 4 \times 3$$ ***Step 1***: Multiplying both sides by 3.

$$2h = 12$$

$$\frac{2h}{2} = \frac{12}{2}$$ ***Step 2:*** Dividing both sides by 2.

$$h = 6$$ Check: $\frac{2 \times 6}{3} = 4$.

Method 2: The backtracking method

$$\frac{2h}{3} = 4$$

$2h = 4 \times 3$ — ***Step 1***: Undo '÷ 3' by multiplying by 3.

$2h = 12$

$h = \frac{12}{2}$ — ***Step 2***: Undo '× 2' in $2h$ by dividing by 2.

$h = 6$ — Check: $\frac{2 \times 6}{3} = 4$

b Method 1: The balancing method

$$\frac{x+1}{2} = 7$$

$\frac{x+1}{2} \times 2 = 7 \times 2$ — ***Step 1:*** Multiplying both sides by 2.

$x + 1 = 14$

$x + 1 - 1 = 14 - 1$ — ***Step 2:*** Subtracting 1 from both sides.

$x = 13$ — Check: $\frac{13+1}{2} = 7$.

Method 2: The backtracking method

$$\frac{x+1}{2} = 7$$

$x + 1 = 7 \times 2$ — ***Step 1***: Undo '÷ 2' by multiplying by 2.

$x + 1 = 14$

$x = 14 - 1$ — ***Step 2***: Undo '+ 1' by subtracting 1.

$x = 13$ — Check: $\frac{13+1}{2} = 7$.

EXERCISE 10.03 ANSWERS ON P. 649

Two-step equations

1 Check which equation has $p = -2$ as its solution. Select the correct answer **A**, **B**, **C** or **D**.

A $2 + 3p = 8$ **B** $5p - 2 = 8$ **C** $-4p + 16 = 8$ **D** $-3p + 2 = 8$

2 Solve each equation, showing all steps. Remember to check your answers. EXAMPLE 3

a $2m + 7 = 19$	**b** $3x - 5 = 13$	**c** $5k + 12 = 52$
d $6w - 17 = 19$	**e** $4h + 21 = 9$	**f** $8d - 5 = -27$
g $-2x + 18 = 4$	**h** $-3a - 9 = -6$	**i** $14c + 2 = 37$
j $12p - 10 = 86$	**k** $9y + 11 = 101$	**l** $2r - 15 = 16$

Foundation Standard Complex

3 Solve each equation.

a $\frac{m}{2} + 4 = 13$ b $\frac{c}{5} + 9 = 12$ c $\frac{d}{7} + 2 = 8$

d $\frac{k}{3} + 15 = 6$ e $\frac{h}{2} + 7 = 17$ f $\frac{x}{4} + 2 = -4$

g $\frac{n}{5} - 9 = 4$ h $\frac{t}{3} - 6 = -2$ i $\frac{a}{8} - 2 = -6$

j $\frac{x}{6} - 16 = -13$ k $\frac{v}{15} - 2 = 4$ l $\frac{b}{3} - 4 = 3$

4 This is Liam's incorrect solution for $7x + 5 = 13$.

C

$7x + 5 = 13$

$7x = 13 - 5$ **A**

$7x = 8$ **B**

$x = \frac{8}{7}$ **C**

$x = 1\frac{1}{8}$ **D**

In which line was the error made? Select the correct answer **A**, **B**, **C** or **D**.

5 Solve each equation.

a $\frac{3x}{2} = 9$ b $\frac{15x}{9} = 15$ c $\frac{2x}{3} = 8$ d $\frac{3N}{5} = 3$

e $\frac{6N}{7} = 18$ f $\frac{5B}{2} = -10$ g $\frac{3x}{4} = -3$ h $\frac{2x}{5} = -4$

i $\frac{5m}{2} = 11$ j $\frac{-x}{3} = 8$ k $\frac{-4x}{5} = 1$ l $\frac{-2x}{3} = -10$

6 Check which value of k is the solution to $\frac{k-12}{9} = 6$. Select the correct answer **A**, **B**, **C** or **D**.

A $k = 27$ **B** $k = 42$ **C** $k = 66$ **D** $k = 162$

7 Solve each equation.

a $\frac{x+1}{3} = 2$ b $\frac{x-3}{2} = 3$ c $\frac{N+2}{5} = 1$ d $\frac{N-3}{4} = 5$

e $\frac{N+8}{5} = 6$ f $\frac{x+1}{2} = -2$ g $\frac{k-5}{2} = -5$ h $\frac{m+2}{3} = 11$

8 Write four equations, one of each type shown in questions **2**, **3**, **5** and **7**, that have the solution $p = 2$.

R C

Foundation Standard Complex

TECHNOLOGY

Guess, check and improve

1 To use a guess, check and improve method to solve the equation $2m + 7 = 19$, pick a number (not too big a number) – let's try the number 4.

2 Substitute the number into the left–hand side (LHS) of the equation, using your calculator, to get:

$2 \times 4 + 7 = 15.$

3 This number is too small, so try a higher number. If the answer was too high, pick a smaller number. Repeat this process until you get the correct answer of 19 (the right-hand side (RHS) of the equation.

4 Now use a spreadsheet to do the checking for you. On the spreadsheet enter the following:

	A	B	C
1	*m*	*2m+7*	*19*
2	1		
3	2		
4	3		
5	4		
6	5		
7	6		
8	7		
9	8		

5 Enter the equation "=2*A2+7" in cell B2 (this is the LHS of the equation).

6 Highlight cells B2 to B9 and fill down.

7 In this column you want the answer to be closest to 19 (which is the RHS of the equation).

8 Enter 19 into cell C2.

9 Highlight cells C2 to C9 and fill down.

10 Find the value of *m* for which columns B and C both say 19. This is the solution to the equation.

11 You can repeat this process entering the LHS of the equation into cell B1 and the RHS of the equation into cell C1 to check your answers to Exercise 10.03.

10.04 Equations with variables on both sides

Worksheet
Equations with unknowns on both sides

Puzzle
Equations with unknowns on both sides

Videos
Equations with variables on both sides

Solving equations

Presentation
Formally solving equations

For equations with variables on both sides, such as $3x + 4 = 2x + 7$, we can only use the **balancing method**, not the backtracking method.

ⓘ Equations with variables on both sides

Perform operations on both sides of the equation to move:

- all the variables onto one side of the equation.
- all the numbers onto the other side of the equation.

Example 5

Solve each equation.

a $3x + 4 = 2x + 7$ **b** $5n - 3 = 2n - 15$ **c** $2d - 8 = -5d - 71$

SOLUTION

a $3x + 4 = 2x + 7$

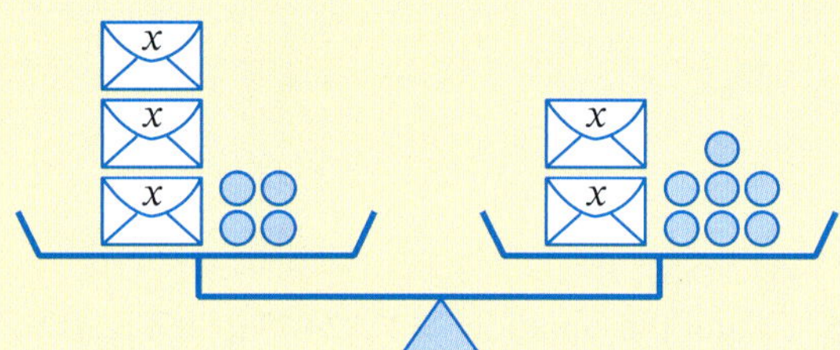

$3x + 4 - 2x = 2x + 7 - 2x$ Subtracting $2x$ from both sides to remove it from the RHS.

$x + 4 = 7$

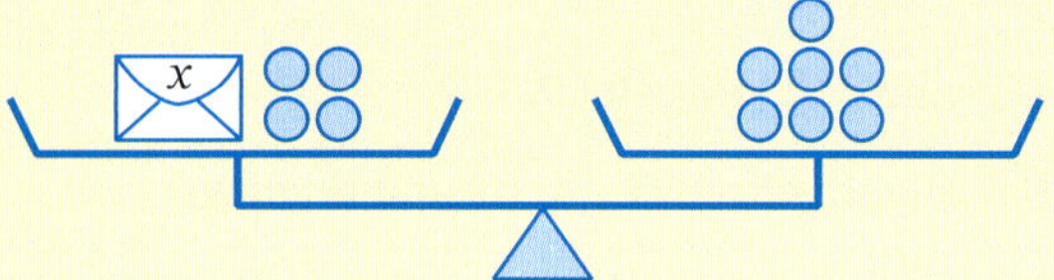

$x + 4 - 4 = 7 - 4$ Subtracting 4 from both sides to remove it from the LHS.

$x = 3$

Check: LHS $= 3 \times 3 + 4 = 13$.

RHS $= 2 \times 3 + 7 = 13$

LHS = RHS.

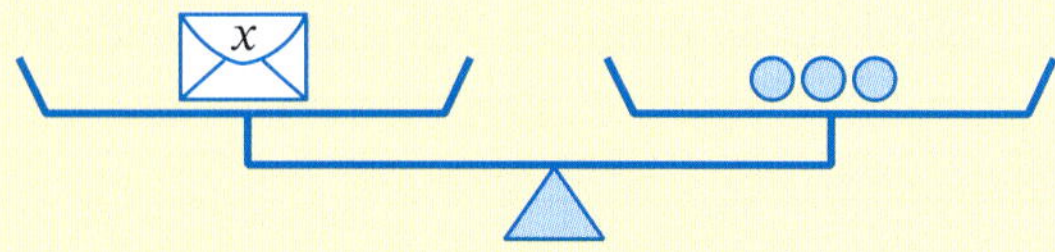

b $5n - 3 = 2n - 15$

$5n - 3 - 2n = 2n - 15 - 2n$ — Subtracting $2n$ from both sides to remove it from the RHS.

$3n - 3 = -15$ — Now this is a two-step equation.

$3n - 3 + 3 = -15 + 3$ — Adding 3 to both sides to remove it from the LHS.

$3n = -12$ — Now this is a one-step equation.

$\frac{3n}{3} = \frac{-12}{3}$ — Dividing both sides by 3.

$n = -4$ — Check:

LHS $= 5 \times (-4) - 3 = -23$.

RHS $= 2 \times (-4) - 15 = -23$.

LHS = RHS.

c $2d - 8 = -5d - 71$

$2d - 8 + 5d = -5d - 71 + 5d$ — Adding $5d$ to both sides to remove it from the RHS.

$7d - 8 = -71$ — Now this is a two-step equation.

$7d - 8 + 8 = -71 + 8$ — Add 8 to both sides to remove it from the LHS.

$7d = -63$ — Now this is a one-step equation.

$\frac{7d}{7} = \frac{-63}{7}$ — Dividing both sides by 7.

$d = -9$ — Check:

LHS $= 2 \times (-9) - 8 = -26$.

RHS $= -5 \times (-9) - 71 = -26$.

LHS = RHS.

EXERCISE 10.04 ANSWERS ON P. 649

Equations with variables on both sides

U F R C

EXAMPLE 5

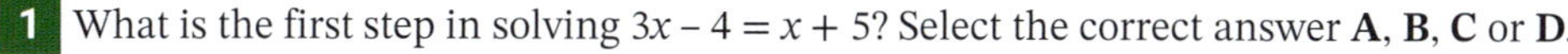

1 What is the first step in solving $3x - 4 = x + 5$? Select the correct answer **A**, **B**, **C** or **D**.

R C

A subtract x from both sides **B** add x to both sides

C add $3x$ to both sides **D** subtract 4 from both sides

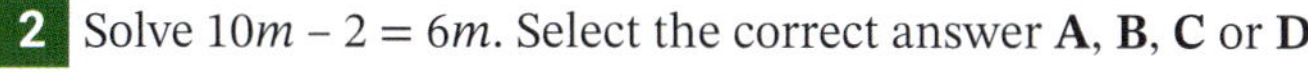

2 Solve $10m - 2 = 6m$. Select the correct answer **A**, **B**, **C** or **D**.

A $-\frac{1}{2}$ **B** $\frac{1}{8}$ **C** $\frac{1}{2}$ **D** 2

Foundation Standard Complex

3 Solve each equation, showing all steps. Remember to check your answers.

a $3a + 6 = a + 18$ **b** $2k + 4 = k + 8$ **c** $3x + 6 = 2x + 9$

d $6p + 4 = p + 19$ **e** $6n + 5 = 2n + 17$ **f** $5q + 6 = 4q + 12$

g $5y + 14 = 3y + 14$ **h** $4a + 7 = 2a + 21$ **i** $2r + 9 = r + 15$

4 Solve $4x - 5 = 2x + 7$. Select the correct answer **A**, **B**, **C** or **D**.

A $x = 1$ **B** $x = 2$ **C** $x = 4$ **D** $x = 6$

5 Solve each equation.

a $5a - 3 = 2a + 6$ **b** $6x - 2 = 3x + 6$

c $3d - 6 = d - 4$ **d** $8p - 15 = 3p - 10$

e $3m + 5 = -m + 26$ **f** $4x + 6 = -6x + 56$

g $4x + 3 = -2x + 7$ **h** $2r + 18 = -5r + 11$

i $3y - 12 = -y + 6$

6 R Here is Bree's solution for $3x - 4 = -2x + 8$.

$3x - 4 = -2x + 8$

$3x + 2x - 4 = 8$ Line 1

$5x - 4 = 8$ Line 2

$5x = 8 - 4$ Line 3

$5x = 4$ Line 4

$x = \frac{5}{4}$ Line 5

In which lines did Bree make mistakes? Select the correct answer **A**, **B**, **C** or **D**.

A Lines 1 and 3 **B** Lines 2 and 5 **C** Lines 1 and 4 **D** Lines 3 and 5

7 R Solve each equation.

a $3d + 4 = d$ **b** $10k = 12 - 8k$ **c** $5 - p = p + 9$

d $7 + x = 6x + 22$ **e** $6k - 11 = k$ **f** $3m + 20 = 7m - 2$

g $4t = 12 - 4t$ **h** $8j - 17 = 10j$ **i** $6 - 3q = 8 - q$

8 R C Write an equation with x on both sides that has the solution $x = 7$.

Foundation Standard Complex

Equations with brackets

ⓘ Equations with brackets (grouping symbols)

Expand the expressions and then solve as usual.

Worksheets
Equations 2
Equations 3 (Extension)
Checking solutions

Videos
Equations with brackets
Solving equations

Puzzles
Equations match
Equations
Equations order activity

Example 6

Solve each equation.

a $3(x + 5) = 9$ **b** $2(3 - y) = 4y + 4$ **c** $5(r - 3) = 2(r + 6)$

SOLUTION

a $3(x + 5) = 9$

$3x + 15 = 9$ Expanding the expression to make it a two-step equation.

Can you think of another way to solve this equation without expanding?

$3x + 15 - 15 = 9 - 15$ Subtracting 15 from both sides.

$3x = -6$

$\frac{3x}{3} = \frac{-6}{3}$ Dividing both sides by 3.

$x = -2$ Check: $3(-2 + 5) = 3 \times 3 = 9$

b $2(3 - y) = 4y + 4$

$6 - 2y = 4y + 4$ Expanding the brackets.

$6 - 2y - 4y = 4y + 4 - 4y$ Subtracting $4y$ from both sides to remove it from the RHS.

$6 - 6y = 4$ Now this is a two-step equation.

$6 - 6y - 6 = 4 - 6$ Subtracting 6 from both sides to remove it from the LHS.

$-6y = -2$

$\frac{-6y}{-6} = \frac{-2}{-6}$ Dividing both sides by (−6).

$y = \frac{1}{3}$ Check:

$\text{LHS} = 2 \times \left(3 - \frac{1}{3}\right) = 5\frac{1}{3}$.

$\text{RHS} = 4 \times \frac{1}{3} + 4 = 5\frac{1}{3}$.

LHS = RHS.

c $5(r - 3) = 2(r + 6)$

$5r - 15 = 2r + 12$ Expanding both sides.

$5\boldsymbol{r} - 15 - 2\boldsymbol{r} = 2\boldsymbol{r} + 12 - 2\boldsymbol{r}$	Subtract $2\boldsymbol{r}$ from both sides to remove it from the RHS.
$3\boldsymbol{r} - 15 = 12$	Now this is a two-step equation.
$3\boldsymbol{r} - 15 + 15 = 12 + 15$	Add 15 to both sides to remove it from the LHS.
$3\boldsymbol{r} = 27$	
$\frac{3r}{3} = \frac{27}{3}$	Dividing both sides by 3.
$\boldsymbol{r} = 9$	Check: LHS $= 5 \times (9 - 3) = 5 \times 6 = 30$. RHS $= 2 \times (9 + 6) = 2 \times 15 = 30$. LHS = RHS.

EXERCISE 10.05 ANSWERS ON P. 649

Equations with brackets

1 Solve each equation, showing all steps. Remember to check your answers.

a $2(x + 3) = 8$ **b** $3(m + 2) = 18$ **c** $5(k + 1) = 35$

d $4(p + 4) = 32$ **e** $10(j + 2) = 50$ **f** $7(d + 4) = 63$

g $3(x - 4) = 15$ **h** $3(x - 4) = 12$ **i** $5(x - 1) = 10$

j $8(x - 5) = 24$ **k** $4(k - 5) = 28$ **l** $2(q - 7) = 26$

2 Solve each equation.

a $2(6 - x) = 6$ **b** $3(4 - p) = 24$ **c** $2(1 - q) = 8$

d $3(16 - h) = 15$ **e** $2(4 + r) = -8$ **f** $3(e - 16) = -15$

g $5(z - 3) = -10$ **h** $-7(y + 2) = 35$ **i** $9(6 - a) = 0$

j $5(d + 3) = -20$ **k** $-2(3 - k) = 12$ **l** $4(6 - c) = -16$

3 Solve $2(12 - 3x) = 30$. Select the correct answer **A**, **B**, **C** or **D**.

A $x = -2$ **B** $x = -1$ **C** $x = 1$ **D** $x = 2$

4 Solve each equation.

a $6(x + 1) = 3x + 3$ **b** $2(r + 2) = r + 5$

c $3(p + 3) = 2p + 2$ **d** $2(x + 1) = x$

e $5(p - 2) = 3p$ **f** $4(e - 2) = 2e + 4$

g $5(y + 2) = 3y + 12$ **h** $3(a - 5) = 5 - 2a$

i $3(2w + 1) = 3w - 15$

□ Foundation ○ Standard ○ Complex

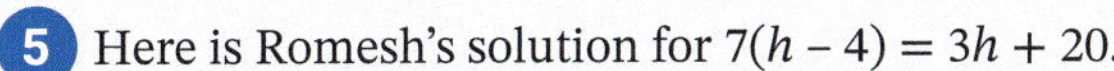

5 Here is Romesh's solution for $7(h-4) = 3h + 20$.

R C

$$7(h-4) = 3h + 20$$

$7h - 4 = 3h + 20$ Line A

$7h - 4 - 3h = 3h + 20 - 3h$

$4h - 4 = 20$ Line B

$4h - 4 + 4 = 20 + 4$

$4h = 24$ Line C

$$\frac{4h}{4} = \frac{24}{4}$$

$h = 6$ Line D

In which line did Romesh make a mistake? Select the correct answer **A**, **B**, **C** or **D**.

6 Solve each equation.

a $5(x-1) = 4(x+2)$ **b** $6(x+2) = 4(x+6)$ **c** $4(k+3) = 3(k-2)$

d $7(y+2) = 4(y+5)$ **e** $3(v+2) = 2(v+5)$ **f** $4(x-2) = 2(x+7)$

g $3(p+1) = 5(p-1)$ **h** $3(2s+1) = 5(s+2)$ **i** $5(d+4) = 2(2d-1)$

7 Write an equation involving brackets whose solution is:

R C

a $c = 5$ **b** $k = -1$ **c** $n = 2$ **d** $q = -4$

☆ **MENTAL SKILLS** (10) ANSWERS ON P. 649 **Maths without calculators**

Quiz
Mental skills 10

Multiplying and dividing by 5, 15, 25 and 50

It is easier to multiply or divide a number by 10 than by 5. So whenever we multiply or divide a number by 5, we can double the 5 (to make 10) and then adjust the first number.

1 Study each example.

a To multiply by 5, halve the number, then multiply by 10.

$$18 \times 5 = 18 \times \frac{1}{2} \times 10 \quad (\text{or } 9 \times 2 \times 10)$$
$$= 9 \times 10$$
$$= 90$$

b To multiply by 50, halve the number, then multiply by 100.

$$26 \times 50 = 26 \times \frac{1}{2} \times 100 \quad (\text{or } 13 \times 2 \times 100)$$
$$= 13 \times 100$$
$$= 1300$$

c To multiply by 25, quarter the number, then multiply by 100.

$$44 \times 25 = 44 \times \frac{1}{4} \times 100 \quad (\text{or } 11 \times 4 \times 25)$$
$$= 11 \times 100$$
$$= 1100$$

d To multiply by 15, halve the number, then multiply by 30.

$$8 \times 15 = 8 \times \frac{1}{2} \times 30 \quad (\text{or } 4 \times 2 \times 15)$$
$$= 4 \times 30$$
$$= 120$$

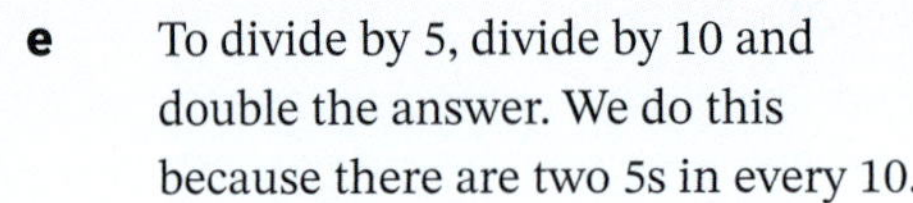

e To divide by 5, divide by 10 and double the answer. We do this because there are two 5s in every 10.

$$\begin{aligned}140 \div 5 &= 140 \div 10 \times 2\\ &= 14 \times 2\\ &= 28\end{aligned}$$

f To divide by 50, divide by 100 and double the answer. This is because there are two 50s in every 100.

$$\begin{aligned}400 \div 50 &= 400 \div 100 \times 2\\ &= 4 \times 2\\ &= 8\end{aligned}$$

g To divide by 25, divide by 100 and multiply the answer by 4. This is because there are four 25s in every 100.

$$\begin{aligned}600 \div 25 &= 600 \div 100 \times 4\\ &= 6 \times 4\\ &= 24\end{aligned}$$

h To divide by 15, divide by 30 and double the answer. This is because there are two 15s in every 30.

$$\begin{aligned}240 \div 15 &= 240 \div 30 \times 2\\ &= 8 \times 2\\ &= 16\end{aligned}$$

2 Now evaluate each expression.

a	32×5	**b**	14×5	**c**	48×5	**d**	18×50
e	52×50	**f**	36×25	**g**	28×5	**h**	12×25
i	12×15	**j**	22×35	**k**	$90 \div 5$	**l**	$170 \div 5$
m	$230 \div 5$	**n**	$1300 \div 50$	**o**	$900 \div 50$	**p**	$300 \div 25$
q	$1000 \div 25$	**r**	$360 \div 45$	**s**	$210 \div 15$	**t**	$360 \div 15$

10.06 Equation problems

Videos
Equation problems
Solving equations

Worksheets
Solving equations review
Angle problems with algebra

Presentation
Writing equations

Puzzle
Writing and solving equations

ⓘ Solving word problems requiring an equation

- Choose your variable.
- Translate the words into an equation.
- Solve the equation.
- Write a sentence that answers the problem.

Example 7

a When a number is doubled and 11 is subtracted, the result is 23. Find the number.

b Five times a number is the same as 6 more than three times the number. What is the number?

SOLUTION

a Let the number be x.

$x \times 2 - 11 = 23$

$2x - 11 = 23$

Translating the words into an equation.

Solving the equation:

$2x - 11 + 11 = 23 + 11$ Adding 11 to both sides.

$2x = 34$

$\frac{2x}{2} = \frac{34}{2}$ Dividing both sides by 2.

$x = 17$ Check: $2 \times 17 - 11 = 23$.

The number is 17.

b Let n represent the number.

$5n = 3n + 6$

$5n - 3n = 3n + 6 - 3n$ Subtracting $3n$ from both sides.

$2n = 6$

$\frac{2n}{2} = \frac{6}{2}$ Dividing both sides by 2.

$n = 3$

The number is 3.

Check:

LHS $= 5 \times 3 = 15$.

RHS $= 3 \times 3 + 6 = 15$.

LHS = RHS.

Example 8

Cooper charges for fixing washing machines using the formula: $\boldsymbol{C} = 45\boldsymbol{h} + 70$, where $\boldsymbol{C}$ is the charge in dollars and $\boldsymbol{h}$ is the number of hours the job takes.

Find:

a the charge for a job that takes 3 hours.

b the number of hours Cooper worked if the charge is $430.

SOLUTION

a Substitute $\boldsymbol{h} = 3$ into the formula.

$$\begin{aligned} C &= 45h + 70 \\ &= 45 \times 3 + 70 \\ &= 205 \end{aligned}$$

The charge is $205.

b Substitute $\boldsymbol{C} = 430$ into the formula and solve the equation.

$$430 = 45\boldsymbol{h} + 70$$

$$45\boldsymbol{h} + 70 = 430$$

$$\begin{aligned} 45h + 70 - 70 &= 430 - 70 \\ 45h &= 360 \\ \frac{45h}{45} &= \frac{360}{45} \\ h &= 8 \end{aligned}$$

The number of hours Cooper worked is 8.

EXERCISE 10.06 ANSWERS ON P. 649

Equation problems

1 Solve each problem by writing an equation and then solving it. You may use diagrams to help you think about the information.

(R) (C)

a Four family tickets for a theme park cost $352. How much does each ticket cost? (Let t represent the price of a family ticket.)

b Twenty cans of drink cost $36. How much does each can cost? (Let c represent the cost of one can.)

c A number is tripled and the result is 99. What is the number? (Let x represent the unknown number.)

d A number has 7 added to it and the result is 23. Find the number. (Let n represent the unknown number.)

2 Select the correct equation **A**, **B**, **C** or **D** for each problem. Then use the equation to solve the problem.

R C

a If 23 is subtracted from a number, the answer is 79. What is the number?

A $79 + N = 23$ **B** $23N = 79$

C $23 - N = 79$ **D** $N - 23 = 79$

b The sum of a number and 18 is 67. What is the number?

A $N + 67 = 18$ **B** $N + 18 = 67$

C $18N = 67$ **D** $N - 18 = 67$

c When a number is multiplied by 23, the answer is 1288. What is the number?

A $1288N = 23$ **B** $N + 23 = 1288$

C $1288 - N = 23$ **D** $23N = 1288$

d When a number is divided by 14, the answer is 42. What is the number?

A $N - 14 = 42$ **B** $\frac{N}{14} = 42$

C $14N = 42$ **D** $14 + N = 42$

e Marika buys a car that costs \$13 499. She pays a deposit of \$4525. How much does she owe?

A $4525N = 13\,499$ **B** $N - 4525 = 13\,499$

C $N + 4525 = 13\,499$ **D** $N = 13\,499 + 4525$

3 For each problem, write an equation and find the unknown number.

R C

a When an unknown number is divided by 5, the result is 15.

b When 12 is subtracted from a number, the answer is 37.

c When a number is subtracted from 52, the answer is 25.

d An unknown number is multiplied by 3, then 8 is subtracted. The result is 40.

e An unknown number is halved, then 7 is added. The result is 13.

f The sum of a number and 4 is doubled. The result is 44.

g The sum of a number and 6 is divided by 3 and the result is 7.

h The product of a number and 5 is decreased by 2. The result is 28.

4 Translate each problem into an equation, then solve the equation to solve the problem.

PS R C

a The student council is holding a disco to raise money. Each ticket bought by students raises \$20, but the costs of running the disco total \$1000. How many tickets must be sold to make a profit of \$2000? (Let n stand for the number of tickets sold.)

b In 8 years Kelly's age will be twice what it is now. How old is Kelly now? (Let n stand for Kelly's age now.)

c Eight sheep have the same mass as three sheep and one cow. If the cow's mass is 500 kg, what is the mass of one sheep? (Let s stand for the mass of one sheep.)

Foundation Standard Complex

d If you multiply Liam's favourite number by 3 and add 1, you get the same answer as if you multiplied the number by 5 and took away 11. What is the number? (Let x represent the number.)

e Mr Yen says, 'If you add 15 to my age and multiply by 7, the answer is 483'. How old is Mr Yen? (Let a stand for his age.)

f The area of a rhombus is calculated by multiplying its diagonals and dividing by 2. If a rhombus has an area of 88 cm^2 and one diagonal is 11 cm, what is the length of the other diagonal? (Let d represent the length of the diagonal.)

5 A temperature in degrees celsius (°C) can be converted to degrees Fahrenheit (°F) using the formula $F = \frac{9C}{5} + 32$.

PS C

a Convert 25°C to °F.

b Convert 108°F to °C.

6 The charge, \$$C$, for hiring a hall for an event is $C = 150 + 2N$, where N stands for the number of people at the event. Find:

PS R

a the charge when 225 people are at the event.

b the number of people at the event when the charge is \$394.

7 The cost, \$$y$, of an online ad on a local news website is $y = 0.8w + 3.5$, where w is the number of words in the ad. Find:

PS R

a the cost of a 13-word ad.

b the number of words in an ad costing \$25.90.

8 The perimeter of this rectangle is 47 cm. Find:

PS R

a the value of x.

b the length of the rectangle.

c the width of the rectangle.

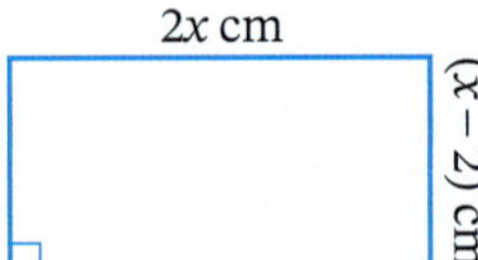

9 The profit, \$$P$, made by a hairdresser is given by $P = 18x - 900$, where x represents the number of customers. Find:

PS R

a the profit made with 195 customers.

b the number of customers if the profit is \$3060.

10 Find the value of x in this isosceles triangle.

R

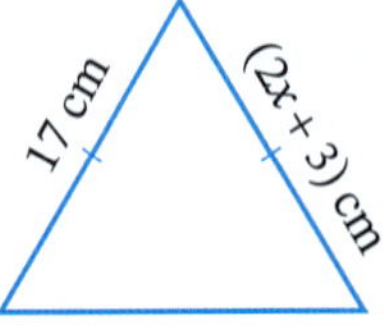

Foundation Standard Complex

INVESTIGATION

Solving $x^2 = c$

1 What is the inverse operation of 'squaring'?

2 The equation $x^2 = 9$ has two solutions. What are the two numbers, x, which when squared give 9?

3 What are the solutions for each equation? What is the pattern found in the answers?

a $x^2 = 25$ b $x^2 = 100$ c $x^2 = 1$

4 Study this example:

$x^2 = 49$

$x = \pm\sqrt{49}$ which means $x = \sqrt{49}$ or $-\sqrt{49}$.

$= \pm 7$ which means $x = 7$ or $x = -7$.

Check: When $x = 7, x^2 = 7^2 = 49$.

When $x = -7, x^2 = (-7)^2 = 49$.

Use the same method to solve each equation and check your answers:

a $x^2 = 81$ b $x^2 = 64$

Extension: Simple quadratic equations $x^2 = c$ 10.07

An equation involving a variable squared, such as $x^2 = 25$ or $3x^2 - 4 = 7$, is called a **quadratic equation**. In this section, we will solve simple quadratic equations of the type $x^2 = c$, where c is a number.

Example 9

Solve each quadratic equation.

a $x^2 = 36$

b $x^2 = 121$

c $x^2 = 40$, writing the solution correct to one decimal place.

d $x^2 = 83$, writing the solution as a surd.

SOLUTION

a $x^2 = 36$ — Finding the square root of both sides.

$x = \pm\sqrt{36}$ — The equation has two solutions, $x = 6$ or $x = -6$.

$= \pm 6$

b $x^2 = 121$ — Finding the square root of both sides.

$x = \pm\sqrt{121}$ — The equation has solutions $x = 11$ or $x = -11$.

$= \pm 11$

YEAR 9 EXTENSION

c $x^2 = 40$

$x = \pm\sqrt{40}$

$= \pm 6.3345\ldots$

$\approx \pm 6.3$

The equation has solutions $\boldsymbol{x} \approx 6.3$ or $\boldsymbol{x} \approx -6.3$.

d $x^2 = 83$

$x = \pm\sqrt{83}$

Leaving the answer as a surd.

$x = \sqrt{83}$ or $-\sqrt{83}$

Note: We have already solved equations of the form in Chapter 1, *Pythagoras' theorem*, but for Pythagoras' theorem we only take the positive solution because the length of a side of a right-angled triangle is always positive.

Simple quadratic equations $x^2 = c$

The simple quadratic equation $x^2 = c$ (where c is a positive number) has two solutions, $x = \pm\sqrt{c}$ (which means $x = \sqrt{c}$ or $x = -\sqrt{c}$).

EXERCISE 10.07 ANSWERS ON P. 649

Simple quadratic equations x² = c

1 Solve each quadratic equation.

a $x^2 = 81$ **b** $x^2 = 144$ **c** $x^2 = 1$

d $x^2 = 169$ **e** $m^2 = 5041$ **f** $u^2 = 1849$

2 Solve each quadratic equation, writing the solution correct to two decimal places.

a $x^2 = 13$ **b** $x^2 = 54$ **c** $x^2 = 88$

d $t^2 = 129$ **e** $h^2 = 946$ **f** $z^2 = 527$

3 Solve each quadratic equation, writing the solution as a surd.

a $x^2 = 41$ **b** $x^2 = 30$ **c** $x^2 = 48$

d $a^2 = 126$ **e** $p^2 = 75$ **f** $b^2 = 509$

4 R Many simple quadratic equations have two solutions (one positive and one negative), but there is one simple quadratic equation that has only **one** solution. What is this equation and what is its solution?

5 R C The equation $x^2 = -100$ has **no solutions**. Explain why.

6 R C Write two more simple quadratic equations that have no solutions.

Foundation Standard Complex

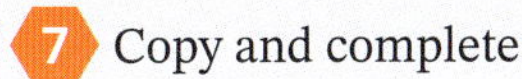

7 Copy and complete:

- (R) **a** $x^2 = c$ has two solutions if c is ______________.
- (C) **b** $x^2 = c$ has no solutions if c is ______________.
- **c** $x^2 = c$ has one solution if c is ______________.

8 Solve each quadratic equation.

(R) **a** $3x^2 = 27$	**b** $2x^2 = 32$	**c** $3x^2 = 75$	**d** $5x^2 - 1 = 19$
e $5x^2 + 1 = 21$	**f** $7 + 3x^2 = 34$	**g** $6x^2 - 8 = 142$	**h** $7x^2 + 9 = 261$

DID YOU KNOW?

Applications of quadratic equations

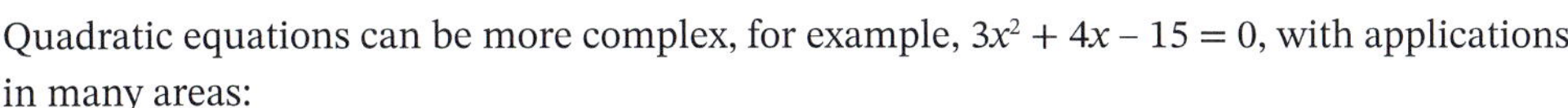

Quadratic equations can be more complex, for example, $3x^2 + 4x - 15 = 0$, with applications in many areas:

- the shape of satellite dishes and reflecting telescopes.
- the path of a thrown object, such as a ball, shotput or javelin.
- the orbits of planets as they move around the Sun.

Research the formula that solves quadratic equations.

Shutterstock.com/Dabarti CGI

10.08 Extension: Equations and formulas

A **formula** is a rule written in algebraic form. It shows the relationship between variables.

For example:

- the formula $A = lw$ gives the area, A, of a rectangle with length l and width w.
- the formula $D = ST$ gives the distance, D, travelled by a car at a speed S for time T.

Worksheet Working with formulas

Solving mathematical problems often involves substituting values in formulas and solving equations.

Example 10

The formula for the perimeter, $\boldsymbol{P}$, of a rectangle of length $\boldsymbol{l}$ and width $\boldsymbol{w}$ is given by $\boldsymbol{P} = 2(\boldsymbol{l} + \boldsymbol{w})$. Find a rectangle's:

a length if its perimeter is 40 cm and its width is 8 cm.

b width if its perimeter is 100 cm and its length is 35 cm.

SOLUTION

a Substitute $\boldsymbol{P} = 40$, $\boldsymbol{w} = 8$:

$$\boldsymbol{P} = 2(\boldsymbol{l} + \boldsymbol{w})$$
$$\mathbf{40 = 2(\boldsymbol{l} + 8)}$$
$$40 = 2\boldsymbol{l} + 16$$
$$2\boldsymbol{l} + 16 = 40$$
$$2l + 16 - 16 = 40 - 16$$
$$2\boldsymbol{l} = 24$$
$$\frac{2l}{2} = \frac{24}{2}$$
$$\boldsymbol{l} = 12$$

b Substitute $\boldsymbol{P} = 100$, $\boldsymbol{l} = 35$:

$$\boldsymbol{P} = 2(\boldsymbol{l} + \boldsymbol{w})$$
$$\mathbf{100 = 2(35 + \boldsymbol{w})}$$
$$100 = 70 + 2\boldsymbol{w}$$
$$70 + 2\boldsymbol{w} = 100$$
$$70 + 2w - 70 = 100 - 70$$
$$2\boldsymbol{w} = 30$$
$$\frac{2w}{2} = \frac{30}{2}$$
$$\boldsymbol{w} = 15$$

EXERCISE 10.08 ANSWERS ON P. 650

YEAR 9 EXTENSION

Equations and formulas

U F PS R C

EXAMPLE 10

1 The volume of a rectangular prism with length, l, width, w and height, h, is given by $V = lwh$. Use this formula to find the length of a rectangular prism with width 4 and height 5 if its volume is 340. Select the correct answer **A**, **B**, **C** or **D**.

A 17 **B** 68 **C** 85 **D** 425

2 The number of toothpicks (T) needed to build a row of N squares is $T = 3N + 1$.

a How many toothpicks are needed to build a row of 4 squares?

b How many squares can be built using 322 toothpicks?

3 PS R The number of hours (H) of sleep that children need depends on their age (A) in years and is given by the formula $H = 17 - \frac{A}{2}$. Find:

a the hours of sleep needed when a child is 14 years old.

b how old a child who needs 8 hours sleep is?

4 PS R Using the perimeter formula for a rectangle, $P = 2(l + w)$, find:

a the perimeter when length = 13 and width = 4.

b the length when perimeter = 36 and width = 7.

c the width when perimeter = 58 and length = 16.

5 PS R The area of a trapezium is given by $A = \frac{1}{2}(a+b)h$, where a and b are the lengths of its parallel sides and h is the perpendicular height between them. Find the value of:

a the area of a trapezium with parallel sides 20 and 13 and height 8.

b the height of a trapezium with parallel sides 22 and 8 and area 90.

c a parallel side of a trapezium with area 64, height 4 and other parallel side 17.

d a parallel side of a trapezium with other parallel side 5, height 12 and area 120.

6 PS R The volume of a pyramid is given by $V = \frac{1}{3}Ah$, where A is the area of the base and h is the height of the pyramid. Find the value of:

a the volume of a pyramid with base area 25 and height 7.

b the base area of a pyramid with volume 100 and height 10.

c the height of a pyramid with volume 225 and base area 75.

7 PS R The simple interest (I) earned on principal $\$P$ invested for n years at an interest rate of r p.a. (where r is a decimal) is given by $I = Prn$.

p.a. means 'per year'

a Find the simple interest earned when \$1250 is invested at 2% p.a. for 5 years (convert 2% to a decimal first).

b Find the principal invested if interest was \$60 and the interest rate was 5% p.a. for 3 years.

Foundation Standard Complex

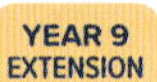

YEAR 9 EXTENSION

c Find the interest rate (as a percentage) if \$700 invested for 3 years earned \$84 in interest.

d Find the number of years that \$1650 was invested if it earned \$346.50 interest at 3% p.a.

8 R **a** Three people meet and each person shakes hands with every other person. How many handshakes were there?

b In how many ways can four people shake hands?

c The formula $H=\frac{P(P-1)}{2}$, where H is the number of handshakes, and P is the number of people, gives the number of shakes between people. In how many ways can 10 people shake hands?

d How many people are needed for 105 handshakes?

INVESTIGATION

Solving inequalities

We have solved equations by doing the same thing to both sides (keeping the equation 'balanced'). Will this method work with inequalities, such as $x + 4 > 10$ or $6x < 13$?

1 Start with an inequality that is true, such as $7 > 4$.

2 Add 5 (or any number you choose) to both sides of the inequality; for example, $7 > 4$ becomes $12 > 9$. Is the new inequality true or false?

3 Subtract 9 (or any number you choose) from each side of the original inequality; for example, $7 > 4$ becomes $-2 > -5$. Is the new inequality true or false?

4 Multiply both sides of the original inequality by 4 (or any positive number you choose); for example, $7 > 4$ becomes $28 > 16$. Is the new inequality true or false?

5 Divide both sides of the original inequality by 2 (or any positive number you choose); for example, $7 > 4$ becomes $3\frac{1}{2} > 2$. Is the new inequality true or false?

6 Multiply both sides of the original inequality by −3 (or any negative number you choose); for example, $7 > 4$ becomes $-21 > -12$. Is the new inequality true or false?

7 Divide both sides of the original inequality by −4 (or any negative number you choose), for example, $7 > 4$ becomes $-1\frac{3}{4} > -1$. Is the new inequality true or false?

8 Which of the six operations used in questions **2** to **7** can be used on inequalities to give a true result?

9 Which of the six operations used in questions **2** to **7** cannot be used with inequalities because they give a false result?

10 Copy and complete each set of inequality statements.

a $6 < 8$

$6 \times 3 < 8 \times$ ___ (multiplying both sides by 3)

$\therefore 18$ ___ 24

Foundation Standard Complex

b $10 > -4$

$10 \div 2$ __ $-4 \div$ __ (dividing both sides by 2)

$\therefore$ ___________

Does the inequality sign (< or >) stay the same when multiplying or dividing by a positive number?

11 a Is it true that $5 < 8$?

b Multiply both sides by −2. Is it true that $-10 < -16$?

c What needs to be reversed to change $-10 < -16$ into a true inequality statement?

d Copy and complete to make a true statement: −10 _______ −16.

12 a Is it true that $18 > -6$?

b Divide both sides by −3. Is it true that $-6 > 2$?

c What needs to be reversed to change $-6 > 2$ into a true inequality statement?

d Copy and complete to make a true inequality statement: −6 _____ 2.

13 Copy and complete: When multiplying or d___________ both sides of an inequality by an___________ number, the inequality sign must be r___________.

Inequalities 10.09

An **inequality** looks like an equation except that the equals sign (=) is replaced by an inequality symbol like > or <.

$2x - 7 = 15$ is an **equation**. There is only one value of x that makes it true.

$2x - 7 < 15$ is an **inequality**. There is a range of values of x that make it true.

Interactive Number line: Integers

Video Inequalities: The world's most populous country

ⓘ Inequality symbols

> 'is greater than'

< 'is less than'

For example, '$x > 3$' is read 'x is greater than 3'. It includes 3 and all the numbers above 3, such as 3.01, 4, 10, 20 000.

Inequality	In words	Meaning
$x > 3$	x is greater than 3.	Values above 3.
$x < 3$	x is less than 3.	Values below 3.

An **inequality** can be solved like an equation.

For example, the solution to $2x - 5 = 11$ is $x = 8$.

The value $x = 8$ is also important for the inequality, $2x - 5 < 11$. The solution to the inequality will either be less than 8 or more than 8. You can pick a number less than or greater than 8 and substitute it into the inequality to see if it will make the inequality true or not.

ⓘ Solving inequalities

Inequalities can be solved algebraically in the same way as equations, using inverse operations. However, when **multiplying** or **dividing** both sides of an inequality by a **negative** number, you must ***reverse* the inequality symbol**.

Note that:

$3 < 4$ '3 is less than 4'.

but

$-3 > -4$ '−3 is greater than −4' (You can look at the number line below to verify this.)

−5 −4 −3 −2 −1 0 1 2 3 4 5

Video
Solving inequalities

Example 11

Solve each equation.

a $m + 4 < 12$ **b** $3y > 18$ **c** $k - 3 < 5$

d $-\frac{t}{8} > 6$ **e** $-3x + 2 < 17$ **f** $\frac{c}{5} + 4 > 7$

SOLUTION

a $m + 4 - 4 < 12 - 4$ Subtracting 4 from both sides.

$m < 8$

Check: Try any number less than 8, try 7

$7 + 4 < 12$

$11 < 12$ correct

b $\frac{3y}{3} > \frac{18}{3}$

$y > 6$

Check: Try any number greater than 6, try 7

$3 \times 7 > 18$

$21 > 18$ correct

c $k - 3 < 5$

$k - 3 + 3 < 5 + 3$ Adding 3 to both sides.

$k < 8$

Check: Try any number less than 8, try 7

$7 - 3 < 5$

$4 < 5$ correct

10.09

d $-\frac{t}{8} > 6$

$-\frac{t}{8} \times 8 > 6 \times 8$ — Multiplying both sides by 8.

$-t > 48$

$\frac{-t}{-1} < \frac{48}{-1}$ — Dividing both sides by (−1) and reversing the inequality symbol from > to <.

$t < -48$

Check: Try any number less than −48, try −56, which is divisible by 8.

$-\frac{-56}{8} > 6$

$\frac{56}{8} > 6$

$7 > 6$

e $-3x + 2 - 2 < 17 - 2$ — Subtracting 2 from both sides.

$-3x < 15$

$\frac{-3x}{-3} > \frac{15}{-3}$ — Dividing both sides by (−3) and reversing the inequality symbol from < to >.

$x > -5$

Check: Try any number greater than −5, try −4.

$-3 \times (-4) + 2 < 17$

$14 < 17$

f $\frac{c}{5} + 4 > 7$

$\frac{c}{5} + 4 - 4 > 7 - 4$ — Subtracting 4 from both sides.

$\frac{c}{5} > 3$

$\frac{c}{5} \times 5 > 3 \times 5$ — Multiplying both sides by 5.

$c > 15$

Check: Try any number greater than 15, try 20.

$\frac{20}{5} + 4 > 7$

$8 > 7$

EXERCISE 10.09 ANSWERS ON P. 650

Inequalities

1 Solve each inequality, showing the working.

a $x + 5 < 12$ **b** $y + 13 > 36$ **c** $m + 9 < 21$ **d** $k + 27 > 54$

e $k + 6 < 2$ **f** $s + 3 > 1$ **g** $n + 9 < -4$ **h** $p + 17 > -3$

i $m + 8 < 0$ **j** $g + 4.5 > 20$ **k** $a + 1\frac{3}{4} < 12$ **l** $k + 2.7 > -5$

2 Solve each inequality.

a $t - 11 < 40$ **b** $w - 7 > 4$ **c** $a - 5 < 2$ **d** $j - 23 > 51$

e $q - 12 < 17$ **f** $f - 42 > 68$ **g** $x - 9 < 24$ **h** $g - 10 > 1$

i $j - 3 < -5$ **j** $w - 9 > -2$ **k** $m - 12 < -27$ **l** $d - 1 > -1$

m $y - 17 < 3.9$ **n** $n - 2.1 > 5.9$ **o** $b - \frac{4}{5} < \frac{1}{4}$ **p** $s - 2\frac{1}{3} > 4\frac{1}{2}$

3 Solve each inequality.

a $6x < 24$ **b** $4g > 16$ **c** $10y < 90$ **d** $5b > 45$

e $12d < 108$ **f** $9c > 81$ **g** $23p < 115$ **h** $2m > 26$

i $5x < -25$ **j** $7q > -42$ **k** $-3c < 54$ **l** $-15x > -120$

m $4p < 37$ **n** $8r > 18$ **o** $3q < -10$ **p** $9w > -12$

4 Solve each inequality.

a $\frac{s}{3} < 5$ **b** $\frac{m}{7} > 6$ **c** $\frac{d}{10} < 12$ **d** $\frac{w}{2} > 29$

e $\frac{a}{3} < -4$ **f** $-\frac{f}{2} > 9$ **g** $\frac{r}{5} < -3$ **h** $-\frac{m}{7} > 1$

i $-\frac{x}{8} < -6$ **j** $\frac{t}{2} > \frac{3}{4}$ **k** $-\frac{n}{4} < -4$ **l** $\frac{x}{3} > -\frac{1}{6}$

5 Solve $3x < -12$. Select the correct answer **A**, **B**, **C** or **D**.

A $x < -9$ **B** $x < -36$ **C** $x < -4$ **D** $x < -15$

6 (R) (C) Write a simple inequality whose solution is:

a $m < 4$ **b** $x > 0$ **c** $r < 7$ **d** $d > -5$

7 Which inequality has the solution $p < -2$? Select the correct answer A, B, C or D.

A $2 + 3p < 8$ B $5p - 2 < 8$ C $-4p + 16 < 8$ D $-3p + 2 < 8$

8 Solve each equation, showing all steps. Remember to check your answers.

a $2m + 7 < 19$ **b** $3x - 5 > 13$ **c** $5k + 12 < 52$

d $6w - 17 > 19$ **e** $4h + 21 < 9$ **f** $8d - 5 > -27$

g $-2x + 18 < 4$ **h** $-3a - 9 > -6$ **i** $14c + 2 < 37$

j $12p - 10 > 86$ **k** $9y + 11 < 101$ **l** $2r - 15 > 16$

▷

☐ Foundation ○ Standard ⬡ Complex

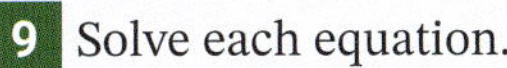

10.09

9 Solve each equation.

a $\frac{m}{2} + 4 < 13$ **b** $\frac{c}{5} + 9 > 12$ **c** $\frac{d}{7} + 2 < 8$

d $\frac{k}{3} + 15 > 6$ **e** $\frac{h}{2} + 7 < 17$ **f** $\frac{x}{4} + 2 > -4$

g $\frac{n}{5} - 9 < 4$ **h** $\frac{t}{3} - 6 > -2$ **i** $\frac{a}{8} - 2 < -6$

j $\frac{x}{6} - 16 > -13$ **k** $\frac{v}{15} - 2 < 4$ **l** $\frac{b}{3} - 4 > 3$

10 Solve each equation.

a $\frac{3x}{2} < 9$ **b** $\frac{15x}{9} > 15$ **c** $\frac{2x}{3} < 8$ **d** $\frac{3N}{5} > 3$

e $\frac{6N}{7} < 18$ **f** $\frac{5B}{2} > -10$ **g** $\frac{3x}{4} < -3$ **h** $\frac{2x}{5} > -4$

i $\frac{5m}{2} < 11$ **j** $-\frac{x}{3} > 8$ **k** $-\frac{4x}{5} < 1$ **l** $\frac{-2x}{3} > -10$

11 Solve $\frac{k-12}{9} > 6$. Select the correct answer **A**, **B**, **C** or **D**.

A $k > 27$ **B** $k > 42$ **C** $k > 66$ **D** $k > 162$

12 Solve each equation.

a $\frac{x+1}{3} < 2$ **b** $\frac{x-3}{2} > 3$ **c** $\frac{N+2}{5} < 1$ **d** $\frac{N-3}{4} > 5$

e $\frac{N+8}{5} < 6$ **f** $\frac{x+1}{2} > -2$ **g** $\frac{k-5}{2} < -5$ **h** $\frac{m+2}{3} > 11$

13 Write 4 equations, one of each type shown in questions **2**, **3**, **5** and **7**, that have the solution $p < 2$.

R C

14 Write 4 equations, one of each type shown in questions **2**, **3**, **5** and **7** that have the same solution as $2q + 1 > 5$.

R C

☐ Foundation ○ Standard ⬡ Complex

POWER PLUS ANSWERS ON P. 650

1 Solve each equation.

a $2(x+1)+2(x-1)=12$

b $2(x+4)-3(x-1)=9$

c $4(2x-1)-5(x-2)=6$

d $2-(3x+5)=4(x+1)$

e $\frac{x+7}{4}=\frac{6(x-1)}{3}$

f $\frac{2(x+1)}{3}=\frac{5(x-2)}{2}$

g $\frac{2x}{3}-\frac{x}{6}=10$

h $\frac{3x}{4}+\frac{9x}{10}=44$

2 For each diagram, write an equation for x and solve it.

a

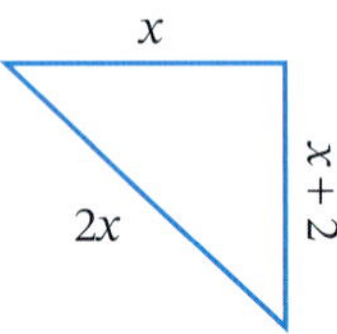

Perimeter = 30

b

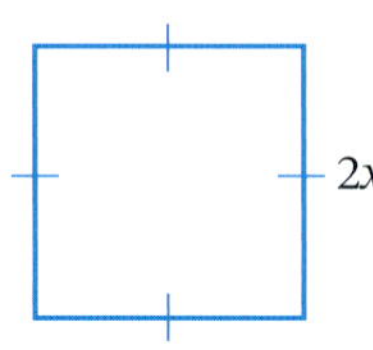

Area = 100

c

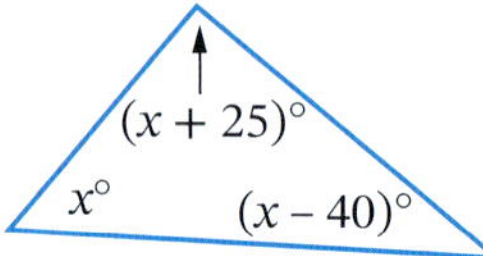

3 Solve each quadratic equation.

a $4x^2=100$

b $5x^2-7=73$

c $\frac{2x^2}{3}=24$

d $3x^2+8=20$

e $(x+4)^2=9$

f $x^2+2x=15$

4 Diophantus was a famous mathematician who was the first to abbreviate his mathematical thoughts using symbols. He is known as the Greek father of algebra and when he died, one of his admirers wrote the following riddle about his life:

Diophantus' youth lasted $\frac{1}{6}$ of his life. He grew a beard after $\frac{1}{12}$ more. After $\frac{1}{7}$ more of his life, Diophantus married; five years later he had a son. The son lived exactly $\frac{1}{2}$ as long as his father, and Diophantus died just four years after his son. All this adds to the years Diophantus lived.

Write this riddle as an equation and solve it to find how long Diophantus lived. (*Hint*: Let x years equal his life.)

10 CHAPTER REVIEW

Language of maths

Quiz
Language of maths 10

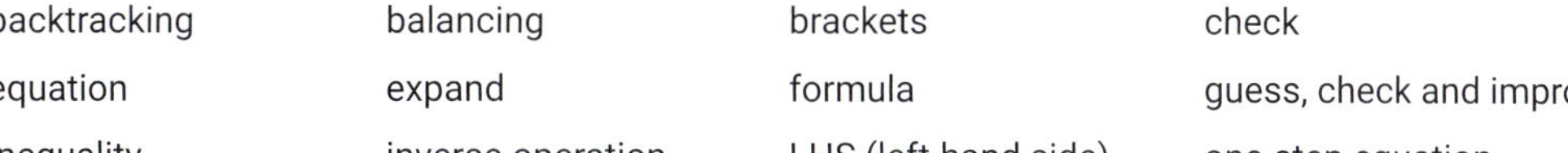

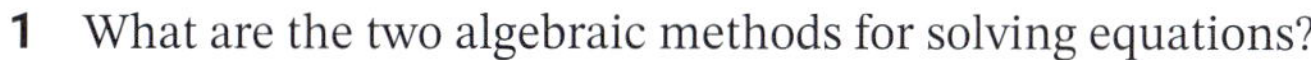

backtracking	balancing	brackets	check
equation	expand	formula	guess, check and improve
inequality	inverse operation	LHS (left-hand side)	one-step equation
quadratic equation	RHS (right-hand side)	solution	solve
substitution	two-step equation	undoing	variable

1 What are the two algebraic methods for solving equations?

2 Which method involves 'undoing' operations?

3 What does the word 'solution' mean?

4 Which word in the list means 'opposite'?

5 Why is the **variable** in an equation sometimes called an **unknown**?

6 What does **RHS** stand for:

  a in congruent triangles?

  b in solving equations?

Topic summary

Worksheet
Mind map: Equations and inequalities

- Which parts of this topic did you find easy? What did you already know?
- Give examples of some problems that might be solved using equations.
- Are there any parts of this topic that you still don't understand? Talk to your teacher about them.
- In what sort of careers would people use equations?

Copy and complete this mind map of the topic, adding detail to its branches and using pictures, symbols and colour where needed. Ask your teacher to check your work.

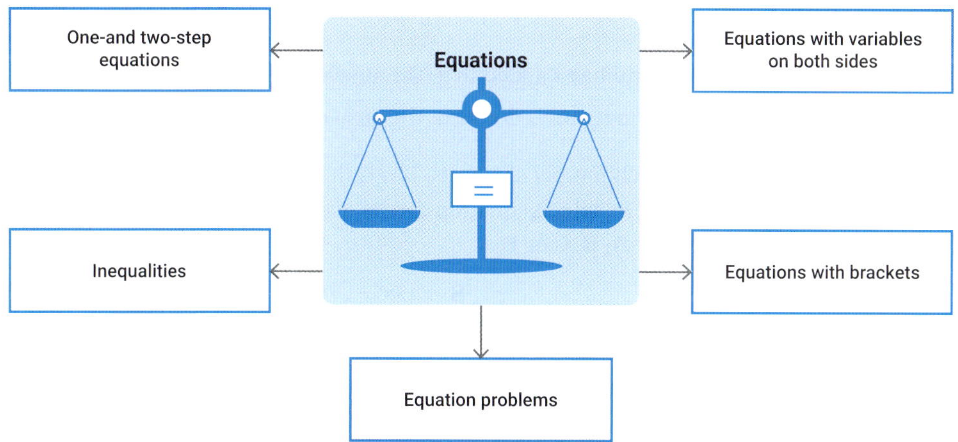

10 TEST YOURSELF

ANSWERS ON P. 650

1 Solve each equation using the 'guess, check and improve' method.

Quiz
Test yourself 10

a $4m - 9 = 7$

b $\frac{k+2}{7} = 2$

c $\frac{d}{3} - 4 = 2$

2 Solve each equation.

a $k + 6 = 13$ **b** $x - 3 = 8$ **c** $a + 3 = 17$

d $a - 12 = 21$ **e** $3x = 12$ **f** $10f = 120$

g $\frac{m}{4} = 7$ **h** $\frac{x}{3} = 8$ **i** $w + 9 = 3$

j $k - 5 = -7$ **k** $-2m = -16$ **l** $\frac{x}{3} = -12$

3 Solve each equation.

a $4p + 3 = 23$ **b** $3m + 17 = 8$ **c** $2x - 12 = 18$

d $\frac{h}{7} + 5 = 16$ **e** $\frac{n}{2} - 8 = -12$ **f** $\frac{a}{5} - 8 = 4$

g $\frac{2y}{5} = 4$ **h** $\frac{x+2}{3} = 4$ **i** $\frac{n-5}{2} = 10$

j $\frac{5x}{2} = -15$ **k** $\frac{k+9}{7} = 5$ **l** $\frac{d-1}{5} = -6$

4 Solve each equation.

a $3x + 4 = x + 6$ **b** $5u - 3 = 2u + 6$ **c** $12h - 8 = 8h + 4$

d $3v - 4 = 7v + 8$ **e** $2x + 9 = 7x - 4$ **f** $9 - 5t = 3t - 15$

10.05

5 Solve each equation.

a $2(n + 3) = 18$ **b** $3(x + 1) = 15$ **c** $10(x - 3) = -10$

d $5(x - 2) = 3x + 4$ **e** $4(y + 1) = y + 18$ **f** $-2(d - 2) = 2d - 20$

g $3(r - 1) = 2(r + 9)$ **h** $7(a + 5) = 3(a + 9)$ **i** $2(2n - 4) = 2(5 - n)$

6 **a** If a number is doubled and has 4 added to it, the answer is 14. What is the number?

b If a number has 4 added to it and is doubled, the answer is 14. What is the number?

7 **a** Keilani is paid \$45 for each jumper she knits. If n is the number of jumpers Keilani knits to earn a total of \$270, which equation can be used to find the value of n? Select the correct answer **A**, **B**, **C** or **D**.

A $\frac{n}{45} = 270$ **B** $45n = 270$

C $n + 45 = 270$ **D** $270 - n = 45$

b Solve the equation to find the value of n.

□ Foundation ○ Standard ⬡ Complex

9780170465601

8 Seven times a number is the same as 9 more than 4 times the same number. Use an equation to find the number.

9 Solve each equation.

a $x^2 = 64$

b $x^2 = 45$, correct to one decimal place.

c $5x^2 = 135$, as a surd.

10 The sum (S) of the angles (in degrees) of a polygon is given by $S = 180(n - 2)$, where n is the number of sides. Find:

a the sum of the interior angles when a polygon has 9 sides.

b the number of sides in a polygon whose angle sum is 1080°.

11 Solve each inequality, showing the working.

a $4p < 37$ b $8r > 18$ c $3q < -10$ d $9w > -12$

e $\frac{s}{3} < 5$ f $\frac{m}{7} > 6$ g $\frac{d}{10} < 12$ h $\frac{v}{2} > 29$

i $3x + 2 < 4$ j $2y - 1 > 7$ k $\frac{2z}{5} + 3 < 5$ i $\frac{q-4}{7} > 2$

Foundation Standard Complex

11

MEASUREMENT, ALGEBRA

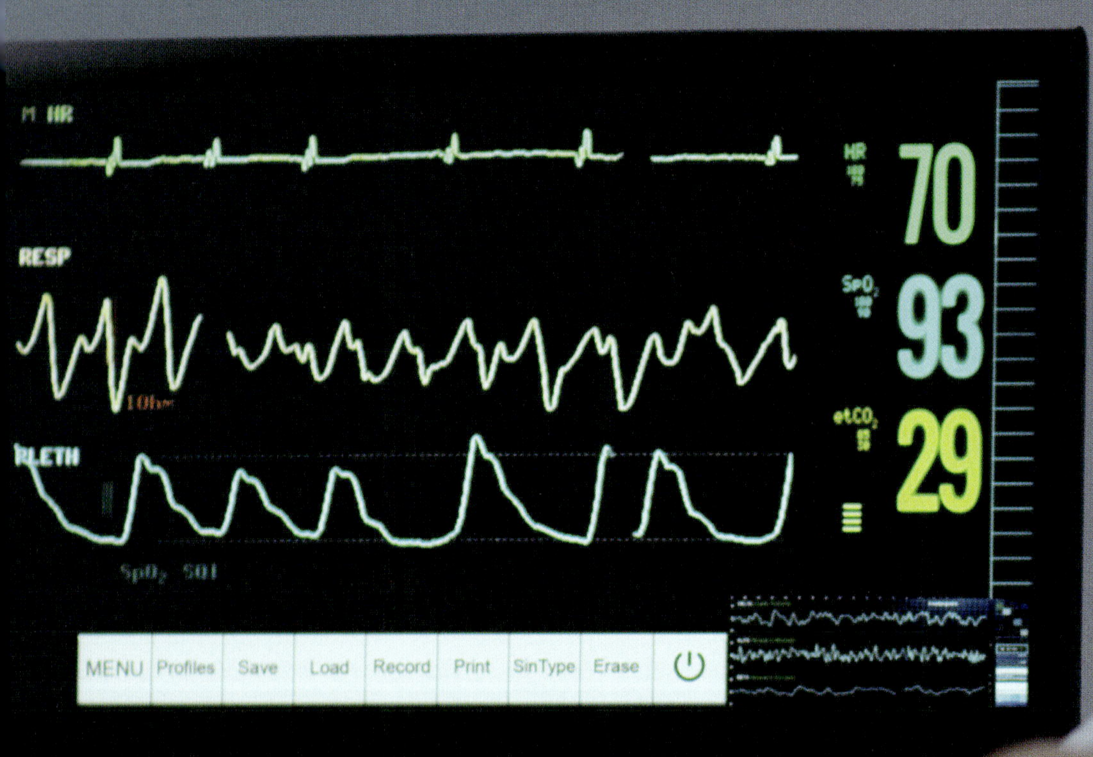

Ratios, rates and time

Ratios and rates are used in many everyday situations to compare things. Ratios are used when mixing hair colour, mixing paint to the chosen colour, paying rent in a share house and constructing buildings from plans. Rates are used when calculating the amount of paint required for a job, measuring heartbeat and preparing a holiday budget. Time skills are needed when working out when to leave home to get to work on time or making travel plans.

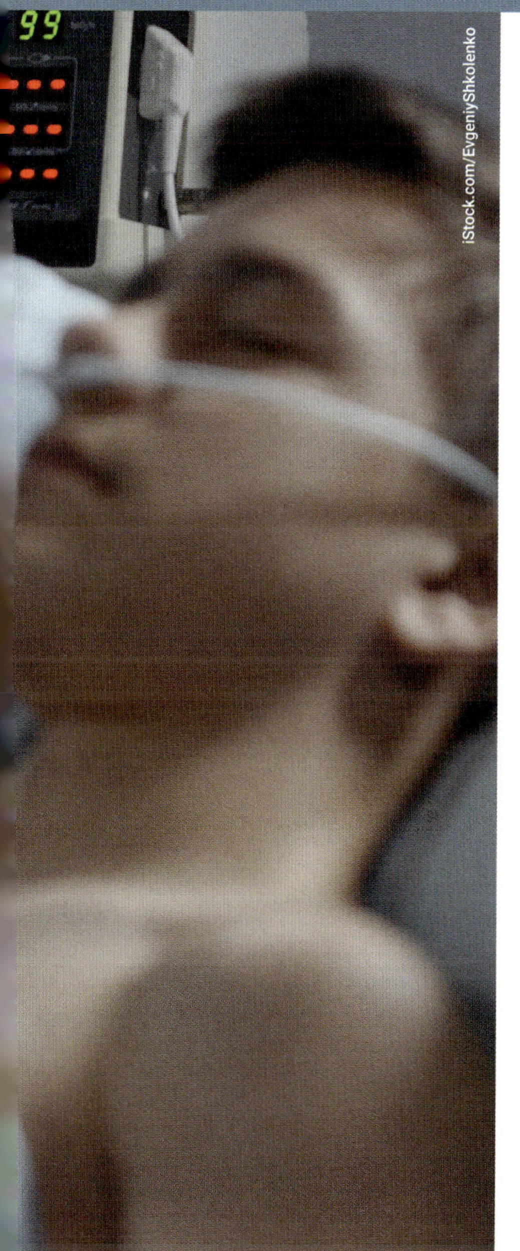

iStock.com/EvgeniyShkolenko

Chapter outline

	U	F	PS	R	C
11.01 Ratios	U	F			C
11.02 Ratio problems	U	F	PS	R	
11.03 Scale maps and plans	U	F	PS	R	C
11.04 Dividing a quantity in a given ratio	U	F	PS	R	
11.05 Rates	U	F		R	C
11.06 Best buys	U	F	PS	R	C
11.07 Rate problems	U	F	PS	R	C
11.08 Speed	U	F	PS	R	C
11.09 Travel graphs	U	F	PS	R	C
11.10 Sketching informal graphs	U	F	PS	R	C
11.11 Time differences	U	F	PS	R	C
11.12 World time zones	U	F	PS	R	C

Proficiencies

U = Understanding
F = Fluency
PS = Problem solving
R = Reasoning
C = Communication

Wordbank

Quiz
Wordbank 11

best buy When comparing different brands or sizes during shopping, this is the item with the lowest unit cost and is the best value for money

per (symbol '/') A word used in rates to mean 'for each'

scaled length A length on a map or plan that represents an actual length, usually much smaller, but in proportion to it

speed A rate that compares distance travelled with time taken

time zone A region of the world where all places experience the same time of day

travel graph A line graph that describes a journey and shows distance travelled over time

unit price The price of one item or unit, such as 1 mL or 1 g

unitary method A method of finding a quantity by finding one part first

Videos (17):

SkillCheck Why do we have ratios? • Time

11.01 Equivalent ratios • Simplifying ratios 2 • Simplifying ratios 1

11.02 Ratio problems 1 • Ratio problems 2

11.03 Map scales • Scale drawings • 24-hour time

11.04 Dividing a quantity in a given ratio

11.06 Unit pricing

11.07 Rate problems 1 • Rate problems 2

11.08 Speed

11.09 Travel graphs

11.11 Time differences • 24-hour time

Twig videos (6):

SkillCheck Ratios: Currency exchange • Ratios: The maths of baking

11.02 The history of the golden ratio

11.08 Speed of the Earth

11.11 Jai Singh (sundial)

11.12 Time zones

PhET interactives (3):

11.01 Ratio and proportion • Proportion playground

11.05 Unit rates

Quizzes (5):

- Wordbank 11
- SkillCheck 11
- Mental skills 11
- Language of Maths 11
- Test Yourself 11

Skillsheets (3):

SkillCheck 24-hour time

11.01 Ratios

11.03 24-hour time

11.11 Units of time • 24-hour time

Worksheets (17):

11.01 Simplifying ratios • Ratio calculations • Ratio skills

11.03 Map of Adelaide • Scale drawings

11.05 Rates

11.07 Rate problems • Ratios and rates review

11.08 Speed • What's my speed?

11.09 The hare and the tortoise

11.11 Time calculations • 12- and 24-hour time • Australian times • Tide chart

11.12 World time zones • Australian times

Mind map: Ratios, rates and time

Puzzle (7):

SkillCheck 12- and 24-hour time

11.01 Simplifying ratios

11.06 Best buys puzzle

11.07 Rate problems

11.08 Speed

11.09 Jane's diary

11.11 Time calculations

Technology (1)

11.04 Applying ratios

Presentations (3):

11.04 Shares in ratios

11.08 Rates: Triangle method

11.11 Time calculations

Spreadsheets (1):

11.04 Applying ratios

Nelson MindTap

To access resources above, visit **cengage.com.au/nelsonmindtap**

11 In this chapter you will:

- ✓ simplify ratios and solve problems involving ratios.
- ✓ interpret and calculate using scales on maps, plans and images.
- ✓ divide a quantity in a given ratio.
- ✓ simplify rates and solve problems involving rates.
- ✓ calculate unit prices to determine 'best buys'.
- ✓ interpret and draw travel graphs, including the slope of the graph.
- ✓ sketch informal graphs of practical situations.
- ✓ solve problems involving time differences, including using 24-hour time.
- ✓ interpret and use international time zones.

SkillCheck

ANSWERS ON P. 650

Videos
Why do we have ratios?
Ratios: Currency exchange
Ratios: The maths of baking
Time

Quiz
SkillCheck 11

1 Copy and complete each conversion.

a	2 m = _______ cm	**b**	3 h = _______ min	**c**	3000 kg = _______ t
d	2.5 L = _______ mL	**e**	380 cm = _______ m	**f**	180 mg = _______ g
g	3 min = _______ s	**h**	8.5 cm = _______ mm	**i**	480 min = _______ h
j	7500 mL = _______ L	**k**	9.15 km = _______ m	**l**	3840 mm = _______ cm

2 Find the highest common factor (HCF) of each pair of numbers.

a 12 and 18 **b** 35 and 21 **c** 16 and 40

3 Copy and complete each pair of equivalent fractions.

a $\frac{7}{10} = \frac{\square}{30}$ **b** $\frac{3}{5} = \frac{12}{\square}$ **c** $\frac{18}{45} = \frac{\square}{5}$

4 Find the lowest common multiple (LCM) of each pair of numbers.

a 2 and 3 **b** 4 and 8 **c** 5 and 3

5 Simplify each fraction.

a $\frac{18}{45}$ **b** $\frac{32}{64}$ **c** $\frac{80}{100}$ **d** $\frac{15}{36}$

6 Evaluate each expression.

a $\frac{2}{5} \times 20$ **b** $\frac{1}{3} \times 15$ **c** $\frac{3}{4} \times 24$ **d** $\frac{5}{6} \times 36$

e 7.6×100 **f** 5.4×10 **g** 0.39×100 **h** 25×10

7 Convert each time to 24-hour time.

a 7:15 a.m. **b** 3:45 p.m. **c** 8:50 p.m. **d** 12:10 a.m.

8 Convert each time to 12-hour time.

a 04:10 **b** 11:05 **c** 14:15 **d** 23:35

Skillsheet
24-hour time

Puzzle
12- and 24-hour time

Ratios 11.01

A **ratio** consists of two or more numbers that compare the parts or shares of things of the same type, in the same units. For example, if a cake recipe uses sugar to flour in a ratio of 1 to 2, written '1 : 2', it means that for every one part of sugar we need two parts of flour.

Each number in a ratio is called a **term** of the ratio.

Equivalent ratios are *equal* ratios and can be found in a similar way to that used for finding **equivalent fractions**.

Skillsheet
Ratios

Puzzle
Simplifying ratios

Video
Equivalent ratios

Worksheets
Simplifying ratios

Ratio calculations

Ratio skills

Interactives
Ratio and proportion

Proportion playground

ⓘ Equivalent ratios

To find an **equivalent ratio**, multiply or divide each term by the same number.

Example 1

Complete each pair of equivalent ratios.

a $6:4 = 18:\_\_\_\_$ **b** $3:4 = \_\_\_\_:24$ **c** $12:8 = 3:\_\_\_\_$

SOLUTION

a To find the missing term, look at the two known matching terms, 6 and 18.

6 is *multiplied* by 3 to give 18, so do the same thing to the 4.

$4 \times 3 = 12$

So, $6:4 = 18:12$.

$\times 3$
$6:4 = 18:12$
$\times 3$

b The two known matching terms are 4 and 24.

4 is *multiplied* by 6 to give 24, so do the same thing to the 3.

$3 \times 6 = 18$

So, $3:4 = 18:24$.

$\times 6$
$3:4 = 18:24$

c The two known matching terms are 12 and 3.

12 is *divided* by 4 to give 3, so do the same thing to the 8.

$8 \div 4 = 2$

So, $12:8 = 3:2$

$\div 4$
$12:8 = 3:2$
$\div 4$

ⓘ Simplifying ratios

- To simplify a ratio, keep dividing both terms by the same number, preferably a large number such as their HCF, until each term is as small as possible.
- If all terms are even, divide by 2 or perhaps 4.
- Otherwise, try dividing by the odd numbers 3, 5 or 7.

Videos
Simplifying ratios 2

Simplifying ratios 1

Example 2

Simplify each ratio.

a $25:40$ **b** $24:16$ **c** $18:30:15$

SOLUTION

a $25:40 = \frac{25}{5}:\frac{40}{5} = 5:8$

Divide both terms by their HCF, 5.

OR enter 25 : 40 as a fraction $\frac{25}{40}$ on the calculator: 25 [fraction key] 40 [=]

[a b/c] is the fraction key on some calculators.

b $24:16 = \frac{24}{8}:\frac{16}{8} = 3:2$ Divide both terms by their HCF, 8.

OR enter 24 : 16 as an improper fraction $\frac{24}{16}$ on the calculator: 24 [▭/▭] 16 [=]

Then, to change the mixed numeral answer to an improper fraction, press: [SHIFT] [S⇔D] or [2nd F] [$a^{b}/_{c}$].

c $18:30:15 = \frac{18}{3}:\frac{30}{3}:\frac{15}{3} = 6:10:5$ Divide all terms by their HCF, 3.

Because there are more than 2 terms in this ratio, the calculator cannot be used here.

Example 3

Video Simplifying ratios 2

Simplify each ratio.

a $\frac{3}{5}:\frac{1}{3}$

b 0.7 : 0.05

SOLUTION

If a ratio has terms that are fractions or decimals, it can be simplified by converting the terms to **whole numbers**.

a For fractions, multiply both terms by a common multiple, preferably the **LCM** of the denominators. The LCM of 5 and 3 is 15.

$$\frac{3}{5}:\frac{1}{3} = \left(\frac{3}{5}\times 15\right):\left(\frac{1}{3}\times 15\right) = 9:5$$

b For decimals, multiply both terms by the appropriate power of 10. In this case, multiply by 100 (move the decimal place 2 places to the right). Simplifying

$$\begin{aligned}0.7:0.05 &= (0.7\times 100):(0.05\times 100)\\ &= 70:5\\ &= 14:1\end{aligned}$$

Example 4

Simplify the ratio of 2 hours to 1 day.

SOLUTION

First, change the values to the same units. 1 day = 24 hours

$$\begin{aligned}2\text{ hours} : 1\text{ day} &= 2\text{ hours} : 24\text{ hours}\\ &= 2:24\\ &= 1:12\end{aligned}$$

EXERCISE 11.01 ANSWERS ON P. 651

Ratios

U F C

1 For each shape, write the ratio of shaded to unshaded parts.

C **a**

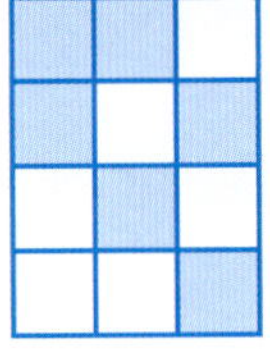

b

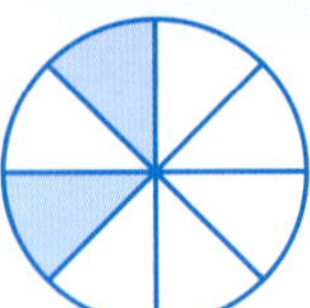

c

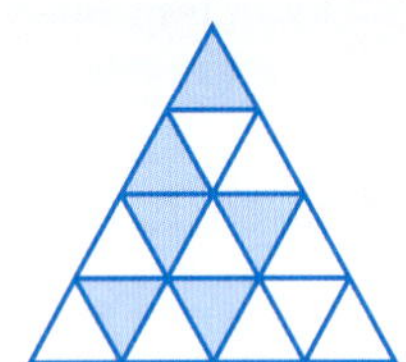

d

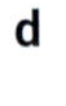

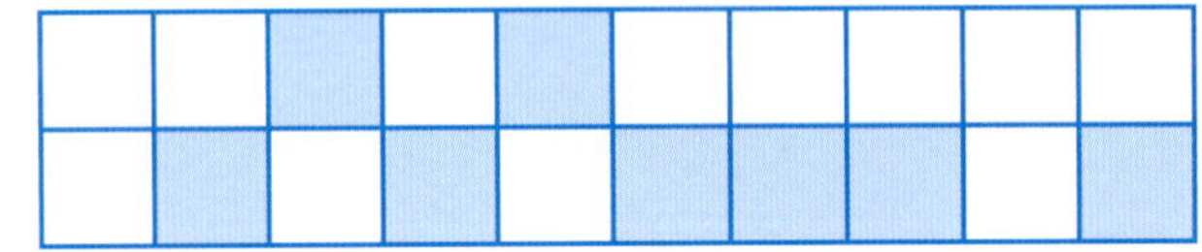

e

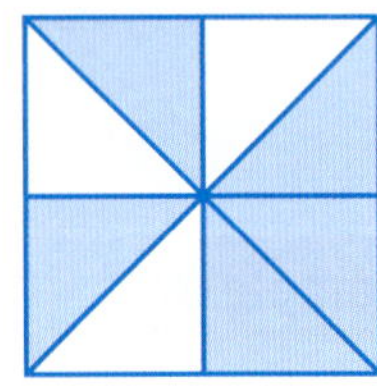

EXAMPLE 1

2 Copy and complete each pair of equivalent ratios.

a 2 : 3 = 8 : ______ **b** 1 : 5 = 2 : ______ **c** 3 : 5 = ______ : 15

d 4 : 7 = ______ : 35 **e** 5 : 8 = 20 : ______ **f** 7 : 12 = 49 : ______

g 5 : 11 = ______ : 66 **h** 3 : 4 = ______ : 100 **i** 2 : 1 = 10 : ______

j ______ : 9 = 20 : 36 **k** 12 : ______ = 3 : 1 **l** 17 : 34 = ______ : 2

m ______ : 45 = 6 : 9 **n** 24 : 12 = 4 : ______ **o** 16 : ______ = 2 : 5

p ______ : 20 = 15 : 60 **q** 24 : 20 = 6 : ______ **r** 50 : 40 = ______ : 20

3 Which of the following ratios is **not** equivalent to the ratio of 32 : 48? Select the correct answer **A**, **B**, **C** or **D**.

A 16 : 24 **B** 4 : 6 **C** 2 : 3 **D** 6 : 8

EXAMPLE 2

4 Simplify each ratio.

a 10 : 100 **b** 12 : 24 **c** 12 : 30 **d** 35 : 49

e 18 : 12 **f** 56 : 24 **g** 1000 : 100 **h** 45 : 99

i 87 : 87 **j** 123 : 321 **k** 51 : 17 **l** 3 : 48

m 8 : 12 : 20 **n** 15 : 20 : 30 **o** 27 : 9 : 36 **p** 14 : 35 : 21 : 49

EXAMPLE 3

5 Simplify each ratio.

a $\frac{1}{3} : \frac{2}{5}$ **b** $\frac{1}{4} : \frac{1}{3}$ **c** $\frac{3}{4} : \frac{2}{3}$ **d** $\frac{1}{2} : \frac{3}{8}$

e $\frac{2}{5} : \frac{3}{10}$ **f** $\frac{4}{5} : \frac{1}{2}$ **g** $\frac{5}{8} : \frac{1}{4}$ **h** $\frac{2}{3} : \frac{1}{2}$

i $\frac{3}{4} : \frac{7}{16}$ **j** $\frac{4}{5} : \frac{1}{2}$ **k** $\frac{5}{6} : \frac{2}{5}$ **l** $\frac{6}{5} : \frac{2}{3}$

m 0.4 : 0.7 **n** 1.3 : 0.8 **o** 0.5 : 0.3 **p** 0.9 : 1.8

▷

☐ Foundation ○ Standard ⬡ Complex

q 0.6 : 0.8 **r** 3.6 : 2.4 **s** 0.05 : 0.2 **t** 0.25 : 0.5

u 0.375 : 0.25 **v** 12 : 8.4 **w** 2.4 : 1.2 : 3.6 **x** 4.5 : 1 : 0.9

6 Simplify each ratio. EXAMPLE 4

a 50 cm to 2 m **b** 300 g to 1.2 kg **c** 5 days to 7 weeks

d 30 min to 2 hours **e** 70 cents to $2.10 **f** 2 years to 6 months

g 15 hours to 2 days **h** 20 mm to 1 m **i** 4 tonnes to 350 kg

j 25 min to 3 hours **k** 18 m to 1 km **l** 8 months to 4 years

m 2 days to 8 hours **n** 75 cents to $5 **o** $2.70 : $12

7 On a farm there are 200 orange and mandarin trees in total. If there are 120 orange trees, what is the ratio of orange to mandarin trees?

8 In a class of 28 students, there are 16 girls. Find, in simplest form, the ratio of:

a girls to boys **b** boys to girls **c** girls to students in the class

9 A store has 30 gas heaters and 20 electric heaters in its warehouse. Find the ratio of:

a gas heaters to electric heaters.

b gas heaters to all heaters.

c all heaters to electric heaters.

10 A man earns $75 000 a year and spends $63 000 a year. Find the ratio of his savings to earnings. Select the correct answer **A**, **B**, **C** or **D**.

A 25 : 21 **B** 21 : 25 **C** 4 : 25 **D** 4: 21

11 Zoe buys goods for $320 and sells them for $380. Find, in simplest form, the ratio of:

a the cost price to the selling price.

b the selling price to the cost price.

c the profit to the selling price.

d the selling price to the profit.

12 The line below is divided into units of length as shown.

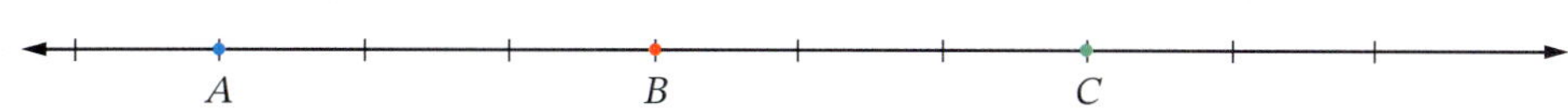

Find each ratio of lengths.

a $AB : BC$ **b** $AC : AB$ **c** $BC : AC$ **d** $AC : BC$

13 A vending machine is filled with bottles and cans. The ratio of the number of bottles to the total number of contents is 3 : 8.

a What fraction of the contents are cans?

b What is the ratio of the number of cans to the number of bottles?

14 0.5 m^3 of cement is added to $\frac{3}{8}$ m^3 of metal to make a mixture. What is the ratio of cement to metal? Select the correct answer **A**, **B**, **C** or **D**.

A 4 : 3 **B** 7 : 8 **C** 3 : 4 **D** 3 : 16

Foundation Standard Complex

15 When comparing the rate at which babies are born in different countries (that have different population sizes), we use a base number of 1000. The annual birth rate in Australia is approximately 12 per 1000 people (a ratio of 12 : 1000). The rate in India is approximately 22 per 1000 people or 22 : 1000. Express each ratio in simplest form.

11.02 Ratio problems

Problems involving ratios can be solved using **equivalent ratios** or the **unitary method**. With the unitary method, we find the size of **one part** first. We have used this method before when calculating with percentages.

Example 5

Videos
Ratio problems 1
Ratio problems 2

The ratio of boys to girls in a class is 2 : 3. If there are 10 boys in the class, how many girls are there?

SOLUTION

Method 1: Equivalent ratios

Write the problem as a pair of equivalent ratios.

Boys : girls = 2 : 3 = 10 : __ (× 5) — 10 boys

Number of girls = 3 × 5 = 15

There are 15 girls in the class.

Method 2: Unitary method — Finding one part first.

Boys : girls = 2 : 3 — 'Unitary' means 'one'

2 parts (boys) = 10

1 part = 10 ÷ 2 = 5

3 parts (girls) = 3 × 5 = 15

There are 15 girls in the class.

Example 6

To make concrete, a builder mixes sand and cement in the ratio of 5 : 4. If a mix of concrete contains 20 kg of cement, find:

a the amount of sand in the mix.

b the total mass of the mix.

SOLUTION

a **Method 1: Equivalent ratios** 20 kg of cement

$$\text{Sand : cement} = 5:4 = \_\_ : 20 \quad (\times 5)$$

Amount of sand = $5 \times 5 = 25$ kg.

Method 2: Unitary method

Sand : cement = 5 : 4

4 parts (cement) = 20 kg

1 part = $20 \div 4 = 5$ kg

5 parts (sand) = $5 \times 5 = 25$ kg.

b Total mass = 20 kg + 25 kg = 45 kg Cement and sand

So, the total mixture was 45 kg.

EXERCISE 11.02 ANSWERS ON P. 651

Ratio problems

U F PS R

1 PS Alison and Elena buy a length of material and divide it between them in the ratio of 2 : 3. If Alison has 5.4 m, what length of material does Elena have? Select the correct answer **A**, **B**, **C** or **D**.

A 2.7 m **B** 3.6 m **C** 8.1 m **D** 13.5 m

2 PS A tiler uses 4 green tiles to every 3 white ones. How many white tiles are used if 100 green tiles are used?

3 PS When making concrete, sand and cement are mixed in the ratio of 4 : 1. If 140 kg of cement has been delivered, what mass of sand is needed?

4 Two lengths of timber are in the ratio of 4 : 7. The longer length is 56 cm. What is the shorter length?

5 PS In a college, the ratio of teachers to students is 1 : 18. If the college has 80 teachers, how many students are there?

6 The speeds of two boats are in the ratio of 7 : 4. The speed of the slower boat is 10 km/h. Find the speed of the faster boat.

7 In a rectangle, the ratio of the width to the length is 5 : 12. The length is 48 cm.

PS **a** Find the width of the rectangle.

R **b** Find the perimeter of the rectangle.

Foundation Standard Complex

8 An alloy contains copper and iron in the ratio of 2 : 5. A quantity of alloy contains 20 kg of copper. What mass of iron does it contain?

9 Toudi, Ash and Felicity share a Lotto prize in the ratio of 11 : 8 : 6. Ash received $720.

PS **a** How much did Toudi and Felicity each receive?

R **b** What was the total prize money shared?

10 In a triangle, the lengths of the sides are in the ratio of 3 : 4 : 5. If the longest side is 45 cm long, find the perimeter of the triangle. Select the correct answer **A**, **B**, **C** or **D**.

PS R **A** 57 cm **B** 72 cm **C** 81 cm **D** 108 cm

11 The masses of two packets of detergent are in the ratio of 3 : 10.

PS **a** If the lighter packet has a mass of 1.5 kg, what is the mass of the larger one?

R **b** If the heavier packet costs $12.50 and the lighter packet costs $3.90, which packet is the cheaper per kilogram and by how much?

12 Ed's Farmers' Market buys fruit and vegetables in the ratio of 7 : 9. The mass of vegetables ordered is 14.4 tonnes. What is the total mass of products ordered?

PS

13 Samantha and James' heights are in the ratio of 7 : 6. If Samantha is 1.75 m tall, how tall is James?

14 In an outback mining town, the ratio of women to men is 2 : 5. If there are 240 women, how many men and women are there in the town altogether? Select the correct answer **A**, **B**, **C** or **D**.

PS R **A** 840 **B** 600 **C** 960 **D** 1680

15 To make Superglue, the contents of Tube A and Tube B are mixed in the ratio of 3 : 1.

PS **a** If 21 mL of Tube A is used, how much of Tube B is needed?

b How much glue is made altogether if 21 mL of Tube B is used?

DID YOU KNOW?

Loose change

All Australian coins are made by mixing copper with other elements.

The $2 and $1 coins were introduced in 1988 to replace notes. They are made of copper, aluminium and nickel in the ratio of 46 : 3 : 1.

The 50c, 20c, 10c and 5c coins are made of copper and nickel in the ratio of 3 : 1.

600 grams of aluminium are used to make some $2 coins. How much copper and nickel is used?

Shutterstock.com/Cre8tive Images

INVESTIGATION

The golden ratio

Ancient Greek mathematicians and artists discovered a 'perfect' rectangle whose dimensions were most naturally pleasing to the eye. Try drawing an ideal rectangle for yourself, one that is neither too thin nor too tall. Such a rectangle has a length that is $\frac{\sqrt{5}-1}{2}$ times its width.

This value, which is approximately 1.618, is called the golden ratio and a rectangle with these dimensions is called a golden rectangle. Is your rectangle's length 1.6 times its width?

The ancient Greeks used the golden ratio in art and architecture. The Parthenon in Athens was built to the dimensions of the golden rectangle. Even today, the golden rectangle is often used in graphic design and publishing, such as in website panels and sidebars.

Your task:

- Find 10 photos from this textbook and measure their lengths and widths.
- Copy and complete this table.

	Description of photos	Page	Pleasing to the eye? (Y/N)	Length (mm)	Width (mm)	Length ÷ width (3 decimal places)
1						
2						
3						
4						
5						
6						
7						
8						
9						
10						

1 How many of the photos were pleasing?

2 How many of these matched the golden ratio?

3 What percentage of the pictures that you found pleasing match the golden ratio?

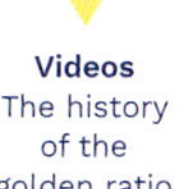

Videos The history of the golden ratio

11.02

11.03 Scale maps and plans

Worksheet
Map of Adelaide

Scale maps and plans are a special application of ratios in real life. Lengths and distances on scale diagrams are in the same ratio as the real lengths and distances.

ⓘ Scale ratios

The scale ratio on a scale diagram is written in the form scaled length : real length, where scaled length is the length on the diagram.

This is similar to **scale factor** $\left(\frac{\text{image length}}{\text{original length}}\right)$ that we learned about in Chapter 8, *Congruent and similar figures*, but written as a ratio.

For example, a scale ratio of 1 : 100 means that the real lengths are 100 times larger than the lengths on the diagram.

Map scales

Map scales are often expressed in the same form as ratios. A scale of 1 cm : 1 km means that 1 cm on the map represents an actual distance of 1 km.

Videos
Map scales

Example 7

Simplify each map scale.

a 1 cm : 1 km

b (scale bar) 0 2 4 6 8 10 km

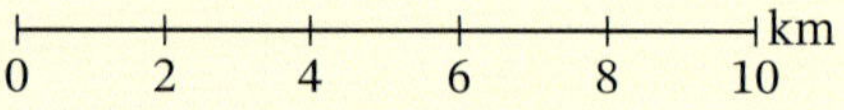

SOLUTION

a 1 cm : 1 km = 1 cm : 1000 m   1 km = 1000 m

= 1 cm : 100 000 cm   1 m = 100 cm

= 1 : 100 000

b The length of the scale from 0 to 10 on the diagram is 5 cm.

So, 5 cm on the map represents 10 km of actual distance.

Scale = 5 cm : 10 km

= 5 cm : 10 × 1000 × 100 cm   Convert 10 km to cm.

= 5 cm : 1 000 000 cm

= 1 : 200 000   Simplifying the ratio.

Map scales

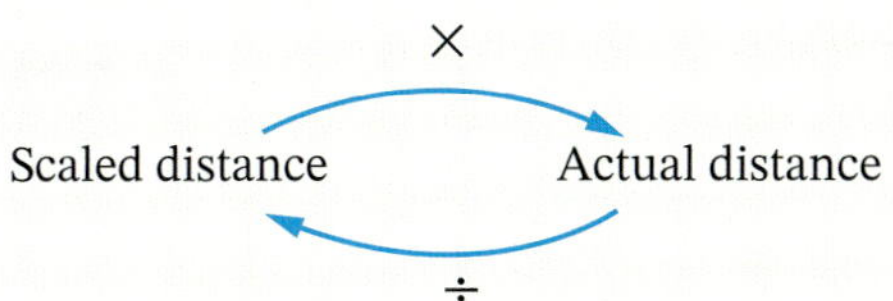

Videos
Map scales

Example 8

A map has a scale of 1 : 25 000.

a What is the actual distance if the scaled distance is 4 cm?

b What is the scaled distance if the actual distance is 3.5 km?

SOLUTION

a Scaled distance = 4 cm — 1 m = 100 cm, 1 km = 1000 m

$$\begin{aligned}\text{Actual distance} &= 4 \times 25\,000 \text{ cm}\\ &= 100\,000 \text{ cm}\\ &= 1000 \text{ m}\\ &= 1 \text{ km}\end{aligned}$$

b — 1 km = 1000 m, 1 m = 100 cm

$$\begin{aligned}\text{Actual distance} &= 3.5 \text{ km}\\ &= 3500 \text{ m}\\ &= 350\,000 \text{ cm}\end{aligned}$$

$$\begin{aligned}\text{Scaled distance} &= 350\,000 \text{ cm} \div 25\,000\\ &= 14 \text{ cm}\end{aligned}$$

Scale drawings

Video
Scale drawings

Worksheet
Scale drawings

Example 9

This square clock is drawn to a scale of 1 : 6. Measure its length and calculate its actual length.

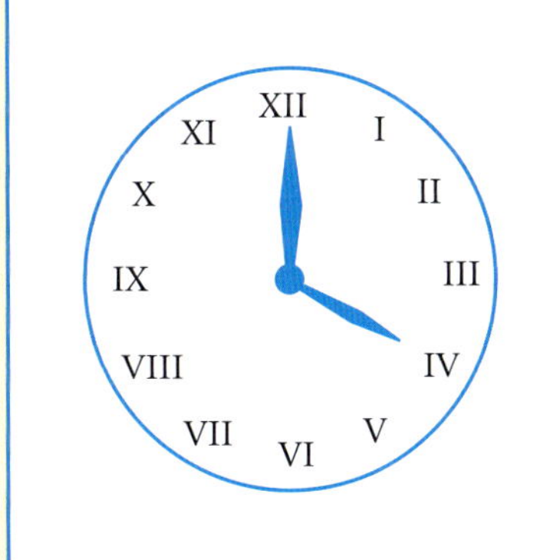

Scale 1 : 6

SOLUTION

Scaled length = 5 cm — by measurement

$$\begin{aligned}\text{Actual length} &= 5 \text{ cm} \times 6\\ &= 30 \text{ cm}\end{aligned}$$

Example 10

This drawing of a screw is drawn to a scale of 5 : 1. Find its actual length.

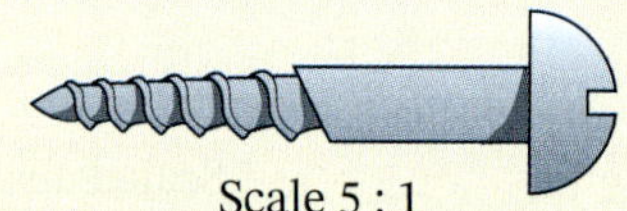

Scale 5 : 1

SOLUTION

A scale of 5 : 1 means that the real screw is 5 times **smaller** than the one drawn.

Scaled length = 4 cm   by measurement

$$\text{Actual length} = 4 \text{ cm} \div 5$$
$$= 0.8 \text{ cm}$$

EXERCISE 11.03 ANSWERS ON P. 651

Scale maps and plans

U F PS R C

EXAMPLE 7

1 Simplify each map scale.

R C

a 1 cm : 5 km **b** 1 mm : 1 m **c** 1 cm : 500 km **d** 1 cm : 25 km

e scale bar: 0, 100, 200, 300 metres

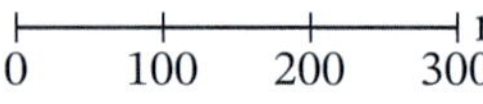

f scale bar: 0, 1, 2, 3, 4, 5 kilometres

g scale bar: 0, 500, 1000, 1500 metres

h scale bar: 0, 1, 2, 3 km

i scale bar: 0, 1, 2, 3, 4 km

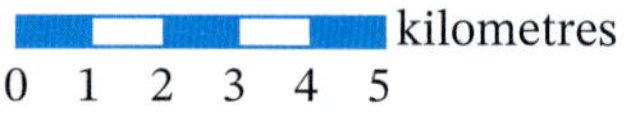

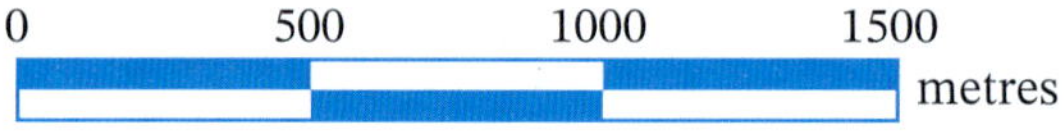

j scale bar: 0, 150, 300, 450, 600, 750 metres

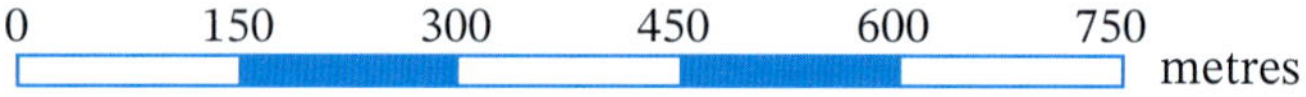

EXAMPLE 8

2 A map has a scale of 1 : 50 000. What distance is represented by 64 mm on the map? Select the correct answer **A**, **B**, **C** or **D**.

R C

A 0.32 km **B** 3.2 km **C** 32 km **D** 320 km

Foundation Standard Complex

3 A street map uses a scale of 1 cm : 200 m.

PS **a** Simplify this ratio.

R **b** Find the actual distance, in kilometres, represented by each scaled distance.

C

- **i** 7 cm
- **ii** 9.5 cm
- **iii** 12.4 cm

c Find the scaled distance, in centimetres, used to represent each actual distance.

- **i** 18 km
- **ii** 1500 m
- **iii** 9.6 km

4 On a map using a scale of 1 : 10 000 000, the world's longest river, the Nile in Egypt, is the length of an average shoe lace (66.7 cm). How many kilometres long is the River Nile?

PS R C

5 The town of Gilgandra is 66 km north of Dubbo. On a map with a scale of 1 : 100 000, what is the scaled distance between the two towns?

PS R C

6 Lord Howe Island is 2.8 km wide. How long would its scaled width be on a map with a scale of 1 : 50 000?

PS R C

7 This map of Nambucca Heads has a scale of 1 : 40 000.

PS R C

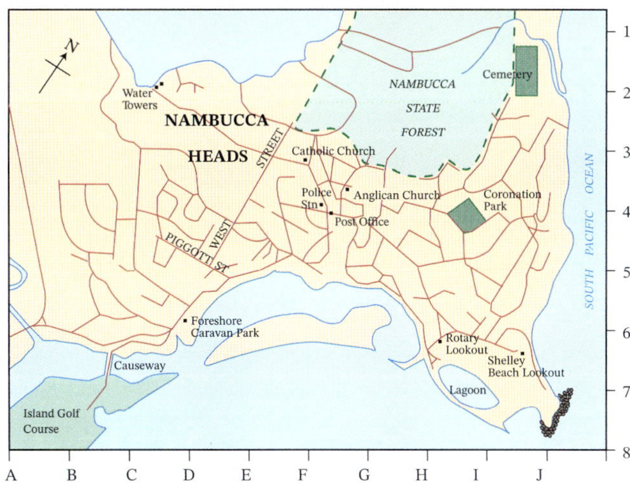

a Find, in metres, the distance between:

- **i** the Water Towers (C2) and the Catholic Church (F3).
- **ii** the Anglican Church (G4) and the centre of Coronation Park (I4).
- **iii** Rotary Lookout (H6) and Shelley Beach Lookout (J6).
- **iv** the post office (F4) and the Foreshore Caravan Park (D6).

b Find the length of:

- **i** West Street (E4)
- **ii** Piggott Street (D5)

c How long is the lagoon (I7)?

d What are the dimensions of the cemetery (J2)?

e How long is the causeway leading to the Island Golf Course (B7)?

f To train for a fun run, Merridy decides to run 8 km 3 times a week. What distance will this be on the map in centimetres? Outline a possible course for her training run, starting and finishing at the Foreshore Caravan Park (D6).

EXAMPLE 8

8 This is a scale plan of a bedroom.

Scale:
1 : 70

By measurement and calculation, find, in metres, the actual:

a length of the bedroom.

b width of the doorway (in cm).

c length of the bed.

d length of the window.

e length of the table.

f area of the bedroom.

9 Measure the length of each scaled-down image below, then use the scale ratio to calculate its actual length in centimetres (metres for the house).

a Fish 1 : 3

b House 1 : 300

Left: Shutterstock.com/Elena Blokhina; Right: iStock.com/pamspix

c Pen 1 : 4

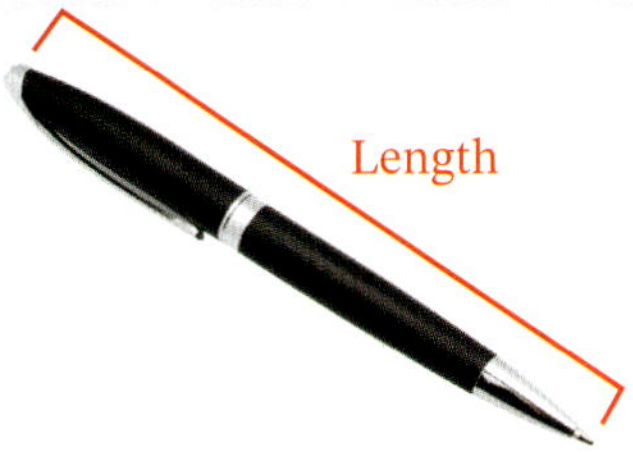

d Tennis racquet 1 : 16

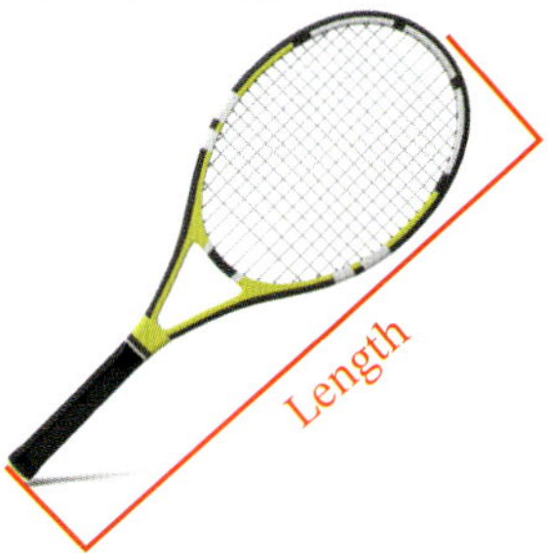

Left: Shutterstock.com/Pakhnyushchy; Right: Shutterstock.com/Gearstd

10 This house plan is drawn to a scale of 1 : 140.

PS

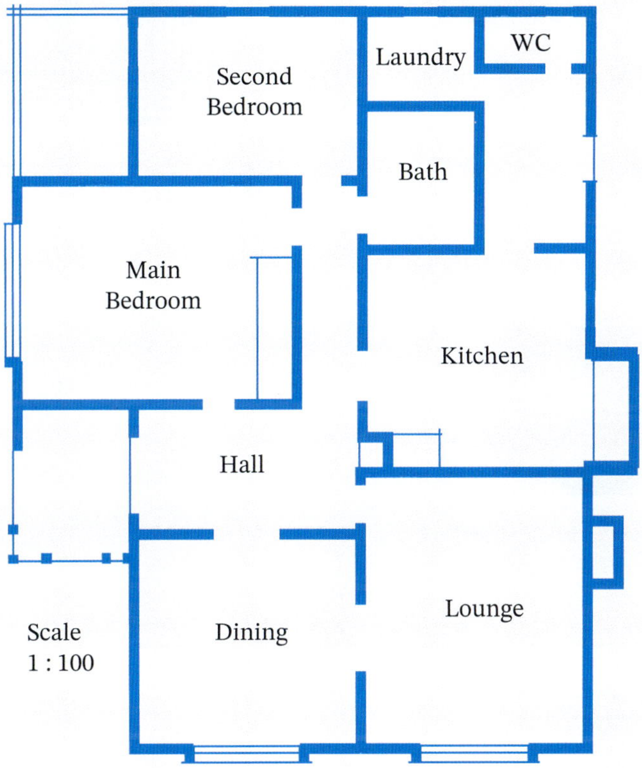

Measure and calculate to the nearest 0.1 m:

R **a** the length of the main bedroom.

C **b** the length of the window in that room.

c the length of the laundry.

d the area of the bathroom.

e the longer side of the lounge room.

f the area of the dining room.

EXAMPLE 10

11 An electronics engineer designs a mobile phone SIM card using a diagram with a scale of 100 : 1. If the scaled drawing is 80 cm long, what is its actual length in millimetres? Select the correct answer **A**, **B**, **C** or **D**.

A 80 **B** 8 **C** 0.8 **D** 0.08

12 Measure the length of each magnified image below, then use the scale ratio to calculate its actual length in millimetres.

R C

a Flea 100 : 1

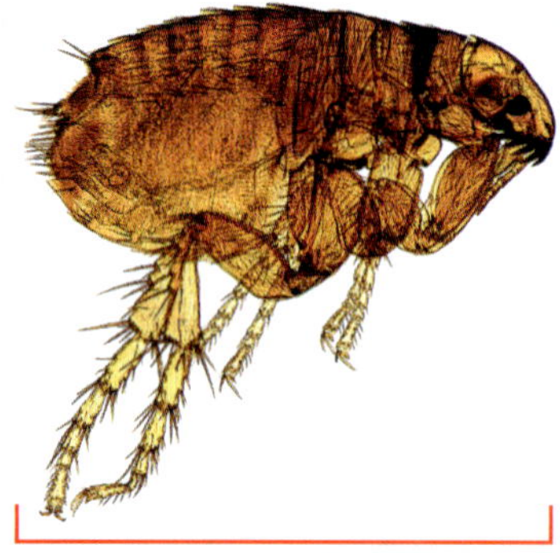

Length

b Microchip 3 : 1

Length

Left: iStock.com/oikim; Right: iStock.com/filonmar

c Nut 4 : 3

Length

d Bacterium 100 : 1

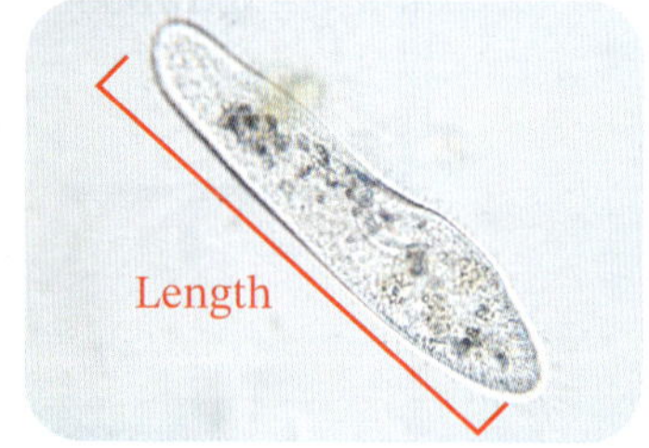

Left: Shutterstock.com/Alexapicso; Right: Dreamstime.com/Puntasit Choksawatdikorn

INVESTIGATION

Scale drawings

1 Measure the dimensions of your bedroom and the dimensions of the items of furniture in it.

2 Using a scale of 1 : 100, draw a scale diagram of your bedroom on graph paper.

3 Draw scale diagrams of each piece of furniture on another piece of graph paper, then cut them out.

4 Rearrange the furniture in your bedroom on the scale diagram to see which arrangement you prefer best.

Foundation Standard Complex

☆ **MENTAL SKILLS** (11) ANSWERS ON P. 651 **Maths without calculators**

Quiz Mental skills 11

Skillsheet 24-hour time

Video 24-hour time

24-hour time

24-hour time	12-hour time
00:00	12 a.m. (midnight)
01:00	1 a.m.
02:00	2 a.m.
03:00	3 a.m.
04:00	4 a.m.
05:00	5 a.m.
06:00	6 a.m.
07:00	7 a.m.
08:00	8 a.m.
09:00	9 a.m.
10:00	10 a.m.
11:00	11 a.m.

24-hour time	12-hour time
12:00	12 p.m. (midday)
13:00	1 p.m.
14:00	2 p.m.
15:00	3 p.m.
16:00	4 p.m.
17:00	5 p.m.
18:00	6 p.m.
19:00	7 p.m.
20:00	8 p.m.
21:00	9 p.m.
22:00	10 p.m.
23:00	11 p.m.

To convert from **24-hour time** to 12-hour time:

- if it begins with '00', then it is the 12 a.m. (midnight) hour.
- if it begins with '12', then it is the 12 p.m. (midday) hour.
- if it is less than 12:00, then it is a.m. (morning).
- if it is 13:00 or more, then it is p.m. (afternoon/evening) time: subtract 12 from the hour.

1 Study each example.

a Convert 18:50 to 12-hour time.

18:00 > 13:00, so it is p.m. time, so subtract 12 from the hour.

18 − 12 = 6

18:50 = 6:50 p.m.

b Convert 04:30 to 12-hour time.

04:30 < 12:00, so it is a.m. time.

04:30 = 4:30 a.m.

c Convert 00:15 to 12-hour time.

00:15 begins with 00, so it is 12 a.m. time

00:15 = 12:15 a.m.

2 Now convert each time to 12-hour time.

a 08:45	**b** 13:20	**c** 17:50	**d** 00:17
e 21:05	**f** 18:32	**g** 11:15	**h** 02:38
i 14:40	**j** 03:20	**k** 16:55	**l** 23:31
m 01:08	**n** 10:18	**o** 20:00	**p** 06:43

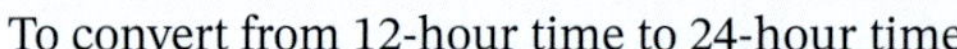

To convert from 12-hour time to 24-hour time:

- if it is the 12 a.m. (midnight) hour, begin with '00'.
- if it is 'a.m.' time or the 12 p.m. (midday) hour, write as is but make sure the hour has two digits (e.g. 02 and 09).
- if it is 1 p.m. or later, then add 12 to the hour.

3 Study each example.

a Convert 4:10 a.m. to 24-hour time.

It is 'a.m.' time, so write as a four-digit number and rename 4 as 04.

4:10 a.m. = 04:10

b Convert 4:10 p.m. to 24-hour time.

It is after 1 p.m., so add 12 to the hour: 4 + 12 = 16.

4:10 p.m. = 16:10

c Convert 12:47 a.m. to 24-hour time.

It is in the 12 a.m. (midnight) hour, so change the 12 to 00.

12:47 a.m. = 00:47

4 Now convert each time to 24-hour time.

a	6:35 p.m.	**b**	8:05 a.m.	**c**	11:45 a.m.	**d**	11:20 p.m.
e	2:21 a.m.	**f**	12:30 p.m.	**g**	3:48 p.m.	**h**	7:11 p.m.
i	9:08 a.m.	**j**	9:50 p.m.	**k**	12:42 a.m.	**l**	7:39 a.m.
m	1:59 a.m.	**n**	10:18 p.m.	**o**	10:46 a.m.	**p**	5:23 p.m.

11.04 Dividing a quantity in a given ratio

Presentation Shares in ratios

Technology Applying ratios

Spreadsheet Applying ratios

Elyse and Henry won a cash prize of $420 for winning an art competition, but instead of dividing the money evenly ($210 each), they decide to divide it in the ratio of 3 : 4, so that Elyse receives three parts while Henry receives four parts. Henry receives more because he did more of the work and paid more when buying the materials for the artwork.

Problems involving dividing a quantity in a given ratio can be solved using the **unitary method** or **fraction method**.

Example 11

Divide a cash prize of \$420 between Elyse and Henry in the ratio of 3 : 4.

SOLUTION

Method 1: Unitary method

Total number of parts = 3 + 4 = 7.

7 parts = \$420

1 part $= \$420 \div 7$

$= \$60$

Elyse's share (3 parts) $= 3 \times \$60$

$= \$180$

Henry's share (4 parts) $= 4 \times \$60$

$= \$240$

So, Elyse receives \$180 and Henry receives \$240.

Check: \$180 + \$240 = \$420.

Method 2: Fraction method

Total number of parts = 3 + 4 = 7.

Elyse's share (3 parts) $= \frac{3}{7} \times \$420$

$= \$180$

Henry's share (4 parts) $= \frac{4}{7} \times \$420$

$= \$240$

Video
Dividing a quantity in a given ratio

11.04

Example 12

Nick, Hong and Rob agreed to divide the profits of their removalist business in the ratio of 2 : 3 : 5. If their profit this year was \$45 000, find the size of each person's share of the profits.

SOLUTION

Method 1: Unitary method

Total number of parts = 2 + 3 + 5 = 10.

10 parts = \$45 000

1 part $= \$45\,000 \div 10$

$= \$4500$

Nick's share $= 2 \times \$4500$

$= \$9000$

Hong's share $= 3 \times \$4500$

$= \$13\,500$

Rob's share $= 5 \times \$4500$

$= \$22\,500$

∴ Nick, Hong and Rob receive \$9000, \$13 500 and \$22 500, respectively.

Check: \$9000 + \$13 500 + \$22 500 = \$45 000.

Method 2: Fraction method

Total number of parts = 2 + 3 + 5 = 10.

Nick's share $= \frac{2}{10} \times \$45\,000$

$= \$9000$

Hong's share $= \frac{3}{10} \times \$45\,000$

$= \$13\,500$

Rob's share $= \frac{5}{10} \times \$45\,000$

$= \$22\,500$

EXERCISE 11.04 ANSWERS ON P. 651

Dividing a quantity in a given ratio

U F PS R

1 Find the total number of parts if the ratio is:

a 2 : 7 **b** 4 : 1 **c** 3 : 4 **d** 2 : 5 : 6

EXAMPLE 11

2 Divide $500 in the ratio:

a 4 : 1 **b** 7 : 3

3 PS R Phuong, Janelle and Ahmet bought a Lotto ticket for $12. They contributed $3, $3 and $6, respectively, to the purchase price. They won $4000 and agree to split the winnings in the same ratio.

a Simplify the ratio of 3 : 3 : 6.

b How much prize money does each person get?

4 George and Amal share the weekly rent of $390 in the ratio of 7 : 6. What is Amal's share?

5 Divide 450 kg in the ratio:

a 4 : 5 **b** 3 : 2

6 Divide 720 cm in the ratio:

a 1 : 3 : 5 **b** 5 : 3 : 4

7 PS R Company directors Rashmi, Rachel and Robert share the company profits in the ratio of 5 : 3 : 3. Which of the following is the amount that Rashmi receives in a year when profits are $121 000? Select the correct answer **A**, **B**, **C** or **D**.

A $11 000 **B** $24 000 **C** $33 000 **D** $55 000

8 PS In Year 8, the ratio of boys to girls is 4 : 5. If there are 225 students in Year 8, find how many girls are there.

9 PS R In Year 9, the ratio of boys to girls is 3 : 2. If there are 125 students in Year 9, how many more boys than girls are there?

10 PS R Adam needs to make 800 g of short-crust pastry. Flour and butter are needed in the ratio of 3 : 1. How much flour is needed?

11 A company posts 1386 letters in a week. The ratio of local to overseas letters is 2 : 7. How many overseas letters are sent in a week?

12 PS R A truck carries fruit and vegetable boxes in the ratio of 5 : 7. If it carries a total mass of 7.5 tonnes, what mass of vegetables does it carry? Select the correct answer **A**, **B**, **C** or **D**.

A 0.625 tonnes **B** 3.125 tonnes **C** 4.375 tonnes **D** 6.873 tonnes

13 An alloy of mass 176 kg is made from copper and zinc in the ratio of 5 : 6. Find the mass of copper in the alloy.

14 At a school, the ratio of students who speak a second language to students who speak only English is 5 : 8. If there are 923 students at the school, how many students speak only English?

15 A 20 m cable is cut into three sections in the ratio of 2 : 3 : 5. Find the length of each section.

16 When making mortar, sand and concrete is mixed in the ratio of 6 : 1. If we need 280 kg of mortar, how much sand will be needed?

PS R

17 In a study of 225 000 people, it was found that the ratio of right-handed people to left-handed people was 13 : 2.

PS R

a How many left-handed people were there?

b How many more right-handed people than left-handed people were there?

18 Angus earns twice as much as Catriona. If the sum of their wages is $210 000, how much does each earn?

R

Rates 11.05

While a ratio compares two or more quantities measured in the same units, a **rate** compares two quantities measured in **different units**.

Worksheet Rates

Interactive Unit rates

A rate shows how one quantity changes with another quantity. We write a rate using a '/' symbol in the form 'something *per* something else'. For example:

- the price of petrol is stated in cents per litre (c/L).
- your heart beats at a rate measured in beats per minute (beats/min).
- the speed of a car is measured in kilometres per hour (km/h).

The word 'per' means 'for each' so we express a rate '*per* single unit'. For instance, travelling at a rate of 50 km/h means travelling 50 km in *each* hour.

Example 13

Write each statement as a simplified rate.

a A factory produces 87 cars in 3 hours.

b Ham costs $50 for 8 kg.

Foundation Standard Complex

SOLUTION

a The production rate would be expressed in cars per hour.

$$\text{Production rate} = \frac{87 \text{ cars}}{3 \text{ hours}}$$
$$= 29 \text{ cars/hour}$$

We can write a rate as a fraction: divide the number of cars by the number of hours.

b The cost would be expressed in dollars per kilogram.

$$\text{Cost} = \frac{\$50}{8 \text{ kg}}$$
$$= \$6.25 / \text{kg}$$

Divide the number of dollars by the number of kilograms.

EXERCISE 11.05 ANSWERS ON P. 651

Rates

1 Write the units suitable for each rate below, in the form _____/_____.

a typing speed
b heart rate
c cost of a mobile phone call
d cost of bananas
e a person's wage
f a runner's speed
g population growth
h the cost of water
i population density of a country

EXAMPLE 13

2 Write each statement as a simplified rate.

C **a** 51 sheep in 3 hours
b $10.75 for 2.5 kg
c 208 students for 8 teachers
d 136 points in 4 games
e 546 words in 6 minutes
f 34 articles in 4 hours
g 72 cars in 14 days
h 5040 boxes in 8 hours
i 259 metres in 7 seconds
j 46 000 bottles in 50 hours
k 7944 revolutions in 6 minutes
l $175 for 5 hours
m 448 km in 8 hours
n $16.50 for 6 kg
o 114 runs in 24 overs
p 243 km using 30 litres
q $126 for 12 hours
r 2520 kg for 60 hectares

3 The cost of sending a 5.5 kg parcel to Malaysia is $88. What is the postage rate?
C Select the correct answer **A**, **B**, **C** or **D**.

A $0.34/kg **B** $16/kg **C** $0.07/kg **D** $484/kg

4. In your own words, explain what is meant by each rate.
   a. a speed of 100 km/h
   b. a traffic flow of 150 cars/h
   c. petrol consumption of 10.3 L/100 km
   d. a farmer keeping 60 sheep/hectare
5. A lift should carry no more than 1600 kg or 20 people. What is this weight allowance, in kg/person?
6. A pulp mill clears 12 600 hectares of forest in 7 years. At what rate in hectares/year does the mill clear the forest?
7. The cost of 53 litres of petrol is $73.67. Express this cost in c/L.
8. A complaints hotline took 2190 calls in one year. Calculate the number of calls per month.

Best buys 11.06

When shopping, it is important to compare the prices of different brands or sizes of items and calculate the **best buy** ('best value for money'). The biggest container does not always provide the best value. This can be done by comparing the **unit price** (cost of one item or unit) of each brand or size and choosing the cheapest one. A unit could be 1 gram or 1 millilitre. Unit price is an example of a rate.

Video Unit pricing

Puzzle Best buys puzzlec

ⓘ Unit price

Unit price = cost ÷ number of items or units

Supermarkets are required by law to display unit prices to allow you to compare brands or sizes of items. If you look carefully at the price tags on their shelves, you will notice unit prices displayed in small print.

Example 14

Which brand of baked beans is the better buy?

$4.60

$1.65

SOLUTION

Calculating the unit price (cost per gram) for each brand:

Bean There = $4.60 ÷ 500 = $0.0092/g.

Mr Beanz = $1.65 ÷ 200 = $0.00825/g.

Mr Beanz has the lower unit price, so it is the better buy.

EXERCISE 11.06 ANSWERS ON P. 652

Best buys

1 Find the better buy for each pair of items.

PS **a** A 2 kg box of sultanas for $8.85 or a 1 kg box for $4.65.

R **b** 6 Notepads for $15.50 or 8 notepads for $20.

C **c** 45 g Chips for $2.10 or 150 g for $6.50.

d 2 Tubs of yoghurt for $1.24 or 7 tubs for $4.30.

e 3 kg of Corn flour for $5.85 or 500 g of corn flour for $1.

f 600 mL Bottle of fruit juice for $3.20 or a 2.25 litre carton for $8.05.

2 Michael purchased 400 g of ham for $3.59, while Kim purchased 300 g for $2.54. Who had the better buy?

PS R C

3 Which size of Fruit Bix cereal is the best buy?

PS  

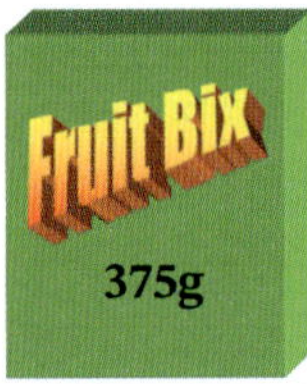

$4.75 **$3.49** **$2.51**

4 Happy Tam cat food can be purchased in four different packages.

PS Pack 1: 24 × 250 g cans for $39.99 Pack 2: 12 × 300 g cans for $24.99

R Pack 3: 9 × 400 g cans for $25.99 Pack 4: 6 × 1 kg cans for $38.50

C Order these packs from best to worst buy. Select the correct answer **A**, **B**, **C** or **D**.

A 4, 2, 1, 3 **B** 3, 2, 1, 4 **C** 4, 1, 2, 3 **D** 2, 1, 3, 4

  Foundation Standard Complex

5 For each item, find which size is the best value for money.

PS R C

a 375 g $1.90; 750 g $3.90

b 2 L $3.25; 375 mL × 6 $3.85

c 440 g $2.21; 735 g $2.78

d 235 g $4.20; 175 g $3.20; 115 g $1.80

Alamy Stock Photo/Oliver Hoffmann
Shutterstock.com/Bahadir Yeniceri
Shutterstock.com/Marek Tr
Shutterstock.com/Gts
Smart Media/Jalna
Smart Media/Jalna
Shutterstock.com/Lipowski Milan

6 For each item, which size gives the best value for money?

PS R C

		Small	Medium	Large
a	Washing powder	450 g for $3.99	600 g for $5.15	1 kg for $8.95
b	Margarine	350 g for $4.15	500 g for $5.99	750 g for $8.50
c	Ice cream	500 mL for $1.39	750 mL for $2.30	2 L for $5.99
d	Peas	400 g for $2.05	500 g for $2.49	600 g for $2.50

11.07 Rate problems

Worksheets
Rate problems

Ratios and rates review

Puzzle
Rates problems

Video
Rate problems 1

Example 15

Lucas types 55 words per minute. How many words can he type in 20 minutes?

SOLUTION

This problem can be solved by writing a pair of equivalent rates in fraction form. On the left, the given rate: in each minute, Lucas types 55 words. On the right, what we're trying to find (the number of words in 20 min):

$$\frac{55 \text{ words}}{1 \text{ min}} = \frac{? \text{ words}}{20 \text{ min}}$$

Number of words $= 55 \times 20$

$= 1100$

Multiply 1 min by 20 to get 20 min, so multiply 55 words by 20 to get the answer.

Problems involving rates can usually be solved by multiplying or dividing. The following strategy may help.

ⓘ Solving rate problems

- Write the units of the rate x/y as a fraction: $\frac{x}{y}$.
- To find the quantity in the numerator, x, **multiply** by the rate.
- To find the quantity in the denominator, y, **divide** by the rate.

Video
Rate problems 2

Example 16

A factory makes pens at the rate of 50 pens per minute.

a How many hours and minutes will it take to produce 10 000 pens?

b How many pens are produced in an 8-hour day at the factory?

SOLUTION

a $$\frac{50 \text{ pens}}{1 \text{ min}} = \frac{10\,000 \text{ pens}}{? \text{ min}}$$

How many times 50 divides into 10 000.

In each minute, the factory makes 50 pens, so to determine how long it will take to make 10 000 pens, we calculate 10 000 ÷ 50.

Time taken = 10 000 ÷ 50

= 200 minutes

$= \frac{200}{60}$ h

$= 3\frac{1}{3}$ h

= 3 h 20 min

By equivalent fractions, 50 goes into 10 000 200 times.

Converting to hours and minutes.

OR

The units of the rate expressed as a fraction are $\frac{\text{pens}}{\text{min}}$.

To find the number of minutes (the denominator), **divide** by the rate.

Time taken = 10 000 ÷ 50

= 200 minutes

= 3 h 20 min

b $\frac{50 \text{ pens}}{1 \text{ min}} = \frac{? \text{ pens}}{8 \text{ h}}$

1 h = 60 min

8 h = 8 × 60 min

= 480 min

$\frac{50 \text{ pens}}{1 \text{ min}} = \frac{? \text{ pens}}{480 \text{ min}}$

Number of pens = 50 × 480

= 24 000

Multiply 1 min by 480 to get 480 min, so multiply 50 pens by 480 to get the answer.

OR

$\frac{\text{pens}}{\text{min}}$: To find the number of pens (the numerator), **multiply** by the rate.

Number of pens in one hour = 60 × 50

= 3000

1 h = 60 min

No. of mins × pens per min

Number of pens in 8 hours = 8 × 3000

= 24 000

EXERCISE 11.07 ANSWERS ON P. 652

Rate problems

U F PS R C

1 Marisol is paid $17.80 per hour. How much will she earn if she works 38 hours in a week?

PS R C

2 A cricket batsman scores at a rate of 32 runs per innings. How many runs he will score in 5 innings?

PS R C

EXAMPLE 16

3 A car uses 14 litres of petrol to travel 147 kilometres.

PS **a** Write this as a rate in km/L.

R **b** How far can the car travel on 20 litres of petrol?

4 A farmer can graze 16 sheep per hectare. If he has 21 hectares set aside for sheep, how many sheep can he graze?

5 A tap drips water at a rate of 18 mL/h. How much water would be wasted in one day?

PS R

6 An aeroplane can carry a total of 450 passengers per flight. How many flights would it take to carry 2700 passengers?

PS R

7 The Wong family needs to buy new carpet for the family home. The carpet costs $125/m and the carpet layers charge $480 to lay it. How much will it cost to carpet the house if the Wongs need 28 metres of carpet?

PS R

8 A farmer uses fertiliser at a rate of 28 kg/ha. How many hectares can she cover if she has 200 kg of fertiliser?

PS R

9 Tatiana is paid $17.50/hour to babysit.

PS **a** How much does she earn for babysitting for 4 hours?

R **b** How long did she babysit to earn $43.75?

10 A plane travelled at 900 km/h. How far did it travel in 20 minutes?

PS R

11 **a** Petrol costs $1.35/L. If Rico had $20 in his pocket, how much petrol could he buy (correct to the nearest 0.1 litre)?

PS R **b** If Rico's car can travel 9 km on 1 litre of petrol, how far can he travel on the petrol he bought with $20?

12 Smokers lose approximately 3% of their lung capacity per year. If Ella with 100% lung capacity starts smoking at 20 years of age, at what age will she have 25% lung capacity? Select the correct answer **A**, **B**, **C** or **D**.

PS R

A 75 **B** 28 **C** 45 **D** 33

13 The temperature in Mittagong at 7 a.m. is 8°C. If the temperature rises at a rate of 3°C/h, find:

PS R

a the temperature at 10 a.m. and at 2 p.m.

b the time at which the temperature will be 23°C.

14 Keira and Xander take turns mowing a rectangular lawn 60 m long and 40 m wide.

PS R

a Find the area of the lawn.

b When Keira cuts the grass, she does it in 1 hour using an old lawn mower. What is her mowing rate in m^2/min?

c Xander uses a new lawn mower and it takes him 40 minutes. What is his mowing rate in m^2/min?

d How long will it take them to mow the lawn if they mowed together?

15 Jayden is filling this tank with a pipe that pumps water at 1.25 litres per second.

PS  C

6750 L

How long will it take to fill the tank:

a in seconds? **b** in minutes? **c** in hours?

16 Mikayla and Jesinta started their holiday on a full tank of petrol. The reading on the odometer (in km) was | 0 | 3 | 4 | 5 | 6 | 8 |

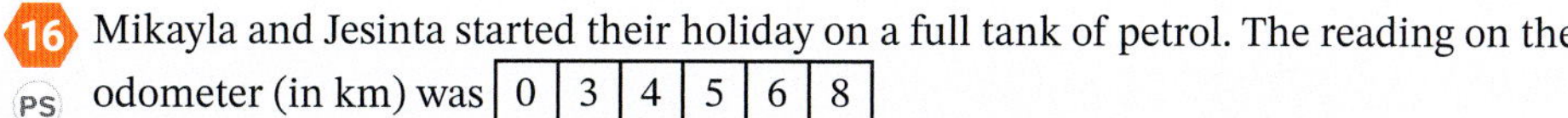

PS R

When the tank became empty, the reading was | 0 | 3 | 5 | 2 | 1 | 8 |

Petrol cost $1.54/litre and they needed to pay $73.92 to fill the tank.

a How many kilometres did they travel on a full tank of petrol?

b How much petrol is there in a full tank?

c How far can their car travel on 1 litre of petrol? Answer to the nearest 0.1 km.

Speed 11.08

Speed is a rate that compares the distance travelled with the time taken. Average speed is calculated by dividing the distance travelled by the time taken.

$$\text{Average speed} = \frac{\text{distance travelled}}{\text{time taken}}$$

Worksheets Speed

What's my speed?

Puzzle Speed

Video Speed of the Earth

Speed can be measured in kilometres/hour (km/h) or metres/second (m/s).

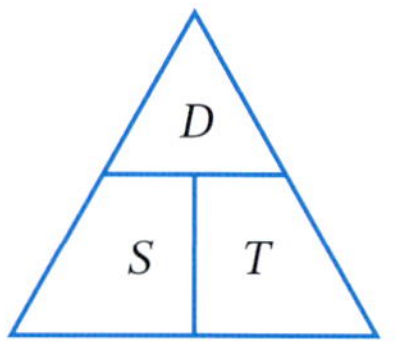

The relationship between distance, speed and time can be remembered using this triangle.

If we cover the quantity we are looking for, the rest of the diagram tells us what to do.

- To find the speed, cover the S and we are left with $\frac{D}{T}$, so speed = distance ÷ time.
- To find the distance, cover the D and we are left with $S \times T$, so distance = speed × time.
- To find the time, cover the T and we are left with $\frac{D}{S}$, so time = distance ÷ speed.

Presentation Rates: Triangle method

Video Speed

Example 17

A bullet train travels 320 km in 1 hour and 20 minutes.

a Calculate its average speed.

b How long will it take to travel 1000 km?

c How far will it travel in 5 hours?

SOLUTION

a $\text{Average speed} = \frac{\text{distance}}{\text{time}}$

$= \frac{320 \text{ km}}{1 \text{h } 20 \text{ min}}$

$= \frac{320 \text{ km}}{1\frac{1}{3} \text{h}}$

$= 240 \text{ km/h}$

$20 \text{ min} = \frac{20}{60} \text{ h} = \frac{1}{3} \text{h}$

b $\text{Time} = \frac{\text{distance}}{\text{speed}}$

$= \frac{1000}{240} \text{ h}$

$= 4\frac{1}{6} \text{ h}$

$= 4 \text{ h } 10 \text{ min}$

By the speed triangle, or because we divide by the rate to find the h in

The train travels 240 km in every hour, so to calculate how long it will take to travel 1000 km, we work out how many times 240 goes into 1000.

Press °′″ or DMS on the calculator.

c $\text{Distance} = \text{speed} \times \text{time}$

$= 240 \times 5$

$= 1200 \text{ km}$

By the speed triangle, or because we multiply by the rate to find the km in

The train travels 240 km in every hour, so to calculate how far it will take to travel in 5 hours, we multiply 240 by 5.

EXERCISE 11.08 ANSWERS ON P. 652

Speed

U F PS R C

EXAMPLE 17

1 Find the average speed in km/h for each statement.

- **a** A horse rider travels a distance of 15 km in 3 hours.
- **b** A jet travels 2400 km in 5 hours.
- **c** A bushwalker walks 25 km in 5 hours.
- **d** A cyclist rides 85 km in 5 hours.
- **e** A motorbike travels 250 km in 2.5 hours.
- **f** A car travels 280 km in 3.5 hours.
- **g** A truck travels 390 km in 3 hours and 15 minutes.
- **h** A car travels 90 km in 45 minutes.
- **i** A motorbike on the race track travels 250 km in 1 hour and 15 minutes.
- **j** A boat travels 15 km in 1 hour and 40 minutes.
- **k** An athlete runs 200 m in 20 seconds.
- **l** A kangaroo travels 1.5 km in 15 minutes.

2 Find the average speed in m/s for each statement.

- **a** An athlete runs 200 m in 20 seconds.
- **b** A swimmer swims 100 m in 64 seconds.
- **c** A bird flies 348 m in 6 seconds.
- **d** A cyclist travels 960 m in 2 minutes.
- **e** A swimmer swims 1500 m in 16 minutes and 40 seconds.

3 Find the distance travelled for each statement.

- **a** A car travels for 8 hours at an average speed of 90 km/h.
- **b** A bushwalker travels for 7 hours and 30 minutes at an average speed of 4 km/h.
- **c** A truck travels for 14 hours at an average speed of 95 km/h.

4 Majid can walk at a speed of 6 km/h. How far can he walk in 30 minutes?

5 A boat travels 24 km in 90 minutes. What is its average speed? Select the correct answer **A**, **B**, **C** or **D**.

A 12 km/h **B** 16 km/h **C** 18 km/h **D** 36 km/h

6 Peter walked to school in 15 minutes, a distance of 1500 metres. Find his speed in:

R C

a metres per minute **b** kilometres per hour

7 Mrs Hassam lives 20 km from the nearest railway station and it takes her 30 minutes to drive there in her car. Find her average speed in km/h.

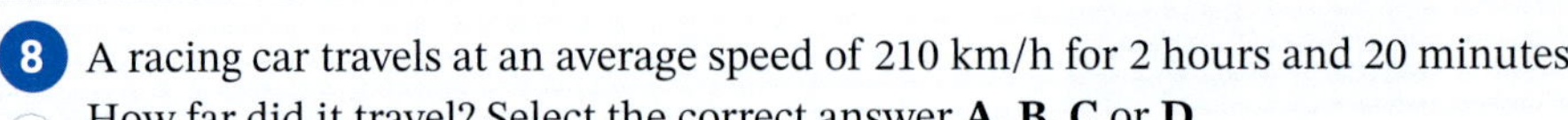

8 A racing car travels at an average speed of 210 km/h for 2 hours and 20 minutes. How far did it travel? Select the correct answer **A**, **B**, **C** or **D**.

PS R

A 90 km **B** 294 km **C** 462 km **D** 490 km

9 A flywheel rotates at a rate of 2400 revolutions per minute. How many revolutions does it make in 30 seconds?

R

10 How long does it take a motor scooter to travel 60 km at a speed of 40 km/h?

R

11 There are other units for measuring speed, such as the **knot** and the **mach**. Find what these special terms mean and where they are used.

C

12 A bus travels at 75 km/h for 5 hours, and then travels a further distance of 200 km in 4 hours. Find the average speed for the whole journey, correct to one decimal place.

PS R

13 A car travels for 3 hours at 60 km/h and then travels a further distance of 100 km at 50 km/h. What is the average speed for the trip?

PS R

14 Kieran swims at a speed of 1.85 m/s. What would be his time (to two decimal places) for a 100 metre sprint?

PS R

15 Nia runs 200 metres in 24 seconds. How far will she run in one minute at the same rate?

PS R

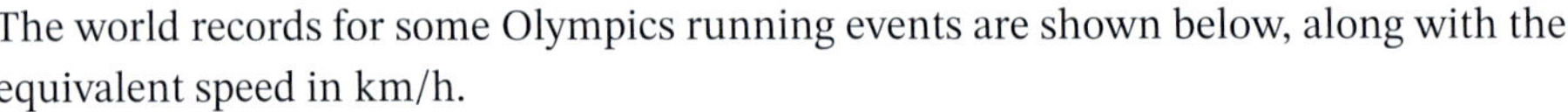

DID YOU KNOW?

Running speeds

The world records for some Olympics running events are shown below, along with the equivalent speed in km/h.

Mens Event	World record	Equivalent speed
100 m	9.58 s	37.58 km/h
200 m	19.19 s	37.52 km/h
400 m	43.03 s	33.47 km/h
800 m	1 min 40.91 s	28.54 km/h
1500 m	3 mins 26 s	26.21 km/h
4 × 100 m relay	36.84 s	39.09 km/h

☐ Foundation ○ Standard ⬡ Complex

Womens Event	World record	Equivalent speed
100 m	10.49 s	34.52 km/h
200 m	21.34 s	33.74 km/h
400 m	47.6 s	30.25 km/h
800 m	1 min 53.28 s	25.42 km/h
1500 m	3 min 50.07 s	23.47 km/h
4 × 100 m relay	40.82 s	35.28 km/h

What do you notice about the speeds as the length of the race increases?

Why would the speeds for the 4 × 100 m relays be faster than the speeds for 400 m?

TECHNOLOGY

Travelling distances and times

Visit a map website such as Google Maps or WhereIs for this investigation to examine travelling distances and times between two locations.

Aidan lives near the Wyong Hill Reserve in NSW and is travelling to the Tuggerah shopping centre to see a movie.

1 Use the website to locate both Wyong Hill Reserve and Tuggerah shopping centre, then find the travelling distance and time for Aidan's trip if he is going by car.

2 Find the travelling distance and time for Aidan's trip if he is travelling on foot (walking).

3 How much further would you have to travel if you were going from the Wyong Hill Reserve to the shops by car instead of walking there?

4 Would you recommend that Aidan travel by car or on foot for the trip? Why?

5 a Calculate the average speed of the car for the trip.

  b Why do you think this average speed is slower than the normal speed limit of 60 km/h?

6 Aidan is meeting Alinta at Tuggerah to see a movie. Alinta lives in Berkeley Vale. Find the travelling distance and time for Alinta's trip if she is travelling by car.

7 If Alinta's actual driving time was 5 minutes, what was her average speed?

8 If Aidan's actual speed was 80 km/h, how many minutes and seconds will his trip take?

9 So will Aidan or Alinta arrive at Tuggerah first? By how many minutes and seconds?

11.09 Travel graphs

Video
Travel graphs

Worksheet
The hare and the tortoise

Puzzle
Jane's diary

A **travel graph** or **distance–time graph** is a line graph that describes a journey, by comparing distance (on the vertical axis) with time (on the horizontal axis). The slope or steepness of the graph indicates the speed.

Travel graphs

On a **travel graph**:

- a horizontal (flat) section on the graph indicates a stop (the traveller is stationary).
- the steeper the line, the greater the speed (more distance covered in less time).
- a section going down, towards the right, indicates a change in direction or that the traveller is returning towards the start.

Example 18

This graph shows Monique's cycling trip.

a At what time did Monique leave home?

b When was Monique's first stop? How far from home was she?

c Find her average speed over the first two hours.

d What time was it when Monique began her journey home?

e How far did she travel all together?

f Find her average speed during the trip home.

g For how long did Monique stop altogether during the trip?

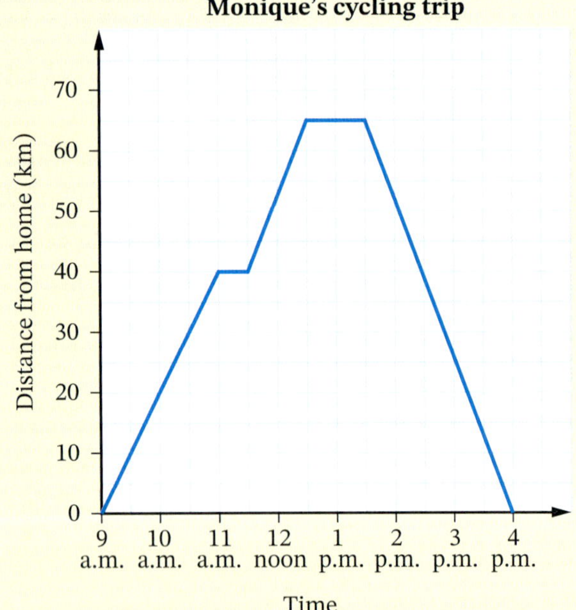

SOLUTION

a Monique left home at 9 a.m. At the start of the graph, when distance = 0.

Where the graph is flat, the distance from home does not change, which means that Monique has stopped.

b Monique first stopped at 11 a.m., 40 km from home. This is where the graph is flat.

Where the graph is flat, the distance from home does not change, which means that Monique has stopped.

c Distance = 40 km, time = 2 hours

Average speed $= \frac{40 \text{ km}}{2 \text{ h}} = 20$ km/h

d Monique started returning home at 1:30 p.m. This is where the graph points downward.

e Monique travelled 65 km, then returned home.

Total distance = 2 × 65 km = 130 km.

f Distance = 65 km, time $= 2\frac{1}{2}$ hours

Average speed $= \frac{65 \text{ km}}{2\frac{1}{2} \text{ h}} = 26$ km / h

g First stop: $\frac{1}{2}$ hour; Second stop: 1 hour.

Total stopping time $= \frac{1}{2} + 1 = 1\frac{1}{2}$ hours

EXERCISE 11.09 ANSWERS ON P. 652

Travel graphs

U F PS R C

1 This travel graph shows Obama's return trip to his friend Joe's house.

R C

a How long did it take Obama to walk to Joe's house?

b How far is Joe's house from Obama's house?

c Find Obama's average speed:

i before his first stop.

ii after his first stop.

d How is a higher speed shown on the graph?

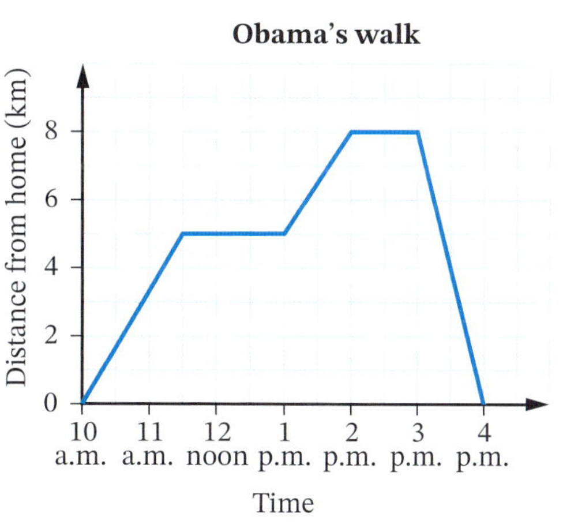

e What was the total distance that Obama travelled?

f Between what times does Obama stop on his walk?

g When did Obama start his trip home?

h How long did it take him to walk home?

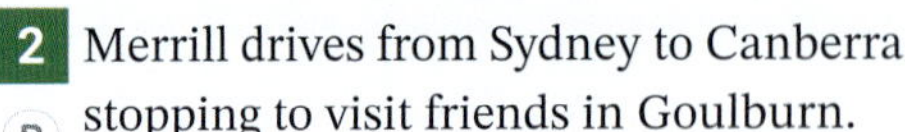

2 Merrill drives from Sydney to Canberra, stopping to visit friends in Goulburn.

R C

a How far is Canberra from Sydney?

b How long does the trip take?

c For how long does Merrill stop at Goulburn?

d Find her average speed for the journey (excluding stops).

e How far it is from Goulburn to Canberra?

f When does Merrill travel faster: before or after Goulburn?

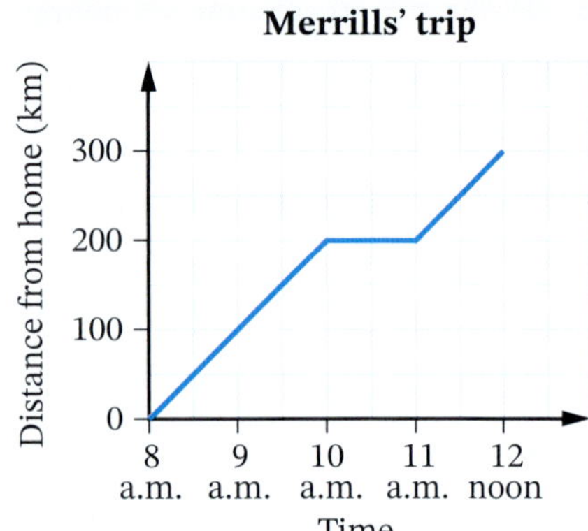

3 This graph shows a cyclist's day trip.

R C

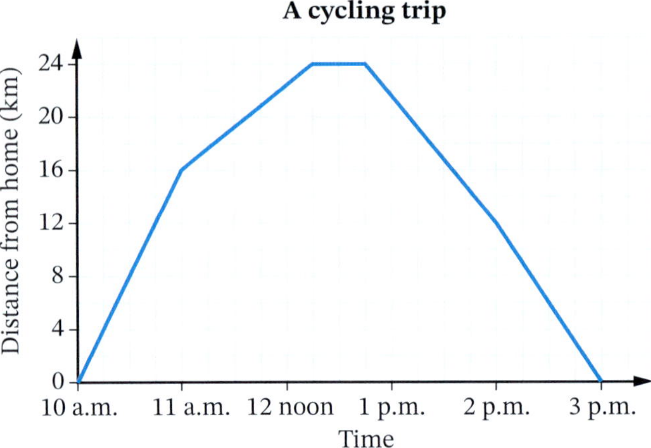

a At what time did the speed of the cyclist

i increase? **ii** decrease?

b When did the cyclist start to return home?

c How far did the cyclist travel altogether on this day?

d How long did the cyclist spend 'on the road'?

e Find the cyclist's average speed for:

i the first hour **ii** 11 a.m. to 12:15 p.m.

iii 12:45 p.m. to 2 p.m. **iv** the entire day

4 Briana travels from Bligh to Macquarie, while Connor travels from Macquarie to Bligh.

R C

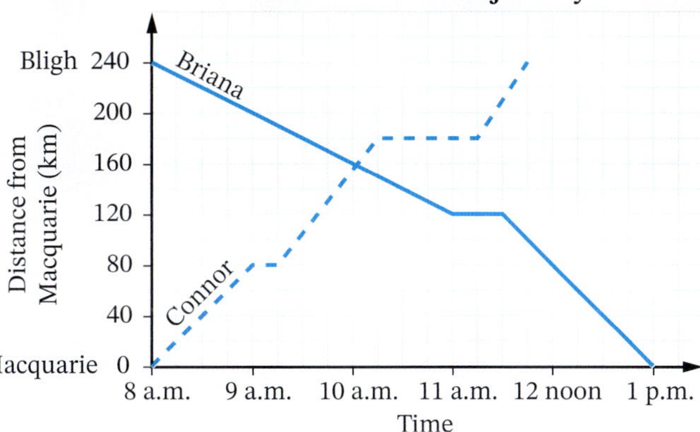

a How far it is between the two towns?

b Who is travelling faster? How can you tell?

c How far is Briana from Macquarie at 11 a.m.?

d How far is Connor from Macquarie at 11 a.m.?

e At what time do Briana and Connor pass each other? How far are they from Macquarie when they pass?

f Is Connor travelling faster before 9 a.m. or after 9:15 a.m.? How does the graph show this?

g Calculate Briana's average speed before he stops.

h For how long did Connor stop altogether on the trip? Select the correct answer **A**, **B**, **C** or **D**.

A 60 minutes **B** 75 minutes **C** 5 minutes **D** 90 minutes

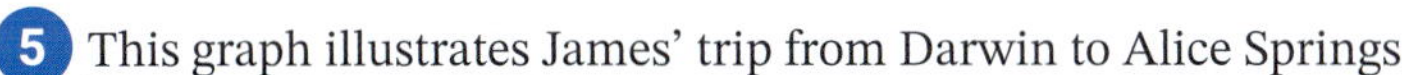

5 This graph illustrates James' trip from Darwin to Alice Springs.

PS R C

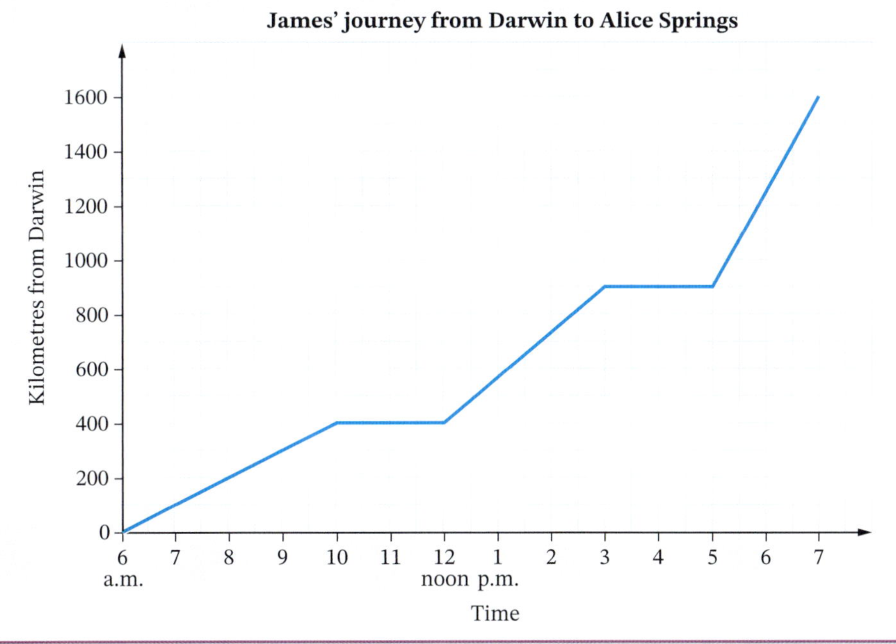

a How far is Alice Springs from Darwin?

b When did James arrive in Alice Springs?

c What was James' average speed in the last 2 hours?

d Which of the following statements is false? Select the correct answer **A**, **B**, **C** or **D**.

A James stopped for 4 hours altogether.

B James was travelling fastest in the last 2 hours.

C James' speed for the first 4 hours was 100 km/h.

D James' average speed for the whole trip was 120 km/h.

e Samantha leaves Darwin one hour after James and travels at a constant speed of 125 km/h towards Alice Springs. Copy the graph and add Samantha's journey to it.

f Find the approximate time when Samantha overtakes James.

6 C This travel graph shows Marnie's cycling journey. Write a story about her ride, based on the information in the graph.

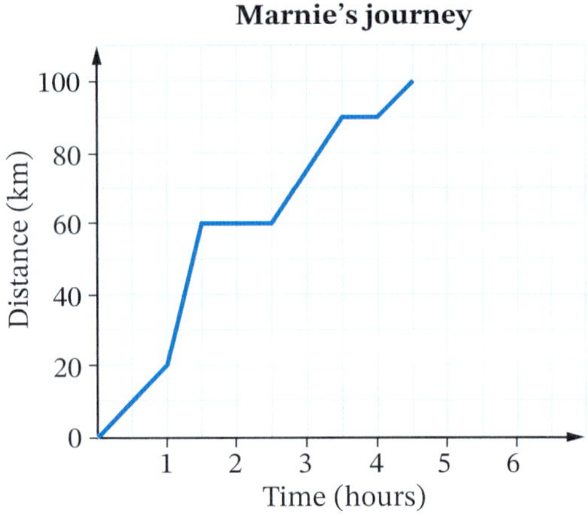

7 R C Match each bushwalker described below with the correct travel graph.

a Santosh maintained the same speed all day.

b Michelle was fast at first, but slowed down.

c Tahlia was the fastest, but then got slower after stopping for lunch.

d Jim was slow at first, but picked up speed.

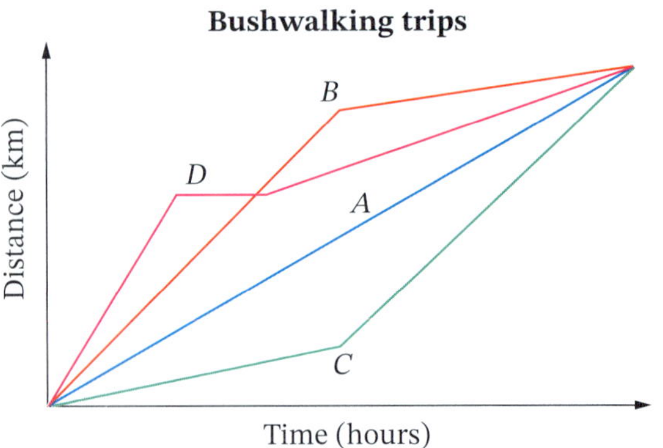

8 Which one of these graphs could **not** be a travel graph? Why not?

R C

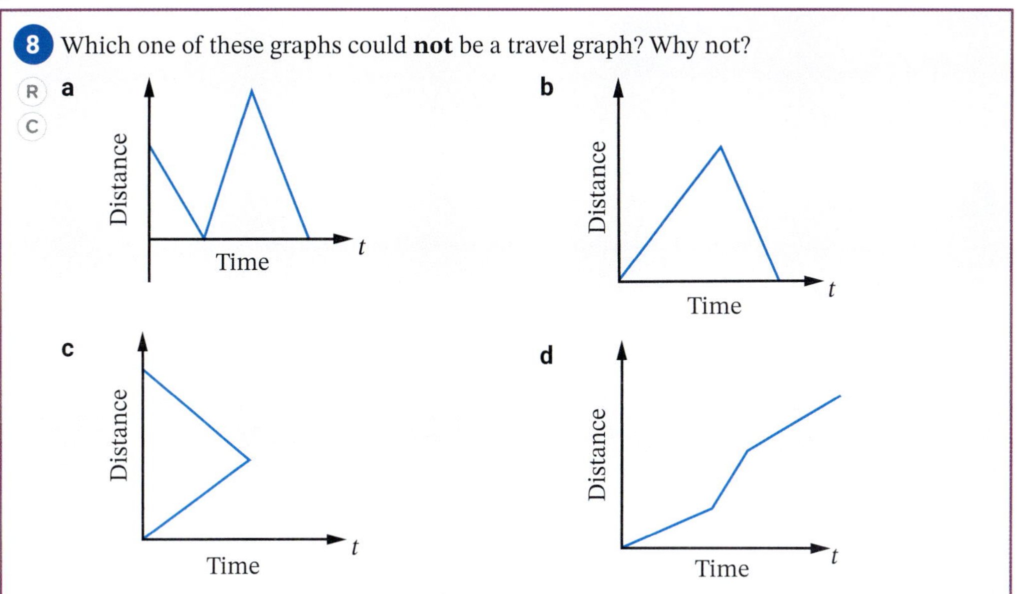

Sketching informal graphs 11.10

While **line graphs** are useful for describing relationships and changes between variables, in most real-life situations, the graphs are more likely to be informal and involve **curves** rather than straight lines.

Example 19

Draw a graph to represent a bath's water level over time.

- Jacob switches on the tap and the water level gradually rises.
- He switches off the tap, gets into the bath and the water level quickly rises.
- Jacob stays in the bath for a while before climbing out, reducing the water level again.
- He unplugs the bath and it drains completely: the water level decreasing quickly at first, then slowly towards the end.

SOLUTION

The horizontal axis should represent time.

The vertical axis should represent height of water.

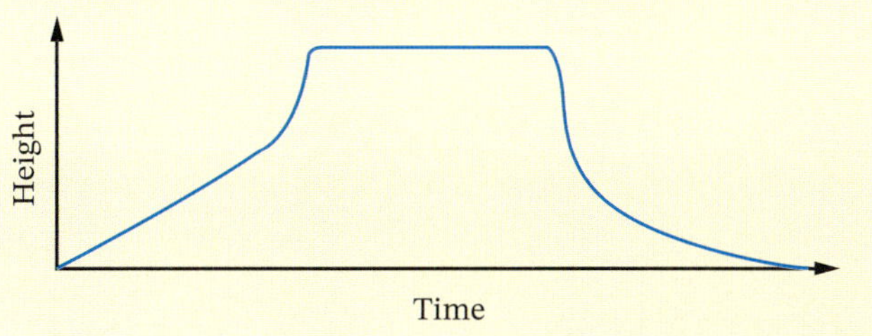

Foundation Standard Complex

EXERCISE 11.10 ANSWERS ON P. 653

Sketching informal graphs

1 Sketch a graph of the noise level of this classroom.

PS R C

- The students enter the classroom and the noise level increases.
- They settle down and work quietly.
- There are small group discussions and then the teacher talks.
- Towards the end of the lesson the noise level increases.
- Then the teacher speaks about the homework for the night and the class is dismissed.

2 The lift in a 6-storey building started at the ground floor, went to the 2nd floor, then the 5th floor, then the 3rd floor, before returning to ground floor. Which graph represents this situation? Select the correct answer **A**, **B**, **C** or **D**.

R C

A

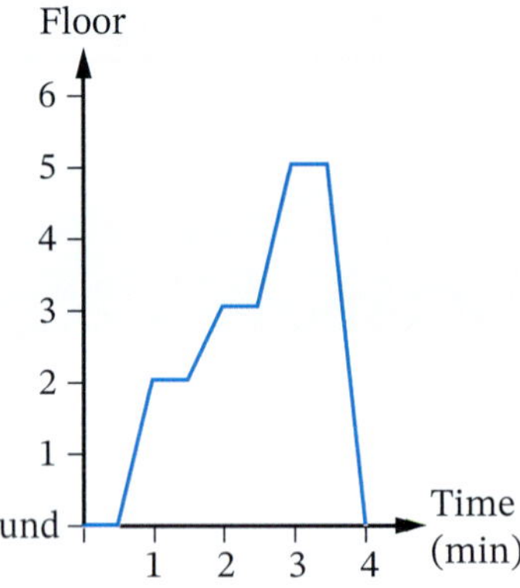

B

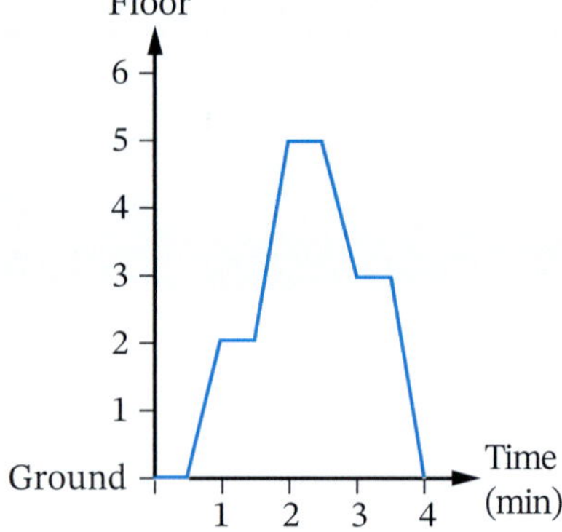

C

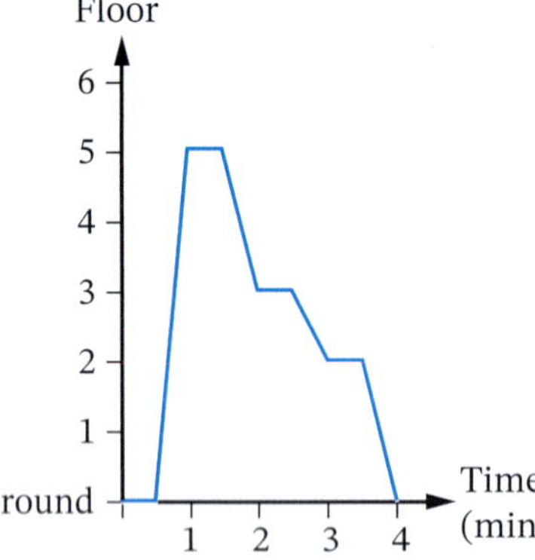

D

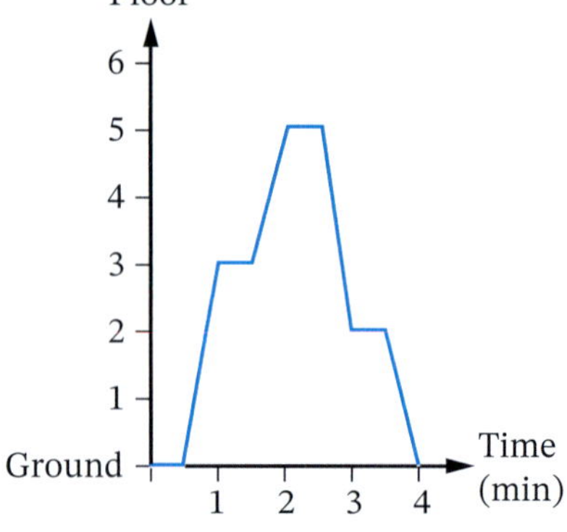

☐ Foundation ○ Standard ⬡ Complex

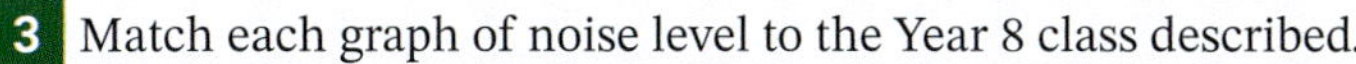

3 Match each graph of noise level to the Year 8 class described.

R C

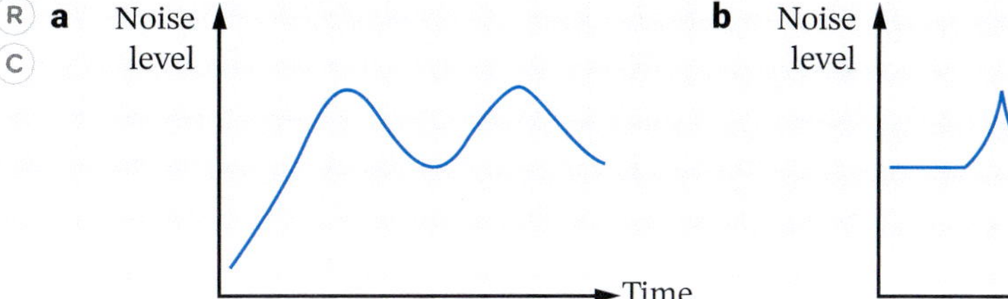

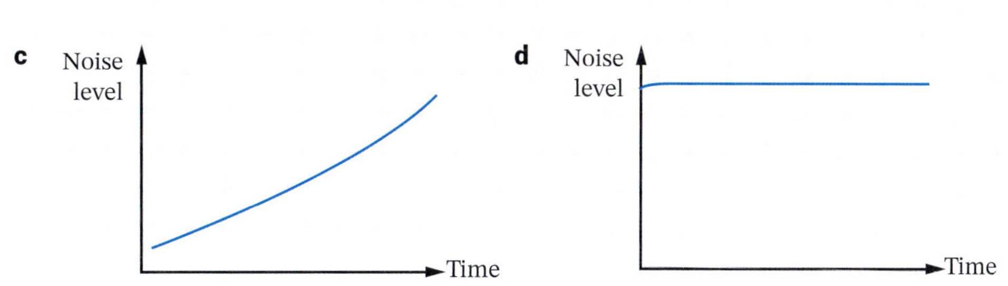

- 8 Green is constantly noisy.
- 8 Yellow is quiet until the teacher leaves the room.
- 8 Brown is regularly told to be quiet by their teacher.
- 8 Red just gets louder and louder.

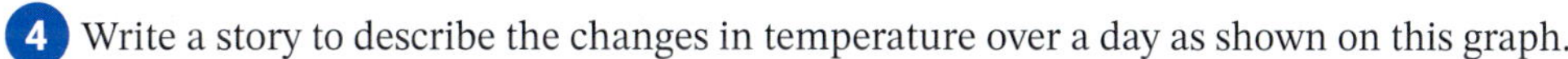

4 Write a story to describe the changes in temperature over a day as shown on this graph.

R C

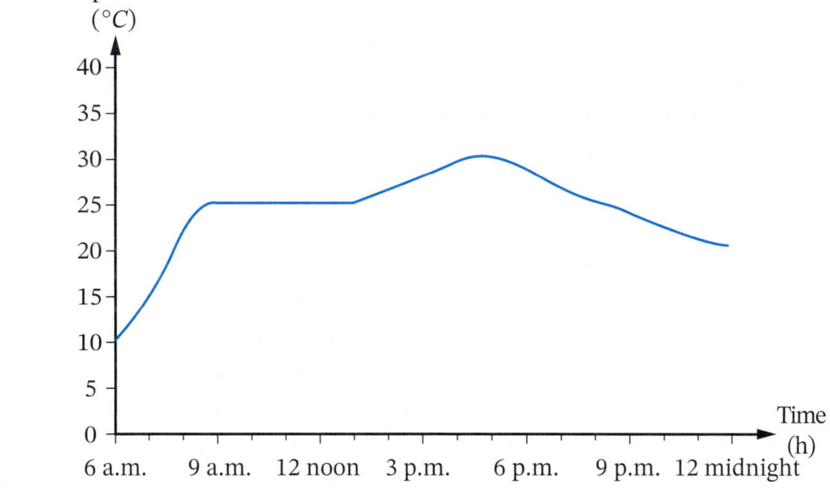

Foundation Standard Complex

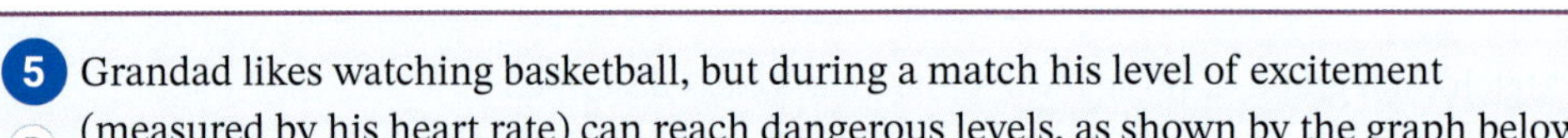

5 Grandad likes watching basketball, but during a match his level of excitement (measured by his heart rate) can reach dangerous levels, as shown by the graph below.

R C

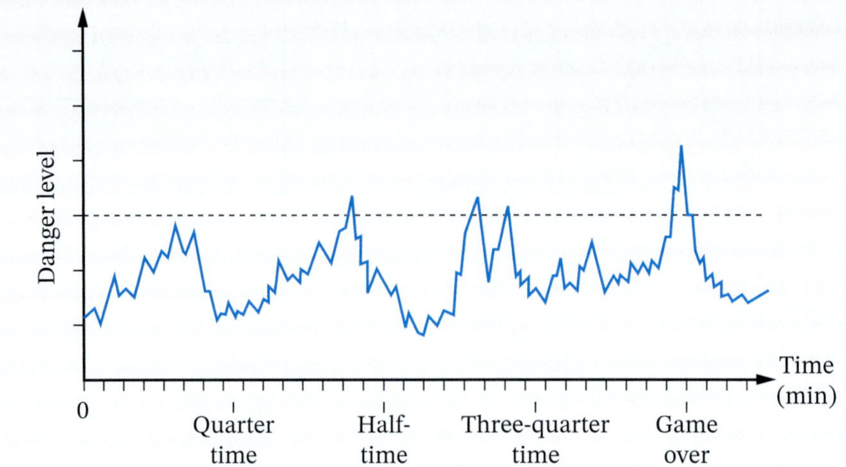

a Describe Grandad's excitement level in the first quarter.

b How many times does he become too excited during the match?

c What happens to his excitement level after the match?

d Describe what might have happened in the final quarter to make him so excited.

6 A triathlete has this training programme for a marathon:

PS R C

- Start at 5 a.m., run 12 km in 2 hours, rest $\frac{1}{2}$ hour.
- Run a further 8 km in the next hour, rest $\frac{1}{2}$ hour.
- Pick up bike and cycle home, arriving at 10 a.m.

  – Draw the graph of this training session.

7 A bowl of soup sits on the kitchen bench cooling. At first it loses heat quickly, but as time passes it loses heat more slowly until it reaches room temperature. Which graph below best illustrates this? Select the correct answer **A**, **B**, **C** or **D**.

R C

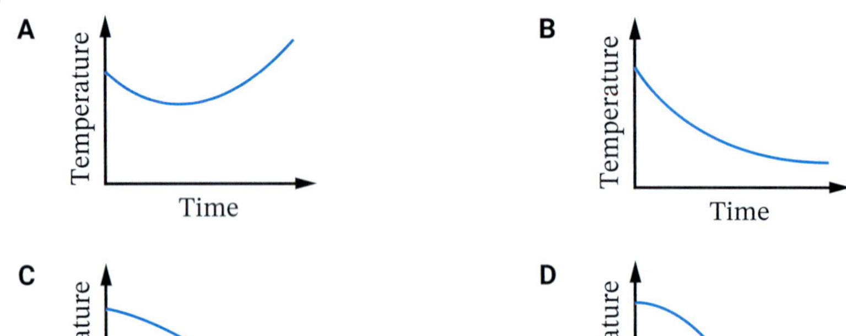

8 Sketch a graph to represent each situation.

PS **a** The temperature in your town over a 24-hour period.

R **b** The volume of petrol in a car. It is filled up, then runs for a while, stops for some time and then continues the trip.

C **c** The height of a plane travelling from Sydney to London with a refuelling stop in Dubai.

d The number of cans of drink in a vending machine if it is filled in the morning and in the afternoon.

e Your hunger level over a day during your waking hours.

Time differences 11.11

Example 20

What is the difference in time between 8:35 a.m. and 3:10 p.m.?

SOLUTION

Method 1

From 8:35 a.m. to 9 a.m. = 25 minutes

From 9 a.m. to 3 p.m. = 6 hours.

From 3 p.m. to 3:10 p.m. = 10 minutes.

25 min, + 6 h, + 10 min
8:35 a.m. 9 a.m. 3 p.m. 3:10 p.m.

$$\begin{aligned}\text{Total time difference} &= 6\text{ h} + 25\text{ min} + 10\text{ min}\\ &= 6\text{ h } 35\text{ min}\end{aligned}$$

Method 2

Convert to 24-hour time first, then use the calculator's [° ' "] or [DMS] key to subtract the times.

$$\begin{aligned}\text{Time difference} &= 3{:}10\text{ p.m.} - 8{:}35\text{ a.m.}\\ &= 15{:}10 - 08{:}35\\ &= 6\text{ h } 35\text{ min.}\end{aligned}$$

15 [° ' "] 10 [° ' "] [−] 8 [° ' "] 35 [° ' "] [=].

Videos
Time differences
24-hour time
Jai Singh (sundial)

Skillsheets
Units of time
24-hour time

Worksheets
Time calculations
12- and 24-hour time
Australian times
Tide chart

Presentation
Time calculations

Puzzle
Time calculations

Foundation Standard Complex

Example 21

Find 7 h 5 min – 3 h 24 min.

SOLUTION

$$\begin{aligned} 7\text{ h }5\text{ min} - 3\text{ h }24\text{ min} &= (7-3)\text{h} + (5-24)\text{min} \\ &= 4\text{ h} + (-19)\text{min} \\ &= 3\text{ h }(60-19)\text{min} \\ &= 3\text{ h }41\text{ min} \end{aligned}$$

OR On a calculator, enter 7 [° ′ ″] 5 [° ′ ″] [−] 3 [° ′ ″] 24 [=]

Example 22

What is the time 3 hours and 20 minutes after 10:42 p.m.?

SOLUTION

3 hours after 10:42 p.m. is 1:42 a.m.

20 minutes after 1:42 a.m. is 2:02 a.m.

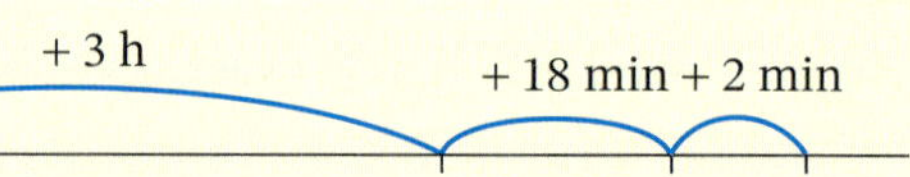

EXERCISE 11.11 ANSWERS ON P. 653

Time differences

U F PS R C

1 What is the difference in time between 11:42 a.m. and 2:13 p.m.? Select the correct answer **A**, **B**, **C** or **D**.

A 2 h 31 min **B** 3 h 55 min **C** 9 h 29 min **D** 11 h 55 min

2 Calculate the time difference between:

a 6:15 p.m. and 8:10 p.m.
b 11:16 a.m. and 12:06 p.m.
c 4:10 a.m. and 8:55 a.m.
d 11:25 p.m. and 3:20 a.m.
e 07:25 and 13:10
f 21:20 and 08:15
g 4:10 a.m. and 12:15 p.m.
h 09:40 and 13:10
i 12:45 and 17:25

3 Simplify each expression.

a 2 h 15 min + 4 h 32 min
b 3 h 25 min + 8 h 27 min
c 6 h 42 min - 3 h 13 min
d 12 h 37 min - 5 h 6 min
e 7 h 12 min + 5 h 18 min
f 1 h 42 min + 6 h 27 min

☐ Foundation ○ Standard ⬡ Complex

g 15 h 57 min - 9 h 48 min

h 6 h 2 min - 4 h 17 min

i 9 h 37 min + 2 h 52 min

j 4 h 49 min + 7 h 18 min

k 8 h 18 min - 3 h 27 min

l 5 h 31 min - 3 h 48 min

4 A film starts at 3:14 p.m. and ends at 5:09 p.m. How long is the film?

5 This part of a timetable is a schedule for a bus travelling between Sydney and Wagga Wagga.

PS R C

Sydney to Wagga Wagga		Wagga Wagga to Sydney	
Sydney	2:30 p.m.	Wagga Wagga	7:15 a.m.
Strathfield	3:00 p.m.	Gundagai	8:25 a.m.
Yagoona	3:20 p.m.	Jugiong	8:54 a.m.
Liverpool	3:45 p.m.	Yass	9:41 a.m.
Mittagong	4:40 p.m.	Goulburn*	10:41 a.m.
Goulburn*	5:40 p.m.	Mittagong	12:10 p.m.
Yass	7:10 p.m.	Liverpool	1:05 p.m.
Jugiong	7:55 p.m.	Yagoona	1:20 p.m.
Gundagai	8:20 p.m.	Strathfield	1:35 p.m.
Wagga Wagga	9:30 p.m.	Sydney	2:05 p.m.

* Thirty-minute meal stop at Goulburn.

a How long does the trip from Sydney to Wagga Wagga take?

b How long would the trip take without a meal break?

c Ali joins the return bus at Jugiong and gets off at Liverpool. How long is his trip?

d Find the time taken from Liverpool to Sydney, and from Sydney to Liverpool on the return trip. Suggest a reason for the difference.

e At what time does the bus from Wagga Wagga arrive in Goulburn?

f Where is the return bus at 1:35 p.m.?

g Renee is waiting at Yagoona at 2:45 p.m. for the bus to Mittagong. How long will she have to wait before the bus arrives and when will she reach Mittagong?

6 What time will it be:

EXAMPLE 22

a 6 hours after 2 p.m.?

b 3 hours after 10 a.m.?

c 20 minutes after 7:15 p.m.?

d 2 hours 32 minutes after 10:45 a.m.?

e 3 hours 29 minutes after 10:35 p.m.?

f 5 hours after 9:32 a.m.?

g 9 hours after 6:17 p.m.?

h 55 minutes after 3:30 p.m.?

i $4\frac{1}{4}$ hours after 4:30 a.m.?

j 5 hours and 25 minutes after 9:45 a.m.?

7 A car rally began at 8:20 a.m. Here are some of the cars and the times they ran. Write the cars in their order of finishing and the time each crossed the finishing line.

R C

Toyota 6:21 (6 h 21 min) Volvo 5:23 Ford 5:44

Nissan 6:01 Subaru 5:59 Peugeot 5:42

8 This is part of the ferry timetable for Woy Woy to Empire Bay.

PS R C

Monday to Friday Ferry departs from Woy Woy					
Departs	Saratoga	Davistown			Arrives
Woy Woy	Veterans Hall	Lintern St.	Central (RSL)	Pine Av.	Empire Bay
6:35 a.m.	6:45 a.m.	6:50 a.m.	7:00 a.m.	–	–
7:45 a.m.	7:55 a.m.	8:00 a.m.	8:10 a.m.	8:10 a.m.	8:15 a.m.
9:00 a.m.	9:10 a.m.	9:15 a.m.	9:25 a.m.	9:25 a.m.	9:30 a.m.
10:45 a.m.	10:55 a.m.	11:00 a.m.	11:10 a.m.	11:10 a.m.	11:15 a.m.
12:30 p.m.	12:40 p.m.	12:45 p.m.	12:55 p.m.	12:55 p.m.	1:00 p.m.
1:50 p.m.	2:00 p.m.	2:05 p.m.	2:15 p.m.	2:15 p.m.	2:20 p.m.
3:30 p.m.	3:40 p.m.	3:45 p.m.	–	–	–
4:50 p.m.	5:00 p.m.	5:05 p.m.	5:15 p.m.	–	–
5:50 p.m.	6:00 p.m.	6:05 p.m.	6:15 p.m.	–	–
6:50 p.m.	7:00 p.m.	7:05 p.m.	7:15 p.m.	7:15 p.m.	7:20 p.m.

- **a** How many trips from Woy Woy arrive at Empire Bay each weekday?
- **b** If Elise catches the 7:45 a.m. ferry from Woy Woy, at what time will she arrive at Empire Bay?
- **c** If Toby catches the 2:05 p.m. ferry from Lintern St, where will he be at 2:15 p.m.?
- **d** Adele needs to be at Central (RSL) before 10 a.m. What is the latest ferry she can catch from Saratoga to be there on time?
- **e** How long does it take to travel from Saratoga to Empire Bay?
- **f** At what time does the last ferry leave Woy Woy in the evening?
- **g** Suppose a new ferry service departs Woy Woy at 11:15 a.m. What time should it arrive at Pine Avenue?

11.12 World time zones

Worksheets
World time zones

Australian times

Video
Time zones

The world is divided into 24 hourly **time zones**. Time is the same throughout each zone. The centre of each time zone is a meridian of longitude (an imaginary line running from the North Pole to the South Pole). Each hourly time zone covers 15° of longitude.

☐ Foundation ○ Standard ○ Complex

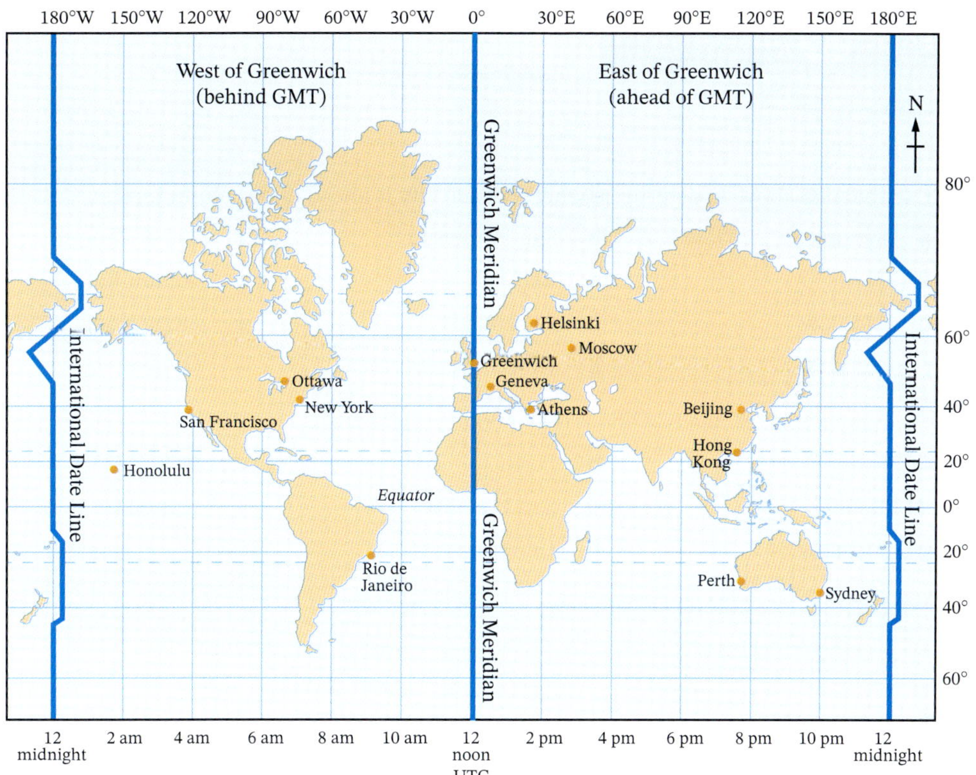

The map above shows how times around the world are related. All time is measured in relation to the time at the Greenwich Observatory (in London), either ahead of or behind **UTC (Coordinated Universal Time)**, also known as **Greenwich Mean Time (GMT)**. Greenwich is pronounced 'Grennitch'. Time zones in Australia are ahead of UTC because Australia is east of Greenwich. Time zones in the USA are behind UTC because the USA is west of Greenwich.

Example 23

For each country, use the map to determine whether it is ahead of or behind UTC.

a South Africa **b** Canada **c** Chile **d** China

SOLUTION

a On the map, South Africa is to the right (east) of the Greenwich Meridian.
South Africa is ahead of UTC.

b Canada is in North America, to the left (west) of the Greenwich Meridian.
Canada is behind UTC.

c Chile is in South America, to the left (west) of the Greenwich Meridian.
Chile is behind UTC.

d China is in Asia, to the right (east) of the Greenwich Meridian.
China is ahead of UTC.

EXERCISE 11.12 ANSWERS ON P. 653

World time zones

1 Use the map above to determine whether each city is ahead of or behind UTC.

R
a Sydney **b** San Francisco **c** Rio de Janeiro
d Perth **e** Beijing **f** Honolulu
g Moscow **h** Athens **i** Hong Kong
j Helsinki **k** New York **l** Ottawa

2 Find the time in each city when it is 12 noon in Greenwich.

a Sydney **b** Perth **c** New York
d Beijing **e** San Francisco **f** Honolulu
g Moscow **h** Geneva

3 What is the time difference between:

a Perth and Sydney? **b** Sydney and Beijing?
c Sydney and Honolulu? **d** Perth and Moscow?
e Sydney and New York? **f** Perth and Beijing?
g San Francisco and New York? **h** Honolulu and Moscow?
i Geneva and Perth? **j** San Francisco and Geneva?

4 If it is 9 p.m. in Sydney, what is the time in:

a London? **b** Perth? **c** New York? **d** Beijing?
e San Francisco? **f** Honolulu? **g** Moscow? **h** Geneva?

5 A cricket match being played in India is telecast live at 3:30 p.m. Brisbane time. What is the local time of the cricket match if Brisbane's time is $4\frac{1}{2}$ hours ahead of India's?

PS R

6 Sabiha, in Tasmania, wants to video call her cousin Habib in Turkey. The time in Turkey is 7 hours behind the time in Tasmania. At what time should Sabiha call to catch Habib when it is 1 p.m. in Turkey?

PS R

7 Sydney is 2 hours behind Auckland. A plane leaves Auckland at 4 p.m. and takes 3 hours to fly to Sydney. What is the local time in Brisbane when the plane lands? Select the correct answer **A**, **B**, **C** or **D**.

PS R

A 3 p.m. **B** 5 p.m. **C** 7 p.m. **D** 9 p.m.

8 Find out what happens if you cross the International Date Line (IDL). Why isn't the IDL straight?

PS R

Foundation Standard Complex

9 This map shows the three time zones for Australia.

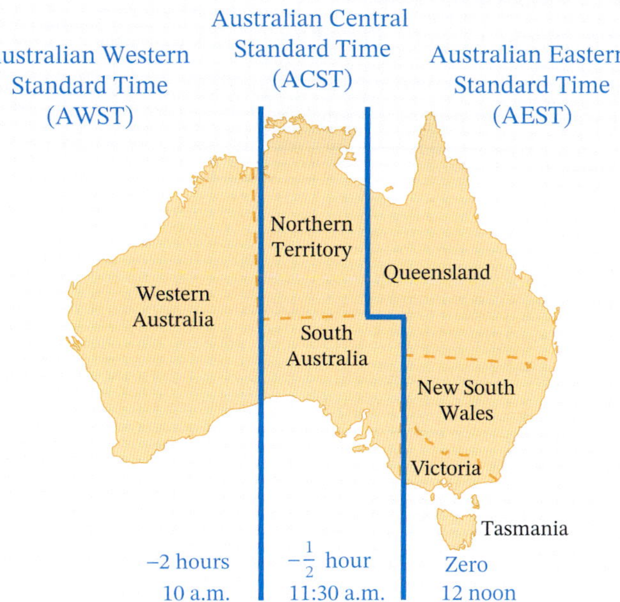

State whether each city is ahead of, behind or has the same time as Adelaide (SA).

a Canberra **b** Hobart **c** Darwin
d Perth **e** Hervey Bay (Qld) **f** Broome (WA)
g Griffith (NSW) **h** Alice Springs (NT) **i** Cairns (Qld)

10 What is the time difference between:

a Sydney and Darwin? **b** Brisbane and Perth?
c Adelaide and Hobart? **d** Melbourne and Adelaide?
e Hobart and Perth? **f** Brisbane and Canberra?

11 If it is 11 a.m. in Brisbane, what time it is in:

a Melbourne? **b** Adelaide? **c** Perth?
d Darwin? **e** Hobart? **f** Canberra?

12 If it is 7:30 p.m. in Perth, what time it is in:

a Melbourne? **b** Sydney? **c** Adelaide?
d Darwin? **e** Hobart? **f** Brisbane?

13 PS R **a** Johanna flies from Perth to Sydney, taking 4 hours. If she leaves Perth at 1 p.m., at what time does she land in Sydney? Give your answer as Sydney local time.

b When Johanna flies home, she leaves Sydney at 1 p.m. At what time does she land in Perth? Give your answer as Perth local time.

Foundation Standard Complex

14 a Find out when daylight saving begins and ends.

PS b Why do we have daylight saving?

R c How does daylight saving affect the different time zones?

C d If it is 6 p.m. in Western Australia (not on daylight saving), what time is it in Victoria on Eastern Standard Daylight Saving Time?

INVESTIGATION

World trip

Plan a trip around the world with at least three stopovers (e.g. Tokyo, Hawaii and Dubai). Use airline timetables so you can give details of departures, arrivals and the length of each flight. Does it matter if you head east or west when you start? What effect does the IDL have on your trip?

DID YOU KNOW?

First Nations seasonal calendars

The First Nations peoples of Australia have acquired knowledge to coexist with their local natural environment over thousands of years. They have done this by observing weather patterns and environmental changes dating all the way back to the Ice Age, and by sharing this knowledge with their community.

Indigenous groups use seasonal calendars to anticipate the coming of seasonal food sources, breeding seasons and when plants are ready to harvest. This detailed understanding is an essential approach to land and resource management.

Within Australia, there are many different seasonal calendars based on different Indigenous nations and localities. These calendars demonstrate a wealth of knowledge about the Australian environments and ecologies (relations of organisms to one another and to their physical surroundings).

The seasonal calendar of the Banbai Nation in NSW identifies 6 distinct seasons that revolve around fire, conditions for fire and the flora and fauna available during that season. Notice that the seasons are of different lengths, ranging from 5 months (November to March) to 1 month (July).

Season	Months	Duration	Climate
Wildfire time	November to March	5 months	Wet and hot, becoming warm
Grass cures	April to mid-May	1.5 months	Dry becoming cool
Burning time	Mid-May to June	1.5 months	Dry and cold, to frosty
Too cold	July	1 month	Freezing and windy
Burning time	August to September	2 months	Cold becoming warm
Risky time	October	1 month	Hot and windy

More details can be found at the Bureau of Meteorology website: www.bom.gov.au/iwk/calendars/banbai.shtml

☐ Foundation ○ Standard ⬡ Complex

INVESTIGATION

First Nations seasonal calendar

Let's conduct some research into the different Indigenous calendars used by our First Nations people.

1 Copy this table.

Indigenous Nation	Location in Australia	Number of seasons in calendar	Indigenous names of seasons	Interesting facts
Banbai	Wattleridge, NSW	6	No Indigenous names have been given to seasons	Wildfire Time season is the longest season in the calendar spanning five months.
D'harawal				
Gariwerd				
Jawoyn				
Kaurna				
Masig				
Maung				
Miriwoong				
Ngoorabul				
Nyoongar				
Tiwi				
Walabunnba				
Wardaman				
Wunambal Gaambera				
Yanyuwa				
Yawuru				
Yirrganygji				

2 Access the BOM website www.bom.gov.au/iwk/calendars/banbai.shtml looking specifically at Indigenous Seasonal Calendars.

3 Select an Indigenous seasonal calendar from the drop-down menu and work your way through the table filling in as many cells as possible. If the seasons in the calendar have been provided with Indigenous names, write them down in the 'Indigenous names' column of the table. The details for the Banbai Nation calendar have already been completed for you.

POWER PLUS ANSWERS ON P. 653

1 Australia's annual birth rate in 2021 was approximately 12.7 per 1000. Given that its population then was 26 200 000, approximately how many babies were born in Australia in 2021?

2 Singapore is a small island, but it has a population of 5 780 000, making it the second most densely populated country in the world. Its population density is 8210 persons/km^2.

a What is the area of Singapore, to the nearest square km?

a Australia is the third least-densely populated country in the world, at a rate of only 3.4 persons/km^2 and an area of 7 682 300 km^2. If Australia was as densely populated as Singapore, what would its population be?

3 A spider moves at 1 cm/s. If the spider is in the back left-hand corner of your classroom, find how long (in minutes) it will take to reach:

a the nearest person
b you
c the teacher's desk

4 Convert each speed to km/h. Round your answers to two decimal places.

a An African cheetah runs at 27 m/s.

b A German peregrine falcon dives at 97 m/s.

c A Tanzanian snake travels at 3.3 m/s.

d A racing cyclist rides at 23 m/s.

5 Work out a formula for converting speed expressed as x m/s to y km/h.

11 CHAPTER REVIEW

Language of maths

best buy	daylight saving	distance	divide
equivalent	International Date Line	km/h	map
per (/)	plan	rate	ratio
scaled length	scale ratio	simplify	speed
stationary	term	time zone	travel graph
unitary method	unit price	UTC (Coordinated Universal Time)	

Quiz
Language of maths 11

1 When comparing the prices and sizes of items to find the **best buy**, do we look for the highest or lowest **unit price**?

2 What type of measurement compares quantities of different types, expressed using two units?

3 What is meant by the **scale ratio** of a map or plan?

4 What does an **average speed** of 65 km/h actually mean?

5 How is a change in speed shown on a travel graph?

6 In which country is the UTC or GMT time zone?

Topic summary

Worksheet
Mind map: Ratios, rates and time

- Give examples of situations where ratios, rates and time are used.
- What did you learn in this chapter?
- What did you find the most difficult about this topic? Discuss any problems with your teacher or with a friend.

Print (or copy) and complete this mind map of the topic, adding detail to its branches and using pictures, symbols and colour where needed. Ask your teacher to check your work.

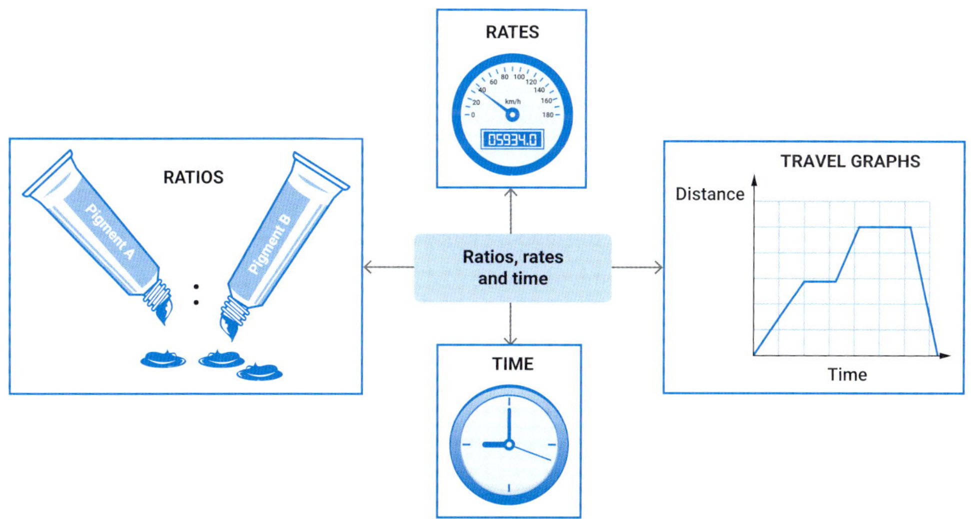

11 TEST YOURSELF

ANSWERS ON P. 653

Quiz Test yourself 11

1 Copy and complete each pair of equivalent ratios.

a 1 : 3 = 4 : ____ **b** 2 : 5 = 6 : ____ **c** 3 : 7 = ____ : 21

d 4 : 9 = ____ : 45 **e** 0.5 : 0.8 = 2 : ____

f 3 : 4 : 5 = ____ : 20 : ____

2 Simplify each ratio.

a 12 : 21 **b** 25 : 75 **c** 6 : 36

d 25 : 45 **e** 18 : 6 : 24 **f** 5 : 25 : 100

g $\frac{1}{3}:\frac{1}{2}$ **h** $\frac{3}{4}:\frac{9}{16}$ **i** $1\frac{1}{2}:2\frac{1}{2}$

j 0.98 : 0.245 **k** 1.5 : 6 **l** 91 : 5.6

m 5 km : 200 metres **n** 20 kg : 3600 g **o** \$25 : \$4.25

p 18 months: 4 years **q** 10 days: 5 weeks **r** 8 h : 3 days

3 **a** The ratio of girls to boys in Year 8 is 4 : 3. If there are 75 boys in Year 8, find how many girls are there.

b Adam and Eve invested in a business in the ratio of 5 : 7. If Eve invested \$63 000, how much did Adam invest?

4 Measure the length of each scale drawing below, and then use the ratio to work out the actual length of the object correct to the nearest centimetre.

a Fish

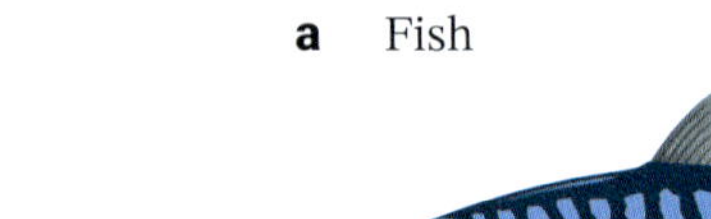

Scale 1 : 10

b Frog

Scale 1 : 4

Shutterstock.com/Yuliia Husar
Shutterstock.com/SGV_Arts

5 On a tourist map of Sydney, the scale is given by 0 ⎯⎯⎯ 500 m.

a Write this scale as a simplified ratio.

b Find the actual distance between the following places given the scaled distance.

i Circular Quay station to the Opera House (2.5 cm).

ii Pyrmont Bridge to Parliament House (4.4 cm).

c Find the scaled distance between the following places given the actual distance.

i Art Gallery of NSW to Sydney Tower (875 m).

ii Circular Quay station to Central station (2.5 km).

☐ Foundation ○ Standard ⬡ Complex

6 Ellen and Portia share the weekly rent of $420 on their apartment in the ratio of 4 : 3. How much does Ellen pay? Select the correct answer **A**, **B**, **C** or **D**.

A $105 **B** $240 **C** $180 **D** $210

7 Write each statement as a rate in simplified form. 11.05

a $10.50 for 3 kg **b** 220 km in 2 hours

c $56.40 for 4 hours **d** 260 runs in 50 overs

8 For each pair, determine which is the better buy.

a A 1 kg box of biscuits for $5.60 or a 700 g carton for $4.00.

b A 350 g can of soup for $4.20 or a 550 g can for $7.15.

c A 1.5 litre bottle of lemonade for $2.45 or a 375 mL can for $0.70.

d 3×40 g chocolate bars for $2.36 or a 500 g chocolate bar for $9.

9 **a** Mince is $3.99/kg. How much does 5 kg of mince cost? 11.07

b Tiwa earns $18.70/h. How much is she paid for 38 hours work?

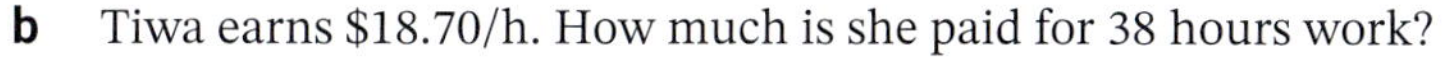

c How many litres of petrol can you buy with $40 if petrol costs $1.35/L? Answer to the nearest 0.1 litre.

d Fertiliser is used at 20 kg/ha. How many hectares can be covered with 144 kg of fertiliser?

10 Find the speed, in km/h, of a cyclist who travels a distance of 45 km in 2 hours and 15 minutes.

11 Find the distance covered by a truck that travels for 5 hours and 30 minutes at an average speed of 45 km/h.

12 Keith and Kent decided to go bushwalking. This travel graph shows their walk.

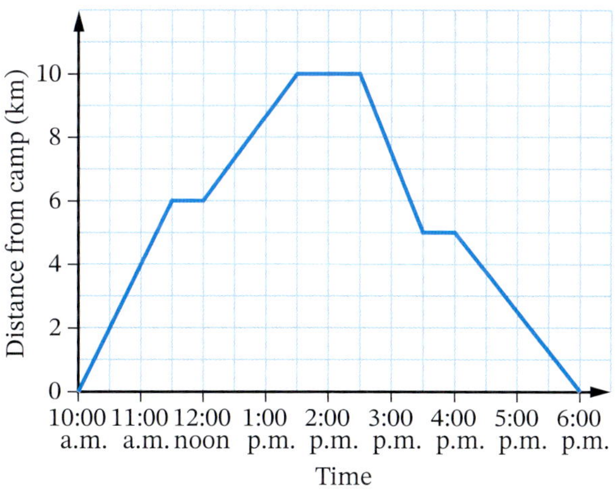

a How far did they travel from camp?

b How many stops did they make?

c Find their average speed between their first and second stops.

d How long did it take them to return home?

e Between what times were they walking the fastest?

☐ Foundation ○ Standard ○ Complex

13 Match each description to its correct graph.

a The water level of the bath stayed constant at 50 cm.

b The bath was filled at a steady rate. The tap was turned off to let the hot bathwater cool. The rest of the bath was filled with cold water.

c The bath was filled with warm water at a steady rate.

A

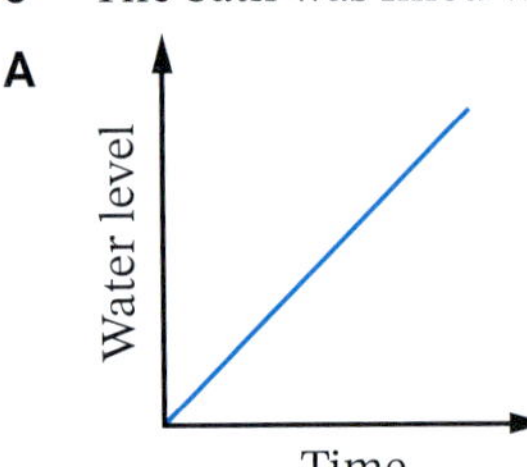

B

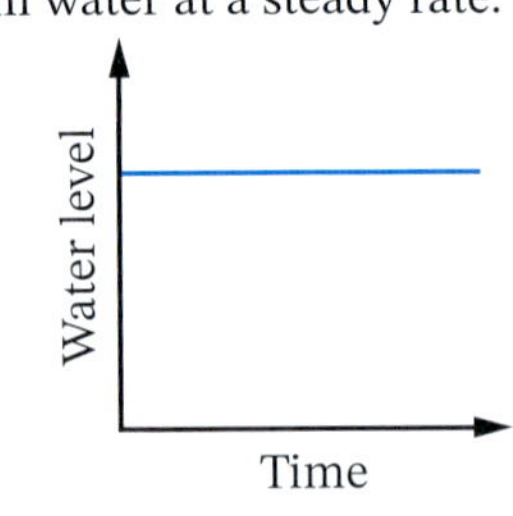

C

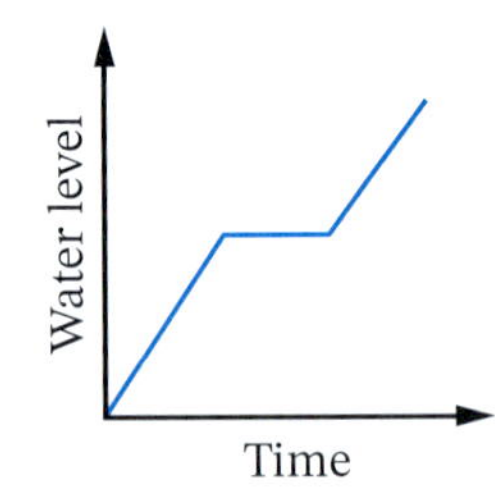

14 Draw a graph to represent the following story. 'The weight of the baby elephant increased quickly at first, but then slowed down to a steady rate'.

11.11 **15** What is the time:

a 3 hours and 20 minutes after 8:05 p.m.?

b 4 hours and 45 minutes before 11:15 p.m.?

11.11 **16** Calculate the time difference between:

a 10:10 a.m. and 6:40 p.m.

b 12:05 p.m. and 3:20 a.m.

c 12:30 and 23:20

11.11 **17** A section of a bus timetable is shown.

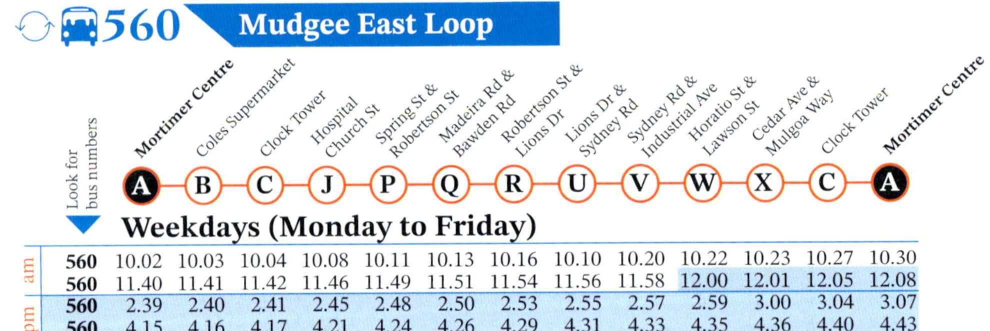

560 Mudgee East Loop

Weekdays (Monday to Friday)

	Look for bus numbers	A Mortimer Centre	B Coles Supermarket	C Clock Tower	J Hospital Church St	P Spring St & Robertson St	Q Madeira Rd & Bawden Rd	R Robertson St & Lions Dr	U Lions Dr & Sydney Rd	V Sydney Rd & Industrial Ave	W Horatio St & Lawson St	X Cedar Ave & Mulgoa Way	C Clock Tower	A Mortimer Centre
am	560	10.02	10.03	10.04	10.08	10.11	10.13	10.16	10.10	10.20	10.22	10.23	10.27	10.30
am	560	11.40	11.41	11.42	11.46	11.49	11.51	11.54	11.56	11.58	12.00	12.01	12.05	12.08
pm	560	2.39	2.40	2.41	2.45	2.48	2.50	2.53	2.55	2.57	2.59	3.00	3.04	3.07
pm	560	4.15	4.16	4.17	4.21	4.24	4.26	4.29	4.31	4.33	4.35	4.36	4.40	4.43

a What do you think the shaded area represents?

b How many bus services run each day?

c How long is the journey from the Mortimer Centre to Robertson St/Lions Dr?

d Tom catches the bus from Spring St at 2:48 p.m. Where does the bus stop after 11 minutes?

e Mary needs to meet a friend at the Clock Tower at 3:10 p.m. Which is the latest bus she can catch from Madeira Rd to be there on time?

Foundation Standard Complex

18 Use the world map on page 539 to find the time difference between: 

a Sydney and Athens.

b Sydney and Hong Kong.

c Perth and Honolulu.

d San Francisco and Helsinki.

19 An international hockey match is being played at 3:30 p.m. local time in The Netherlands. Shegufta lives in Cairns, QLD, where the time is 8 hours ahead of The Netherlands. At what time should she be ready to watch the game live on TV in Cairns? 

12

ALGEBRA, SPACE

Graphing linear functions

The famous French philosopher and mathematician René Descartes (1596–1650) is famous for discovering the connections between algebra and geometry. His concept of coordinate geometry connected algebra and geometry using the graphs of straight lines and curves. In computer modelling, the use of coordinates allows us to locate things more precisely, especially in three-dimensional (3D) space. Number plane graphs are used to track changes over short and long periods of time, such as the height and speed of an aeroplane.

Chapter outline

	Proficiencies
12.01 Tables of values	U F
12.02 Finding the rule	U F R C
12.03 Finding rules for number patterns	U F PS R C
12.04 The number plane	U F R C
12.05 The 3D number space	U F R C
12.06 Graphing number patterns	U F R C
12.07 Graphing linear functions	U F R C
12.08 Comparing linear functions	U F R C
12.09 Solving linear equations graphically	U F PS
12.10 Applying linear functions	U F PS U F
12.11 Graphing linear inequalities	U F U F

U = Understanding
F = Fluency
PS = Problem solving
R = Reasoning
C = Communication

Wordbank

consecutive numbers Numbers that follow each other in order, for example, 2, 3, 4

constant term The term in an equation that is a number only and does not contain a variable, for example, the 5 in $y = 2x + 5$

horizontal Going across, sideways, flat

linear Involving a line

linear function A function whose graph is a straight line

number plane A coordinate grid system based on the x- and y-axes, also called a Cartesian plane

table of values A table of ordered pairs of numbers, usually following a formula and which can be graphed on a number plane

vertical Going up and down, at a right angle to the horizontal

y-intercept The y-value at which a line crosses the y-axis on the number plane

Quiz
Wordbank 12

Videos (6):

12.02 Finding the rule
12.05 The number plane
12.06 Graphing number patterns
12.07 Linear functions • Graphing linear equations
12.08 Linear functions
12.11 Graphing linear inequalities

***Twig* video (1):**

12.05 Coordinate geometry: Descartes

***PhET* interactives (4):**

12.02 Function builder: Basics • Function builder
12.07 Graphing lines
12.08 Graphing slope-intercept

Skillsheet (1):

12.04 The number plane

Quizzes (5):

- Wordbank 12
- SkillCheck 12
- Mental skills 12
- Language of maths 12
- Test yourself 12

Worksheets (12):

12.01 Tables of values
12.03 Patterns and rules
12.04 The number plane • Plane grid 2 • Number plane review • Number plane grid paper • A page of number planes • Plotting points and lines • Number plane writing activity
12.06 Number plane grid paper • A page of number planes • Graphing tables of values
12.07 Number plane grid paper • A page of lines • A page of number planes
12.09 Number plane grid paper • A page of number planes
12.11 Number plane grid paper • A page of number planes

Mind map: Graphing linear functions

Puzzles (3):

12.04 Coordinates code puzzle
12.07 Matching linear equations • Graphing functions
12.08 Matching linear equations

Technology (2):

12.04 Plotting points
12.07 Graphing straight lines: Finding intercepts

Presentations (1):

12.04 Graph coordinates

Spreadsheets (2):

12.04 Points plotter
12.07 Linear functions plotter

Nelson MindTap

To access resources above, visit **cengage.com.au/nelsonmindtap**

12 In this chapter you will:

- ✓ determine the formula for a table of values and set of points on a number plane.
- ✓ extend your understanding of the number plane and coordinates to the 3D number space.
- ✓ complete a table of values for a linear function and graph the linear function on a number plane.
- ✓ identify and describe a number pattern, then represent it algebraically and graphically on a number plane.
- ✓ compare and contrast the graphs of different linear functions
- ✓ solve linear equations graphically.
- ✓ apply linear functions to real-life situations and problems.
- ✓ graph simple linear inequalities on a number plane.

Quiz SkillCheck 12

SkillCheck

ANSWERS ON P. 654

1 For each equation, find the missing number.

a $2 \times 3 + \_\_\_ = 10$ **b** $2 \times (-1) + \_\_\_ = 3$ **c** $3 \times 1 - \_\_\_ = 1$

d $-2 \times 2 + \_\_\_ = -2$ **e** $4 \times 3 - \_\_\_ = 8$ **f** $-1 \times (-1) - \_\_\_ = -4$

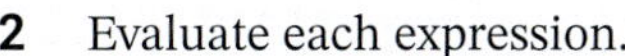

2 Evaluate each expression.

a $3x - 2$ if $x = 2$ **b** $\frac{1}{2}x + 3$ if $x = 6$ **c** $2x - 1$ if $x = -1$

d $-x + 1$ if $x = 0$ **e** $-2x - 1$ if $x = 1$ **f** $10 - x$ if $x = 3$

3 Solve each equation.

a $3x - 2 = 7$ **b** $-2x - 1 = 3$ **c** $-x + 1 = 4$

d $2x - 1 = -5$ **e** $3x + 2 = x - 1$ **f** $x - 2 = 4 - x$

Tables of values 12.01

In order to graph straight lines, we need points. In Year 7, you were given a **table of values** to plot on the number plane, but where do the values come from? We will now use a rule or formula to create a table of values.

Worksheet
Tables of values

Example 1

Complete this table of values using the formula $v = 2u + 1$.

u	-2	-1	0	1	2	6	9
v							

SOLUTION

Substitute each value of u in the table into the formula $v = 2u + 1$ to find the value of v.

- When $u = -2$, $v = 2 \times (-2) + 1 = -3$
- When $u = -1$, $v = 2 \times (-1) + 1 = -1$
- When $u = 0$, $v = 2 \times 0 + 1 = 1$
- When $u = 1$, $v = 2 \times 1 + 1 = 3$
- When $u = 2$, $v = 2 \times 2 + 1 = 5$
- When $u = 6$, $v = 2 \times 6 + 1 = 13$
- When $u = 9$, $v = 2 \times 9 + 1 = 19$

Completing the table, we have:

u	-2	-1	0	1	2	6	9
v	-3	-1	1	3	5	13	19

EXERCISE 12.01 ANSWERS ON P. 654

Tables of values

U F

1 Complete each table of values given the formula.

a $y = x - 1$

x	3	7	-1	-8	4	5
y					3	

b $q = 3p$

p	8	-3	2	5	10	-4
q				15		

Foundation Standard Complex

c $k = h \div 2$

h	12	8	-2	-6	0	7
k	6					

d $y = x + 7$

x	-2	11	7	0	13	-9
y		18				

e $b = 2a$

a	8	-1	-5	0	4	2
b			-10			

f $y = x - 3$

x	10	-4	7	5	-8	11
y	7					

2 Given the rule $y = -x - 4$, what is the value of y when $x = -3$?
Select the correct answer **A**, **B**, **C** or **D**.

A -1 **B** -7 **C** 7 **D** 12

3 Complete each table of values given the formula.

a $p = 5m - 1$

m	3	10	1	4	8	-6
p		49				

b $q = 3p - 2$

p	4	-1	7	10	2	-6
q			19			

c $t = 4r + 6$

r	2	0	-3	9	-5	4
t	14					

d $z = 5y + 2$

y	4	8	7	-5	6	-2
z		42				

e $y = 2x - 3$

x	0	1	2	3	4	5
y						

f $b = \frac{1}{2}a - 2$

a	-2	-1	0	1	2	3
b						

g $k = 3h + 5$

h	-1	0	1	2	3	4
k						

h $y = 1 - x$

x	-2	-1	0	1	2	3
y						

4 Which table of values matches the formula $y = 3x - 1$?
Select the correct answer **A**, **B**, **C** or **D**.

A

x	1	2	3	4	5
y	2	3	4	5	6

B

x	1	2	3	4	5
y	2	5	8	11	14

C

x	1	2	3	4	5
y	4	7	10	13	16

D

x	1	2	3	4	5
y	0	3	6	9	12

5 Which table of values follows the rule $c = \frac{h+2}{2}$? Select **A**, **B**, **C** or **D**.

A

h	0	2	4
c	1	3	5

B

h	0	2	4
c	1	2	5

C

h	0	2	4
c	0	1	3

D

h	0	2	4
c	1	2	3

☐ Foundation ○ Standard ○ Complex

Finding the rule

12.02

We can often find the formula for a table of values. If the values in the top row are **consecutive** (increase by 1 each time), look for a pattern in the values in the bottom row.

Interactives
Function builder: Basics
Function builder

Video
Finding the rule

Example 2

Find the formula for each table of values.

a

p	2	3	4	6	7	8
q	1	2	3	5	6	7

b

m	−2	−1	0	1	2	3
n	−6	−3	0	3	6	9

c

r	1	2	3	4	5	6
t	1	3	5	7	9	11

d

d	1	2	3	4	5	6
e	8	12	16	20	24	28

SOLUTION

a What has been done to each value in the top row to get the value in the bottom row?

p	2	3	4	6	7	8
q	1	2	3	5	6	7

The pattern is:

$2 - 1 = 1$

$3 - 1 = 2$

$4 - 1 = 3$, and so on.

The formula is $q = p - 1$.

b

m	−2	−1	0	1	2	3
n	−6	−3	0	3	6	9

The pattern is:

$3 \times (-2) = -6$

$3 \times (-1) = -3$

$3 \times 0 = 0$, and so on.

The formula is $n = 3m$.

c The formula for this table of values is not so obvious. It involves **two** operations: a multiplication and either an addition or a subtraction.

r	1	2	3	4	5	6
t	1	3	5	7	9	11

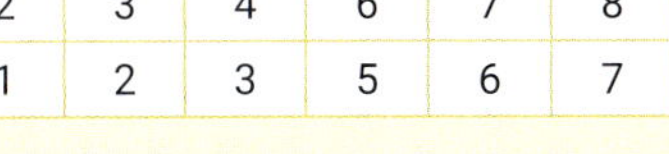

If the values in the top row are consecutive, the bottom row helps us find the **multiplier.**

The values in the bottom row go up by 2 each time, so the multiplier is 2. This means the formula must have $2 \times r$ in it.

$2 \times 1 - 1 = 1$

$2 \times 2 - 1 = 3$

$2 \times 3 - 1 = 5$, and so on.

The formula is $t = 2r - 1$.

d

d	1	2	3	4	5	6
e	8	12	16	20	24	28

4 4 4 4 4

The values in the top row are **consecutive**, the values in the bottom row go up by 4, so the multiplier is 4, and the equation is of the form $e = 4d$ ____.

Choose any ordered pair from the table to find the missing number, say (2, 12).

When $d = 2$, $e = 4 \times 2 + 4 = 12$, so the missing number is 4.

So the equation is $e = 4d + 4$.

(Checking that this is also true for another ordered pair (1, 8).

When $d = 1$, $e = 4 \times 1 + 4 = 8$)

EXERCISE 12.02 ANSWERS ON P. 654

Finding the rule

U F R C

EXAMPLE 2

1 Copy and complete the formula for each table of values.

R C

a $y =$ ___ x

x	0	1	2	3
y	0	3	6	9

b $y =$ ___ x

x	0	1	2	3
y	0	-2	-4	-6

c $y = x +$ ______

x	-1	0	1	2
y	3	4	5	6

d $y = x -$ ______

x	0	1	2	3
y	-4	-3	-2	-1

e $y = 10 -$ ______ x

x	-1	0	1	2
y	11	10	9	8

f $y = x \div$ ______

x	0	1	2	3
y	0	$\frac{1}{2}$	1	$1\frac{1}{2}$

Foundation Standard Complex

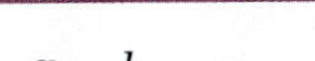

g $k =$ ______

p	1	2	3	4
k	5	6	7	8

h $y =$ ______

x	−1	0	1	2
y	−4	−3	−2	−1

i $j =$ ______

h	4	5	9	11
j	8	10	18	22

j $p =$ ______

m	0	5	30	45
p	0	1	6	9

2 Find the formula for each table, then complete the last two columns.

R C

a

f	1	2	3	4	7	8
h	1	4	7	10		

b

m	1	2	3	4	6	9
p	2	7	12	17		

c

m	0	1	2	3	5	8
b	3	6	9	12		

d

h	3	4	5	6	9	11
k	8	10	12	14		

e

r	0	1	2	3	4	7
s	1	4	7	10		

f

a	2	3	4	5	10	11
b	2	4	6	8		

g

m	0	1	2	3	5	6
n	5	8	11	14		

h

c	3	4	5	6	9	10
d	31	41	51	61		

i

w	3	4	5	6	7	8
x	14	19	24	29		

j

y	5	6	7	8	9	10
z	4	6	8	10		

k

a	1	2	3	4	7	9
m	5	9	13	17		

l

z	4	5	6	7	9	12
t	1	3	5	7		

3 What is the formula for this table? Select the correct answer **A**, **B**, **C** or **D**.

p	3	4	5	6
m	8	11	14	17

A $m = 2p + 1$ **B** $m = 3p - 1$ **C** $m = 2p + 4$ **D** $m = 3p + 1$

4 What is the formula for this table? Select **A**, **B**, **C** or **D**.

x	1	2	3	4	5
y	−5	−7	−9	−11	−13

A $y = -2x - 5$ **B** $y = 2x - 5$ **C** $y = -2x - 3$ **D** $y = 2x - 3$

5 Find the formula for each table, then complete the table.

R C

a

x	0	1	2	3	9
y	−1	−3	−5	−7	

b

A	0	1	2	3	10
C	0	0.5	1	1.5	

c

s	−1	0	1	2	7
t	18	13	8	3	

d

t	−1	0	1	2	5
v	30	20	10		

e

x	0	1	2	3	6
D	−0.5	−1	−1.5		

12.03 Finding rules for number patterns

Worksheet Patterns and rules

Example 3

For this number pattern: 2, 5, 8, 11, 14, ...

a find the 8th term.

b find a formula for the nth term, T.

SOLUTION

a This number pattern is increasing by 3 each time. By continuing the pattern:

2, 5, 8, 11, 14, 17, 20, 23

8th term = 20 + 3 = 23

b Write the number pattern as a table of values.

Term of pattern, n	1	2	3	4	5
Number, T	2	5	8	11	14

The number pattern goes up by 3 each time, so the multiplier for the formula is 3, and the equation is of the form $T = 3n$ ____.

Choose any ordered pair from the table to find the missing number, say (1, 2).

When $n = 1$, $T = 3 \times 1 - 1 = 2$, so the missing number is −1.

$3 \times 1 - 1 = 2$

$3 \times 2 - 1 = 5$

$3 \times 3 - 1 = 8$, and so on

So the general formula for the nth term is $T = 3n - 1$.

Example 4

Examine this geometrical pattern.

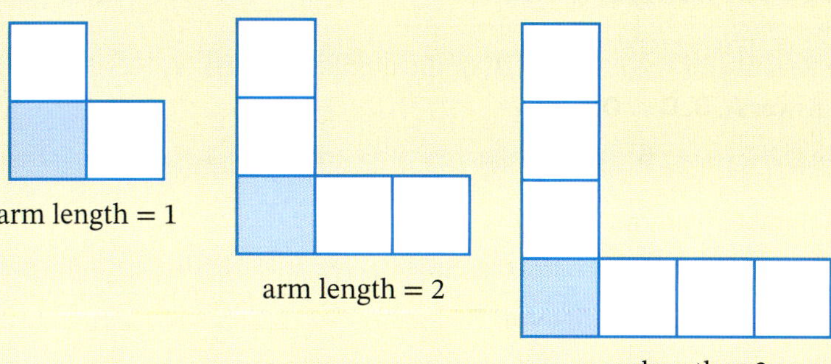

a Copy and complete this table for the pattern.

Arm length, a	1	2	3	4	5	6
Number of tiles, t						

b Write the rule for this pattern in words.

c Write the rule as a formula.

d How many tiles would be needed to make a design with an arm length of:

i 18? **ii** 60?

SOLUTION

a

Arm length, a	1	2	3	4	5	6
Number of tiles, t	3	5	7	9	11	13

b The number of tiles increases by 2 each time. This means that the multiplier is 2, and the equation is of the form $t = 2a$ ____.

Choose (1, 3) from the table to find the missing number.

When $a = 1$, $t = 2 \times 1 + 1 = 3$.

In words, the rule is: The number of tiles is 2 times the arm length plus 1.

c As a formula, the rule is $t = 2a + 1$.

d **i** When $a = 18$, $t = 2 \times 18 + 1 = 37$,

37 tiles are needed for a design with an arm length of 18.

ii When $a = 60$, $t = 2 \times 60 + 1 = 121$,

121 tiles are needed for a design with an arm length of 60.

EXERCISE 12.03 ANSWERS ON P. 654

Finding rules for number patterns

U F PS R C

1 Find the formula for the nth term, T, of the number pattern –1, –3, –5, –7, ... Select the correct answer **A**, **B**, **C** or **D**.

R C

A $T = n - 2$ **B** $T = 1 - 2n$ **C** $T = 2n - 1$ **D** $T = n - 3$

2 For each number pattern:

R C

i find the 8th term.

ii find a formula for the nth term, T.

iii use the formula to find the 8th term and the 20th term.

a 1, 3, 5, 7, 9, ... **b** 2, 4, 6, 8, 10, ...

c 8, 10, 12, 14, 16, ... **d** 7, 10, 13, 16, 19, ...

e 7, 15, 23, 31, 39, ... **f** 2, 7, 12, 17, 22, ...

g 19, 17, 15, 13, 11, ... **h** 1, 4, 9, 16, 25, ...

3 **a** Here are the first 2 T-shapes in a geometrical pattern. Draw the next 3 T-shapes.

R C

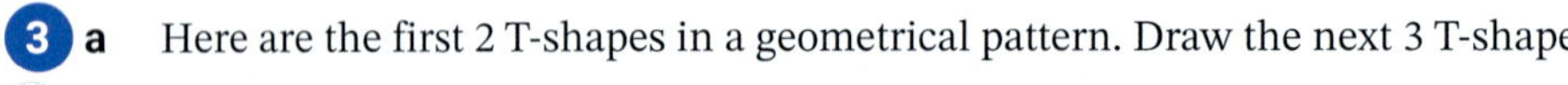

arm length = 1

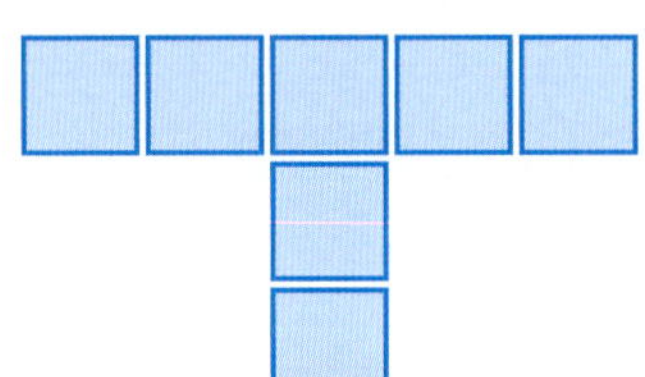

arm length = 2

b Copy and complete the table.

Arm length, a	1	2	3	4	5	8	11
Number of tiles, t	4						

c Write the rule for the pattern in words.

d Write the rule as a formula.

e How many tiles are needed to build a T-shape with arm length of:

i 15? **ii** 33?

4 **a** Here are the first two flight shapes, made from toothpicks. Draw the next 3.

R C

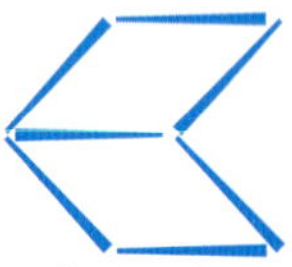

flight shape 1

flight shape 2

▷

b Copy and complete the table.

Number of flight shape, *f*	1	2	3	4	6	9	10
Number of toothpicks, *t*	7						

c Write the rule for the pattern in words.

d Write the rule as a formula.

e How many toothpicks are needed to build:

i flight shape 25? **ii** flight shape 40?

5 R C **a** The first two rockets are shown below. Draw the next three rockets.

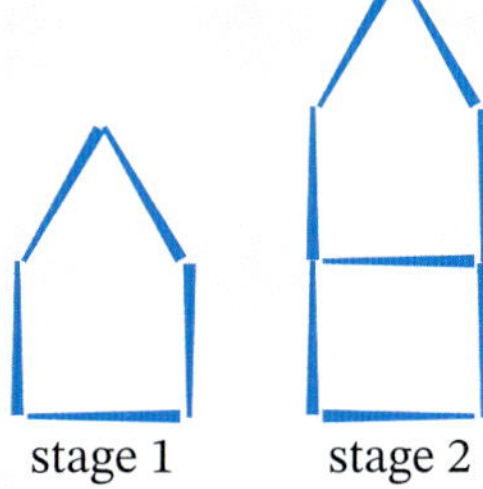

b Copy and complete the table.

Number of stage, *s*	1	2	3	4	5	8	11
Number of toothpicks, *t*							

c Write the rule for the pattern in words.

d Write the rule as a formula.

e How many toothpicks are needed to build:

i stage 12? **ii** stage 45?

6 R C For each matchstick pattern on the next page:

i draw the next three terms.

ii copy and complete this table.

Number of shapes (*n*)	1	2	3	4	5
Number of matchsticks (*T*)					

iii find the formula for the pattern.

iv find the number of matchsticks required to make 12 shapes.

a

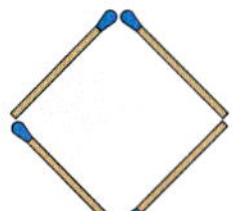

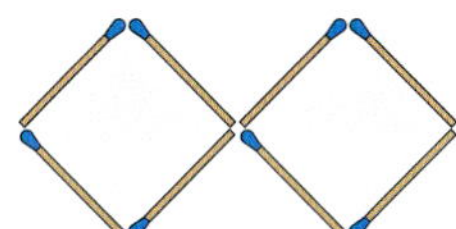

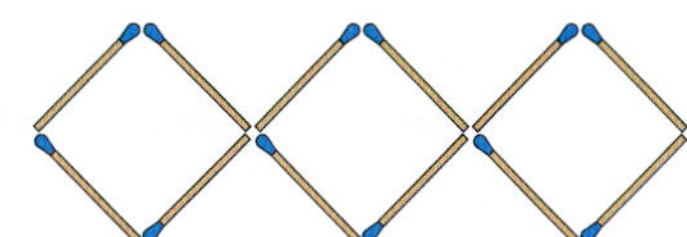

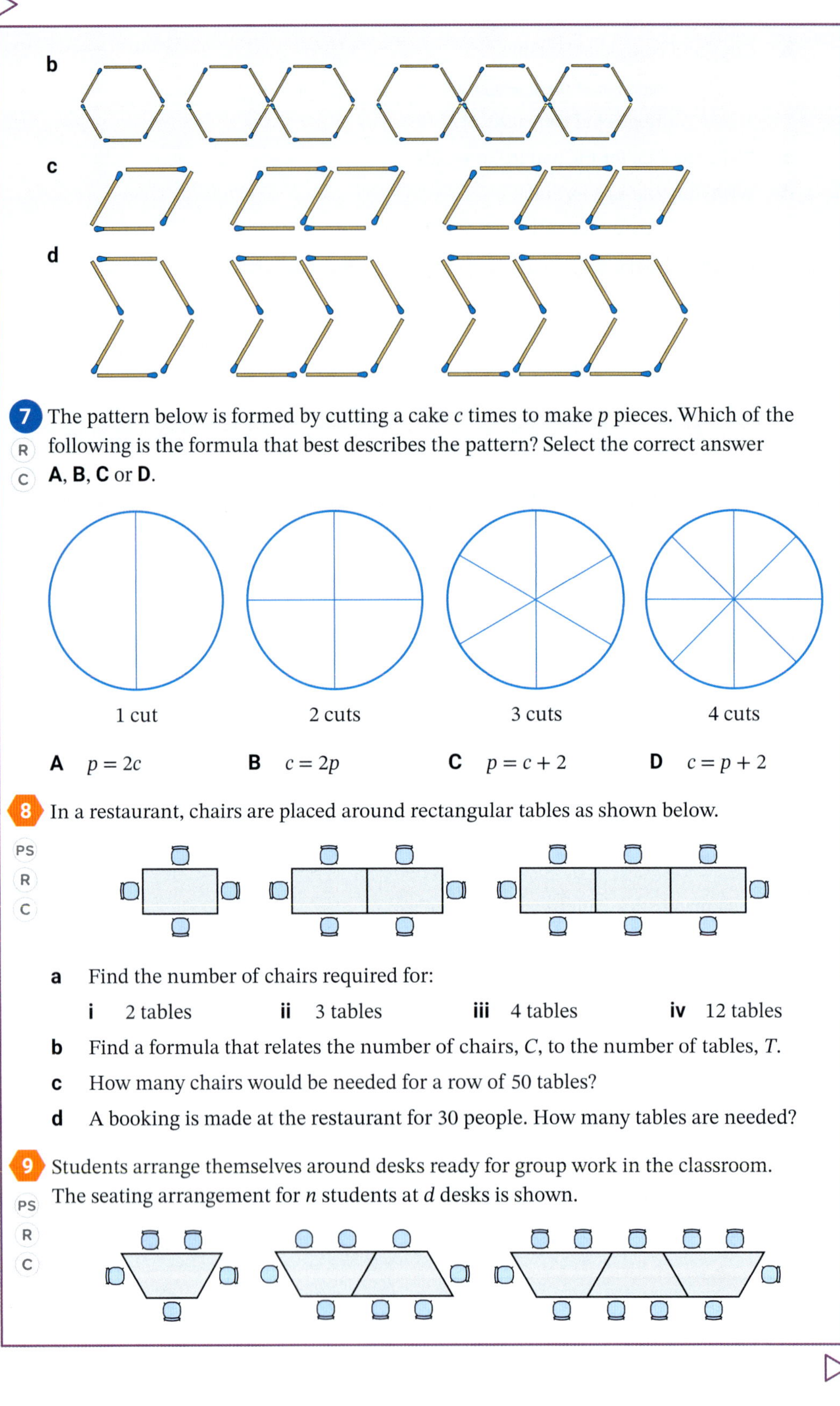

7 R C The pattern below is formed by cutting a cake c times to make p pieces. Which of the following is the formula that best describes the pattern? Select the correct answer **A**, **B**, **C** or **D**.

A $p = 2c$ **B** $c = 2p$ **C** $p = c + 2$ **D** $c = p + 2$

8 PS R C In a restaurant, chairs are placed around rectangular tables as shown below.

a Find the number of chairs required for:

i 2 tables **ii** 3 tables **iii** 4 tables **iv** 12 tables

b Find a formula that relates the number of chairs, C, to the number of tables, T.

c How many chairs would be needed for a row of 50 tables?

d A booking is made at the restaurant for 30 people. How many tables are needed?

9 PS R C Students arrange themselves around desks ready for group work in the classroom. The seating arrangement for n students at d desks is shown.

Foundation Standard Complex

- **a** Find the number of students who can be seated around a row of:
  - **i** 2 desks
  - **ii** 3 desks
  - **iii** 4 desks
  - **iv** 10 desks
- **b** Find a formula that relates the number of students, n, to the number of desks, d.
- **c** Find the number of students that could be seated around 18 desks.
- **d** Class 8Y has 29 students. How many desks are needed for them?

The number plane

12.04

A **number plane** is a grid made from a horizontal number line called the ***x*-axis** and a vertical number line called the ***y*-axis**. The number plane is also called a **Cartesian plane**, named after René Descartes (pronounced 'Ren-ay Day-cart'), a French mathematician and philosopher who developed the idea.

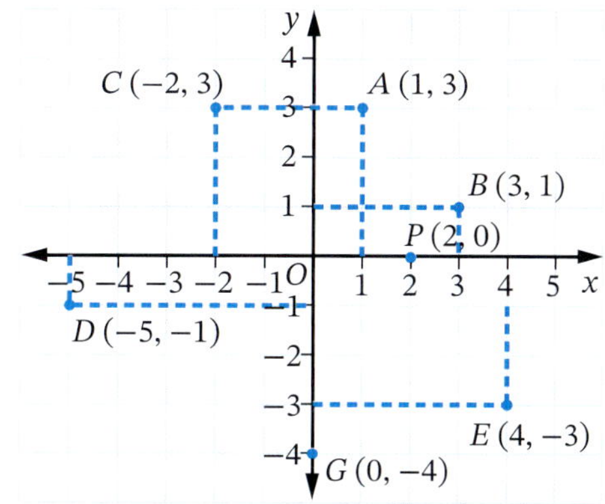

Points on the number plane are located using a pair of **coordinates** or an **ordered pair**. Note that order is important with coordinates: $A(1, 3)$ is not the same as $B(3, 1)$.

Videos
The number plane
Coordinate geometry: Descartes

Skillsheet
The number plane

Worksheets
The number plane
Plane grid 2
Number plane review
Number plane grid paper
A page of number planes
Plotting points and lines
Number plane writing activity

Puzzle
Coordinates code puzzle

ⓘ Points on the number plane

In any ordered pair, the first number is called the ***x*-coordinate** and the second number is called the ***y*-coordinate**.

The point (0, 0) is called the **origin**.

To locate a point, always start from (0, 0): the x-coordinate tells you how far to move *across*, the y-coordinate tells you how far to move *up* or *down*.

The number plane is divided evenly into 4 regions called **quadrants**.

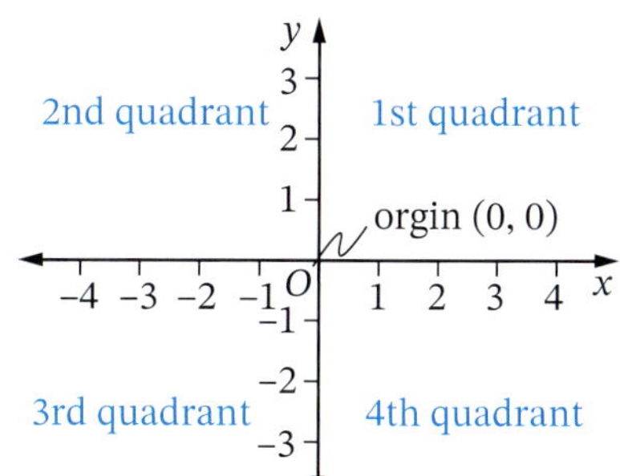

Presentation Graph coordinates

Technology Plotting points

Spreadsheet Points plotter

Example 5

a Write the coordinates of point D and state what quadrant it is in.

b Write the point with coordinates $(-4, -3)$ and state what quadrant it is in.

c Write the coordinates of point Q.

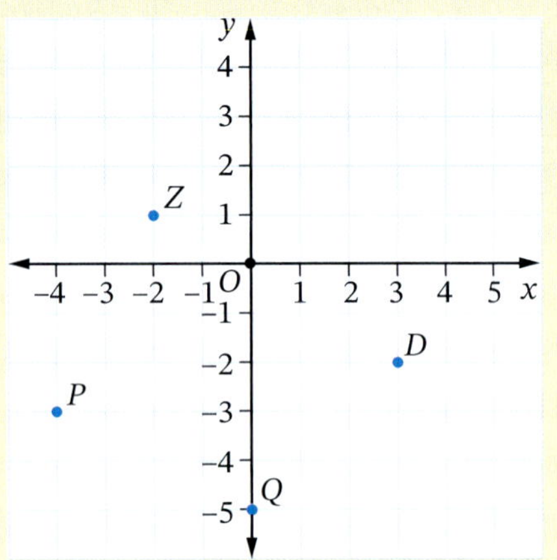

SOLUTION

a $D(3, -2)$ is in the 4th quadrant.

b $(-4, -3)$ is the point P, which is in the 3rd quadrant.

c Point Q is 0 on the x-axis and -5 on the y-axis, so its coordinates are $(0, -5)$.

EXERCISE 12.04 ANSWERS ON P. 655

The number plane

U F R C

Questions **1** to **5** refer to this diagram.

1 Name the point with each pair of coordinates.

a $(2, 1)$ **b** $(-3, 1)$

c $(5, 2\frac{1}{2})$ **d** $(5, -2)$

e $(0, 3)$ **f** $(0, -4)$

g $(2, 0)$ **h** $(-1\frac{1}{2}, 0)$

i $(-4, 2)$ **j** $(4, -1)$

2 Write the coordinates of each point.

C **a** A **b** B **c** C

d D **e** E **f** F

g G **h** H **i** I **j** J

3 In which quadrant does each point lie?

C **a** N **b** T **c** E **d** L **e** C

4 List all of the points that are on the x-axis and write their coordinates.

C

5 List all of the points that are on the y-axis and write their coordinates.

C

Foundation Standard Complex

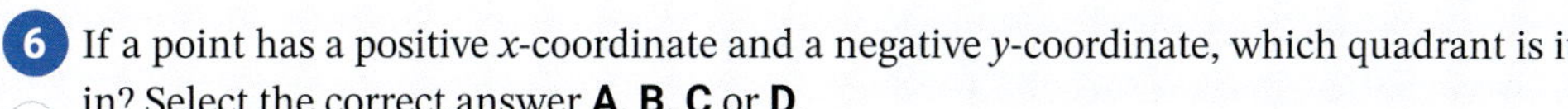

6 If a point has a positive x-coordinate and a negative y-coordinate, which quadrant is it in? Select the correct answer **A**, **B**, **C** or **D**.

R

A 1st quadrant **B** 2nd quadrant **C** 3rd quadrant **D** 4th quadrant

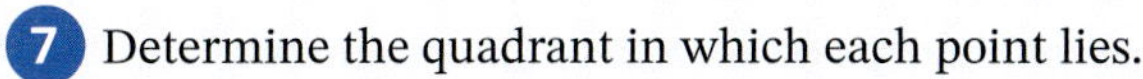

7 Determine the quadrant in which each point lies.

R C

a (3, –5) **b** (–2, –4) **c** (–8, 1) **d** (5, 4)

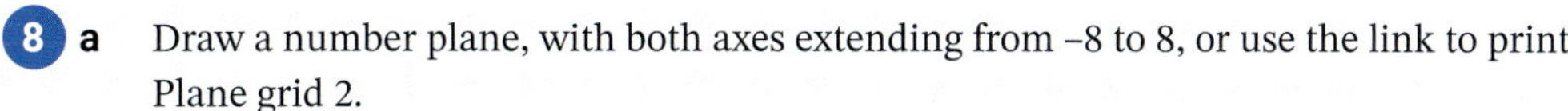

8 **a** Draw a number plane, with both axes extending from –8 to 8, or use the link to print Plane grid 2.

b Plot and label these points.

$A(3, 0)$ $B(-2, 2)$ $C(4, 5)$ $D\left(3\frac{1}{2}, -5\right)$ $E(-1, 7)$

$F(0, -4)$ $G(-6, 0)$ $H\left(\frac{1}{2}, -7\frac{1}{2}\right)$ $I(-2, 6)$ $J(7, 0)$

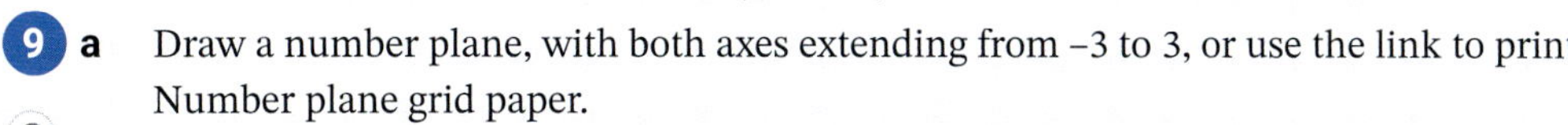

9 **a** Draw a number plane, with both axes extending from –3 to 3, or use the link to print Number plane grid paper.

C

b Mark these points.

$A(0, 1)$, $B(2, -1)$, $C(-1, 2)$, $D(-2, 3)$, $E(3, -2)$ $F(1, 0)$.

c What do you notice about the points? Check this with your ruler.

d Points that lie on the same straight line are said to be **collinear**. Write down the coordinates of any four points on a number plane that are collinear.

Worksheets
Plane grid 2
Number plane grid paper

DID YOU KNOW?

Polar coordinates

The number plane is not the only way to represent the position of a point. The **polar coordinates** of a point give its position in terms of a distance from the origin and angle measured from the x-axis.

We say that (r, θ) are the polar coordinates of the point P, where r is the distance P is from the origin O and θ is the angle between Ox and OP. A polar grid shows both distance and angle. For example, the grid below shows the coordinates of P as (2, 45°) and Q as (3, 225°).

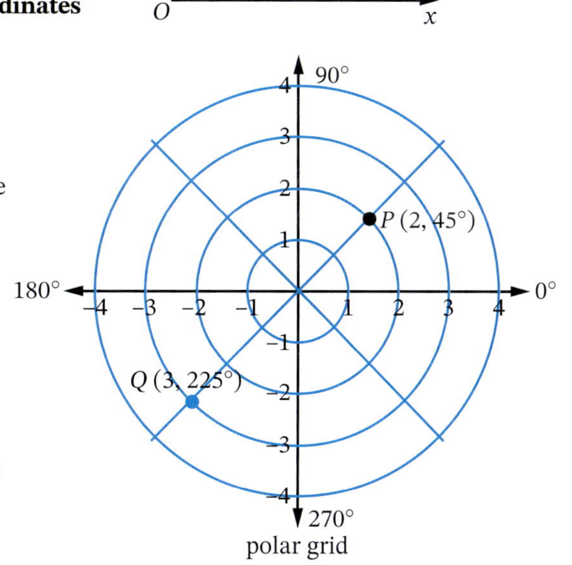

polar grid

Some of the real-life uses of polar coordinates include avoiding collisions between vessels and other ships, calculating the flow of ground water and guiding industrial robots.

On polar grid paper, plot the points $L(2, 90°)$, $M(4, 135°)$ and $N(0, 270°)$

Quiz
Mental skills 12

MENTAL SKILLS 12 ANSWERS ON P. 655 **Maths without calculators**

The unitary method with percentages

The unitary method is used when you are only given a *percentage* of an amount and you need to find the amount. It is called the unitary method because we find 1% of the amount first, then multiply that by 100 to find the whole (100%).

1 Study each example.

a If 8% of a number is 24, what is the number?

8% of the number = 24

$\therefore$ 1% of the number = $24 \div 8 = 3$

$\therefore$ 100% of the number = $3 \times 100 = 300$.

The number is 300. *Check*: $8\% \times 300 = 24$

b If 15% of an amount is \$90, what is the whole amount?

15% of the amount = \$90

$\therefore$ 1% of the amount = \$90 ÷ 15 = \$6

$\therefore$ 100% of the amount = \$6 × 100 = \$600.

The amount is \$600. *Check*: 15% × \$600 = \$90

2 Find the whole amount if:

a 5% of the amount is \$35

b 11% of the amount is \$88

c 20% of the amount is 80

d 6% of the amount is 42

e 90% of the amount is \$270

f 15% of the amount is \$60

g 40% of the amount is 100

h 120% of the amount is \$360

i 25% of the amount is \$75

j 8% of the amount is 40

12.05 The 3D number space

The number plane or Cartesian plane is a 'plane' because it is a flat surface in 2 dimensions, and points on the number plane grid are located and described by an ordered pair of coordinates (x, y) that refer to the x-axis and y-axis.

In 3 dimensions, it is possible to add a **z-axis** pointing 'out of' the number plane, perpendicular to both the x- and y-axes. The number plane then becomes a 3D 'number space' or **Cartesian space**, and points in the number space are located by an 'ordered triple' of 3D coordinates (x, y, z) that refer to the x-, y- and z-axes.

All 3 axes meet at the centre of the number space, the origin $O(0, 0, 0)$.

While the number plane is divided into 4 equal quadrants, the number space is divided into 8 equal **octants.** The 1st octant closest facing us as indicated on the diagram has the points where all of the values for (x, y, z) are positive, while the 8th octant, opposite the 1st octant and completely hidden below the x–y number plane, has the points where all of the values for (x, y, z) are negative.

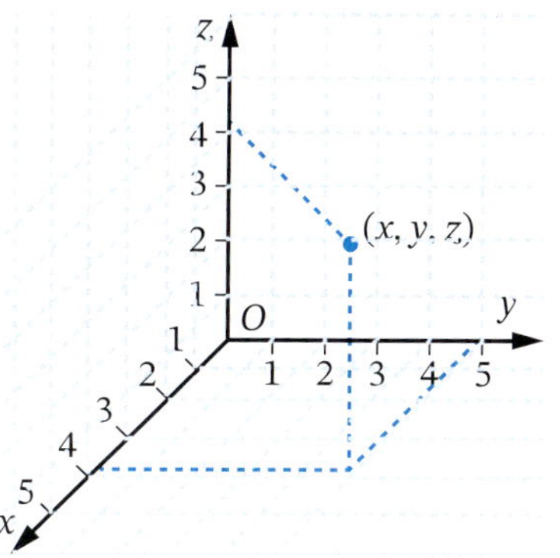

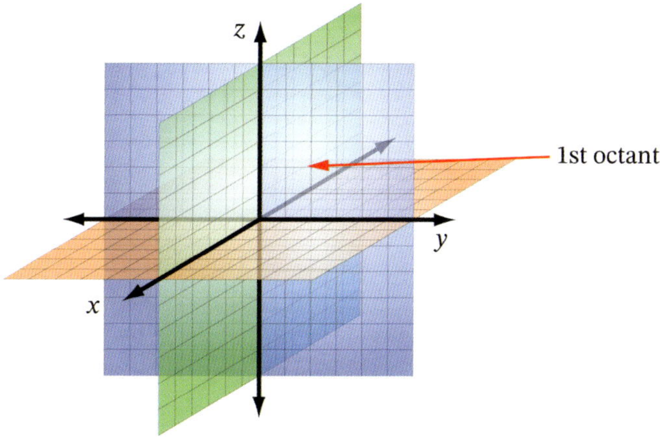

ⓘ Points in 3D space

In any ordered triple (x, y, z), the first number is called the ***x*-coordinate**, the second number is called the ***y*-coordinate** and the third number is called the ***z*-coordinate.**

The point (0, 0, 0) is the **origin**.

To locate a point, start from (0, 0, 0): the x-coordinate tells you how far to move *across*, the y-coordinate tells you how far to move *up* or *down,* and the z-coordinate tells you how far to move *above* or *below* the number plane.

When plotting points in 3D space, it is important to take note of where the x-, y- and z-axes are.

Example 6

a Write the coordinates of point P.

b Draw a set of axes with the point Q, with coordinates (3, 4, 1).

c Search online for 'graph 3D coordinates' to find a 3D drawing package to plot the point (2, 3, 5).

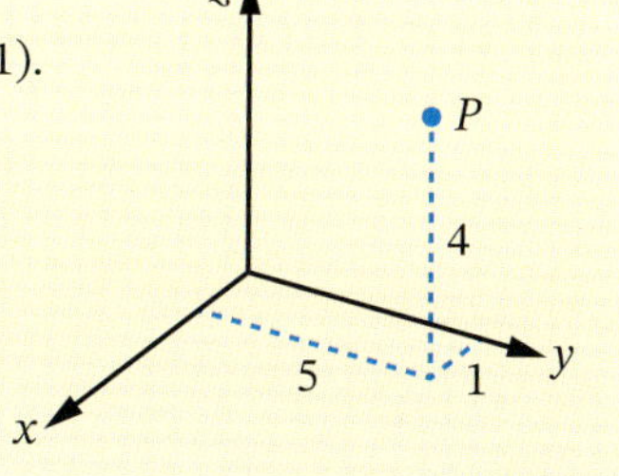

SOLUTION

a $P(1, 5, 4)$

b $(3, 4, 1)$ is point Q. **c** $(2, 3, 5)$

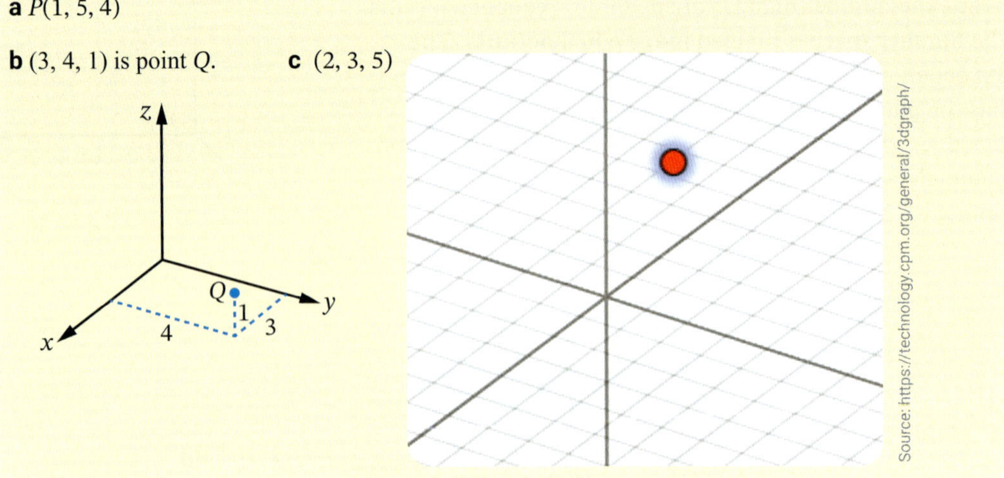

Example 7

Position coordinates of locations on Earth can also be described by 3D coordinates: (latitude, longitude, height). The location of a drone that is above Singapore at a height of 3000 m can be shown using 3D coordinates. Find the latitude and longitude of Singapore and hence state the location of the drone.

SOLUTION

Search online for 'Singapore latitude and longitude' to find the position coordinates (1.35°N, 103.82°E). So the 3D coordinates of the drone is (1.35°N, 103.82°E, 3000).

Note: The more decimal places of accuracy in the latitude and longitude, the closer you can get to a particular location.

Example 8

Draw a cube of side length 3 units in the first octant with one vertex on the origin.

SOLUTION

The vertices would be $(0, 0, 0)$, $(3, 0, 0)$, $(0, 3, 0)$, $(0, 0, 3)$, $(3, 3, 0)$, $(3, 0, 3)$, $(0, 3, 3)$ and $(3, 3, 3)$.

Note: that the directions of the x- and y-axes are different with this software (GeoGebra).

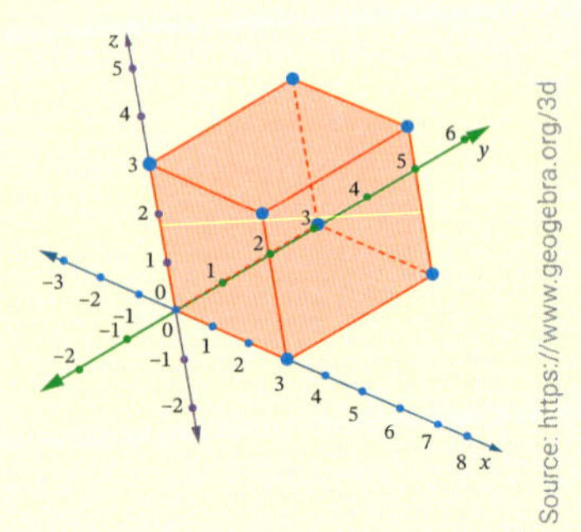

EXERCISE 12.05 ANSWERS ON P. 655

12.05

The 3D number space

EXAMPLE 6

1 **a** Write the coordinates of point P.

b Draw a set of axes with the point Q, with coordinates (0, 2, 4).

c Use an online 3D drawing package to represent the point (0, 2, 4).

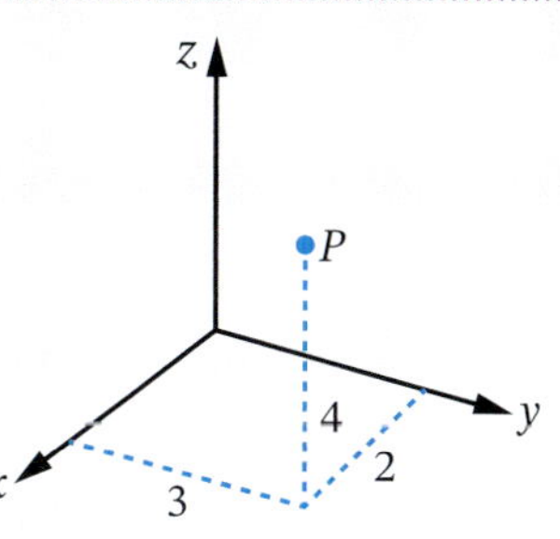

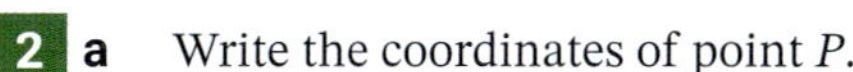

2 **a** Write the coordinates of point P.

b Draw a set of axes with the point Q, with coordinates (5, 0, 3).

c Use an online 3D drawing package to represent the point (5, 0, 3).

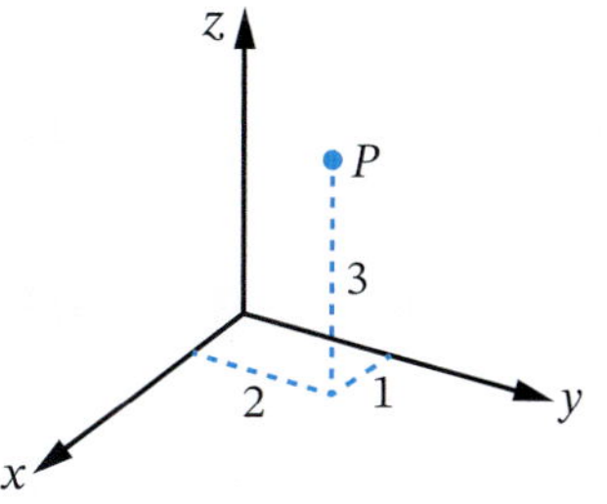

3 Write down the location of a drone that is:

a flying 1500 m directly over the London Eye.

b flying 600 m directly above the Sydney Harbour Bridge.

c flying 3000 m directly above the Bell Tower in Perth.

d flying 1 km directly over the Australia Zoo in Queensland.

For Questions 4 to 7, you may do them on paper or use an online 3D drawing package.

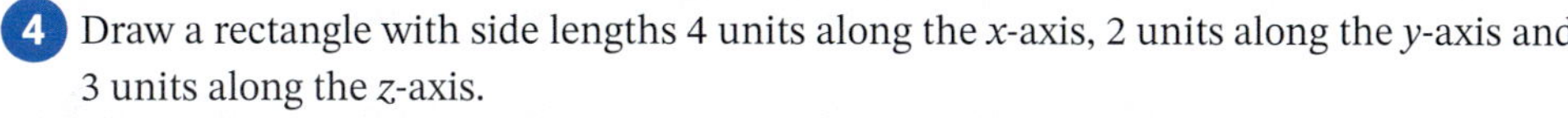

4 Draw a rectangle with side lengths 4 units along the x-axis, 2 units along the y-axis and 3 units along the z-axis.

5 Draw a rectangle with side lengths 3 units along the x-axis, 5 units along the y-axis and 4 units along the z-axis.

6 Draw a square-based pyramid with side lengths 6 units along the x-axis, 6 units along the y-axis and a height of 6 units.

7 Write down the coordinates of a square-based prism with side lengths of 2 units along the x-axis, 2 units along the y-axis and 4 units along the z-axis.

Foundation | Standard | Complex

DID YOU KNOW?

Legal maximum heights for drones

There are laws in different places about how high drones are allowed to fly, due to the technology used in manufacturing them.

Drone type	Maximum altitude
Hand-held	600 m
Close	1500 m
NATO type	3000 m
Tactical	5500 m
MALE	9000 m
HALE	9100 m
Hypersonic	15 200 m

iStock.com/akiyoko

12.06 Graphing number patterns

Video Graphing number patterns

Worksheets Number plane grid paper

A page of number planes

Graphing tables of values

Example 9

Consider this pattern of matchsticks used to make triangles.

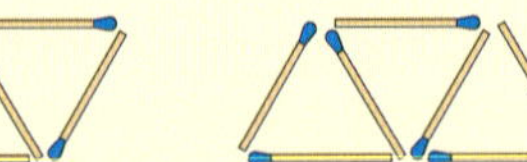

a Copy and complete the table of values for this pattern.

No. of triangles, x	1	2	3	4	5
No. of matchsticks, y					

b Find the formula for this table of values.

c Graph this table of values.

SOLUTION

a

No. of triangles, x	1	2	3	4	5
No. of matchsticks, y	3	5	7	9	11

b Bottom row of y-values increase by 2, so the multiplier in the formula is 2, and the equation is of the form $y = 2x$ ____.

Choose (2, 5) from the table to find the missing number.

When $x = 2, y = 2 \times 2 + 1 = 5$.

The formula is $y = 2x + 1$.

c Each column in the table forms an ordered pair.

No. of triangle, x	1	2	3	4	5
No. of matchsticks, y	3	5	7	9	11

(1, 3) (2, 5) (3, 7) (4, 9) (5, 11)

Graph each ordered pair on a number plane.

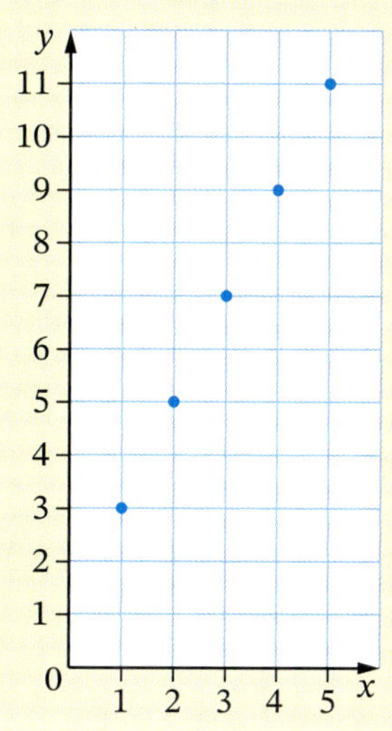

Note that the points are **collinear**, that is, they lie on a straight line. When this happens, we say that the relationship or pattern is **linear**.

- **linear** = involving a line.
- **collinear points** = points that lie on a straight line.

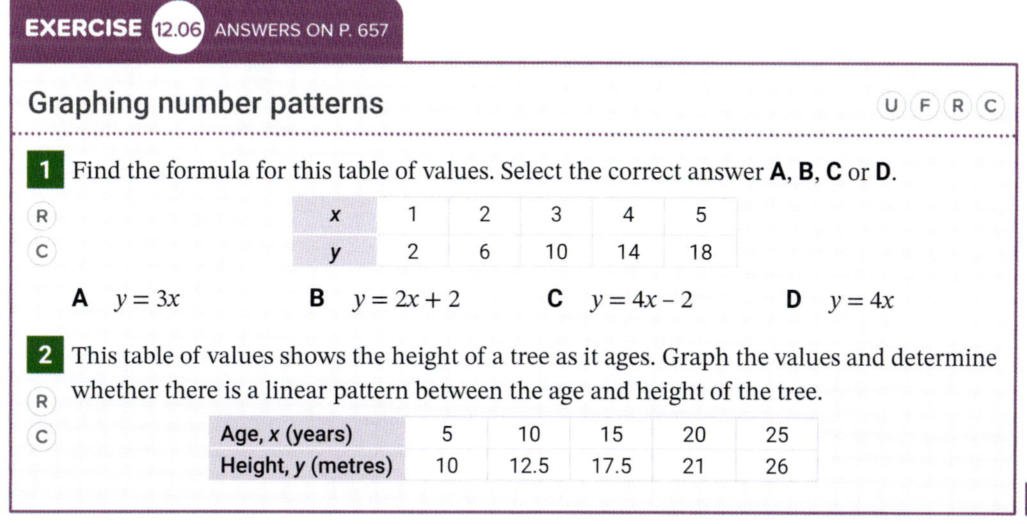

EXERCISE 12.06 ANSWERS ON P. 657

Graphing number patterns

U F R C

1 Find the formula for this table of values. Select the correct answer **A**, **B**, **C** or **D**. EXAMPLE 9

R C

x	1	2	3	4	5
y	2	6	10	14	18

A $y = 3x$ **B** $y = 2x + 2$ **C** $y = 4x - 2$ **D** $y = 4x$

2 This table of values shows the height of a tree as it ages. Graph the values and determine whether there is a linear pattern between the age and height of the tree.

R C

Age, x (years)	5	10	15	20	25
Height, y (metres)	10	12.5	17.5	21	26

Foundation Standard Complex

3 For each pattern below:

i copy and complete this table.

No. of shapes, x	1	2	3	4	5
No. of matchsticks, y					

ii find the formula for the table of values.

iii graph the table of values.

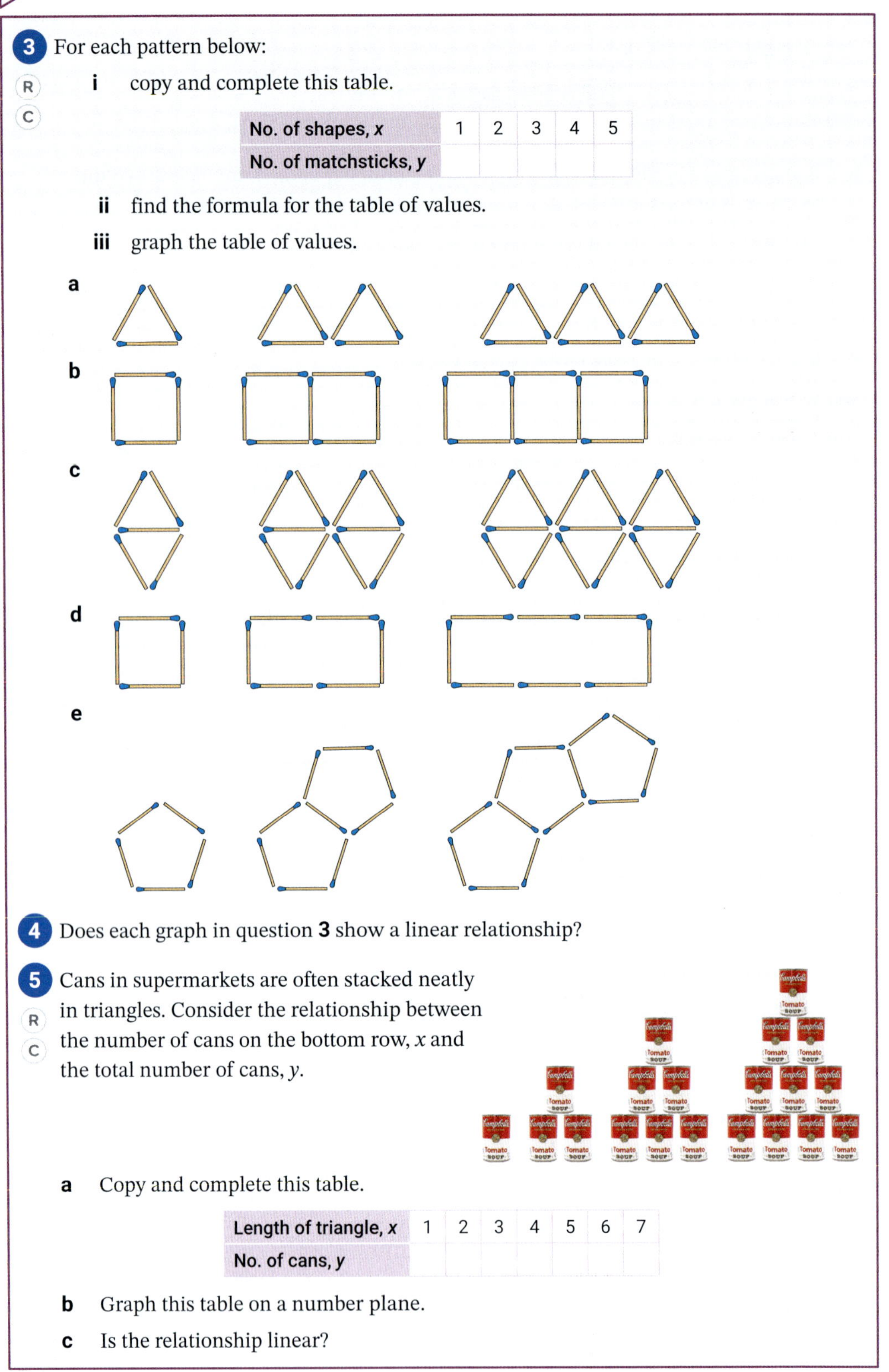

4 Does each graph in question **3** show a linear relationship?

5 Cans in supermarkets are often stacked neatly in triangles. Consider the relationship between the number of cans on the bottom row, x and the total number of cans, y.

a Copy and complete this table.

Length of triangle, x	1	2	3	4	5	6	7
No. of cans, y							

b Graph this table on a number plane.

c Is the relationship linear?

Foundation Standard Complex

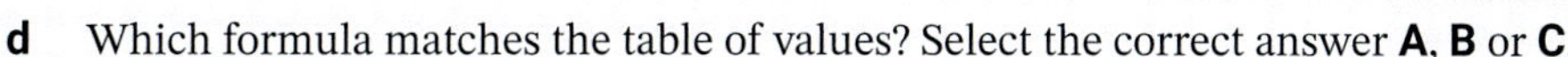

d Which formula matches the table of values? Select the correct answer **A**, **B** or **C**.

A $y = x(x - 1)$ **B** $y = \frac{x+1}{2}$ **C** $y = \frac{x(x+1)}{2}$

6 The following pattern is made with buttons.

R
C

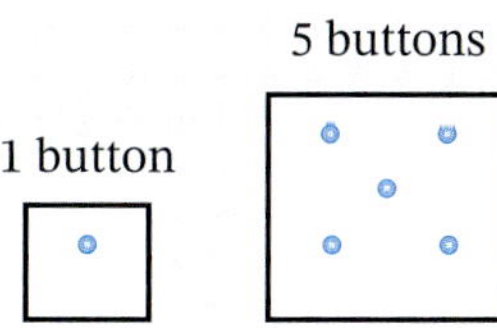

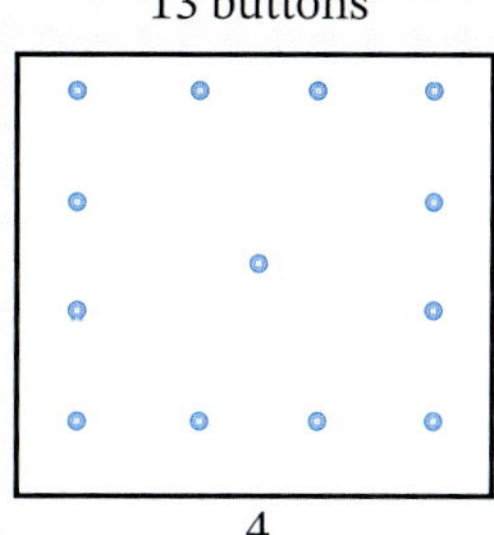

a Complete a table of values showing the relationship between the length of the square (S) and the number of buttons (B).

b Find a rule for S in terms of B.

c Use your rule to find the number of buttons required to make a pattern for a square of length:

i 10 **ii** 24

d Kovo has 106 buttons. What is the length of the biggest complete square that she can build?

Graphing linear functions 12.07

A **linear function** is an equation or formula whose graph on a number plane is a **straight line**.

ⓘ Graphing linear functions

- Complete a table of values.
- Graph the table of values on a number plane.
- Rule a line through the points and label the line with the equation.

Video Linear functions

Worksheets Number plane grid paper

A page of lines

Puzzles Matching linear equations

Graphing functions

Interactive Graphing lines

Foundation Standard Complex

Technology
Graphing straight lines: Finding intercepts

Spreadsheet
Linear functions plotter

Example 10

Graph $y = x + 1$ on a number plane.

SOLUTION

Complete a table of values. Choose x-values close to 0 for easy calculation and graphing.

x	-2	-1	0	1	2
y	-1	0	1	2	3

Graph the table of values on a number plane.
Rule a straight line through the points, place arrows at each end and label the line with its equation.

Note:

- Every pair of values that follows the linear function $y = x + 1$ lies on this line, for example, (3, 4), (–4, –3), $(1\frac{1}{2}, 2\frac{1}{2})$, (5, 6), (100, 101), $(-\frac{1}{2}, \frac{1}{2})$.
- Every point on the line follows the linear function $y = x + 1$.
- There are an infinite number of points that satisfy the rule.

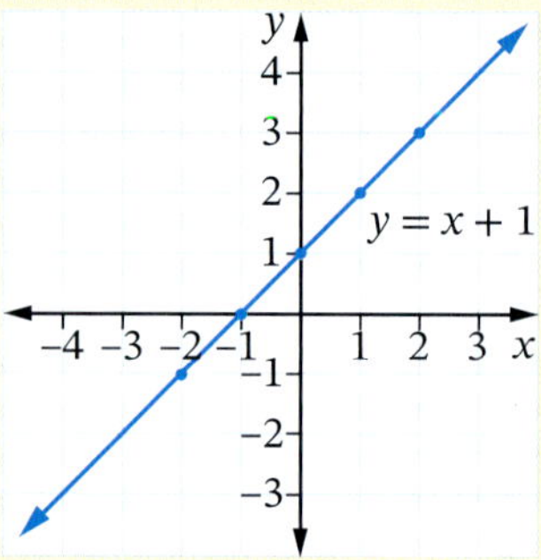

For these reasons, the line is infinite so we must draw arrows at each end.

Video
Graphing linear equations

Example 11

a Graph $y = 2x - 3$ on a number plane.
b Test whether the point (5, 8) lies on the line $y = 2x - 3$.

SOLUTION

a

x	-1	0	1	2
y	-5	-3	-1	1

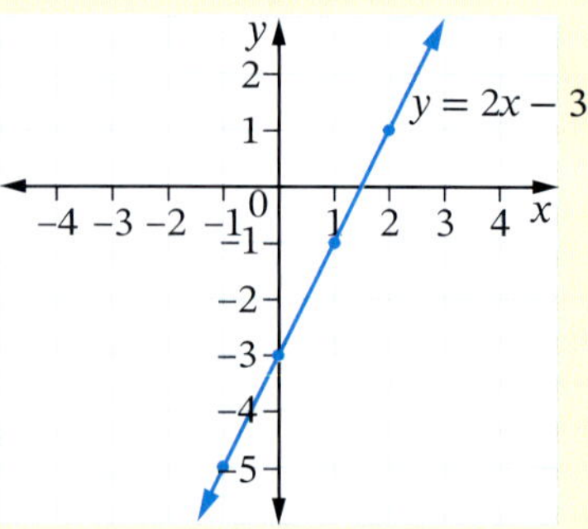

b We could extend the line $y = 2x - 3$ to see if (5, 8) lies on it, but a simpler way is to check whether (5, 8) follows or **satisfies** the function $y = 2x - 3$. If it does, then when $x = 5$, $y = 8$.

When $x = 5$, $y = 2 \times 5 - 3 = 7 \neq 8$.

$\therefore$ (5, 8) does not lie on the line. (In fact, (5, 7) does)

EXERCISE 12.07 ANSWERS ON P. 657

Graphing linear functions

Worksheets
Number plane grid paper
A page of number planes

EXAMPLE 11

1 (C) For each linear function, copy and complete the table of values and graph the function on a number plane.

a $y = x + 3$

x	−1	0	1	2
y				

b $y = x - 2$

x	0	1	2	3
y				

c $y = 2x$

x	−1	0	1	2
y				

d $y = \frac{x}{2}$

x	−1	0	1	2
y				

e $y = 6 - x$

x	0	1	2	3
y				

f $y = x$

x	−1	0	1	2
y				

EXAMPLE 12

2 Graph each linear function on a number plane.

a $y = 2x + 1$ **b** $y = 2x - 1$ **c** $y = 3x$
d $y = 3x - 2$ **e** $y = 2x + 4$ **f** $y = 3x - 5$

3 (C) Look at the graphs you drew in question **2**. Why are they called **increasing graphs**?

4 Graph each linear function on a number plane.

a $y = -x$ **b** $y = -x + 5$ **c** $y = 4 - x$
d $y = -x - 3$ **e** $y = 1 - 3x$ **f** $y = -2x$

5 (C) Look at the graphs you drew in question **4**. Why are they called **decreasing graphs**?

6 (R) (C) By looking at each linear function, predict whether its graph will be increasing or decreasing.

a $y = 5x - 4$ **b** $y = x + 9$ **c** $y = 3 - x$
d $y = 2x$ **e** $y = -x + 5$ **f** $y = -3x - 1$

7 (R) Test whether each point lies on the graph of the given linear function by:

i substituting its coordinates into the linear function.
ii examining the graphs of the lines you drew in questions **1** and **2**.

a $(4, 7)$, $y = x + 3$ **b** $(-1, -1)$, $y = x - 2$
c $(5, 1)$, $y = 6 - x$ **d** $(-3, -5)$, $y = 2x + 1$
e $(4, 10)$, $y = 3x - 2$ **f** $(-2, 1)$, $y = 2x + 4$

8 On which line does the point $(1, 3)$ lie? Select the correct answer **A**, **B**, **C** or **D**.

A $y = -x + 3$ **B** $y = x + 4$ **C** $y = 2 - 2x$ **D** $y = 2x + 1$

☐ Foundation ◯ Standard ⬡ Complex

TECHNOLOGY

Graphing linear functions

We are going to use a spreadsheet to graph $y = 2x - 3$.

	A	B
1	x	y
2	−4	
3	−3	
4	−2	
5	−1	
6	0	
7	1	
8	2	
9	3	
10	4	
11	5	
12		

1 Enter these values into a spreadsheet.

2 In cell B2, enter the formula **=2*A2−3** to evaluate $2x - 3$ and use **Fill Down** to copy the formula down to cell B11.

3 Highlight columns A and B, then click **Insert** and **Scatter** (with **Smooth Lines and Markers**).

4 Click the bottom-right corner of the graph border to enlarge the graph.

5 Use your graph to predict the value of y when:

a $x = 5$ **b** $x = -6$ **c** $x = -1.5$

6 Use your graph to predict the value of x when:

a $y = 11$ **b** $y = -9$ **c** $y = 14$

INVESTIGATION

Special lines

1 Draw a number plane with both axes extending from −4 to 4 and graph each table of values on it.

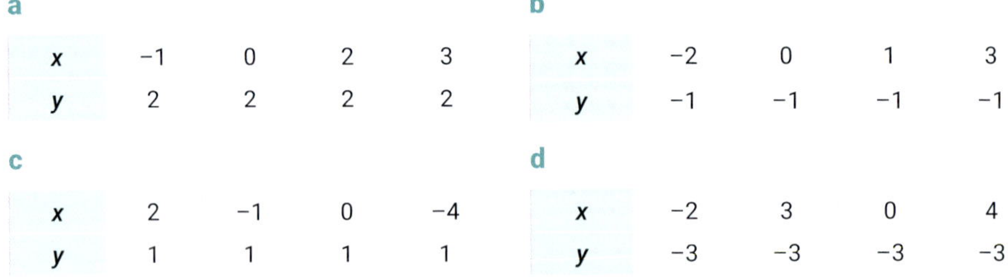

a

x	−1	0	2	3
y	2	2	2	2

b

x	−2	0	1	3
y	−1	−1	−1	−1

c

x	2	−1	0	−4
y	1	1	1	1

d

x	−2	3	0	4
y	−3	−3	−3	−3

2 **a** What do all of the lines in question **1** have in common?

b What do all of the tables of values in question **1** have in common?

c The formula for the table of values and line in question **1a** is $y = 2$. It does not have an x in it because the value of y does not depend on x, but is always equal to 2. What are the formulas for the other 3 lines in question **1**?

3 Draw another number plane and graph each table of values on it.

a

x	1	1	1	1
y	−3	−2	0	2

b

x	−2	−2	−2	−2
y	−2	0	1	3

c

x	3	3	3	3
y	−3	2	−2	0

d

x	−3	−3	−3	−3
y	−4	−1	2	4

4 a What do all of the lines in question 3 have in common?

b What do all of the tables of values in question 3 have in common?

c The formula for the table of values and line in question 3a is $x = 1$. What are the formulas for the other 3 lines in question 3?

TECHNOLOGY

Comparing linear functions

In this activity, we will use a spreadsheet to compare the graphs of $y = x$, $y = 2x$ and $y = 3x$.

1 Set up this spreadsheet.

	A	B	C	D
1	x	y = x	y = 2x	y = 3x
2	-2			
3	-1			
4	0			
5	1			
6	2			
7				

2 In cell B2 enter **=A2** to calculate $y = x$, then use **Fill Down** to copy this formula down to B6.

3 In cell C2 enter **=2*A2** to calculate $y = 2x$, then **Fill Down** to C6.

4 In cell D2, write a formula to calculate $y = 3x$, then **Fill Down** to D6.

5 Highlight all cells. Click **Insert** and **Scatter (Smooth Lines and Markers)** to graph all 3 lines on the same axes.

6 Compare the 3 lines. What is the same and what is different? Describe any patterns and important features. Write your answers on your spreadsheet or in your workbook.

7 Set up a similar spreadsheet for each set of 3 linear functions shown below. Use appropriate formulas to complete each column and graph each set of 3 lines. Compare the 3 lines in each set and describe the pattern and important features of each set.

a

G	H	I	J
x	y = -x	y = -2x	y = -5x
-2			
-1			
0			
1			
2			

b

L	M	N	O
x	y = x + 1	y = x + 2	y = x - 4
-2			
-1			
0			
1			
2			

c

Q	R	S	T
x	y = -x + 1	y = -x + 2	y = -x - 4
-2			
-1			
0			
1			
2			

12.08 Comparing linear functions

Video Linear functions

Puzzle Matching linear equations

Interactive Graphing slope-intercept

A **linear function** such as $y = 2x + 1$ is made up of 2 terms:

- the **variable term**, or the term with the x
- the **constant term**, the number without the x.

The number in front of the x is called the **coefficient** of x. For example,

$$y = 2x + 1$$

The coefficient of x is 2. The constant term is 1.

Example 12

The lines $y = 2x + 1$, $y = 2x - 1$ and $y = 2x - 3$ are graphed.

a How are the lines similar?

b How are the linear equations similar?

c The y-intercept of a line is the value at which the line crosses the y-axis. Find the y-intercept of each line in the diagram.

d How are the linear equations different?

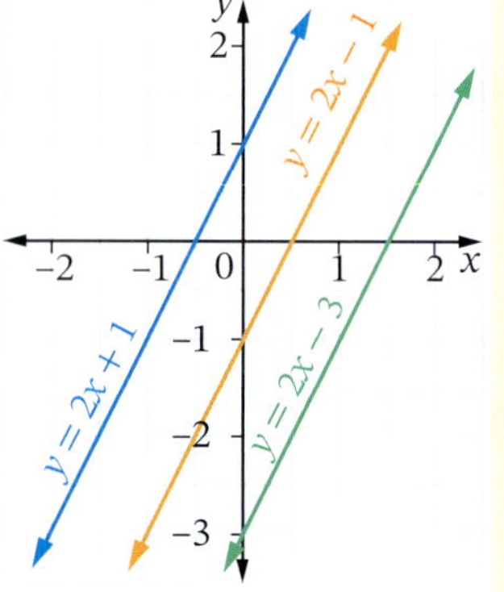

SOLUTION

a The lines point in the same direction and have the same slope, they are parallel.

b The linear equations all have the same coefficient of x, which is 2.

c $y = 2x + 1$: y-intercept $= 1$

$y = 2x - 1$: y-intercept $= -1$

$y = 2x - 3$: y-intercept $= -3$

d The linear equations have different constant terms: 1, −1, −3.

Example 13

The lines $y = x + 2$, $y = -x + 2$ and $y = 2x + 2$ are graphed.

a Find the y-intercept of each line.

b How are the linear equations similar?

c How are the lines different?

d How are the linear equations different?

e How does the coefficient of x in a linear equation affect its graph?

f How does the constant term in a linear equation affect its graph?

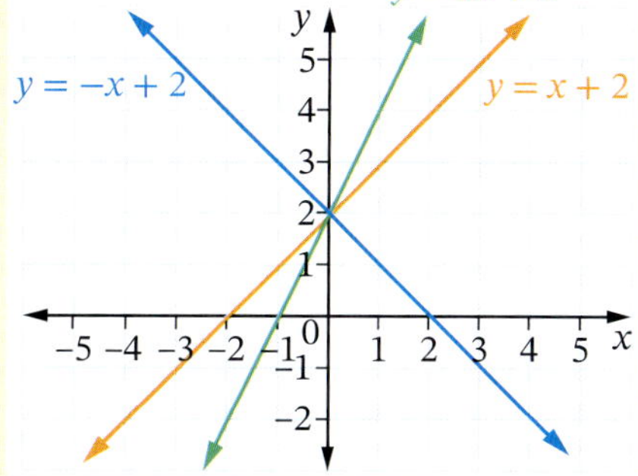

SOLUTION

a $y = -x + 2$: y-intercept $= 2$

$y = 2x + 2$: y-intercept $= 2$

$y = x + 2$: y-intercept $= 2$

b The linear equations all have the same constant term, 2.

c The lines point in different directions: they are not parallel.

d The linear equations have different coefficients of x: −1, 2 and 1.

e The coefficient of x affects the direction of the graph.

f The constant term gives the y-intercept of the graph.

ⓘ Features of a linear function

- In a linear function, the **coefficient of** x gives the **direction** of the line.
- If 2 or more lines have the same coefficient of x in their equations, then they are parallel.
- The **constant term** gives the y-**intercept** of the line.

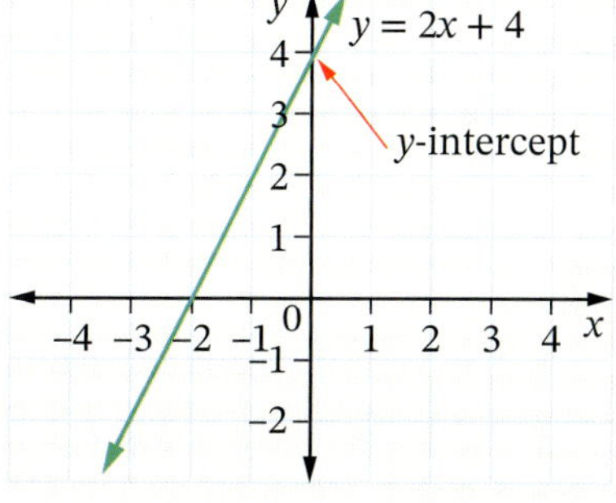

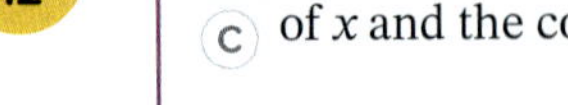

Comparing linear functions

U F R C

The exercise is best completed using dynamic geometry or graphing software.

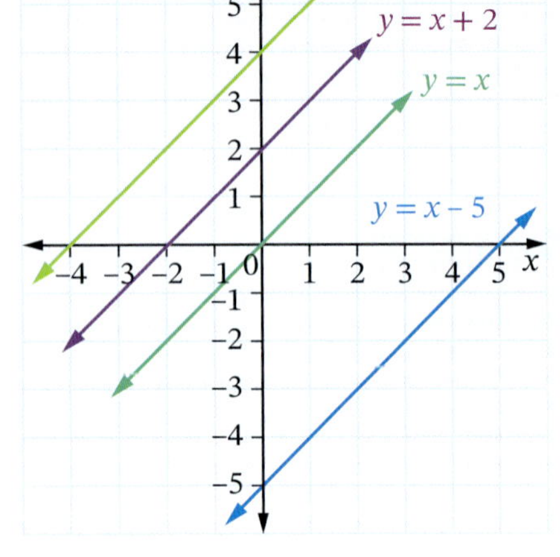

EXAMPLE 12

1 For each linear function, state the coefficient of x and the constant term. (C)

a $y = 4x + 1$ **b** $y = \frac{1}{2}x - 3$

c $y = x - 5$ **d** $y = -x + 6$

2 The lines $y = x + 4$, $y = x + 2$, $y = x$ and $y = x - 5$ are graphed. (R) (C)

a How are the lines similar?

b How are the linear equations similar?

c How are the lines different?

d How are the linear equations different?

EXAMPLE 13

3 The lines $y = \frac{1}{4}x$, $y = x$, $y = 3x$ and $y = -x$ are graphed. (R) (C)

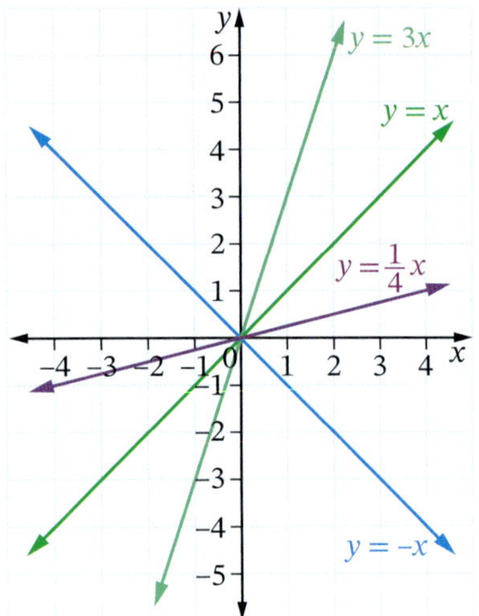

a How are the lines similar?

b How are the linear equations similar?

c How are the lines different?

d How are the linear equations different?

e Which line is the steepest? What is the coefficient of x in its equation?

f Which line is the least steep? What is the coefficient of x in its equation?

g How does the size of the coefficient of x in a linear equation affect the steepness of its line?

4 Graph $y = 3x$, $y = 3x - 1$ and $y = 3x + 2$ on the same number plane. (R) (C)

a What is the same about all 3 lines?

b What is the same about all 3 linear equations?

c What is different about the lines?

d What is different about the linear equations?

5 Graph $y = x + 1$, $y = 3x + 1$ and $y = \frac{1}{2}x + 1$ on the same number plane. (R) (C)

a What is the same about all 3 lines?

b What is the same about all 3 linear equations?

c What is different about the lines?

Foundation | Standard | Complex

d What is different about the linear equations?

e What is the coefficient of x in the equation with the steepest line?

6 Graph $y = -x - 2$, $y = -2x - 2$ and $y = -\frac{1}{2}x - 2$ on the same number plane.

R C

a How are the lines similar?

b How are the linear equations similar?

c How are the lines different?

d How are the lines different to the lines in question **5**?

e What is the coefficient of x in the equation with the least steepest line?

7 The graphs in question **5** are all increasing, while the graphs in question **6** are all decreasing. What feature of their equations indicates this?

R C

8 Which equation below when graphed gives a decreasing line with a y-intercept of 5? Select the correct answer **A**, **B**, **C** or **D**.

R C

A $y = 5x$ **B** $y = 5 - x$ **C** $y = 5x - 5$ **D** $y = -5x$

9 Copy and complete each statement.

R C

a The equations of parallel lines have the same ___________.

b Linear equations with the same constant term have graphs that have the same. ___________

c Linear equations with a negative coefficient of x have graphs that are ___________.

d The equations of lines with the same y-intercept have the same ___________.

e The equations of lines that are decreasing have a ___________ coefficient of x.

10 Graph each **non-linear equation** after completing the table of values given.

R C

a $y = x^2$

x	-3	-2	-1	0	1	2	3
y							

b $y = \sqrt{x}$

x	0	1	2	4	5	7	9
y							

c $y = x^3$

x	-2	-1.5	-1	-0.5	0	0.5	1	1.5	2
y									

d $y = \frac{1}{x}$

x	0.5	1	1.5	2	2.5	3	3.5	4	4.5
y									

11 **a** Why do you think the equations graphed in question **10** are called non-linear equations?

R C

b How are the non-linear equations algebraically different to the linear equations?

Foundation Standard Complex

12.09 Solving linear equations graphically

Equations can be solved **algebraically** using the balancing and backtracking methods. However, we can also solve an equation **graphically** by first graphing it on the number plane.

Example 14

Solve the equation $2x - 3 = 5$ graphically.

SOLUTION

Graph $y = 2x - 3$ on a number plane (the LHS of the equation).

x	−1	0	1	2
y	−5	−3	−1	−1

For each point in the table and on the line, the y-coordinate is the value of $2x - 3$. So to solve $2x - 3 = 5$, we need to find the point whose y-coordinate is 5.

Draw a dotted horizontal line at $y = 5$ (shown in red on the diagram) and read off the coordinates for the point on $y = 2x - 3$ crossed by this line.

This point is (4, 5), which means that the solution to $2x - 3 = 5$ is $x = 4$.

Check: $2 \times 4 - 3 = 5$.

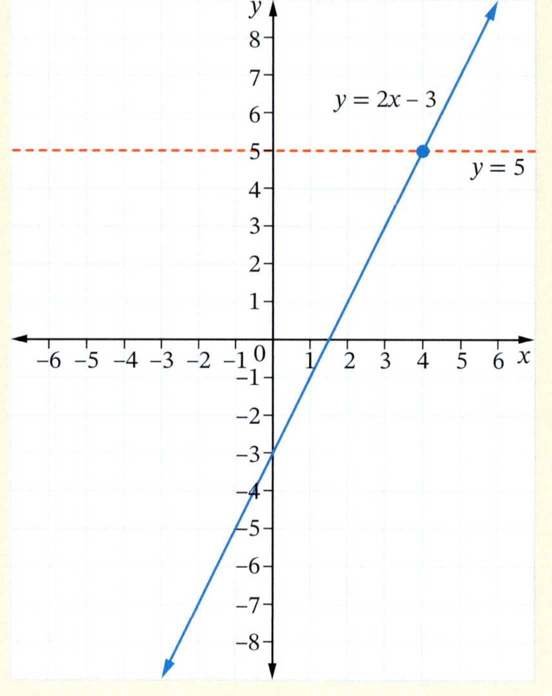

EXERCISE 12.09 ANSWERS ON P. 661

Solving linear equations graphically

U F

This exercise is best completed using dynamic geometry software or a graphing website.

Worksheets
Number plane grid paper
A page of number planes

1 Use the graph in Example **13** above to solve each equation below. Check your solutions.

a $2x - 3 = 3$ **b** $2x - 3 = -7$ **c** $2x - 3 = 0$

2 Graph $y = 2x + 1$ and use it to solve each equation below graphically.

a $2x + 1 = 7$ **b** $2x + 1 = 10$ **c** $2x + 1 = -5$

3 Solve each equation in question **2** algebraically.

4 Graph $y = 2x - 1$ and use it to solve each equation below graphically.

a $2x - 1 = -5$ **b** $2x - 1 = -1$ **c** $2x - 1 = 4$

☐ Foundation ◯ Standard ⬡ Complex

5 Graph $y = 3x - 2$ and use it to solve each equation graphically.

a $3x - 2 = 7$ **b** $3x - 2 = -11$ **c** $3x - 2 = -2$

6 Graph $y = -x + 1$ and use it to solve each equation graphically.

a $-x + 1 = -3$ **b** $-x + 1 = 4$ **c** $-x + 1 = -1$

7 Graph $y = -2x - 1$ and use it to solve each equation graphically.

a $-2x - 1 = -9$ **b** $-2x - 1 = 0$ **c** $-2x - 1 = 3$

8 Graph $y = \frac{1}{2}x + 3$ and use it to solve each equation graphically.

a $\frac{1}{2}x + 3 = 5$ **b** $\frac{1}{2}x + 3 = 1$ **c** $\frac{1}{2}x + 3 = 2$

Applying linear functions 12.10

Linear functions can be used to model situations such as the cost of a taxi trip, the price of a tradesperson completing a job, the amount of water left in a leaking tank, and the temperature of food at different times cooling after cooking.

Example 15

The graph below shows the cost of a taxi ride based on the distance travelled.

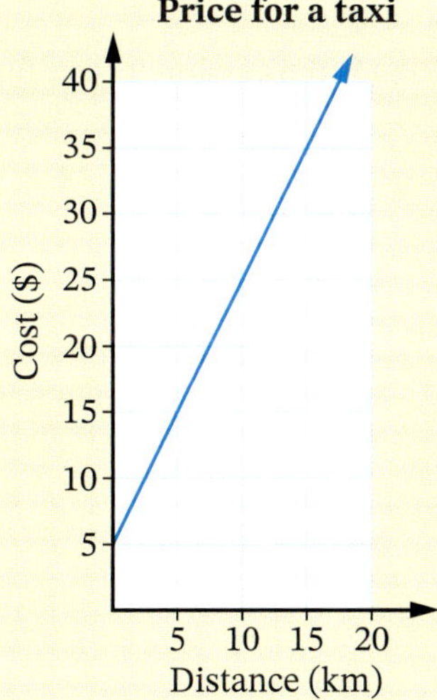

a How much was the flagfall, the amount it costs for the taxi, before you travel anywhere?

b How much was the taxi fare for a 15 km trip?

c What distance taxi trip costs $25?

Foundation Standard Complex

SOLUTION

a The price when no kilometres have been travelled is the y-intercept of the graph, so \$5.

b

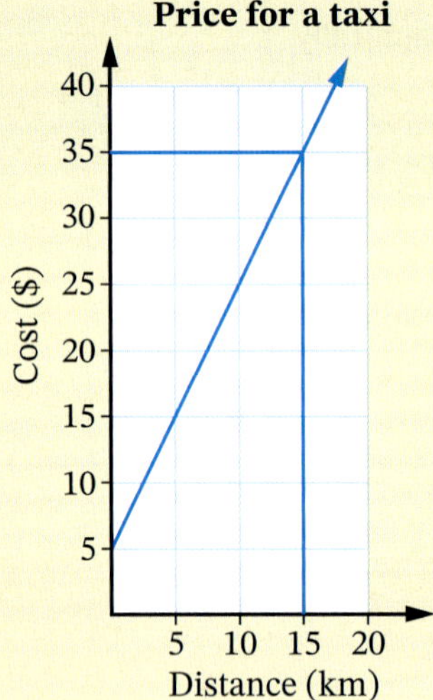

At 15 km, the cost is \$35.00.

c

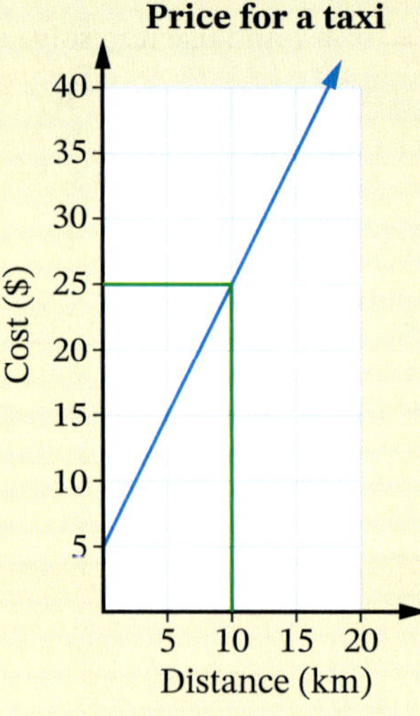

Cost \$25, 10km travelled.

Example 16

a Copy and complete table of values for the equation for the graph is $C = 2d + 5$, where C is the cost of the taxi fare and d is the distance travelled.

d	0	5	10	15
c				

a How much was the flag fall (that is the amount it costs for the taxi, before you travel anywhere)?

b What is the cost of a taxi trip of 15 km?

c What distance taxi trip costs \$25?

SOLUTION

a These values are the points on the graph in Example 14, but if you were to calculate them from the equation then a simple substitution is all that is required:

$d = 0, C = 2 \times 0 + 5 \quad = 5.$

$d = 5, C = 2 \times 5 + 5 \quad = 15.$

$d = 5, C = 2 \times 10 + 5 = 25.$

$d = 5, C = 2 \times 15 + 5 = 35.$

b $C = 2 \times 15 + 5 = \$35.00$

c $C = 2d + 5$

$25 = 2d + 5$ (–5 from each side of the equation)

$20 = 2d$ (divide each side of the equation by 2)

$10 = d$

The distance is 10 km.

EXERCISE 12.10 ANSWERS ON P. 661

Applying linear functions

EXAMPLE 15

1 The graph shows the cost of a taxi ride based on the distance travelled. (R)

a How much was the flagfall?

b How much was the taxi fare for a 10 km trip?

c What distance taxi trip costs $17?

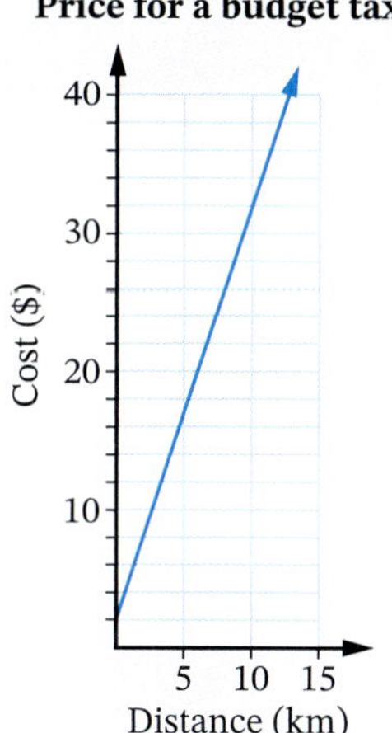

2 The graph shows the cost of a plumber.

a How much was the callout fee for the plumber?

b How much would it cost for the plumber to replace washers in all the taps in the house, if it took 3 h to complete the job?

c If a job cost $250 how long did it take to complete?

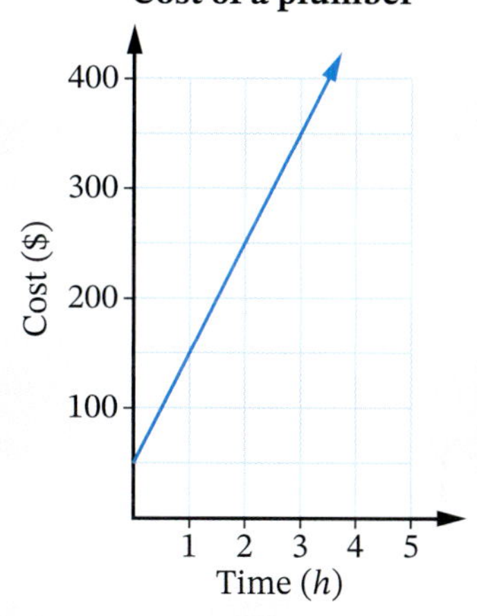

3 The graph shows the cost for *Thrifty Electricians.* (R)

a How much was the callout fee for the electrician?

b How much would it cost for the electrician to rewire the meter box if it took 4 hours to complete the job?

c If a job cost $520 how long did it take to complete?

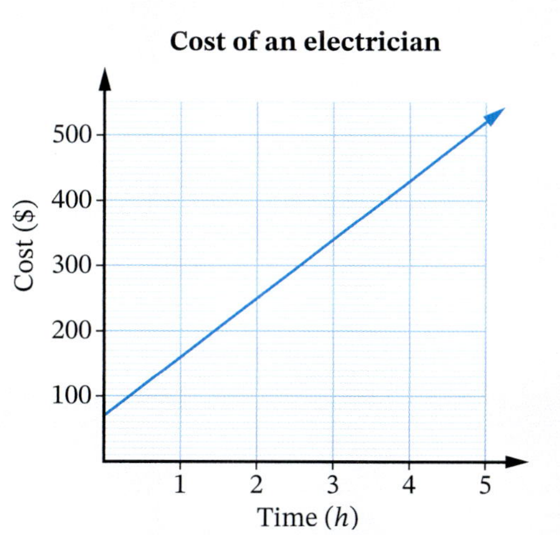

Foundation Standard Complex

4 The graph shows the amount of water left in a leaking tank over time.

a How much water was in the tank initially?

b How much water was in the tank after 6 h?

c Explain what happens to the water level after 10 h?

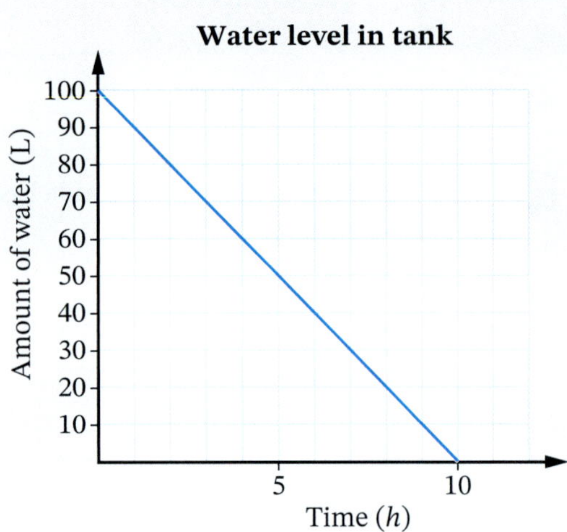

5 The graph shows the temperature of a cake left out to cool over time.

a What was the temperature of the cake initially?

b What was the temperature of the cake after 15 min?

c How long would it take for the cake to cool to 70°?

d For the values shown in the graph, is there a time when the cake is 0°? Suggest a reason for this.

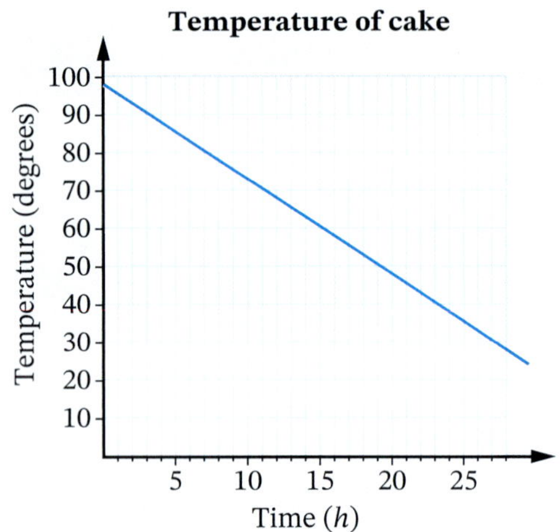

6 The cost of a taxi fare is calculated using the equation $C = 4d + 7$, where C is the cost of the taxi and d is the distance travelled.

a Copy and complete the table of values below:

d	0	5	10	15
c				

b How much was the flag fall (that is the amount it costs for the taxi, before you travel anywhere)?

c How much was the taxi fare for a 10 km trip?

d What distance taxi trip costs \$27?

7 The cost for a plumber to complete a job is calculated using the equation $C = 80h + 90$, where C is the cost of the job and h is the number of hours the job takes to complete.

a Copy and complete the table of values below:

d	0	1	2	3	4	5
c						

b How much was the call out fee for the plumber?

c How much was the cost of a job that takes 3 h?

d How long did a job take if it cost $410.00?

8 The cost for an electrician to complete a job is $C = 85h + 120$, where C is the cost of the job and h is the time taken to complete the job (in hours).

a Copy and complete the table of values below:

d	0	1	2	3	4	5
c						

b How much was call out fee for the electrician?

c How much did it cost for a job that took 4 h?

d How long did a job take if it cost $290?

9 The amount of water left in a leaking tank is given by the equation $L = 200 - 4h$, where L is the amount of water left (in litres) and h is the time since the tank was filled (in hours).

a Copy and complete the table of values.

d	0	5	10	15	20	25
c						

b How much water was in the bucket when it was completely filled?

c How much water was in the tank after 18 h?

d How long has the tank been leaking if there is 32 L left in the tank?

Graphing linear inequalities 12.11

The graph of a linear function or equation such as $y = x + 1$ on a number plane is a straight line. Every point on the line has coordinates that follow or satisfy the rule $y = x + 1$.

So what does the graph of a **linear inequality** such as $y < x + 1$ look like? It is the region on one side of the line, where every point in the region has coordinates that satisfy the rule $y < x + 1$.

ⓘ Graphing linear inequalities

- Complete a table of values for the linear *equation*.
- Graph the table of values as a *dotted line* on the number plane.
- Test a point on one side of the line to see if it makes the inequality true.
- If it does, shade that side; if not, shade the other side of the line.

Video
Graphing linear inequalities

Example 17

Graph $y < x + 1$ on a number plane.

SOLUTION

First graph $y = x + 1$ after completing a table of values.

x	-2	-1	0	1	2
y	-1	0	1	2	3

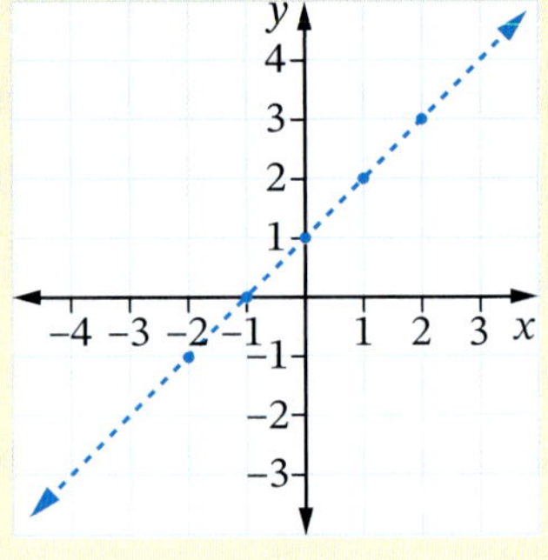

Graph the table of values on a number plane, but make the line ($y = x + 1$) dotted because it will not be included in the graphed region of $y < x + 1$.

Test a point on one side of the line for $y < x + 1$.

The point (0, 0) lies below the line, so see if this point makes the inequality true (or *satisfies* the inequality): $0 < 0 + 1$

$0 < 1$, which is true, so shade that side of the line and label it '$y < x + 1$'

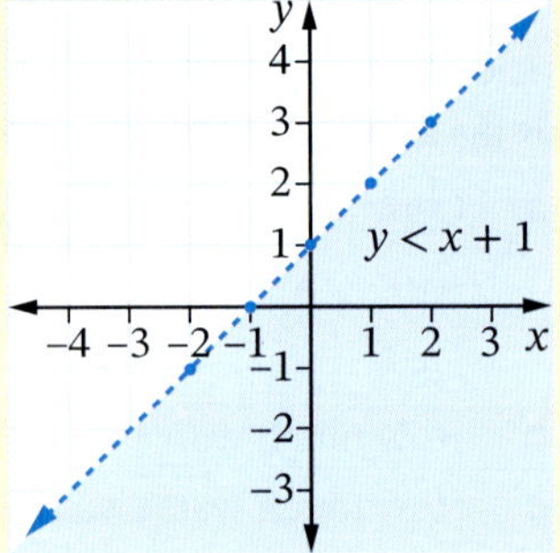

Inequalities involving '$y <$' (less than) tend to be *below* the line

Example 18

Graph $y > 2x - 3$ on a number plane.

SOLUTION

First, graph $y = 2x - 3$ as a dotted line.

x	-1	0	1	2
y	-5	-3	-1	1

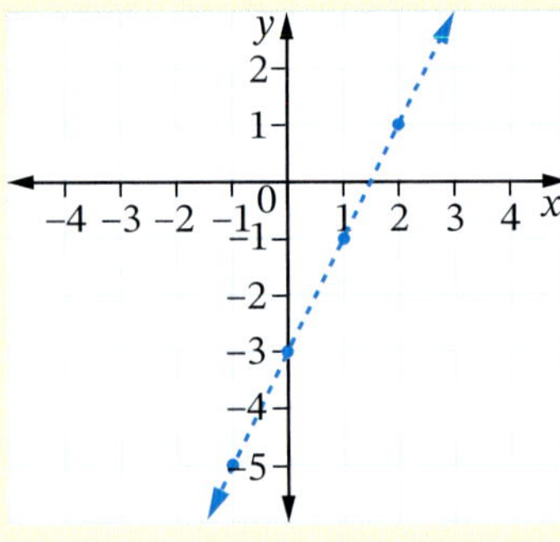

Test a point on one side of the line for $y > 2x - 3$.

The point (1, 1) lies above the line, so see if this point satisfies the inequality:

$1 > 2 \times 1 - 3$

$1 > -1$, which is true, so (1, 1) satisfies the inequality, so shade that side of the line and label it '$y > 2x - 3$'.

Inequalities involving '$y >$' (greater than) tend to be *above* the line.

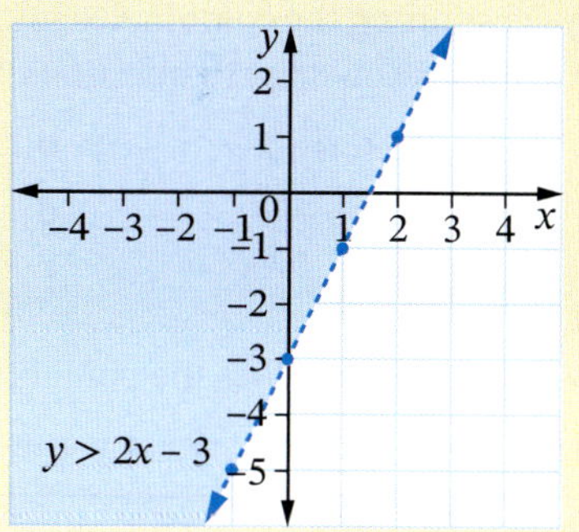

EXERCISE 12.11 ANSWERS ON P. 661

Graphing linear inequalities

U F R C

EXAMPLE 17

Worksheets: Number plane grid paper; A page of number planes

1 For each linear inequality, copy and complete the table of values for the *equation* and then graph the inequality on a number plane. (C)

a $y < x + 3$

x	−1	0	1	2
y				

b $y > x - 2$

x	0	1	2	3
y				

c $y < 2x$

x	−1	0	1	2
y				

d $y > \frac{x}{2}$

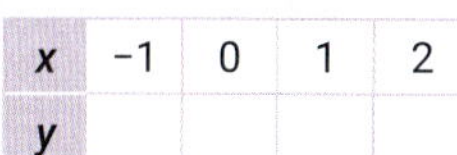

x	−1	0	1	2
y				

e $y < 6 - x$

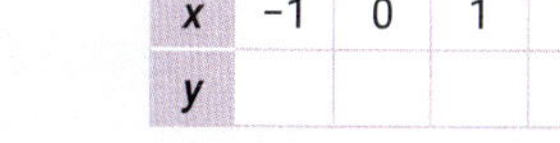

x	0	1	2	3
y				

f $y < x$

x	−1	0	1	2
y				

EXAMPLE 18

2 Graph each linear inequality on a number plane. (C)

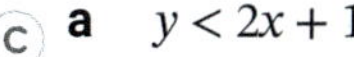

a $y < 2x + 1$ **b** $y > 2x - 1$ **c** $y < 3x$

d $y > 3x - 2$ **e** $y < 2x + 4$ **f** $y > 3x - 5$

3 Graph each linear inequality on a number plane. (C)

a $y < -x$ **b** $y > -x + 5$ **c** $y < 4 - x$

d $y > -x - 3$ **e** $y < 1 - 3x$ **f** $y > -2x$

4 By looking at each linear inequality, predict whether its graph will be shaded above or below the line. (R)

a $y < 5x - 4$ **b** $y > x + 9$ **c** $y < 3 - x$

d $y > 2x$ **e** $y < -x + 5$ **f** $y > -3x - 1$

☐ Foundation ○ Standard ⬡ Complex

5 Test whether each point makes the inequality true by substituting it into the inequality by substituting the coordinates into the linear inequality AND examining the graphs of the regions you drew in questions **1** and **2.**

a $(4, 4),\ y < x + 3$

b $(-1, -1),\ y > x - 2$

c $(5, 3),\ y < 6 - x$

d $(-3, -6),\ y > 2x + 1$

e $(4, 8),\ y < 3x - 2$

f $(-2, -1),\ y > 2x + 4$

6 Which inequality does the point (1, 3) satisfy? Select the correct answer **A**, **B**, **C** or **D**.

A $y < -x + 3$ **B** $y > x + 4$ **C** $y < 2 - 2x$ **D** $y < 3x + 1$

POWER PLUS ANSWERS ON P. 663

1 a George receives twice as much as Bill in pocket money. This can be represented by $y = 2x$, where y is George's pocket money and x is Bill's amount of money. Draw the graph of this line on a number plane where the x-axis runs from 0 to 12 and the y-axis from 0 to 24.

b The total of George and Bill's pocket money is $24. Write an equation that represents this.

c Draw the graph of your equation in part **b** on the same set of axes as part **a**.

d Use your graphs to find George's and Bill's pocket money.

2 Graph each equation accurately and state whether it is linear or non-linear.

a $y = x^3 - 2$

b $x + y = 5$

c $y = x^2 - 1$

d $y = 4 - x^2$

e $x - y = 4$

f $y = \frac{4}{x}$

Foundation Standard Complex

12 CHAPTER REVIEW

Language of maths

3D	coefficient	collinear	consecutive
constant term	coordinates	formula	graphically
horizontal	inequality	linear	number plane
number space	octant	origin	quadrant
satisfy	solution	table of values	variable
vertical	*x*-axis	*y*-axis	*z*-axis

Quiz
Language of maths 12

1 What is the difference between **linear** and **collinear**?

2 In which quadrant of the number plane are the *x*- and *y*-coordinates of a point both negative?

3 What are **consecutive** numbers?

4 What is the 'answer' to an equation called?

5 What does it mean if the coordinates of a point **satisfy** the equation of a line?

6 For the linear inequality $y > 4x + 1$, what does '>' stand for?

Topic summary

Worksheet
Mind map: Graphing linear functions

- Write in your own words what you have learnt about linear equations and their graphs.
- What parts of this topic did you have difficulty with? Discuss them with a friend.
- Where might you use the skills acquired in this chapter?

Print (or copy) and complete this mind map of the topic, adding detail to its branches and using pictures, symbols and colour where needed. Ask your teacher to check your work.

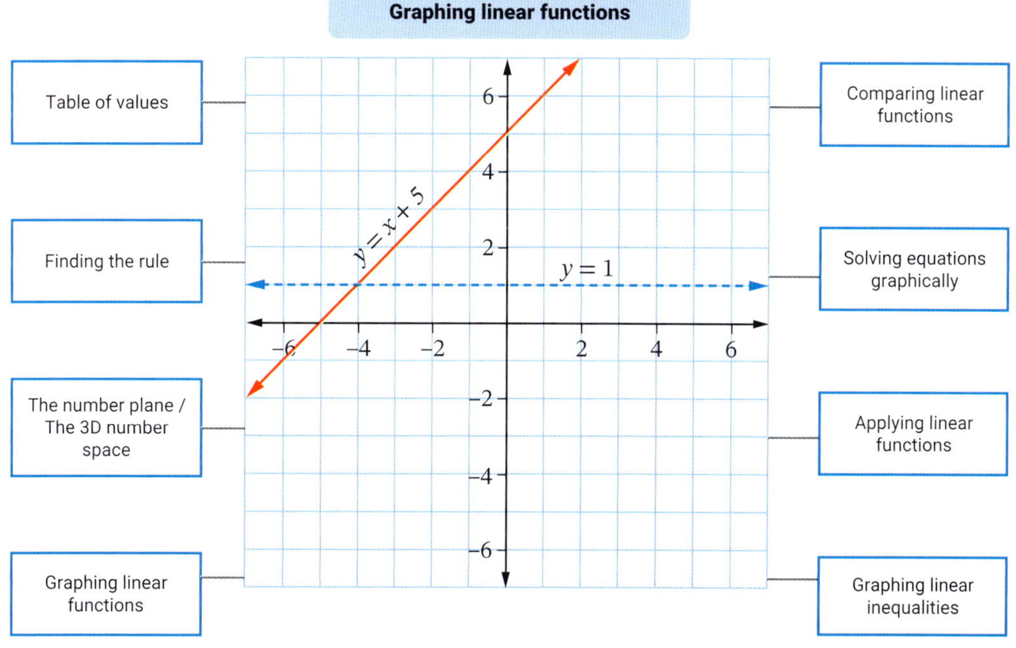

12 TEST YOURSELF

ANSWERS ON P. 664

Quiz
Test yourself 12

12.01

1 Copy and complete each table of values.

a $y = x + 3$

x	−1	0	1	2	3
y					

b $q = 2p - 7$

p	−2	−1	0	1	2
q					

c $d = 5 - c$

c	−1	0	1	2	3
d					

12.02

2 Find the formula for each table of values.

a

x	−1	0	1	2	3
y	−3	−2	−1	0	1

b

m	1	2	3	4	5
p	2	5	8	11	14

c

m	0	1	2	3	4
n	10	8	6	4	2

12.03

3 For this number pattern: 6, 10, 14, 18, 22, ...

a find the 7th term.

b find a formula for the nth term, T.

12.03

4 This pattern of rhombuses is made up of matchsticks.

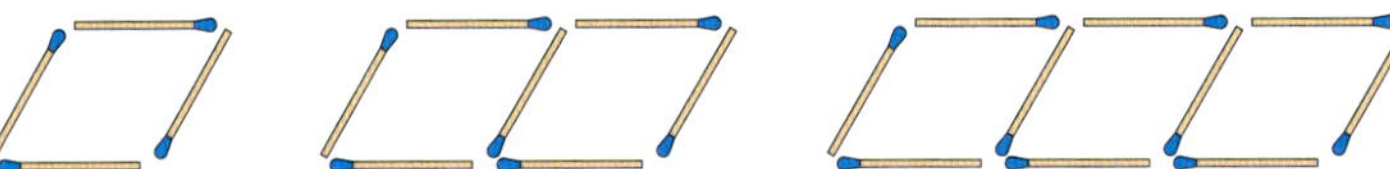

a Copy and complete the table.

No. of rhombuses, x	1	2	3	4	5	8	10
No. of matchsticks, y	4						

b Write the rule for the pattern in words.

c Write the rule as a formula.

d How many matchsticks are needed to build a pattern of:

i 12 rhombuses? **ii** 15 rhombuses?

12.04

5 State the coordinates of the points A, B, C and D.

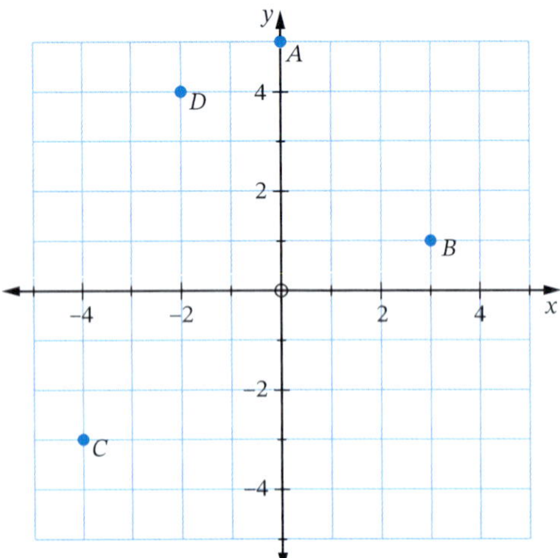

Foundation Standard Complex

6 Plot each point on a number plane. 12.04

a $P(1, -4)$ **b** $Q(-3, -1)$ **c** $R(-1, 2)$

d $S(0, 4)$ **e** $T\left(2\frac{1}{2}, -2\right)$ **f** $U(-2, 0)$

7 Which point from question **6**: 12.04

a is on the y-axis? **b** is in the 2nd quadrant?

c is on the x-axis? **d** is in the 4th quadrant?

8 **a** Write the coordinates of point P. 12.05

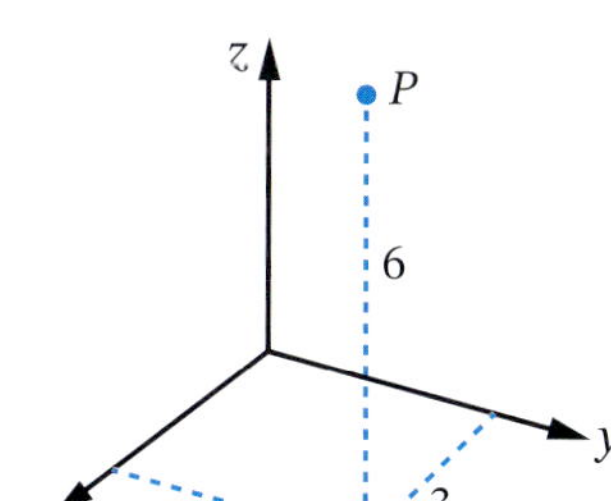

b Draw a set of axes with the point Q with coordinates (1, 4, 2).

c Use an online 3D drawing package to represent the point (3, 2, 1).

9 **a** Copy and complete this table for the pattern of tables and chairs. 12.06

Number of tables (x)	1	2	3	4	5
Number of chairs (y)					

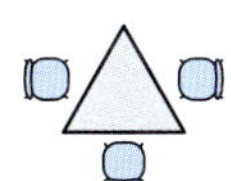 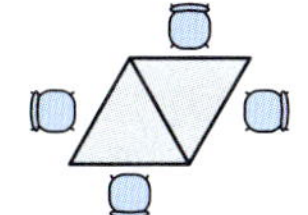 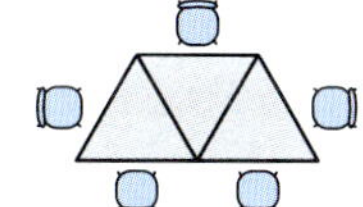

b Find the formula for y.

c Graph the table of values.

10 Graph each linear equation on a number plane after completing a table of values. 

a $y = 2x - 1$ **b** $y = -x + 3$ **c** $y = \frac{1}{2}x - 2$

11 Test whether each point lies on the line $y = 3x + 2$. 12.07

a (1, 2) **b** (−1, 4) **c** (−2, −4) **d** (0, 2)

12 **a** Graph $y = -x + 2$, $y = x + 2$ and $y = 2x + 2$ on the same number plane. 

b What is the same about the 3 lines?

c What is the same about the 3 linear equations?

d What is different about the 3 lines?

13 **a** Graph $y = 3x - 4$ and use it to solve $3x - 4 = 5$ graphically. 

b Graph $y = 2x + 1$ and use it to solve $2x + 1 = -2$ graphically.

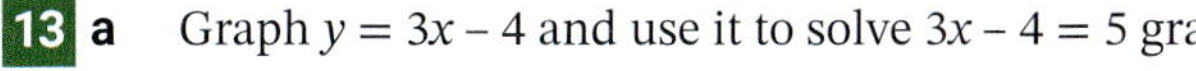

Foundation Standard Complex

12.10 **14** This graph shows the cost for *Deluxe Electricians*.

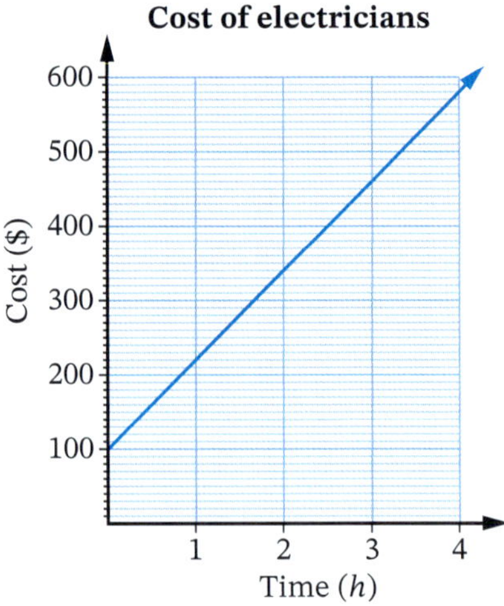

a How much was the callout fee for the electrician?

b How much would it cost for the electrician to fix a faulty oven, if it took 3 hours to complete the job?

c If a job cost $300, how long did it take?

d Given the cost of the electrician can be calculated using the equation $C = 120h + 100$, copy and complete the table of values below:

h	0	1	2	3
C				

15 **a** Graph the inequality $y < 3x + 4$ on a number plane.

b Does the point (1, 5) satisfy the inequality $y < 3x + 4$?

Practice set 4 ANSWERS ON P. 665

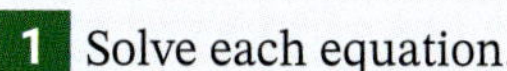

1 Solve each equation. 10.02

a $2a + 7 = 9$ **b** $3m - 2 = 7$ **c** $3x + 5 = -10$

d $\frac{t}{3} - 5 = 1$ **e** $\frac{k}{7} + 1 = 12$ **f** $8 - 2x = 4$

g $\frac{2x}{3} = 6$ **h** $\frac{5x}{4} = 10$ **i** $\frac{a+3}{2} = 9$

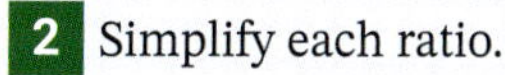

2 Simplify each ratio. 11.01

a 2 : 6 **b** 96 : 34

c 2 m to 15 cm **d** 10 minutes to 2 hours

e $\frac{2}{5} : \frac{1}{3}$ **f** 0.7 : 2.8

3 For each pair of sizes, find the better buy.

a Chips: 300 g for \$5.20 or 400 g for \$7.20

b Pasta: 1.5 kg for \$3.30 or 750 g for \$1.80

4 Complete this table of values using the formula $y = 2x + 9$.

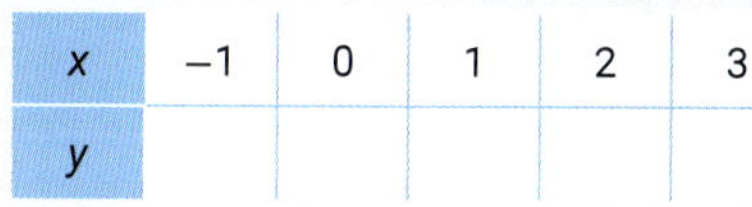

x	−1	0	1	2	3
y					

YEAR 9 EXTENSION

5 Solve $3x^2 = 48$. Select the correct answer **A**, **B**, **C** or **D**.

A $x = -2$ **B** $x = 2$ **C** $x = \pm 2$ **D** $x = \pm 4$

6 Solve each equation.

a $5q - 10 = 2q + 8$

b $4x + 5 = 8x - 7$

7 Solve each equation.

a $2(y + 3) = 10$

b $5(d - 4) = 2(d + 5)$

8 How many hours and minutes it is from:

a 4:30 a.m. to 9:55 a.m.?

b 15:05 to 22:10?

9 The ratio of boys to girls at a conference is 4 : 3. If there are 15 girls, find the number of boys. 11.02

☐ Foundation ○ Standard ⬡ Complex

10 **a** If 5 cm on a map represents 1 km in actual distance, what length would be used to represent 5 km?

b Using a scale of 1 : 500, what actual length, in metres, would a scaled length of 4 cm represent?

11 5 bricks have the same mass as one brick and 4 kg. What is the mass of one brick? Use an equation to solve this problem.

12 **a** A train travels 264 km in 8 hours. What is its average speed?

b A bus travels at 45 km/h for 5 hours. How far does it travel?

12.02

13 Find the formula for this table of values.

x	2	3	4	5	6	7
y	6	9	12	15	18	21

12.03

14 Here are the first 3 shapes in a pattern made of toothpicks.

a Copy and complete this table.

Number of triangles t	1	2	3	4	5	8
Number of toothpicks, p	3					

b Write the rule for the pattern in words.

c Write the rule as a formula.

d How many toothpicks are needed to build 20 triangles?

12.04

15 Write the coordinates of each point marked on the number plane.

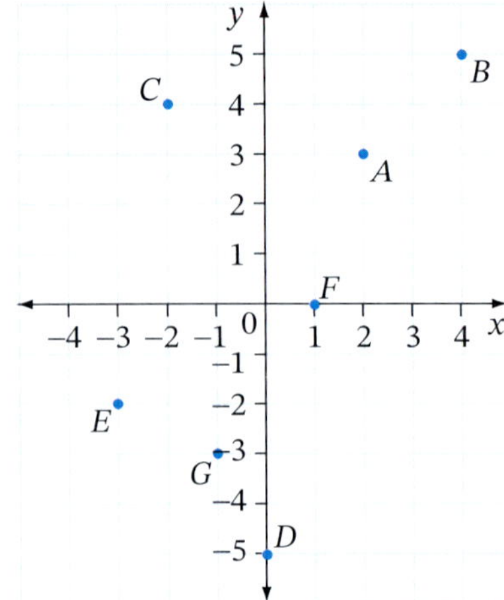

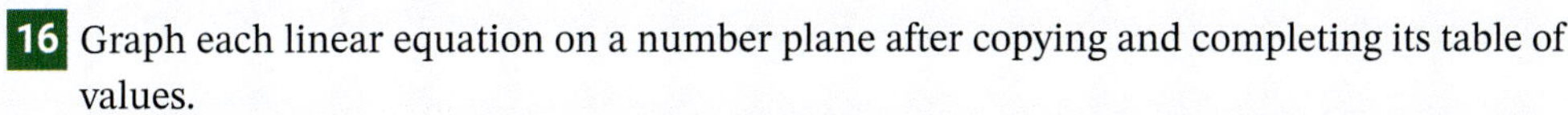

16 Graph each linear equation on a number plane after copying and completing its table of values.

a $y = x + 3$

x	−1	0	1	2
y				

b $y = 4x - 1$

x	−1	0	1	2
y				

17 Using the formula for the area of a triangle $A = \frac{1}{2}bh$, find h, the height of the triangle with area 63 and base length 10.5.

18 A truck is loaded with fruit and vegetables in the ratio of 5 : 9. The total mass is 4.2 tonnes. What is the mass of the fruit carried in the truck?

19 In 20 hours, a train travels 800 km. How far would it travel in 11 hours at the same rate?

20 Test whether the point (3, 7) lies on the line with the equation:

a $y = 2x + 4$

b $y = 3x - 2$

21 Water flows into a tank at a rate of 15 litres per hour. How long will it take to fill a tank that holds 750 litres?

22 Graph the line $y = 2$ on a number plane.

23 Graph the line $y = 2x - 3$ and use it to solve the equation $2x - 3 = 5$ graphically.

24 Graph the linear inequality $y > x + 2$ on a number plane.

25 The time in Melbourne is 10 hours ahead of UTC time in London. What is the time in Melbourne when it is 2 p.m. in London?

26 Solve each inequality.

a $2x < 8$ **b** $x - 4 > 8$ **c** $3x + 4 < 19$ **d** $-x - 6 > -10$

☐ Foundation ○ Standard ⬡ Complex

11.09 **27** Simon's cycling trip is described by this travel graph.

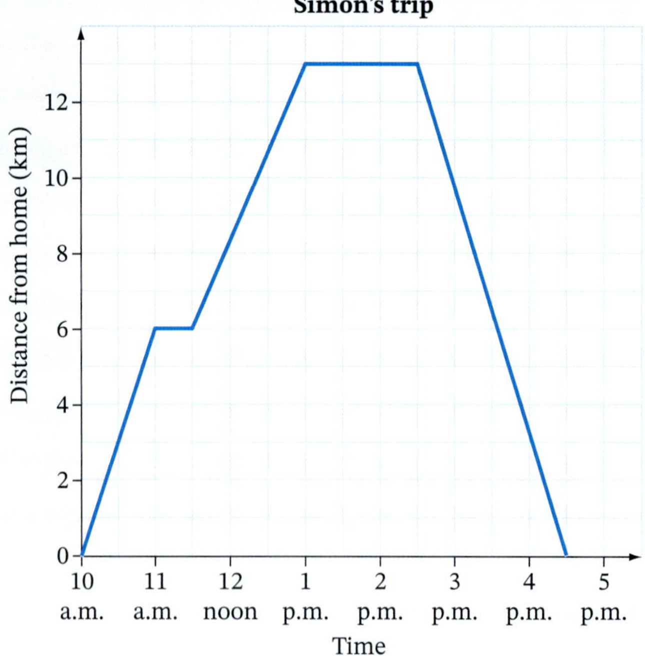

a When, and for how long, did Simon stop for lunch?

b How far did he cycle in total?

c What was Simon's speed on the return journey home?

28 Solve $\frac{1+2x}{3}=-1$.

29 Leo and Chris share the cost of their internet bill in the ratio of 3 : 5. What fraction does Chris pay? Select the correct answer **A**, **B**, **C** or **D**.

A $\frac{2}{5}$ **B** $\frac{5}{8}$ **C** $\frac{5}{2}$ **D** $\frac{1}{5}$

30 Dounya takes 24.5 seconds to walk 70 metres, while Anna walks 160 metres in 1 minute. Answer correct to one decimal place.

a What is the average speed of each person?

b If each person walks 2 km at their normal speed, who finishes first, and by how much?

☐ Foundation ○ Standard ⬡ Complex

General practice

ANSWERS ON P. 666

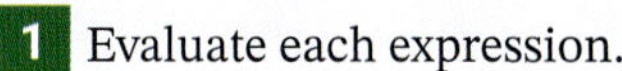

1 Evaluate each expression.

a $\sqrt{169}$ **b** $\sqrt{256}$ **c** $\sqrt{289}$

d $\sqrt[3]{27}$ **e** $\sqrt[3]{125}$ **f** $\sqrt[3]{5832}$

2 Write an algebraic expression for:

a 2 times x, plus y. **b** y times z, minus x. **c** the sum of x, y and z.

d the product of x and y decreased by 4 times z.

3 Simplify each fraction.

a $\frac{3}{12}$ **b** $\frac{18}{24}$ **c** $\frac{35}{100}$

4 Convert each improper fraction into a simplified mixed numeral.

a $\frac{36}{5}$ **b** $\frac{67}{10}$ **c** $\frac{74}{12}$

5 If $m = 2$, $n = -1$ and $p = 3$, evaluate each expression.

a mnp **b** $2p - m$ **c** $3m - n^2$

6 Simplify each expression.

a $3a \times 2r$ **b** $-7p - 2p + 6p$ **c** $7u \times 5u$

d $9x - 3 + 12 - 4x$ **e** $6p - 3q + 2p - 4q$ **f** $\frac{24ut}{4mt}$

7 Thay scores 32 points out of his team's score of 128. What percentage is this?

8 Write as a rate:

a 450 m in 15 seconds **b** 82 points in 5 games **c** 3.5 kg for $28

9 These are scores in a singing contest:

5 4 8 6 1 5 8 5 4 2 5.

Find:

a the mean (to one decimal place) **b** the median

c the mode **d** the range

10 **a** Write the coordinates of point P.

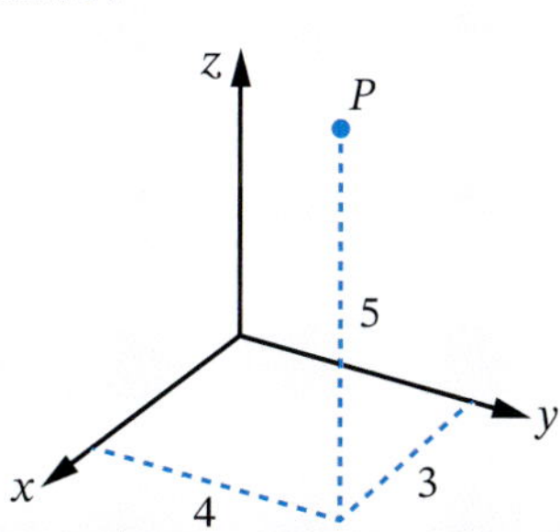

b Plot the point Q with coordinates (2, 1, 5) in a number space.

☐ Foundation ○ Standard ⬡ Complex

11 Find the value of each variable, giving reasons.

a
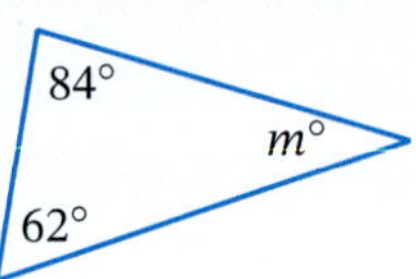

b
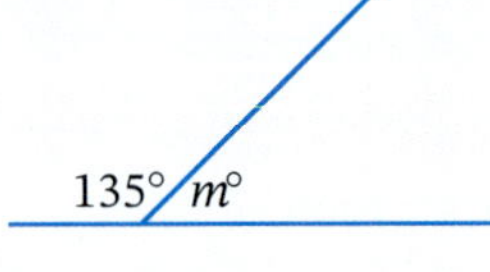

c
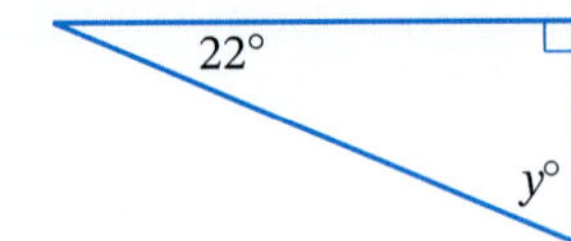

d
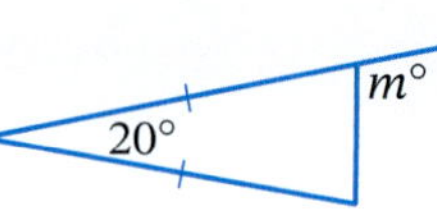

e
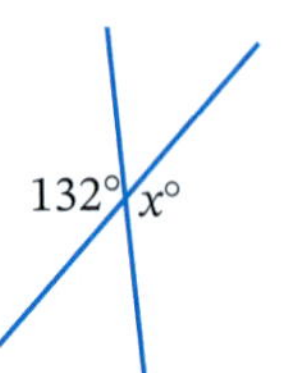

f
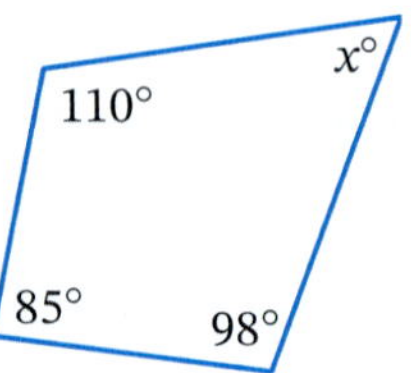

g
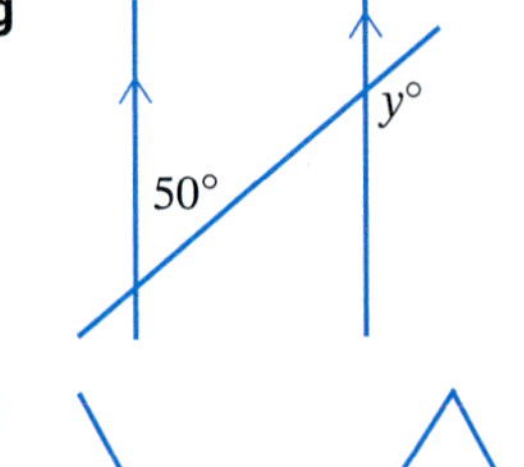

h
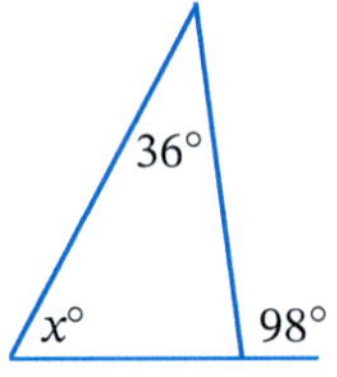

i
m° 56°

CHAPTER 1

12 Find the value of each variable, correct to one decimal place.

a
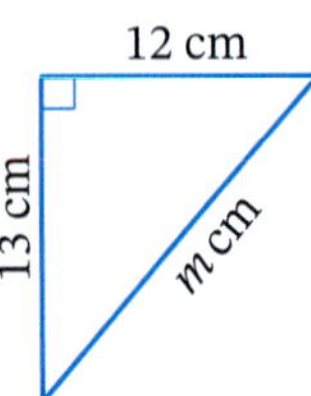

b
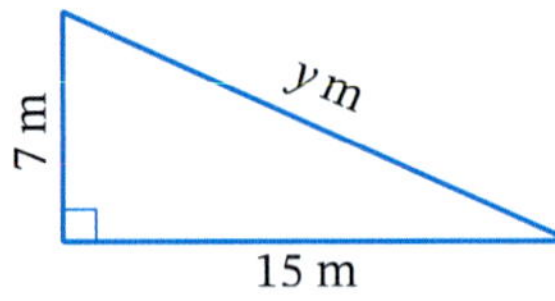

c
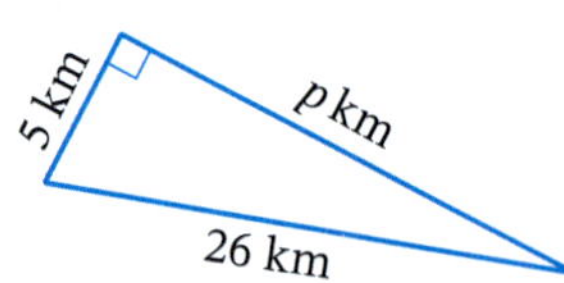

d
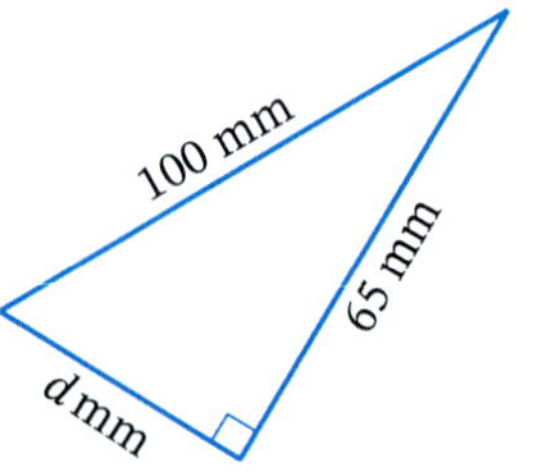

13 Use index notation to simplify each expression.

a $6^5 \div 6^3$ **b** $2^3 \times 2^4$ **c** $3^9 \div 3^7$ **d** 5×5^2

e $(2^3)^4$ **f** $(2^3 \times 5)^2$ **g** $3^2 \div 3^2$ **h** 6×4^0

☐ Foundation ○ Standard ⬡ Complex

14 The probability of choosing an Ace from a pack of cards is $\frac{4}{52}$. CHAPTER 9

a What is the complementary event to choosing an Ace?

b What is the probability of this complementary event?

15 Find the area of each quadrilateral. CHAPTER 5

a

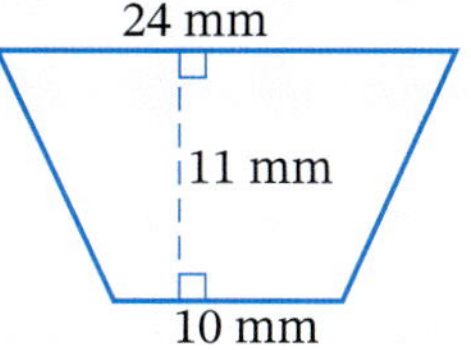

b

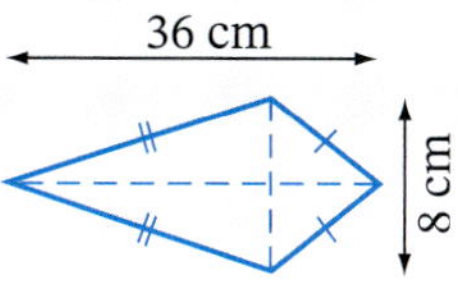

c

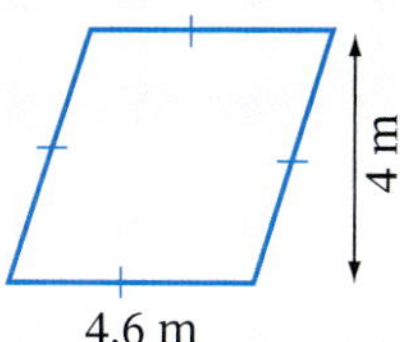

16 Classify each triangle below: CHAPTER 4

i by sides **ii** by angles

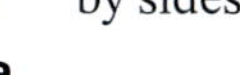

a

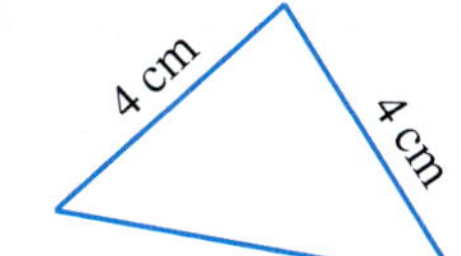

b

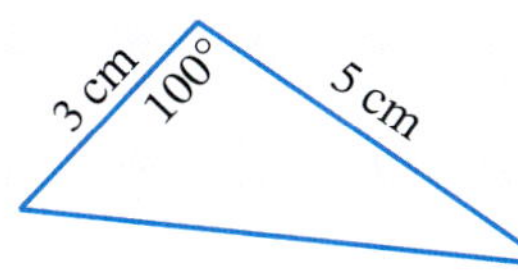

c

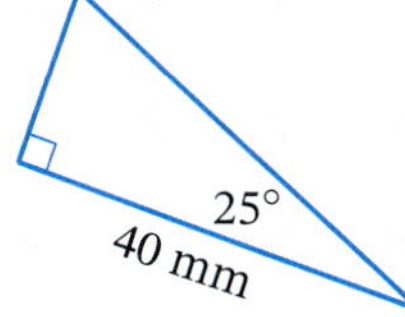

17 Use a factor tree to write 200 as a product of its prime factors. CHAPTER 2

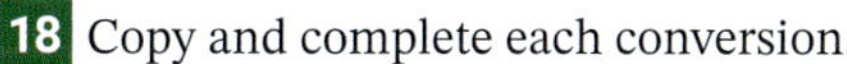

18 Copy and complete each conversion. CHAPTER 5

a $8 \text{ m}^2 = \_\_\_\_ \text{ cm}^2$ **b** $4900 \text{ m}^2 = \_\_\_\_ \text{ ha}$ **c** $650 \text{ m}^3 = \_\_\_\_ \text{ cm}^3$

19 Evaluate each expression. CHAPTER 6

a $\frac{2}{3}+\frac{3}{4}$ **b** $\frac{2}{3}-\frac{1}{5}$ **c** $\frac{7}{10}\times\frac{2}{3}$ **d** $\frac{4}{9}\div\frac{5}{6}$

e $1\frac{3}{8}+\frac{1}{4}$ **f** $4-\frac{5}{7}$ **g** $5\div\frac{4}{9}$ **h** $3\frac{2}{5}\times 1\frac{4}{5}$

20 Convert 15% into a simple fraction. CHAPTER 6

21 Solve each equation. CHAPTER 10

a $5a-13=12$ **b** $5d+6=2d-9$ **c** $4(2r+1)=28$

22 Plot the following points in a number space and state the solid shape formed by those points: (0, 0, 0), (0, 4, 0), (0, 4, 4), (0, 0, 4) and (2, 2, 2) CHAPTER 12

23 Test whether each triangle is right-angled. CHAPTER 1

a

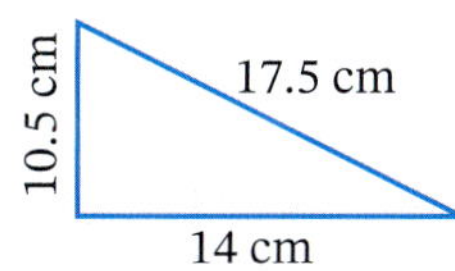

b

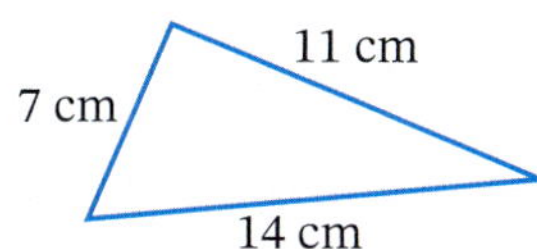

Foundation Standard Complex

24 These 2 triangles are congruent.

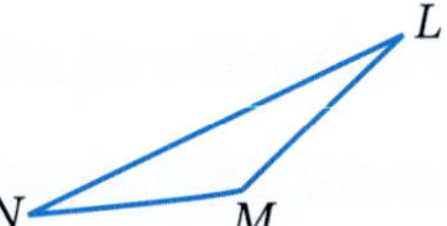

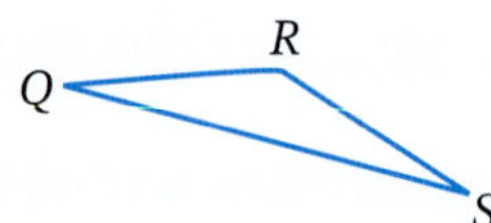

- **a** Which side in $\triangle QRS$ matches side NM?
- **b** Which angle in $\triangle LMN$ matches $\angle S$?
- **c** Copy and complete: $\triangle LMN \equiv$ ______

25 Gaurish earns \$22.40 per hour. How much will he earn if he works 42 hours in a week?

26 Find the formula for each table of values.

a

x	4	5	6	7	8
y	16	19	22	25	28

b

x	3	3	3	3	3
y	–1	2	4	–2	0

27 When catering for events, *Dine with Us* allows 1.2 L of drink per person.

- **a** If each glass contains 150 mL, how many glasses of drink is this per person?
- **b** *Dine with Us* are catering at an exhibition where 500 guests are expected. How many litres of drink are required?

28 Graph each linear equation on the same number plane.

- **a** $y = 3$
- **b** $x = -1$

29 Expand and simplify each expression.

- **a** $3(a + 5)$
- **b** $4(5 - 2k)$
- **c** $3(2p - 5) - 6p$

30 Simplify each ratio.

- **a** 42 : 49
- **b** 16 kg : 800 g
- **c** 10 mm : 1 m

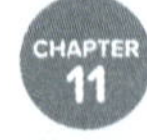

31 A machine makes transistors, of which 2 in 35 are faulty. If the machine has made 16 faulty transistors, how many good ones would you expect?

32 Find the width of a TV screen if its diagonal is 125 cm and its height is 62 cm.

33 What time will it be:

- **a** 4 hours 25 minutes after 2:15 p.m.?
- **b** 6 hours 40 minutes before 8:35 a.m.?

34
- **a** Graph $y = 2x - 1$ on a number plane.
- **b** Test whether (5, 9) lies on the line.

35 This graph shows the cost for a job completed by *Crafty Carpenters*.

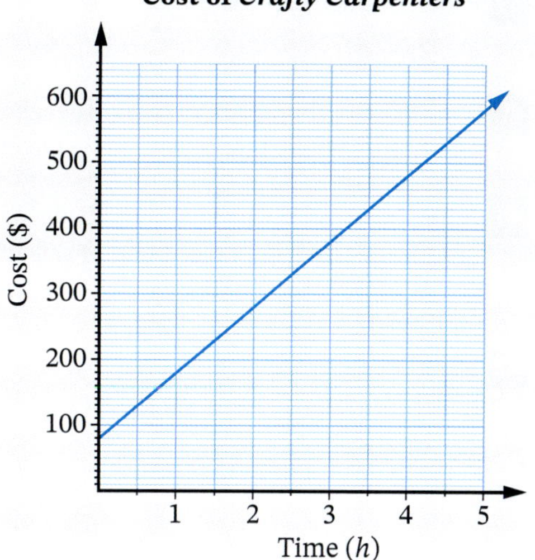

a How much was the callout fee for the carpenter?

b How much would it cost for the carpenter to replace a countertop if the job took 4.5 hours?

c The formula for the graph is $C = 100t + 80$, where C is the cost in dollars and t is the time is hours. If the carpenter charged \$880, for how long did she work?

36 A department store buys jeans for \$60 and sells them for \$96. Find:

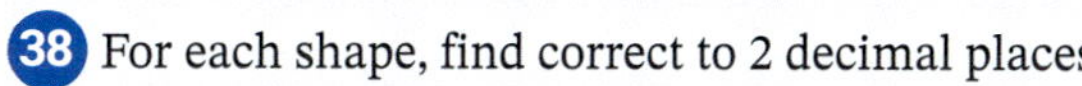

a the store's profit.

b the percentage profit.

37 Name the shape that matches each description.

a A quadrilateral with equal diagonals.

b A triangle with 3 equal sides.

c A quadrilateral with opposite sides equal and 2 axes of symmetry.

d A triangle with 2 equal angles.

38 For each shape, find correct to 2 decimal places:

i the perimeter **ii** the area

a

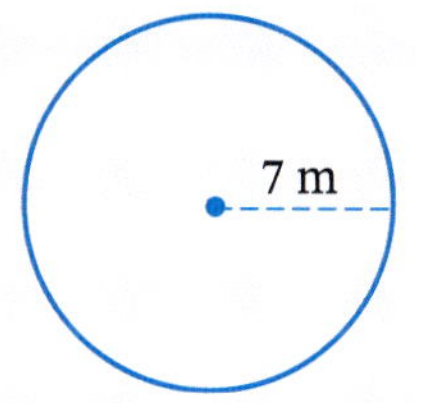

b

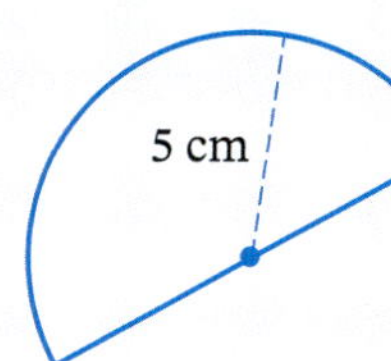

c

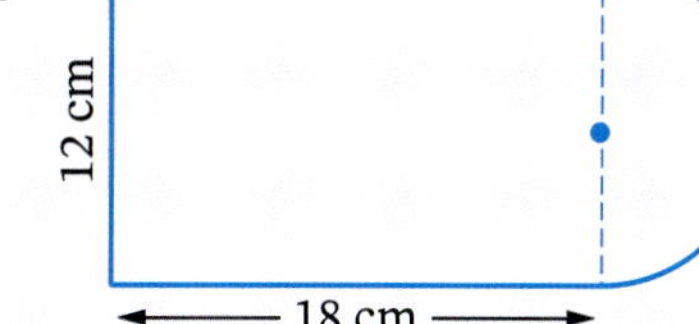

d

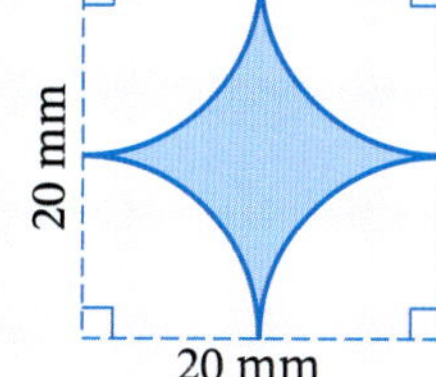

☐ Foundation ○ Standard ○ Complex

39 Factorise each expression.

a $2y + 8$ **b** $5a - 15ab$ **c** $-8ab + 2b$

40 A savings account pays 4% interest each year. If Haroon puts $5000 in his account, how much does he get interest after one year?

CHAPTER 9

41 A bag has 7 yellow badges, 3 red badges and 4 blue badges. If one badge is selected at random, what is the probability of selecting a:

a yellow badge? **b** blue badge? **c** badge that is not red?

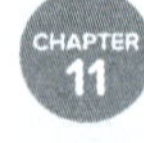

42 In an orchard, the ratio of orange trees to lemon trees is 7 : 4. If there are 2800 lemon trees, how many orange trees are there?

43 A ramp for a shopping centre is built as shown. Find the length of the ramp, correct to two decimal places.

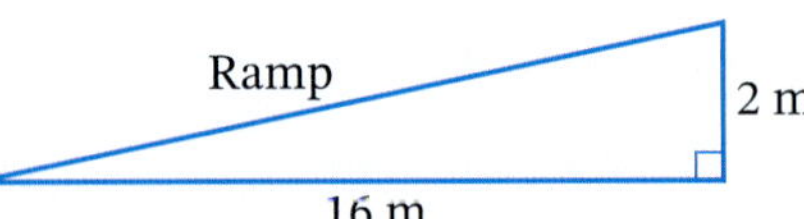

44 Luis uses 8 litres of paint to cover 32 square metres of wall. What area can he cover with 20 litres of paint?

45 The ages of patients waiting at a doctor's surgery are shown on this stem-and-leaf plot.

Stem	Leaf
1	3 8
2	1 9
3	5
4	2
5	0 4 7
6	3 5
7	

Find:

a the range.

b the median.

c the mean, correct to one decimal place.

46 **a** Draw a rectangle and its diagonals.

b Shade 2 pairs of congruent triangles.

c What does this show about the diagonals of a rectangle?

47 Each shape is made up of 2 congruent triangles combined. For each shape, state which test proves that they are congruent.

a

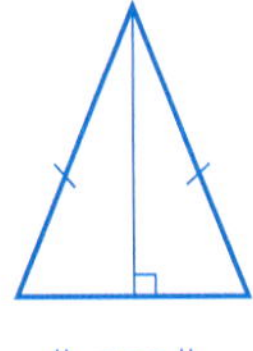

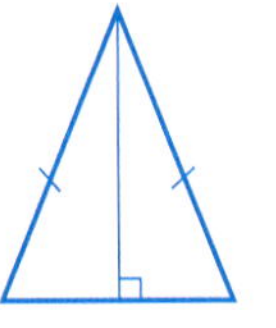

b

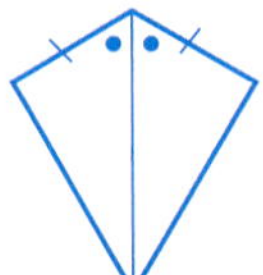

c

d

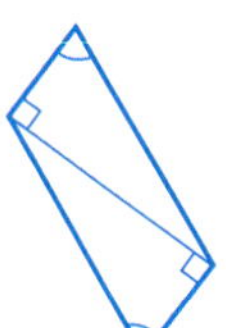

Foundation Standard Complex

48 The scale of a plan is 1 : 2000. What distance, in centimetres, on the plan would represent a real distance of 50 m? CHAPTER 11

49 Which is the best buy for strawberries? Select the correct answer **A**, **B** or **C**. CHAPTER 11

A 600 g for \$8.60 **B** 500 g for \$7.00 **C** 1 kg for \$15

50 Find the volume of each solid. CHAPTER 5

a

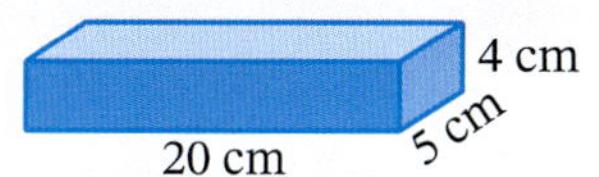

b

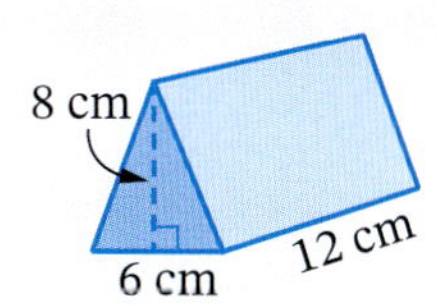

c

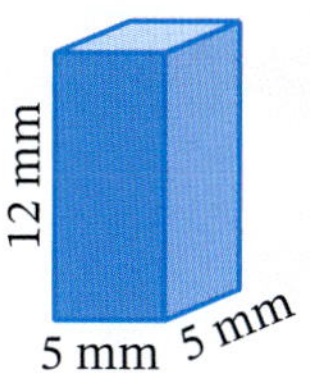

d

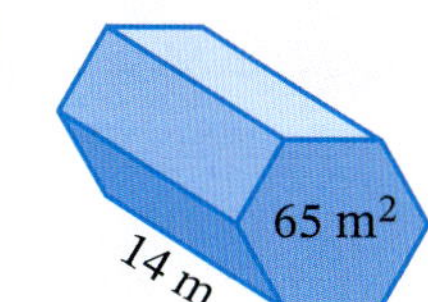

e

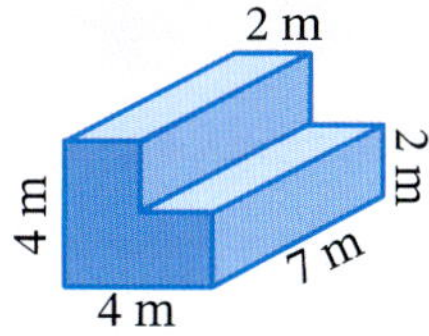

51 Anne earned \$42 for delivering newspapers. If she was given a 20% pay rise, how much does she earn now?

52 **a** Complete a frequency table for this data, including an '*fx*' column.

53	50	50	52	47	49	49	50	48	50
52	49	47	51	51	52	50	53	50	52
50	48	52	49	50	49	51	45	52	50
51	52	51	50	52	50	51	52	49	51

b Find the mean of this data.

c Show this data in a frequency histogram and polygon.

53 A survey was held to find whether a sample of 30 Australians had travelled to Japan or Vietnam. The results are shown below.

Japan 9 Vietnam 18 Japan and Vietnam 5

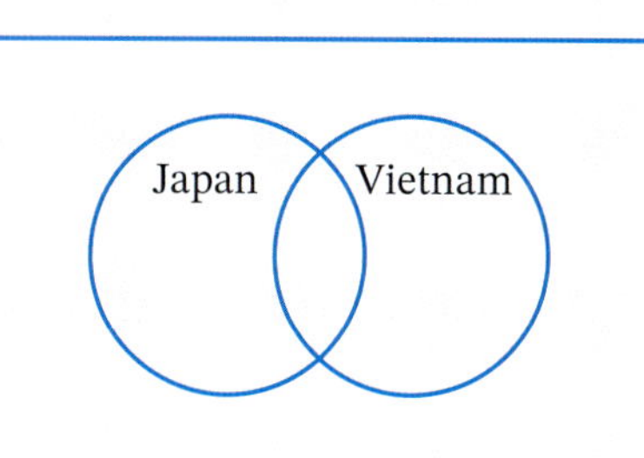

a Copy and complete the Venn diagram to show this information.

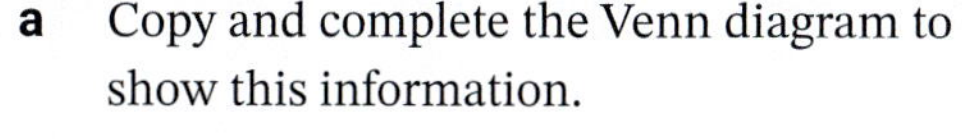
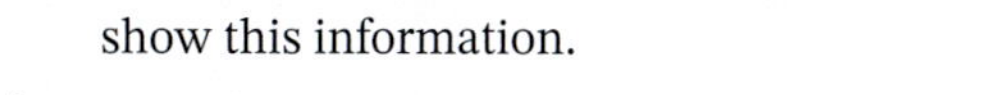

b How many people did not visit either country?

c How many people had visited Japan but not Vietnam?

d What is the probability that a person selected randomly from this sample has not been to Japan?

Foundation Standard Complex

54 Adrian's fish tank is a rectangular prism 60 cm by 25 cm and filled to a depth of 13 cm. Find the volume of water in the tank in:

a cm^3 **b** litres

55 If the perimeter of this isosceles triangle is 24 cm, find the value of x.

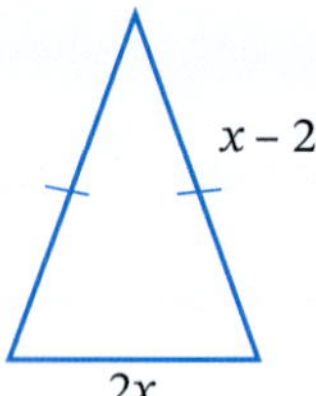

56 If Omar can run 400 metres in 66 seconds, calculate his speed in m/s, correct to one decimal place.

57 What is the whole amount if:

a 35% of it is \$210? **b** 68% of it is 306 kg?

58 Mel went on a bike ride and her trip is shown in the travel graph.

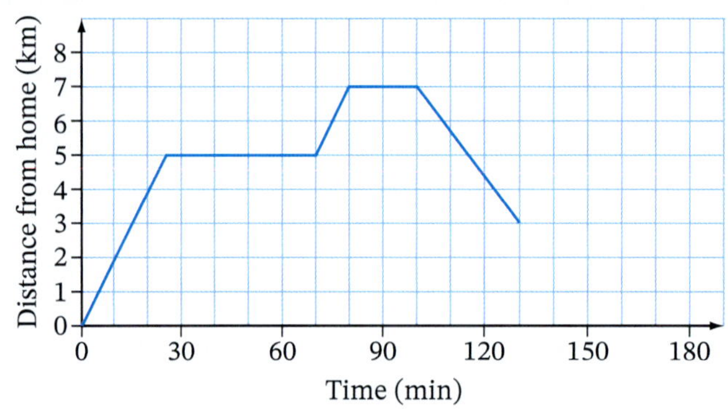

a How far from home was Mel after 20 minutes?

b When and where she was furthest from home?

c For how long did Mel rest on her second stop?

d When Mel was travelling home, what was her average speed in km/h?

59 Solve the linear equation $4 - 2x = 6$ graphically.

60 Graph the inequality $y < 2x - 4$ on a number plane.

61 Izak paid \$88 for a shirt at a '20% off' sale. What was the original price of the shirt?

☐ Foundation ○ Standard ⬡ Complex

Calculator skills

Basic keys

Key	Use	Key	Use
+, −, ×, ÷	Basic operations	(−) or +/−	Enters negative numbers
=	Equals sign, gives the answer	▭/▭ or a b/c	Enters fractions
·	Decimal point	( )	Enters parentheses (brackets)
DEL	Delete previous entry	x^2	Squares a number
Ans	Retrieves previous answer	√	Finds the square root of a number
← ↑ → ↓	Moves cursor around the screen	x^3	Cubes a number
MODE or SHIFT or 2ndF	Accesses other operations	∛	Finds the cube root of a number

1 Basic operations

Example 1

PERFORMING A CALCULATION

Question	Calculator steps	Answer
34 + 6 − 16	34 + 6 − 16 =	24
7.4 × 9.1	7.4 × 9.1 =	67.34
234 ÷ 1.5	234 ÷ 1.5 =	156
4569 ÷ (1.7 + 4.3)	4569 ÷ (1.7 + 4.3) =	761.5

Example 2

CORRECTING A WRONG ENTRY

Calculating 34 + 6 − 16

a Wrong entry: 34 + 8 — To delete the 8 and continue, enter DEL 6 − 16 =

b Wrong entry: 34 + 6 + 16 = — To correct this, press ← repeatedly to go back to the second +, press DEL and then −, then press =

EXERCISE 1 ANSWERS ON P. 668

Basic operations

1 Calculate each expression.

- **a** 362×14
- **b** $26.3 - 11.8$
- **c** $810 \div 18$
- **d** 5.3×1.08
- **e** $51.1 \div 14.6$
- **f** $19.3 + 8.8 - 1.2$
- **g** $2.4 \div 0.2 + 11$
- **h** $5.6 \times (0.1 + 0.2)$
- **i** $(56 - 24) \div (1.2 - 1)$
- **j** $12.6 \times 3.8 - 18.4$
- **k** $(16.07 + 9.02) - (38.3 - 5.126)$
- **l** $(24.8 + 3.06 + 9.4) \div 3$

2 Bronte purchased a loaf of bread for $2.75, a packet of cornflakes for $4.55 and 3 kg of tomatoes at $3.95 per kg.

- **a** Find the total cost.
- **b** Find the change Bronte will receive from a $20 note.

3 Mark started the month with $1650 in his bank account. He made deposits of $380 and $725.50 and withdrew $880. How much was left in Mark's account?

4 An unloaded truck has a mass of 6550 kg. It is loaded with 8 boxes, each weighing 124 kg. What is the total mass of the truck and the boxes?

5 Salma needs to cut a 9.6 m length of rope into 4 equal pieces. Find the length of each piece.

6 Hamish posted 25 Christmas cards to his friends. If the stamps were 75 c each, how much did he spend on stamps?

7 An orchard has 1288 trees arranged in 56 rows. How many trees are in each row?

8 A train takes 15 hours to travel 1128 km. What is its average speed in km/h?

9 Chloe bought 60 L of petrol at 142.9 c/L. How much did she pay for the petrol?

10 Nelson College has the following numbers of students in each year level.

Year 7	Year 8	Year 9	Year 10	Year 11	Year 12
122	128	123	131	117	94

The whole school is going on an excursion. If each bus can carry 55 people, how many buses will be needed?

11 The manager of a sports store purchased 32 tennis racquets for $79.60 each and sold them for $120.80 each.

- **a** Find the profit made on each racquet.
- **b** Find the total profit.

12 A batter scored the following runs over 5 cricket innings.

27 4 123 66 47

Find the batter's average score.

9780170465601

2 Integers

Example 3

Question	Calculator steps		Answer
	Casio	Sharp	
-4 + 9	(–) 4 + 9 =	+/– 4 + 9 =	5
17 × (-2)	17 × (–) 2 =	17 × +/– 2 =	-34
18 -(-3) × 4	18 – (–) 3 × 4 =	18 – +/– 3 × 4 =	30

EXERCISE 2 ANSWERS ON P. 668

Integers

1 Calculate each expression.

a $-5+12$ **b** $-6-3$ **c** -3×4

d $16\div(-2)$ **e** $12-(-3)$ **f** $-19+4$

g $-25\times(-3)$ **h** $-55\div(-11)$ **i** $8-12+(-1)$

j $-3+7-2-(-5)$ **k** $6\times(-2)\times 8$ **l** $-18\div(-2)\times 5$

m $-3\times 7-4$ **n** $(-3+9)\times 4$ **o** $80-(-2)\times 8$

p $12\times(-3)+50$ **q** $64\div(-8)\times 2$ **r** $27-(-3)\times(-4)$

s $18\times(-5+3)$ **t** $-14-3\times(-8)$

3 Powers and roots

Look for these keys on your calculator: x^2 x^3 $\sqrt{\ }$ $\sqrt[3]{\ }$

You may need to use the SHIFT or key.

Example 4

Question	Calculator steps	Answer
16^2	16 x^2 =	256
11^3	11 x^3 =	1331
$\sqrt{196}$	$\sqrt{\ }$ 196 =	14
$\sqrt[3]{4913}$	$\sqrt[3]{\ }$ 4913 =	17

EXERCISE 3 ANSWERS ON P. 668

Powers and roots

1 Calculate each expression.

a 14^2 b 23^2 c 6^3 d 1.2^3

e 5.13^2 f 101^2 g 2.04^2 h 19^3

i $\sqrt{676}$ j $\sqrt{21904}$ k $\sqrt[3]{729}$ l $\sqrt{6^2 \times 7^2}$

2 Evaluate each expression, correct to one decimal place.

a $\sqrt{24}$ b $\sqrt{101}$ c $\sqrt{11 \times 2}$ d $\sqrt[3]{100}$

3 Calculate each expression.

a $5^2 - 3^2$ b $16^2 - 10^2$ c $\sqrt{6} \times \sqrt{6}$ d $\sqrt{25 \times 4}$

4 Fractions

Casio calculators have two ways of entering fractions: MATH mode or LINE mode. MATH mode allows you to enter the numerator and denominator into two blank spaces on the calculator's screen, while LINE mode makes the fraction key act like a vinculum (fraction bar). For MATH mode, use the arrow keys to move the cursor to the spaces to enter the numerator and denominator. Press SHIFT MODE to change between MATH I/O and LINE I/O (input/output). To ensure that mixed numeral answers are not converted to improper fractions, press SHIFT MODE and the down arrow to choose a b/c.

Example 5

CONVERTING FRACTIONS

Question	Calculator steps		Answer
	Casio LINE mode	Sharp	
Simplify $\frac{16}{20}$	16 [fraction] 20 [=]	16 [a b/c] 20 [=]	$\frac{4}{5}$
Change $1\frac{3}{4}$ to an improper fraction	1 [fraction] 3 [fraction] 4 [=] [SHIFT] [S⇔D]	1 [a b/c] 3 [a b/c] 4 [=] [2ndF] [a b/c]	$\frac{7}{4}$
Change $\frac{19}{4}$ to a mixed numeral	19 [fraction] 4 [=]	19 [a b/c] 4 [=]	$4\frac{3}{4}$
Change $\frac{3}{4}$ to a decimal	3 [fraction] 4 [=] [S⇔D] or 3 [÷] 4 [=]	3 [a b/c] 4 [=] [a b/c] or 3 [÷] 4 [=]	0.75
Change 1.7 to a fraction	1.7 [=]	1.7 [=] [a b/c]	$1\frac{7}{10}$

S⇔D converts an answer between standard fraction and decimal forms.

SHIFT S⇔D or 2ndF a b/c converts an answer between mixed numeral and improper fraction forms.

Example 6

OPERATIONS WITH FRACTIONS

Question	Calculator steps		Answer
	Casio LINE mode	Sharp	
$\frac{1}{2}+\frac{1}{3}$	1 [fraction] 2 [+] 1 [fraction] 3 [=]	1 [a b/c] 2 [+] 1 [a b/c] 3 [=]	$\frac{5}{6}$
$\frac{5}{7}-\frac{2}{3}$	5 [fraction] 7 [−] 2 [fraction] 3 [=]	5 [a b/c] 7 [−] 2 [a b/c] 3 [=]	$\frac{1}{21}$
$\frac{3}{5}\times\frac{2}{7}$	3 [fraction] 5 [×] 2 [fraction] 7 [=]	3 [a b/c] 5 [×] 2 [a b/c] 7 [=]	$\frac{6}{35}$
$\frac{4}{5}\div\frac{2}{3}$	4 [fraction] 5 [÷] 2 [fraction] 3 [=]	4 [a b/c] 5 [÷] 2 [a b/c] 3 [=]	$1\frac{1}{5}$

Example 7

MIXED OPERATIONS WITH FRACTIONS

Question	Calculator steps	Answer
$\frac{13+7}{9\times 2}$	[(] 13 [+] 7 [)] [÷] [(] 9 [×] 2 [)] [=] or [(] 13 [+] 7 [)] [fraction] (or [a b/c]) [(] 9 [×] 2 [)] [=]	1.1111... $1\frac{1}{9}$

EXERCISE 4 ANSWERS ON P. 669

Fractions

1 Simplify each fraction.

a $\frac{20}{32}$ **b** $\frac{40}{85}$ **c** $\frac{45}{72}$

d $\frac{112}{126}$ **e** $\frac{84}{100}$ **f** $\frac{48}{192}$

2 Convert each mixed numeral into an improper fraction.

a $1\frac{7}{8}$ **b** $5\frac{2}{9}$ **c** $10\frac{3}{4}$

3 Convert each improper fraction into a mixed numeral.

a $\frac{20}{9}$ **b** $\frac{44}{5}$ **c** $\frac{18}{8}$

4 Convert each fraction into a decimal.

a $\frac{1}{2}$ **b** $\frac{4}{5}$ **c** $\frac{7}{8}$

d $\frac{9}{25}$ **e** $\frac{34}{200}$ **f** $\frac{5}{9}$

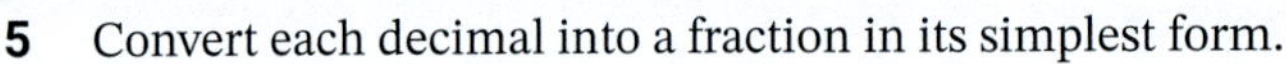

5 Convert each decimal into a fraction in its simplest form.

a 0.35 **b** 2.04 **c** 0.001

d 0.365 **e** 1.06 **f** 0.325

6 Simplify each expression.

a $\frac{3}{5}+\frac{1}{4}$ **b** $\frac{4}{7}+\frac{1}{2}$ **c** $\frac{2}{3}-\frac{3}{5}$ **d** $\frac{1}{3}\times\frac{4}{7}$

e $\frac{5}{7}\div\frac{15}{28}$ **f** $\frac{5}{6}-\frac{5}{8}$ **g** $\frac{9}{10}+\frac{2}{3}$ **h** $\frac{7}{8}\times\frac{5}{12}$

i $\frac{7}{8}\div\frac{1}{2}$ **j** $24\div1\frac{1}{4}$ **k** $5\frac{1}{4}+1\frac{2}{5}$ **l** $9\frac{7}{8}-1\frac{1}{3}$

m $5\frac{1}{4}\times2\frac{3}{7}$ **n** $4\frac{2}{3}\div\frac{2}{3}$ **o** $\frac{8}{9}\div6$ **p** $10\frac{1}{6}\times1\frac{9}{10}$

7 Simplify each expression.

a $\frac{12-4\times2}{18+14}$ **b** $\frac{6^2}{100\div4+23}$ **c** $\frac{9\times9-2}{\sqrt{30-5}}$ **d** $\frac{20\div2+8}{3\times9\times4}$

8 Mr Pettis is at his office for $7\frac{1}{2}$ hours each day. He takes $\frac{3}{4}$ of an hour for lunch each day. How long does he work:

a in 1 day? **b** in 5 days?

Answers

CHAPTER 1

SkillCheck

1 a 16 b 144 c 900 d 4.41
e 106.09 f 25 g 169 h 18.92

2 a 4 b 8 c 10 d 14
e 11 f 7 g 1.1 h 1.5

3 a 25 cm b 28 cm c 76 mm

4 a 40 cm^2 b 16 cm^2 c 100 mm^2

5 a 10.3 b 4.7 c 7.7 d 0.9

6 a 7.6 b 10.8 c 12.7

Exercise 1.01

1 a 14 b 30 c 8
d 25 e 11 f 19

2 a 3.46 b 6.71 c 31.64
d 18.03 e 12.37 f 14.39
g 9.90 h 29.80 i 4.90
j 10.49 k 17.23 l 26.25

3 a 10.91 b 8.60 c 12.53
d 12.04 e 7.42 f 9.85
g 4.46 h 10.29 i 31.49

4 A

5 $\sqrt{98}$, $\sqrt{160}$, $\sqrt{52}$, $\sqrt{77}$, $\sqrt{18}$, $\sqrt{200}$

Exercise 1.02

1 a p b z c LN

2 b 13 cm c 169 d 169
e 13, 169, 5, 12, 169

3 B

4 b

	Shorter sides		Hypotenuse	a^2	b^2	$a^2 + b^2$	c^2
	a	b	c				
a	6	8	10	36	64	36 + 64	100
b	2.5	6	6.5	6.25	36	6.25 + 36	42.25
c	1.5	2	2.5	2.25	4	2.25 + 4	6.25
d	4	7.5	8.5	16	56.25	16 + 56.25	72.25

5 Teacher to check.

Exercise 1.03

1 D

2 a $p^2 = m^2 + n^2$ b $t^2 = r^2 + s^2$
c $LK^2 = JL^2 + JK^2$ d $5^2 = d^2 + 3^2$
e $12.5^2 = 10^2 + 7.5^2$ f $x^2 = 3^2 + 3^2$

3 a $b = 10$ b $d = 25$ c $x = 15$

4 a $m = \sqrt{865}$ b $b = \sqrt{52}$ c $x = \sqrt{569}$
d $p = \sqrt{2}$ e $d = \sqrt{19.4}$

5 a $m = 29.4$ b $b = 7.2$ c $x = 23.9$
d $p = 1.4$ e $d = 4.4$

6 D

7 a $b = 7.81$ b $p = 7.94$ c $z = 3.02$

8 12.5 cm

9 a 40 b 39 c 41
d 28 e 11 f 8
g 45 h 6, 8 i 9, 12

10 a 1700 m b 600 m
c Iluka might have been needed water from the waterhole, or might have met someone there, or there may have been terrain or obstacles directly between the campsite and meeting place preventing him from going that way. Other answers are possible.

Exercise 1.04

1 a $x = 12$ b $y = 16$ c $y = 8$
d $a = 15$ e $d = 18$ f $x = 12$

2 a $x = \sqrt{8}$ b $y = \sqrt{84}$ c $m = \sqrt{13\,321}$
d $y = \sqrt{13}$ e $x = \sqrt{94\,176}$ f $x = \sqrt{128}$

3 a $x = 2.8$ b $y = 9.2$ c $m = 115.4$
d $y = 3.6$ e $x = 306.9$ f $x = 11.3$

4 B

5 a $k = 10.39$ b $q = 25.16$ c $a = 28.28$

6 30 cm 7 500 m

8 Teacher to check.

Mental skills 1

2 a 625 b 3025 c 2025 d 7225
e 13 225 f 56.25 g 9025 h 38 025
i 2.25 j 4225 k 24 025 l 60 025

4 a 441 b 10 201 c 961 d 8281
e 26.01 f 6561 g 3721 h 40 401
i 1.21 j 16.81

Exercise 1.05

1 a H b S c H
d S e H f S

2 a $f = 85$ b $b = 4$ c $s = 55$
d $r = 4.5$ e $l = 29$ f $k = 38$

3 a $g = \sqrt{130}$ b $t = \sqrt{160}$ c $x = \sqrt{735}$

4 a $q = 62.2$ b $a = 21.0$ c $p = 4.7$

5 B

6 a $i = 12.59$ b $y = 27.97$ c $v = 54.38$

7 a $d = \sqrt{74}$ b $x = \sqrt{60}$ c $x = \sqrt{288}$
d $p = \sqrt{85}$ e $x = \sqrt{164}$ f $p = \sqrt{560}$
g $x = \sqrt{16.65}$ h $x = \sqrt{624}$ i $y = \sqrt{786.25}$

Exercise 1.06

1 a right-angled b right-angled
c not right-angled d right-angled
e not right-angled f right-angled
g not right-angled h right-angled
i not right-angled j right-angled
k not right-angled l right-angled
Teacher to check diagrams.

2 a

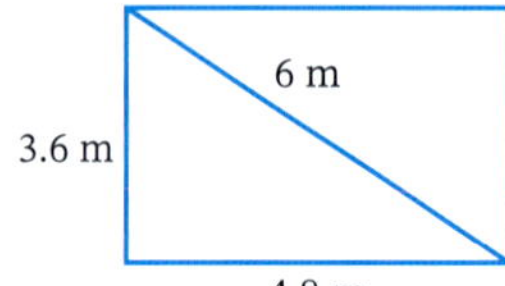

b Yes, $6^2 = 36$ and $4.8^2 + 3.6^2 = 36$

Exercise 1.07

1 **a**, **b**, **c**, **e**, **g**, **h** are Pythagorean triads.

2 D **3–4** Teacher to check.

5 a 12, 13
b $a = 5, b = 12, c = 13$
$a^2 + b^2 = 5^2 + 12^2 = 25 + 144 = 169$
$c^2 = 13^2 = 169$
hence $a^2 + b^2 = c^2$, so it is a Pythagorean triad.

6 a 7, 24, 25 b 11, 60, 61
c 15, 112, 113 d 4, 7.5, 8.5
e 9, 40, 41 f 19, 180, 181
g 10, 49.5, 50.5 h 51, 1300, 1301

Exercise 1.08

1 3.61 m **2** 92.2 km **3** B

4 $\sqrt{72}$ **5** 9.7 m **6** 68.7 cm

7 a 48 cm b 31.38 cm c 170 cm
d 12.92 m e 28.94 m

8 $h = 10.39$ cm **9** 6.32 m **10** 16 m

11 a 192 cm^2 b 360 mm^2 c 84 mm^2
d 22 m^2 e 123.0 cm^2

12 38 m **13** 23 cm

14 4.58 m or 458 cm

Power plus

1 a 22.6 b 10.6 c 27.7
d 5.4 e 10.3 f 2.8

2 a 16.2 m b 16.9 m

3 a 21.2 cm b 26.0 cm

4 18.4 cm

Test yourself 1

1 a 14.25 b 16.16 c 5.74 d 6.21

2 $\sqrt{104}, \sqrt{96}, \sqrt{12}, \sqrt{45}, \sqrt{88}$

3 a $BC, BC^2 = AB^2 + AC^2$
b $d, d^2 = e^2 + f^2$
c $s, s^2 = h^2 + k^2$

4 a $c = \sqrt{41}$ b $p = \sqrt{1325}$ c $x = \sqrt{914}$

5 a $m = 0.99$ b $f = 10.58$ c $r = 4.87$

6 a $z = 145.0$ b $y = 7.0$ c $v = 42.6$

7 a right-angled b right-angled

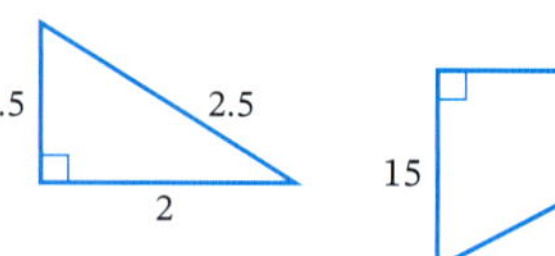

c not right-angled

8 **a**, **c**, **d** are Pythagorean triads.

9 206 mm

10 a 9.5 m b 152.2 cm

CHAPTER 2

SkillCheck

1 a Missing numbers are −6, −4, −3, −1, 1, 2, 4, 6, 7, 9
b Missing numbers are −35, −30, −25, −15, −10, 5

2 −7, −5, −2, 0, 1, 6, 7, 11

3 12, 8, 5, 4, 2, −1, −3, −9

4 a 7 b 8 c 1

5 a $\frac{3}{100}$ b $\frac{7}{1000}$ c $\frac{49}{100}$ d $\frac{9}{10}$

6 a 3 b 1 c 2 d 5

7 a 1, 3, 5, 15 b 15, 30, 45, 60, 75

8 a prime b prime
c composite d composite

Exercise 2.01

1 a 331 b 157 c 1587 d 255
e 421 f 413 g 368 h 734
i 612 j 35 k 276 l 47

2 a 119 b 112 c 233
d 95 e 250 f 242

3 A

4 a 36 b 54 c 24 d 62
e 28 f 43 g 81 h 146
i 36 j 18 k 68 l 14
m 43 n 172 o 26 p 99

5 a 233 b 1375 c 85

6 **a** 24 **b** 16 **c** 35 **d** 54
e 36 **f** 40 **g** 64 **h** 21
i 24 **j** 36 **k** 27 **l** 28

7 **a** 7 **b** 5 **c** 4 **d** 9
e 9 **f** 4 **g** 3 **h** 9
i 4 **j** 9 **k** 3 **l** 5

8 **a** 800 **b** 90 **c** 700 **d** 160
e 100 **f** 1000 **g** 1500 **h** 240
i 180 **j** 320 **k** 1700 **l** 1400

9 **a** 160 **b** 3500 **c** 18 000 **d** 300
e 210 **f** 1800 **g** 36 000 **h** 2400
i 30 **j** 6 **k** 5 **l** 8
m 20 **n** 100 **o** 7 **p** 3

10 C

11 **a** 170 **b** 78 **c** 228 **d** 112
e 128 **f** 264 **g** 324 **h** 116
i 54 **j** 122 **k** 109 **l** 26

12 **a** 280 **b** 400 **c** 132 **d** 260
e 210 **f** 63 **g** 192 **h** 324

13 **a** 110 **b** 70 **c** 1400 **d** 540
e 1100 **f** 2600 **g** 16 **h** 18
i 8 **j** 40 **k** 82 **l** 14

14 1140 words

15 **a** 306 **b** 459 **c** 216 **d** 576
e 396 **f** 451 **g** 768 **h** 264

16 **a** 330 **b** 182 **c** 88 **d** 396
e 45 **f** 54 **g** 26 **h** 27

17 C **18** $34

Exercise 2.02

1 **a** −2 **b** 5 **c** 2 **d** 6
e 4 **f** −3 **g** 4 **h** 0
i 6 **j** 3 **k** 2 **l** 0
m −5 **n** −3 **o** −1 **p** −8

2 **a** −2 **b** −6 **c** −7 **d** −4
e −8 **f** −10 **g** −6 **h** −12
i 12 **j** 5 **k** 12 **l** 18
m −1 **n** 3 **o** 0 **p** 1

3 9°C **4** C

5 **a** −2 **b** 0 **c** −11 **d** 4
e −13 **f** 6 **g** −1 **h** 2
i −2 **j** −7 **k** −15 **l** −5

6 B

7 **a** $5 + (-9) = -4$ **b** $-2 + 3 = 1$
c $-4 + (-5) = -9$ **d** $-4 + (-7) = -11$
e $3 - 6 = -3$ **f** $-6 - 1 = -7$
g $-2 - (-7) = 5$ **h** $-4 - 2 = -6$
i $-8 - (-8) = 0$

8 Jackson Creek **9** 30 cm

10 Teacher to check.

11 **a** True **b** False

12 **a** 0 **b** $2a$ **c** $2a$ **d** 0

13 **a** + **b** E **c** + **d** −
e E **f** − **g** + **h** E

Exercise 2.03

1 **a** −20 **b** −18 **c** 32 **d** −45
e −70 **f** 30 **g** 9 **h** −24
i −21 **j** 100 **k** −36 **l** 40
m −44 **n** −60 **o** 36 **p** 63
q −55 **r** 81 **s** −100 **t** 42

2 **a** C **b** D **c** B

3 **a**

×	−3	5
−3	9	−15
5	−15	25

b

×	−5	5
5	−25	25
−5	25	−25

c

×	6	−4
−4	−24	16
6	36	−24

d

×	−1	1
−1	1	−1
1	−1	1

e

×	7	−3
4	28	−12
−6	−42	18

f

×	−3	−6
8	−24	−48
2	−6	−12

4 **a** (−2) **b** (−4) **c** 5
d −7 **e** −7 **f** −7

5 **a** 28 **b** −90 **c** −40
d 75 **e** 20 **f** −8

6 B

7 **a** + **b** + **c** − **d** + **e** +

8 True

9 You get the negative of the integer squared.

10 Teacher to check.

11 **a** a^2 **b** a^2

Exercise 2.04

1 **a** −9 **b** 5 **c** −7 **d** −6
e 5 **f** −7 **g** −5 **h** −2
i 6 **j** −8 **k** −8 **l** 7
m −5 **n** 20 **o** −9 **p** −6

2 **a** A **b** B

3 **a**

÷	−6	−3	3	6
−6	1	$\frac{1}{2}$	$-\frac{1}{2}$	−1
−3	2	1	−1	−2

b

÷	8	−8	32	−32
−2	−4	4	−16	16
4	2	−2	8	−8

4 a −9 b −4 c 6 d −7
e 4 f −4 g 9 h −7

5 a −2 b 3 c −2
d 30 e −5 f −1
g −20 h −2 i −70

6 a (−6) b (−5) c 9
d 32 e −24 f 100

7 D

8 a + b + c −

9 False

10 a The answer is 1. b The answer is −1.

11 a 1 b −1

12 Teacher to check.

Mental skills 2A

2 a 3.5 b 2.4 c 0.12 d 0.36
e 0.8 f 0.027 g 0.2 h 8.8
i 0.24 j 0.012 k 1.8 l 0.028

4 a 66.3 b 6630 c 6.63 d 0.663
e 6.63 f 663 g 0.663 h 663
i 6630 j 66.3 k 0.663 l 0.0663

Exercise 2.05

1 a 5 b 30 c 22 d −32
e 26 f 8 g −12 h 23
i −1 j −5 k −19 l 0

2 a −2 b 1 c 31 d 10
e 28 f 3 g −42 h −8
i 16 j 19 k 24 l 22

3 a $\frac{1}{2}$ b 2 c $\frac{9}{10}$ d −6
e −2 f $\frac{1}{4}$ g 1 h 3

4 B

5 a She has not followed the correct order of operations: need to multiply first, before adding.
b Ms Ferme should explain the correct order of operations.
c (−9 + 13) × 4

6 a 8 b 5 c 3
d (−2) e (−1) f −20
g (−6) h 6 i 20

7 a 8 − (3 + 7) = −2
b (40 − 10) × 5 = 150
c 27 ÷ (9 ÷ 3) = 9
d 8 + (4 − 3) × 2 = 10
e [8 + (4 − 3)] × 2 = 18
f (6 + 4) × (−1) − 1 = −11
g (13 + 3) ÷ (4 − 6) = −8
h 100 ÷ (10 + 10) + 5 = 10
i 100 ÷ (10 + 10 + 5) = 4

8 Teacher to check.

Exercise 2.06

1 B

2 a 4.035, 4.05, 4.3, 4.35
b 12.173, 17.12, 17.2103, 17.231
c 14.08, 14.8, 14.801, 14.81
d 0.032, 0.2, 0.295, 0.3

3 a 1.621, 1.612, 1.61, 1.16
b 3.305, 3.053, 3.05, 3.035

4 a 3 b 2 c 1 d 0
e 3 f 3 g 1 h 2

5 a 3.9 b 4.1 c 0.3 d 7.3
e 15.1 f 3.0 g 2.0 h 16.2

6 C

7 a 68.91 b 107.06 c 3.60 d 4.71
e 3.20 f 33.00 g 19.73 h 11.25

8 a 9.704 b 13.168 c 0.083 d 53.094

9 a \$460.40 b \$460 c \$460.4
d \$460.40 e \$460.395

10 a 164.91 b 25.36 c 72.6 d 91.61
e 15.7 f 77.27 g 18.44 h 2.68
i 66.68 j 51.61 k 4.64 l 2.391

11 C 12 7.6°C 13 0.8 m

14 a \$10.05 b \$9.95

15 a 3.04 b 12.74 c 5.45 d 0.889
e 4.326 f Teacher to check.

Exercise 2.07

1 a 26 b 4.05 c 11.5 d 27.3
e 0.21 f 0.088 g 0.02 h 1.26
i 28.35 j 88.97 k 0.806 l 0.3672

2 B 3 \$40.16

4 a 6.2 b 16.3 c 2.61 d 0.139
e 316.5 f 69.55 g 370 h 58.05
i 0.58 j 5.2 k 170 l 1.2

5 B

6 a 43.44 m b 1.81 m

7 a 5.91 b 14 c 5.66

8 \$67.50

9 a 8 b 2 c 0.04
d 0.9 e 13.9 f 25.92

10 16.2 km 11 \$14.10

12 a 41.04 b 0.829 c 328.24
d 4.7292 e 2.51706

Exercise 2.08

1 a 0.4 b 0.375 c 0.75 d 1
e 0.5 f 0.75 g 0.6 h 0.25
i 0.625 j 0.875 k 0.5 l 1

2 They are equivalent fractions.

3 a $0.\dot{6}$ b $4.\dot{2}\dot{7}$ c $0.\dot{2}\dot{8}$ d $3.8\dot{3}$
e $9.\dot{6}0\dot{7}$ f $0.\dot{1}51\dot{9}$ g $0.0\dot{5}\dot{2}$ h $12.23\dot{4}$

4 a 0.858333... b 0.1656565...
c 0.272727... d 0.461538461538...

5 **a** $0.\dot{1}$ **b** $0.1\dot{6}$ **c** $0.8\dot{3}$
d $0.\dot{1}4285\dot{7}$ **e** $0.\dot{6}$ **f** $0.\dot{2}8571\dot{4}$
g $0.\dot{2}$ **h** $0.\dot{4}2857\dot{1}$ **i** $0.\dot{5}7142\dot{8}$
j $0.\dot{4}$ **k** $0.\dot{6}$ **l** $0.\dot{5}$
m $0.\dot{8}5714\dot{2}$ **n** $0.\dot{7}$ **o** $0.\dot{7}1428\dot{5}$

6 $0.\dot{1}4285\dot{7}$, $0.\dot{2}8571\dot{4}$, $0.\dot{4}2857\dot{1}$, $0.\dot{5}7142\dot{8}$, $0.\dot{7}1428\dot{5}$, $0.\dot{8}5714\dot{2}$. It's the same digits, in order, but starting with a different digit each time.

7 **a** $3.\dot{6}$ **b** $1.7\dot{1}$ **c** $531.\dot{1}\dot{8}$ **d** $3.2\dot{6}$

Mental skills 2B

2 **a** 0.5 **b** 90 **c** 8 **d** 0.9
e 90 **f** 0.7 **g** 8 **h** 30
i 0.08 **j** 3.5 **k** 4 **l** 0.3

4 **a** 16 **b** 160 **c** 1.6 **d** 1.6
e 1.6 **f** 16 **g** 160 **h** 0.16
i 0.016 **j** 160 **k** 0.16 **l** 0.0016

Exercise 2.09

1 **a** 7^3 **b** 4^5 **c** 5^{10} **d** $(-8)^7$

2 **a** 64 **b** 8 **c** −216
d 81 **e** 7 **f** −32
g 1000 **h** 256 **i** 1 771 561
j 18 954 **k** 972 **l** 2529
m 39 **n** 4094

3 **a** 3 **b** 3 **c** 2
d 6 **e** 3 **f** 5

4 B **5** 100^3 **6** 2^{10}

7 **a** 36
b 36
c Yes, they both equal 36.

8 **a** 400 **b** 400 **c** Yes

9 **a** $3^2 \times 8^2$
b $a^2 \times b^2$

10 **a** $6^2 \times 3^2 = 324$
b $2^2 \times 11^2 = 484$
c $(2 \times 8)^2 = 2^2 \times 8^2 = 256$
d $(3 \times 5)^2 = 3^2 \times 5^2 = 225$
e $(10 \times 3)^2 = 10^2 \times 3^2 = 900$
f $(4 \times 7)^2 = 4^2 \times 7^2 = 784$

11 **a** 28 **b** 16 **c** 17 **d** 33
e 2 **f** 7 **g** 13 **h** 6

12 **a** 6.1 **b** 9.7 **c** 22.4 **d** 1.9
e −7.9 **f** 44.7 **g** 1.0 **h** 10.3

14 D **15** D **16** 3 and 4

17 **a** 5.7 **b** 10.2 **c** 8.2

18 $\sqrt{6}$ $\sqrt{62}$ $\sqrt{14}$ $\sqrt[3]{181}$ $\sqrt[3]{49}$ $\sqrt[3]{674}$

19 **a** 10 **b** 10
c The answer is the same.

20 **a** 9 **b** 9
c The answer is the same.

21 **a** $\sqrt{16} \times \sqrt{9}$ **b** $\sqrt{a} \times \sqrt{b}$

22 **a** $2 \times 4 = 8$
b $\sqrt{9} \times \sqrt{49} = 3 \times 7 = 21$
c $\sqrt{36} \times \sqrt{4} = 6 \times 2 = 12$
d $\sqrt{81} \times \sqrt{25} = 9 \times 5 = 45$

Exercise 2.10

1 **a** **b**

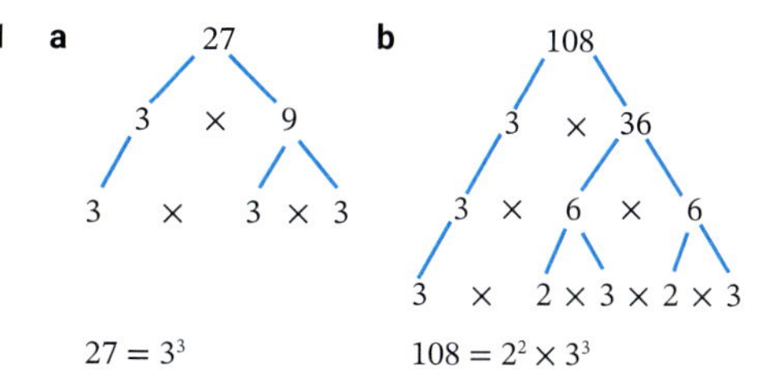

$27 = 3^3$ $108 = 2^2 \times 3^3$

c

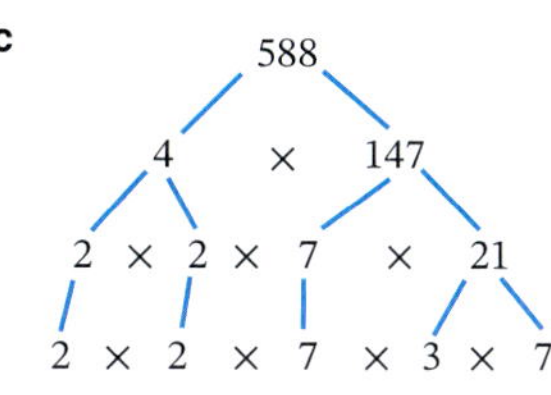

$588 = 2^2 \times 7^2 \times 3$

d

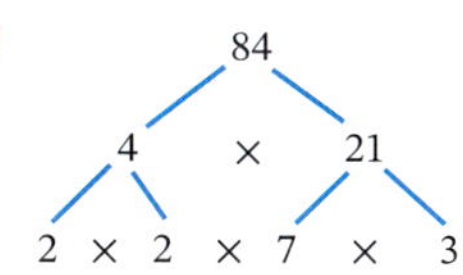

$84 = 2^2 \times 3 \times 7$

e

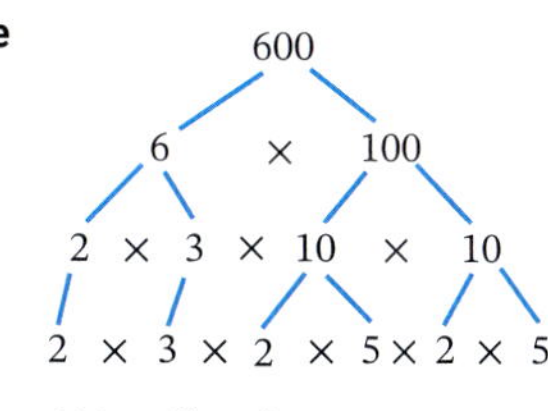

$600 = 2^3 \times 5^2 \times 3$

13

Number, x	1	2	3	4	5	6	7	8	9	10	11	12
Number squared, x^2	1	4	9	16	25	36	49	64	81	100	121	144
Number cubed, x^3	1	8	27	64	125	216	343	512	729	1000	1331	1728

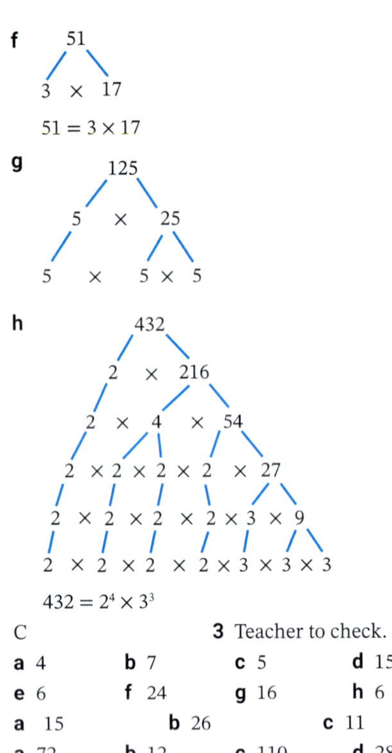

2 C **3** Teacher to check.

4 **a** 4 **b** 7 **c** 5 **d** 15
e 6 **f** 24 **g** 16 **h** 6

5 **a** 15 **b** 26 **c** 11

6 **a** 72 **b** 12 **c** 110 **d** 28
e 30 **f** 24 **g** 45 **h** 110

7 **a** 90 **b** 252 **c** 80

8 **a** 45 **b** 21 **c** 16
d 16 **e** 9 **f** 18

9 Teacher to check.

10 **a** 4 **b** 532

Exercise 2.11

1 **a** 6^5 **b** 4^7 **c** 2^{11} **d** 8^{10}
e 11^8 **f** 7^{10} **g** 5^7 **h** 6^{16}
i 2^9 **j** 5^9 **k** 2^{14} **l** 3^{16}

2 D

3 **a** $5^6 = 15\,625$ **b** $7^8 = 5\,764\,801$
c $3^{13} = 1\,594\,323$ **d** $4^6 = 4096$
e $9^7 = 4\,782\,969$ **f** $6^9 = 10\,077\,696$
g $8^{10} = 1\,073\,741\,824$ **h** $7^4 = 2401$

4 **a** 2^4 **b** 3^2 **c** 7^6 **d** 2^5
e 5^2 **f** 10^4 **g** 11^7 **h** 6^7
i 2 **j** 16^6 **k** 35^{11} **l** 20

5 D

6 **a** $2^4 = 16$ **b** $5^2 = 25$
c $4^3 = 64$ **d** $10^5 = 100\,000$
e $8^3 = 512$ **f** $6^8 = 1\,679\,616$
g 11 **h** $3^6 = 729$

7

Number, n	n^2	n^3	n^4	n^5	n^6
2	4	8	16	32	64
3	9	27	81	243	729
4	16	64	256	1024	4096
5	25	125	625	3125	15 625

8 **a** 32 **b** 81 **c** 125 **d** 3125
e 4096 **f** 1944 **g** 64 **h** 3125
i 9 **j** 2 **k** 4 **l** 1

9 **a** 5^4 **b** 3^7 **c** 2^3

10 **a** a^{10} **b** x^6 **c** b^5 **d** y^5
e d^4 **f** x^2 **g** a **h** r^4

Exercise 2.12

1 **a** 2^{18} **b** 5^6 **c** 7^{20} **d** 8^{14}
e 3^{24} **f** 9^{24} **g** 6^{14} **h** 11^{25}
i 8^{30} **j** 4^{45} **k** $(-2)^{28}$ **l** $(-7)^{18}$

2 D

3 **a** $5^8 = 390\,625$ **b** $2^{15} = 32\,768$
c $3^{18} = 387\,420\,489$ **d** $6^{12} = 2\,176\,782\,336$
e $7^6 = 117\,649$ **f** $4^{12} = 16\,777\,216$
g $8^0 = 1$ **h** $4^9 = 262\,144$

4 **a** 1 **b** 1 **c** 1 **d** 1
e 1 **f** 1 **g** 1 **h** 6
i 1 **j** 7 **k** 1 **l** 3
m 1 **n** 2 **o** 6 **p** 9

5 B

6 1, 2, 8, 16, 64, 128, 256, 512, 2048, 4096

7 **a** 64 **b** 4096 **c** 16 **d** 512
e 1 **f** 256 **g** 1 **h** 128

8 **a** $(6^2)^4 = 6^8$ **b** $(5^3)^3 = 5^9$
c $(10^6)^3 = 10^{18}$, other possible answers

9 **a** a^8 **b** x^{36} **c** d^{10} **d** n^{18}
e 1 **f** 4 **g** 1 **h** 0

Exercise 2.13

1

Fraction	Decimal	Terminating or recurring? *See example 2*	Prime factors of denominator
$\frac{1}{2}$	0.5	Terminating	2
$\frac{1}{3}$	$0.\dot{3}$	Recurring	3
$\frac{1}{4}$	0.25	Terminating	2
$\frac{1}{5}$	0.2	Terminating	5
$\frac{1}{6}$	$0.1\dot{6}$	Recurring	2, 3
$\frac{1}{7}$	$0.\dot{1}4285\dot{7}$	Recurring	7
$\frac{1}{8}$	0.125	Terminating	2
$\frac{1}{9}$	$0.\dot{1}$	Recurring	3
$\frac{1}{10}$	0.1	Terminating	2, 5
$\frac{1}{11}$	$0.\dot{0}\dot{9}$	Recurring	11
$\frac{1}{12}$	$0.08\dot{3}$	Recurring	2, 3

Fraction	Decimal	Terminating or recurring? *See example 2*	Prime factors of denominator
$\frac{1}{13}$	$0.\dot{0}7692\dot{3}$	Recurring	13
$\frac{1}{14}$	$0.0\dot{7}1428\dot{5}$	Recurring	2, 7
$\frac{1}{15}$	$0.0\dot{6}$	Recurring	3, 5
$\frac{1}{16}$	0.0625	Terminating	2

2 Fractions with denominators with prime factors of 2 or 5 are terminating, otherwise they are recurring.

3

Fraction	Decimal
$\frac{1}{20}$	0.05
$\frac{1}{50}$	0.02
$\frac{1}{100}$	0.01
$\frac{1}{500}$	0.002
$\frac{1}{1000}$	0.001
$\frac{1}{2000}$	0.0005
$\frac{1}{5000}$	0.0002

4 As the denominator of the fraction gets larger, the decimal gets smaller and closer to zero.

5

Fraction	Decimal
$\frac{19}{20}$	0.95
$\frac{49}{50}$	0.98
$\frac{99}{100}$	0.99
$\frac{499}{500}$	0.998
$\frac{999}{1000}$	0.999
$\frac{1999}{2000}$	0.9995
$\frac{4999}{5000}$	0.9998

6 As the value of n increases, the decimal value of $\frac{n-1}{n}$ gets closer and closer to 1.

7

n	$\frac{1}{n}+\frac{1}{n+1}$	Fraction
1	$\frac{1}{1}+\frac{1}{2}$	$\frac{3}{2}$
2	$\frac{1}{2}+\frac{1}{3}$	$\frac{5}{6}$
3	$\frac{1}{3}+\frac{1}{4}$	$\frac{7}{12}$
4	$\frac{1}{4}+\frac{1}{5}$	$\frac{9}{20}$
5	$\frac{1}{5}+\frac{1}{6}$	$\frac{11}{30}$
6	$\frac{1}{6}+\frac{1}{7}$	$\frac{13}{42}$
7	$\frac{1}{7}+\frac{1}{8}$	$\frac{15}{56}$
8	$\frac{1}{8}+\frac{1}{9}$	$\frac{17}{72}$
9	$\frac{1}{9}+\frac{1}{10}$	$\frac{19}{90}$
10	$\frac{1}{10}+\frac{1}{11}$	$\frac{21}{110}$
11	$\frac{1}{11}+\frac{1}{12}$	$\frac{23}{132}$

8 When adding $\frac{1}{n}$ and $\frac{1}{n+1}$, in the answer the numerator is the sum of the denominators $n + (n + 1)$ and the denominator is the product of the denominators $n \times (n + 1)$.

Power plus

1 **a** 9 **b** $2\sqrt{2}$ **c** $3\sqrt{3}$ **d** $5\sqrt{2}$
e $6\sqrt{2}$ **f** $3\sqrt{5}$ **g** $6\sqrt{5}$ **h** $8\sqrt{2}$
i $2\sqrt{7}$ **j** $10\sqrt{2}$ **k** $4\sqrt{3}$ **l** $7\sqrt{5}$

2 **a** 2^{-1} **b** $\frac{1}{2}$ **c** $\frac{1}{2}$ **d** $\frac{1}{2^2}=\frac{1}{4}$

3 **a** Yes **b** No
c For odd values of n. **d** 0

4 **a** 24 000 **b** 455 000 **c** 0.0933
d 0.0638 **e** 8 700 000 **f** 0.005 82
g 0.126 **h** 2690 **i** 314 000 000

5 **a** 1.2×10^4 **b** 3.45×10^8 **c** 7×10^{-3}
d 4×10^3 **e** 5×10^{-4} **f** 4.1×10^{-4}
g 1.92×10^6 **h** 3.61×10^{-4} **i** 6.37×10^{-8}

6 Teacher to check.

Test yourself 2

1 **a** 1400 **b** 210 **c** 160 **d** 131
e 87 **f** 385 **g** 260 **h** 480
i 276 **j** 800 **k** 276 **l** 42
m 9 **n** 2800 **o** 333 **p** 90
q 48 **r** 725 **s** 160

2 **a** 116 **b** 94 **c** 22 **d** 93

3 **a** 2 **b** 7 **c** −11 **d** 0

4 9237 m **5** 23°C

6 **a** −14 **b** 24 **c** 49 **d** −80

7 **a** −9 **b** −4 **c** −12 **d** 3

8 **a** 57 **b** 29 **c** 3
d 62 **e** $\frac{1}{5}$ **f** 5

9 0.471, 0.47, 0.417, 0.147

10 a 0.5 b 13.11 c 98.427

11 12.2 m

12 a 30.5 b 19.72 c 10.4 d 3.3
e 0.923 f 6.25 g 2.3925 h 11.6

13 15 bottles

14 a $0.\dot{2}$ b 0.875 c $0.1\dot{6}$

15 a 343 b 161 051 c 625
d 20 e 3 f −5

16 5.6

17 a 2, 2, 4, 100, 400 b 4, 2, 7, 14

18 a 6 b 140

19 $132 = 2^2 \times 3 \times 11$, $88 = 2^3 \times 11$,
HCF $= 2 \times 2 \times 11 = 44$

20 $12 = 2^2 \times 3$, $45 = 3^2 \times 5$,
LCM $= 2^2 \times 3 \times 3 \times 5 = 180$

21 $784 = 2^4 \times 7^2$, $\sqrt{784} = 2^2 \times 7 = 28$

22 a 4^7 b 6^{11} c 12^8 d 9^4
e 2^4 f 5^3 g 3 h 6^{12}

23 a 3^8 b 7^{10} c 6^9 d 3^{15}

24 a 1 b 1 c 4 d 8
e 1 f 1 g 1 h 0

25

n	$\frac{1}{n} - \frac{1}{n+1}$	Fraction
1	$\frac{1}{1} - \frac{1}{2}$	$\frac{1}{2}$
2	$\frac{1}{2} - \frac{1}{3}$	$\frac{1}{6}$
3	$\frac{1}{3} - \frac{1}{4}$	$\frac{1}{12}$
4	$\frac{1}{4} - \frac{1}{5}$	$\frac{1}{20}$
5	$\frac{1}{5} - \frac{1}{6}$	$\frac{1}{30}$
6	$\frac{1}{6} - \frac{1}{7}$	$\frac{1}{42}$
7	$\frac{1}{7} - \frac{1}{8}$	$\frac{1}{56}$
8	$\frac{1}{8} - \frac{1}{9}$	$\frac{1}{72}$
9	$\frac{1}{9} - \frac{1}{10}$	$\frac{1}{90}$
10	$\frac{1}{10} - \frac{1}{11}$	$\frac{1}{110}$
11	$\frac{1}{11} - \frac{1}{12}$	$\frac{1}{132}$

26 In the subtraction $\frac{1}{n} - \frac{1}{n+1}$, in the answer the numerator is 1 and the denominator is the product of the denominators $n \times (n + 1)$.

CHAPTER 3

SkillCheck

1 a −1 b 4 c −10 d 5
e −18 f 3 g −13 h 32
i 36 j 8 k −1 l 18

2 a $7p$ b ab c k^2 d n^3
e $2mp$ f dkp g $6y$ h $3qt$
i $-d$ j $3q$ k $4a$ l x

3 a + b × c − d ÷
e − f + g + h −

4 a 10 b 24 c 6 d 10
e 16 f 7 g 25 h 20
i 4 j 4 k 20 l 42

5 a 13 b 18 c 16
d 3 e 21 f 8

Exercise 3.01

1 a $5w$ b m^2 c $3abc$ d $6k$
e $\frac{m}{8}$ f $3f$ g $16m$ h $21wx$
i $9p$ j $4t^2$ k d l $16bcd$
m $\frac{26}{n}$ n $9h^2$ o $20x$ p a
q $8q$ r $12cn$ s $-6d^2r$ t $3a + 2c$
u a^3x^2

2 a $\frac{13}{2a}$ b $9 + 4m$ c $\frac{e-6}{5}$
d $4t - 8$ e $y^2 + z^2$ f $\frac{k}{9} - n$
g $\frac{p}{n+9}$ h $22 - \frac{e}{2}$ i $11 + 11u^2$

3 A 4 Teacher to check.

5 B

6 a $8 \times s \times t$ b $-2 \times y \times y$
c $8 \div f$ d $r \times r + t \times t$
e $4 \times b \times b \times d$ f $5 \times x \times y - 2 \times a$
g $(x - 2) \div 3$ h $q \times q \times q$
i $16 - 2 \times d \times d$ j $-4 \times d \div 3$
k $h \times h \times i \times i$ l $4 \times j - 9 \div d$

7 D

Exercise 3.02

1 C

2 a 8 b 12 c 4 d 10
e 30 f 12 g 0 h 8
i 5 j 19 k −3 l 3

3 a −150 b 25 c −27 d 2
e −15 f 75 g −21 h 6
i 18 j −10 k −90 l $\frac{1}{2}$

4 B

5 a −13 b −3 c 10 d −5
e $-\frac{1}{4}$ f 1 g $\frac{1}{3}$ h 5

6 C 7 $a = 42$ m^2 8 \$212

9 78.8 F　**10** 14 hours　**11** $5900

12 **a** **i** 31 minutes　**ii** 56 minutes

b 8:10 a.m.

Exercise 3.03

1 **a** $5n$　**b** $n+6$　**c** $\frac{n}{10}$　**d** $n+1$

e $\frac{n}{2}$　**f** $n-20$　**g** n^2　**h** $20-n$

i $n-1$　**j** $2n$　**k** $\frac{6}{n}$　**l** $n-8$

m $-4n$　**n** $\sqrt{n}$

2 **a** $a+b+c$　**b** ab　**c** $b-c$

d $ac+b$　**e** $3b+a$　**f** $\frac{c}{b}$

g $a-b$　**h** $5+bc$　**i** $b-c$

j $\frac{ab}{c}$　**k** $\sqrt{a+c}$　**l** $ab-7c$

3 D

4 **a** $3y+8$　**b** $\frac{p}{3}-6$　**c** $4M+27$

d $\frac{5k}{2}$　**e** $\frac{x}{6}+13$　**f** $\frac{B}{2}-2$

g $\frac{R}{9}+10$　**h** $21n$

5 **a** x plus y minus z

b 5 times K plus L

c x times y minus 3

d W divided by 5

e 3 times x, minus 1

f y minus x, then divided by 10

g 2 times N, plus 21

h 7 minus e

6 **a** $t-b$　**b** $\frac{n}{5}$　**c** $y-3p$　**d** $2r+9$

e $\frac{a}{12}$　**f** $4x$　**g** $\frac{160}{r}$　**h** $5x+3y$

Exercise 3.04

1 **a** $x+2$　**b** $2x+3$

c $2+2x$　**d** $3x+4$

e $x+1+x+1$ or $2x+2$

f $2+x+2+2x$ or $3x+4$

g $1+x+3$ or $x+4$

h $2x+2+2x+1$ or $4x+3$

i $x+3+x$ or $2x+3$

j $4x+5$

2 **a** $x+y+2$　**b** $2x+2y+1$　**c** $x+2y$

d $2x+2y+3$　**e** $x+2y+2$　**f** $2x+3y+4$

g $x+y$　**h** $2x+3y+1$

3 **a** 　**b**

c 　**d**

e

f

4 B

5 **a** D　**b** A　**c** F

d B　**e** C　**f** E

6 **a** D　**b** F　**c** A

d B　**e** C　**f** E

7 **a** $m+2p$　**b** $m+p$　**c** $3p+5$

d $2m+2p+1$　**e** $m+2p$

8 C

9 **a** $4a+5d$　**b** $5h+9r$　**c** e

d $4k-3$　**e** $2x+6z$　**f** $2p+s$

g $4u+12$　**h** $4n-1$　**i** $3i$

Mental skills 3A

2 **a** 720　**b** 486　**c** 825　**d** 128

e 560　**f** 216　**g** 189　**h** 432

i 560　**j** 616　**k** 135　**l** 675

Exercise 3.05

1 **a** $4p, 2p, p$　**b** $m^2, 3m^2$

c $5ac, 7ca$　**d** $vw, 2wv$ and $5v, 9v$

e $6pq, 2qp, 3pq$　**f** $2d, 3d$

g $4mn, 2nm, mn$　**h** $x^2y, 4x^2y$

i $5ba, ab$

2 C

3 **a** $11k$　**b** $7mn$　**c** $2xy$　**d** $9abc$

e $6x^2$　**f** $5ef^2$　**g** $5d$　**h** $5mk$

i $4xy$　**j** $-4de$　**k** $3k^2$　**l** $3w^2$

4 C

5 **a** $5k+4j$　**b** $9ab+2$　**c** $7m^2+2m$

d $10s^2-7st$　**e** $10mn$　**f** $9k+3$

g $2y^2$　**h** $12d-4e$　**i** $4-2x$

j $8bg$　**k** $9-2q$　**l** $5a+2c$

6 **a** $3x-3y$　**b** $4ab+3bc$　**c** $9k^2+11$

d $6bc-9a$　**e** x^2-7x+7　**f** $4p^2$

g $12y+14$　**h** $7r+3s$　**i** $7b+6g$

j $6x^2-13x+6$　**k** $52a$　**l** $7q-7$

m $-6y$　**n** 12

7 C

8, 9 Teacher to check.

Exercise 3.06

1 D

2 **a** $15y$　**b** $16m$　**c** $4km$　**d** $20t$

e $12f$　**f** $16xy$　**g** $30mp$　**h** $18bd$

i $12d^2$　**j** $120mn$　**k** $48y^2$　**l** $27r^2$

m $72ab$　**n** x^2y^2　**o** $42jk$　**p** $10r^2s^2t$

q $18ac^2d$　**r** $h^2j^3k^2$　**s** $24p^2q^2$　**t** $6m^2n^2$

3 **a** $-8bd$　**b** $-24m^2$　**c** $-90kl$

d $-60p$　**e** $-63k$　**f** $75b^2$

g $-12m^2n^2$　**h** $-12a^2b^2c$　**i** $40r^2$

j $-216d^2e$　**k** $144x^2$　**l** $-40m^2$

4 B

5 **a** y^4　**b** a^3　**c** $2c^2$　**d** t^5

e $7q^3$　**f** k^3l^2　**g** m^3n^2　**h** $-5p^4k^3$

i $-3y^3$　**j** $-d^2e^2f$　**k** $4q^2r^4$

6 A

7 **a** $4m$ **b** $9e$ **c** $-7p$ **d** $9x$
e $4w$ **f** 8 **g** -5 **h** $\frac{3}{4}$
i $2p$ **j** -7 **k** $\frac{p}{5}$ **l** $-\frac{7n}{m}$

8 **a** $2m^2$ **b** $-3e$ **c** $-5p$ **d** $7x$
e $-2w$ **f** $4m$ **g** $-8ky$ **h** $\frac{3}{4}$
i $5de$ **j** a **k** $-8a$ **l** $-2m$

9 D

10 **a** -3 **b** -3 **c** $-15y^3$ **d** $6e$
e $6x^2$ **f** $8mn$ **g** 3 **h** $3a$
i $\frac{a}{5c}$ **j** $2d$ **k** $\frac{5p}{m}$ **l** 1
m $8x$ **n** $9r + 14$ **o** $18p$

11, 12 Teacher to check.

Exercise 3.07

1 D

2 **a** m^7 **b** k^9 **c** y^{10} **d** x^8
e n^8 **f** q^{10} **g** b^{16} **h** s^{15}
i $42k^6$ **j** $20y^4$ **k** $3d^9$ **l** $55e^4$
m $h^{10} n^7$ **n** $-30c^6 d^5$ **o** $f^8 g^{11}$ **p** $-10p^7 q^2$

3 A

4 **a** y **b** p^3 **c** q^3 **d** n^5
e k^2 **f** x^2 **g** 1 **h** m^5
i q **j** $9g^4$ **k** $3p^4$ **l** a^5
m $11xy$ **n** x^3 **o** $7k^3 l^5$ **p** $\frac{5e^2}{7f}$

5 B

6 **a** y^{16} **b** x^{10} **c** y^9 **d** m^{18}
e k^{21} **f** t^{40} **g** $9m^2$ **h** $8m^{12}$
i $x^3 y^3$ **j** $p^{12} q^4$ **k** $81h^{16}$ **l** $k^{15} z^5$
m $16k^4 j^4$ **n** $16c^6 d^2$ **o** $125x^6 y^{24}$ **p** $64m^{24} p^{12}$

7 **a** 1 **b** 1 **c** 1 **d** 5
e 1 **f** 1 **g** 1 **h** h^2
i $\frac{b^6}{2}$ **j** $16x^2$ **k** 1 **l** $13x$
m $3m + 1$ **n** $3b^2$ **o** 32 **p** $\frac{1}{2}$

8 **a** $\frac{1}{3pq}$ **b** $2mn$ **c** $\frac{2}{3y}$

Mental skills 3B

2 **a** 176 **b** 363 **c** 261 **d** 405
e 682 **f** 707 **g** 1818 **h** 3564
i 152 **j** 540 **k** 2142 **l** 588

Exercise 3.08

1 **a** $5m + 15$ **b** $8m - 16$ **c** $2j + 20$
d $3a + 6$ **e** $4h - 28$ **f** $6x + 6y$
g $36 - 9b$ **h** $11y - 66$ **i** $8k + 20$
j $15m - 6$ **k** $18x + 24$ **l** $20a + 50b$

2 **a** $-3n - 12$ **b** $-4v + 20$ **c** $-7d - 21$
d $-10k + 20$ **e** $-2d - 2e$ **f** $-25 + 5z$
g $-6h - 6$ **h** $-m + 1$ **i** $-45 - 18x$
j $-2f - 3$ **k** $-3u + 3v$ **l** $-32 + 8k$

3 B

4 **a** $xy + 6x$ **b** $pq - 2p$ **c** $a^2 + 4a$
d $3b^2 - b$ **e** $d^2 + 3de$ **f** $8m + 10n$
g $15k^2 - 10k$ **h** $21j^2 + 7ijk$ **i** $4x^2 + xy$
j $-8p + 6q$ **k** $k^2 - 12k$ **l** $-4m + m^2$
m $8n^2 + 20mn$ **n** $2x - 14x^2$ **o** $-2f^2 - 4f$
p $-5j + 7k$

5 C **6, 7** Teacher to check. **8** D

9 **a** $6a + 26$ **b** $6k + 6$ **c** $3q - 4m$
d $6p - 22$ **e** $4 - 2y$ **f** $17 - 7t$
g $12h + 5$ **h** $-8x + 8$ **i** $18n + 4$

10 Teacher to check.

11 **a** $5a + 14$ **b** $8x + 4$ **c** $17k + 18$
d $22t + 11$ **e** $n + 7$ **f** $4d + 2$
g $22y + 23$ **h** $-2p - 7$ **i** $m^2 + 4m + 3$
j $7d + e$ **k** $6f^2 + 4$ **l** $-38e^2 + e$

12 **a** $x + 5$ **b** $r - 3$ **c** $8k - 3$
d $2a + 7$ **e** $y + 9$ **f** $2t - 7$

Exercise 3.09

1 **a** 1, 3, 5, 15 **b** 1, 2, 4, 8
c 1, 17 **d** 1, 5, 25
e 1, 5, 11, 55 **f** 1, 2, 3, 6, 7, 14, 21, 42

2 **a** 3 **b** 2 **c** 6 **d** 4 **e** 11
f 5 **g** 3 **h** 15 **i** 16

3 Teacher to check. **4** C

5 **a** $3x$ **b** $4j$ **c** $2x$ **d** $7k$
e 5 **f** $4m$ **g** $16c$ **h** de
i 9 **j** 3 **k** $2m$ **l** $4y$
m a **n** 32 **o** $5w$ **p** $2xy$
q d **r** $12t$

6 C

Exercise 3.10

1 **a** 2 **b** 4 **c** x **d** y

2 **a** $x + 2$ **b** $p - 4$ **c** $4k - 5$ **d** $j - 4k$
e $x^2 + y^2$ **f** 5 **g** 6 **h** 2
i 2 **j** $3x + 1$ **k** $y + z$ **l** $b - c$
m $q - 1$ **n** $b + d$ **o** $a + 2$ **p** h
q n **r** y **s** k **t** f

3 **a** $4(x + 2)$ **b** $3(m + 2)$ **c** $5(a - 2)$
d $5(3m - 2a)$ **e** $4(2x - y)$ **f** $4(3m - 4w)$
g $h(k - 1)$ **h** $x(2y - m)$ **i** $p(12 + 5r)$
j $x(x + 5)$ **k** $m(4 - m)$ **l** $y(y - 11)$

4 C

5 **a** $m + n$ **b** $2y - 1$ **c** $2p + 1$
d $n - p$ **e** $y - 4$ **f** $q + r$
g $7 - y$ **h** $4k + 1$ **i** $3x - 4z$

6 D

7 **a** $7a(b + 2)$ **b** $2x(1 - 3y)$ **c** $3m(1 - 3w)$
d $5x(2y - 5w)$ **e** $4m(2a - 3p)$ **f** $3y(3x + 2m)$
g $3x(1 + 3y)$ **h** $ab(1 + c)$ **i** $xy(z + a)$
j $6m(3 - 4n)$ **k** $ac(2 + 3b)$ **l** $pq(r + m)$
m $r(s + 7)$ **n** $4(s^2 + 4)$ **o** $p(p - 2)$
p $2k(3k^2 - 1)$ **q** $j(3 + j)$ **r** $n(7 - n)$

s $7e(1 + 3e)$ t $6q(1 - 2q)$ u $a^2(a^2 + 1)$
v $xy(x + y)$ w $3m(mn + 3)$
x $3abc(4abc + 1)$

Exercise 3.11

1 a $k + 1$ b $n + p$ c $r - 2$ d $k - 9$
e $-a$ f $-y$ g -6 h $-f$

2 a $-4(x - 3)$ b $-3(m + 3)$
c $-10(x - 2)$ d $-5(3x - 2m)$
e $-2(3x + 4m)$ f $-5(x + 6)$
g $-a(1 + b)$ h $-4(xyz - 2abc)$
i $-4(5 - x)$ j $-d(x + y)$
k $-m(2 - 5p)$ l $-8(3x + 2y)$

3 B

4 a $-8x(2y + 3k)$ b $-4x(5 - 3y)$ c $-4m(2w - 3)$
d $5e(1 - 3f)$ e $-ab(1 + c)$ f $xy(a + w)$
g $-6c(2 + 3w)$ h $-rt(2 - 7s)$ i $5h(4f + 1)$
j $-19w(b + 2)$ k $-abc(1 - d)$ l $-abc(1 + d)$

5 a 10, 10 b Yes

6 A

7 a $a^3bc(a - b)$ b $2\pi r(r + h)$

Exercise 3.12

1 a

	x	7
x	x^2	$7x$
2	$2x$	14

$x^2 + 9x + 14$

b

	x	1
x	x^2	x
5	$5x$	5

$x^2 + 6x + 5$

2 D

3 a $x^2 + 5x + 4$ b $x^2 + 9x + 18$
c $x^2 + 9x + 20$ d $x^2 + 4x + 4$
e $x^2 + 8x + 12$ f $x^2 + 13x + 30$
g $x^2 + 9x + 8$ h $x^2 + 14x + 49$
i $x^2 + 3x - 4$ j $x^2 - x - 6$
k $x^2 - 3x - 4$ l $x^2 + x - 30$
m $x^2 - 9x + 20$ n $x^2 + 6x - 27$
o $x^2 - 8x + 15$ p $x^2 - 14x + 49$

4 a i $x^2 - 9$ ii $x^2 - 16$
iii $x^2 - 4$ iv $x^2 - 49$
b $x^2 - 8^2 = x^2 - 64$

5 a $2x^2 + 6x + 4$ b $3x^2 + 5x - 2$
c $2x^2 + 5x + 3$ d $5x^2 - 13x - 6$
e $4x^2 - 11x + 6$ f $10x^2 - 11x - 6$
g $3x^2 - 10x + 3$ h $4x^2 - 5x - 21$

6 a $a^2 + 10a + 25$ b $x^2 + 4x + 4$
c $y^2 + 14y + 49$ d $x^2 - 6x + 9$
e $4a^2 + 4a + 1$ f $9a^2 - 30a + 25$

7 a $a^2 + 2ab + b^2$ b $a^2 - b^2$

8 A

Power plus

1 a $-4p - 9q$ b $-6m - 3n - 7mn$
c $4j - k + 4$ d $5b^2$
e $3x^2 - 6x$ f $5xy + 2yz + 4xz$
g $-4d^2 + 4e^2 + 16$ h $13f^2 + 6f + 1$

2 a i $4x + 6$ ii $x^2 + 3x$
b i $4p$ ii p^2
c i $x + y + 16$ ii $\frac{xy}{2}$
d i $2m + n + 8$ ii $\frac{n}{2}(2m+3)$ or $mn+\frac{3m}{2}$
e i $3x + 2y$ ii $x^2 + x$
f i $2a + 2b$ ii $ad + bc - cd$

3 a $x^2 + 2x - 15$ b $2am - 3m$
c $4y - 9$ d $6a - 2b + 14c$
e $-5p^2 - 7pq + 10pr$ f $4k^2 - 5k - 5$
g $27 - 20f + 30g$ h $-2x^2 + 7x + 115$
i $x^2 + 8x + 15$ j $p^2 + 9p - 10$

4 a -2 b -24 c 10 d 9
e 8 f 4 g 7 h 4
i -1 j 3 k 1.2 l 0

5 a $p(9q + 5 - 2p)$ b $11(4k - 3l - m)$
c $x(2 - 7x + 3y)$ d $(y + 2)(x + 3)$
e $(b - 3)(a - 7)$ f $(k + 1)(4 - j)$

6 a m b $\frac{2n}{5}$ c k d $\frac{2d}{5}$
e $\frac{3f}{2}$ f $\frac{7c}{10}$ g $\frac{6}{x}$ h $\frac{7b+5a}{ab}$

Test yourself 3

1 a $6p$ b $-r$ c $4dg$
d $5t$ e $20m$ f $5k$
g $6y^2$ h $\frac{16}{a}$ i $20cd$

2 a $c \times d$ b $3 \times y \times y$
c $w \div 6$ d $c \times c + y$

3 B

4 a 3 b -10 c 19
d 24 e -15 f 29

5 a $y + 4$ b $6b - 3$ c $\frac{p}{q}$

6 a $12k$ b $17d$ c $17m$
d $18m$ e $3ac$ f $9cd$
g $10x + 10y$ h $15r + 15s$ i $28h + 7g$

7 a $10x$ b $14bd$ c $66m$ d $24pq$
e $32w^2$ f $75j$ g $36m^2$ h $60p^3$

8 a $25x$ b $\frac{4b}{d}$ c $11m$ d $8rs$
e 5 f $7d^2$ g $7e$ h $13b$
i $2x$ j $4mn^2$ k $4e$ l $5ae$

9 a $6n + 12$ b $-10b - 35$ c $9j^2 - 3j$
d $2n + 18$ e $19v - 15$ f $9y - 18$

10 A

11 a 6 b $3p$ c 7

12 a $3(m + 3)$ b $5(h - 7)$ c $3a(2 + 3b)$
d $2h(5 - 6m)$ e $xy(1 + z)$ f $d(c - e)$
g $a(a - 6)$ h $8j(k + 3j)$

13 a $-2(1 + 2q)$ b $-4(f + 4)$ c $-3(4 - 5d)$
d $-b(a - 1)$ e $-8b(2 + 3d)$ f $-k(4k + 1)$
g $-x(x - 7)$ h $-hi(h + i)$

PRACTICE SET 1

1 a 343 b 1225 c 256
d 20 e 14 f 1.5

2 a 140 b 135 c 120
d 480 e 180 f 15

3 a $x = 5$ b $y = 41$ c $x = 6.4$
d $m = 24$ e $a = 17.7$ f $p = 33.5$

4 a 8 b −30 c −20 d 5
e 28 f −6 g −2 h 2
i −16 j −28 k 6 l −15

5 a 18 b 23 c −3
d −11 e 19 f −99
g $4\frac{3}{4}$ h 8 i 2

6 a 13 b 9 c 10 d 10

7 a $2k + 4j$ b $4p$ c $11p + 5q$ d $8x + 6y$

8 a −4.23 b 22.39

9 a 7.2 b 0.45 c 0.42 d 18.4

10 a $k + 5$ b $20k$ c $\frac{k}{2}$
d $3k + 10$ e $k - 12$ f k^2

11 a $10p$ b 6 c $-12x^2y$
d $24ab$ e 2 f $2a$

12 a $180 = 2^2 \times 3^2 \times 5$ b $875 = 5^3 \times 7$

13 a 4 b 18

14 a 2^7 b 7^{11} c 5^5 d 8
e 3^8 f 11^{15} g 1 h 1

15 a $\sqrt{48}$ cm b 6.93 cm

16 Right-angled.

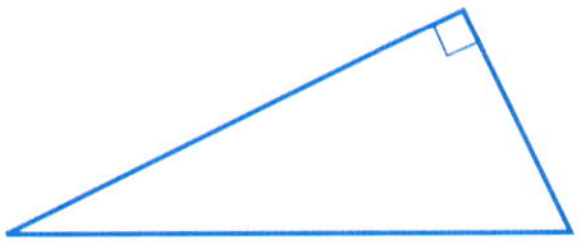

17 a 36 b 54

18 a $4x + 8$ b $6m - 24$ c $-14k - 7$
d $-3p - q$ e $6m + 4$ f $4p$

19 a $2(m + 2)$ b $d(d - 1)$ c $6(p + 2q)$
d $-12(x + 2)$ e $a(2b + 3)$ f $-2x(x - 3)$

CHAPTER 4

SkillCheck

1 Teacher to check.

2 a acute b obtuse c reflex
d obtuse e right f revolution
g reflex h straight

3 a vertically opposite
b right
c reflex

4 a $x = 60$ b $x = 123$ c $x = 162$

5 a rectangle

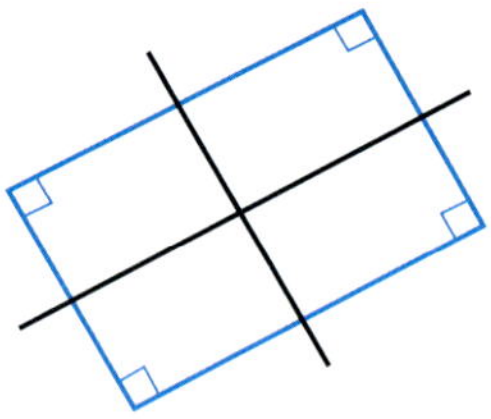

b isosceles triangle

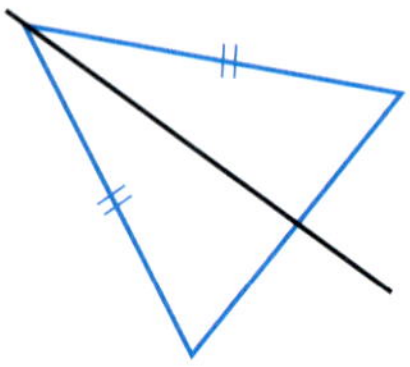

c rhombus

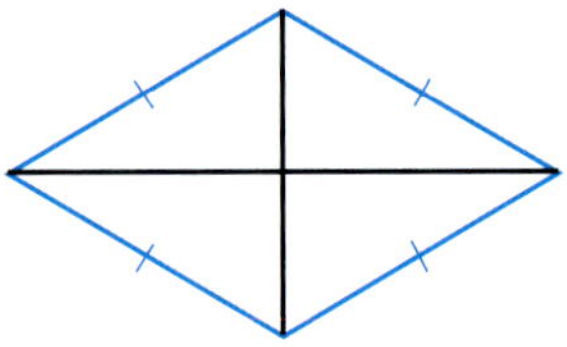

d equilateral triangle

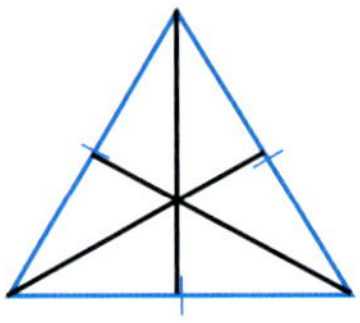

6 a 2 b 1 c 2 d 3

7 D

8

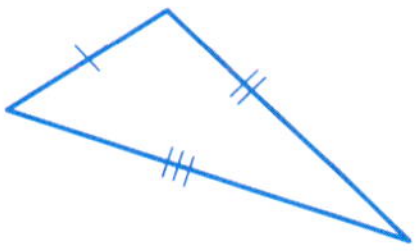

9 a

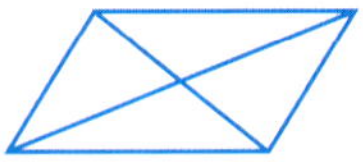

b No

Exercise 4.01

1 a complementary b reflex
c supplementary d obtuse
e straight f vertically opposite
g acute h adjacent
i angles at a point

2 a 74° b 57° c 19° d 88°

3 a 155° b 31° c 73° d 132°

4 D 5 C

6 **a** $x = 70$ (angles in a right angle)
b $a = 112$ (angles on a straight line)
c $a = 110$ (vertically opposite angles), $b = 70$ (angles on a straight line)
d $y = 285$ (angles at a point)
e $a = 19$ (angles on a straight line)
f $x = 53$ (adjacent angles)
g $a = 52$ (angles on a straight line), $b = 128$ (vertically opposite angles), $c = 52$ (angles on a straight line)
h $x = 165$ (angles at a point)
i $x = 270$ (angles at a point)
j $a = 45$ (vertically opposite angles), $b = 135$ (angles on a straight line), $c = 135$ (vertically opposite angles)
k $k = 58$ (angles in a right angle)
l $n = 70$ (angles on a straight line)
m $w = 156$ (angles at a point)
n $q = 62$ (angles on a straight line)
o $y = 40$ (angles at a point)
p $p = 45$ (equal angles in a right angle)
q $p = 62$ (angles on a straight line)
r $a = 23$ (adjacent angles)

7 **a** $m = 235$ (angles at a point)
b $m = 270$ (angles at a point)
c $a = 50$ (angles on a straight line)
d $x = 170$ (angles at a point)
e $a = 80$ (angles on a straight line), $b = 65$ (vertically opposite angles), $c = 115$ (angles on a straight line)
f $a = 45$ (angles on a straight line)
g $p = 90$ (angles on a straight line)
h $m = 30$ (angles on a straight line)

8 C

Exercise 4.02

1 **a** alternate **b** co-interior
c corresponding **d** co-interior
e alternate **f** co-interior

2 C

3 **a** $e°$ **b** C **c** $e°$ or A or $g°$
d $f°$ **e** D or $h°$ or $f°$ or B

4 **a** $a = 70$ (alternate angles on parallel lines)
b $d = 45$ (co-interior angles on parallel lines)
c $a = 61$ (alternate angles on parallel lines)
d $m = 120$ (corresponding angles on parallel lines)
e $p = 89$ (co-interior angles on parallel lines)
f $m = 112$ (co-interior angles on parallel lines)

5 (Other reasons possible)
a $a = 61$ (alternate angles on parallel lines), $b = 119$ (angles on a straight line)
b $c = 70$ (co-interior angles on parallel lines), $d = 70$ (vertically opposite angles)
c $a = 65$ (co-interior angles on parallel lines, $b = 115$ (alternate angles on parallel lines
d $g = 50$ (alternate angles on parallel lines), $h = 60$ (alternate angles on parallel lines), $m = 70$ (angles on a straight line)
e $p = 100$ (angles on a straight line), $q = 80$ (alternate angles on parallel lines)
f $y = 89$ (vertically opposite angles and corresponding angles on parallel lines)
g $w = 108$ (co-interior angles on parallel lines), $x = 108$ (angles on a straight line), $y = 108$ (corresponding angles on parallel lines)
h $a = 100$ (alternate angles on parallel lines [by extending the middle parallel line])
i $n = 250$ (co-interior angles on parallel lines)

6 D

7 **a** Yes (alternate angles are equal)
b Yes (corresponding angles are equal)
c No (co-interior angles do not add to 180°)
d Yes (co-interior angles add to 180°)

8 **a** 255 **b** 52

Exercise 4.03

1 **a** scalene, obtuse-angled
b equilateral, acute-angled
c scalene, acute-angled
d isosceles, right-angled
e isosceles, obtuse-angled
f scalene, obtuse-angled
g scalene, right-angled
h isosceles, acute-angled
i isosceles, right-angled
j isosceles, obtuse-angled
k equilateral, acute-angled
l scalene, right-angled

2 **a**
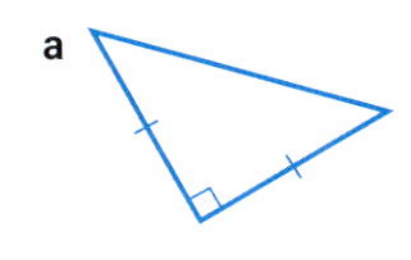
b
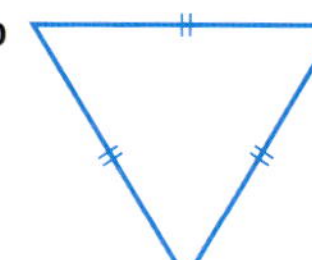
c
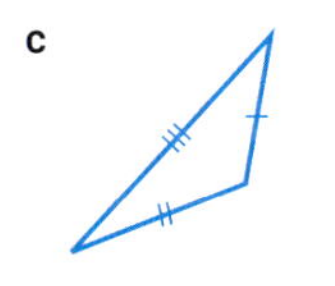
d
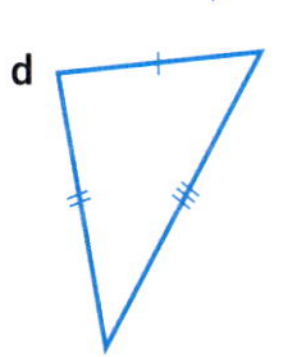
e
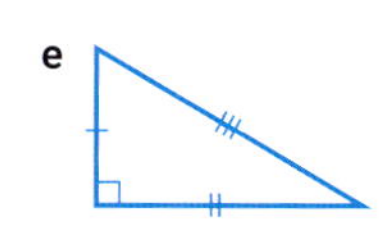
f
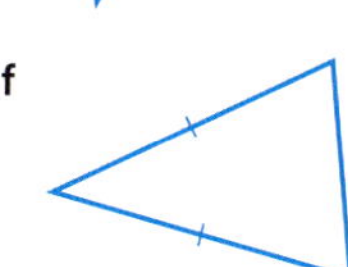

3 A

4 No, all angles are 60° in equilateral triangles so they are not obtuse.

5 **a** b, d, e, h, i, j, k **b** b, k

6 **a** $x = 42$ (angles in an isosceles triangle)
b $y = 7$ (angles in an isosceles triangle)
c $x = 60$ (angles in an equilateral triangle)

d $l = 4.8$ (angles in an isosceles triangle)
e $a = 17.2$, $b = 17.2$ (angles in an equilateral triangle)
f $r = 15$ (angles in an isosceles triangle)

7 No, the sum of all angles cannot be greater than 180°.

8 C

9

Equilateral triangle	3	3
Isosceles triangle	1	No rotational symmetry
Scalene triangle	0	No rotational symmetry

Mental skills 4A

2 **a** 70% **b** 66% **c** 45% **d** 88%
e 75% **f** 75% **g** 80% **h** 80%
i 55% **j** 35% **k** 30% **l** 80%
m 90% **n** 45% **o** 52%

4 **a** 25% **b** 68% **c** 17% **d** 60%
e 10% **f** 33.3% **g** 59% **h** 70.2%
i 84% **j** 70% **k** 42.8% **l** 5.5%
m 91% **n** 78.25% **o** 31.4%

Exercise 4.04

1 **a** convex **b** convex **c** non-convex
d non-convex **e** convex **f** convex

2 **a** square **b** trapezium **c** kite
d parallelogram

3 **a** square, rectangle
b trapezium
c square, rhombus
d parallelogram, rectangle, square, rhombus
e parallelogram, rectangle, square, rhombus
f kite, square, rhombus

4 rhombus

5

Quadrilateral	Number of axes of symmetry	Order of rotational symmetry
Rectangle	2	2
Parallelogram	0	2
Trapezium	0	No rotational symmetry
Rhombus	2	2
Square	4	4
Kite	1	No rotational symmetry

6 **a** trapezium **b** rhombus

7 B

8 **a** 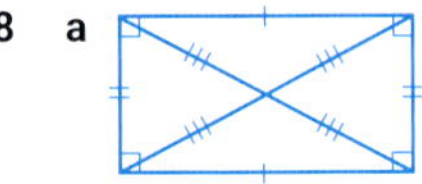 **b** 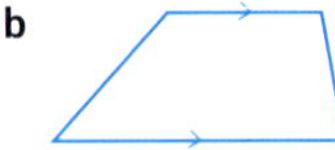

c 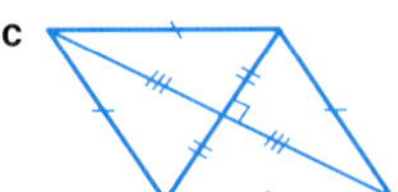 **d** 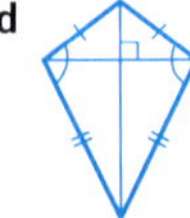

9 **a** False **b** True **c** True **d** True
e False **f** True **g** False

Exercise 4.05

1 **a** parallel
b equal; angles; right angles
c opposite; sides; equal; bisect
d sides; opposite; parallel; opposite; equal; right; bisect
e equal; parallel; right angles; equal; bisect
f equal; right angles; sides; equal; right angles

2 **a** Parallelogram, rectangle, rhombus, square
b Square, rhombus, kite
c Parallelogram, rectangle, rhombus, square
d Trapezium
e Parallelogram, rectangle, rhombus, square
f Parallelogram, rectangle, rhombus, square
g Parallelogram, rectangle, trapezium, kite
h Rectangle, square
i Rectangle, square
j Rhombus, square

3 **a** $a + 70 = 180$ (co-interior angles on parallel lines); $a = 110$, $b = 70$ (opposite angles of a parallelogram)

4 **a** $m = 123$ (opposite angles of a parallelogram)
b $a = 154$ (co-interior angles on parallel lines), $b = 26$ (opposite angles of a rhombus or co-interior angles on parallel lines)
c $x = 156$ (co-interior angles on parallel lines), $y = 156$ (opposite angles of a parallelogram and vertically opposite angles, or co-interior angles on parallel lines and vertically opposite angle)
d $l = 45$ (diagonals bisect the angles of a square)
e $a = 49$ (angle sum of a triangle), $b = 90$ (diagonals of a rhombus bisect each other at right angles)
f $x = 87$ (corresponding angles on parallel lines), $y = 93$ (co-interior angles on parallel lines)
g $m = 75$ (co-interior angles on parallel lines), $n = 75$ (equal angles of an isosceles trapezium, or by symmetry)
h $a = 147$ (co-interior angles on parallel lines), $b = 33$ (opposite angles of a parallelogram or co-interior angles on parallel lines)
i $a = 5$, $b = 7$ (opposite sides of a parallelogram)
j $a = 4$, $c = 7$ (diagonals of a parallelogram bisect each other)
k $a = 70$, $b = 92$ (co-interior angles on parallel lines)

5 B **6** D **7** A **8** A

9 **a** No, teacher to check **b** No, teacher to check
c Yes, teacher to check

Exercise 4.06

1 **a** $a = 45$ **b** $b = 60$ **c** $c = 50$
d $d = 60$ **e** $e = 18$ **f** $f = 33$
g $g = 120$ **h** $h = 80$ **i** $i = 45$

2 **a** $a = 127$ **b** $b = 140$ **c** $c = 60$
d $d = 28$ **e** $e = 100$ **f** $f = 120$
g $g = 50$ **h** $h = 136$ **i** $i = 120$

3 **a** $a = 89$ **b** $b = 62$ **c** $c = 109$
d $d = 90$ **e** $e = 33$ **f** $f = 245$
g $g = 90$ **h** $h = 90$ **i** $i = 140$

4 C 5 A 6 A 7 C

8 $x = 68$ (angle sum of an isosceles triangle), $y = 34$ (exterior angle of an isosceles triangle)

Exercise 4.07

1 **a** hexagon, irregular, convex
b quadrilateral, irregular, convex
c pentagon, irregular, non-convex
d triangle, irregular, convex
e decagon, regular, convex
f heptagon, irregular, convex
g octagon, regular, convex
h nonagon, irregular, non-convex

2

Polygon	Number of sides	Sum of angles inside polygon
hexagon	6	720°
heptagon	7	900°
octagon	8	1080°
nonagon	9	1260°
decagon	10	1440°

3 **a** 2340° **b** 3240° **c** 4140° **d** 17 640°

4 **a** $a = 95$ **b** $b = 110$ **c** $d = 108$ **d** $c = 160$
e $g = 140$ **f** $x = 90$ **g** $e = 135$

5 **a** 90° **b** 60° **c** 120° **d** 135°
e 144° **f** 108° **g** 150°

6 **a** 14 **b** 34 **c** 26 **d** 53 **e** 125

7 **a** Yes, 10 **b** No

Mental skills 4B

2 **a** $\frac{3}{4}$ **b** $\frac{7}{25}$ **c** $\frac{3}{10}$ **d** $\frac{7}{50}$ **e** $\frac{3}{50}$
f $\frac{17}{20}$ **g** $\frac{8}{25}$ **h** $\frac{49}{100}$ **i** $\frac{14}{25}$ **j** $\frac{9}{10}$
k $\frac{18}{25}$ **l** $\frac{13}{20}$ **m** $\frac{1}{5}$ **n** $\frac{6}{25}$ **o** $\frac{53}{100}$

4 **a** $\frac{19}{25}$ **b** $\frac{1}{10}$ **c** $\frac{4}{5}$ **d** $\frac{9}{20}$ **e** $\frac{22}{25}$
f $\frac{14}{25}$ **g** $\frac{3}{4}$ **h** $\frac{31}{100}$ **i** $\frac{17}{25}$ **j** $\frac{1}{20}$
k $\frac{3}{5}$ **l** $\frac{27}{50}$ **m** $\frac{3}{50}$ **n** $\frac{49}{100}$ **o** $\frac{41}{50}$

Power plus

1 **a** $x = 121$ **b** $x = 88$ **c** $x = 96$
d $x = 110$ **e** $x = 6$ **f** $x = 34$
g $x = 67$ **h** $x = 155$ **i** $x = 117$
j $x = 108$ **k** $x = 30$ **l** $x = 140$

Test yourself 4

1 **a** obtuse **b** acute **c** reflex

2 **a** $x = 50$ (angles on a straight line)
b $a = 50$ (angles on a straight line), $b = 130$ (vertically opposite angles or angles on a straight line)
c $y = 5$ (angles in a right angle)
d $n = 72$ (vertically opposite angles)
e $w = 130$ (angles at a point)

3 **a** corresponding **b** alternate
c co-interior

4 **a** $m = 100$ (alternate angles on parallel lines)
b $x = 95$ (co-interior angles on parallel lines)
c $x = 80$ (straight angle on parallel lines); $y = 80$ (corresponding angles on parallel lines)
d $m = 88$ (alternate angles on parallel lines); $p = 88$ (corresponding angles on parallel lines)

5 **a** scalene, right-angled
b scalene, obtuse-angled
c isosceles, obtuse-angled
d equilateral, acute-angled
e isosceles, acute-angled
f isosceles, acute-angled
g isosceles, right-angled
h scalene, obtuse-angled
i isosceles, obtuse-angled

6 **a**

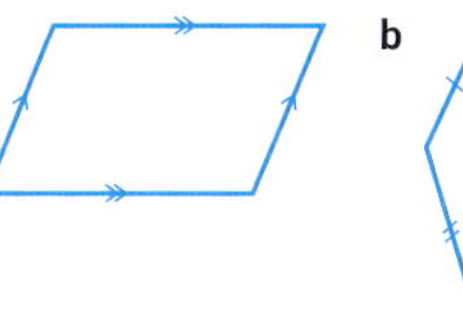

b

c

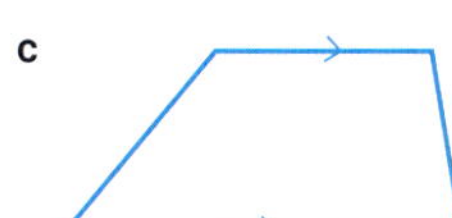

7 **a** rhombus, square
b kite
c rectangle, square
d parallelogram, rhombus, rectangle, square
e parallelogram, rhombus, rectangle, square

8 **a** $a = 50$ **b** $a = 42$
c $y = 30$ **d** $y = 53$
e $x = 34$ **f** $m = 106$

9 **a** $m = 63$ **b** $x = 135$
c $y = 123$ **d** $d = 70$
e $a = b = 115$ **f** $a = 56, b = 68$
g $a = 48, b = 66$

10 **a** $x = 42$ **b** $m = 66$ **c** $y = 130$

11 A

CHAPTER 5

SkillCheck

1 a 200 b 3.5 c 2.5
d 4.6 e 4000 f 5.2
g 0.6 h 8.2 i 400

2 a 12 cm b 19.2 m c 86 m
d 112 mm e 21 cm f 13 cm

3 a 4 m² b 30 cm² c 56 m²
d 91.84 cm² e 58 mm² f 140 m²

4 a diameter b segment c sector
d arc e radius f chord
g circumference

Exercise 5.01

1 a 84 mm b 33.2 cm c 69.3 mm
d 158 cm e 43.3 m f 12.4 km

2 a 31.5 m b 34 m c 45 cm
d 60 km e 66 cm f 64 m
g 164 mm h 58 cm i 92 m

3 a 12 cm b 34 m
c 15 cm d 22 m

4 C
5 D

6 a 56 mm b 84 mm
c 92 mm d 80 mm

7 A
8 7 m
9 4.5 m
10 4 cm
11 16 cm
12 18 cm
13 36 m

Exercise 5.02

1 a 21 cm b 32 m c 270 mm d 90 cm

2 a 21.99cm b 12.57 m
c 282.74 mm d 47.12 cm

3 a 7π cm b 4π m c 90π mm d 15π cm

4 a 30 m b 78 mm c 54 cm d 36 cm

5 a 31.4 m b 81.7 mm c 56.5 cm d 37.7 cm

6 a 10π m b 26π mm c 18π cm d 12π cm

7 7.54 m

8 a 527.8 cm b 150

9 21 300 km
10 100.5 mm
11 236 mm

12 a 15.71 cm b 43.98 cm
c 56.55 cm d 106.81 cm

13 C
14 D
15 7.5 m
16 120 minutes (or 2 hours)
17 Ali, 25 m

Exercise 5.03

1 a m² b cm² c m²
d km² e cm² f cm²

2 A

3 a 1 b 7.5 c 0.45
d 0.034 e 0.06 f 43.2
g 0.000 95 h 0.76 i 0.048
j 300 000 k 85 000 l 6 900 000
m 5400 n 7 000 000 o 4800
p 450 000

4 D
5 0.0025 km²
6 3 002 800 ha
7 1348.05 km²

8 a 0.875 km² b 87.5 ha

9 a 0.0076 b 70 000 000 c 8 000 000
d 8.5 e 69 000 f 152 000
g 0.85 h 49

10 32.9%

11 a 8.81 km² b 6.2

Exercise 5.04

1 a 24 m² b 36 cm² c 48 cm²
d 3 cm² e 53.3 mm² f 37.21 m²
g 54.4 cm² h 85.56 km² i 340 mm²
j 6 m² k 84 cm² l 136 m²

2 a 10 800 cm² b 42 140 mm² c 547 200 m²
d 552 mm² e 243.6 mm² f 528 000 m²

3 D
4 C
5 5 cm
6 3 m
7 7 cm
8 Teacher to check.
9 60 m

10 a 15.6 cm² b 66.4 mm²
c 8.4 m² d 30 km²

Exercise 5.05

1 A

2 a 150 m² b 51 m² c 88 mm²
d 156 m² e 950 mm² f 36 cm²
g 189 m² h 154 cm² i 71 m²
j 201.19 km² k 88 000 cm² l 280 mm²
m 20 m²

3 C
4 B
5 1874 cm²

Mental skills 5A

2 a 0.7 b 0.6 c 0.07
d 0.45 e 0.98 f 0.4

4 a 0.61 b 0.04 c 0.9
d 0.38 e 0.271 f 0.007

Exercise 5.06

1 a 148 mm² b 1080 m² c 97.6 m²
d 33.02 mm² e 472 cm² f 30 cm²
g 862.5 mm² h 162 cm² i 680 cm²

2 a 8200 m² b 6560 m²

3 630 000 mm²
4 D

5 a 546 cm² b 1772 cm²

6 D
7 4 m

Exercise 5.07

1 a 1.5 cm² b 850 mm²
c 8.16 m² d 210 cm²

2 a 240 m² b 80 cm² c 315 mm²
d 1344 m² e 1.8 m²

3 B

4 a 300 mm^2 b 172.5 mm^2 c 90 mm^2
d 360 mm^2 e 120 mm^2 f 72 mm^2
g 66.5 mm^2 h 510 mm^2 i 112 mm^2

5 a 10 cm and 6 cm b 30 cm^2
c 90 cm^2

Mental skills 5B

2 a $\frac{3}{5}$ b $\frac{1}{4}$ c 0.105
d 3.68 e $\frac{4}{10}$ f $\frac{5}{6}$
g 2.817 h 0.503 i $\frac{1}{5}$

3 a 0.082, 0.81, 0.821 b 3.5, 3.51, 3.513, 3.55
c $\frac{1}{6}, \frac{2}{5}, \frac{2}{3}$ d 0.007, 0.07, 0.7, 0.707
e $\frac{2}{10}, \frac{3}{8}, \frac{1}{2}$ f $\frac{1}{4}, \frac{4}{10}, \frac{3}{5}$

5 a $0.25, \frac{1}{5}, \frac{1}{6}$, 16% b $0.4, \frac{1}{3}$, 0.28, 27%
c 50%, $\frac{1}{8}$, 6%, 0.05 d $\frac{3}{4}$, 0.639, 55%, $\frac{2}{5}$
e 69%, $\frac{2}{3}$, 0.609, 0.6 f 22.5%, $\frac{2}{9}$, 17%, 0.105

Exercise 5.08

1 a 28.3 cm^2 b 452.4 mm^2 c 0.2 m^2
d 19.6 cm^2 e 3421.2 mm^2

2 a $9\pi\text{ cm}^2$ b $144\pi\text{ mm}^2$ c $0.0625\pi\text{ m}^2$
d $6.25\pi\text{ cm}^2$ e $1089\pi\text{ mm}^2$

3 a 13.85 cm^2 b 38.48 cm^2 c 78.54 cm^2

4 a $4.41\pi\text{ cm}^2$ b $12.25\pi\text{ cm}^2$ c $25\pi\text{ cm}^2$

5 539 m^2 6 3690.67 cm^2 7 $196\pi\text{ cm}^2$

8 D 9 6.2 m^2 10 12.03 cm

11 A 12 A

13 a 160.97 cm^2 b 221.54 m^2 c 85.81 mm^2
d 46.27 cm^2 e 21.46 cm^2 f 16.08 cm^2

14 a i 153.9 m^2 ii 95.0 m^2 iii 58.9 m^2
b i 314.2 m^2 ii 254.5 m^2 iii 59.7 m^2

15 a 9168 m^2 b 7326 m^2
c 1842 m^2 d \$44 668.50

Exercise 5.09

1 a $\frac{1}{2}$ b $\frac{1}{6}$ c $\frac{1}{3}$ d $\frac{13}{18}$

2 a 50.0 mm b 37.5 cm
c 137.1 cm d 12.2 m

3 D

4 a 153.94 mm^2 b 87.27 cm^2
c 1073.24 cm^2 d 6.87 m^2

5 a $P = 14.3$ cm, $A = 12.6\text{ cm}^2$
b $P = 128.5$ mm, $A = 981.7\text{ mm}^2$
c $P = 9.3$ cm, $A = 5.3\text{ cm}^2$
d $P = 18.0$ cm, $A = 19.2\text{ cm}^2$
e $P = 61.4$ mm, $A = 235.6\text{ mm}^2$
f $P = 2.0$ m, $A = 0.2\text{ m}^2$
g $P = 106.7$ mm, $A = 641.4\text{ mm}^2$
h $P = 82.6$ m, $A = 351.9\text{ m}^2$
i $P = 177.7$ mm, $A = 1954.8\text{ mm}^2$

6 D

7 a 6° b 10.26 cm^2

8 9.62 cm^2

Exercise 5.10

1 a m^3 b cm^3 c cm^3
d m^3 e cm^3 f m^3

2 C

3 a 5 000 000 b 1 600 000
c 2000 d 4 000 000 000
e 0.006 f 8 200 000
g 7 000 000 h 0.0096
i 4 j 0.16
k 0.25 l 12 000
m 180 000 n 0.2

4 0.272 m^3 5 D 6 A

7 a E b G c F d C
e A f B g D

8 $2\,460\,000\text{ mm}^3$

Exercise 5.11

1 a Yes b No c No
d Yes e No f Yes

2 a i 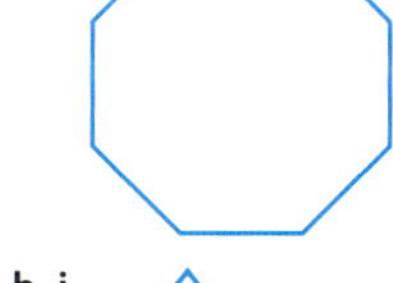 ii octagonal prism

b i 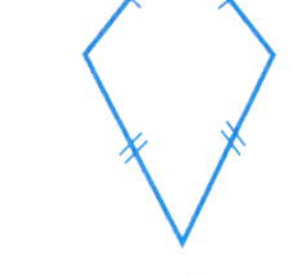 ii kite prism

c i  ii pentagonal prism

3 a  b 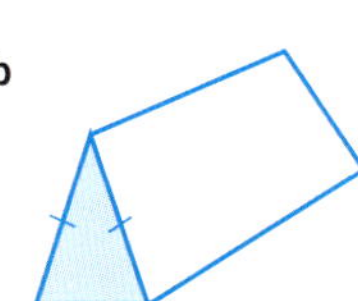

c

d 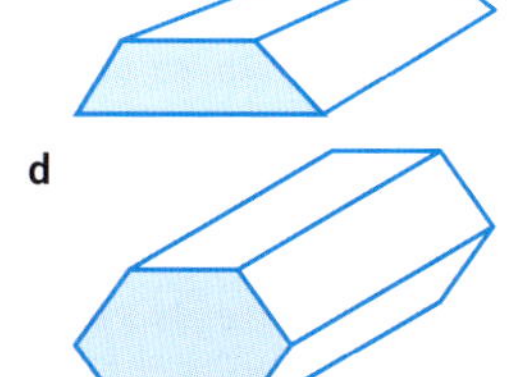

4 **a** 210.8 m^3 **b** $4.096\,875 \text{ m}^3$ **c** 486.42 cm^3

5 **a** 135 cm^3 **b** 64 m^3 **c** 6120 mm^3
d 1.728 m^3 **e** $21\,952 \text{ mm}^3$ **f** 258.06 cm^3

6 **a** 90 cm^3 **b** $15\,400 \text{ mm}^3$ **c** 7.8125 m^3

7 B

8 **a** $22\,500 \text{ mm}^3$ **b** 928 cm^3
c $13\,200 \text{ cm}^3$ **d** 288 cm^3

9 **a** 0.306 m^3 **b** 0.306 m^3 Teacher to check.

10 A 11 5 m 12 7 m

13 **a** 220 cm^3 **b** $11\,220 \text{ mm}^3$ **c** $24\,000 \text{ m}^3$
d $29\,450 \text{ mm}^3$ **e** 82 m^3

14 B 15 Teacher to check.

Exercise 5.12

1 **a** 153.93 m^2 **b** 1847.26 m^3

2 C

3 **a** 1133.0 cm^3 **b** $17\,671.5 \text{ mm}^3$
c 1.8 m^3 **d** 1606.0 cm^3
e $17\,583.1 \text{ cm}^3$ **f** $301\,592.9 \text{ mm}^3$
g 499.6 m^3

4 **a** 4.25 cm **b** 4.75 cm **c** 678.58 cm^3

5 B 6 8 cm

7 **a** 4021.2 m^3 **b** $732\,944.7 \text{ m}^3$
c 1005.3 mm^3 **d** 786.1 m^3
e 380.8 m^3 **f** $3\,685\,088.2 \text{ mm}^3$

8 4 m

Exercise 5.13

1 **a** mL **b** ML **c** L
d mL **e** L **f** kL

2 **a** 5 **b** 3400 **c** 1600
d 4 **e** 2.98 **f** 7 100 000
g 0.875 **h** 8200 **i** 800
j 1.85 **k** 5400 **l** 900
m 6000 **n** 3.5 **o** 1 200 000

3 A 4 A

5 **a** $121\,500 \text{ cm}^3$ **b** 121.5 L

6 1 800 000 L

7 **a** $70\,200\,000 \text{ cm}^3$ **b** 70 200 L

8 **a** 23.4 L **b** 10

9 Teacher to check. 10 1728 L

11 **a** 1.32 m^3 **b** 1320 L

12 424 mL

13 **a** 18.850 m^3 **b** 18 849.6 L

14 **a** 0.72 L **b** 17.28 L

15 Teacher to check.

16 Seventy-five water trucks will need to be ordered

Power plus

1 **a** 35.4 cm **b** 1228.3 mm **c** 57.1 cm

2 **a** 7.9 m^2 **b** 14.4 m

3 **a** 2.09 cm^2 **b** Teacher to check
c 1.73 cm^2 **d** 0.36 cm^2 **e** 2.81 cm^2

4 **a** 4.4 m^2 **b** 47.1 cm^2 **c** 39.3 m^2

Test yourself 5

1 C

2 **a** 76 mm **b** 256 cm **c** 31.4 km

3 **a** 12.6 cm **b** 62.8 mm
c 44.0 cm **d** 88.0 cm

4 **a** 2 **b** 70 000 **c** 0.8
d 30 000 000 **e** 7 **f** 7.7
g 0.0007 **h** 72 000 **i** 0.058

5 B

6 **a** 56.7 m^2 **b** 51.6 mm^2 **c** 345 cm^2
d 49 km^2 **e** 69.3 m^2 **f** 775.5 cm^2

7 **a** 450 mm^2 **b** 133 cm^2
c 6.3 m^2 **d** 86 cm^2

8 **a** 288.51 mm^2 **b** 17.16 m^2 **c** 4126.5 cm^2

9 **a** 189 m^2 **b** 1560 mm^2 **c** 418 cm^2

10 **a** **i** $4\pi \text{ cm}^2$ **ii** 12.6 cm^2
b **i** $100\pi \text{ mm}^2$ **ii** 314.2 mm^2
c **i** $49\pi \text{ cm}^2$ **ii** 153.9 cm^2
d **i** $196\pi \text{ mm}^2$ **ii** 615.8 mm^2

11 **a** 20.38 cm **b** 40.85 cm
c 26.85 cm **d** 56.55 cm

12 **a** 25.13 cm^2 **b** 116.55 cm^2
c 37.70 cm^2 **d** 208.23 cm^2

13 **a** 8 200 000 **b** 3400 **c** 2
d 1 000 000 000 **e** 8 000 000 000 **f** 45

14 **a** 391.392 cm^3 **b** 2200 mm^3 **c** 672 m^3
d $35\,200 \text{ mm}^3$ **e** $25\,350 \text{ cm}^3$

15 **a** 452.39 mm^3 **b** 543.43 m^3
c $154\,893.81 \text{ cm}^3$

16 2.5 L

17 **a** 3000 **b** 2.5 **c** 6500
d 0.0072 **e** 0.12 **f** 0.035

18 **a** 5.7 L **b** 4000 L

19 **a** 8.75 m^2 **b** 35 m^3 **c** 35 000 L

CHAPTER 6

SkillCheck

1 **a** 3 **b** 8 **c** 20 **d** 4
e 6 **f** 3 **g** 40 **h** 15

2 **a** $\frac{17}{100}$ **b** $\frac{3}{100}$ **c** $\frac{51}{100}$
d $\frac{63}{100}$ **e** $\frac{89}{100}$

3 **a** 70% **b** 5% **c** 61%
d 29% **e** 88% **f** 93%

4 **a** 13% **b** 73% **c** 8%

5 **a** 87% **b** 27% **c** 92%

6 65%

7 1, 2, 4, 5, 10, 20, 25, 50, 100

8 **a** \$40 **b** 6 kg **c** \$5
d 10 kg **e** \$3 **f** 2 kg

Exercise 6.01

1 a $1\frac{1}{2}$ b $3\frac{2}{3}$ c $2\frac{1}{4}$
d $2\frac{2}{5}$ e 5 f $4\frac{3}{11}$

2 a $\frac{7}{2}$ b $\frac{13}{3}$ c $\frac{23}{4}$ d $\frac{23}{3}$
e $\frac{39}{5}$ f $\frac{65}{9}$ g $\frac{71}{7}$ h $\frac{93}{10}$

3 Teacher to check.

4 Mixed numeral, because it is greater than 1, while a proper fraction is less than 1.

5 a $\frac{1}{2}$ b $\frac{3}{4}$ c $\frac{3}{8}$ d $2\frac{7}{10}$
e $\frac{11}{15}$ f $\frac{5}{12}$

6 a $\frac{1}{2}$ b $\frac{1}{3}$ c $\frac{6}{13}$ d $\frac{3}{4}$
e $\frac{3}{5}$ f $\frac{4}{7}$ g $\frac{2}{3}$ h $\frac{3}{5}$
i $\frac{4}{7}$ j $\frac{4}{7}$ k $\frac{3}{5}$ l $\frac{9}{10}$

7 A

8 a $1\frac{1}{2}$ b 3 c $4\frac{1}{2}$ d $1\frac{4}{5}$

Exercise 6.02

1 a $\frac{4}{5}$ b $\frac{1}{2}$ c $\frac{5}{8}$ d $1\frac{3}{8}$

2 a $\frac{7}{10}$ b $\frac{5}{21}$ c $\frac{17}{20}$ d $\frac{1}{4}$
e $\frac{13}{24}$ f $\frac{1}{30}$ g $1\frac{5}{18}$ h $\frac{11}{24}$
i $1\frac{1}{8}$ j $\frac{1}{7}$ k $3\frac{5}{9}$ l $2\frac{3}{5}$

3 $\frac{17}{40}$ 4 D

5 a $\frac{5}{12}$ b $\frac{17}{30}$

6 $\frac{1}{2}$ 7 C 8 A

9 a $1\frac{9}{10}$ b $3\frac{5}{8}$ c $16\frac{17}{24}$ d $5\frac{1}{45}$
e $1\frac{21}{40}$ f $1\frac{19}{20}$ g $\frac{5}{6}$ h $1\frac{7}{10}$

10 a $2, 2\frac{1}{2}, 3\frac{1}{2}, 1$ b $1, \frac{5}{6}, 1\frac{1}{3}, 0, \frac{1}{2}$

Exercise 6.03

1 a $\frac{1}{3}$ b $\frac{10}{21}$ c $\frac{1}{64}$ d $\frac{3}{4}$
e $\frac{9}{20}$ f $4\frac{1}{2}$ g $12\frac{7}{20}$ h $3\frac{1}{9}$

2 D

3 a $\frac{3}{14}$ b $\frac{8}{15}$ c $\frac{2}{3}$ d $1\frac{1}{2}$
e 16 f 9 g $\frac{1}{8}$ h $\frac{1}{15}$
i $3\frac{1}{2}$ j $3\frac{2}{11}$ k $2\frac{4}{5}$ l $3\frac{9}{22}$

4 $\frac{3}{32}$

5 a 1 (or −1) b 0 c 1 d $\frac{1}{3}$

6 a $\frac{7}{8}$ b $\frac{2}{3}$ c $\frac{2}{21}$ d $\frac{7}{10}$

7 a Increase: dividing by a proper fraction is the same as multiplying by its reciprocal, an improper fraction, whose value is greater than 1, so the number will increase.
b Increase: multiplying by an improper fraction, whose value is greater than 1, will make the number increase.

Exercise 6.04

1 a $\frac{3}{5}$ b $\frac{3}{4}$ c $\frac{31}{100}$ d $\frac{2}{25}$
e $\frac{3}{10}$ f $\frac{17}{20}$ g $\frac{99}{100}$ h $\frac{3}{100}$
i $1\frac{3}{5}$ j $1\frac{7}{20}$ k $\frac{1}{4}$ l $2\frac{1}{2}$

2 C

3 a 0.18 b 0.82 c 0.02 d 0.5
e 1.2 f 0.511 g 0.79 h 0.125
i 0.163 j 0.04 k 0.187 l 0.0525

4 a 17% b 70% c 26%
d 55% e 62.5% f 96%
g $66\frac{2}{3}\%$ h 125% i 140%
j 67.5% k 6.25% l $44\frac{4}{9}\%$

5 a 38% b 55% c 96% d 62.5%
e 8% f 5.4% g 60% h 0.3%
i 190% j 40.5% k 126% l 11.4%

6 a $\frac{13}{20}$, 65% b $\frac{3}{5}$, 60% c $\frac{1}{5}$, 0.2
d $\frac{21}{25}$, 0.84 e \$0.5, 50% f 0.125, 12.5%
g $\frac{9}{25}$, 0.36 h 0.625, 62.5% i $\frac{73}{100}$, 73%
j $0.\dot{3}$, $33\frac{1}{3}\%$ k $\frac{2}{3}$, $0.\dot{6}$

7 a 75% b $\frac{4}{25}$ c $\frac{2}{3}$
d 18% e $\frac{1}{4}$ f 55%
g $\frac{19}{20}$ h 60% i $\frac{1}{6}$

8 a 0.75, 78%, $\frac{4}{5}$, $\frac{9}{11}$ b 22%, $\frac{1}{4}$, 0.29, $\frac{7}{20}$.
c 57%, $\frac{3}{5}$, 0.605, 0.62

ANSWERS

9 C

10 a 0.47, $\frac{9}{20}$, 43%, $\frac{2}{5}$ b 0.88, 86%, $\frac{21}{25}$, 0.08
c $\frac{19}{20}$, 91%, 0.905, $\frac{9}{10}$

Mental skills 6A

2 a 19 b $7.50 c 87.5 d $20.20
e $3.76 f 40 g $0.925 h 89.6
i $270 j $0.38 k $152.76 l $7.25
m 315.4 n $1.07 o 42.6 p $2431.76

4 a 10 b 124 c $490 d $1.72
e 14.4 f $2540 g 78 h $1.16
i $9 j $16.80 k $920 l 64

6 a 100 b $0.60 c 2.5 d $1.35
e $1.84 f $4.20 g 40 h 6.5
i $0.48 j $6.90 k $3.60 l 42

Exercise 6.05

1 a 24 b 7 c 4
d 10 e 42 f 10
g 750 m h 8 h i 400 mL
j 125 kg k 10 months l 35 min

2 C

3 a $55 b 5 kg c 19.5 L
d 126 cm e $67.50 f 285 g
g 473.76 m h 286.72 km i 152.29 mL
j $15.62 k $89.68 l 45.539 kg
m 104.5 n 453 o $2613
p 126 m q 595 g r $460

4 0.675 kg

5 a $105 b $315

6 234

7 a 96 cm b 350 kg
c 144 min d 75 mL
e 11 250 g f 900 m
g 84 h h 30 min
i 27 000 mg

8 63 9 $0.86 10 273

11 D

12 17 13 24 318 14 1400

15 a $90 b $500

Exercise 6.06

1 a $\frac{1}{2}$ b $\frac{19}{25}$ c $\frac{87}{100}$
d $\frac{2}{3}$ e $\frac{1}{4}$ f $\frac{3}{8}$

2 a 50% b 76% c 87%
d $66\frac{2}{3}$% e 25% f $37\frac{1}{2}$%

3 62.5%

4 a $\frac{2}{5}$ b 36%

5 No

6 a $\frac{172}{189}$ b 9%

7 a 32% b $\frac{9}{67}$ c 13.4%

8 5%

9 a i $169 ii $205 iii $249
b i 11.8% ii 12.7% iii 12.0%
c ii $179 oven

10 A

11 a $\frac{1}{12}$ b $\frac{1}{4}$ c $\frac{7}{10}$ d $\frac{23}{100}$
e $\frac{1}{8}$ f $\frac{1}{6}$ g $\frac{3}{20}$ h $\frac{19}{70}$
i $\frac{1}{8}$ j $\frac{3}{8}$ k $\frac{8}{15}$ l $\frac{4}{35}$

12 a $8\frac{1}{3}$% b 25% c 70% d 23%
e $12\frac{1}{2}$% f $16\frac{2}{3}$% g 15% h $27\frac{1}{7}$%
i $12\frac{1}{2}$% j $37\frac{1}{2}$% k $53\frac{1}{3}$% l $11\frac{3}{7}$

13 C

14 a Broncos 124.9%, Bulldogs 112.0%, Cowboys 84.7%, Dragons 81.3%, Eels 115.6%, Knights 86.9%, Panthers 76.1%, Rabbitohs 97.2%, Raiders 86.9%, Roosters 61.8%, Sea Eagles 184.7%, Sharks 117.6%, Storm 230.9%, Tigers 77.6%, Titans 79.9% and Warriors 151.5%
b Storm, Sea Eagles, Warriors and Broncos

Exercise 6.07

1 a $157.50 b 480 c 73.2 km
d $2650 e 152 kg f 13.3 L

2 a $176 b 102 c 101.2 kg
d $243.75 e 1240 L f $1522.90

3 $232.20

4 a $405 b $3105

5 $1024.10 6 171.36 cm 7 D

8 $18 691.50 9 $145.20 10 798

11 $1 160 000 12 B 13 $80.10

14 a $540 b $378 c $92 d 19.6%

15 650 16 $423.30 17 510

18 Less than the original price, $673.20

Exercise 6.08

1 a 8 kg b 45 mm c 6 L d $250

2 C

3 a 18 m b 9 g c 3 h
d 1 L e $20 f 12 km
g 5 mL h 6 days i 75 kg

4 a 6 tonne b 4 h
c 30 sheep d $9

5 a 7 b $40 c 48 min
d $63 e 150 cm f 80 mL

6 B

7 a $9 b 5 kg c 20 kL

8 a $100 b 154 c $200
d 42 e $270 f 54

9 a $279 b 96 c $70
d 57 e $300 f 500

10 a 24 b 10

Mental skills 6B

2 a $308 b 60 c 30 d $450
e 96 f $90 g $60.50 h 150
i 210 j $84 k $300 l 98

4 a $360 b 40 c 88 d $22.50
e 450 f $63 g $80 h 50
i $50.40 j $87.50 k 225 l $105

Exercise 6.09

1 a $700 b $560 c $4600
d $1900 e 400 cm f 280 kg
g 3 m h 150 min i $400
j $360 k 400 kg l 110 min

2 $1200 3 79 800 4 $82 176

5 $15 250 6 D 7 24 000

8 $3000 9 $300 10 $435

Exercise 6.10

1 a i $18 700 ii Loss
b i $543 000 ii Profit
c i $799 ii Loss
d i $6800 ii Profit

2 a i $17 ii 20%
b i $185 ii 44.6%
c i $1.45 ii 52.7%

3 a i $5 ii 26.3%
b i $1250 ii 46.3%
c i $21 ii 26.6%

4 C

5 a $150 b 12%

6 a $1800 b 10.6%

7 a $234 b $71.40 c $468
d $1267 e $6.25 f $16.20

8 23.31% 9 44.4%

10 a $2090, $22 990 b $181, $1991
c $0.30, $3.30 d $14.50, $159.50
e $18, $198 f $2.70, $29.70

11 a $72, $720 b $18, $180 c $84, $840
d $1.70, $17 e $3.15, $31.50 f $1.52, $15.18

12 a $1440 b $1584 c $950.40
d $86.40 e Loss of $35 f 4.0%

Exercise 6.11

1 D

2 a $120 b $720 c $262.40
d $6 e $1706.25 f $3777.75
g $393.68 h $588 i $42.36
j $2428.38

3 a $1350 b $2250 c $11 200
d $108 000 e $101.50 f $6255
g $4981.62 h $540 i $36.85
j $374.40

4 a $27 b $11.38 c $18.75
d $1020 e $3.96 f $225.50
g $988.17 h $186.43 i $12.25
j $1364.18

5 a $735 b $7735

6 a $10 500 b $40 500 c $675

7 a $1620 b 4 years

Exercise 6.12

1 27 2 81.25% 3 80%

4 $41.25 5 60% 6 234 points

7 348 8 37.5% 9 40%

10 12.5% 11 $8625 12 11.4%

13 $262.50 14 C 15 $4369.95

16 $44 561.25

Power plus

1 $96.80

2 a $1627.50 b $1765.84

3 a $499 200 b $519 168 c $583 993.39

4 a Less b 20%

5 a $357 000 b $303 500 c $186 400

6 a Less b 50%

Test yourself 6

1 a $\frac{3}{4}$ b $\frac{7}{12}$ c $\frac{7}{9}$

2 a $3\frac{3}{4}$ b $4\frac{2}{5}$ c $4\frac{1}{3}$

3 a $\frac{9}{2}$ b $\frac{11}{3}$ c $\frac{34}{5}$

4 $\frac{5}{6}, \frac{3}{4}, \frac{7}{12}, \frac{1}{2}$

5 a $\frac{11}{15}$ b $\frac{1}{14}$ c $1\frac{5}{12}$
d $6\frac{7}{10}$ e $6\frac{1}{6}$ f $2\frac{7}{8}$

6 $\frac{3}{8}$

7 a $\frac{7}{18}$ b $\frac{8}{15}$ c $\frac{15}{28}$
d $\frac{27}{40}$ e $\frac{2}{5}$ f $3\frac{7}{12}$

8 16 500

9 a $\frac{37}{50}$ b $\frac{29}{100}$ c $\frac{13}{20}$ d $1\frac{57}{100}$

10 a 0.12 b 0.162 c 0.02 d 0.054

11 a 87.5% b 24% c $22\frac{2}{9}$% d 82.5%

12 a 56% b 61.3% c 70% d 4.8%

13 0.7,0.725,$\frac{3}{4}$,77%

14 a 6 b 16 c $242
d 8 months e 108 L f 210 cm

15 a 24.31 kg b 162.69 kg

16 a i $\frac{4}{5}$ ii 80%

b i $\frac{3}{4}$ ii 75%

c i $\frac{11}{20}$ ii 55%

17 a $\frac{17}{20}$ b 85%

18 48.1% 19 \$327 20 \$67 275

21 a 11 tonnes b 12 cm c 9 min

22 \$147 111.11 23 20%

24 a \$8.70 b \$87.00

25 \$42.50 26 57.6%

PRACTICE SET 2

1 a $\frac{1}{2}$ b $\frac{1}{3}$ c $\frac{3}{5}$

2 a $y = 65$ (angles on a straight line).

b $k = 135$ (vertically opposite angles), $l = 45$ (angles on a straight line).

c $x = 313$ (angles at a point).

d $q = 62$ (angles in a right angle).

3 a 48 cm b 35 m c 65 cm

4 a 60 cm^2 b 110 cm^2 c 60 cm^2 d 880 mm^2

5 a i scalene ii right-angled

b i scalene ii acute-angled

c i isosceles ii acute-angled

d i equilateral ii acute-angled

6 a $m = 60$ b $x = 50$ c $k = 65$ d $m = 60$

7 a $\frac{5}{2}$ b $\frac{7}{4}$ c $\frac{22}{5}$

8 a i $\frac{1}{4}$ ii 0.25

b i $\frac{2}{5}$ ii 0.4

c i $\frac{17}{20}$ ii 0.85

d i $\frac{1}{8}$ ii 0.125

9 a $d = 78$ (alternate angles on parallel lines).

b $f = 112$ (corresponding angles on parallel lines).

c $t = 72$ (co-interior angles between parallel lines).

10 a 34% b 75% c 21.5% d 62.5%

11 a parallelogram b rectangle

c square d kite

12 a 248.62 m^2 b 22.68 cm^2 c 90.2 m^2

13 a $5\frac{1}{2}$ b $6\frac{1}{3}$ c $3\frac{1}{2}$

14 a $1\frac{1}{5}$ b $\frac{11}{15}$ c $\frac{7}{20}$ d $4\frac{1}{10}$

e $1\frac{1}{3}$ f $\frac{3}{10}$ g $\frac{3}{5}$ h $1\frac{3}{5}$

15 a 137 m^2 b 112.5 cm^2

16 a 31.42 cm b 21.99 cm

17 a 78.54 cm^2 b 38.48 cm^2

18 a $x = 130$ (exterior angle of a triangle).

b $m = 50$ (exterior angle of a triangle).

c $n = 35$ (exterior angle of a triangle).

d $p = 105$ (angle sum of a quadrilateral).

e $k = 40$ (angle sum of a quadrilateral).

f $x = 90$ (angle sum of a quadrilateral).

19 \$2.30 20 26.928 L

21 a 4752 cm^3 b 640 m^3

22 a 36 000 cm^2 b 8.75 cm^2

23 $\frac{3}{5}$, 63%, 0.67, $\frac{6}{7}$

24 a true b false c true d false

25 a \$0.30 b 528 kg c \$4.80 d 15 min

26 a 40% b 35% c 50%

27 3281

28 a 15.42 cm b 91.42 mm c 80.55 cm

29 a 14.14 cm^2 b 557.08 mm^2 c 339.29 cm^2

30 A 31 589.0 cm^2 32 \$1360

33 a \$499 b 45.4%

34 a 62.1 m^2 b 55.2 m^2

35 a 942.5 cm^3 b 14 726.2 mm^3

CHAPTER 7

SkillCheck

1

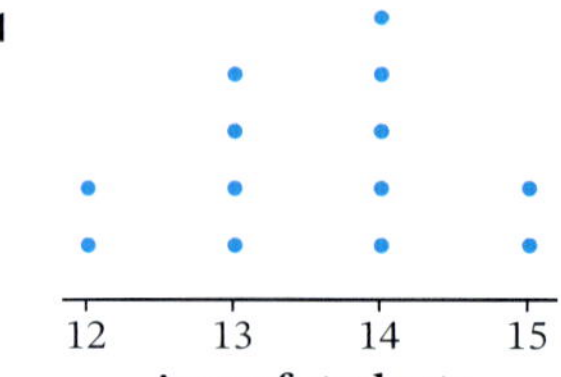

2 a

Stem	Leaf
3	0
4	7 8 8 9
5	4 9 9
6	3 4 8 8 8

b 30

3 a

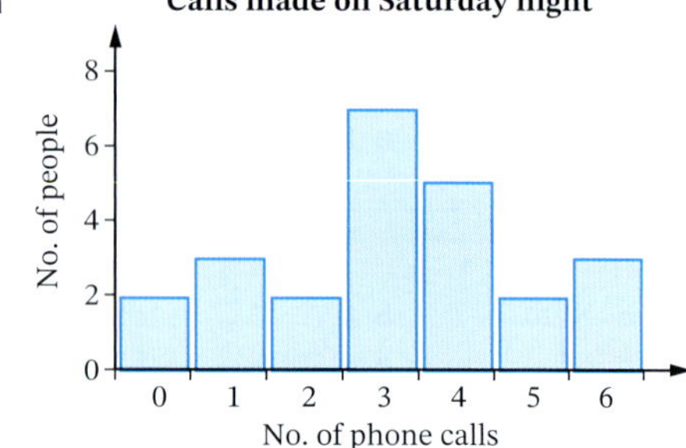

b 24 c 3

d 0, 2 and 5 e 6

4 a 53 b 10 c 240

Exercise 7.01

1

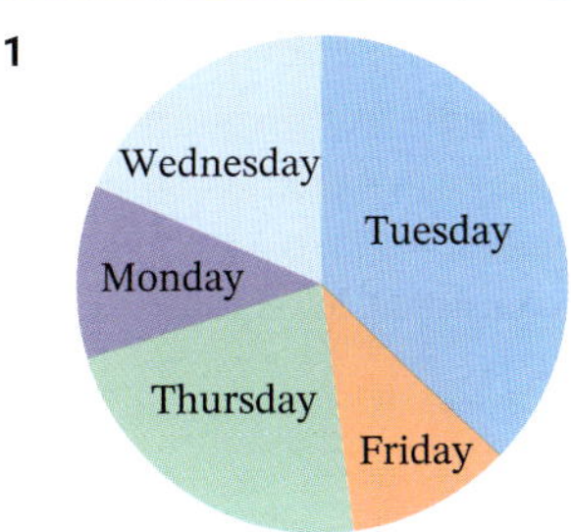

Day of sport training
Sector graph with angles: Mon = 36°, Tues = 156°, Wed = 48°, Thurs = 84°, Fri = 36°

2

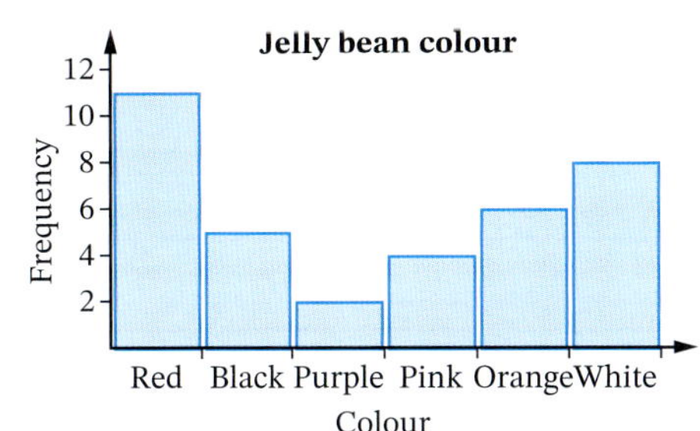

3 Hours of sleep

6	7	8	9	10	11

6 = 0.75 cm, 7 = 2.25 cm , 8 = 2.75 cm, 9 = 1.5 cm, 10 = 1.25 cm, 11 = 0.5 cm

4 D

5 **a**

Score	Tally	Frequency
1	||	2
2		0
3	|	1
4	|||	3
5	|||	3
6		0
7	~~||||~~ |	6
8	~~||||~~ ||	7
9	||||	4
10	||	1
	Total	27

b 10 **c** 3 **d** 8

e

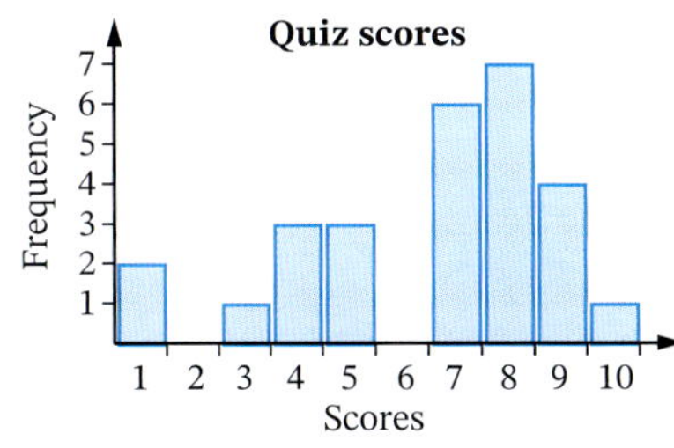

6 **a**

Colour	Frequency
Black (B)	10
Silver (S)	3
White (W)	6
Blue (U)	3
Red (R)	5
Total	27

b

Silver, Blue

Black		White		Red

c Black

7 **a**

No. of siblings	Tally	Frequency
0	~~||||~~	5
1	~~||||~~ ~~||||~~	10
2	~~||||~~ |||	8
3	~~||||~~ ||||	9
4	~~||||~~	5
5	|||	3
	Total	40

b 5 **c** 1

d Sector graph with angles: 0 = 45°, 1 = 90°, 2 = 72°, 3 = 81°, 4 = 45°, 5 = 27°

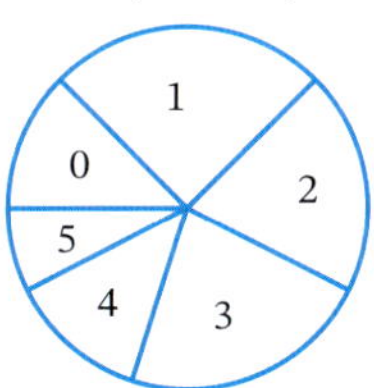

No. of siblings

Exercise 7.02

1 **a** 1.6 **b** 1 and 2

2 **a** 6.75, 6 **b** 4.71, 2 **c** 4.88, 1 and 7
d 70.44, 96 **e** 47.25, no mode

3 See answers to question **2**.

4 **a** Hyundai **b** spade **c** Korean

5 Data are categorical, not numerical.

6 **a** 7.71 **b** 7.0
c Mean, every value is involved in calculating it.

7 C

8 **a** Carrozza 26.3, Binns 33.3
b Binns family has higher values, outlier of 93.
c Binns family has higher mean.

9 **a** 8Y 6.1; 8Z 6.1
b 8Z's marks are more spread out.
c 4

10 2940

11 **a** \$1812.50 **b** \$2750

12 Teacher to check.

Exercise 7.03

1 B **2** D

3 **a** **i** 8 **ii** 7
b **i** 47.5 **ii** 10
c **i** 12 **ii** 4
d **i** 51 **ii** 99
e **i** 57 **ii** 0
f **i** 82.35 **ii** 37.4

4 **a** −$0.57
b median $9.50, mode −$25
c No. His winnings are small and his loss big. Not worth the effort.

5 **a** no mode **b** $686 000 **c** $771 845.45
d Median, because it's not affected by the outlier of $1 670 000.

6 **a** Aldo 168.1, Peter 175.7
b Aldo 35, Peter 6
c **i** Aldo, higher mean.
ii Peter is more consistent, lower range.
d 194

7 C

8 **a** Daniel 29.68, Cristian 29.08
b Daniel 5, Cristian 4.5
c Cristian, he is consistent and fast.

9 6, 7, 8 or 9 **10** Teacher to check.

Mental skills 7

2 **a** **b** **c** **d** **e** **f** **g** **h**

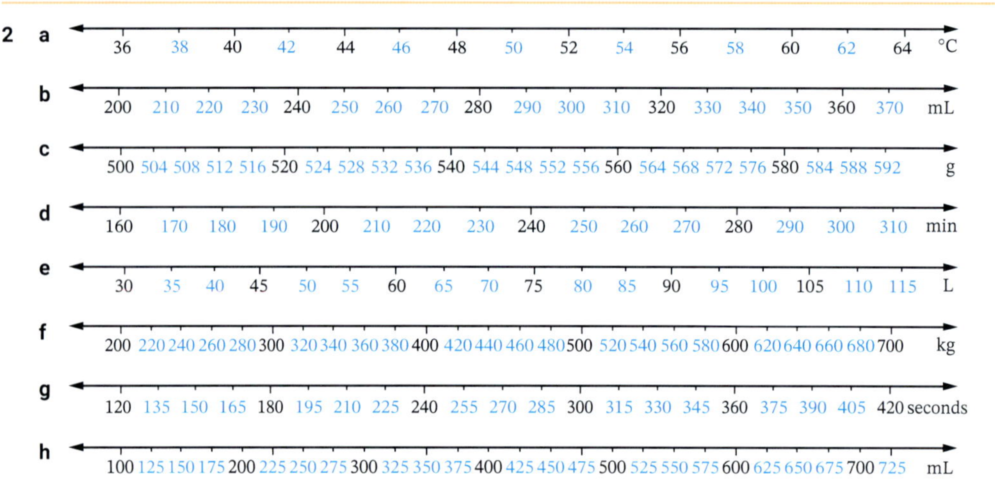

Exercise 7.04

1 **a** 34 **b** 2 **c** 2.1 **d** 2 **e** 8

2 **a** 24 **b** 11 **c** 8.6 **d** 5 **e** 9

3 **a**

Judge's score, x	Frequency, f	fx
1	2	2
3	1	3
4	3	12
5	3	15
7	6	42
8	7	56
9	4	36
10	1	10
Total	27	176

b 8 **c** 6.5 **d** 7 **e** 9

4 **a** 5 **b** 2 **c** 2.5 **d** 2.8

5 **a** 37 **b** 1.47 **c** 0.1 **d** 1.48 **e** 1.48

6 A

7 **a** 7 **b** 3 **c** 3 **d** 3.4

8 **a** 12 **b** 15 **c** 14.5

Exercise 7.05

1 **a** **i** 4 **ii** 11 **iii** 10 **iv** 9.8
b **i** 4 **ii** 24 **iii** 22 **iv** 22.3
c **i** 6 **ii** 36 **iii** 37 **iv** 37.6
d **i** 0.3 **ii** 1.8 **iii** 1.8 **iv** 1.8

2 **a**

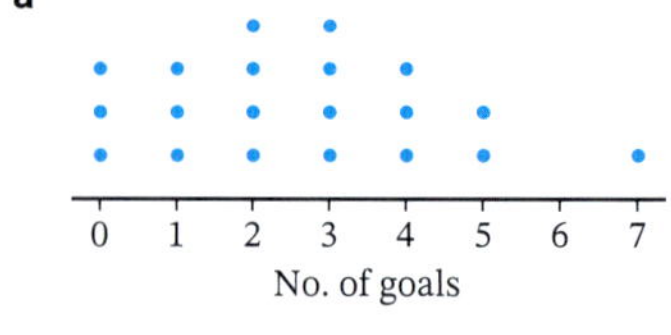

b 22 **c** 7 **d** 2.5 **e** 2.6

3 **a** 18 **b** 8 **c** 22 and 25
d 22.3 **e** 71% **f** B

4 **a**

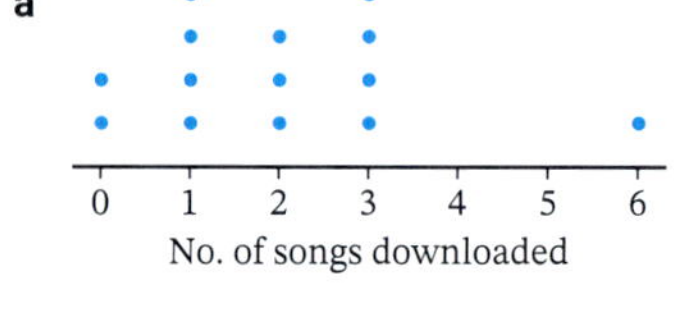

b 14 **c** 6 **d** 2 **e** 2 **f** 14

5 a

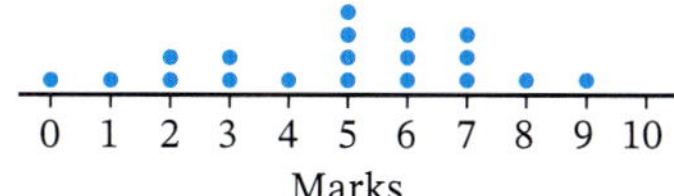

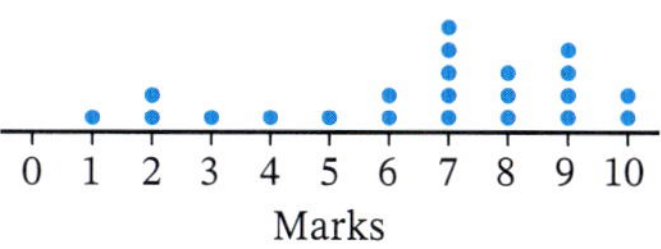

b 7 c 5 d 9

e

x	f	fx
0	1	0
1	1	1
2	2	4
3	2	6
4	1	4
5	4	20
6	3	18
7	3	21
8	1	8
9	1	9
Total	19	91

$\bar{x} \approx 4.8$

f 8 Yen, marks clustered around higher values 7 to 9.

6 a i 39 ii 33 iii 33 iv 30.1
b i 26 ii 94 iii 88 iv 86.3
c i 40 ii 131 iii 124.5 iv 121.5
d i 44 ii 17 iii 17 iv 21.7

7 a

Stem	Leaf
7	6
8	
9	2 5 7 8
10	1 1 5 5 6 8
11	2 4 7
12	4

b 76 c 101, 105 d 124
e 9, 10, 11 f 66.7%

8 a

Stem	Leaf
13	7
14	5
15	4 5 6 7 8 8 9
16	0 0 1 3 3 4 4 5 6 7 7 7 9
17	15

b 16 c 24 d 2
e 38 f 162

9 a 19 b 1
c Fuller's Ridge Rd, times are clustered around the lower values.
d D

10 a 16 b 8 c 30
d 9, 13 and 30 e 30
f Narrabri Lions
g Wee Waa Reds, it has a range of 28 compared to Narrabri Lion's 26.

Exercise 7.06

1 a

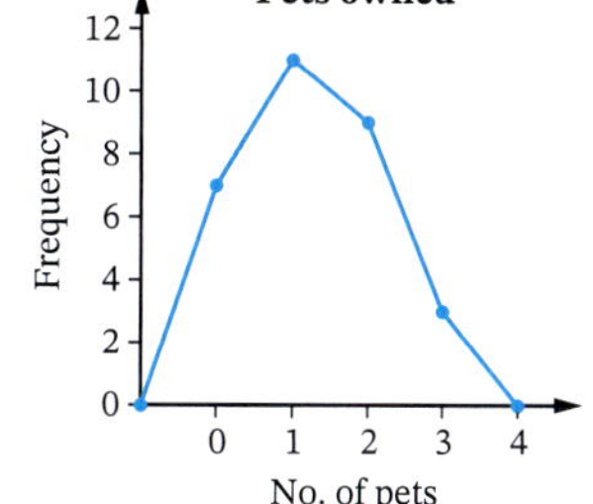

b 30 c 7 d 30%
e 1 f B

2 a

x	f
1	3
2	5
3	7
4	4
5	5
6	4
7	2
	30

b

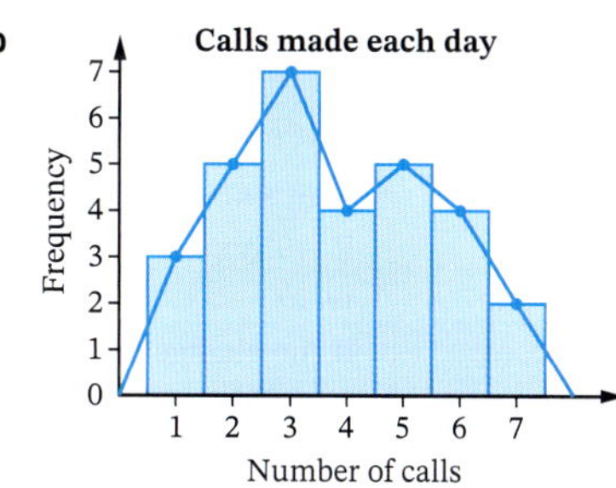

c 7 d 113 e 8 f 3 g 3.8

3 a 25 b 1 c 1 d 4

e

x	f	fx
1	13	13
2	5	10
3	3	9
4	1	4
5	3	15
	25	51

$\bar{x} = 2.04$

f 1

4 **a**

x	f
0	3
1	4
2	6
3	4
4	2
5	1
	20

b

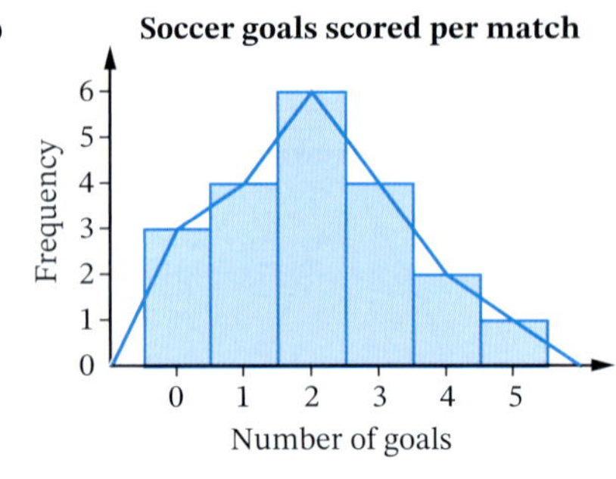

c 20 **d** 5 **e** 2

f 2.05 **g** 7 **h** 15%

5 **a** 5, 6, 9, 12, 4, 8, 44 **b** 4

c 5 **d** 155 **e** 154.64

f 11 **g** 155 **h** 24

Exercise 7.07

1 **a** **i** 37 **ii** 49 **iii** 52.5 **iv** 55.3

b Mean, because there are no outliers.

2 **a** **i** 24 **ii** 32 **iii** 37 **iv** 36.4

b **i** increase **ii** no change

iii decrease **iv** decrease

3 **a** 30 **b** B

c Grades are categorical data, not quantitative (numerical).

4 **a** **i** \$427 000 **ii** \$91 500 **iii** \$135 000

b There is no mode, and the mean is affected by the outlier of \$484 000.

c Mean \$103 727, median \$90 000, yes.

5 **a** **i** 192.4 cm **ii** 193 cm

iii no mode **iv** 23

b **i** decrease **ii** no change

iii There will be a mode of 181.

iv no change

6 **a** D **b** Sunjeet 32.8, Chris 33.7

c **i** Sunjeet **ii** Chris

d Sunjeet: his scores are consistently higher.

7 **a**

Forbes		Cobar
9 6 2 0	3	0 1 3 6 7
4	4	1 7 9 9
8 5 4	5	1 1 8
9 5 3 1	6	0 5
6 5 1	7	2 3 7 9
4 0 0	8	8
1 0	9	7

b

	Mean	Mode	Median	Range
Forbes	62.7	80	64	61
Cobar	56.2	49, 51	51	67

c Forbes, it has a higher mean, mode and median.

d 97, mean

8 **a**

	Mean	Mode	Median	Range
Christina	20.9	19, 25	21.5	31
Mary	21.9	17, 20	20	21

b Mary, lower range **c** Mary

d Mary, higher mean **e** 4

f **i** no **ii** yes

9 **a**

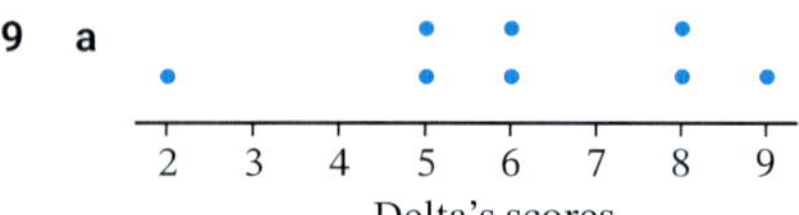

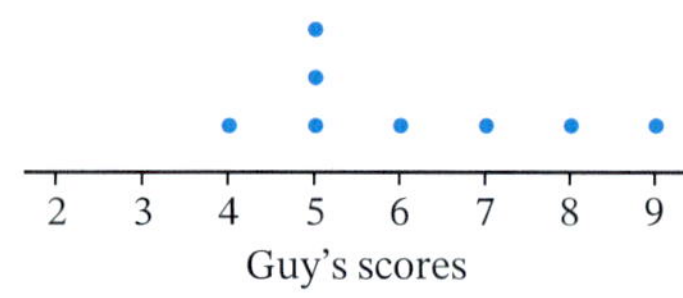

b Delta: range 7, median 6; Guy: range 5, median 5.5

c Other than the outlier of 2, Delta allocates higher marks, while Guy's marks are spread out.

Exercise 7.08

1 **a** sample **b** census **c** sample **d** sample

e sample **f** census **g** census **h** census

2 **a** sample, survey **b** sample, survey

c census, observation **d** sample, survey

e sample, survey **f** census, measurement

g sample, measurement **h** census, survey

3 C **4** Teacher to check.

5 **a** Not random, students volunteered themselves for the survey rather than be chosen.

b Random.

c Not random, many people do not have their phone numbers listed in the book.

d Random.

e Not random, station focuses on rock music only, and listeners call in to vote rather than be randomly selected.

6 Teacher to check.

Exercise 7.09

1 **a** What we mean by happy is not clear?

b Too many possible answers.

c Too personal and possibly unclear.

d Uses emotion to manipulate possible answers.

e Too personal.

f What is included in take-away food is not clear.
g Not all computer brands are covered by the choices given.
Teacher to check alternative questions.

2–4 Teacher to check.

Exercise 7.10

Teacher to check all answers.

Power plus

1 **a** 74.2 **b** 77.2
c Increased by 3. **d** Does the same to the mean?

2 92%

3 No, but the values will be centred around 20 (teacher to check).

4 **a** 40 **b, c** Teacher to check.

5 Yes. The sum of the four sisters' wages is \$8000, so each of the other two sisters earn much more (the sum of their wages is \$6700.).

Test yourself 7

1 **a**

x	f
1	2
2	4
3	7
4	5
5	1
6	1
	20

b 3 **c** 5 **d** 10%
e Sector graph with angles: $1 = 36°$, $2 = 72°$, $3 = 126°$, $4 = 90°$, $5 = 18°$, $6 = 18°$.

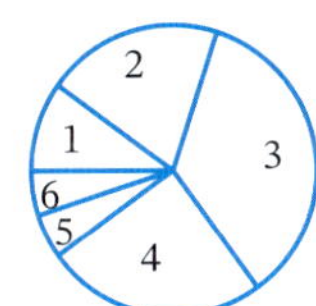

No. of computers owned

2 **a** **i** 8.78 **ii** 6 **iii** 10 **iv** 6
b **i** 3.5 **ii** 1 **iii** 3.5 **iv** 7
c **i** 22.6 **ii** no mode **iii** 23 **iv** 5

3 **a** 0, 7, 12, 12, 8, 5, 6, totals 24, 50
b 24 **c** 6 **d** 1 **e** 2 **f** 2.08

4 **a** **i** 4 **ii** 35 **iii** 33 **iv** 34.6
b **i** 35 **ii** 144 **iii** 148 **iv** 140.25

5 **a**

x	f
1	2
2	3
3	6
4	5
5	9
6	5
	30

b House visits

c 14 **d** 5 **e** 4

6 **a** 37 **b** 36 **c** 37.9
d mean **e** 11
f **i** increase **ii** no change
iii no change **iv** no change

7 **a** census **b** sample **c** sample **d** census

8, 9 Teacher to check.

10 **a** Too broad, 'What is the area of your house in square metres?'
b Suggestive/biased towards republican answer, 'Should Australia become a republic?'

CHAPTER 8

SkillCheck

1 **a**

b

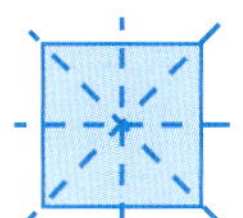

c

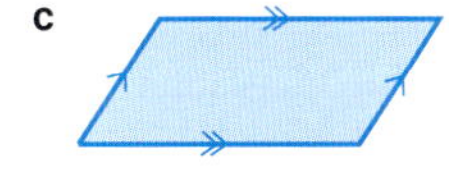

d

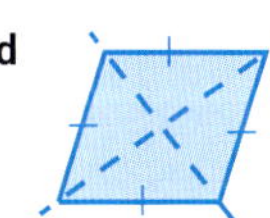

e

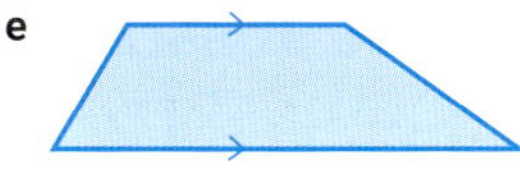

f

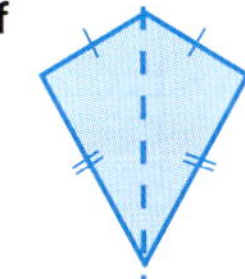

2 **a** rectangle, square
b parallelogram, rectangle, rhombus, square
c parallelogram, rectangle, rhombus, square
d trapezium
e rhombus, square
f kite, rhombus, square

3 **a** translation **b** reflection or rotation
c rotation **d** reflection
e rotation **f** translation

Exercise 8.01

1 **a** reflection–reflection, or reflection–translation or rotation–translation

b rotation–translation or reflection–reflection

c rotation–reflection

2 **a**

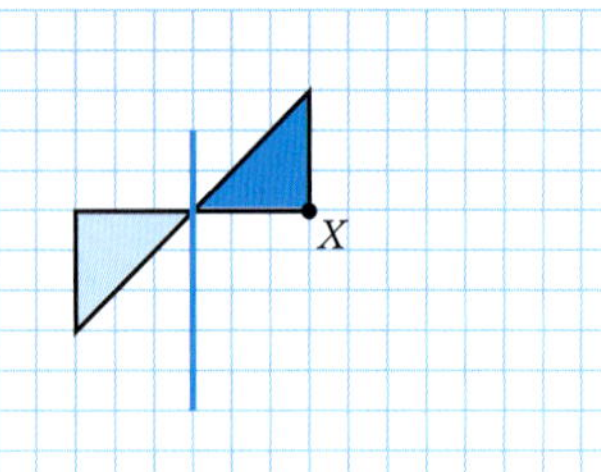

b

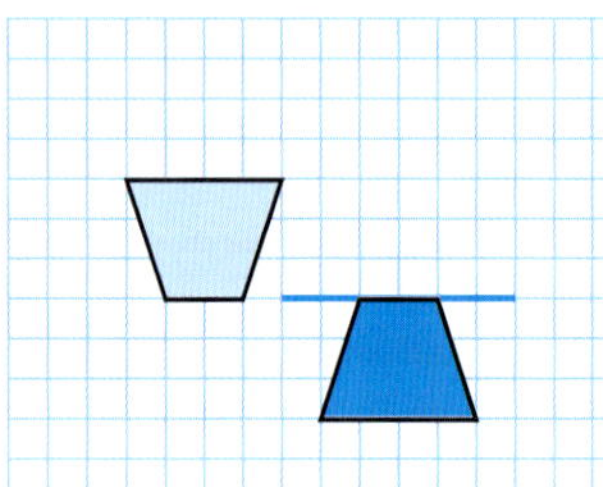

c

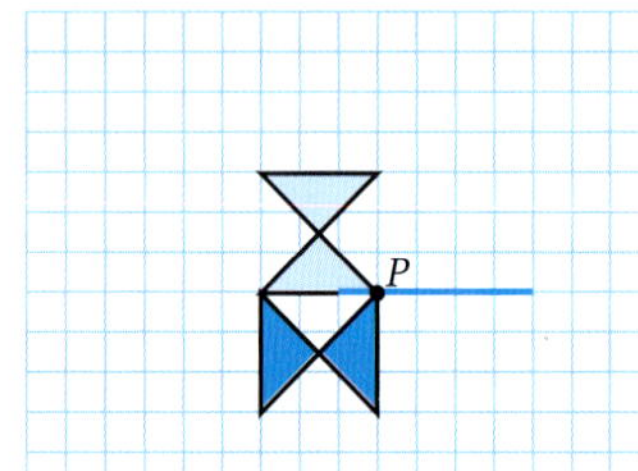

3 **a i** rotation, rotation **ii** translation

b i reflection, reflection **ii** rotation

c i rotation, reflection **ii** reflection

4 **a**

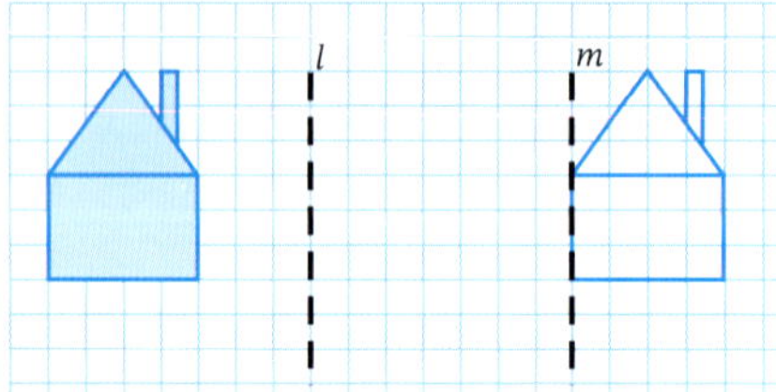

b translation

5 **a**

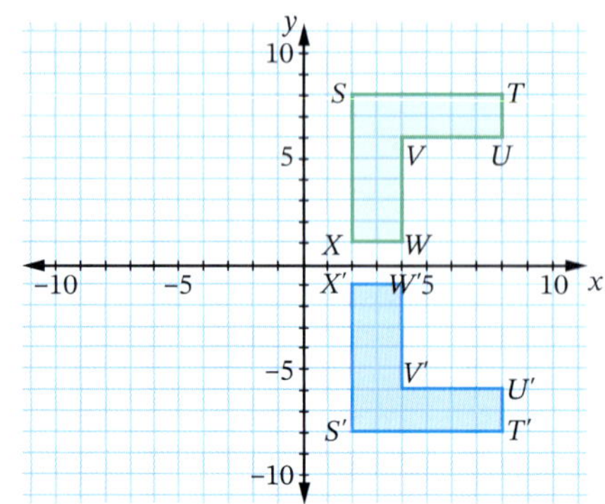

b $S(2, 8) \rightarrow S'(2, -8)$, $T(8, 8) \rightarrow T'(8, -8)$, $U(8, 6) \rightarrow U'(8, -6)$, $V(4, 6) \rightarrow V'(4, -6)$, $W(4, 1) \rightarrow W'(4, -1)$, $X(2, 1) \rightarrow X'(2, -1)$

The x-coordinate of each vertex does not change while the y-coordinate changes sign.

6 **a** Rotation 180° clockwise about $S(0, 0)$.

b $S(0, 0) \rightarrow S'(0, 0)$, $T(3, 0) \rightarrow T'(-3, 0)$, $U(3, -4) \rightarrow U'(-3, 4)$, $V(5, -4) \rightarrow V'(-5, 4)$, $W(5, 0) \rightarrow W'(-5, 0)$, $X(8, 0) \rightarrow X'(-8, 0)$, $Y(8, -7) \rightarrow Y'(-8, 7)$, $Z(0, -7) \rightarrow Z'(0, 7)$

The x- and y-coordinates of each vertex change sign.

7 **a**

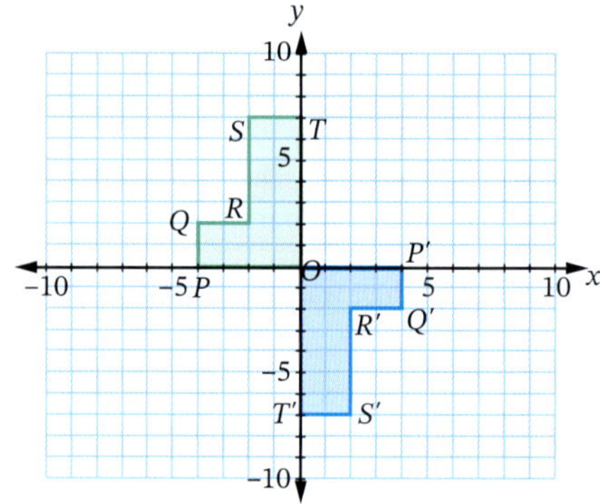

b $Q(-4, 2) \rightarrow Q'(4, -2)$; The x- and y-coordinates both change sign.

8 **a** Translation 8 units right and 2 units down.

b $W(-6, 8) \rightarrow W'(2, 6)$; $X(-2, 1) \rightarrow X'(6, -1)$; $Y(-6, 1) \rightarrow Y'(2, -1)$

The x-coordinate of each vertex increases by 8 units while the y-coordinate decreases by 2 units.

Exercise 8.02

1 A and H, B and K, E and L, F and P

2 **a** D and J **b** C **c** H

3 Other answers possible.

a $ABCD \equiv JKLI$ **b** $\triangle MNO \equiv \triangle PRQ$

c $NOPQ \equiv RSTU$ **d** $\triangle ACB \equiv \triangle EDF$

e $WXYZ \equiv KLMJ$ **f** $JKLMNO \equiv STUPQR$

4 Other answers possible.

a $\angle F$, DE **b** $\angle P$, RQ

c $\angle Q$, OP **d** $\angle R$, WV

e $\angle IHG$, EF

f $\angle QRS$ (or $\angle PON$), ON (or RS)

g $\angle EDH$ (or $\angle FGH$), NJ **h** $\angle YXW$, ZU

5 A

6 **a i** $AB = DE$, $BC = EF$, $AC = DF$

ii $\angle A = \angle D$, $\angle B = \angle E$, $\angle C = \angle F$

b i $LM = RQ$, $MN = QP$, $LN = RP$

ii $\angle L = \angle R$, $\angle M = \angle Q$, $\angle N = \angle P$

c i $AB = DF$, $AC = DE$, $CB = EF$

ii $\angle A = \angle D$, $\angle B = \angle F$, $\angle C = \angle E$

d i $PQ = CD$, $QR = DE$, $PR = CE$

ii $\angle P = \angle C$, $\angle Q = \angle D$, $\angle R = \angle E$

7 

8 

Exercise 8.03

1 Teacher to check.

2 **a** $\triangle MNO$ **b** 60°

3 **a** $\triangle HIJ$ **b** 2 cm

4 **a** different **b** congruent **c** different
d different **e** different **f** congruent
g congruent

Mental skills 8A

2 **a** 12:25 p.m. **b** 1:10 a.m. **c** 10:50 p.m.
d 10:55 p.m. **e** 06:10 **f** 00:10
g 9:10 a.m. **h** 3:15 a.m. **i** 11:00
j 23:05 **k** 12:20 a.m. **l** 11:35 a.m.

4 **a** 6:05 p.m. **b** 6:40 a.m. **c** 12:10 p.m.
d 2:50 a.m. **e** 12:45 **f** 03:55
g 10:50 p.m. **h** 12:15 p.m. **i** 15:45
j 04:00

Exercise 8.04

1 B

2 **a** SAS **b** AAS **c** AAS **d** SSS
e RHS **f** SAS **g** AAS **h** SSS
i RHS **j** SAS

3 **a** B **b** $\triangle ABC \equiv \triangle DEF$

4 **a** $p = 2, n = 20$ **b** $x = 15, y = 3$
c $x = 58$ **d** $x = 4, y = 60$

5 **a** AAS, $x = 6$ **b** AAS, $x = 6$

6 **a** Vertically opposite angles.
b SAS
c $\triangle PQR \equiv \triangle STR$ **d** $y = 28$
e $\angle S$
f Alternate angles are equal.

Exercise 8.05

1 **a** SSS **b** $\angle C$ **c** $\angle CDB$
d
e SAS
f XC **g** $\angle DXC$ **h** 90°
i The diagonals intersect at right angles.

2 **a** alternate angles on parallel lines.
b $\angle QSR$ **c** AAS
d RQ **e** SR
f opposite sides are equal.
g $\angle R$
h opposite angles are equal.

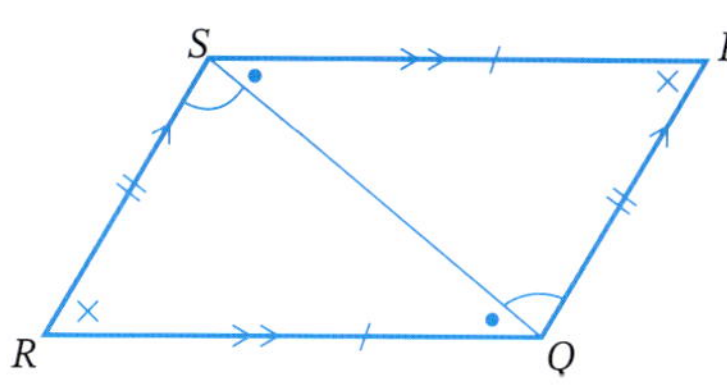

3 **a** $\angle XRQ$ **b** AAS **c** XQ
d XR **e** Diagonals bisect each other.

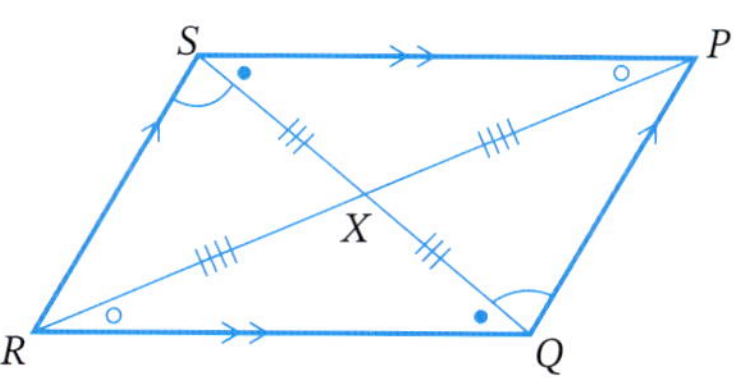

4 **a** SSS
b 90°, four equal angles at a point (in a revolution) add up to 360°.
b, d, e, f

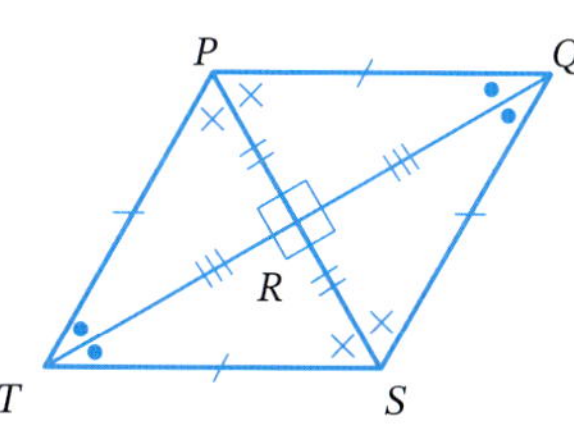

c The diagonals intersect at right angles.
g The diagonals bisect the angles of the rhombus.

5 **a** SAS **b** XZ
c Diagonals are equal. **e** SSS
f 90° **g** Diagonals bisect at right angles.
h 45°

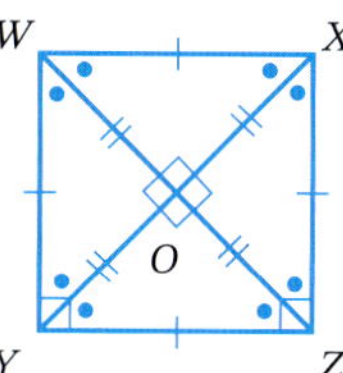

i Diagonals bisect the angles of a square.

Exercise 8.06

1 **a** $ABCD \mathrel{|||} IJKL$ **b** $\triangle STU \mathrel{|||} \triangle XVW$
c $ABCD \mathrel{|||} JKLI$ **d** $UVWXY \mathrel{|||} MNJKL$
e $\triangle HGF \mathrel{|||} \triangle JIK$

2 **a** Similar

i $\angle A = \angle E, \angle B = \angle F, \angle C = \angle G, \angle D = \angle H$

ii AB and EF, AD and EH, BC and FG, DC and HG

iii 3 : 2

b Not similar **c** Similar

i $\angle J = \angle T, \angle K = \angle P, \angle L = \angle Q, \angle M = \angle R, \angle N = \angle S$

ii JK and TP, KL and PQ, LM and QR, NM and TS, JN and TS

iii 5 : 7

d Not similar **e** Not similar

f Not similar **g** Not similar

h Similar

i $\angle S = \angle U, \angle P = \angle T, \angle R = \angle V, \angle Q = \angle W$

ii SR and UV, SP and UT, RQ and VW, PQ and TW

iii 3 : 2

3 **a** $x = 44$ **b** $y = 60$ **c** $m = 150$

d $n = 100$ **e** $p = 55$ **f** $q = 32$

4 **a** **i** Yes **ii** No **iii** Yes

b **i** All circles are similar.

ii Measured length and width, ratio is different.

iii All semi-circles are similar.

5 Teacher to check.

6 **a** True **b** False **c** False **d** True

Exercise 8.07

1 **a** 2 **b** 3 **c** $\frac{3}{4}$

d 1.5 **e** 2.5 **f** $\frac{1}{3}$

2 **a** AA **b** SSS **c** SAS

d SAS **e** SSS **f** AA

3 If two corresponding angles are same then the third pair of corresponding angles will also be the same using the sum of angles in a triangle angle relationship.

4 **a** Yes using AA **b** Yes using SSS

c Yes using SAS **d** No

e No **f** Yes using AA

g Yes using SSS **h** No

5 **a** 4 cm **b** 2 cm **c** 9 cm **d** 4.8 cm

e 10 cm **f** 9 cm **g** 7 cm **h** 52 cm

6 **a** 15 cm **b** 10 mm **c** 9 cm **d** 15 cm

e 5 m **f** 8 mm **g** 27 cm **h** 7.65 m

7 **a** Similar using AA, $x = 4, y = 0.75$

b Similar using AA, $x = 10, y = 12$

c Similar using RHS, $x = 72$cm

d Similar using AA, $x = 4.8, y = 12$

8 10.5 m

9 1.8 m

10 2.8 m

11 72.2 m

Mental skills 8B

2 **a** 8 h 30 min **b** 5 h 40 min **c** 3 h 25 min

d 8 h 15 min **e** 11 h 25 min **f** 1 h 40 min

g 5 h 10 min **h** 5 h 45 min **i** 7 h 55 min

j 7 h 40 min

Power plus

1 **a** WX; XY, equal sides of a kite; WY

b AC, $\triangle ABC$ is isosceles; DC, the midpoint of BC; common; SSS

c JK; opposite sides of a parallelogram; GK, opposite sides of a parallelogram; GKJ, opposite angles of a parallelogram; GHJ, JKG, SAS

2 **a** all angles 90°/equal **b** all sides equal

c all sides equal

3 **a** parallelogram, rectangle, rhombus, square

b kite, rhombus, square

c rectangle, square

d rectangle, square

e parallelogram, rectangle, rhombus, square

f parallelogram, rectangle, rhombus, square

Test yourself 8

1

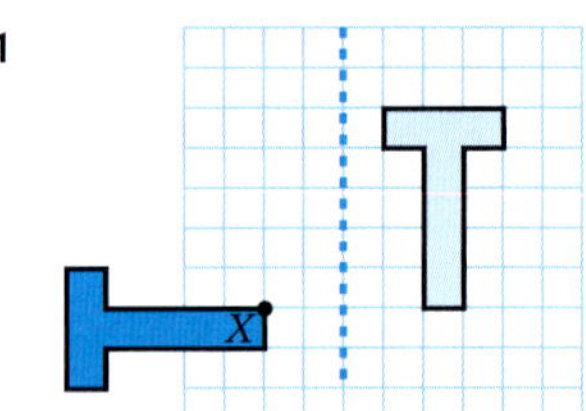

2 **a** Reflection across the y-axis.

b $C(5, -4) \rightarrow (-5, -4)$, the x-coordinate changes sign while the y-coordinate stays the same.

3 **i** **a** $AB = UT$ or $BC = TS$ or $AC = US$

b $\angle A = \angle U$ or $\angle B = \angle T$ or $\angle C = \angle S$

c $\triangle ABC \equiv \triangle UTS$

ii **a** $DE = WX$ or $FE = VX$ or $DF = WV$

b $\angle D = \angle W$ or $\angle E = \angle X$ or $\angle F = \angle V$

c $\triangle DEF \equiv \triangle WXV$

iii (other answers possible)

a $GH = OM$ or $IH = NM$ or $GI = ON$

b $\angle G = \angle O$ or $\angle H = \angle M$ or $\angle I = \angle N$

c $\triangle GHI \equiv \triangle OMN$

iv (other answers possible)

a $JK = RQ$ or $KL = QP$ or $JL = RP$

b $\angle J = \angle R$ or $\angle K = \angle Q$ or $\angle L = \angle P$

c $\triangle JKL \equiv \triangle RQP$

4 **b** no

5 **a** SAS **b** RHS **c** AAS

6 **a** SSS

b four equal matching angles at R add up to 360°, so each angle is 90°.

c $\angle PTR$, $\angle STR$, $\angle SQR$

d **i** false **ii** true **iii** true

7 a $PQ = XY$ or $QR = XW$ or $SR = ZW$ or $PS = YZ$
  b $\angle P = \angle Y$ or $\angle Q = \angle X$ or $\angle R = \angle W$ or $\angle S = \angle Z$
  c $PQRS \parallel\!| YXWZ$
8 4
9 AA; SSS; SAS; RHS
10 a 3 m b 45 cm
  c $t = 9$ cm; $m = 5.3$ cm d 100
11 $\triangle BCD \parallel\!| \triangle BEA$ using SAS and scale factor of 1.5.
12 The scoreboard is 25 m high.

CHAPTER 9

SkillCheck

1 a 0.85 b 0.375 c 0.02 d 0.3
2 a {head, tail} b {1, 2, 3, 4, 5, 6}
  c {female, male} d {red, amber, green}
  e {win, lose, draw} f {pass, fail}
3 a 70% b 56% c $66\frac{2}{3}\%$ d 24%
4 a $\frac{2}{3}$ b $\frac{3}{5}$ c $\frac{5}{8}$
5 C
6 a $\frac{7}{25}$ b $\frac{1}{50}$ c $\frac{16}{25}$ d $\frac{4}{5}$
7 Answers will vary between students.

Exercise 9.01

1 a i {red, blue} ii $\frac{1}{2}$
  b i {red, blue, yellow, green} ii $\frac{1}{4}$
  c i {red, blue} ii $\frac{1}{2}$
  d i {red, white} ii $\frac{1}{5}$
  e i {red, white, black, green} ii $\frac{1}{4}$
  f i {red, green, yellow} ii $\frac{1}{2}$
2 a 2, equally likely b 3, not equally likely
  c 26, not equally likely d 2, equally likely
  e 10, equally likely f 2, not equally likely
3 a {1, 3, 5} b {A, E, I, O, U}
  c {5, 6, 7, 8, 9} d {April, August}
  e {NSW, Qld, SA, Tas., Vic., WA}
  f {Years 7, 8, 9, 10, 11, 12}
4 a $\frac{2}{20} = \frac{1}{10}$ b $\frac{3}{20} = 0.15$ c $\frac{11}{20} = 55\%$
  d $\frac{15}{20} = \frac{3}{4}$ e $\frac{7}{20} = 0.35$ f $\frac{13}{20} = 65\%$
5 a $\frac{3}{16}$ b $\frac{8}{16} = \frac{1}{2}$
  c $\frac{14}{16} = 0.875$ d $\frac{10}{16} = 62.5\%$
6 C

7 a {win, lose, draw}
  b No, teams have different abilities.
8 D
9 a 3 b {5, 6, 7, 8}
  c

Outcome	5	6	7	8
Probability as a fraction	$\frac{1}{4}$	$\frac{1}{3}$	$\frac{1}{6}$	$\frac{1}{4}$
Probability as a decimal	0.25	$0.\dot{3}$	$0.1\dot{6}$	0.25
Probability as a percentage	25%	$33\frac{1}{3}\%$	$16\frac{2}{3}\%$	25%

10 C
11 a C b A c E d H
  e D f G g B h F
12 a 52 b Yes c 4
  d 13 e 2
  f i $\frac{1}{2}$ ii $\frac{1}{4}$ iii $\frac{1}{52}$
   iv $\frac{1}{13}$ v $\frac{5}{13}$ vi $\frac{3}{26}$
13 a low b 9
14 Teacher to check.
15 a 1 b 5 c 9 d 10

Exercise 9.02

1 a tossing heads.
  b sunny day tomorrow.
  c selecting a brown chocolate.
  d rolling 1, 2, 3, 4 or 5.
  e losing or having a draw (not winning).
  f selecting a diamonds, clubs or spades card (not selecting hearts).
  g born in spring, summer or autumn (not being born in winter).
  h being 15 or over.
2 A
3 a $\frac{1}{6}$ b $\frac{5}{6}$ c $\frac{1}{2}$ d $\frac{1}{2}$
  e $\frac{5}{6}$ f $\frac{2}{3}$
4 a $\frac{1}{3}$ b $\frac{2}{3}$ c $\frac{5}{6}$ d 1
5 a $\frac{45}{56}$ b D
6 a $\frac{1}{6}$ b $\frac{5}{6}$
7 0.8
8 $\frac{199}{200}$
9 a $\frac{3}{26}$ b $\frac{23}{26}$ c 1
  d $\frac{11}{26}$ e $\frac{9}{13}$ f $\frac{1}{2}$
10 a $\frac{12}{13}$ b 39% c 0.85

11 **a** 0.65 **b** 0.35
c Red 16, Blue 14, White 10
d 0

12 **a** $\frac{3}{4}$ **b** $\frac{1}{2}$ **c** $\frac{1}{2}$ **d** 0

13 15% **14** C

15 **a** $\frac{7}{16}$ **b** $\frac{9}{16}$

16 **a** $\frac{10}{11}$ **b** $\frac{7}{11}$ **c** $\frac{9}{11}$ **d** $\frac{8}{11}$

17 **a** $\frac{2}{5}$ **b** $\frac{3}{5}$ **c** $\frac{4}{5}$ **d** $\frac{3}{5}$

18 **a** $\frac{1}{4}$ **b** $\frac{3}{4}$ **c** $\frac{1}{2}$

Exercise 9.03

1 **a** 50 **b** 21 **c** 42
d 33 **e** 8 **f** A

2 **a** Yes, mutually exclusive **b** 6
c 40 **d** 46 **e** 0

3 **a** No **b** 52 **c** 4
d **i** 0.808 **ii** 0.731 **iii** 1 **iv** 0.192
e right-handed male

4 **a** 440 **b** 368 **c** 15 **d** A

5 **a**

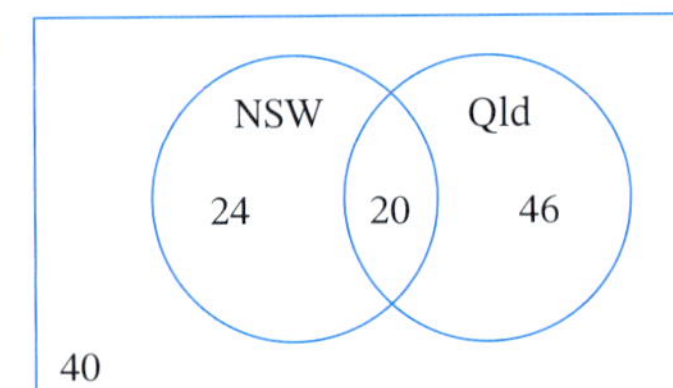

b 90 **c** 70 **d** $\frac{12}{65}$ **e** $\frac{4}{13}$

6 **a**

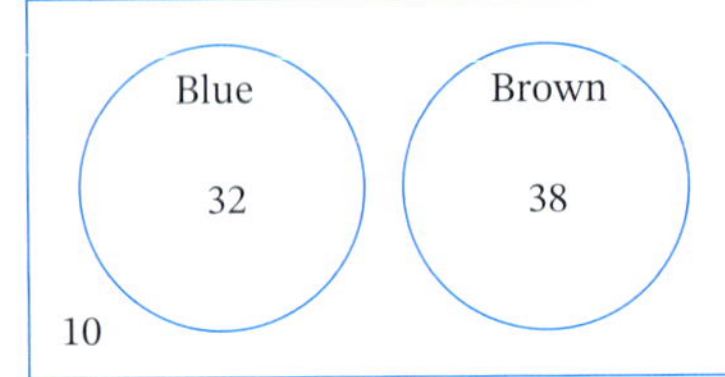

b Yes, cannot be both blue-eyed and brown-eyed.
c 80 **d** 70 **e** 0 **f** 52.5%

7 **a**

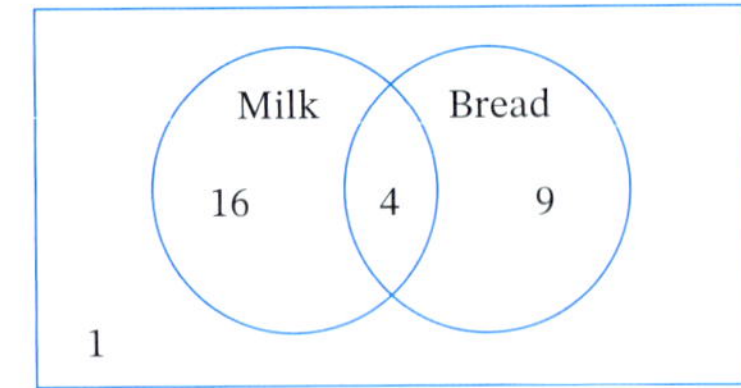

b 4 **c** 16

d **i** $\frac{3}{10}$ **ii** $\frac{1}{3}$ **iii** $\frac{29}{30}$ **iv** $\frac{5}{6}$

8 **a**

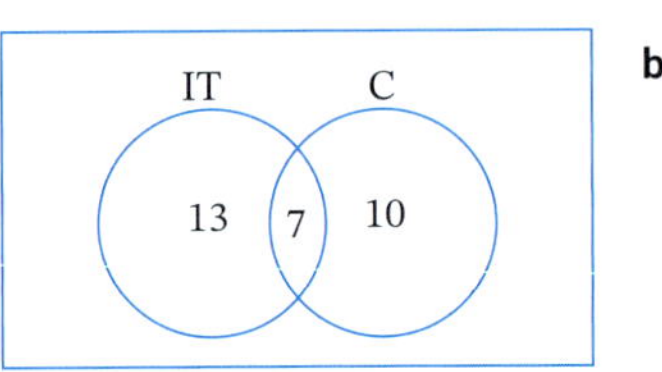

b $\frac{17}{30}$

Exercise 9.04

1 **a** 27 **b** 18 **c** 2 **d** $\frac{4}{9}$
e **i** $\frac{5}{27}$ **ii** $\frac{9}{27}=\frac{1}{3}$

2 **a** 336 **b** 192 **c** 96
d **i** $\frac{11}{21}$ **ii** $\frac{4}{21}$ **iii** $\frac{17}{21}$

3 **a** 56; 101, 146; 101, 243
b 146 **c** 56
d **i** $\frac{101}{344}$ **ii** $\frac{36}{43}$ **iii** $\frac{187}{344}$ **iv** $\frac{243}{344}$

4 **a** 125; 70, 110; 165, 310 **b** 165
c **i** $\frac{11}{31}$ **ii** $\frac{24}{31}$ **iii** $\frac{25}{62}$
d D

5

	Swimming	Not swimming	
Bowling	11	20	31
Not bowling	34	5	39
	45	25	70

6 **a**

	GPS navigation	No GPS navigation	
Four doors	25	42	67
Not four doors	3	12	15
	28	54	82

b **i** $\frac{14}{41}$ **ii** $\frac{67}{82}$ **iii** $\frac{25}{82}$ **iv** $\frac{27}{41}$

7 **a**

	Needs glasses	Does not need glasses	
Male	22	60	82
Female	13	85	98
	35	145	180

b **i** $\frac{49}{90}$ **ii** $\frac{7}{36}$ **iii** $\frac{13}{180}$ **iv** $\frac{167}{180}$
v $\frac{2}{3}$ **vi** $\frac{107}{180}$

Mental skills 9

2 **a** 16 000 **b** 210 **c** 9900 **d** 600
e 16 000 **f** 400 **g** 490 **h** 52
i 1500 **j** 1360 **k** 195 **l** 17 600

4 **a** 8 **b** 50 **c** 8 **d** 22
e 4 **f** 50 **g** 9 **h** 9.3
i 8.5 **j** 4.23 **k** 9.6 **l** 40.8

9780170465601

Exercise 9.05

1 **a** **i** 2

ii 6

b **i** {H1, H2, H3, H4, H5, H6, T1, T2, T3, T4, T5, T6} where H = head, T = tail

		Die					
		1	**2**	**3**	**4**	**5**	**6**
Coin	**H**	H1	H2	H3	H4	H5	H6
	T	T1	T2	T3	T4	T5	T6

Coin	Die	Outcomes
H	1	H1
	2	H2
	3	H3
	4	H4
	5	H5
	6	H6
T	1	T1
	2	T2
	3	T3
	4	T4
	5	T5
	6	T6

c 12 **d** **i** $\frac{1}{4}$ **ii** $\frac{1}{6}$ **iii** $\frac{5}{12}$

2 **a**

		1st die					
	+	**1**	**2**	**3**	**4**	**5**	**6**
2nd die	**1**	2	3	4	5	6	7
	2	3	4	5	6	7	8
	3	4	5	6	7	8	9
	4	5	6	7	8	9	10
	5	6	7	8	9	10	11
	6	7	8	9	10	11	12

b 36

c **i** $\frac{1}{12}$ **ii** $\frac{5}{36}$ **iii** $\frac{1}{12}$

iv 0 **v** $\frac{1}{2}$ **vi** $\frac{2}{9}$

d **i** 2.7% **ii** 16.66%

e **i** 7 **ii** 2, 12

3

1st digit	2nd digit	Outcomes
4	5	45
	7	47
	8	48
5	4	54
	7	57
	8	58
7	4	74
	5	75
	8	78
8	4	84
	5	85
	7	87

b 12

c **i** 25% **ii** 50% **iii** 50% **iv** 25%

4 **a**

		1st course			
		C	**R**	**W**	**Y**
2nd course	**B**	CB	RB	WB	YB
	P	CP	RP	WP	YP
	S	CS	RS	WS	YS

b 12 **c** **i** $\frac{1}{2}$ **ii** $\frac{1}{6}$ **iii** $\frac{1}{2}$ **iv** $\frac{1}{12}$

a

		Chairperson				
		H	**R**	**A**	**K**	**T**
Secretary	**H**	–	RH	AH	KH	TH
	R	HR	–	AR	KR	TR
	A	HA	RA	–	KA	TA
	K	HK	RK	AK	–	TK
	T	HT	RT	AT	KT	–

H = Henryk, R = Ramy, A = Anastacia, K = Krystal, T = Tom

b **i** $\frac{1}{20}$ **ii** $\frac{2}{5}$ **iii** $\frac{1}{5}$ **iv** $\frac{1}{10}$

6 **a**

Saturday Sunday Outcomes

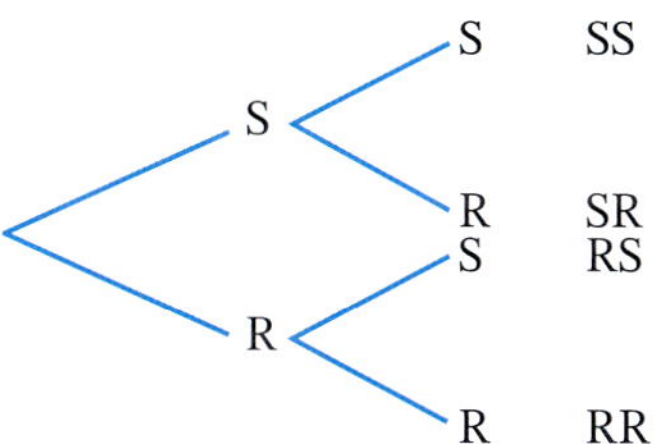

b **i** $\frac{1}{4}$ **ii** $\frac{1}{2}$ **iii** $\frac{1}{2}$ **iv** $\frac{3}{4}$

7 **a** **i** {HHH, HHT, HTH, HTT, THH, THT, TTH, TTT}

1st toss 2nd toss 3rd toss Outcomes

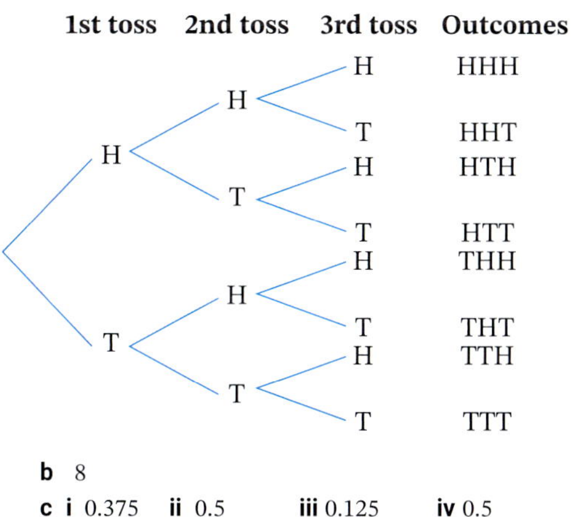

b 8

c **i** 0.375 **ii** 0.5 **iii** 0.125 **iv** 0.5

Exercise 9.06

1 a $\frac{17}{28}$ b $\frac{2}{7}$ c $\frac{6}{7}$

2 a {U, E, C, S}

b No, some occur more than others.

c i $\frac{3}{7}$ ii $\frac{2}{7}$ iii $\frac{5}{7}$ iv 0

3 a {black, red, blue}

b i $\frac{1}{3}$ ii $\frac{1}{3}$ iii 0

c choosing a black or blue sock, $\frac{2}{3}$

4 a 20% b 40% c 80% d 100%

5 D

6 a $\frac{1}{100}$ b $\frac{3}{5}$ c $\frac{1}{2}$ d $\frac{17}{20}$

7 $\frac{1}{2}$

8 a $\frac{8}{15}$ b $\frac{3}{5}$ c $\frac{7}{15}$ d $\frac{2}{5}$

9 a $\frac{17}{30}$ b $\frac{1}{10}$ c $\frac{2}{3}$ d $\frac{13}{30}$

10 a

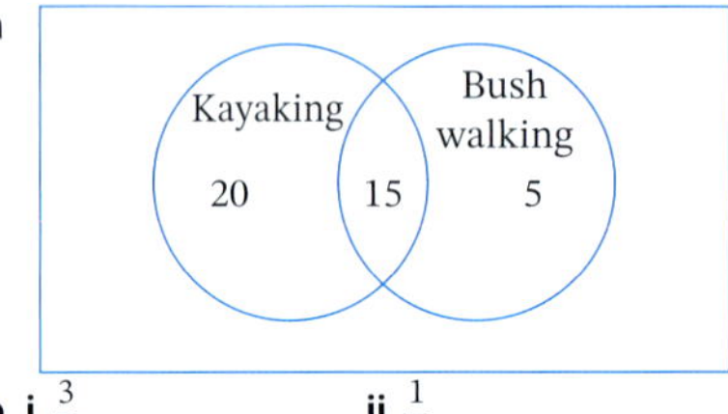

b i $\frac{3}{8}$ ii $\frac{1}{2}$

11 0.9

12 a 232; 140; 228, 272, 500

b i $\frac{57}{125}$ ii $\frac{7}{25}$ iii $\frac{1}{5}$

iv $\frac{32}{125}$ v $\frac{23}{25}$ vi $\frac{83}{125}$

13 a 0.3 b 5 red, 2 green, 3 purple

c 50 red, 20 green, 30 purple

d No difference, the probabilities are the same.

Exercise 9.07

1 a $\frac{1}{4}$ b 125

2 Teacher to check.

3 a $\frac{1}{2}$ b 40, less than

c $\frac{11}{20}$ d $\frac{1}{5}$

e 16, the observed frequency was less than expected.

f $\frac{11}{80}$

4 a 50 b $\frac{8}{25}$ c $\frac{9}{50}$

d i 400 ii 820

5 c i $\frac{1}{6}$ ii $\frac{2}{3}$ iii $\frac{1}{2}$

d i 12 ii 48 iii 36

g i 167

6 46%

7 a Yes

b Yes, 6 comes up much more often.

c $13\frac{3}{4}$% d 11

8 a i 0.06 ii 0.22 iii 0.44

iv 0.76 v 0.34 vi 0.68

b i 8 ii 5, 9 iii 2

iv 10 v 10 vi 6, 7

Exercise 9.08

Answers may vary: teacher to check.

Power plus

1

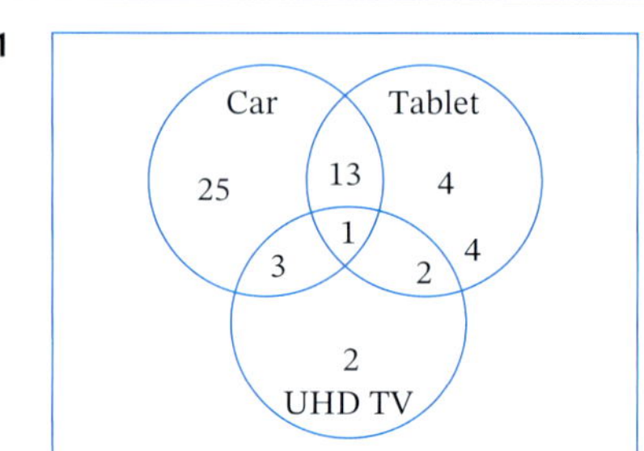

2 a

		1st die					
	×	1	2	3	4	5	6
2nd die	1	1	2	3	4	5	6
	2	2	4	6	8	10	12
	3	3	6	9	12	15	18
	4	4	8	12	16	20	24
	5	5	10	15	20	25	30
	6	6	12	18	24	30	36

b 18 c They occur different numbers of times.

d 6, 12 e 1, 9, 16, 25, 36

f i $\frac{1}{9}$ ii $\frac{1}{18}$ iii $\frac{2}{9}$

g 4

3 No, the probabilities of the 3 events are not equal.

4 a {Paul and Summet, Paul and Tash, Paul and Nadine, Sumeet and Tash, Sumeet and Nadine, Tash and Nadine}

b $\frac{1}{2}$

Test yourself 9

1 B

2 a $\frac{1}{6}$ b $\frac{2}{3}$ c $\frac{1}{2}$ d $\frac{1}{3}$

3 a 5, 6, 7, 8

b i 25% ii 100% iii 25% iv 0%

4 a Choosing a card that is not an Ace, $\frac{12}{13}$.

b Rolling a number that is not a factor of 6, $\frac{1}{3}$.

c Buying a ticket that does not win, $\frac{999}{1000}$.

5 26%

6 **a** $\frac{6}{25}$ **b** $\frac{17}{40}$ **c** $\frac{67}{200}$ **d** $\frac{133}{200}$

7 **a** 73 **b** 16 **c** 50

d **i** $\frac{44}{73}$ **ii** $\frac{66}{73}$ **iii** $\frac{7}{73}$

8 **a** 70 **b** 43 **c** 63

d **i** $\frac{1}{10}$ **ii** $\frac{1}{2}$ **iii** $\frac{11}{35}$ **iv** $\frac{27}{70}$

9

	Tall	Not tall	
Blond	3	5	8
Not blond	10	12	22
	13	17	30

10 **a** M = Marcus, W = Washi, P = Piyush, S = Souraya, D = Diane, A = Amia

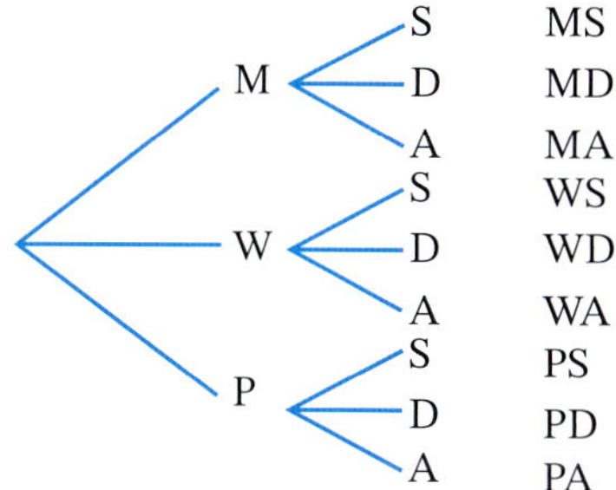

b $\frac{1}{9}$ **c** $\frac{4}{9}$

11 **a** 0.125 **b** 0.25 **c** 0.75 **d** 0.625

12 **a** $31\frac{1}{4}\%$ **b** $33\frac{1}{3}\%$ **c** 27

PRACTICE SET 3

1 A and J, C and F, D and G

2 **a** 25% **b** 75% **c** 75%

3 **a**

Mark	Tally	Frequency
6	𝍸	5
7	𝍸 ‖	7
8	𝍸	5
9	𝍸 𝍸	10
10	‖‖	3

b 30

c

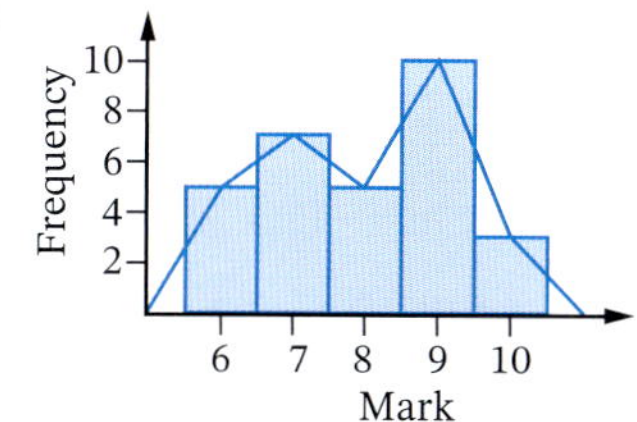

4 **a** $\triangle ABC \equiv \triangle GHI$

b $AB = GH$ or $BC = HI$ or $AC = GI$, $\angle A = \angle G$ or $\angle B = \angle H$ or $\angle C = \angle I$

5 **a**

Score	Frequency
1	4
2	9
3	8
4	3
5	10
6	6
Total	40

b

Result of roll of die

1	2	3	4	5	6

6 **a** $ED = ZY$ or $DC = YX$ or $CB = XW$ or $BA = WV$ or $AE = VZ$

b $\angle A = \angle V$ or $\angle E = \angle Z$ or $\angle D = \angle Y$ or $\angle C = \angle X$ or $\angle B = \angle W$

c $ABCDE \mathbin{|||} VWXYZ$

7 **a** $\frac{1}{2}$ **b** $\frac{5}{6}$ **c** $\frac{1}{3}$

8

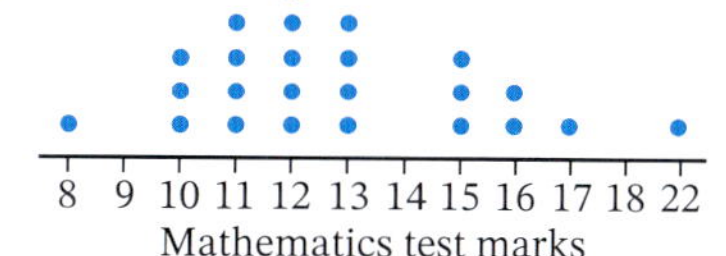

a 12 **b** 12 **c** 22 **d** 10–13

9 **a**

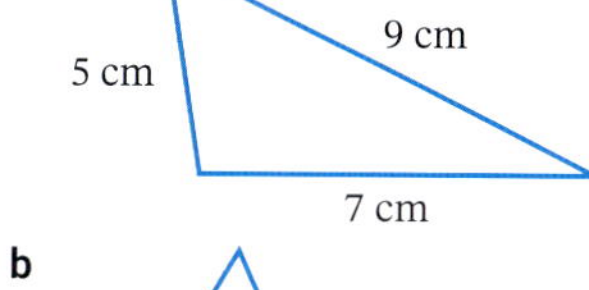

b

5 cm

65°

4 cm

10 **a** 14.4 **b** $x = 8; y = 20$

11 **a** 92.5 **b** 90 **c** 92 **d** 8

12 $\frac{7}{10}$

13 3

14 **a**

x	f	fx
28	6	168
29	1	29
30	6	180
31	2	62
32	3	96
33	2	66
34	4	136
	24	737

b $\frac{7}{24}$

c i 30.7 **ii** 30 **iii** 28, 30 **iv** 6

15 a

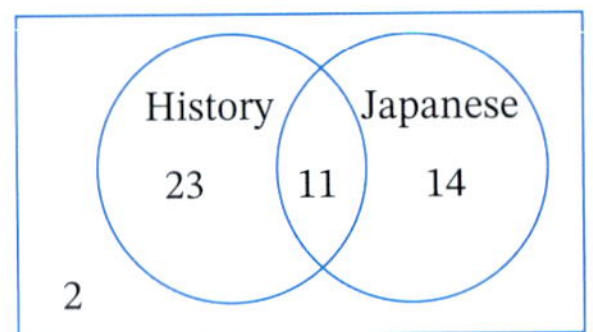

b 14 **c** $\frac{1}{25}$

16 a SSS, SAS, AAS, RHS **b** AAS

17 a $\frac{1}{8}$

b expected = 15; it is smaller than the observed frequency

c 0.19 (19%) **d** $\frac{1}{4}$

e expected = 30 ; it is larger than the observed frequency

f 0.175 (17.5%)

18 a

	Plays weekend sport	Does not play weekend sport	Total
Male	49	23	72
Female	39	45	84
Total	88	68	156

b 156 **c** 88

d i $\frac{7}{13}$ **ii** $\frac{31}{78}$

19 a SAS **b** RHS

20 a

Maths		Science
3	2	4
5 5	3	
8 3	4	0 0 3
8 8 2	5	4
	6	0 2 9
	7	3
7	8	8
6	9	9

b

	Median	Range
Maths	50	73
Science	57	64

c Maths, lower median.

21 $\triangle ABC ||| \triangle DBE$ using AA; $x = 24$

22 a $\frac{3}{64}$ **b** $\frac{5}{16}$ **c** $\frac{55}{64}$ **d** $\frac{37}{64}$

23 a Sample – quick, simple, inexpensive.

b, c, d Teacher to check.

24 40 m

25 a

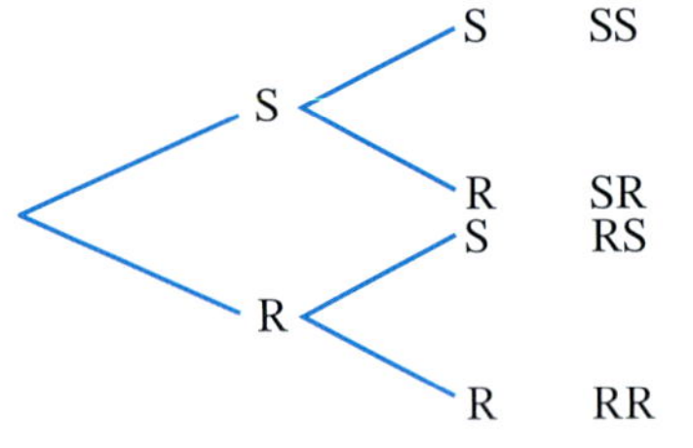

b i $\frac{1}{2}$ **ii** $\frac{1}{4}$ **iii** $\frac{3}{4}$ **iv** $\frac{3}{4}$

CHAPTER 10

SkillCheck

1 a 7 **b** 7 **c** 8 **d** 27
e 6 **f** 23 **g** 5 **h** 3
i 10 **j** 5 **k** 8 **l** 20

2 a −3 **b** −5 **c** −12 **d** −2
e 3 **f** −11 **g** 36 **h** 3
i −10 **j** −8 **k** −9 **l** −72

3 a 17 **b** 17 **c** 4 **d** 4

4 a −6 **b** 7 **c** 6 **d** −14

5 a n **b** $11x$ **c** $4r$ **d** $8t$

6 a $2m + 6$ **b** $3x - 6$
c $4k + 28$ **d** $5d - 5$
e $6a - 18$ **f** $4b + 32$
g $10q - 30$ **h** $45j + 9$

7 a subtracting **b** multiplying
c dividing **d** adding

Exercise 10.01

1 a $x = 3$ **b** $x = 9$ **c** $a = 5$ **d** $m = 20$
e $b = 10$ **f** $c = 4$ **g** $k = 6$ **h** $d = 13$
i $m = 15$ **j** $x = 5$ **k** $y = 15$ **l** $n = 25$

2 B

3 a $x = 5$ **b** $p = 7$ **c** $k = 5$ **d** $x = 4$
e $x = 3$ **f** $x = 8$ **g** $m = 28$ **h** $a = 12$
i $d = 2$ **j** $n = 1$ **k** $r = 15$ **l** $k = 11$
m $p = -2$ **n** $r = 10$ **o** $y = 15$ **p** $x = 2$

4 B

Exercise 10.02

1 a $x = 7$ **b** $y = 23$ **c** $m = 12$
d $k = 27$ **e** $k = -4$ **f** $s = -2$
g $n = -13$ **h** $p = -20$ **i** $m = -8$
j $g = 15.5$ **k** $a = 10\frac{1}{4}$ **l** $k = -7.7$

2 a $t = 51$ **b** $w = 11$ **c** $a = 7$
d $j = 74$ **e** $q = 29$ **f** $f = 110$
g $x = 33$ **h** $g = 11$ **i** $j = -2$
j $w = 7$ **k** $m = -15$ **l** $d = 0$
m $y = 20.9$ **n** $n = 8$ **o** $b = 1\frac{1}{20}$
p $s = 6\frac{5}{6}$

3 **a** $x=4$ **b** $g=4$ **c** $y=9$
d $b=9$ **e** $d=9$ **f** $c=9$
g $p=5$ **h** $m=13$ **i** $x=-5$
j $q=-6$ **k** $c=-18$ **l** $x=8$
m $p=9\frac{1}{4}$ **n** $r=2\frac{1}{4}$ **o** $q=-3\frac{1}{3}$
p $w=-1\frac{1}{3}$

4 **a** $s=15$ **b** $m=42$ **c** $d=120$
d $w=58$ **e** $a=-12$ **f** $f=-18$
g $r=-15$ **h** $m=-7$ **i** $x=48$
j $t=1\frac{1}{2}$ **k** $n=16$ **l** $x=-\frac{1}{2}$

5 C **6** Teacher to check.

Exercise 10.03

1 D

2 **a** $m=6$ **b** $x=6$ **c** $k=8$
d $w=6$ **e** $h=-3$ **f** $d=-2\frac{3}{4}$
g $x=7$ **h** $a=-1$ **i** $c=2\frac{1}{2}$
j $p=8$ **k** $y=10$ **l** $r=15\frac{1}{2}$

3 **a** $m=18$ **b** $c=15$ **c** $d=42$
d $k=-27$ **e** $h=20$ **f** $x=-24$
g $n=65$ **h** $t=12$ **i** $a=-32$
j $x=18$ **k** $v=90$ **l** $b=21$

4 D

5 **a** $x=6$ **b** $x=9$ **c** $x=12$
d $N=5$ **e** $N=21$ **f** $B=-4$
g $x=-4$ **h** $x=-10$ **i** $m=4\frac{2}{5}$
j $x=-24$ **k** $x=-1\frac{1}{4}$ **l** $x=15$

6 C

7 **a** $x=5$ **b** $x=9$ **c** $N=3$
d $N=23$ **e** $N=22$ **f** $x=-5$
g $k=-5$ **h** $m=31$

8 Teacher to check.

Exercise 10.04

1 A **2** C

3 **a** $a=6$ **b** $k=4$ **c** $x=3$
d $p=3$ **e** $n=3$ **f** $q=6$
g $y=0$ **h** $a=7$ **i** $r=6$

4 D

5 **a** $a=3$ **b** $x=2\frac{2}{3}$ **c** $d=1$
d $p=1$ **e** $m=5\frac{1}{4}$ **f** $x=5$
g $x=\frac{2}{3}$ **h** $r=-1$ **i** $y=4\frac{1}{2}$

6 D

7 **a** $d=-2$ **b** $k=\frac{2}{3}$ **c** $p=-2$
d $x=-3$ **e** $k=2\frac{1}{5}$ **f** $m=5\frac{1}{2}$
g $t=1\frac{1}{2}$ **h** $j=-8\frac{1}{2}$ **i** $q=-1$

8 Teacher to check.

Exercise 10.05

1 **a** $x=1$ **b** $m=4$ **c** $k=6$
d $p=4$ **e** $j=3$ **f** $d=5$
g $x=9$ **h** $x=8$ **i** $x=3$
j $x=8$ **k** $k=12$ **l** $q=20$

2 **a** $x=3$ **b** $p=-4$ **c** $q=-3$
d $h=11$ **e** $r=-8$ **f** $e=11$
g $z=1$ **h** $y=-7$ **i** $a=6$
j $d=-7$ **k** $k=9$ **l** $c=10$

3 B

4 **a** $x=-1$ **b** $r=1$ **c** $p=-7$
d $x=-2$ **e** $p=5$ **f** $e=6$
g $y=1$ **h** $a=4$ **i** $w=-6$

5 A

6 **a** $x=13$ **b** $x=6$ **c** $k=-18$
d $y=2$ **e** $v=4$ **f** $x=11$
g $p=4$ **h** $s=7$ **i** $d=-22$

7 Teacher to check.

Mental skills 10

2 **a** 160 **b** 70 **c** 240 **d** 900
e 2600 **f** 900 **g** 140 **h** 300
i 180 **j** 770 **k** 18 **l** 34
m 46 **n** 26 **o** 18 **p** 12
q 40 **r** 8 **s** 14 **t** 24

Exercise 10.06

1 **a** \$88 **b** \$1.80 **c** 33 **d** 16

2 **a** D, 102 **b** B, 49 **c** D, 56
d B, 588 **e** C, \$8974

3 **a** 75 **b** 49 **c** 27 **d** 16
e 12 **f** 18 **g** 15 **h** 6

4 **a** 150 **b** 8 **c** 100 kg
d 6 **e** 54 **f** 16 cm

5 **a** 77°F **b** $42\frac{2}{9}$°C

6 **a** \$600 **b** 122

7 **a** \$13.90 **b** 28

8 **a** $x=8\frac{1}{2}$ **b** 17 cm **c** $6\frac{1}{2}$ cm

9 **a** \$2610 **b** 220

10 $x=7$

Exercise 10.07

1 **a** $x=\pm 9$ **b** $x=\pm 12$ **c** $x=\pm 1$
d $x=\pm 13$ **e** $m=\pm 71$ **f** $u=\pm 43$

2 **a** $x\approx\pm 3.61$ **b** $x\approx\pm 7.35$ **c** $x\approx\pm 9.38$
d $t\approx\pm 11.36$ **e** $h\approx\pm 30.76$ **f** $z\approx\pm 22.96$

3 **a** $x=\pm\sqrt{41}$ **b** $x=\pm\sqrt{30}$

c $x = \pm\sqrt{48}$ or $\pm 4\sqrt{3}$ **d** $a = \pm\sqrt{126}$ or $\pm 3\sqrt{14}$
e $p = \pm\sqrt{75}$ or $\pm 5\sqrt{3}$ **f** $b = \pm\sqrt{509}$

4 $x^2 = 0$, with solution $x = 0$.

5 It is not possible to find the square root of a negative number.

6 For example, $x^2 = -1$ and $x^2 = -12$.

7 **a** positive **b** negative **c** zero

8 **a** $x = \pm 3$ **b** $x = \pm 4$ **c** $x = \pm 5$ **d** $x = \pm 2$
e $x = \pm 2$ **f** $x = \pm 3$ **g** $x = \pm 5$ **h** $x = \pm 6$

Exercise 10.08

1 A

2 **a** 13 **b** 107

3 **a** 10 hours **b** 18 years old

4 **a** 34 **b** 11 **c** 13

5 **a** 132 **b** 6 **c** 15 **d** 15

6 **a** $58\frac{1}{3}$ **b** 30 **c** 9

7 **a** \$125 **b** \$400 **c** 4% p.a. **d** 7 years

8 **a** 3 **b** 6 **c** 45 **d** 15

Exercise 10.09

1 **a** $x < 7$ **b** $y > 23$ **c** $m < 12$ **d** $k > 27$
e $k < -4$ **f** $s > -2$ **g** $n < -13$ **h** $p > -20$
i $m < -8$ **j** $g > 15.5$ **k** $a < 10\frac{1}{4}$ **l** $k > -7.7$

2 **a** $t < 51$ **b** $w > 11$ **c** $a < 7$ **d** $j > 74$
e $q < 29$ **f** $f > 110$ **g** $x < 33$ **h** $g > 11$
i $j =< -2$ **j** $w > 7$ **k** $m < -15$ **l** $d > 0$
m $y < 20.9$ **n** $n > 8$ **o** $b < 1\frac{1}{20}$ **p** $s > 6\frac{5}{6}$

3 **a** $x < 4$ **b** $g > 4$ **c** $y < 9$ **d** $b > 9$
e $d < 9$ **f** $c > 9$ **g** $p < 5$ **h** $m > 13$
i $x < -5$ **j** $q > -6$ **k** $c < -18$ **l** $x > 8$
m $p < 9\frac{1}{4}$ **n** $r > 2\frac{1}{4}$ **o** $q < -3\frac{1}{3}$ **p** $w > -1\frac{1}{3}$

4 **a** $s < 15$ **b** $m > 42$ **c** $d < 120$ **d** $w > 58$
e $a < -12$ **f** $f > -18$ **g** $r < -15$ **h** $m > -7$
i $x < 48$ **j** $t > 1\frac{1}{2}$ **k** $n < 16$ **l** $x > -\frac{1}{2}$

5 C

6 Teacher to check.

7 D

8 **a** $m < 6$ **b** $x > 6$ **c** $k < 8$
d $w > 6$ **e** $h < -3$ **f** $d > -2\frac{3}{4}$
g $x > 7$ **h** $a < -1$ **i** $c < 2\frac{1}{2}$
j $p > 8$ **k** $y < 10$ **l** $r > 15\frac{1}{2}$

9 **a** $m < 18$ **b** $c > 15$ **c** $d < 42$
d $k > -27$ **e** $h < 20$ **f** $x > -24$
g $n < 65$ **h** $t > 12$ **i** $a < -32$
j $x > 18$ **k** $v < 90$ **l** $b > 21$

10 **a** $x < 6$ **b** $x > 9$ **c** $x < 12$
d $N > 5$ **e** $N < 21$ **f** $B > -4$
g $x < -4$ **h** $x > -10$ **i** $m < 4\frac{2}{5}$
j $x > -24$ **k** $x > -1\frac{1}{4}$ **l** $x < 15$

11 C

12 **a** $x < 5$ **b** $x > 9$ **c** $N < 3$
d $N > 23$ **e** $N < 22$ **f** $x > -5$
g $k < -5$ **h** $m > 31$

13, 14 Teacher to check.

Power plus

1 **a** $x = 3$ **b** $x = 2$ **c** $x = 0$
d $x = -1$ **e** $x = 2\frac{1}{7}$ **f** $x = 3\frac{1}{11}$
g $x = 20$ **h** $q = 26\frac{2}{3}$

2 **a** $x = 7$ **b** $x = 5$ **c** $x = 65$

3 **a** $x = 5$ or -5 **b** $x = 4$ or -4 **c** $x = 6$ or -6
d $x = 2$ or -2 **e** $x = -7$ or -1 **f** $x = 3$ or -5

4 $\frac{1}{6}x + \frac{1}{12}x + \frac{1}{7}x + 5 + \frac{1}{2}x + 4 = x$, 84 years

Test yourself 10

1 **a** $m = 4$ **b** $k = 12$ **c** $d = 18$

2 **a** $k = 7$ **b** $x = 11$ **c** $a = 14$
d $a = 33$ **e** $x = 4$ **f** $f = 12$
g $m = 28$ **h** $x = 24$ **i** $w = -6$
j $k = -2$ **k** $m = 8$ **l** $x = -36$

3 **a** $p = 5$ **b** $m = -3$ **c** $x = 15$
d $h = 77$ **e** $n = -8$ **f** $a = 60$
g $y = 10$ **h** $x = 10$ **i** $n = 25$
j $x = -6$ **k** $k = 26$ **l** $d = -29$

4 **a** $x = 1$ **b** $u = 3$ **c** $h = 3$
d $v = -3$ **e** $x = 2\frac{3}{5}$ **f** $t = 3$

5 **a** $n = 6$ **b** $x = 4$ **c** $x = 2$
d $x = 7$ **e** $y = 4\frac{2}{3}$ **f** $d = 6$
g $r = 21$ **h** $a = -2$ **i** $n = 3$

6 **a** 5 **b** 3

7 **a** B **b** $n = 6$

8 $x = 3$

9 **a** $x = \pm 8$ **b** $x = \pm 6.7$ **c** $x = \pm\sqrt{27}$

10 **a** 1260° **b** 8

11 **a** $p < 9.25$ **b** $r > 2.25$ **c** $q < -3.33$
d $w > -1.33$ **e** $s < 15$ **f** $m > 42$
g $d < 120$ **h** $v > 58$ **i** $x < \frac{2}{3}$
j $y > 4$ **k** $z < 5$ **l** $q > 18$

CHAPTER 11

SkillCheck

1 **a** 200 **b** 180 **c** 3 **d** 2500
e 3.8 **f** 0.18 **g** 180 **h** 85
i 8 **j** 7.5 **k** 9150 **l** 384

2 **a** 6 **b** 7 **c** 8

3 **a** 21 **b** 20 **c** 2

4 **a** 6 **b** 8 **c** 15

5 a $\frac{2}{5}$ b $\frac{1}{2}$ c $\frac{4}{5}$ d $\frac{5}{12}$

6 a 8 b 5 c 18 d 30
e 760 f 54 g 39 h 250

7 a 07:15 b 15:45 c 20:50 d 00:10

8 a 4:10 a.m. b 11:05 a.m.
c 2:15 p.m. d 11:35 p.m.

Exercise 11.01

1 a 6 : 6 or 1 : 1 b 2 : 6 or 1 : 3
c 6 : 10 or 3 : 5 d 8 : 12 or 2 : 3
e 5 : 3

2 a 12 b 10 c 9 d 20 e 32
f 84 g 30 h 75 i 5 j 5
k 4 l 1 m 30 n 2 o 40
p 5 q 5 r 25

3 D

4 a 1 : 10 b 1 : 2 c 2 : 5 d 5 : 7
e 3 : 2 f 7 : 3 g 10 : 1 h 5 : 11
i 1 : 1 j 41 : 107 k 3 : 1 l 1 : 16
m 2 : 3 : 5 n 3 : 4 : 6 o 3 : 1 : 4 p 2 : 5 : 3 : 7

5 a 5 : 6 b 3 : 4 c 9 : 8 d 4 : 3
e 4 : 3 f 8 : 5 g 5 : 2 h 4 : 3
i 12 : 7 j 8 : 5 k 25 : 12 l 9 : 5
m 4 : 7 n 13 : 8 o 5 : 3 p 1 : 2
q 3 : 4 r 3 : 2 s 1 : 4 t 1 : 2
u 3 : 2 v 10 : 7 w 2 : 1 : 3 x 45 : 10 : 9

6 a 1 : 4 b 1 : 4 c 5 : 49 d 1 : 4
e 1 : 3 f 4 : 1 g 5 : 16 h 1 : 50
i 80 : 7 j 5 : 36 k 9 : 500 l 1 : 6
m 6 : 1 n 3 : 20 o 9 : 40

7 3 : 2

8 a 4 : 3 b 3 : 4 c 4 : 7

9 a 3 : 2 b 3 : 5 c 5 : 2

10 C

11 a 16 : 19 b 19 : 16 c 3 : 19 d 19 : 3

12 a 1 : 1 b 2 : 1 c 1 : 2 d 2 : 1

13 a $\frac{5}{8}$ b 5 : 3

14 A

15 Australia 3 : 250, India 11 : 500

Exercise 11.02

1 C 2 75 3 560 kg

4 32 cm 5 1440 6 17.5 km/h

7 a 20 cm b 136 cm

8 50 kg

9 a Toudi \$990, Felicity \$540
b \$2250

10 D

11 a 5 kg b Heavier packet by 10 c/kg

12 25.6 tonne 13 1.5 m 14 A

15 a 7 mL b 84 mL

Exercise 11.03

1 a 1 : 500 000 b 1 : 1000 c 1 : 50 000 000
d 1 : 2 500 000 e 1 : 10 000 f 1 : 200 000
g 1 : 25 000 h 1 : 40 000 i 1 : 50 000
j 1 : 10 000

2 B

3 a 1 : 20 000
b i 1.4 km ii 1.9 km iii 2.48 km
c i 90 cm ii 7.5 cm iii 48 cm

4 6670 km 5 66 cm 6 5.6 cm

7 a i 880 m ii 680 m
iii 440 m iv 960 m
b i 880 m ii 440 m
c 360 m d 120 m by 280 m
e 160 m f 20 cm, Teacher to check

8 a 3.9 m b 77 cm c 2.0 m
d 1.2 m e 1.6 m
f 3.9 m × 2.9 m = 11.31 m^2

9 a 12.6 cm b 7.2 m
c 13.5 cm d 67.2 cm

10 a 5.0 m b 2.5 m
c 2.0 m d 2.0 m × 2.5 m = 5 m^2
e 5.0 m f 4.1 m × 3.8 m = 15.58 m^2

11 B

12 a 0.45 mm b 11 mm
c 26.25 mm d 0.37 mm

Mental skills 11

2 a 8:45 a.m. b 1:20 p.m. c 5:50 p.m.
d 12:17 a.m. e 9:05 p.m. f 6:32 p.m.
g 11:15 a.m. h 2:38 a.m. i 2:40 p.m.
j 3:20 a.m. k 4:55 p.m. l 11:31 p.m.
m 1:08 a.m. n 10:18 a.m. o 8 p.m.
p 6:43 a.m.

4 a 18:35 b 08:05 c 11:45 d 23:20
e 02:21 f 12:30 g 15:48 h 19:11
i 09:08 j 21:50 k 00:42 l 07:39
m 01:59 n 22:18 o 10:46 p 17:23

Exercise 11.04

1 a 9 b 5 c 7 d 13

2 a \$400, \$100 b \$350, \$150

3 a 1 : 1 : 2
b Phuong \$1000, Janelle \$1000, Ahmet \$2000

4 \$180

5 a 200 kg, 250 kg b 270 kg, 180 kg

6 a 80 cm, 240 cm, 400 cm
b 300 cm, 180 cm, 240 cm

7 D 8 125 9 25

10 600 g 11 1078 12 C

13 80 kg 14 568

15 4 m, 6 m, 10 m 16 240 kg

17 a 30 000 b 165 000

18 \$140 000, \$70 000

Exercise 11.05

1 a words/min b beats/min c c/min
d \$/kg e \$/h f m/s or km/h
g persons/year h c/kL i persons/km^2

2 **a** 17 sheep/h **b** $4.30/kg
c 26 students/teacher **d** 34 points/game
e 91 words/min **f** 8.5 articles/h
g 5.1 cars/day **h** 630 boxes/h
i 37 m/s **j** 920 bottles/h
k 1324 revolutions/min **l** $35/h
m 56 km/h **n** $2.75/kg
o 4.75 runs/over **p** 8.1 km/L
q $10.50/h **r** 42 kg/ha

3 B

4 **a** Can travel 100 km in one hour at this speed.
b 150 cars will pass in an hour at this rate.
c 10.3 L can be used over a distance of 100 km.
d For each hectare of land, there are 60 sheep.

5 80 kg/person **6** 1800 ha/year

7 139 c/L **8** 182.5

Exercise 11.06

1 **a** 2 kg box **b** 8 notepads **c** 150 g
d 7 tubs **e** 3 kg **f** 2.25 L carton

2 Kim **3** 750 g box **4** C

5 **a** 375 g **b** 2 L **c** 735 g **d** 115 g

6 **a** medium **b** large **c** small **d** large

Exercise 11.07

1 $676.40 **2** 160 runs

3 **a** 10.5 km/L **b** 210 km

4 336 **5** 432 mL **6** 6

7 $3980 **8** $7\frac{1}{7}$ ha or 7.1 ha

9 **a** $70 **b** 2.5 h

10 300 km

11 **a** 14.8 L **b** 133.2 km

12 C

13 **a** 17°C, 29°C **b** 12 p.m.

14 **a** 2400 m^2 **b** 40 m^2/min
c 60 m^2/min **d** 24 min (at 100 m^2/ min)

15 **a** 5400 s **b** 90 min **c** 1.5 h

16 **a** 650 km **b** 48 L **c** 13.5 km

Exercise 11.08

1 **a** 5 km/h **b** 480 km/h **c** 5 km/h
d 17 km/h **e** 100 km/h **f** 80 km/h
g 120 km/h **h** 120 km/h **i** 200 km/h
j 9 km/h **k** 36 km/h **l** 6 km/h

2 **a** 10 m/s **b** 1.5625 m/s **c** 58 m/s
d 8 m/s **e** 1.5 m/s

3 **a** 720 km **b** 30 km **c** 1330 km

4 3 km **5** B

6 **a** 100 m/min **b** 6 km/h

7 40 km/h **8** D **9** 1200

10 1.5 h **11** Teacher to check.

12 63.9 km/h **13** 56 km/h

14 54.05 s **15** 500 m

Exercise 11.09

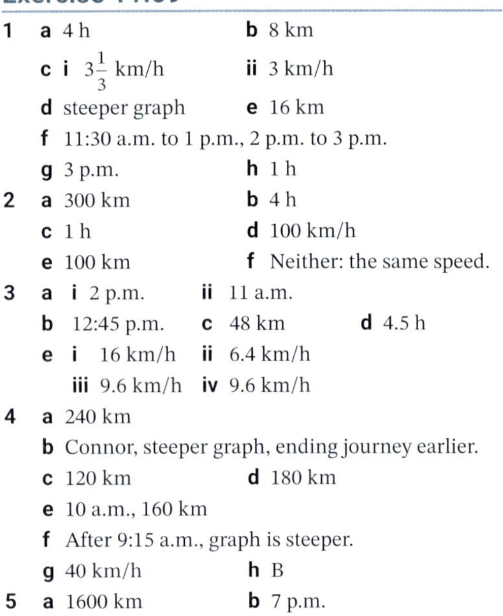

1 **a** 4 h **b** 8 km
c **i** $3\frac{1}{3}$ km/h **ii** 3 km/h
d steeper graph **e** 16 km
f 11:30 a.m. to 1 p.m., 2 p.m. to 3 p.m.
g 3 p.m. **h** 1 h

2 **a** 300 km **b** 4 h
c 1 h **d** 100 km/h
e 100 km **f** Neither: the same speed.

3 **a** **i** 2 p.m. **ii** 11 a.m.
b 12:45 p.m. **c** 48 km **d** 4.5 h
e **i** 16 km/h **ii** 6.4 km/h
iii 9.6 km/h **iv** 9.6 km/h

4 **a** 240 km
b Connor, steeper graph, ending journey earlier.
c 120 km **d** 180 km
e 10 a.m., 160 km
f After 9:15 a.m., graph is steeper.
g 40 km/h **h** B

5 **a** 1600 km **b** 7 p.m.
c 350 km/h **d** D

e

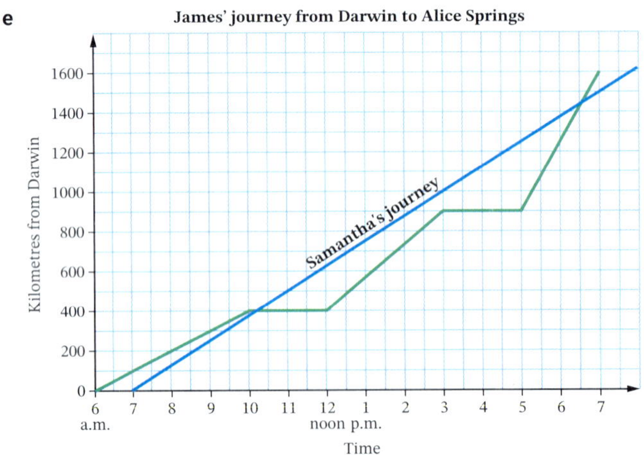

f 10:10 a.m.

6 Teacher to check.

7 **a** A **b** B **c** D **d** C

8 C, not possible to be at two different places/distances at the same time.

Exercise 11.10

1

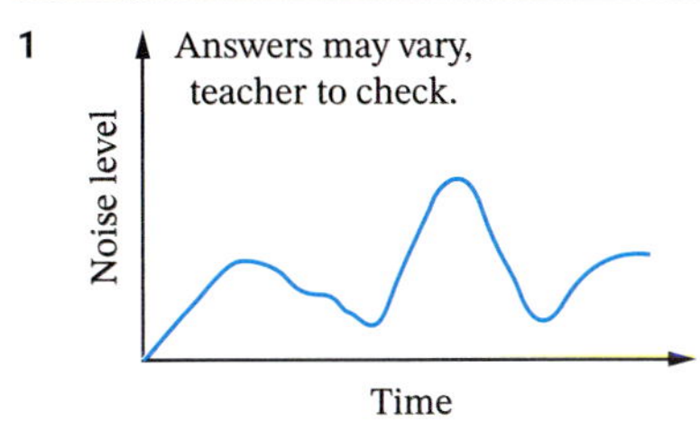

2 B

3 **a** 8 Brown **b** 8 Yellow
c 8 Red **d** 8 Green

4 Teacher to check.

5 **a** Grew steadily, then dropped.
b 4
c Returns to normal.
d His team was losing and it was a close match. Answers may vary, teacher to check.

6

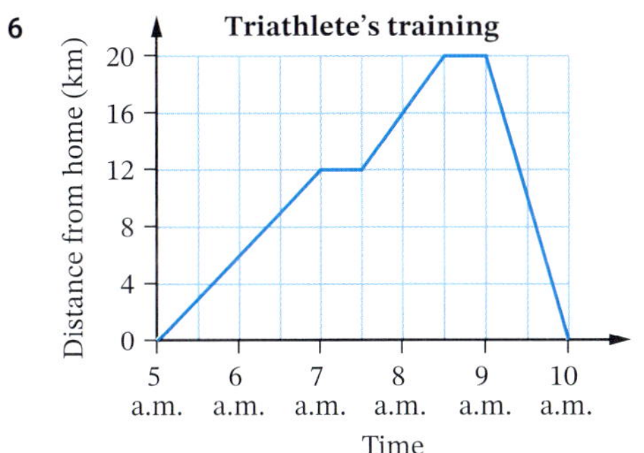

7 B **8** Teacher to check.

Exercise 11.11

1 A

2 **a** 1 h 55 min **b** 50 min **c** 4 h 45 min
d 3 h 55 min **e** 5 h 45 min **f** 10 h 55 min
g 8 h 5 min **h** 3 h 30 min **i** 4 h 40 min

3 **a** 6 h 47 min **b** 11 h 52 min **c** 3 h 29 min
d 7 h 31 min **e** 12 h 30 min **f** 8 h 9 min
g 6 h 9 min **h** 1 h 45 min **i** 12 h 29 min
j 12 h 7 min **k** 4 h 51 min **l** 1 h 43 min

4 1 h 55 min

5 **a** 7 h **b** 6 h 30 **c** 4 h 11 min
d 1 h, 1 h 15 min, faster trip due to less traffic earlier in the day.
e 10:41 a.m. **f** Strathfield
g 35 min, 4:40 p.m.

6 **a** 8 p.m. **b** 1 p.m. **c** 7:35 p.m.
d 1:17 p.m. **e** 2:04 a.m. **f** 2:32 p.m.
g 3:17 a.m. **h** 4:25 p.m. **i** 8:45 a.m.
j 3:10 p.m.

7 Volvo 1:43 p.m., Peugot 2:02 p.m., Ford 2:04 p.m., Subaru 2:19 p.m., Nissan 2:21 p.m. and Toyota 2:41 p.m.

8 **a** 6 **b** 8:15 a.m.
c Central (RSL) or Pine Ave.
d 9:10 a.m. **e** 20 min
f 6:50 p.m. **g** 11:40 a.m.

Exercise 11.12

1 **a** ahead **b** behind **c** behind **d** ahead
e ahead **f** behind **g** ahead **h** ahead
i ahead **j** ahead **k** behind **l** behind

2 **a** 10 p.m. **b** 8 p.m. **c** 7 a.m. **d** 8 p.m.
e 4 a.m. **f** 2 a.m. **g** 3 p.m. **h** 1 p.m.

3 **a** 2 h **b** 2 h **c** 20 h **d** 5 h
e 15 h **f** 0 h **g** 3 h **h** 13 h
i 7 h **j** 9 h

4 **a** 11 a.m. **b** 7 p.m. **c** 6 a.m. **d** 7 p.m.
e 3 a.m. **f** 1 a.m. **g** 2 p.m. **h** 12 noon

5 11 a.m. **6** 8 p.m.

7 B **8** Teacher to check.

9 **a** ahead **b** ahead **c** same
d behind **e** ahead **f** behind
g ahead **h** same **i** ahead

10 **a** $\frac{1}{2}$ h **b** 2 h **c** $\frac{1}{2}$ h
d $\frac{1}{2}$ h **e** 2 h **f** 0 h

11 **a** 11 a.m. **b** 10:30 a.m. **c** 9 a.m.
d 10:30 a.m. **e** 11 a.m. **f** 11 a.m.

12 **a** 9:30 p.m. **b** 9:30 p.m. **c** 9 p.m.
d 9 p.m. **e** 9:30 p.m. **f** 9:30 p.m.

13 **a** 7 p.m. **b** 3 p.m.

14 **a**, **b**, **c** Teacher to check **d** 9 p.m.

Power plus

1 332 740

2 **a** 704 km^2 **b** 63 071 683 000

3 Teacher to check.

4 **a** 97.2 km/h **b** 349.2 km/h
c 11.88 km/h **d** 82.8 km/h

5 $y = 3.6x$

Test yourself 11

1 **a** 4 : 12 **b** 6 : 15 **c** 9 : 21
d 20 : 45 **e** 2 : 3.2 **f** 15 : 20 : 25

2 **a** 4 : 7 **b** 1 : 3 **c** 1 : 6 **d** 5 : 9
e 3 : 1 : 4 **f** 1 : 5 : 20 **g** 2 : 3 **h** 4 : 3
i 3 : 5 **j** 4 : 1 **k** 1 : 4 **l** 65 : 4
m 25 : 1 **n** 50 : 9 **o** 100 : 17 **p** 3 : 8
q 2 : 7 **r** 1 : 9

3 **a** 100 **b** $45 000

4 **a** 67 cm **b** 12 cm

5 **a** 1 : 25 000
b **i** 625 m **ii** 1.1 km or 1100 m
c **i** 3.5 cm **ii** 10 cm

6 B

7 a \$3.50/kg b 110 km/h
c \$14.10/h d 5.2 runs/over

8 a 1 kg box b 350 g can
c 1.5 L bottle d 500 g bar

9 a \$19.95 b \$710.60 c 29.6 L d 7.2 ha

10 20 km/h 11 247.5 km

12 a 10 km b 3 c $2\frac{2}{3}$ km/h
d $3\frac{1}{2}$ h e 2:30 p.m. to 3:30 p.m.

13 a B b C c A

14

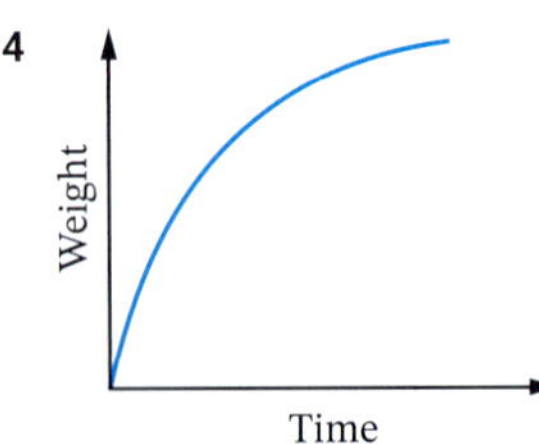

15 a 11:25 p.m. b 6:30 p.m.

16 a 8 h 30 min b 15 h 15 min c 10 h 50 min

17 a p.m. time b 4 c 9 min
d Horatio St e 2:50 p.m.

18 a 8 h b 2 h c 18 h d 10 h

19 11:30 p.m.

CHAPTER 12

SkillCheck

1 a 4 b 5 c 2
d 2 e 4 f 5

2 a 4 b 6 c −3
d 1 e −3 f 7

3 a $x = 3$ b $x = -2$ c $x = -3$
d $x = -2$ e $x = -1\frac{1}{2}$ f $x = 3$

Exercise 12.01

1 a 2, 6, −2, −9, 4 b 24, −9, 6, 30, −12
c 4, −1, −3, 0, $3\frac{1}{2}$ d 5, 14, 7, 20, −2
e 16, −2, 0, 8, 4 f −7, 4, 2, −11, 8

2 A

3 a 14, 4, 19, 39, −31 b 10, −5, 28, 4, −20
c 6, −6, 42, −14, 22 d 22, 37, −23, 32, −8
e −3, −1, 1, 3, 5, 7 f $-3, -2\frac{1}{2}, -2, -1\frac{1}{2}, -1, -\frac{1}{2}$
g 2, 5, 8, 11, 14, 17 h 3, 2, 1, 0, −1, −2

4 B 5 D

Exercise 12.02

1 a $y = 3x$ b $y = -2x$ c $y = x + 4$
d $y = x - 4$ e $y = 10 - x$ f $y = x \div 2$
g $k = p + 4$ h $y = x - 3$ i $j = 2h$
j $p = m \div 5$

2 a $h = 3f - 2$, 19, 22 b $p = 5m - 3$, 27, 42
c $b = 3m + 3$, 18, 27 d $k = 2h + 2$, 20, 24
e $s = 3r + 1$, 13, 22 f $b = 2a - 2$, 18, 20
g $n = 3m + 5$, 20, 23 h $d = 10c + 1$, 91, 101
i $x = 5w - 1$, 34, 39 j $z = 2y - 6$, 12, 14
k $m = 4a + 1$, 29, 37 l $t = 2z - 7$, 11, 17

3 B 4 C

5 a $y = -2x - 1$, −19 b $c = \frac{1}{2}A$, 5
c $t = -5s + 13$, −22 d $v = -10t + 20$, 0, −30
e $D = -\frac{1}{2}x - \frac{1}{2}, -2, -3\frac{1}{2}$

Exercise 12.03

1 B

2 a i 15 ii $T = 2n - 1$ iii 15, 39
b i 16 ii $T = 2n$ iii 16, 40
c i 22 ii $T = 2n + 6$ iii 22, 46
d i 28 ii $T = 3n + 4$ iii 28, 64
e i 63 ii $T = 8n - 1$ iii 63, 159
f i 37 ii $T = 5n - 3$ iii 37, 97
g i 5 ii $T = -2n + 21$ iii 5, −19
h i 64 ii $T = n^2$ iii 64, 400

3 a

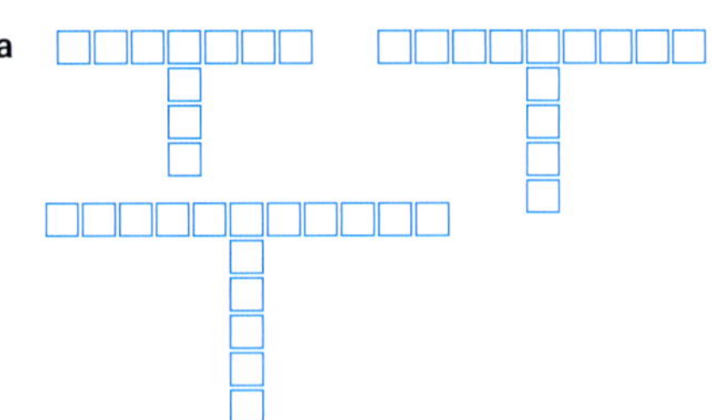

b 7, 10, 13, 16, 25, 34
c The number of tiles equals 3 times the arm length plus 1.
d $t = 3a + 1$
e i 46 ii 100

4 a

b 12, 17, 22, 32, 47, 52
c The number of toothpicks equals 5 times the number of the flight shape plus 2.
d $t = 5f + 2$
e i 127 ii 202

5 **a**

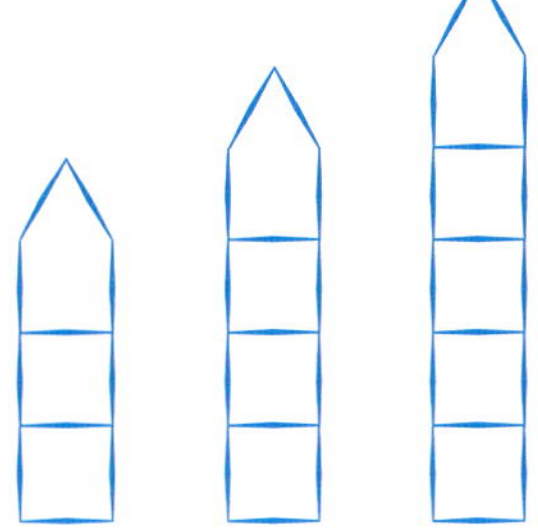

b 5, 8, 11, 14, 17, 26, 35

c The number of toothpicks equals 3 times the number of stages plus 2.

d $t = 3s + 2$

e **i** 38 **ii** 137

6 **a** **i**

ii 4, 8, 12, 16, 20 **iii** $T = 4n$ **iv** 48

b **i**

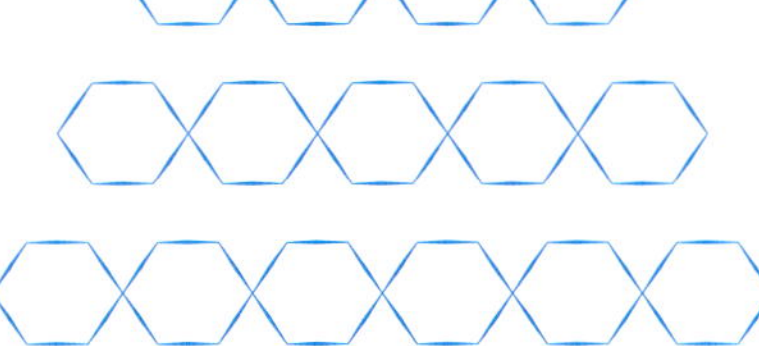

ii 6, 12, 18, 24, 30 **iii** $T = 6n$ **iv** 72

c **i**

ii 4, 7, 10, 13, 16 **iii** $T = 3n + 1$ **iv** 37

d **i**

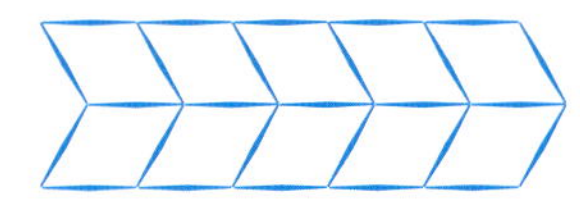

ii 6, 10, 14, 18, 22 **iii** $T = 4n + 2$ **iv** 50

7 A

8 **a** **i** 6 **ii** 8 **iii** 10 **iv** 26

b $C = 2T + 2$ **c** 102 **d** 14

9 **a** **i** 8 **ii** 11 **iii** 14 **iv** 32

b $n = 3d + 2$ **c** 56 **d** 9

Exercise 12.04

1 **a** P **b** N **c** K **d** T **e** M **f** S **g** Q **h** R **i** L **j** U

2 **a** (4, 1) **b** (−2, 4) **c** (−3, −2) **d** (1, −3) **e** (4, 4) **f** (−5, 0) **g** (0, 1) **h** (0, −2) **i** (5, 0) **j** (3, −4)

3 **a** 2nd **b** 4th **c** 1st **d** 2nd **e** 3rd

4 $F(-5, 0)$, $R\left(-1\frac{1}{2}, 0\right)$, $O(0, 0)$, $Q(2, 0)$, $I(5, 0)$

5 $M(0, 3)$, $G(0, 1)$, $O(0, 0)$, $H(0, -2)$, $S(0, -4)$

6 D

7 **a** 4th **b** 3rd **c** 2nd **d** 1st

8

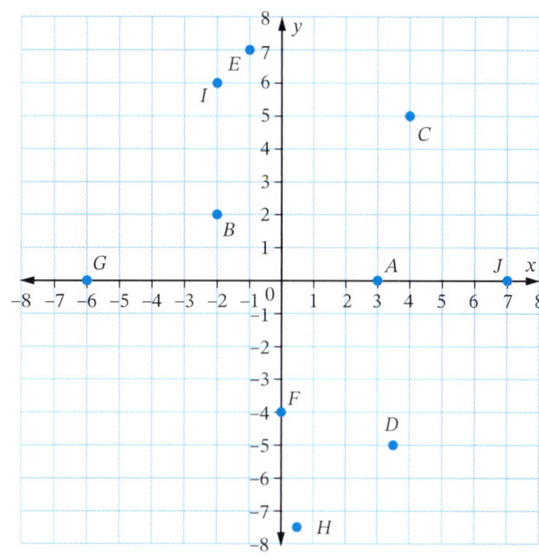

9 **b**

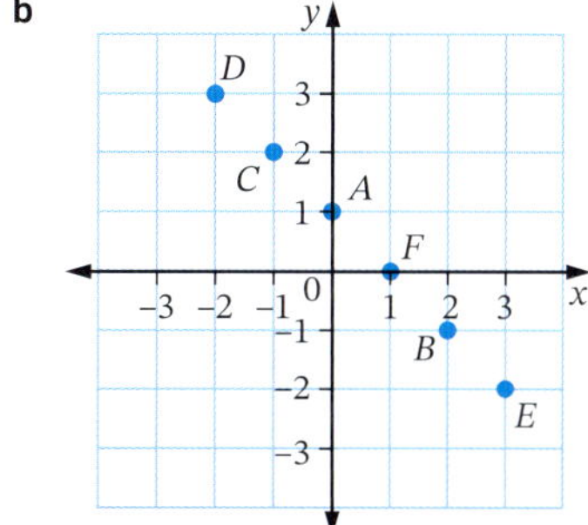

c They all lie on the same line.

d Teacher to check.

Mental skills 12

2 **a** \$700 **b** \$800 **c** 400 **d** 700 **e** \$300 **f** \$400 **g** 250 **h** \$300 **i** \$300 **j** 500

Exercise 12.05

1 **a** (3, 2, 4)

b

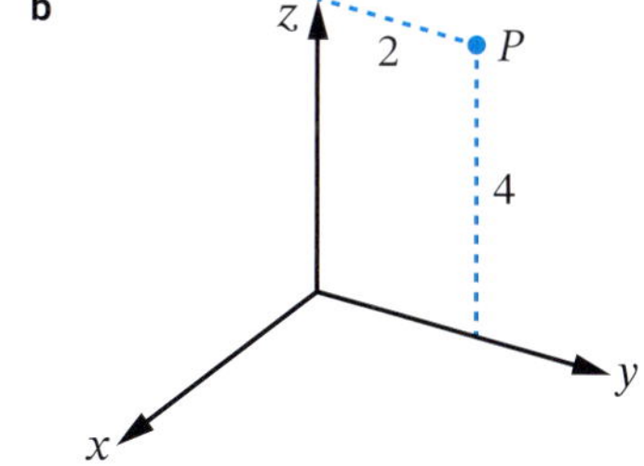

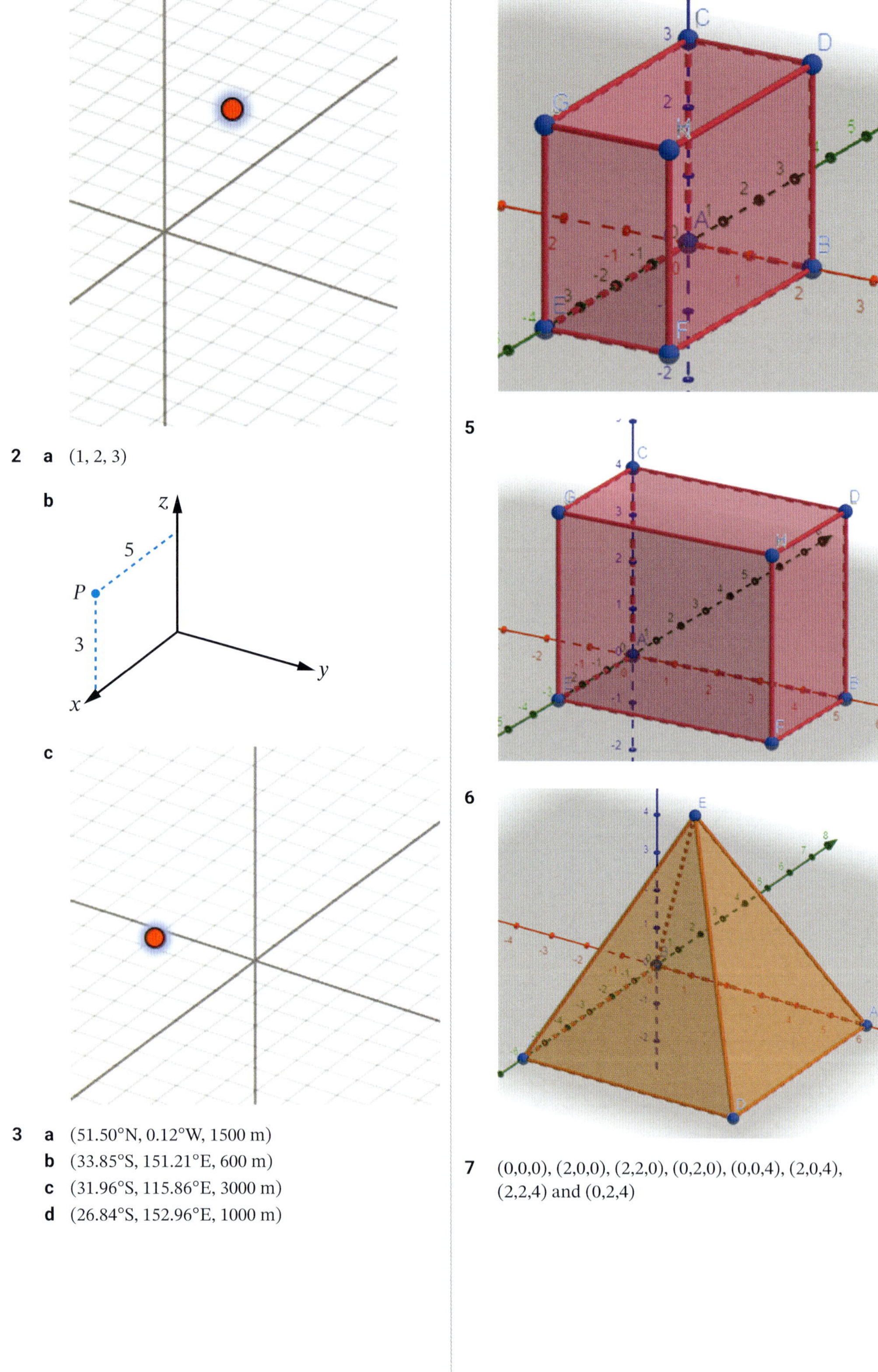

c

2 **a** (1, 2, 3)

b

c

3 **a** (51.50°N, 0.12°W, 1500 m)

b (33.85°S, 151.21°E, 600 m)

c (31.96°S, 115.86°E, 3000 m)

d (26.84°S, 152.96°E, 1000 m)

4

5

6

7 (0,0,0), (2,0,0), (2,2,0), (0,2,0), (0,0,4), (2,0,4), (2,2,4) and (0,2,4)

Exercise 12.06

1 C

2

There is a rough linear pattern but it is not perfect.

3 a i 3, 6, 9, 12, 15 **ii** $y = 3x$

iii

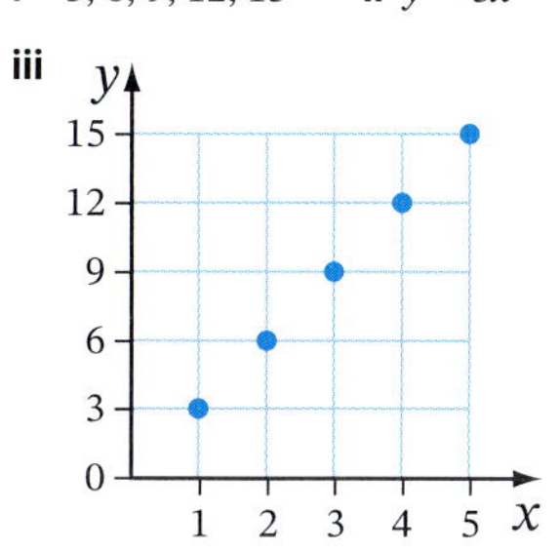

b i 4, 7, 10, 13, 16 **ii** $y = 3x + 1$

iii

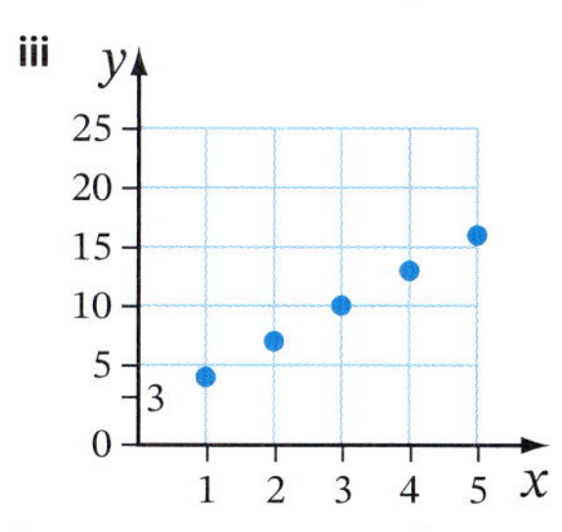

c i 5, 10, 15, 20, 25 **ii** $y = 5x$

iii

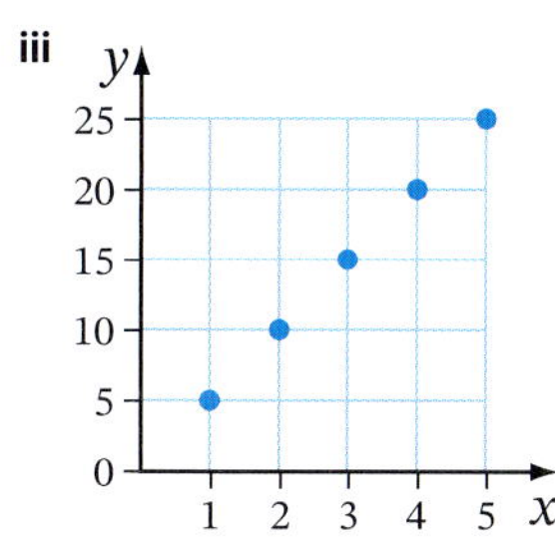

d i 4, 6, 8, 10, 12 **ii** $y = 2x + 2$

iii

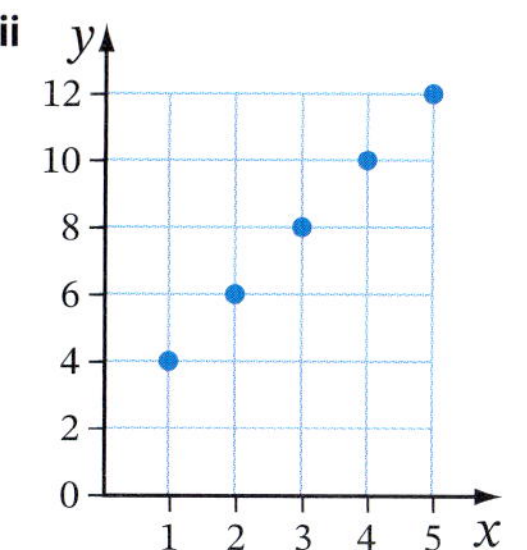

e i 5, 9, 13, 17, 21 **ii** $y = 4x + 1$

iii

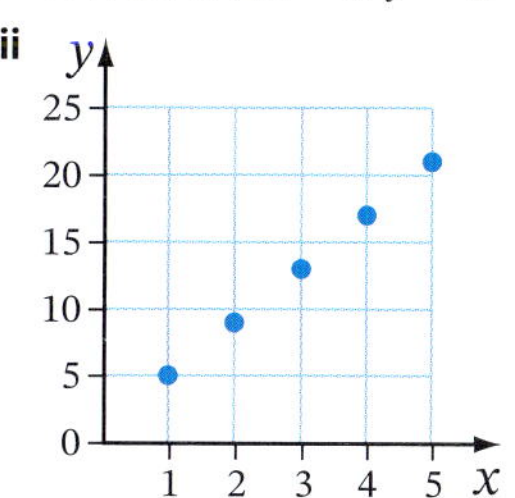

4 Yes

5 a 1, 3, 6, 10, 15, 21, 28

b

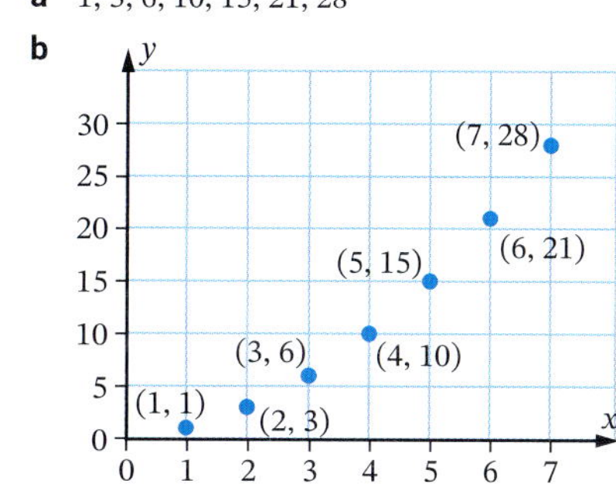

c No **d** C

6 a

S	1	2	3	4	5	6
B	1	5	9	13	17	21

b $B = 4S - 3$

c i 37 **ii** 93

d 27

Exercise 12.07

1 a 2, 3, 4, 5

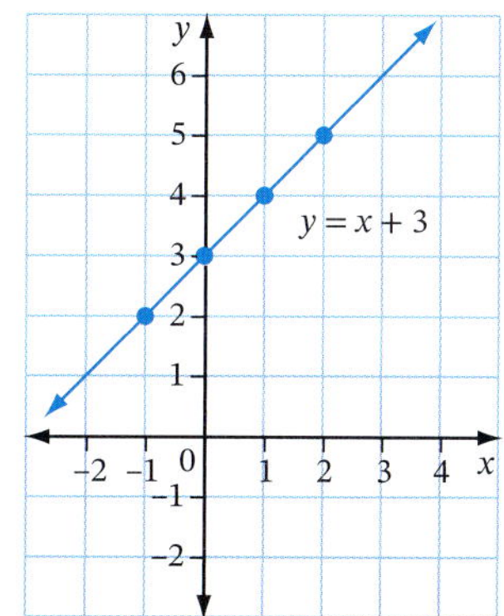

b $-2, -1, 0, 1$

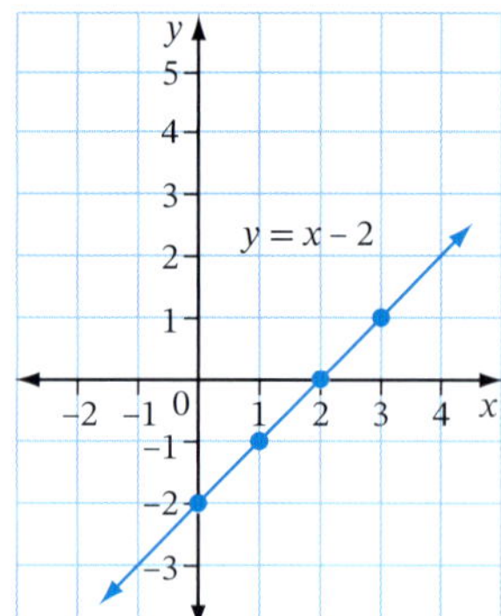

c $-2, 0, 2, 4$

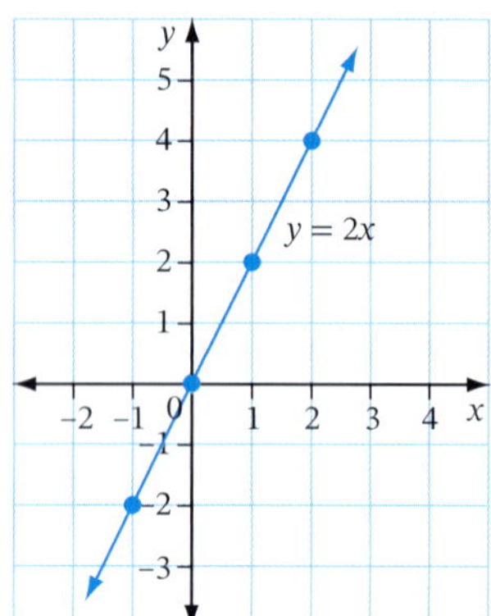

d $-\frac{1}{2}, 0, \frac{1}{2}, 1$

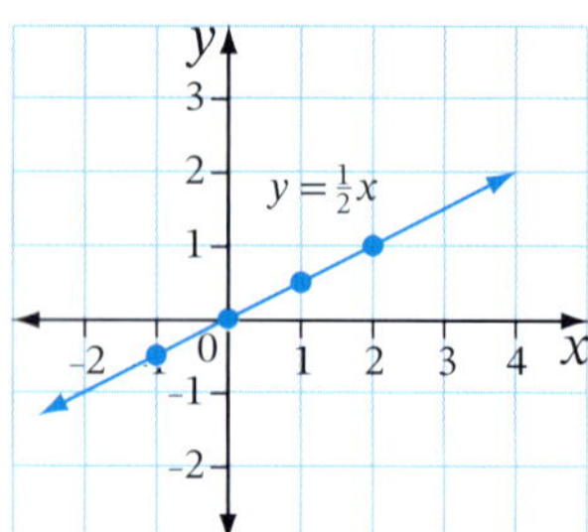

e $6, 5, 4, 3$

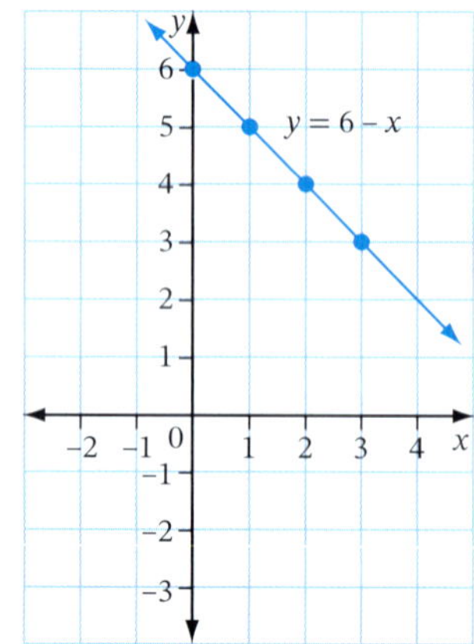

f $-1, 0, 1, 2$

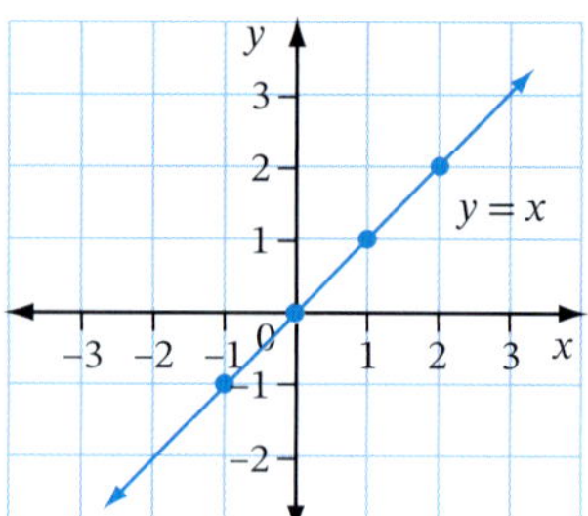

2 **a**

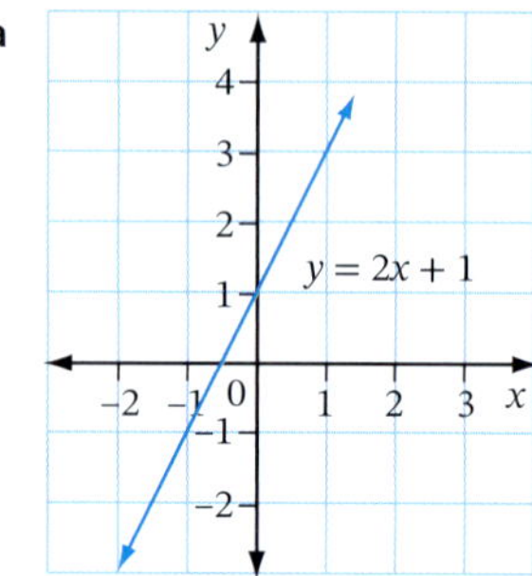

b

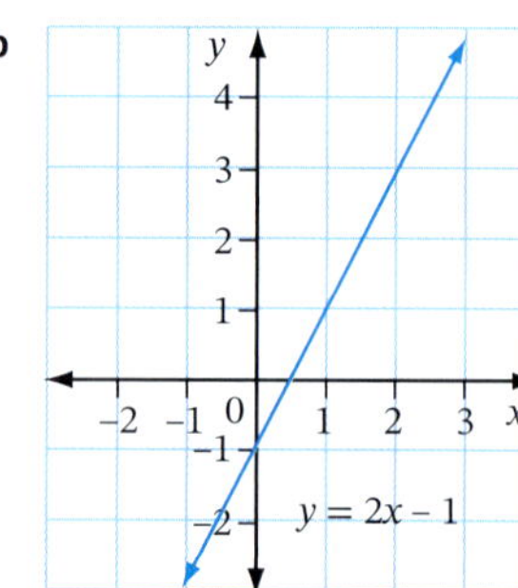

c

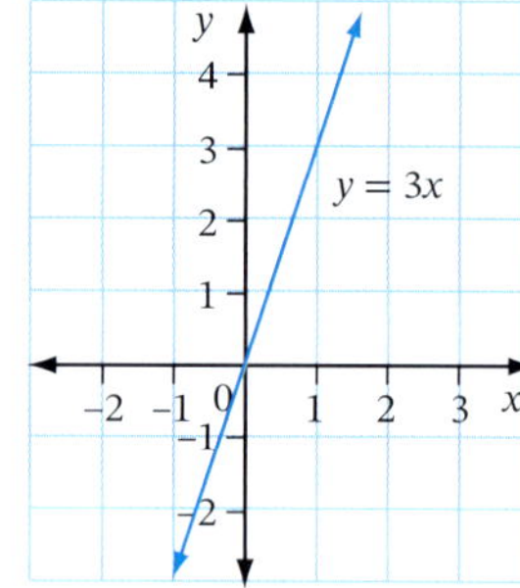

d

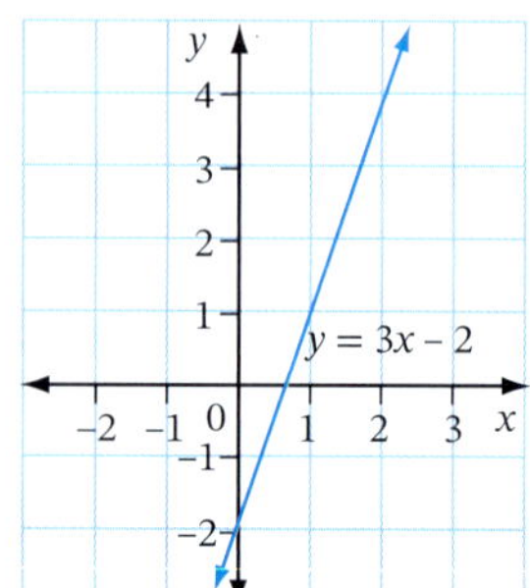

e

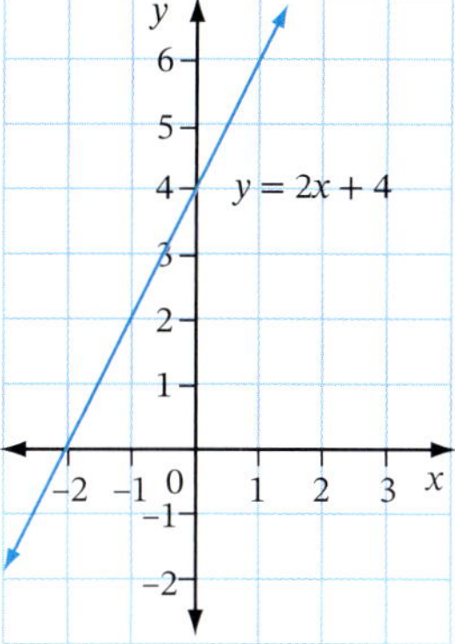

f

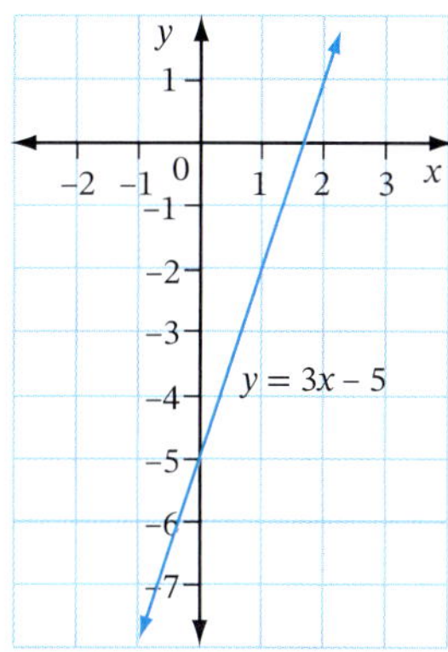

3 From left to right, the line is rising, going up.

4 **a**

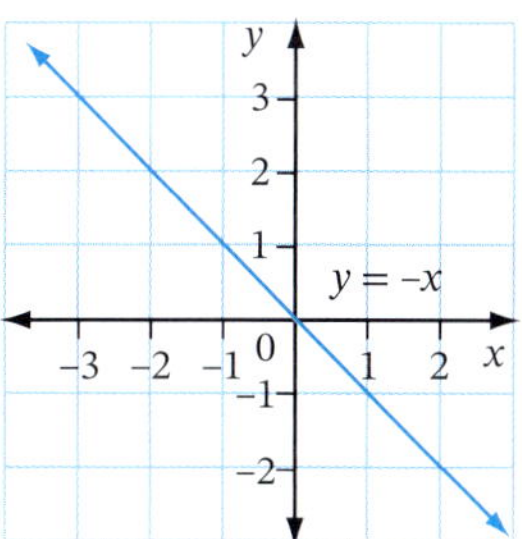

b

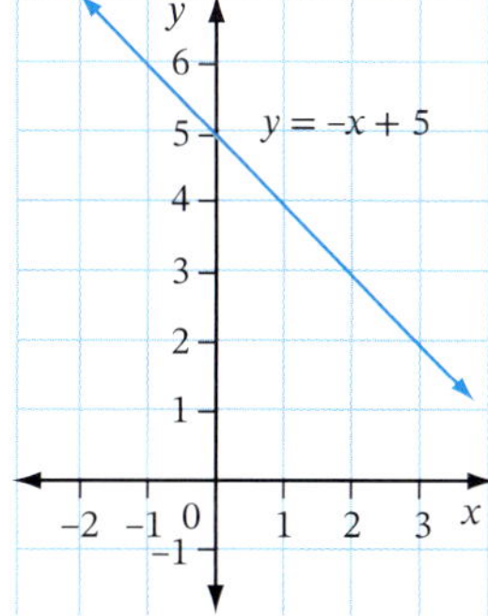

c

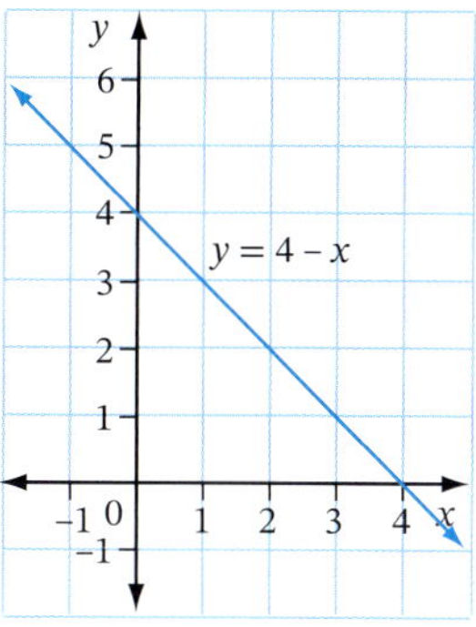

d

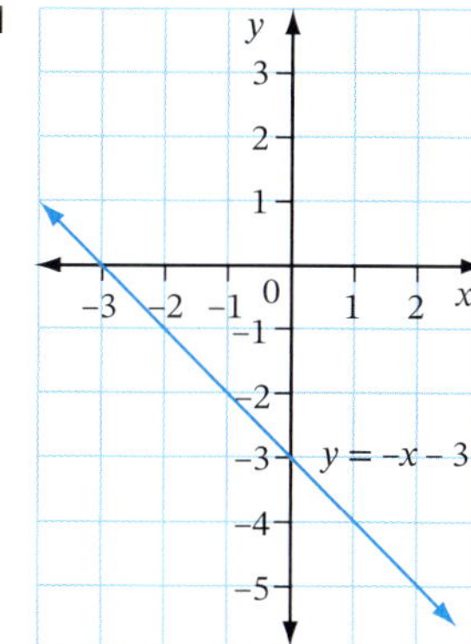

e

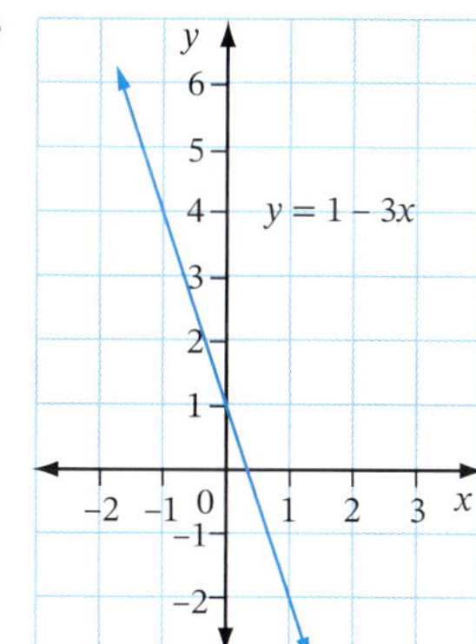

f

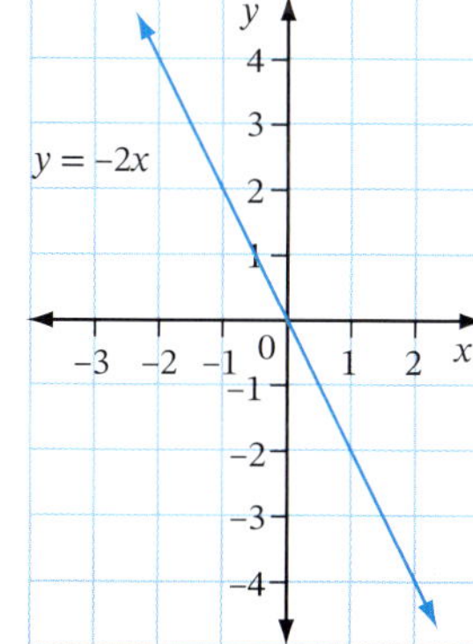

5 From left to right, the line is falling, going down.

6 **a** increasing **b** increasing **c** decreasing
d increasing **e** decreasing **f** decreasing

7 **a** yes **b** no **c** yes
d yes **e** yes **f** no

8 D

Exercise 12.08

1 **a** 4, 1 **b** $\frac{1}{2}$, −3 **c** 1, −5 **d** −1, 6

2 **a** point in the same direction, parallel
b coefficient of x is 1
c different y-intercepts
d different constant terms

3 **a** all pass through (0, 0), y-intercept of 0
b constant term of 0
c point in different directions
d different coefficients of x
e $y = 3x$, 3 **f** $y = \frac{1}{4}x$, $\frac{1}{4}$
g The greater the size of the coefficient, the steeper the line.

4

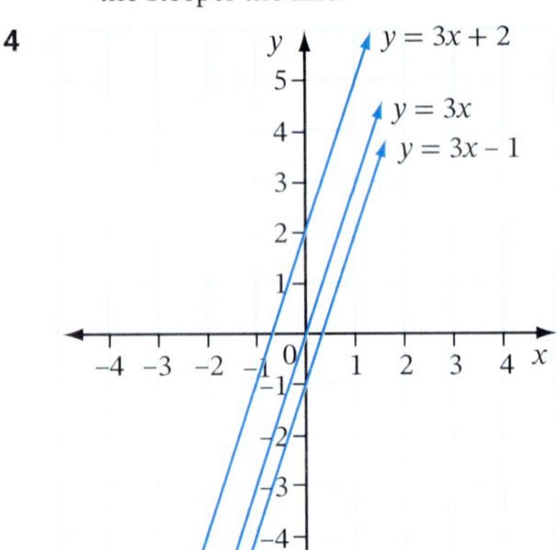

a parallel **b** coefficient of x is 3
c y-intercepts **d** constant terms

5

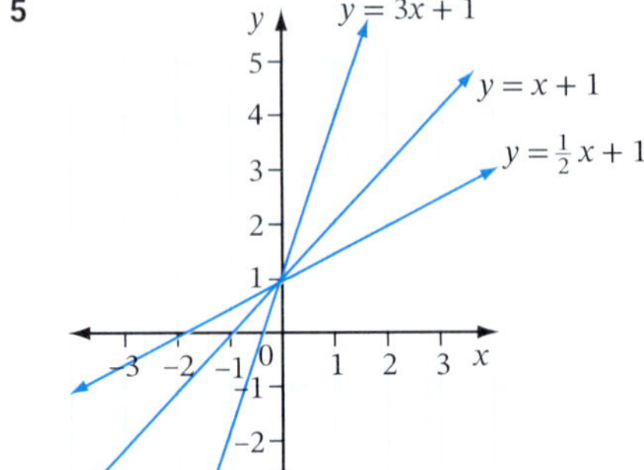

a y-intercept of 1
b constant term of 1
c different steepnesses, directions
d coefficients of x
e 3

6

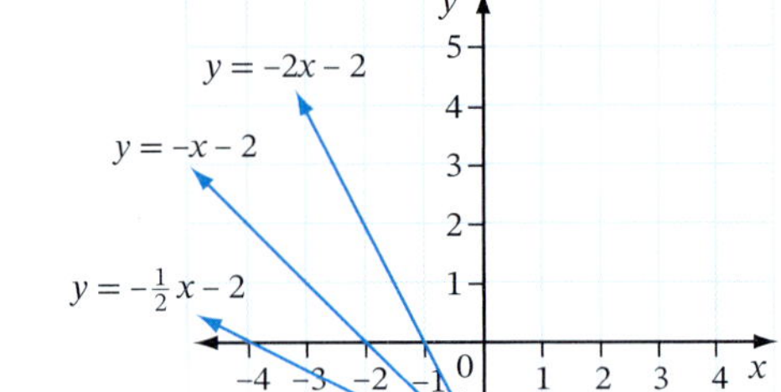

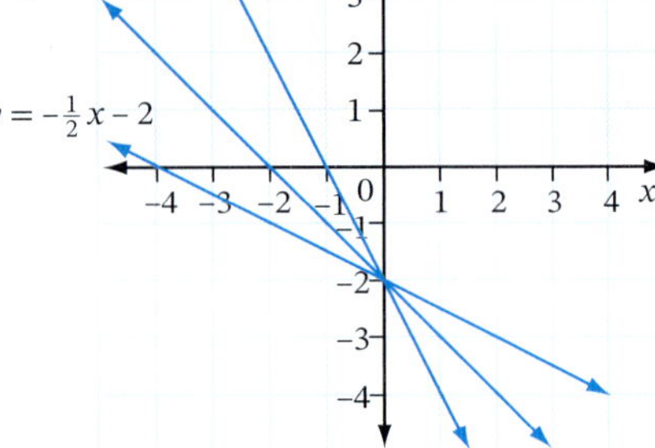

a y-intercept of −2
b constant term of −2
c different steepnesses, directions
d decreasing lines, pointing downwards
e $-\frac{1}{2}$

7 the sign of the coefficients of x 8 B

9 **a** coefficient of x **b** y-intercept
c decreasing **d** constant term
e negative

10 **a** 9, 4, 1, 0, 1, 4, 9

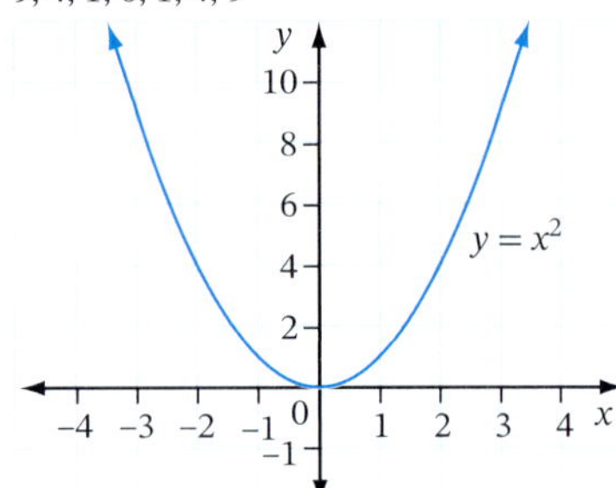

b 0, 1, 1.41, 2, 2.24, 2.65, 3
$y = \sqrt{x}$

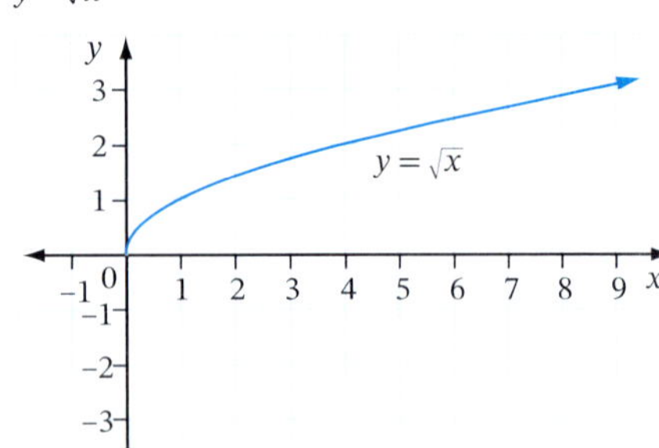

c −8, −3.375, −1, −0.125, 0, 0.125, 1, 3.375, 8

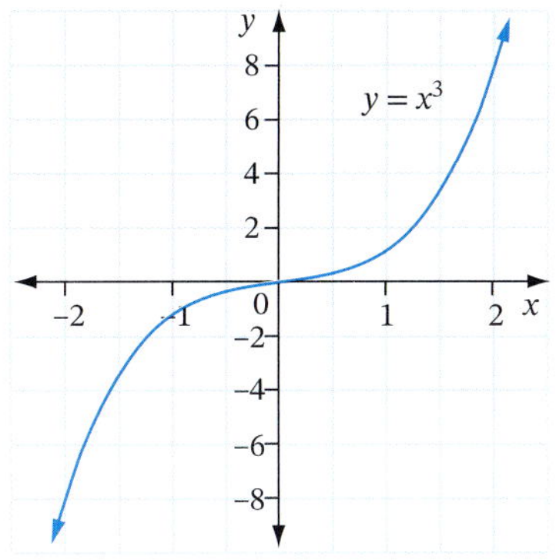

d 2, 1, 0.$\dot{6}$, 0.5, 0.4, 0.$\dot{3}$, 0.29, 0.25, 0.$\dot{2}$

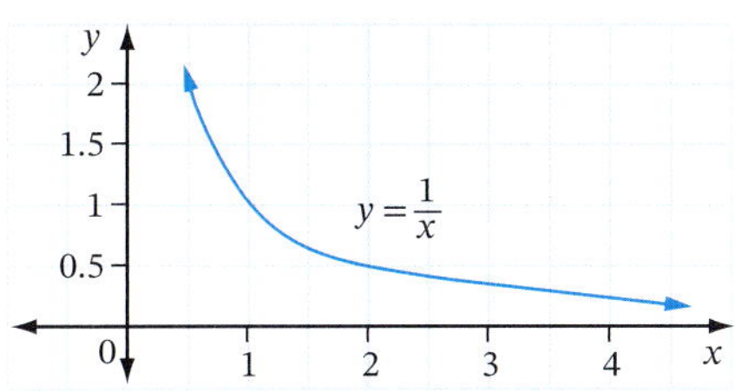

11 **a** Their graphs are not lines, but curves.
b x is raised to a power, or square root of x, or appears in the denominator of a fraction.

Exercise 12.09

1	**a** $x = 3$	**b** $x = -2$	**c** $x = 1\frac{1}{2}$
2, 3	**a** $x = 3$	**b** $x = 4\frac{1}{2}$	**c** $x = -3$
4	**a** $x = -2$	**b** $x = 0$	**c** $x = 2\frac{1}{2}$
5	**a** $x = 3$	**b** $x = -3$	**c** $x = 0$
6	**a** $x = 4$	**b** $x = -3$	**c** $x = 2$
7	**a** $x = 4$	**b** $x = -\frac{1}{2}$	**c** $x = -2$
8	**a** $x = 4$	**b** $x = -4$	**c** $x = -2$

Exercise 12.10

1	**a** $2	**b** $32	**c** 5h
2	**a** $50	**b** $350	**c** 2 h
3	**a** $70	**b** $430	**c** 5 h
4	**a** 100 L	**b** 40 L	

c The tank is empty so there are 0 L of water in the tank.

5 **a** 98° **b** 60° **c** 11 min
d The cake would not cool to 0° in the time shown, and after that, unless it was put in a freezer and frozen.

6 **a** $C = 4d + 7$

D	0	5	10	15
C	7	27	47	67

b $7
c $C = 4 \times 10 + 7 = \$47$
d 5km

7 **a** $C = 80h + 90$

d	0	1	2	3	4	5
C	90	170	250	330	410	490

b $90.00
c $330.00
d 4 h

8 **a** $C = 85h + 120$

D	0	1	2	3	4	5
C	120	205	290	375	460	545

b $120.00
c $460.00
d 2 h

9 **a** $L = 200 - 4h$

D	0	5	10	15	20 ×	25
C	200	180	160	140	120	100

b 200L
c $L = 200 - 4 \times 18 = 128$ L
d 42 hours

Exercise 12.11

1 **a** 2, 3, 4, 5

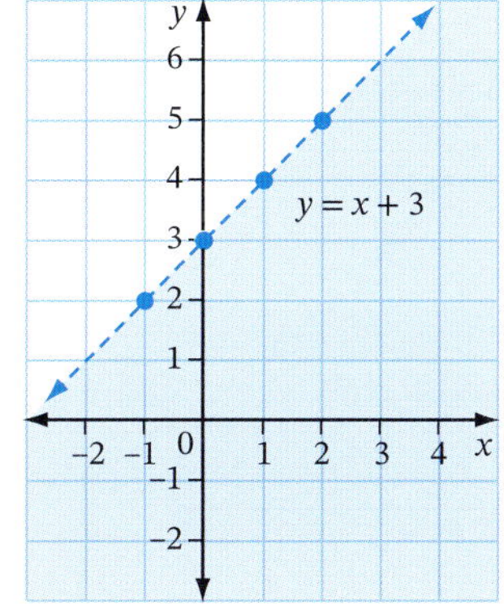

b −2, −1, 0, 1

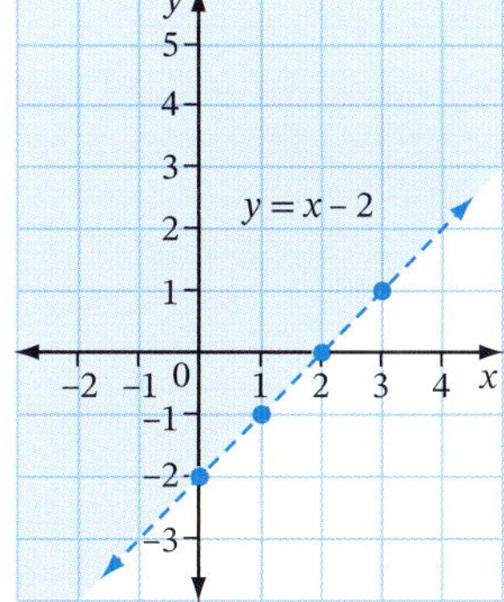

c −2, 0, 2, 4

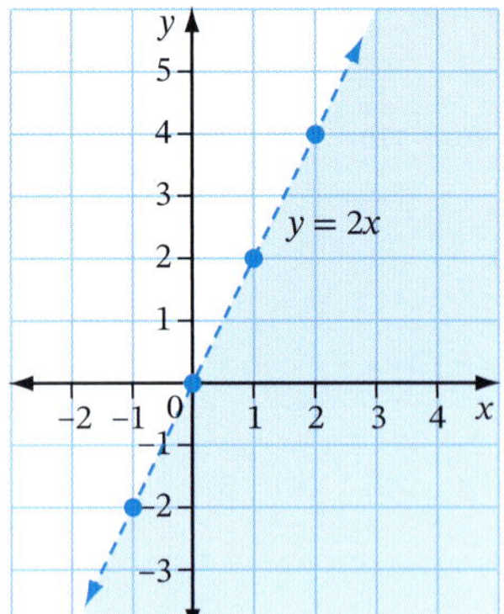

d $-\frac{1}{2}, 0, \frac{1}{2}, 1$

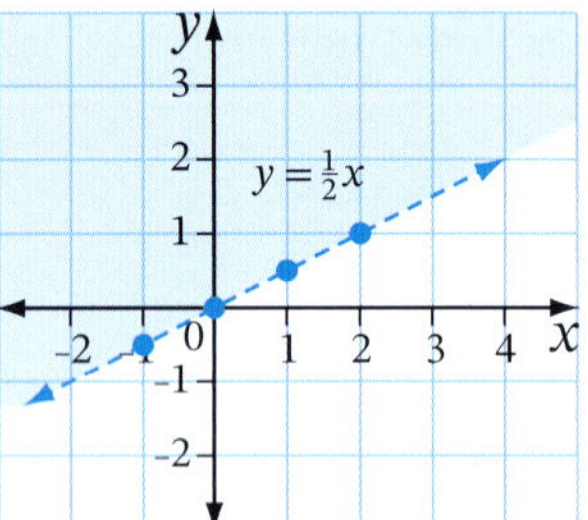

e 6, 5, 4, 3

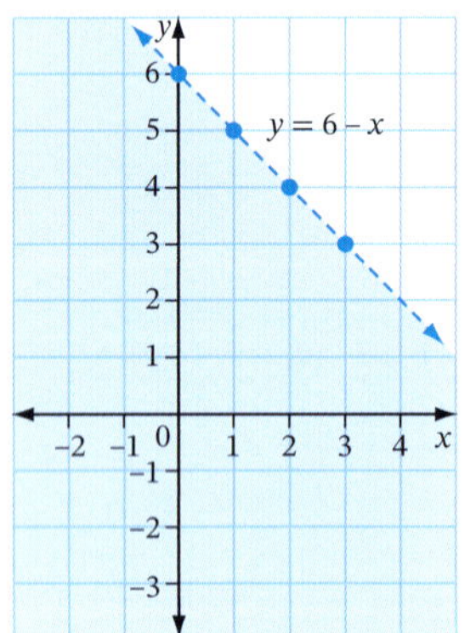

f −1, 0, 1, 2

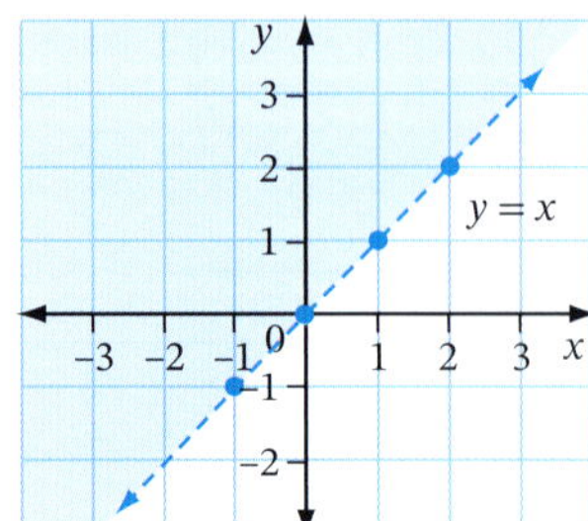

2 **a**

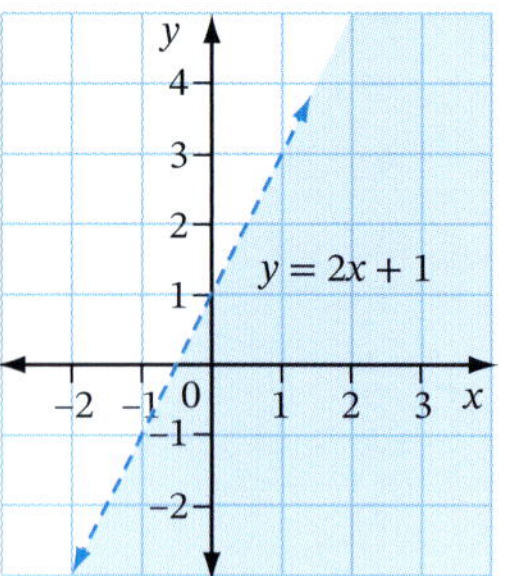

b

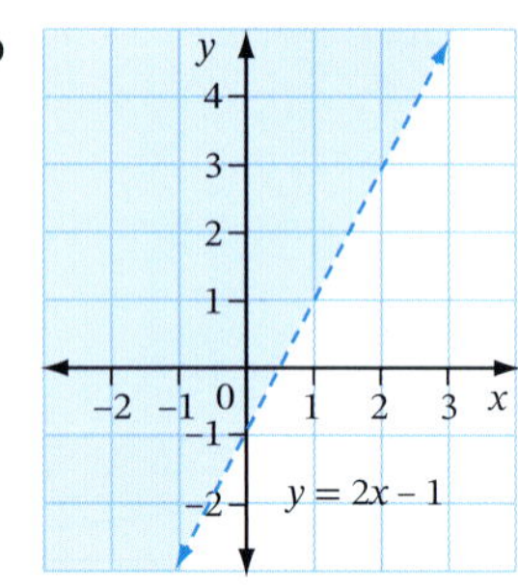

c

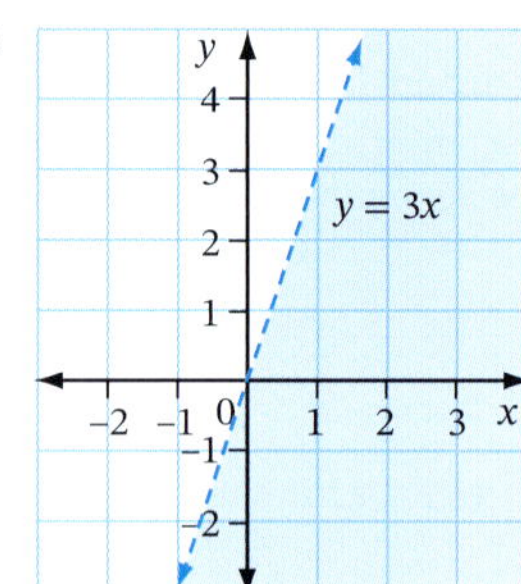

d

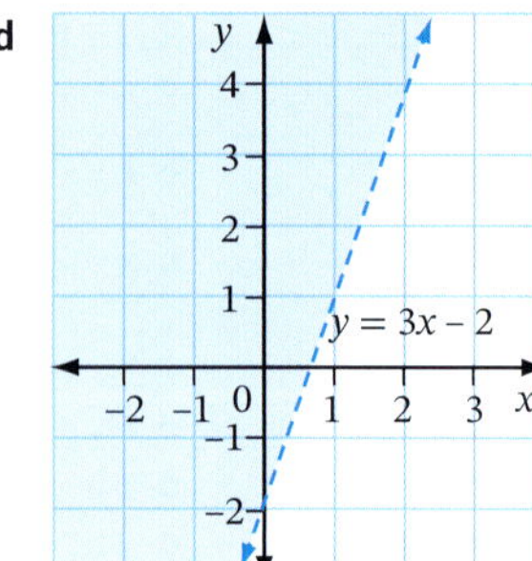

e

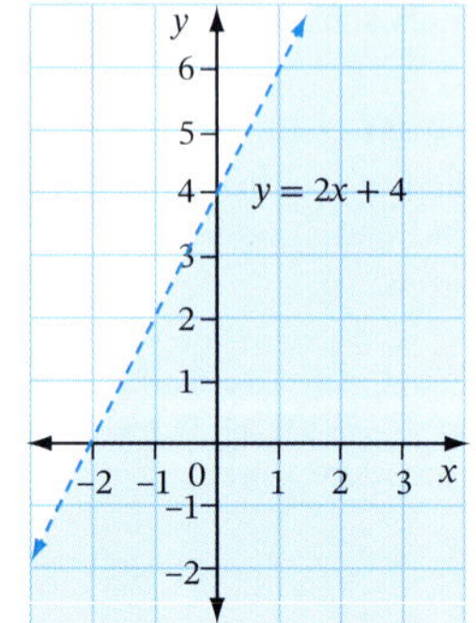

f

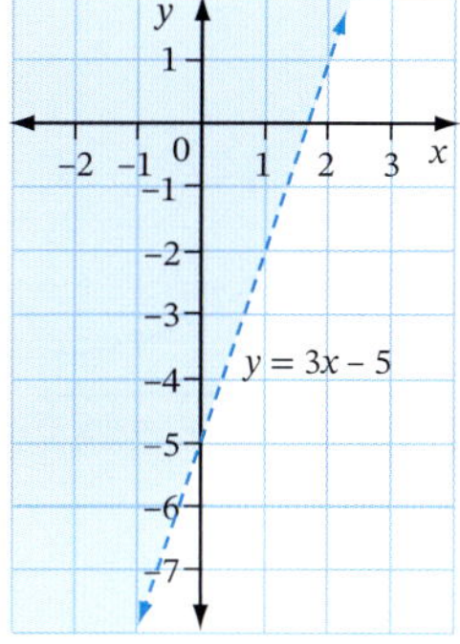

3 a

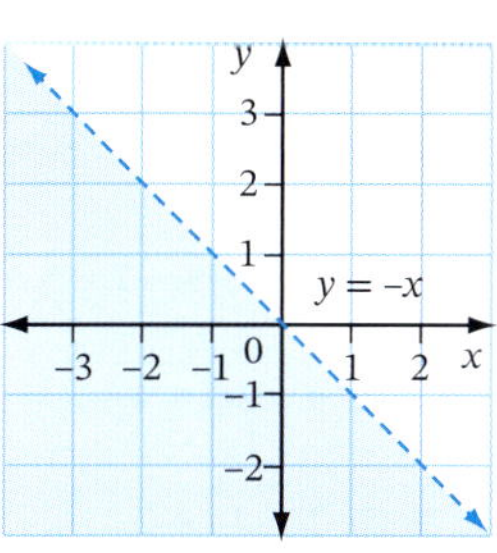

b

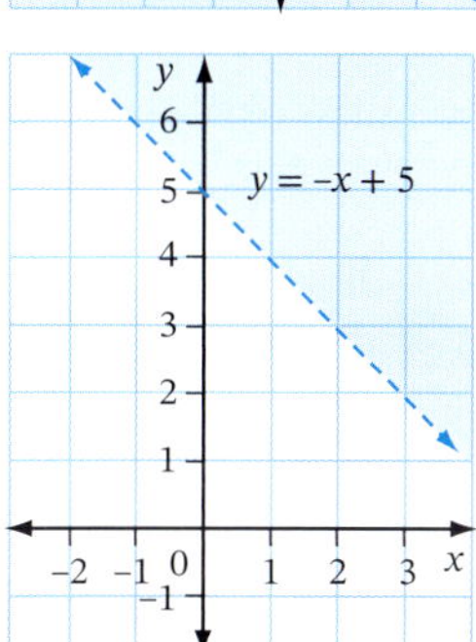

c

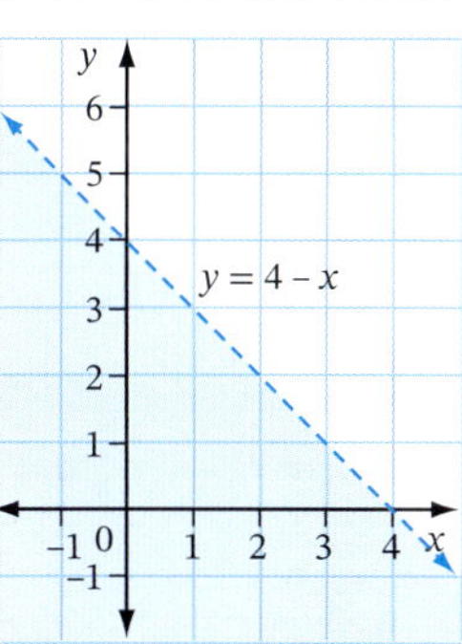

d

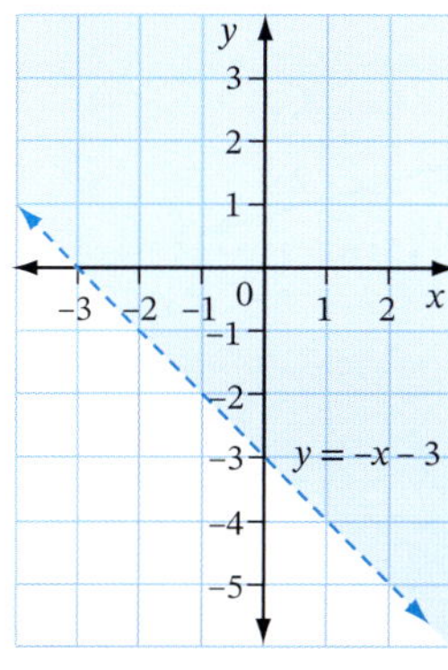

e

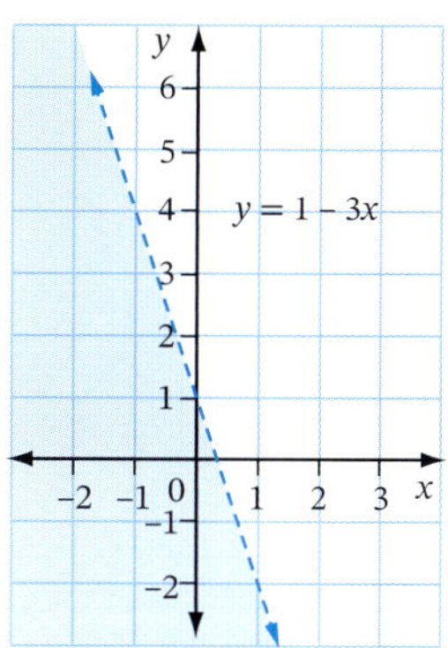

f

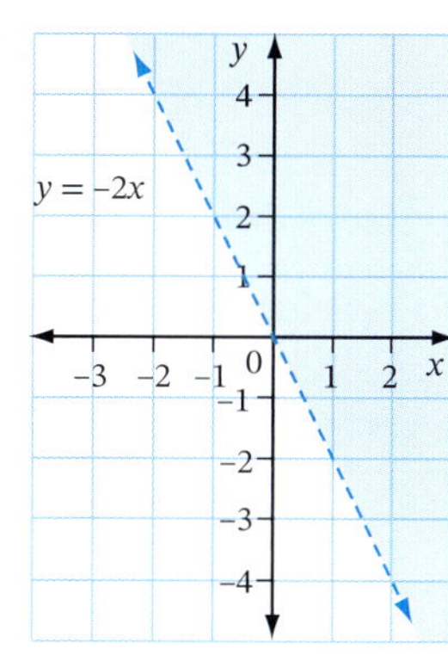

4 **a** below **b** above **c** below
d above **e** below **f** above

5 **a** yes **b** yes **c** no
d no **e** yes **f** no

6 D

Power plus

1 **a, c**

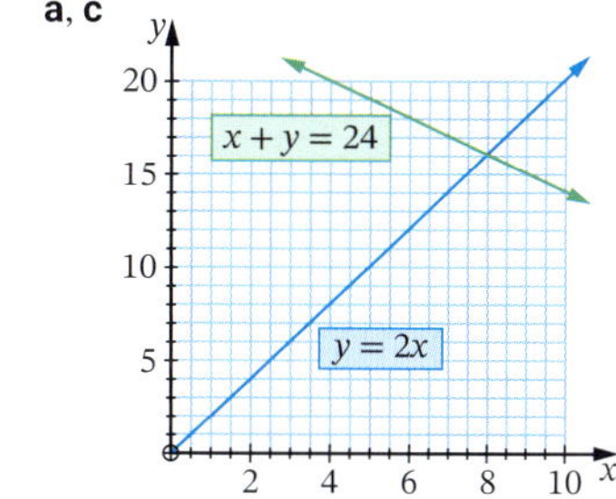

b $x + y = 24$ **d** George $16, Bill $8

2 **a** non-linear

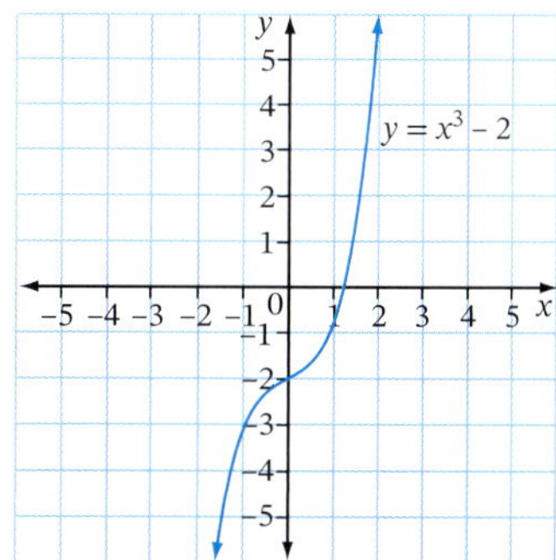

b linear

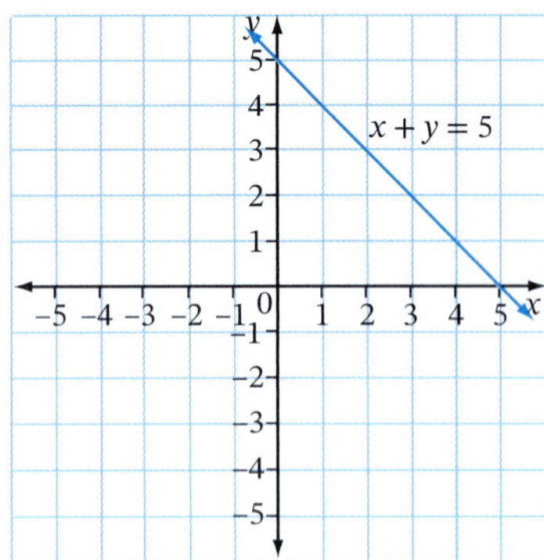

c non-linear

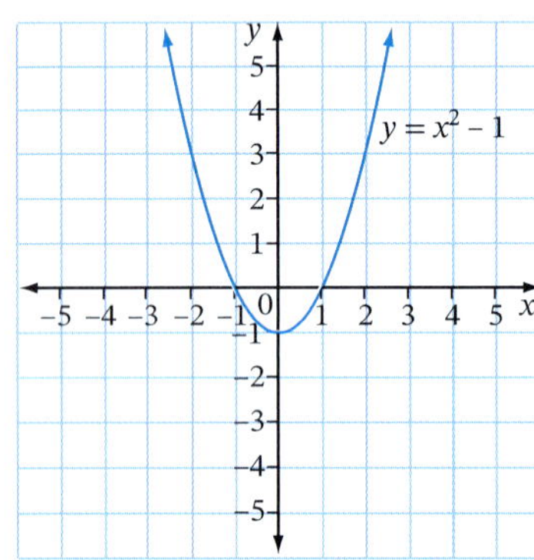

d non-linear

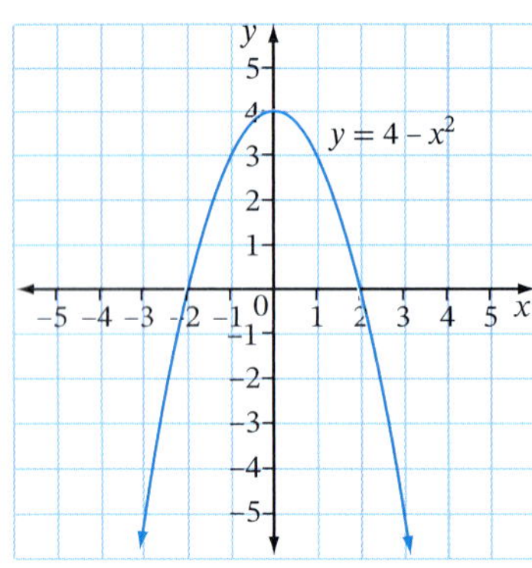

e linear

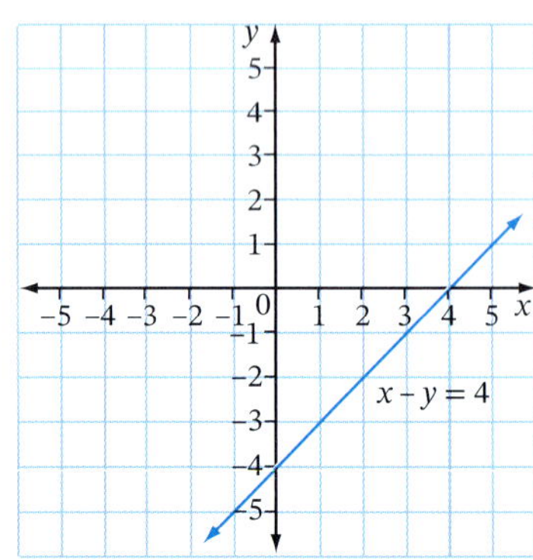

f non-linear

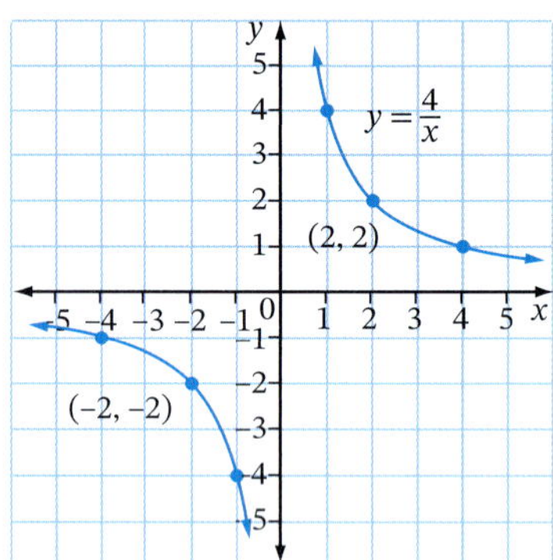

Test yourself 12

1 **a** 2, 3, 4, 5, 6 **b** −11, −9, −7, −5, −3
c 6, 5, 4, 3, 2

2 **a** $y = x - 2$ **b** $p = 3m - 1$
c $n = -2m + 10$

3 **a** 30 **b** $T = 4n + 2$

4 **a** 4, 7, 10, 13, 16, 25, 31
b Number of matchsticks equals 3 times the number of rhombuses plus 1.
c $y = 3x + 1$
d **i** 37 **ii** 46

5 $A(0, 5)$, $B(3, 1)$, $C(-4, -3)$, $D(-2, 4)$

6

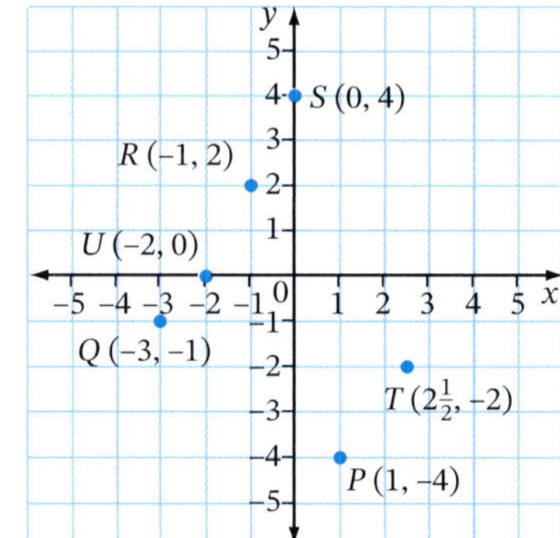

7 **a** S **b** R **c** U **d** T and P

8 **a** $P(3, 4, 6)$
b

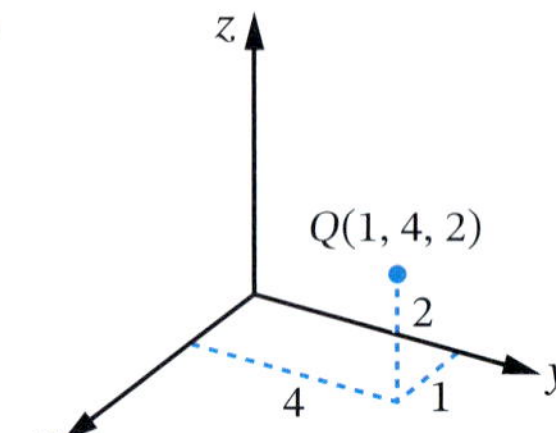

c

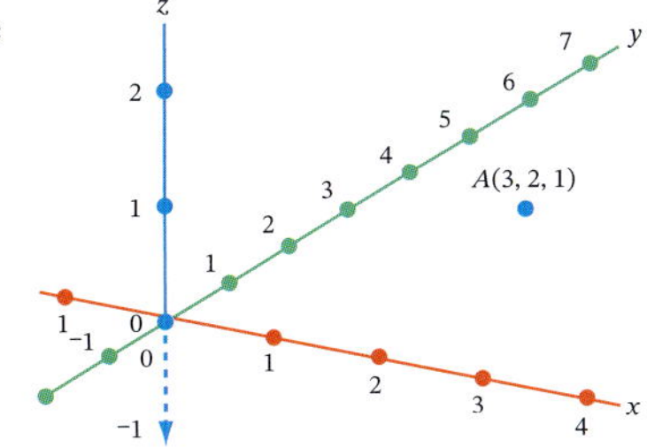

9 **a** 3, 4, 5, 6, 7 **b** $y = x + 2$

c

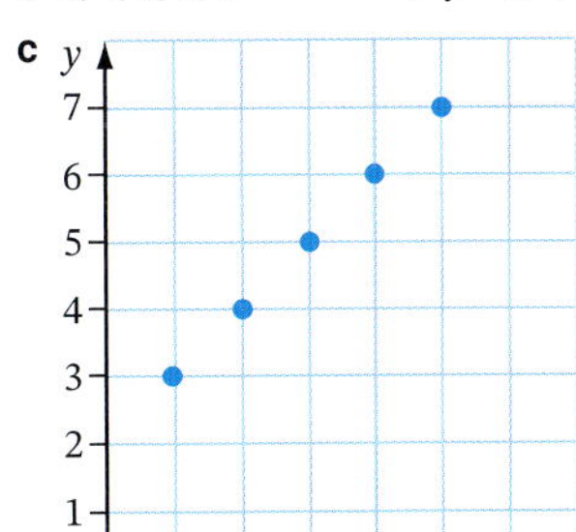

10 **a**

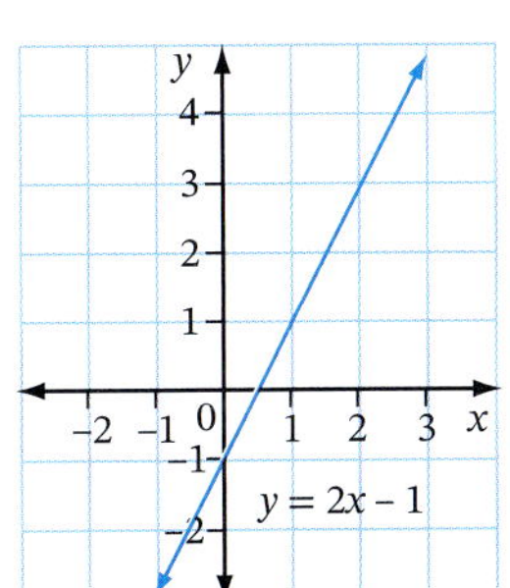

b

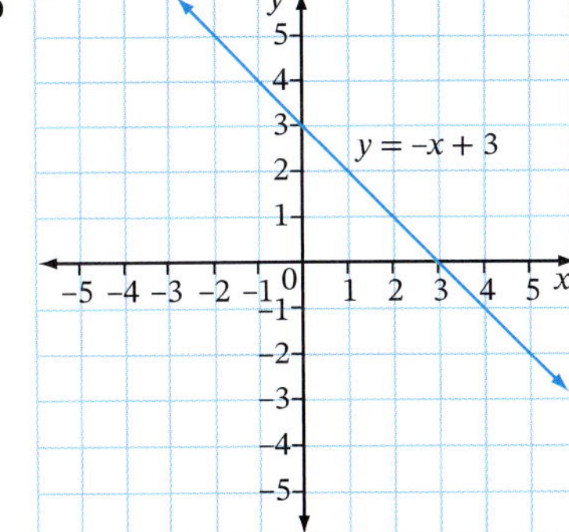

c

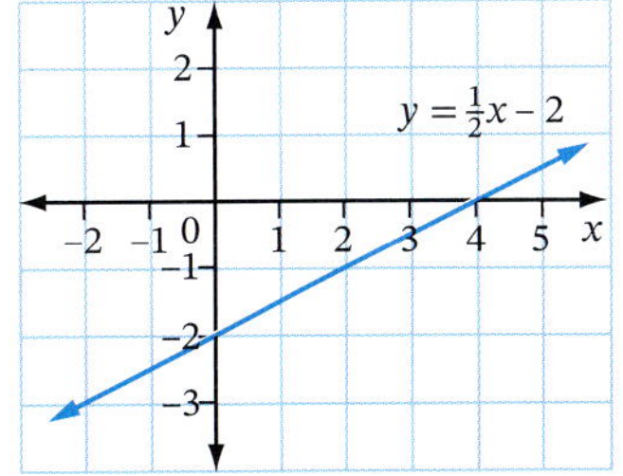

11 **a** No **b** No **c** Yes **d** Yes

12

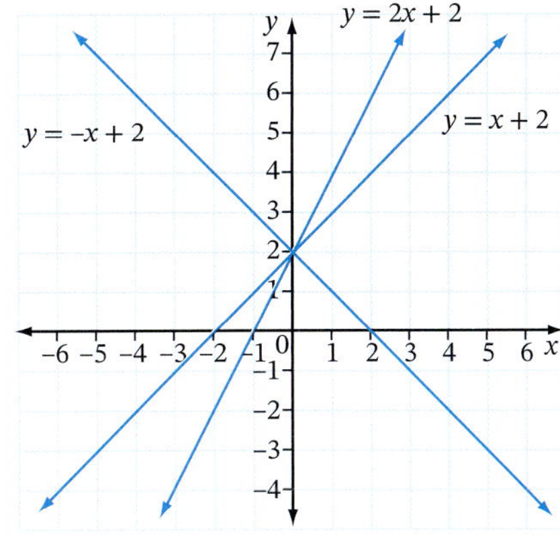

b y-intercept of 2

c constant term of 2

d different steepnesses, directions and coefficients of x

13 **a** $x = 3$ **b** $x = -1\frac{1}{2}$

14 **a** \$100 **b** \$460

c 1 h and 40 min (1.67 h)

d $C = 120h + 100$

h	0	1	2	3
C	100	220	340	460

15 **a**

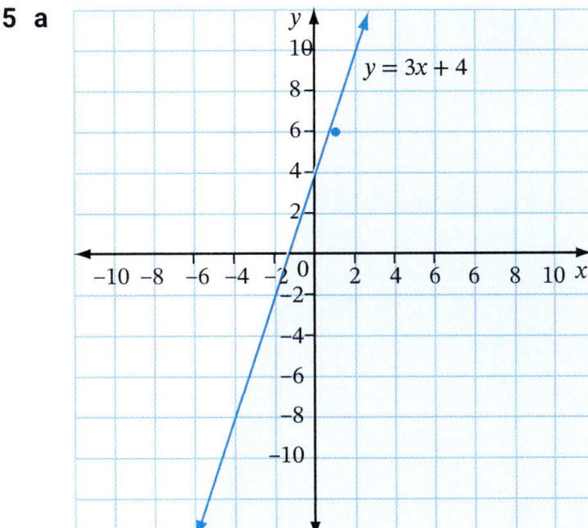

b Yes.

$y < 3x + 4$

$5 < 3 + 4$

$5 < 7$, which is true. (1, 5) is in the shaded region so it satisfies the inequality.

PRACTICE SET 4

1 **a** $a = 1$ **b** $m = 3$ **c** $x = -5$

d $t = 18$ **e** $k = 77$ **f** $x = 2$

g $x = 9$ **h** $x = -8$ **i** $a = 15$

2 **a** 1 : 3 **b** 48 : 17 **c** 40 : 3

d 1 : 12 **e** 6 : 5 **f** 1 : 4

3 **a** 300 g for \$5.20 **b** 1.5 kg for \$3.30

4 7, 9, 11, 13, 15 **5** D

6 **a** $q = 6$ **b** $x = 3$

7 **a** $y = 2$ **b** $d = 10$

8 **a** 5 h 25 min **b** 7 h 5 min

9 20

10 **a** 25 cm **b** 20 m

11 1 kg

12 **a** 33 km/h **b** 225 km

13 $y = 3x$

14 **a** 5, 7, 9, 11, 17

b The number of toothpicks is equal to the double the number of triangles plus 1.

c $p = 2t + 1$ **d** 41

15 $A(2, 3)$, $B(4, 5)$, $C(-2, 4)$, $D(0, -5)$, $E(-3, -2)$, $F(1, 0)$, $G(-1, -3)$.

16 **a** 2, 3, 4, 5

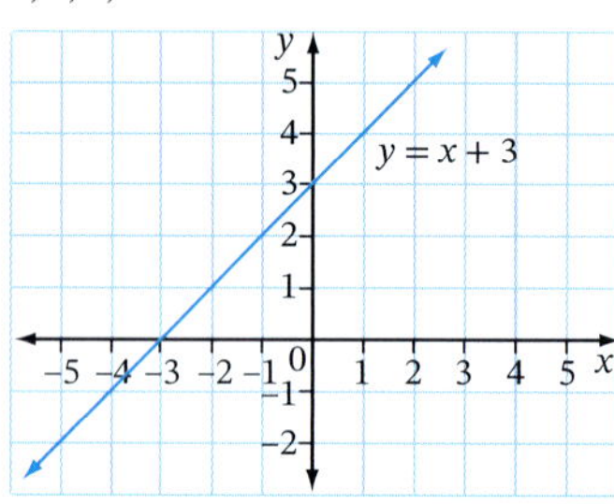

b −5, −1, 3, 7

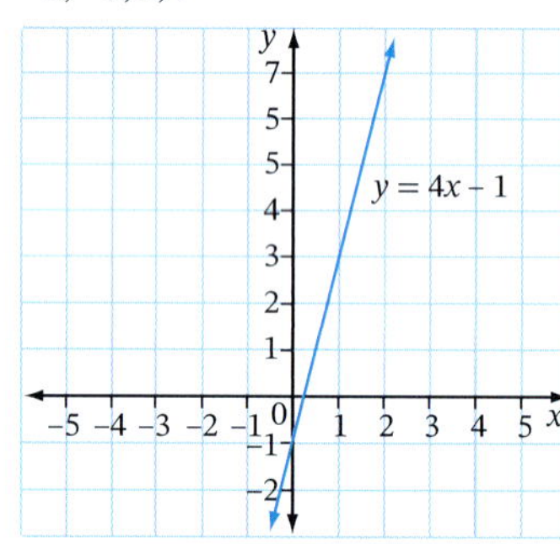

17 $h = 12$ **18** 1.5 tonnes **19** 440 km

20 **a** No **b** Yes

21 50 h

22

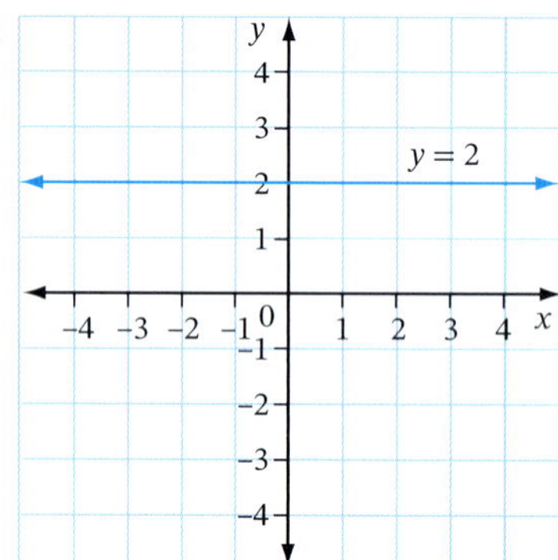

23 $x = 4$

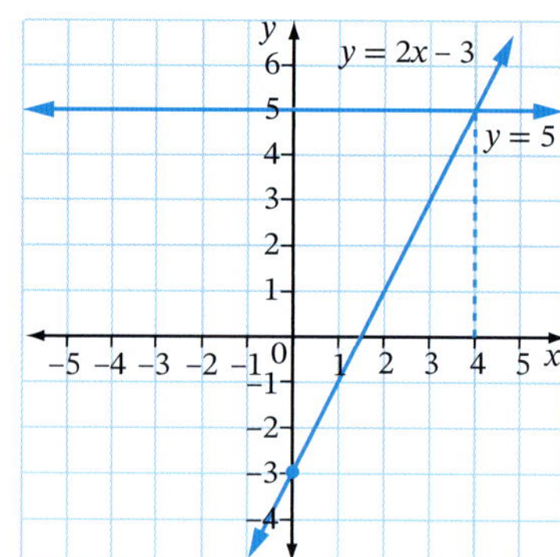

24

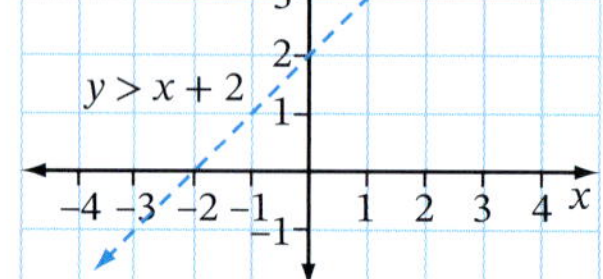

25 12 midnight

26 **a** $x < 4$ **b** $x > 12$ **c** $x < 5$ **d** $x < 4$

27 **a** 1 p.m., for 1.5 hours **b** 26 km

c 6.5 km/h

28 $x = -2$ **29** B

30 **a** Dounya 2.9 m/s, Anna 2.7 m/s

b Dounya by 51.1 s

GENERAL PRACTICE

1 **a** 13 **b** 16 **c** 17

d 3 **e** 5 **f** 18

2 **a** $2x + y$ **b** $yz - x$

c $x + y + z$ **d** $xy - 4z$

3 **a** $\frac{1}{4}$ **b** $\frac{3}{4}$ **c** $\frac{7}{20}$

4 **a** $7\frac{1}{5}$ **b** $6\frac{7}{10}$ **c** $6\frac{1}{6}$

5 **a** −6 **b** 4 **c** 5

6 **a** $6ar$ **b** $-3p$ **c** $35u^2$

d $5x + 9$ **e** $8p - 7q$ **f** $\frac{6u}{m}$

7 25%

8 **a** 30 m/s **b** 16.4 points/game

c \$8/kg

9 **a** 4.8 **b** 5 **c** 5 **d** 7

10 **a** $m = 34$ (angle sum of a triangle).

b $m = 45$ (angles on a straight line).

c $y = 68$ (angle sum of a triangle).

d $m = 100$ (isosceles triangle, exterior angle of a triangle or angles on a straight line).

e $x = 132$ (vertically opposite angles).

f $x = 67$ (angle sum of a quadrilateral).

g $y = 130$ (different reasons possible: corresponding/alternate angles on parallel lines and angles on a straight line).

h $x = 62$ (exterior angle of a triangle).

i $m = 62$ (angle sum of an isosceles triangle, corresponding angles on parallel lines).

11 a (3, 4, 5)

b

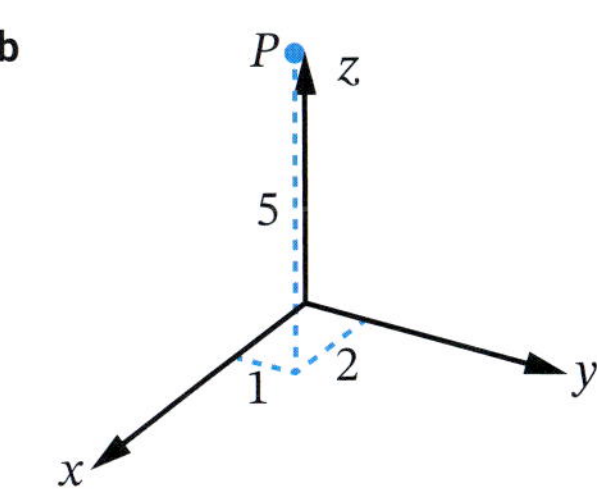

12 a $m = 17.7$ **b** $y = 16.6$

c $p = 25.5$ **d** $d = 76.0$

13 a 6^2 **b** 2^7 **c** 3^2 **d** 5^3

e 2^{12} **f** $2^6 \times 5^2$ **g** $3^0 = 1$ **h** 6

14 a Not choosing an Ace **b** $\frac{48}{52} = \frac{12}{13}$

15 a 187 mm² **b** 144 cm² **c** 18.4 m²

16 a i isosceles **ii** acute-angled

b i scalene **ii** obtuse-angled

c i scalene **ii** right-angled

17 $200 = 2^3 \times 5^2$

18 a 80 000 **b** 0.49 **c** 650 000 000

19 a $1\frac{5}{12}$ **b** $\frac{7}{15}$ **c** $\frac{7}{15}$ **d** $\frac{8}{15}$

e $1\frac{5}{8}$ **f** $3\frac{2}{7}$ **g** $11\frac{1}{4}$ **h** $6\frac{3}{25}$

20 $\frac{3}{20}$

21 a $a = 5$ **b** $d = -5$ **c** $r = 3$

22 a

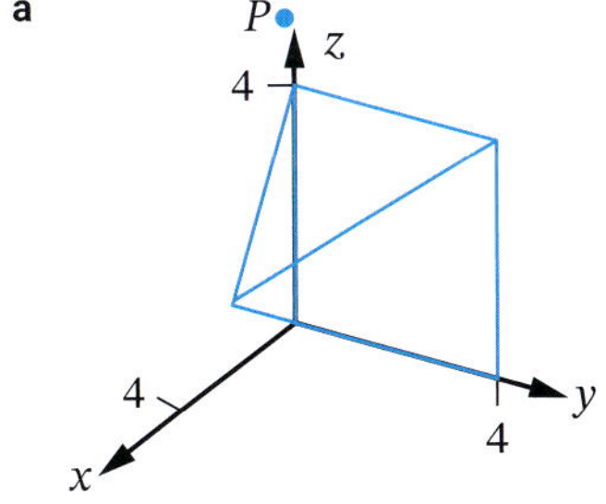

b Square pyramid

23 a Yes **b** No

24 a QR **b** $\angle L$ **c** $\triangle SRQ$

25 $940.80

26 a $y = 3x + 4$ **b** $x = 3$

27 a 8 glasses **b** 600 L

28 a

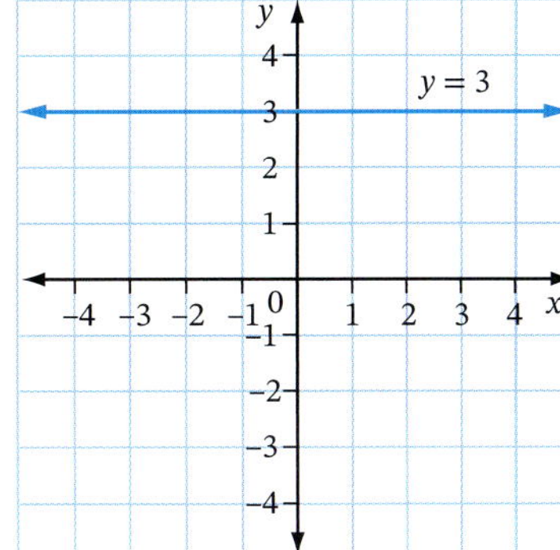

b

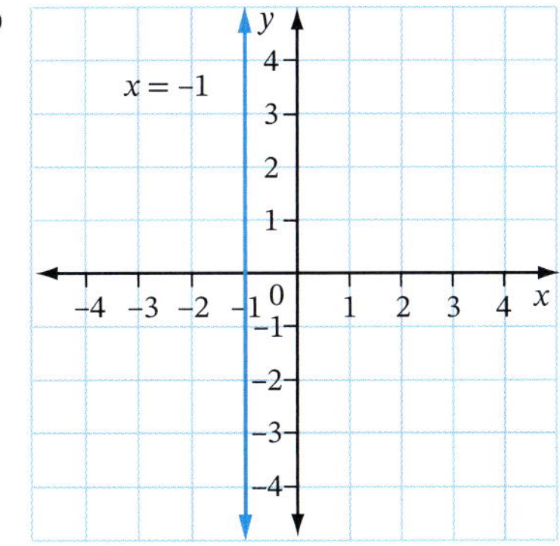

29 a $3a + 15$ **b** $20 - 8k$ **c** -15

30 a 6 : 7 **b** 20 : 1 **c** 1 : 100

31 280 **32** 109 cm

33 a 6:40 p.m. **b** 1:55 a.m.

34 a

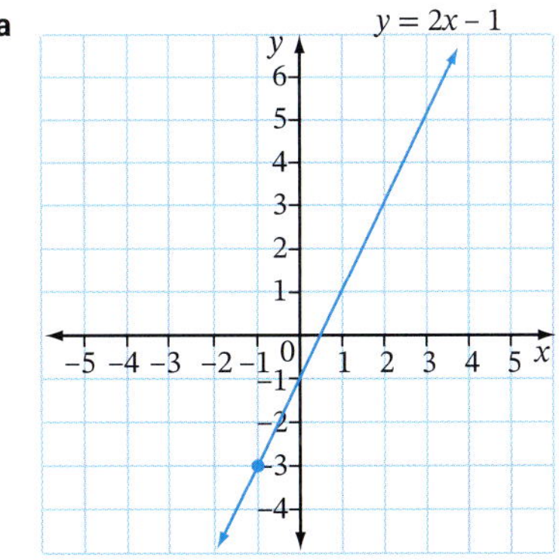

b yes

35 a $80 **b** $530 **c** 8 hours

36 a $36 **b** 60%

37 a rectangle (or square)

b equilateral triangle

c rectangle (or rhombus)

d isosceles triangle

38 a i 43.98 m **ii** 153.94 m²

b i 25.71 cm **ii** 39.27 cm²

c i 66.85 cm **ii** 272.55 cm²

d i 62.83 mm **ii** 85.84 mm²

39 a $2(y + 4)$ **b** $5a(1 - 3b)$ **c** $-2b(4a - 1)$

40 $200

41 **a** $\frac{1}{2}$ **b** $\frac{2}{7}$ **c** $\frac{11}{14}$

42 4900 **43** 16.12 m **44** 80 m^2

45 **a** 62 **b** 42 **c** 42.5

46 **a, b**

c Diagonals are equal and bisect each other.

47 **a** RHS **b** SAS **c** SSS **d** AAS

48 2.5 cm **49** B

50 **a** 400 cm^3 **b** 288 cm^3 **c** 300 mm^3 **d** 910 m^3 **e** 84 m^3

51 $50.40

52 **a**

Score	Tally	*f*	*fx*
45	|	1	45
47	||	2	94
48	||	2	96
49	𝍸 |	6	294
50	𝍸 𝍸 |	11	550
51	𝍸 ||	7	357
52	𝍸 ||||	9	468
53	||	2	106
	Total	40	2010

b 50.25

c

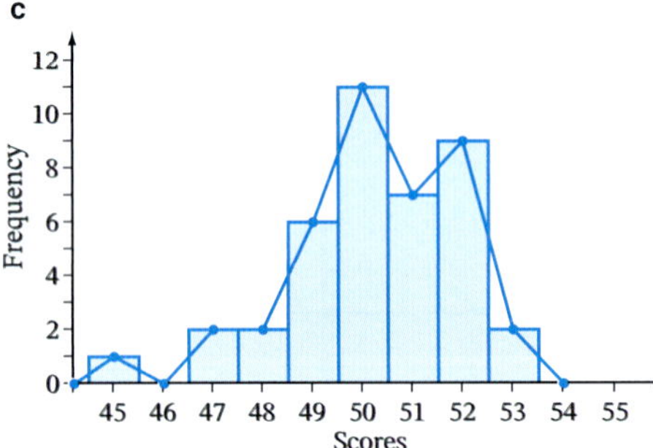

53 **a**

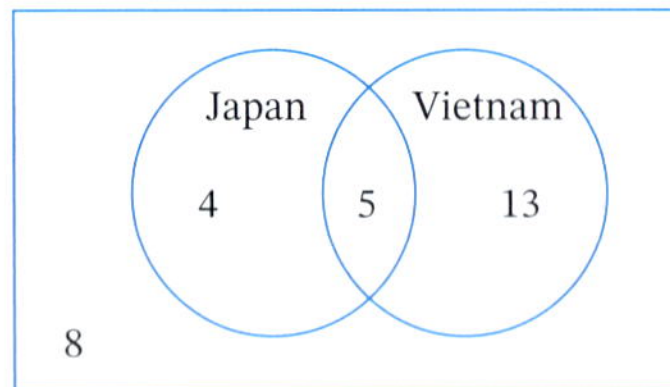

b 8 **c** 4 **d** $\frac{7}{10}$

54 **a** 19 500 cm^3 **b** 19.5 L

55 $x = 7$ **56** 6.1 m/s

57 **a** $600 **b** 450 kg

58 **a** 4 km **b** 80 min, 7 km **c** 20 min **d** 8 km/h

59 $x = -1$

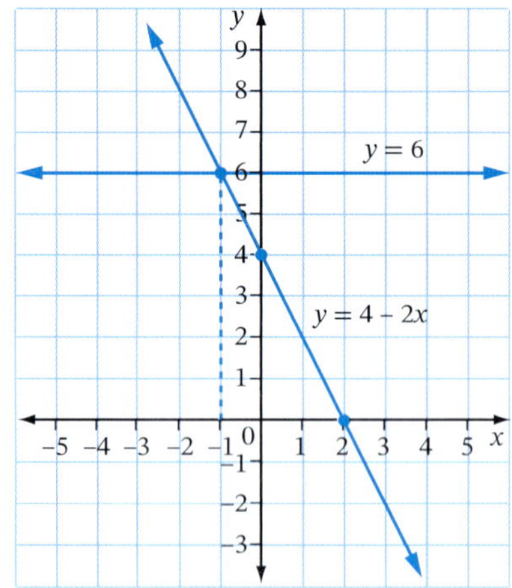

60

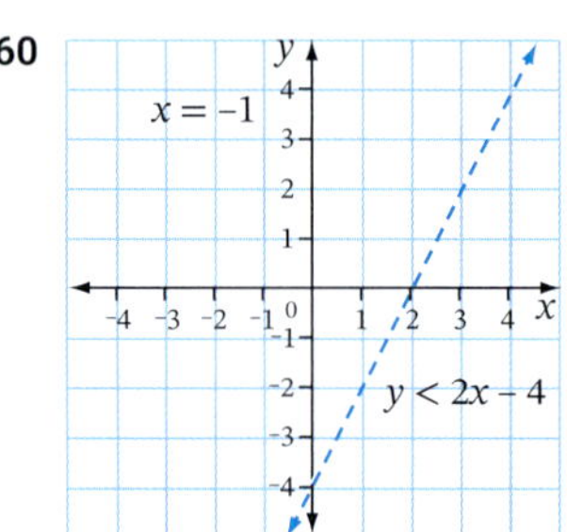

61 $110

CALCULATOR SKILLS

Exercise 1

1 **a** 5068 **b** 14.5 **c** 45 **d** 5.724 **e** 3.5 **f** 26.9 **g** 23 **h** 1.68 **i** 160 **j** 29.48 **k** −8.084 **l** 12.42

2 **a** $19.15 **b** 85 cents

3 $1875.50 **4** 7542 kg

5 2.4 m **6** $18.75

7 23 trees **8** 75.2 km/h

9 $85.74 **10** 13 buses

11 **a** $41.20 **b** $1318.40

12 53.4

Exercise 2

1 **a** 7 **b** −9 **c** −12 **d** −8 **e** 15 **f** −15 **g** 75 **h** 5 **i** −5 **j** 7 **k** −96 **l** 45 **m** −25 **n** 24 **o** 96 **p** 14 **q** −16 **r** 15 **s** −36 **t** 10

Exercise 3

1 **a** 196 **b** 529 **c** 216 **d** 1.728 **e** 26.3169 **f** 10 201 **g** 4.1616 **h** 6859 **i** 26 **j** 148 **k** 9 **l** 42

2 **a** 4.9 **b** 10.0 **c** 4.7 **d** 4.6

3 **a** 16 **b** 156 **c** 6 **d** 10

Exercise 4

1 **a** $\frac{5}{8}$ **b** $\frac{8}{17}$ **c** $\frac{5}{8}$

d $\frac{8}{9}$ **e** $\frac{21}{25}$ **f** $\frac{1}{4}$

2 **a** $\frac{15}{8}$ **b** $\frac{47}{9}$ **c** $\frac{43}{4}$

3 **a** $2\frac{2}{9}$ **b** $8\frac{4}{5}$ **c** $2\frac{1}{4}$

4 **a** 0.5 **b** 0.8 **c** 0.875

d 0.36 **e** 0.17 **f** $0.\dot{5}$

5 **a** $\frac{7}{20}$ **b** $2\frac{1}{25}$ **c** $\frac{1}{1000}$

d $\frac{73}{200}$ **e** $1\frac{3}{50}$ **f** $\frac{13}{40}$

6 **a** $\frac{17}{20}$ **b** $1\frac{1}{14}$ **c** $\frac{1}{15}$ **d** $\frac{4}{21}$

e $1\frac{1}{3}$ **f** $\frac{5}{24}$ **g** $1\frac{17}{30}$ **h** $\frac{35}{96}$

i $1\frac{3}{4}$ **j** $19\frac{1}{5}$ **k** $6\frac{13}{20}$ **l** $8\frac{13}{24}$

m $12\frac{3}{4}$ **n** 7 **o** $\frac{4}{27}$ **p** $19\frac{19}{60}$

7 **a** $\frac{1}{8}$ **b** $\frac{3}{4}$

c $15\frac{4}{5}$ or 15.8 **d** $\frac{1}{6}$ or $0.1\dot{6}$

8 **a** $6\frac{3}{4}$ hours **b** $33\frac{3}{4}$ hours

Glossary and index

24-hour time Time of day written using 4 digits (instead of a.m. or p.m.) and the hours 0–23. For example, 18:20 is the 24-hour time for 6:20 p.m. (p. 509)

AA The angle–angle (equiangular) test for similar triangles. (p. 384)

AAS The angle–angle–side test for congruent triangles. (p. 369)

acute angle A 'sharp' angle between 0° and 90°, for example, the marked angle in the diagram. (p. 144)

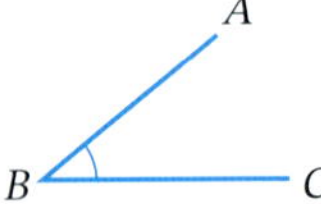

acute-angled triangle A triangle with all 3 angles acute. (p. 152)

algebraic expression A number written in algebraic form using variables, for example, $2xy + 4y - 5$. (p. 97)

algebraic term See **terms of an expression**.

alternate angles A pair of angles between 2 lines crossed by a transversal, on opposite sides of the transversal; for example, the 2 angles marked in the diagram. (p. 148)

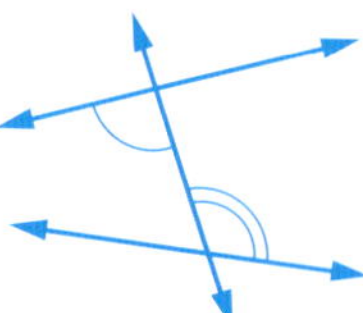

angle sum The total of the sizes of the angles in a shape. The angle sum of a triangle is 180°. (p. 163)

annulus A ring shape between 2 different-sized circles with the same centre. (p. 213)

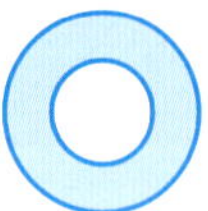

arc Part of the circumference of a circle. (p. 214)

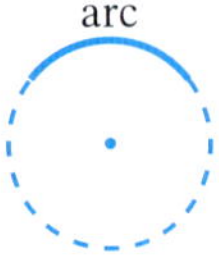

area The amount of surface enclosed by a shape, measured in square units. (p. 189)

ascending order Going up, increasing, from smallest to largest (1–2–3). The opposite of **descending order**. (p. 60)

at least Referring to the smallest number, for example, 'at least 2' means 2, 3, 4, ... , that is, '2 or more'. (p. 405)

average See **measure of central tendency**. (p. 300)

backtracking method A method of solving equations by 'undoing', or performing inverse (opposite) operations in reverse order. (p. 455)

balancing method A method of solving equations by performing inverse (opposite) operations on both sides. (p. 455)

base (in index notation) When a number is raised to a power, the number raised is the base. In the expression 3^5, the 3 is called the base. (p. 71)

base (of a shape) The bottom side of a flat shape such as a triangle. (p. 192)

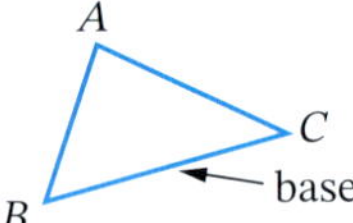

best buy When comparing different brands or sizes of items during shopping, the best buy is the item with the lowest unit price and is the best value for money. (p. 515)

bias In statistics, something that causes a sample to not truly represent the population. (p. 333)

binomial expression An algebraic expression with 2 terms, for example, $x + 9$, $2y - 12$. (p. 130)

braces/brackets See **grouping symbols**.

capacity The amount of material (usually liquid) that a container can hold, measured in millilitres (mL), litres (L), kilolitres (kL) and megalitres (ML). See also **volume**. (p. 228)

Cartesian plane Another name for **number plane**. (p. 563)

categorical data Non-numerical data that can be classified into categories, such as hair colour, favourite radio station or postcode. Data that is not numerical. (p. 300)

census A survey of the entire **population** of people or items, not just a survey of a **sample**. (p. 333)

certain Must happen; has a probability of 1% or 100%. (p. 404)

circumference The perimeter of a circle or the length of that perimeter. $C = \pi d$ or $C = 2\pi r$, where C is the circumference, π is pi, d is the diameter and r is the radius. (p. 185)

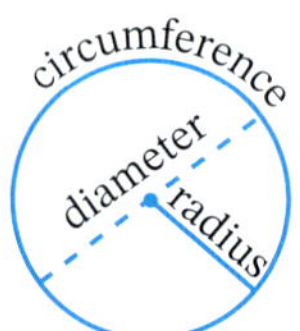

cluster A group of data values that are bunched or close together. (p. 318)

coefficient The number 'in front of' a variable in an equation; a multiplier. For example, in $y = 3x - 7$, the coefficient of x is 3. (p. 578)

co-interior angles A pair of angles between 2 lines crossed by a transversal, on the same side of the transversal (co-interior means 'together inside'), for example, the 2 angles marked in the diagram. (p. 148)

collinear points Points that lie in a straight line. (p. 571)

compass Geometrical instrument for constructing circles and lines of equal length.

complementary angles Two angles whose sum is 90°. The angles 30° and 60° are complementary. (p. 144)

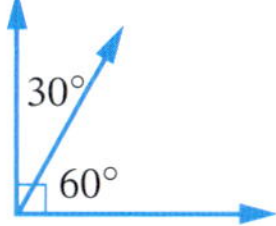

complementary event All the outcomes that are *not* the event; the 'opposite' event. For example, the complementary event to rolling 1 on a die is rolling a number that is not 1. (pp. 408, 410)

composite shape A shape made up of 2 or more basic shapes. (p. 195)

congruent Identical, exactly the same. The symbol '≡' means 'is congruent to' or 'is identical to'. (p. 356)

congruent figures Identical figures, having the same shape and size. (p. 356)

congruence test One of 4 tests for proving that 2 triangles are congruent: SSS, SAS, AAS and RHS. (p. 369)

consecutive numbers Any series of integers that follow each other in order, for example, 8, 9 and 10. (p. 103)

constant term The term in an equation that is a number only and does not contain a variable. For example, in $y = 3x - 7$ the constant term is -7. (p. 578)

converse The 'reverse' of a rule, the rule written 'back-to-front' or 'turned around'. The converse of Pythagoras' theorem is that if the square of the longest side of a triangle is equal to the sum of the squares of the other 2 sides, then the triangle is right-angled. (p. 22)

convex quadrilateral A quadrilateral whose vertices all point outwards. All diagonals lie within the shape, and all angles are less than 180°. (p. 156)

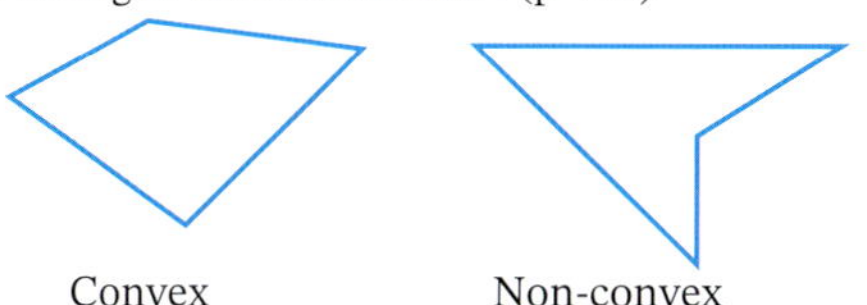

Coordinated Universal Time See **UTC**.

coordinates An ordered pair of numbers used to locate a point or position, for example, $A(2, 5)$ tells us that the point A is located 2 units to the right of origin and 5 units up from the origin. (p. 563)

corresponding angles A pair of angles in matching positions when 2 lines are crossed by a transversal. They are on the same side of the transversal and the lines, for example, the 2 angles marked on the diagram. (p. 148)

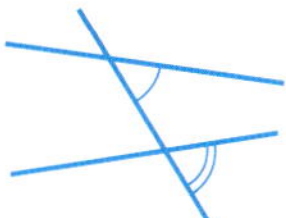

cost price The price it costs a retailer (shop) to buy (to resell). The cost price is usually less than the **selling price**. (p. 272)

cross-section A 'slice' of a solid cut across it rather than along it. (p. 219)

cube (of a number) The number raised to the power of 3. For example, 7 cubed $= 7^3 = 7 \times 7 \times 7 = 343$. (p. 72)

cube root (of a number) The value which, if cubed, gives the number. For example, $\sqrt[3]{8} = 2$ because $2^3 = 2 \times 2 \times 2 = 8$. (p. 72)

cubic metre A metric unit of volume, the volume of a 1 m cube. (p. 217)

cylinder A can-shaped solid with ends that are circles. (p. 225)

data Information, a collection of facts. (p. 293)

data set A collection of data about the same subject. (p. 291)

decimal places The places after the decimal point in a number. For example, 3.1416 has 4 decimal places.

denominator The number below the line in a fraction. The denominator of $\frac{2}{3}$ is 3.

descending order Going down, decreasing, from largest to smallest (3–2–1). The opposite of **ascending order.** (p. 60)

diagonal An interval joining 2 non-adjacent vertices of a shape.

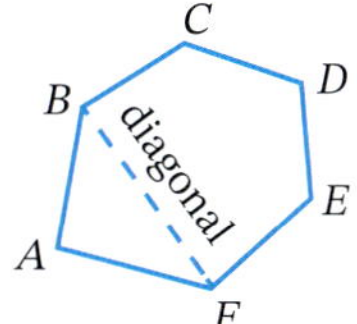

frequency polygon A line graph that shows the frequencies of numerical data. It can be made by joining the midpoints of the tops of the columns of a histogram. The graph looks like a mountain. (p. 325)

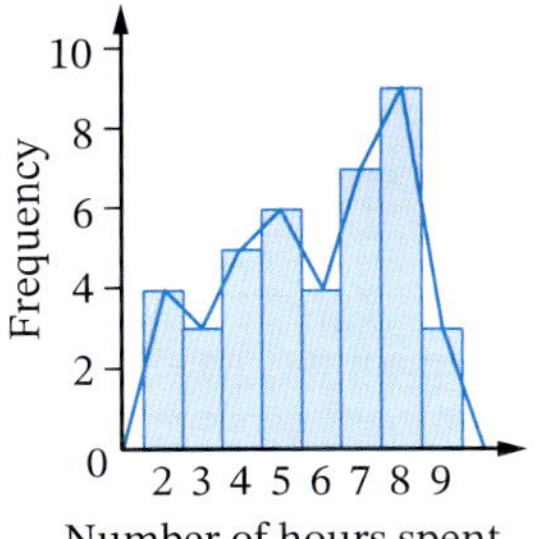

frequency (distribution) table A table listing the frequency of each value in a set of data, with columns for Score (x), Frequency (f) and sometimes Tally and fx. (p. 296)

goods and services tax (GST) A 10% tax added to the original price of an item or service. (p. 273)

greatest common divisor (GCD) See highest common factor. (p. 76)

Greenwich Mean Time (GMT) See **UTC**.

grouping symbols The collective name for parentheses ( ), (square) brackets [ ] and braces {}.

GST See **goods and services tax.**

hectare A large metric unit of area, equivalent to 10 000 m^2 or to the area of a square 100 m × 100 m. (p. 189)

height See **perpendicular height** (p. 192)

highest common factor (HCF) Also called **greatest common divisor (GCD)**. The largest factor shared by 2 or more numbers or algebraic terms. For example, the HCF of 36 and 8 is 4 and the HCF of $6xy$ and $12y^2$ is $6y$. (p. 125)

horizontal Going across, sideways, flat. (p. 551)

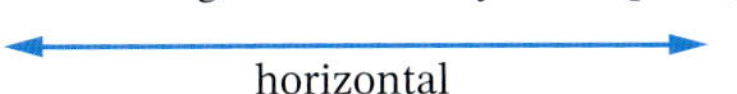

hypotenuse The longest side of a right-angled triangle, opposite the right angle. (p. 7)

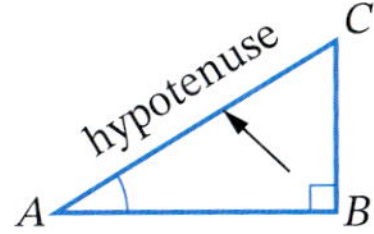

image A transformed shape after it has been translated, reflected, rotated, enlarged or reduced. (p. 351)

impossible Cannot happen, no chance, has a probability of 0. (p. 404)

improper fraction A fraction whose numerator is greater than or equal to its denominator, such as $\frac{7}{4}$. (p. 244)

included angle The angle between 2 given sides of a shape. For example, the included angle for sides AC and CB in this triangle is $\angle C$. (p. 362)

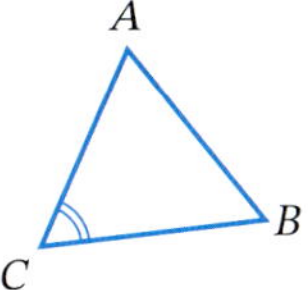

index (plural: **indices**, pronounced 'in-da-sees') See **power**. (p. 71)

index notation A way of writing repeated multiplication using indices (powers), in the form *an*. Index notation for 2 × 2 × 2 × 2 is 2^4. (p. 71)

inequality A mathematical statement that 2 quantities are unequal. For example, $8 + 2 < 14$ or $3x - 3 > 5$. (p. 481)

integer A number that is a positive or negative whole number or zero. The numbers $\{..., -3, -2, -1, 0, 1, 2, 3, ...\}$ are integers. (p. 47)

inverse operation The opposite or reverse operation, used when solving equation. For example, the inverse operation to adding is subtracting, the inverse operation to dividing is multiplying. (p. 458)

irrational number A number such as π or $\sqrt{2}$ that cannot be expressed as a fraction. In decimal form, its digits run endlessly without repeating. (pp. 72, 185)

isosceles triangle A triangle with 2 equal sides. (p. 152)

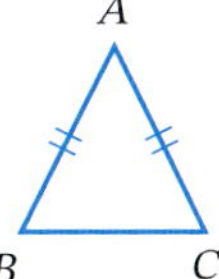

kilo- A prefix meaning 1000, represented by the symbol k. For example, 1 kilogram is 1000 grams.

kite A quadrilateral with 2 pairs of equal adjacent sides. (p. 156)

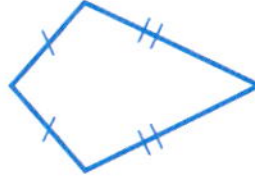

LHS The left-hand side (of an equation). (p. 463)

like terms Algebraic terms that have exactly the same variables. For example, $5xy$ and $2xy$ are like terms, $3xy$ and $4x^2$ are not like terms. (p. 111)

likely A high chance it will happen, probable. (p. 404)

line graph A graph made up of a line or several line intervals, often showing how a quantity is changing over time. (p. 531)

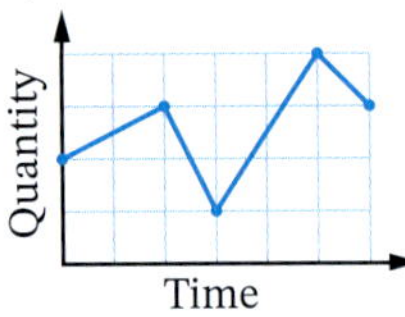

linear Involving a line. (p. 571)

linear equation (or **linear function**) A formula whose graph is a straight line, or an equation involving a variable that is not raised to a power, such as $2x + 9 = 17$. (pp. 573, 578)

long division A method for dividing numbers with 2 or more digits, that is, a number greater than 10. (p. 42)

loss The money lost when selling an item at a lower price, when the selling price is less than the cost price. The opposite of **profit**. (p. 272)

lowest common multiple (LCM) The smallest multiple, that is, shared by 2 or more numbers. For example, the LCM of 4 and 10 is 20. (p. 77)

mean The average of a set of data, represented by $\bar{x}$, calculated by dividing the sum of the data values by the number of values. (p. 300)

measure of central tendency An average, middle or typical value of a set of data. The 3 measures of location are the **mean**, **median** and **mode**. (p. 300)

median The middle value when the values of a data set are arranged in order. If the number of values is even, then the median is the average of the 2 middle values. (pp. 300, 303)

mega- A prefix meaning 1 million (1 000 000), represented by the symbol M. A megalitre is 1 million litres.

micro- A prefix meaning one-millionth, represented by the Greek letter μ. One microsecond is the one-millionth of a second.

milli- A prefix meaning one-thousandth $\left(\frac{1}{1000}\right)$, represented by the symbol *m*. One millimetre is one-thousandth of a metre.

mixed numeral A number written as a whole number and a fraction, for example, $5\frac{3}{4}$. (p. 244)

mode The most common or frequent value(s) in a set of data. (p. 300)

multiple (of a number) The product of the number and a whole number. For example, the multiples of 6 are 6, 12, 18, 24, ...

mutually exclusive events Events or categories that have no items in common. (p. 414)

negative number A number less than 0, written with a negative sign (−), for example, −3, −10 and −7.

number plane A coordinate grid system based on 2 number lines that cross at right angles: a horizontal line called the *x*-axis and a vertical line called the *y*-axis. Also called **Cartesian plane**. (p. 563)

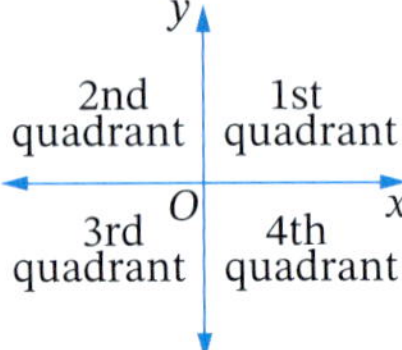

number space A coordinate grid system in 3D based on 3 number lines that cross at right angles: the *x*-axis, *y*-axis and *z*-axis. Also called **Cartesian space**. (p. 563)

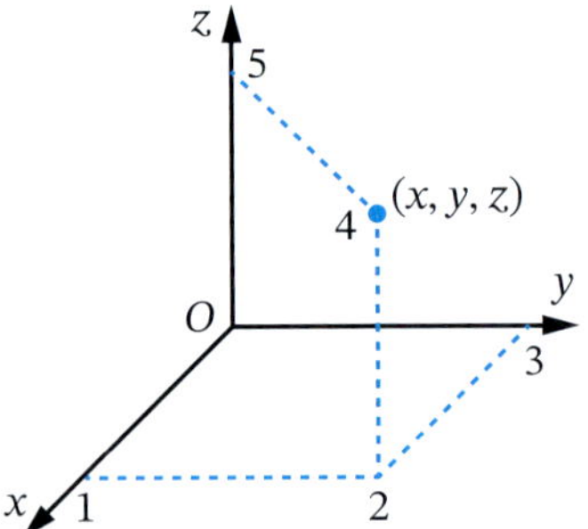

numerator The number above the line in a fraction. The numerator of $\frac{2}{3}$ is 2.

numerical data Data can be measured or counted using values, such as a person's height or number of goals scored. Data that is not **categorical**. (p. 300)

observed frequency The actual number of times an event occurs over repeated trials in an experiment (p. 427)

obtuse angle A 'wide' angle greater than 90°, but less than 180°. (p. 144)

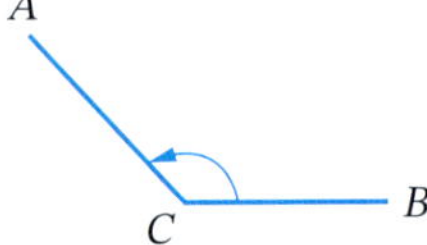

obtuse-angled triangle A triangle with one obtuse angle (between 90° and 180°). (p. 152)

order of operations The order in which a mixed expression such as $(17 - [3 + 1]) \times 2$ is evaluated. Work from left to right, first performing any operations in grouping symbols, then multiplication and division, and finally addition and subtraction. (p. 57)

ordered pair A pair of numbers (x, y) that can be used as coordinates to plot a point on the number plane. (p. 563)

origin The point $O(0, 0)$ at the centre of the number plane, where the x-axis and y-axis cross. (p. 563)

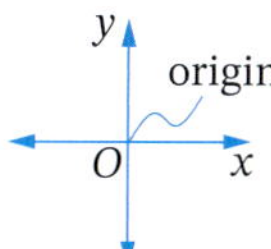

outcome In probability, the result of a situation or experiment. For example, when rolling a die, one possible outcome is rolling a 4. (p. 404)

outlier An extreme data value that is very different from the other values in a set. (p. 318)

P

parallel lines Lines point in the same direction and do not intersect. $AB \parallel CD$ means 'AB is parallel to CD.' (p. 148)

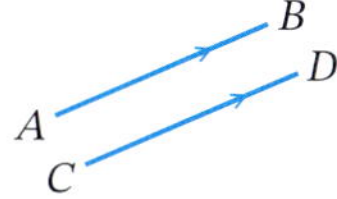

parallelogram A quadrilateral in which the opposite sides are parallel. (p. 156)

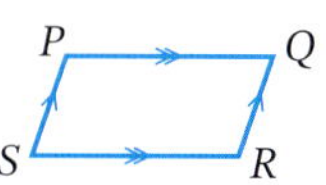

parentheses (Pronounced 'pa-ren-th-sees') See **grouping symbols**.

per annum (p.a.) Per year. (p. 276)

percentage A fraction with denominator 100 that is written in a special way. For example, 7% means $\frac{7}{100}$. (p. 251)

perimeter The distance around the outside of a shape. The sum of the lengths of its sides. (p. 182)

perpendicular height The height of a shape, measured at right angles to the base. (p. 192)

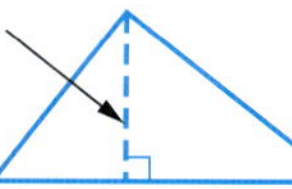

perpendicular lines Lines intersect to form a right angle. $AB \perp CD$ means AB is perpendicular to CD.

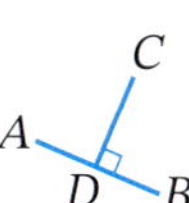

pi (π) A special number, approximately 3.1416, used in calculating circle measurements. Pi is a constant value found by dividing the circumference of any circle by the diameter. (p. 185)

polygon Any flat shape made up of straight sides. (p. 168)

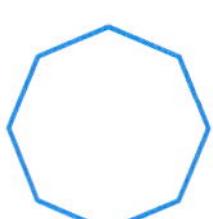

population In statistics, all of the items being studied, the entire group. (p. 332)

positive number A number greater than 0, for example, 11, 6 and 4.

power (or index) The number of times a base is multiplied by itself. In 2^5, the power is 5. Also called the **exponent**. (p. 71)

prime number A number that has only 2 factors, 1 and itself. For example, 2 and 7 are prime numbers, but 15 is not.

principal An amount of money invested or borrowed, on which interest is calculated. (p. 276)

prism A solid shape with identical cross-sections with straight sides, such as this hexagonal prism. (p. 219)

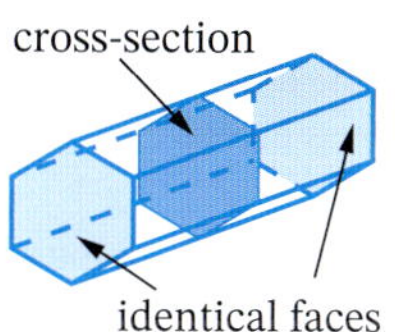

probability The chance of an event occurring, measured as a fraction, decimal or percentage between 0 and 1. (p. 404)

product The result of multiplication. The product of 7 and 3 is 21. (p. 51)

profit The amount made when selling an item at a higher price, when the selling price is more than the cost price. The opposite of **loss**. (p. 272)

pronumeral Another name for **variable**. (p. 97)

proper fraction A fraction whose numerator is less than its denominator, such as $\frac{3}{8}$. (p. 246)

protractor A geometrical instrument that measures the size of an angle in degrees.

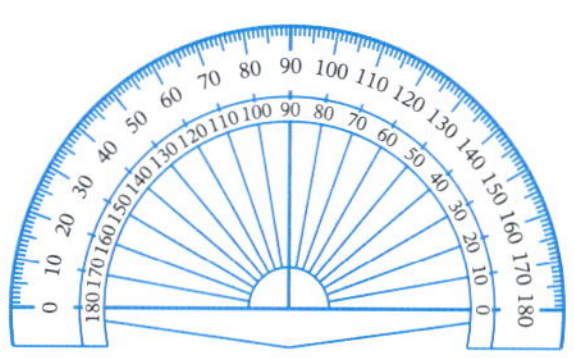

Pythagoras' theorem The formula $c^2 = a^2 + b^2$ for a right-angled triangle, where c is the length of the hypotenuse and a and b are the lengths of the other 2 shorter sides. (pp. 7, 9)

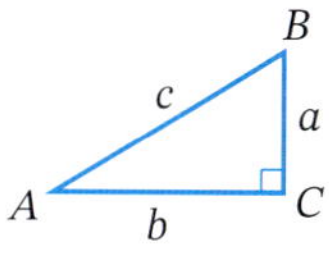

Pythagorean triad A set of 3 numbers that follow Pythagoras' theorem, such as 3, 4, 5. (p. 25)

Q

quadrant (of a number plane) A quarter of the number plane created by the x-axis and y-axis crossing at right angles. (p. 563)

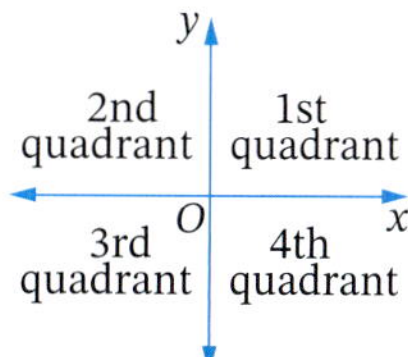

quadratic equation An equation in which the highest power of the variable is 2, that is, a variable squared, for example, $3x^2 - 6 = 69$. (p. 475)

quadrilateral Any polygon with 4 sides. (p. 156)

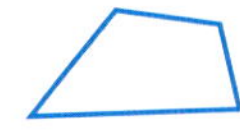

quotient The result of a division. If 36 is divided by 3, the quotient is 12. (p. 53)

R

radius (plural: **radii**) An interval joining the centre of a circle to the circumference, or the length of that interval. The radius is half of the **diameter**. (p. 185)

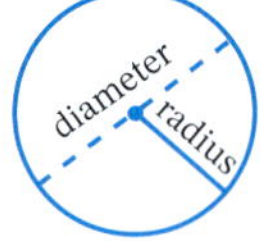

random In probability, describing a situation where every possible outcome has an equal chance, or is equally likely. (p. 404)

random sampling In statistics, selecting a sample in which every person or item in the population has an equal chance of being selected. A sample should be random to be truly representative of the population. (p. 333)

range In a set of data, the difference between the highest and lowest values. (p. 300)

rate A relationship between 2 quantities measured in different units. For example, a speed of 107 km/h compares distance travelled (in kilometres) with time (in hours). (p. 513)

ratio A relationship between quantities measured in the same units. For example, the ratio of 3 teachers to 40 students is 3: 40 (read '3 to 40'). (p. 493)

reciprocal The reciprocal of any number is found by first writing the number as a fraction and then swapping the numerator with the denominator. The reciprocal of 5 is $\frac{1}{5}$ and the reciprocal of $\frac{2}{3}$ is $\frac{3}{2}$. The product of any number and its reciprocal is 1. (p. 249)

rectangle A quadrilateral with 4 right angles. (p. 156)

recurring (or repeating) decimal A decimal with one or more digits that repeat endlessly. For example 0.1666 ... is abbreviated as $0.1\dot{6}$. (p. 68)

reflection The process of 'flipping' a shape across a line to give a mirror image that is back-to-front. (p. 351)

reflex angle A 'bent-back' angle greater than 180° but less than 360°. (p. 144)

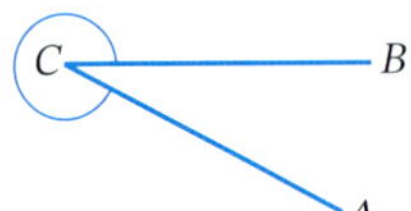

regular polygon A polygon that has all sides equal and all angles equal. For example, this regular pentagon has 5 equal sides and 5 equal angles. (p. 169)

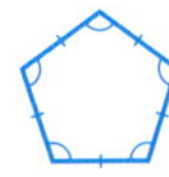

relative frequency The frequency of an event in repeated trials of a chance experiment, written as a fraction of the total frequency, found using the formula $P(E) = \frac{\text{frequency of } E}{\text{total frequency}}$. (p. 431)

revolution An angle of 360°. (p. 144)

rhombus A quadrilateral with 4 equal sides. (p. 156)

RHS The right-hand side (of an equation). (p. 463)

RHS The right angle–hypotenuse–side test for congruent triangles or similar triangles. (pp. 369, 384)

right angle A 90° angle, a 'square corner', a quarter-turn. (p. 7)

right-angled triangle A triangle with one 90° angle. (p. 7)

rotation The process of 'spinning' or turning a shape about or around a fixed point in a certain angle and direction (clockwise or anti-clockwise). (p. 351)

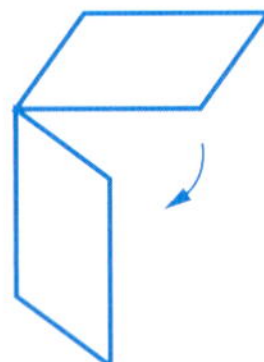

S

sample In statistics, a group of people or items selected from a population for study. (pp. 332, 333)

sample space In a probability situation, the set of all possible outcomes. (pp. 403, 404)

SAS The side–angle–side test for congruent triangles or similar triangles. (p. 369)

satisfy an equation When a variable (or variables) can be substituted into an equation to show that it is correct or true. For example, $x = 5$, $y = 8$ satisfies the equation $y = 3x - 7$. (p. 574)

scale factor The amount by which a figure is enlarged (or reduced), or the ratio of matching sides in a pair of similar figures: Scale factor = $\frac{\text{image length}}{\text{original length}}$. (p. 382)

scaled length The length on a scale drawing, map or plan, not the actual length. (p. 502)

scalene triangle A triangle with no equal sides. (p. 152)

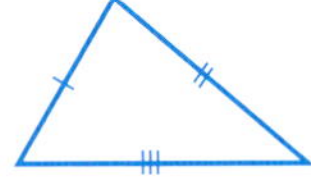

sector A region of a circle cut off by 2 radii, shaped like a slice of pizza. (p. 214)

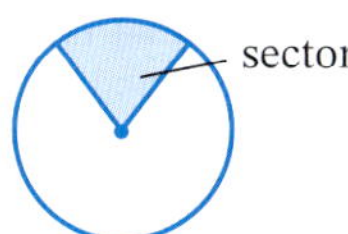

sector graph Also called a pie chart, a circular graph that is divided into proportionately sized sectors to represent parts of a whole. (p. 292)

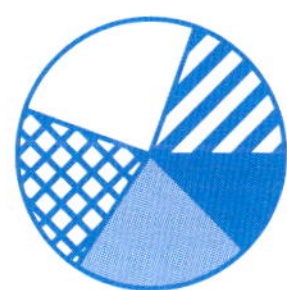

selling price The price at which an item is sold by the retailer (shop). The shop buys the item for the **cost price**. (p. 272)

semicircle Half a circle. (p. 211)

similar In proportion, referring to shapes whose matching sides are in the same ratio. The symbol '|||' means 'is similar to'. (p. 377)

similar figures Two or more figures whose matching sides are in the same ratio, having the same shape but not necessarily the same size. One figure is an enlargement or reduction of the other. See also **scale factor**. (p. 382)

similarity test One of 4 tests for proving that 2 triangles are similar: SSS, SAS, AA and RHS. (p. 384)

simple interest Interest is calculated as a percentage of the original principal. (p. 276)

simulation An imitation of a real event, often used in probability when the real event is difficult or impractical to conduct. (p. 436)

solution In algebra, the value that makes an equation true. (p. 453)

speed A rate that compares distance travelled with time taken. Usually measured in kilometres per hour (km/h) or metres per second (m/s). (p. 521)

$$\text{Average speed} = \frac{\text{distance travelled}}{\text{time taken}}.$$

square A quadrilateral with 4 equal sides and 4 right angles. (p. 156)

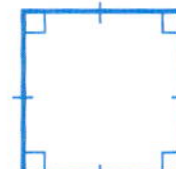

square (of a number) The number multiplied by itself. For example, $7^2 = 7 \times 7 = 49$. (p. 5)

square metre A unit for measuring area, equal to the area of a square with length 1 m (p. 189)

square root (of a number) The positive value which, if squared, gives the number. For example, $\sqrt{25} = 5$ because $5^2 = 5 \times 5 = 25$. (pp. 5, 72)

SSS The side–side–side test for congruent triangles or similar triangles. (p. 369)

stationary Not moving, stopped. (p. 526)

stem-and-leaf plot A 'number graph' that lists all the data values in groups. Each value is split into a 'stem' and a 'leaf'. This stem-and-leaf plot shows 12 test marks, from 42 to 82. (p. 319)

Stem	Leaf
4	2 5
5	0 2 8
6	6 7
7	3 5 7 7
8	2

Key: 5|8 stands for 58

straight angle An angle of 180°. (p. 144)

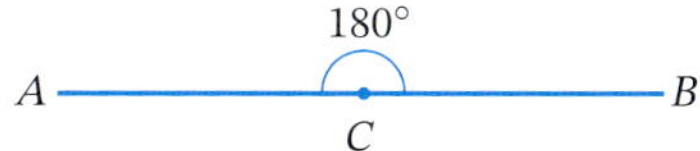

sum The result of an addition. The sum of 5 and 7 is 12. (p. 48)

superimpose To place one figure on top of another congruent figure so that sides and angles match. (p. 356)

supplementary angles Two angles whose sum is 180°. The angles 70° and 110° are supplementary. (p. 144)

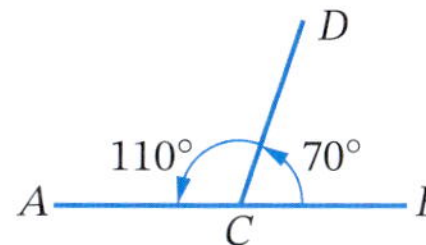

surd A square root (or other root) whose exact value cannot be found, such as $\sqrt{10}$ or $\sqrt[3]{7}$. (pp. 5, 72)

survey A study involving collecting information, facts and opinions. (p. 337)

table of values A table of ordered pairs of numbers, usually following a formula which can be graphed on a number plane. The table of values below is for the formula and linear equation $y = 2x - 3$. (p. 553)

x	−1	0	1	2
y	−5	−3	−1	1

term (of a pattern) A number or diagram in a pattern. (pp. 111, 493)

term (of an expression) A part of an algebraic expression. For example, $b^2 + 6b - 9$ has 3 terms: b^2, $6b$ and -9. (p. 111)

term (of a ratio) A number in a ratio, for example, in 14 : 9, the terms are 14 and 9. (p. 493)

terminating decimal A decimal whose digits do not repeat endlessly, for example, 0.125. (p. 68)

time zone A region of the world where all places experience the same time of day, for example, 10 a.m. (p. 538)

transformation In geometry, the process of moving or changing a shape. (p. 351) See also **reflection**, **rotation** and **translation**.

translation The process of 'sliding' a shape a certain distance and direction. (p. 351)

transversal A straight line that intersects 2 or more lines at different points. (p. 148)

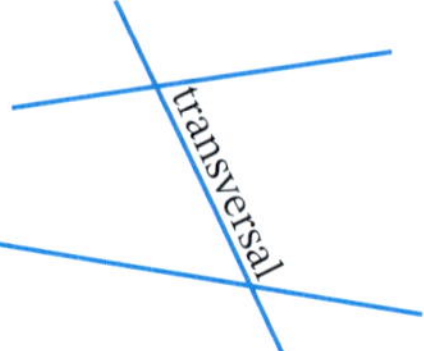

trapezium A quadrilateral with one pair of opposite sides parallel. (p. 156)

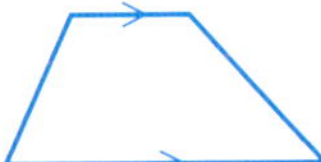

travel graph A line graph that describes a journey and shows the distance travelled over time. (p. 526)

tree diagram A diagram of branches for listing all of the possible outcomes in a multi-step chance experiment. (p. 425)

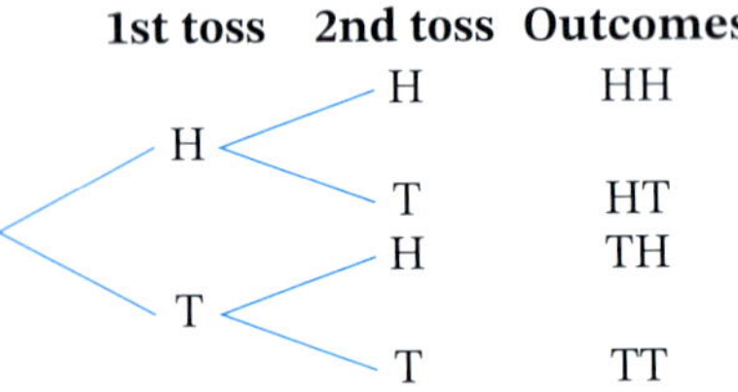

trial One go or run of a repeated probability experiment, for example, one roll of a die. (p. 404)

two-way table A table that shows the number of items belonging to overlapping categories. (p. 419)

	Can swim	Cannot swim
Boys	13	2
Girls	9	3

unit price The cost of one item or unit, found by dividing the cost of the item by the number of items or units. (p. 515)

unitary method A method for finding a quantity by finding the size of one part or 1% first. (pp. 270, 498, 510)

unknown Another name for **variable** because its value is usually not known, especially in an equation where the value can be found.

unlikely A low chance it will happen, improbable, probably won't happen. (p. 404)

UTC (Coordinated Universal Time) The time zone from which all time around the world is measured. Also called Greenwich Mean Time (GMT), it is the time measured at the Greenwich Observatory in London, England. (p. 539)

variable A symbol, usually a letter of the alphabet, stands for a number. Also called a **pronumeral** or **unknown**. (p. 97)

Venn diagram A diagram of circles (usually overlapping) for grouping items into categories. (p. 413)

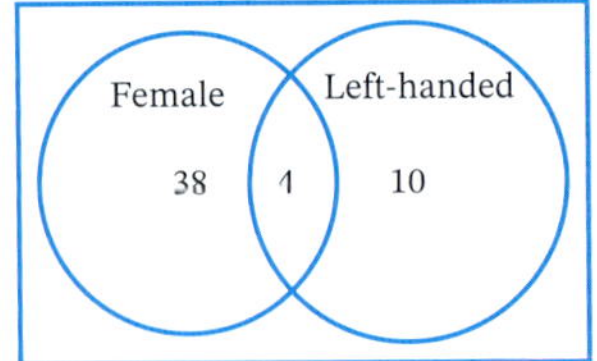

vertex (plural: **vertices**) A corner of a shape or angle. (p. 156)

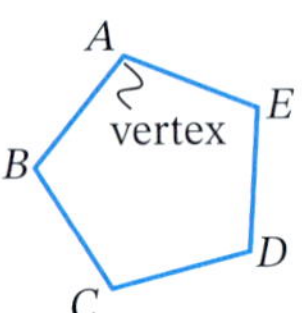

vertical Going up and down, at a right angle to the **horizontal**. (p. 51)

vertically opposite angles A pair of opposite and equal angles formed when 2 lines cross. (p. 144)

volume The amount of space taken up by a solid object, measured in cubic units. (p. 217)

***x*-axis** The horizontal axis of a number plane (running across, see diagram below). (p. 563)

***y*-axis** The vertical axis of a number plane (running up and down). (p. 563)

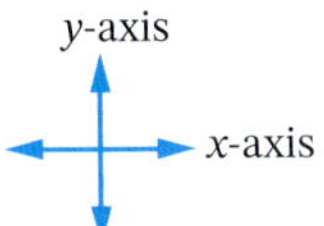

***y*-intercept** The *y*-value at which a line cuts the *y*-axis. (p. 578)